Earthquakes and Coseismic Surface Faulting on the Iranian Plateau

DEVELOPMENTS IN EARTH SURFACE PROCESSES, 17
SERIES EDITOR -J.F. SHRODER JR

For previous volumes refer
http://www.sciencedirect.com/science/bookseries/09282025

Developments in Earth Surface Processes

Volume 17

Earthquakes and Coseismic Surface Faulting on the Iranian Plateau

A Historical, Social and Physical Approach

Manuel Berberian

Ocean County College

and

Onduni Grung Scientific Enterprise

Toms River, New Jersey, USA

ELSEVIER

AMSTERDAM • BOSTON • HEIDELBERG • LONDON • NEW YORK • OXFORD

PARIS • SAN DIEGO • SAN FRANCISCO • SINGAPORE • SYDNEY • TOKYO

Elsevier
The Boulevard, Langford Lane, Kidlington, Oxford, OX5 1 GB, UK
Radarweg 29, PO Box 211, 1000 AE Amsterdam, The Netherlands

First edition **2014**

British Library Cataloguing in Publication Data
A catalogue record for this book is available from the British Library

Library of Congress Cataloging-in-Publication Data
A catalog record for this book is available from the Library of Congress

For information on all **Elsevier** publications
visit our web site at store.elsevier.com

ISBN: 978-0-444-63292-0
ISSN: 0928-2025

Contents

Part II
Dynamic Phenomena Associated with the Earthquakes on the Iranian Plateau

Foreword

Manuel Berberian, an Iranian-Armenian geologist and earthquake seismologist, is responsible for much of what we know about the geology and tectonics of Iranian earthquakes. In addition to field study of Iranian earthquakes and surface faulting, Berberian began studying the archaeology and history of Iran to use that rich record to expand the knowledge of Iranian earthquakes and active faults and their hazards to society. He became a pioneer in Iranian plate tectonics, active tectonics, and seismicity.

Berberian's current work is a continuation of his long-term commitment to using science and engineering to make his native country a safer place for its citizens. Manuel and I are working on improving understanding of the earthquake hazard to the megacity of Tehran. His manuscript on **Earthquakes and Coseismic Surface Faulting on the Iranian Plateau: A Historical, Social, and Physical Approach** is likely to be a major contribution to a better understanding of the hazards facing Iran.

Robert Yeats
Professor Emeritus of Geology and Geophysics
Oregon State University
Corvallis, OR

In their pioneering 1982 account, *A History of Persian Earthquakes,* Nicolas Ambraseys and Charles Melville established meticulous procedures for merging archival records and inscriptions documenting former earthquakes, instrumental seismic data, and field observations of ground rupture and shaking damage in Iran. The result was a benchmark history of 389 earthquakes from the earliest times to 1979, that superseded all previous works on the seismicity of Iran, and which for three decades has been upheld as an interdisciplinary masterpiece that few thought could be surpassed.

In the present work, Manuel Berberian not only adds to and updates the seismic record but brings new insights to the study of earthquakes in Iran. The past thirty years has seen the development of paleoseismology which, through the exhumation of surface faults has extended the earthquake record into pre-history. Digital topography and advances in structural geology have extended our spatial understanding of the structures on which earthquakes occur, and the past three decades have also seen advances in digital seismology, and space geodesy (GPS and Insar) that have permitted a glimpse at the subsurface processes underlying and accompanying earthquakes in Iran.

The industry contained within this new work on the seismicity of Iran is abundantly clear in the detailed studies of individual earthquakes. In a lifetime of dedicated study, Berberian has more than doubled the number of archival earthquakes described by Ambraseys and Melville and has critically evaluated previous accounts to reconcile historical data with known geological structures and newly available historical sources. Entries include a careful evaluation of commonly quite scant data, supported in places by extensive field observations by the author, incurred amid remarkable hardships caused by wars, persecution, and regime changes in Iran.

The importance of the current work to the people of Iran cannot be underestimated. In the past 100 years, the death toll per capita from earthquakes has been the highest in the world, due to an unfortunate combination of geographical earthquake productivity and a societal indifference to the earthquake resistance of the villages and towns of Iran. Despite the recognition of seismic risks to its population and the loss of almost 164,000 lives in the past several decades, the leadership of Iran has yet to mandate ubiquitous earthquake resistance to its dwellings and civic structures. It is to be hoped that the present work will provide Iran a factual fabric on which to formulate wise decisions concerning its inescapable seismic future. If its construction industry continues to ignore Iran's seismic past, many thousands more deaths from earthquakes will assuredly occur.

Roger Bilham
Professor of Geological Sciences
University of Colorado
Boulder, CO

Editorial Foreword

This book is an exhaustive analysis of seismicity on the Iranian Plateau, and it provides a wealth of information that Dr. Manuel Berberian has compiled from his vast experience in the region. His knowledge of and access to the extensive ancient documents of Persia have allowed him to compile a superb record of ancient earthquakes. He has been working on this project for much of his life, with the intention of creating a descriptive catalogue of earthquakes in Iran from ancient times to the present day, and the data informing this book were collected during the last 42 years of his field work and research in the region.

The Iranian plateau has a record of documented earthquakes that is many times longer than the records of most other regions, and that comprehensive record has added to the world's seismic database. Such a long-term record combines individual human memory with geological data that extends even farther into the past, and as a result, it is invaluable because it enables an increased understanding of a process with a time scale so far out of synchronicity or harmony with short-term human endeavors. If more well-executed interpretations of the geological record are added to this understanding through the interpretation of trenches excavated across the profuse stratigraphy of buried fault scarps, scientists will continue to elucidate the mechanics and timing of the seismic activities at these important sites.

Dr. Berberian was born in an ethnic and religious minority community in Iran, which he always thought was the wrong country for parents who barely escaped the Armenian massacre of 1915, in a region known for disgustingly harsh treatment of its minorities. Nonetheless, in spite of these apparent handicaps and detriments, and with English being his fifth language, he has dedicated his life's work to informing and possibly even saving people from the huge and pervasive seismic risks presented by the active fault zones in which they live. He feels that, if he can save even one life, then that accomplishment would perhaps be a baby step forward. Yet, he also notes that one life is infinitesimal when compared to the 164,000 lives lost during the past century. Dr. Berberian's hope is that governments in the developing world will wake up and do something better for their citizens, and in this book, he shows that it is certainly possible.

I was also pleased to see the information and opinions provided by Dr. Berberian about the basic fatalistic incompetence of the existing Iranian government, apparently because of cultural–religious–political issues that, in combination, have caused the unnecessary deaths of thousands of innocent people. Dr. Berberian has also related that the same situation applies in neighboring Afghanistan, almost all of the Middle East, and all of the so-called developing countries around the world. In particular, he has noted to me that, because

Afghanistan is an extremely poor country, such problems might be more expected, but Iran is not so poor. Thus, one might think that, in a country as wealthy as Iran, conditions of seismic safety have been greatly improved, but they have not. This lack of government action is a tragedy of considerable proportions that is constantly unfolding, as the next big seismic event in Iran will surely generate additional hundreds to thousands of needless deaths. Dr. Berberian's exhaustive analysis of seismicity on the Iranian Plateau does provide a rather strong indictment of the Iranian government's incompetence to handle a common natural hazard that exists in the region. In contrast to many other developed countries that pay attention to seismic hazards and engineer their way around many such problems, Iran instead seems to stay mired in corruption, lethargy, and incompetence, so that its earthquake hazard is apparently largely ignored. This certainly calls into question the idea that the Islamic Iranian Revolution was intended to help the people of Iran. The problem of earthquake safety might be addressed more properly in future, however, because, to their credit, the Iranian clerics of today seem to regard education as a universal good (Axworthy, 2013; Secor, 2014). Perhaps with the publication of this work, they will be able to find more state mechanisms to employ in overcoming their cultural–religious fatalism and resistance. Instead of continuing the old patterns, they might begin to improve and apply seismic zoning and even build more seismically resistant buildings.

Because Dr. Berberian is not a native speaker of English, he has done a superb job of pulling all this material together in one place, while translating obscure sources. Through this book, we can see that this is his life's work. He is intensely devoted to it, and he knows more about the topic than probably any other single person in the world. Thus, this book is an important scientific contribution in which I am happy to have been involved. The book's many excellently rendered graphics and photographs of seismic damages and fault-related geomorphologies add much to the understanding of the active faults in Iran. Many pages of this book are also dedicated to coseismic surface faulting and blind-fault-related folding at those faults, adding considerable scientific value and understanding of the phenomena. This book has the potential to become a well-known primary source on an essential topic for a tremendously active seismic region. For these reasons, this is an important work about a part of the world in which he and I have had a long interest, and the book should do well because it is such a thorough compendium of that important material.

John F. Shroder, Jr.
Editor in Chief
Developments in Earth Surface Processes

Preface

Out beyond the ideas of wrong-doing and right-doing,
There is a field. I will meet you there.

Maulavi Rumi (1207–1273)

The sacred Avestan and the Pahlavi texts,[1] as well as later Persian and Armenian literature, are rich in the portrayal of human beings in relation to the forces of nature during the time that nature was considered the arena in which the divine became visible.

According to the ancient Zoroastrian complex cosmogony, addressed briefly in the extant Avestā and the Pahlavi texts rooted in traditions dating back to at least 1200 BCE, the sacred planet Earth[2] was the third good creation of "Ahurā Mazdā/Ohrmazd" (lit., "the Wise Lord") after the "sky" and the "water." Over time, and as a result of the third assault on the good creation by the evil spirit (*Ahriman*), the demons (*Div*; the same word as in "divine" and "deity") rushed in, the Earth quaked, and the essence of the mountains was created in the Earth. As the first mountain range, the *Alborz*,[3] grew, all the mountains remained in "motion." Like the growth of a tree, mountains grew up to the clouds and down underground from the original flat surface of the Earth; the mountain roots were connected to each other (cf., the modern concept of orogeny [mountain building] in tectonics).

The Earth and all that is on it, such as "water," "fire," "soil," and "air" (the four primordial elements of ancient Zoroastrian Iran, ca. 1200 BCE), is considered a sacred manifestation of Ahurā Mazdā; therefore, all elements are "sacred" and precious, homage is rendered to every angel to whose care these are entrusted, and thanks are offered to the Court of the great Creator. At the end of time, the active planet Earth will be flattened by earthquakes and erosion; no longer will there be mountain peaks or valleys.

The ancient view of an *active sacred planet* (with creation and evolution of mountain belts by orogenic forces and movements, shaking of the Earth caused by evil spirits and storm-demons, and an ultimate apocalyptic erosion and leveling process of the mountains by earthquakes) is a remarkable

☆"To view the full reference list for the book, click here."

1. Bondeheshn (Bundahishn), Menog-e-Kherad, and Selections of Zād Sparam.
2. Zem, Ārmaiti; Mother Earth.
3. Pahlavi: *Harburz*; Avestan: *Harā Berezaiti;* lit., the "Lofty Watchpost," the mythical mountain in Yasht 19.1 & Greater Bondeheshn. IX.1.

observation that has survived since ca. 1200 BCE (see Boyce, 1989 for dating the period). These expressions and visions of antiquity were conceived during the time that people lived a free and independent life in open nature. They reveal an early awareness of living in a seismically active region; such knowledge must have been achieved by the close surveillance of successive changes on the active Earth by people who lived in reasonable harmony with nature, some of whom suffered from the frequent destruction caused by earthquakes. These people observed a mechanism inside the active and living Earth, which they defined as an evil spirit and demon, that caused fundamental deformations that continuously shaped and changed the surface of the Earth (geological processes in modern geology).

Similar observations rooted in ancient Iranian culture can be found at later times by master poets Ferdowsi and Rumi. For example, in the beginning of the Persian masterpiece *Shāhnāmeh* (1010), Ferdowsi Tusi wrote:

Mountains grew and water appeared;
The plants grew upward as well;
The Earth did not have mountains earlier;
It was a black and dark center.

Later, *Maulanā Jalal U'ddin Rumi* (1207–1273),[4] writing about Qāf Mountain,[5] mentioned that, despite looking stable, the connected mountains around the world are active and remain in motion:

When He says 'Enough!' my vein rests,
I am (apparently) at rest, but actually I am in rapid motion.
At rest, like the (medicinal) ointment, and very active (efficacious),
At rest, like the intellect, while the speech (impelled) by it is moving.

The extant Avestā and the Pahlavi texts are viewed here as treatises on spiritual secrets and an early understanding of nature in general—and the active sacred planet Earth, specifically, in antiquity—and not as religious texts. The investigation, which I present in the following pages of this book (Part One) is therefore centered upon very ancient myths and folk epics and addresses the psychology of the shaking of the sacred active planet Earth by evil spirits/demons as well as its social impacts on, and psychology of, the commingled inhabitants of Greater Iran[6] and Armenia on the Iranian plateau throughout several millennia. It covers all the human aspects of earthquakes and Iranian and Armenian behavior and thoughts from antiquity to the present day, which has mostly survived by word of mouth and been transmitted from one generation to another by village bards. Part 1 of the book, consequently,

4. Mathnavi-e Ma'navi, Book IV:9; Nicholson (1926).
5. The mythical Alborz Mountain; refers to the present Elbrus (Alborz) Mountain in the Caucasus.
6. Land of Iran (syn., Irānshahr (Country of Iran)); the Iranian plateau with the Ārmenian Highlands and the Caucasus; the Greater Iran during the Achaemenid Empire (550-330 BCE).

embarks upon the difficult task of understanding the myths, legends, folk epics, early interpretations, and facts that have survived in Iran despite the mixture of the blood of its inhabitants with different cultures and religions during successive incursions and conquests (conquests apparently became an inherited industry of the Middle East).

Rural traditions in Iran and Greater Armenia are full of myths, legends, national epics, folklore and stories about earthquakes—the dynamic and physical phenomena associated with them and their social, cultural, and religious inferences since antiquity. Actual geological phenomena of a different era were referred to or explained as epical or mythological events. These events fluctuated between reality and mythology; thus, it is impossible and inappropriate to incorporate these references into chronological contexts. In view of the limited reliability of some of the sources, which refer to ancient myths, legends, epics, and oral traditions, it is impossible to reach any firm conclusion about the date, the site, and/or in some cases, the nature of the events. In order to understand the present Iranian psyche regarding earthquakes and the Iranian's acceptance of doctrine of fatalism despite large-scale destruction and unfathomable death tolls, we need to review Part 1 of this study.

Although relics of the myths and legends are still traceable in the beliefs, customs, and folk narratives, modern scholars have not carried out any systematic and in-depth myth/folktale field collections, or survey and analysis of the early Arab-, Armenian-, and later European-travelers' reports, on this specific issue. The rapid urbanization of the counties in Iran and Armenia since the mid-twentieth century has brought about the speedy disappearance of ancient traditions, popular beliefs and songs, customs, habits, folktales, and myths. I was encouraged by the absence of a regional systematic study on Earth-related folklore to create a preliminary database for further in-depth research and study by the young generations about life on the Greater Iranian plateau.

I have tried to review different aspects of earthquakes and active faulting from the Pleistocene geomorphological changes to ancient myths to present examples. An understanding of the young geomorphological records and ancient metaphoric references to earthquakes embedded in our myths, legends, folklore, epics, and linguistic traces referring to prehistoric events may help us update our natural hazards study programs as well as raise social awareness of the potential earthquakes and active faulting that could cause heavy loss of life and property. This book is not a catalog of earthquakes. It is aimed at improving our knowledge on humanitarian crises and human security as well as socioenvironmental relations at local, national, regional, and international levels by analyzing basic information of earthquakes and coseismic surface ruptures within its historical scientific and social context.

Earthquakes on the Greater Iranian plateau have been intertwined with and influenced the culture, religion, philosophy, history, and political perceptions of the region's inhabitants since antiquity; they have reverberated in people's affairs with severe repercussions. Although the economic damage of devastating earthquakes to the region and society is incalculable, they have also had a catastrophic role in shaping the course of development and then decline of cities and villages that eventually faded into obscurity. Earthquakes and invasions have been the catalyst for decline of numerous places in the region through high death tolls and great destruction of human infrastructure and resources. Therefore, there is a growing need for a better understanding of regional earthquakes, active faults and folds, and seismic hazards. Due to the scarcity of firsthand data from past events, the interdisciplinary geological records of the Quaternary, coupled with the extant myths and archeological data, have been used to review earthquake-triggered hazards, especially coseismic surface faulting of the past, and raise awareness of the importance of the extant fragmentary documents in the reconstruction of the past events. This becomes vital when we realize that in most developing countries prone to earthquakes, most inhabitants have been migrating to large cities and mega-cities built in the vicinity of active fault zones holding a high risk of earthquakes.

I believe that an earthquake scientist working in areas with ancient culture and civilization, like the greater Iranian plateau and Armenia, should have basic regional knowledge of mythology, archeology, theology, history, literature, and culture. Hence, when required, I have applied archeological, historical, religion, literature and social data for better analyses of the discussed entries. I have tried to develop a body of social beliefs, structure, and activity in reference to earthquakes and coseismic surface faulting in the pursuit of social welfare. Although I do not consider myself an expert in all these fields, I do know Earth sciences. This book, therefore, presents the scientific and social aspects of earthquake geology and active tectonics in an exploration of the impacts of earthquakes on human development and evolution throughout a very long regional history.

In the course of my investigation, I was not able to consult solely primary references. I have emphasized quoted texts extracted from the firsthand sources (when available) in *italics*, nesting them within double quotation marks in the main body of the text. My intention is to emphasize that this is source data without changes and/or corruptions, just translation. Where necessary, explanations are added within square brackets.

I must also point out some personal issues that have definitely influenced the outcome of all my publications, including the present work. First, the reader should consider the limitations within which I have worked in a politically volatile, developing country with little access to proper libraries, computer facilities, technical publications, professionals, and mentors. Secondly, as with my all previous works, research for this book was not supported by grants, organizations, or personal moral encouragement. This book was not the subject of different dissertations prepared by research students working under my supervision. Bear in mind that English is not my first language; it is the fifth language that I speak (in chronological order of learning and proficiency).

Finally, working and living as part of an ethnic and religious minority in the Middle East is psychologically, physically, and financially difficult. Nonetheless, I have great affection for the enchanting Iranian plateau and the Greater Armenian highlands and the genuinely delightful people in the modern Iranian and Armenian villages there. I have dedicated my entire life to the once-sacred land of the great teacher (and not prophet), Zartosht [Zoroaster], and the paradise of masters such as Giumars, Hāyk, Feraydun, Vāhāgen, Kāveh, Ārash, Rostam, Jāmāsb, Cyrus the great, Movses Khorenātsi, Mesrop Māshdots, Vārtān Māmigoniān, Ferdowsi, Biruni, Khayyām, Maulavi Rumi, Āndrānik Ozāniān, Pedros Āndreāsiān, Yeprem Khān Armani (Dāvidiān), Qa'em Maqām, Amirkabir, Dr, Mosaddeq, Esq., and many more. Sorrowfully, their paradise and utopia has been lost eternally.

This book grew out of my love of and passion for the geology, active tectonics, archaeology, and history of the Iranian plateau as well as my single-handed, lifelong earthquake research into the glorious mountains and enchanting deserts as well as the textural archives of history, literature, religion, culture, and archaeology since at least 1971. My study starts with the social aspects of earthquakes in ancient times; in-depth analyses of the physical aspects of the seismicity and seismic sources of the Iranian plateau follow. A growing, voluminous database for these series, collected and enlarged since 1971, has traveled with my Ārmeno-Irānicā self as I have wandered for approximately 27,000 km: from Tehrān to Cambridge, UK, back to Tehrān, then to New York City, NY, San Diego, CA, and finally to New Jersey. I have pursued a neverending search for a lost paradise in Diaspora while trying to minimize the earthquake destruction in the urban and rural areas.

The present work provides a valuable context for interpreting prehistoric, historic, and contemporary seismicity and coseismic faulting and folding and their effects on environment, human lives, soul, and culture as well as attendant economic losses. I hope this "infant database" will provide a better understanding of earthquake impacts on people and infrastructure, seismic hazards, and risk on the Iranian plateau with its attendant complicated psychological, sociological, and epidemiological side effects. This work cannot possibly cover the entire breadth of the multidisciplinary subjects that it addresses. Given the tremendous toll in human lives and economic losses, it is appropriate to better understand the long-term effects of earthquakes and earthquake faulting. The greatest price has always been paid by the poor villagers living on one of the richest petroleum resources of the planet Earth.

The systematic collection and analysis of data on the impacts of natural disasters are necessary to mitigate and control the adverse chronic impacts of disasters in general. Governments of developing countries should incorporate planning for disaster control and risk minimization in their national plans. Unfortunately, statistics on different aspects of earthquakes in the region are not available to the public. My numerous requests to different governmental organizations have gone unanswered. Therefore, although the available data are incomplete, my aim is to construct a basis to build on when and if more information becomes available.

One of the motivations behind my 43-year-long, single-handed effort in this field is to get the authorities in developing countries to take earthquake hazards seriously in order to create a culture of prevention and to develop solutions to disasters now rather than handing the problem over to the next generation. As a result of the urbanization, growth in size, and vulnerability of populations, the human losses and destruction of infrastructure from natural hazards have increased in recent decades. Sharing the issues discussed in this work among the regional policymakers, civil servants, scientists, engineers, and social workers is essential and may encourage resilience within the at-risk communities of vulnerable populations in all earthquake-prone countries.

ORGANIZATION OF THE BOOK

The present work covers the social aspects of Iranian earthquakes and the mechanics of surface faulting as a terrifying and invisible phenomenon that strikes with no warning. It has influenced people's daily lives, helping their myths, legends, folklore, beliefs, religions, cultures, and literature evolve. In each case, I have tried to bring the extant, scattered data regarding earthquakes and surface faulting together for present and further analyses.

The text is organized into two parts:

1. Earthquake Hazard Warning in Oral Traditions and Literature (Chapters 1–8), which deals with the human aspects of earthquakes; and
2. Dynamic Phenomena Associated with Earthquakes (Chapters 9–17), which mainly covers the active faulting and folding that occurs during earthquakes.

Each part is divided into chapters that cover specific issues; however, some entries may fall into a few overlapping chapters.

EARTHQUAKE ORIGIN TIME

Since the local time that an earthquake occurs plays an important role in the search-and-rescue operation and number of casualties, this (when available) has been extracted and added to tables showing historical (pre-1900) earthquake data. The primary standard of Coordinated Universal Time (UTS; successor of the Greenwich Mean Time, GMT) is used for post-1900 earthquakes. Depending on the year and season, the difference between local time and UTC varies from 3.5 to 4.5 h.

EARTHQUAKE MAGNITUDE

The original equivalent surface-wave magnitudes of the historical (pre-1900) earthquakes mentioned throughout the text are taken from Ambraseys and Melville (1982), Ambraseys and Adams (1989), Berberian (1994), and Ambraseys and Bilham (2003a,b). Because of a lack of primary data and the exaggeration of figures, we cannot assign an exact number for most of the historic earthquakes. Thus, magnitudes as well as intensities have been revised in this study (Table 17.11). The meizoseismal areas of some events (and hence the intensity and equivalent surface-wave magnitude), such as the 856, 958, 1209, 1336, 1721, and other earthquakes, were exaggerated in most publications and have been corrected here. These preinstrumental magnitudes were derived from macroseismic information embedded in written accounts calibrated against instrumental M_s values. Therefore, they represent poorly constrained magnitude estimates. Instrumental period magnitudes are mainly taken from ISS, ISC, Ambraseys (2001), and Engdahl et al. (2006). Until future volumes become available, the reader is referred to Berberian (1994, 1997, 2005) for variable estimates and complete list of events (see also Table 17.12).

EARTHQUAKE CENTROID DEPTH

Depths reported in this work (CD_W; averaged center of the ruptured plane) are centroid depths derived from body waveform modeling (Berberian et al., 1992, 1999, 2000a,b, 2001; Baker, 1993; Baker et al., 1993; Priestley et al., 1994; Gao and Wallace, 1995; Maggi et al., 2000a,b; Jackson et al., 2002; Walker et al., 2003, 2004, 2005a,b; Talebian and Jackson, 2004; Talebian et al., 2006a; Nissen et al., 2007; Tatar et al., 2007; Peyret et al., 2008; Roustaei et al., 2010; Copley et al., 2013). Also, see the references for individual earthquakes cited in the text as well as Table 17.12.

ACTIVE FAULTS AND MEIZOSEISMAL AREAS OF EARTHQUAKES

In order to predict future large-magnitude earthquakes—including their surface and blind faulting—comprehend earthquake-fault hazard, and minimize earthquake risk, we need to understand active crustal deformation and locate and study the patterns in previous earthquake faults. A multidisciplinary approach has been used to examine past earthquakes and their causative faults. Active faults have been drawn from aerial photographs (Worldwide Aerial Surveys, Inc., Imperial Army of the Government of Iran, Tehran, 1956, and National Cartographic Center, Tehran), satellite imagery, and the author's numerous field investigations since 1971; some of them have been presented in the previously published and internal reports.

With large errors in instrumental epicenters and focal depths, and in the absence of radar interferometry and pixel-matching techniques for most earthquakes as well as definitive observed and mapped coseismic surface faulting, it is oftentimes very difficult to assign historic earthquakes to specific faults. Therefore, firsthand macroseismic data from archeological, epigraphic, literary, seismological, remote sensing, and field investigation sources are used as often as possible to locate and constrain the region of intense shaking in the near-field of each event within the estimated isoseismal of the highest intensity. Hence, the meizoseismal area of each event is schematically presented by an ellipse with variable quality. In some cases, the elongated damage zone could be an artifact of topography and uneven distribution of population. Nonetheless, limited intensity data resulting in unconstrained meizoseismal areas makes it difficult and unwise to draw broader conclusions regarding the interpretation of causative faults and further active tectonics. The question in each case is how well constrained or significant the constructed meizoseismal areas are. This is the only tool that remains with which to study the historic and recent earthquakes on the Iranian plateau.

TRANSLITERATION

The Persian, Armenian, Arabic, and Turkic geographical names and words are written as they are pronounced and in the script of their country of origin, with direct and simplified transliteration into English. Diacritical marks and special characters are used to differentiate vowel "A" [short; e.g., "ant"] from "Ā" [long; e.g., "Armenian"]. The Persian spoken linking particle [ézāfé/afzudeh], which is never written in the script, is used to join proper names rendered by "-e" or "ye."

Character		Equivalent	Example
ء	hamza	′	
ا	Alif	a	Ant
آ		ā	Ārmenian
ب	bāa′	b	
پ	pe′	p	
ت	tāa′	t	
ث	thāa′	th	Th′aālebi; ′āthār
ج	Jim	J	
چ	Ché	ch	
ح	hāa′	H	

Character		Equivalent	Example
خ	khāa′	kh	
د	Dāl	d	
ذ	dhāl	dh	Rendering a dental fricative
ر	rāa′	r	
ز	Zāy	z	
ژ	Zhé	zh	
س	Sin	s	
ش	Shin	Sh	
ص	ṣād	ṣ	
ض	ḍād	ḍ	
ط	ṭāa;	ṭ	
ظ	ẓāa′	ẓ	
ع	≈ayn	≈a; ‘u	‘Abbās; ‘Umar
غ	ghayn	gh	Rendering a voiced grittinal
ف	fāa′	f	
ق	qāf	q	Qanāt
ك	kāf	k	
گ	gāf	g	
ل	lām	l	
م	mim	m	
ن	nun	n	
و	wāw	w	
ه	Hāa′		
ی	Yāa′	y	
ة	Tā marbuta	a; at	

Short vowels:
Fatha: a
ḍamma: u/o
kasra: é

Persian poems are translated literally and without poetic format. Where appropriate, translations are added in the brackets to emphasize the meaning and message embedded in the names.

CHRONOLOGY

The first year of the traditional Armenian chronological system was 551. In 1312, parallel to the traditional Armenian calendar, the Council of the Armenian bishops of Ādānā adopted the Julian calendar ["Old Style," with 10 days discrepancy with the Gregorian calendar (New Style)], which was implemented until 1917 (Badalyan, 1970; iranicaonline.org).

The Islamic Lunar [H; Hijra] and the Persian solar [Sh./Khayyāmi] calendars are converted into the Christian calendar or so-called Common Era [CE] using Wüstenfeld and Mahler (revised by Qureshi, 1982) to avoid religiously determined calendar designations. Where the conversion is given, the Muslim and/or Iranian years are separated from the Common Era by a slash [1292/1875]; the Common Era has a larger digit. The Muslim lunar date [H.] is given up to the end of the nineteenth century, when the Common Era date is given. The Persian solar date [Sh./Khayyāmi] is only given since 1900 (if at all). These dates are given to help locate historic accounts in Arabic and/or Persian sources.

Unlike the Christian or Common Era and planetary weekdays, which begin at midnight, the Islamic and Jewish weekdays begin at sunset. Therefore, the *shab-e-Ādineh* in Persian, or *laila al-Jum'a* in Arabic [lit., "the night of Friday"] means the night preceding Friday, i.e., Thursday evening/night on the civil calendar. *Jum'a shab* means "Friday night." Furthermore, Muslim writers, following the ancient Zoroastrian tradition, consider times during the day relative to sunrise or daytime prayers: the Zohr [*nimruz*, "noon"] and the 'Asr [*pasin*, "afternoon"] prayers. Times at night are considered relative to nighttime prayers: the Maghrib ["sunset"; immediately after the sunset], 'Ashā ["nightfall"; end of evening twilight], or Fajr ["dawn"; start of morning twilight, prior to sunrise] prayers. These terminologies have been converted to local time in the tables. Dates are rendered in the tables as "year.month.day," and in the text as "day month year."

FINAL NOTE

This work includes systematic field studies of earthquakes since 1971 organized by myself alone while I was working at the Tectonics and Seismotectonics Department of the Geological Survey of Iran which I personally established; I had a limited field budget and low salary and was simultaneously teaching at the Universities of Tehran and Tarbiat Modarres. What followed was a long period of desk studies without sponsorship, research grants, or support from any organization from Tehran, Cambridge, UK, or the United States. The loss of precious lives and the profound impact of earthquakes on the social and economic life of survivors have always been the dark chapters of this work, as was living and working in an Iran under siege, with rationed food and fuel, during the 8 years of the disastrous Islamic Iran–Iraq war. Despite all these circumstances, I have tried to keep earthquake research alive in Iran and to collect field data prior to their disappearance.

Carrying out field work was particularly difficult during wartime in light of shortages of food, automobile fuel, and daily bombardments of the cities, especially Tehran.

In my work, I have tried to concentrate on the Iranian plateau. I am most familiar with this region, as it is where I have spent 45 years of my life—the last 43 dedicated to studying different aspects of earthquakes and coseismic surface faulting, through both bright and dark times. The data and discussion provided in this work will definitely increase our ability to identify active faults: their patterns, geomorphology, earthquakes, and risk minimization in other regions of the world. The conclusions also provide useful insights for earthquake engineers, planners, and earthquake-fault risk management. I also discuss how we failed to identify beforehand the active geomorphological indicators of the 1978 Tabas-e Golshan, 1990 Rudbār, 2003 Bam, 2010–2011 South Rigān, and 2012 South Ahar earthquake surface ruptures.

In a work like this, composed from an abundance of published papers and books, unpublished documents, internal reports, and press documents in Persian, Armenian, Assyrian, English, Arabic, and Turkic languages, there is no doubt that many errors will be found. Furthermore, I might have overlooked other published data or missed some references. I would be most grateful, therefore, for any corrections, or notices of omissions or misquoting, that could help eventually produce a complete database that is as accurate and reliable as possible.

The cases discussed in this work will have a profound and direct application in earthquake hazard assessment and risk-minimization efforts on the Iranian plateau as well as other active regions of the world. Knowledge of past earthquakes and active faulting should help the authorities to extrapolate information from the long-term earthquake-faulting history and its effects on social life and the economy. My hope is that this first stage in a long, challenging process will rekindle and enhance the interests of governments throughout the world in natural and environmental risk-minimization project, with special attention given to earthquakes.

I am confident if a more sophisticated instrument is built,
A few minutes after the needle's anomalous move,
The earthquake will be felt.
And if the system is connected to a big bell
[an alarm system], it can be heard by all the people,
and their lives will be saved.

Yusef Telegerāfchi Kermāni (1909), in Berberian (2013)

Manuel Berberian
Ocean County College
and
Onduni Grung Scientific Enterprise
Toms River, New Jersey, USA
2014

Acknowledgments

What is Past is Prologue.

William Shakespeare, *The Tempest*, Act 2, Scene 1 (1610)

This book grew out of my love and passion for the Iranian Plateau, its inhabitants, and their ancient civilization. My hope is that it will help reduce the average number of earthquake casualties each year, which is about 1500. I dedicate this work to the memory of the pioneering efforts of professor Setrāk Ābdāliān (1894–1963), Nasollāh Khādem (1910–1999), Safi Asfiā' (1916–1999), Heinrich Martin Huber (1917–1992), Jovan Stöcklin (1921–2008), and Nicolas Neocles Ambraseys (1929–2012); to John Sebastian Tchalenko, Ārek Megerdichian, and 'Ali Akbar Mo'infar; to the more than 164,000 Iranians who have lost their precious lives in earthquakes since 1900 in a country rich in oil and natural gas resources; to the more than 25,000 Armenians who perished during the 1988 Spitāk earthquake; and to all those who seek freedom, happiness, 'we the people', and a republic.

I thank my wife, Rose, and my son, Sam, who tolerated the mess I created at our homes in Tehran, Cambridge (UK), and the United States with maps, books, reports, papers, and notes spread out in almost every room (including the kitchen); they were also tolerant of the time and expense I have spent on my research work since 1971. My family and I have paid a high price for my work, values, and beliefs; we have suffered during the calamitous 8-year Islamic Iran-Iraq war (with daily bombardment of the major cities, especially Tehran where we lived) as well as a long period of unjust treatment and discrimination as an ethnic and religious minority. I am therefore grateful for their unrelenting support, love, and dedication during all these years.

I am grateful to Nicolas Ambraseys, Emmanuela Guidoboni, George A. Bournoutian, the Very Reverent Father Krikor Maksoudian, Basil Collins, Robert Thomson, Murād Menenshāin, Frances Pritchet, Howard Schwartz, Håkan Wahlquist and Anne Murray (the Sven Hedin Foundation, Stockholm; http://svenhedinfoundation.org), Lars Larsson (the Sven Hedin Project; svenhedin.com), Oxford Journals (Oxford University Press), Mehrdād Bahār, Fereidun Joneidi, Hossein Nasr, Yahyā Zokā', and Ezatollāh Negahbān for permission to use their publications. Prince Mickey Kādjār (Qājār) kindly helped translate the French texts. Fereidun Biglari, Ahmad Kabiri, Jesse Aratunian (Balal), and Kāmrān Ansāri assisted in my proxy search by providing information during numerous occasions from Tehran. My sister Mary in New York city, NY, helped find Armenian sources. I also thank Robert Yeats (Oregon State University, Corvallis, OR) for numerous discussions on the earthquakes and active faulting of different countries.

Thanks are also due to Roger Bilham (University of Colorado, Boulder, CO), Susan Dennis (acquisitions editor, Elsevier, Oxford, UK), Derek Coleman (senior editorial project manager, Elsevier, Amsterdam, the Netherlands), John Shroder (University of Nebraska at Omaha, NE, editor, Elsevier), Mohanapriyan Rajendran (Elsevier, Chennai), and Elsevier B.V. for publishing this work.

This book was not supported by any grant or organization; all expenses, including my time and the drafting of figures, were paid from my family's pocket. Catherine Lane drafted the figures.

Finally, I cherish the memory of my parents' struggle that unjustly faced unreal circumstances, punished by 'the crime of silence', and had to go through the hell in their lifetime.

"If you are coming to see me,
I am beyond the Nowhereland."

Sohrab Sepehri, 1964

Manuel Berberian
New Jersey, USA
2014

Introduction

The past is never dead,
It's not even past.

William Faulkner, *The Sound and the Fury*, Act 1, Scene III, 1929

How does the crust of the Iranian plateau respond to the active tectonic processes resulting from the Arabian–Eurasian collision and northward motion and indentation during earthquakes?

How does strain temporally and spatially accumulate and release over various time scales on the Iranian plateau?

Is the strain accumulation and release along the active faults constant or nonconstant in time and space on the Iranian plateau?

What are the characteristics of the earthquakes and coseismic surface rupture patterns on the Iranian plateau?

How does the active faulting accommodate the present-day velocity field, and do the fault patterns on the Iranian plateau change over time?

When and where have the medium- to large-magnitude earthquakes occurred on the Iranian plateau, and what were the seismic parameters and locations of past earthquakes gathered by social, archeological, epigraphic, literary, seismological, remote sensing, and field sources?

Why, unlike in Central Iran and the Alborz, and despite having an almost identical seismogenic thickness, haven't the faults in the Zāgros fold-and-thrust belt generated large-magnitude ($M_w > 7.0$) earthquakes?

Why are the major urban and rural areas on the Iranian plateau built on or in close proximity to active surface faults or blind thrusts?

Why have the medium- and large-magnitude earthquakes on the Iranian plateau, even the most recent event (the 2003 M_w 6.6 Bam), been catastrophic and brought complete devastation to rural and urban areas, excessive infrastructure damage, and unfathomable casualties such as those documented only in very poor countries such as Afghanistan or Haiti?

What are the exclusive geodynamic, active tectonics, and geopolitical processes that cause earthquakes on the Iranian plateau to become national and regional catastrophes?

☆"To view the full reference list for the book, click here"

Are there any lessons to be learned from the most comprehensive earthquake history in the world that has captured the events of the Iranian plateau for the past two millennia?

As geologically analogous to other seismic parts of the world, such as California, New Zealand, Japan, China, Afghanistan, Pakistan, the Caucasus, Asia Minor, Venezuela, the Andes, and the Atlas Mountains, can the active tectonics data of the Iranian plateau be utilized in other seismically active countries?

Finally, can this study contribute to our scientific understanding of earthquakes and active faulting in the region as well as in other active parts of the world?

Hoping to reduce seismic risk in this unfortunate part of the world, this book tries to answer some of these questions yet raises more questions than it resolves. The impetus for writing this book was the absence of a single publication covering the coseismic surface ruptures and patterns of historical seismicity as well as the archeoseismicity and active tectonics on the Iranian plateau. This book is the result of a lifetime devotion to understanding the historical earthquake fault and seismicity patterns of the Iranian plateau, as well as an unfortunately unsuccessful attempt to reduce their risks.

HISTORICAL PERSPECTIVE

It should be noted that the history of earthquake geology, seismo- and active tectonics, and seismic risk minimization on this part of the planet goes back only to 1951, when Professor Setrāk Ābdālian, an Armenian migrant naturalized Iranian citizen (1894–1963), wrote about earthquake risk in the capital city of Tehrān (Abdalian, 1951). He began to practice earthquake geology in the field by studying coseismic ground deformation associated with the 12 February 1953 M_w 6.5 Torud earthquake in the northern part of the Central Desert (Abdalian, 1953) and prepared a preliminary seismotectonics map of Iran in 1961 and 1963 (Abdalian, 1961, 1962, 1963a). The 1953 Torud earthquake study was carried out a long time before the establishment of the Institute of Geophysics, University of Tehran (1957), and the Geological Survey of Iran (1962). Ābdālian's 1953 earthquake study was conducted during political/petro-turmoil in the country during the last months of the premiership of Dr. Mosaddeq, Esq., whose democratically elected government was overthrown six months after the earthquake on 19 August 1953 coup d'état (Abrahamian, 2013). Ābdālian also studied the 1 September 1962 M_w 7.0 Bu'in earthquake and published his field observations just before his death in 1963 in Tehrān (Abdalian, 1963b).

After the 2 July 1957 M_w 7.1 Band-e Pay earthquake in central Alborz Mountains V. Vrolyk (1957), chief of the Central Civil Protection Service at Algiers, was sent on a mission to study the disaster. The study was followed by C.D.W. Savage (1957) and a year later by Hagiwara and Naito (1959). In 1958, Francesco Peronaci (Geophysical Institute, Rome) and T. Hagiwara and

T. Naito (1959) studied the 1958 earthquake along the Zāgros Main Recent fault. Peronaci (1958b) published a list of Iranian earthquakes that occurred from 23 January 1909 through 13 December 1957. He also prepared a seismicity map of the country divided into five magnitude classes of a: 7.7–8.8, b: 7.0–7.7, c: 6.0–6.9, d: 5.3–5.9, and e: 5.3. No faults were added to his seismicity map.

A few days after the 1962 M_w 7.0 Bu'in earthquake—which killed 12,200 people—at the request of the Imperial Government of Iran, the United Nations Educational, Scientific, and Cultural Organization (UNESCO) sent two scientific and engineering missions to study the earthquake: (i) Nicolas Ambraseys and (ii) J. Despeyroux and M. Lescuyer. Ambraseys arrived at Tehrān on 25 September 1962 and on 29 September departed for the epicentral area for a 20-day field study; he was accompanied by architect D.A. Qāssemlu (Construction Department of the University of Tehrān). J.P. Rothé and J. Gostoli (Strasbourg) also prepared a preliminary seismological study of the event.

As an aftermath of the 1962 earthquake, Dr. E. Michael Fornier d'Albe, director of UNESCO, convened an educational workshop at the University of Tehrān with participation of Professor Tachu Naito (Waseda University, Japan; 1886–1970), Professor Takuji Kobori (Kyoto University, Japan; 1920–2007), Professor Setrāk Ābdālian (University of Tehran), and many others. During this meeting, a preliminary report of the first Iranian code for seismic-resistant building design was discussed. The code, which was based on the San Francisco, California, and U.S. Uniform Building Code, was prepared by 'Ali Akbar Mo'infar, Ārék Megerdichian, Kuros Āmuzégār, and Javād Sālehi, among others, and was approved and implemented later in 1969 as the first Iranian Standard ISIRI, Code No. 519. At this time, there was absolutely no knowledge of active faults, seismic history, or morphotectonics on the Iranian plateau. Except for the Zāgros fold-and-thrust mountain belt of the southwest and south Iran (which was systematically mapped by the Anglo-Persian and then Anglo-Iranian Oil Company for oil and natural gas explorations), the country was not geologically known or mapped.

After the 31 August 1968 M_w 7.1 Dasht-e Bayāz earthquake, with a death toll of over 10,000, the second UNESCO scientific and engineering mission (N.N. Ambraseys, J.S. Tchalenko, S.B. Bubnov, T.S. Tassios, S. Crampin, G. Anderson, and M. Shahidi) arrived to study the disaster area. For the first time, the National Cartographic Center of the Imperial Government of Iran took postearthquake aerial photographs of the damage zone at a scale of 1:7500 which provided a valuable tool for detailed mapping of the coseismic surface ruptures (unfortunately, such steps have not been taken since then). During this time, a joint project among the Imperial College of Science, University of London, and Technical Bureau of the Plan and Budget Organization of the Imperial Government of Iran was begun. Accordingly, Mohammad Hassan Banisadr, Rezā Aryan (both from the Plan and Budget Organization), and Mortezā Seyed Nabavi (Institute of Geophysics, University of Tehran) studied in London.

Ambraseys led additional UNESCO mission field trips after the earthquakes of Karnāveh (1970, with J.S. Tchalenko), Kārzin (1972, with J.S. Tchalenko), Nāghān (1977; a CENTO mission), and Dartangal (1979, with M. Arsovski). Much of Ambraseys's field work in Iran during those 16 years was carried out in the company of structural engineer 'Ali Akbar Mo'infar (Plan and Budget Organization of Iran), who made the field trips possible. In 1973, Mo'infar established the first strong motion network of Iran (which in 1981 was transferred to the Housing and Urban Development Research Center, Ministry of Roads and Urban Development). John Tchalenko also travelled to Iran in 1968 (Dasht-e Bayāz earthquake), 1970 (Karnāveh earthquake), 1971 (Tehrān-Alborz active tectonics, with Mohammad Hossein Irānmanesh and me from the Geological Survey of Iran), 1972 (Kārzin earthquake, and the Khidbas bedrock section of the Dasht-e Bayāz fault mapping with me), and 1973 (the 1930 Salmās coseismic surface fault rupture mapping with me).

In 1971, I became frustrated by the lack of studies of seismotectonics, active faulting, and tectonics in Iran and established the Research Department of Tectonics and Seismotectonics at the newly UN-established Geological Survey of Iran. The aim was to study and improve our knowledge of active faulting, coseismic surface ruptures, earthquake history, and tectonics of the Iranian plateau, subjects that had long been neglected or left to foreign experts despite Iran's being an earthquake-prone country. At that time, there was no knowledge of active or paleo-tectonics, seismic faulting, earthquake history, or plate tectonics of the Iranian plateau. Subsequent systematic field investigations, studies, and contributions provided by the new (and understaffed) Department became invaluable in advancing our understanding of active tectonic processes as well as in helping design numerous national projects. As Robert Yeats (2012) wrote, my work is responsible for much that we know about the earthquake geology and tectonics of Iran. In addition to field study of the Iranian earthquakes and tectonics, I began studying archeology and the ancient history of Iran in order to expand our knowledge of Iranian earthquakes and active faults and their hazards.

During the 1974 site selection studies for nuclear power plants that was conducted by international consulting engineers, I was hired by Dames and Moore International, S.R.L. (for site selection studies along the Persian Gulf), and D'Appolonia Engineers (for the southern Caspian Sea in northern Iran). These comprehensive field studies generated important data on historical and modern earthquakes and active faulting and tectonics. On 5 December 1974, while waiting for my flight to the Persian Gulf area, the roof of the Mehrābād International Airport at Tehrān collapsed under normal winter snow load, killing 17 and injuring tens of passengers. Barely escaping a meeting with death at the age of 29, I heard the "silence of God" and experienced a sneak preview of the "danse macabre"—an experience I would repeat during the 1980–1988 Iran–Iraq war (but this time with my wife and our baby boy during heavy bombardment of Tehran).

In January 1975, the Earthquake Risk Assessment Section, working under the Nuclear Safety Division of the Atomic Energy Organization of the Imperial Government of Iran (AEOI), was established to help site selection investigations for 20 nuclear power plants. ‘Ali Akbar Nowroozi and Arsalān Mohājer-Ashjai became involved in the seismological aspects of this project. They investigated a few earthquakes that had taken place in the late 1970s. The Section also installed and operated three earthquake monitoring networks in Bushehr (at a radius of 100 km around the nuclear power plant) and around the cities of Esfahān (central Iran) and Tabriz (NW Iran). Such activities were halted during the regime change in 1979, and experts fled the country.

After five years of research, in 1976 I published a series of basic color maps on the active faulting, pre-1900 and post-1900 seismicity, and seismotectonics of the country along with a 519-page report in English in a series entitled, “Contribution to the Seismotectonics of Iran.” Due to lack of computer facilities, I painstakingly plotted by hand almost 1800 epicenters from 1900 to 1975 on a 1:2,500,000-scale, fault map of the county. (If I had been fully aware of the large errors in the teleseismic epicenters, I would not have wasted my time on the painstaking work of plotting the epicenters.)

The coseismic surface fault mapping and aftershock sequence recording by portable seismographs and strong motion recorders of the 16 September 1978 M_w 7.3 Tabas-e Golshan earthquake, which killed 20,000 people, was conducted in September and October 1978. At the time, widespread social unrest was unfolding throughout the country: well-organized demonstrations and protests were being carried out by the mullahs and the mobs in nearly every village in the epicentral area. Despite the difficulties, I managed to map the coseismic surface rupture and record the aftershocks. However, I was forced to abandon the expedition, collect all the seismographs and strong motion recorders, and immediately depart from the camp tents because of serious threats made on my life from local religious leaders (though their mobs). The seismographs and strong motion recorders were confiscated at the airport. They were finally released to me, but the films of the strong motion recorders had been extracted from the instruments and exposed to light! Although it was very difficult and dangerous, I am glad that I was able to collect and salvage some data.

Despite the rationing of automobile gasoline and food during the devastating 1980–1988 Iran–Iraq war, I tried to continue systematic mapping of the surface ruptures associated with the 11 June 1981 M_w 6.6 Golbāf and 28 July 1981 M_w 7.0 Sirch earthquakes as well as to map the active faults and earthquake-fault hazard studies of the Greater Tehran and Qazvin quadrangles (the majority of the active faults are now covered by buildings constructed since 1980). During this very dark time, I studied in the field and collected macroseismic data for the 1983 M_w 5.5 and 5.4 Bāijān, 1983 M_w 5.5 Charazeh, 1985 M_w 6.1 and 6.2 Nomal, 1988 M_w 5.9 Tashān, and 1988 M_w 5.5 and 5.8 Doshman Ziāri earthquakes.

The two difficult years after the war, with an economy in shambles and rationed fuel and food, I mapped the coseismic surface faults of the 1989 M_w 5.8 South Golbāf and 1990 M_w 7.3 Rudbār earthquakes. When reminiscing about the research work that I have carried out since September 1978, I see that the studies were conducted under surreal, unconventional, and abnormal circumstances with numerous hindrances. Had these investigations not been carried out, the data on the surface ruptures as well as macroseismic information would have been lost.

As the founder of the Tectonics and Seismotectonics Department of the Geological Survey of Iran, in 1994 I invited my colleague James Jackson of the University of Cambridge to sign onto a joint research project to study the active tectonics of the Iranian plateau and train young Iranian as well as British geologists and research students in the field. This cooperation resulted in a better understanding of seismicity, active tectonics, and surface geomorphological features associated with active faulting of the Iranian plateau. I am pleased that I took this final step for the benefit of the science and the nation and that I have tried to save the people of Iran from earthquakes.

SYNOPSIS: ROOT CAUSES OF THE SEISMIC PAST, PRESENT, AND FUTURE

The Iranian plateau has been frequently struck by catastrophic earthquakes along its active faults—with unfathomable death tolls and destruction—that have been historically documented (a complete list of the pre-1900 and 1900–2013 earthquakes and earthquakes associated with surface fault break are presented in Tables 14.9–14.11 (Chapter 14) and 17.10–17.12 (Chapter 17)). These earthquakes have destroyed agricultural and industrial lifelines and infrastructure, disrupting rural and urban areas. Some historically documented disasters have resulted in repetitive damage throughout the long history of the Iranian and Armenian civilizations (Hakobyan, 1951, 1956; Berberian, 1976a, 1977a, 1981a, 1991, 1994, 1997a, 2005; Ambraseys and Melville, 1982; Babayan, 2006). Throughout the long history of Iran and Armenia, earthquakes have spent national resources, forced the decline of ancient settlements, and resulted in man-made catastrophes in the region.

Traces of the effects of earthquakes on people's minds and souls have survived in the present-day oral traditions, superstitions, mentality, habits, customs, linguistic traces left in the names of mountains and plains (Chapter 1), and historic literature (Chapter 6). These traces are firmly rooted in the rural people, who still stick to their traditional beliefs and are closer to nature because of living in greater harmony with their environment. It is impossible to understand the rural people and their everyday fears of disease, drought, crop failure, famine, evil eyes, flood, storms, thunder and lightning, and earthquakes unless one studies their curious beliefs and practices, which have crept in through numerous religious faiths and long-term occupations throughout the long history of their ancient civilizations.

Direct reference to the earthquakes on the Iranian plateau—including the Greater Iran and Greater Armenian Highlands—is nonexistent or seldom seen in the extant Zoroastrian Avestan and the Pahlavi (Classic Middle Persian) texts (Chapter 2). In ancient times, demons and monsters, and later gods, were judged to be responsible for having created earthquakes (Chapters 2 and 3). Hence, the traditional rationalization of ancient natural phenomena improvised ancient earthquake theories (Chapter 4), folklores, legends, and traditions that evolved into myths (Chapters 2–5), epic literature (Chapter 6), poems with chronogrammatic verses preserving the dates of earthquakes and their impacts (Chapter 7), and inscriptions preserved in historical monuments (Chapter 8).

Although rare, there are more references to earthquakes in the Armenian, Persian, and Arabic chronicles written in the past two millennia. With very few exceptions, historical earthquakes in the chronicles and treatises are generally referred to in short and apocalyptic phrases buried in long texts of literary, philosophical, geographic, historic, religious, dynastic, and political information dedicated to the patrons, governors, and kings (Hakobyan, 1951, 1956; Heuckroth, and Karim, 1970; Nabavi, 1972; Ambraseys, 1974a; Berberian, 1976a, 1977b, 1991, 1994, 1997a, 2005; Ambraseys and Melville, 1982; Berberian et al., 1983, 1985, 1996, 2000a; Guidoboni and Traina, 1995; Ambraseys and Bilham, 2003a,b; Babayan, 2006).

Furthermore, little discussion or understanding about the mechanisms, causes, nature, or interpretative theories of earthquakes occurs in the ancient Iranian and Armenian treatises. The earthquake has been generally considered as a predetermined and inevitable calamity—divine punishment due to the heavy sins of the people that should be appealed through prayers, pleas, vows, and sacrificial offerings (Chapter 3). Nonetheless, occasional rational observations and viewpoints have been expressed by a few individuals; they are addressed in Chapter 4.

The diffuse plate boundary of the Iranian plateau, with complex interactive earthquake sequences along active strike-slip and reverse faults, is analogous to other intracontinental active regions such as Southern California, New Zealand, Venezuela, and Argentina. However, unlike those places, the 2000-year-long historical seismic data for the Iranian plateau show temporal and spatial clustering of intense seismic activity of large-magnitude earthquakes within very short periods of time. Throughout this book, I use the patterns of historical seismicity and coseismic faulting (Chapter 16), as well as their spatial and temporal distributions, to show that the Iranian plateau data are valuable for providing unique, critical, and adequate information regarding other active areas on the planet.

Like the Iranian plateau, Southern California is cut by many active strike-slip and reverse faults, with active blind faulting and surface folding. The main difference is the throughgoing San Andreas fault, which has generated earthquakes of $M_w \sim 8.0$; this has no counterpart on the Iranian plateau. Historical earthquakes have occurred off the San Andreas fault with $M_w \geq 7.0$ that resemble the character of Iranian plateau earthquakes, including the 28 June 1992 M_w 7.3 Landers earthquake (Sieh et al., 1993), the 21 July 1952 M_w 7.3 Arvin-Tehachapi, Kern County, earthquake (Buwalda and St. Amand, 1955; Stein and Thatcher, 1981), and the 4 November 1927 M_s 7.0 Lompoc earthquake at the western end of the Transverse Ranges of Southern California (Helmberger et al., 1992; Satake and Somerville, 1992). However, active faults in and adjacent to the heavily populated regions of the Transverse Ranges, northernmost Peninsular Ranges, and the Los Angeles Basin have generated only three moderate-magnitude earthquakes, all with $M<7.0$. These are the 10 March 1933 M 6.4 Long Beach earthquake (Harding, 1973; Yeats, 1973; Hauksson, 1987) and the two San Fernando Valley earthquakes of 9 February 1971 M_w 6.5 Sylmar (Sharp, 1975) and 17 January 1994 M_w 6.7 Northridge (Mori et al., 1995; Tsutsumi and Yeats, 1999). The unique pattern of historical earthquakes on the Iranian plateau (Chapter 16) can be applied to California, New Zealand, Venezuela, Argentina, and other earthquake-prone countries.

The several thousand years of earthquake history of the Iranian plateau (Chapters 10–17) suggest that earthquakes of $M\geq7.0$ are to be expected in the Los Angeles metropolitan region, similar to those that have occurred in northwest Nelson (New Zealand), northwest Argentina, Japan, Pakistan, Afghanistan, Caucasus, Asia Minor, and Greece (Berberian and Yeats, 1999, 2001; Yeats, 2012). The unique and nearly complete data presented in this study, especially the historic (pre-1900) as well as modern earthquakes and coseismic faulting (Tables 14.9–14.11 and 17.10–17.12) can be used in empirical-relation analyses combined with the other world data, considering that the constrained centroid depths by waveform modeling are factored in.

Since 1900, the Iranian plateau has experienced about 17 earthquakes with $M\geq7.0$ (at least approximately one $M\geq7.0$-magnitude earthquake every 6.6 years); more than 100 earthquakes with $6.0<M<6.9$ (one M 6.0–6.9 earthquake every 1.1 years); and more than 180 earthquakes with $5.5<M<5.9$ (one M 5.5–5.9 in every 7.5 months), culminating in a very large death toll averaging 1577 persons/year (Figure 1). Since 1900, earthquakes in Iran have killed more than 164,000 people (Berberian, 2005; Table 17.12).

During this period, the largest earthquakes seem to be related to the Makrān subduction zone (27 November 1945 M_w 7.9 and 16 April 2013 M_w 7.7) of southeast Iran. The latter event was an intermediate-depth slab earthquake, along the Iran–Pakistan border (Barnhart et al., 2013), associated with the subduction of oceanic crust underneath the Makrān accretionary prism.

The largest intracontinental strike-slip earthquake on the Iranian plateau had a magnitude of M_w 7.4 (23 January 1909, Silākhor, >8000 killed), whereas the largest reverse fault earthquake had a magnitude of M_w 7.3 (16 September 1978, Tabas-e Golshan, ~20,000 killed). No large-magnitude

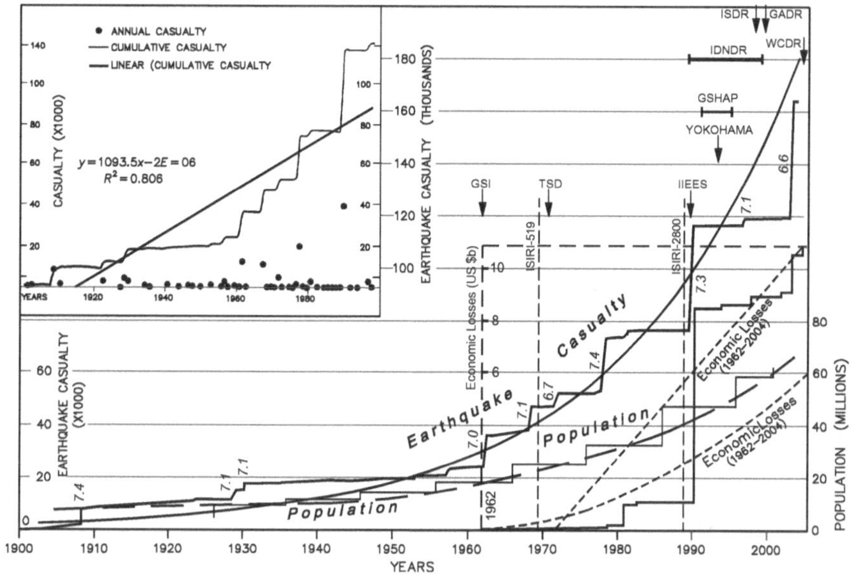

FIGURE 1 Cumulative diagram of the explosion of Iran's large earthquake casualties and population since 1900, with the available minimum uninsured economic losses provided in original values since the 1 September 1962 M_w 7.0 Bu'in earthquake (see Table 17.12). The italicized number on the cumulative death toll diagram denotes the earthquake moment magnitude. Important Iranian (ISIRI Code Nos. 519 and 2800) and International earthquake risk minimization milestones are added to the diagram with arrows. IRISI-519: 1969 Iranian Code for Seismic Resistant Design of Buildings. IRISI-2800: 1988 Iranian Code for Seismic Resistant Design of Buildings. GADR: Global Alliance for Disaster Reduction (2001 onward). GSHAP: UN/IDNDR/ICSUC Global Seismic Hazard Assessment Program (1992–1996). GSI: Geological Survey of Iran (Tehran, 1962 onward). IDNDR: United Nations International Decade of Natural Disaster Reduction (1900–2000). IIEES: International Institute of Earthquake Engineering and Seismology (Tehran, 1989 onward). ISDR: International Strategy for Disaster Reduction (1999). TSD: Tectonics–Seismotectonics Department of the Geological Survey of Iran (Tehran, established by Berberian in 1971). WCDR: World Conference on Disaster Reduction (January 18–22, 2005). Yokohama: Yokohama Strategy and Plan of Action for a Safe World, Guidelines for Natural Disaster Prevention, Preparedness, and Mitigation and its Plan of Action, UN (1994). Between 1993 and 2001, about 600,000 dwellings (5% of Iran's total housing stock) were severely damaged or destroyed, causing US$10 billion in damage. From 1999 to 2001, government subsidies to homeowners affected by disasters exceeded US$1 billion (1.3% of GPD). *Modified from Berberian (2005). For population data see the text.*

earthquake has affected the metropolitan area of the capital Tehrān or the large provincial capital cities such as Tabriz, Ardebil, Zanjān, Qazvin, Karaj, Qom, Rasht, Sāri, Gorgān, Bojnurd, Mashhad, Birjand, Kermān, Zāhedān, Kermānshāh, Bushehr, Ahvāz, Shirāz, and Bandar 'Abbās—that is, at least since 1900. The seismic hazards and risks are high—and increasing—in these areas, with increases of population and urban growth. The destruction of Tehrān and other provincial capital cities, with unreinforced crucial public buildings (schools, hospitals, fire and police stations, detention centers), would paralyze the government and the nation.

The oasis city of Bam (with no recorded seismic history) and its inhabitants were buried in a matter of seconds during the 26 December 2003 M_w 6.6 earthquake; the death toll ranged from 32,000 to 43,200. A similar case is the oasis city of Tabas-e Golshan during the 16 September 1978 M_w 7.3 earthquake, which killed 20,000; the city had no recorded history of seismicity during the past millennium (Berberian 1979a,b, 1982, 2005). The four towns of Rudbār, Manjil, Harzévil, and Lowshān were totally demolished during the 20 June 1990 M_w 7.3 earthquake, and about 40,000 people lost their lives (Berberian et al. 1992; Berberian and Walker, 2010). The 2003 M_w 6.6 Bam urban earthquakes was: (i) 26–36 times more fatal than any previous M_w 6.6 rural event in Iran; (ii) 6.4–8.6 times more fatal than the 1972 M_w 6.7 Kārzin rural earthquake; and (iii) as fatal as the 1990 M_w 7.3 Rudbār urban–rural earthquake (Figure 1).

These losses cannot be justified in light of the existing scientific and engineering knowledge and expertise in disaster management, which is available had anyone thought to use it. These are examples of the failure of earthquake risk reduction in a rich developing country. Such disasters could have been averted by governments since 1970 if the credible earthquake-resistant building codes had been implemented and enforced; if four decades of government-permitted but unsupervised construction had not been the norm; and if the authorities had taken responsibility for the safety of their citizens. These disastrous cases also clearly prove that the Iranian code for seismic-resistant building design (ISIRI Code Nos. 519 dated 1969 and 2800 dated 1988; BHRC, 1988, 1999) has little effect on the country (Figures 1 and 2; Table 17.12). It is particularly painful to consider that hundreds of millions of people living on the Iranian plateau (including the Caucasus, Turkmenistan, Afghanistan, Pakistan, Iran, Iraq, and Asia Minor) are trapped in a "vicious circle" of earthquake vulnerability. Unfortunately, little has been done to reduce the risk of earthquakes in the urban and rural areas of developing countries (cf., Bilham, 2009; Ambraseys and Bilham, 2011; GP-DRR, 2013).

Earthquakes in the Iranian plateau show a nonuniform distribution concentrated within the active folded-faulted mountain belts surrounding the relatively aseismic, less-deformed rigid and stable blocks (Chapter 9). Since the time of the 1 September 1962 M_w 7.0 Bu'in earthquake in Iran, there have been about eight earthquakes of $M_w \geq 7.0$, and the minimum economic losses of the Iranian earthquakes have exceeded US$ 10.6 billion (Berberian, 2005; Figures 1–3; Table 17.12). These destructive earthquakes represent a mix of urban and rural events in different geological environments, with levels of documentation regarding actual loss estimates, socioeconomic, and financial impacts varying from one event to another. Unfortunately, such documentation is not available for most of the Iranian, Armenian, Pakistani, and Afghan earthquakes.

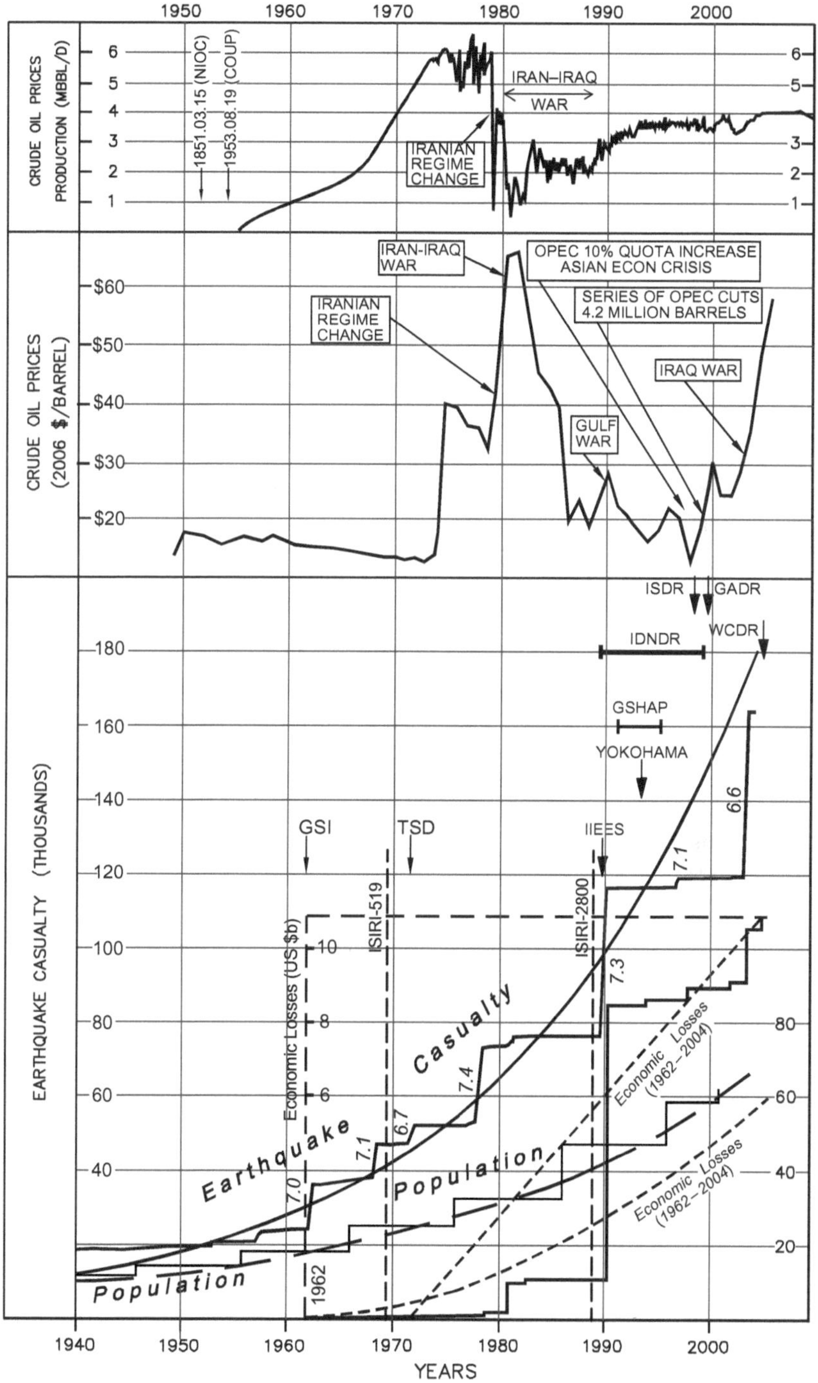

FIGURE 2—CONT'D

Like many urban areas in developing countries, earthquake risks have increased significantly in Irān since the mid-1900s (Figures 1 and 3). Until possibly the early 1960s, the rate of investment in large urban and industrial developments in Iran was minimal, at best. Although earthquakes had taken more than 24,750 lives during the first 60 years of the twentieth century, damage in economic terms had remained relatively low. However, with the massive investments of the last five decades in new urban and industrial centers, future earthquakes in the region are likely to result in serious economic loss and possibly government and political change. There has been a 3.5-fold increase in the Iranian population since 1956, and a twofold increase since 1976 (Bharier, 1968, 1972; Zandjani, 1977, 1992; Zandjani and Raḥmani, 1989; Hourcade, 1994; SCI, 2004–2013; see Chapter 17). Population growth has been concentrated in the earthquake-prone mega-city of Tehrān and in the large provincial capital cities throughout the country. At the present time, an estimated 12–15 million people are living in the vicinity and on top of active faults of Tehran, dramatically increasing the seismic risk (Berberian and Yeats, 2014).

The Iranian urban population increased from 5.953 million in 1956 to 9.794 million in 1966, 15.854 million in 1976, 26.844 million in 1986, and 31.836 million in 1991. The number of cities has increased from 199 in 1956 to 514 in 1991 (Hourcade, 1994; iranicaonline.org). The continuous high rates of population growth, longer life expectancy, and migration from the rural areas to the cities have a direct effect on the increasing risk of earthquakes in the cities. Unlike Japan, Chile, or California, the risk is increasing rapidly because of another reason: the aseismic building codes have not been enforced by the authorities in these cities, buildings are built on top of active faults (Plates 1 and 2), and there has never been accountability for the destruction and deaths following an event, even after the city of Bam was leveled to the ground during the 2003 earthquake. Buildings are poorly constructed, and the corrupt building industry and inspectors (if any) in the cities have approved constructions built with low-quality materials and poor workmanship in major earthquake-prone cities with a documented history of devastating earthquakes (see Chapters 11–17).

During the late nineteenth and the early twentieth century, concern over earthquakes and reducing their effects increased in Japan [after the 11 November

FIGURE 2—CONT'D Comparison of Iranian crude oil production and revenues with international crude oil prices, as well as the time trends of earthquake death tolls. Despite astronomical oil revenues, Iran has not developed or implemented any plans to create earthquake hazard minimization or warning alarm systems (see also Berberian, 2013). Modified from U.S. Energy Information Administration, Independent Statistics and analysis (eia.gov/cabs/iran; tonto.eia.gov), Ziff Energy Group (ziffenergy.com), WTRG Economics (wtrg.com/oil), and mongabay.com. Top diagram: Iran's crude oil production and revenues from petroleum, with total oil supplies in million barrels per day (2006 US$). NIOC: National Iranian Oil Company; Coup: 1953 Coup d'état. Middle diagram: International oil prices (in 2006 US$), with important national and international crises. Lower Diagram: Time trend of earthquake death tolls since 1940 (see Figure 1 for complete picture).

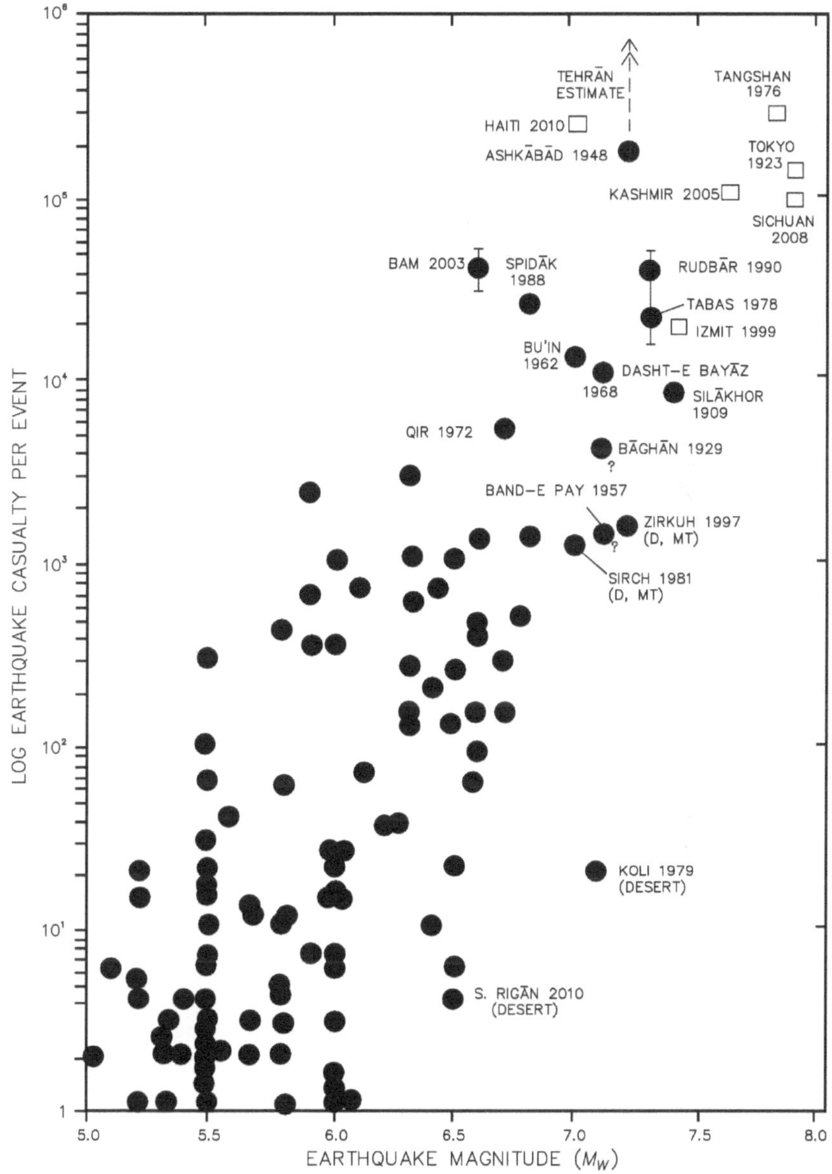

FIGURE 3 Recorded earthquake casualties (log) versus magnitude (M_w) for medium- to large-magnitude earthquakes in Iran from 1900 to 2013. Death tolls of some early earthquakes are unknown and are not included here. Major urban earthquakes outside Iran are shown by blank squares. D: Desert with remote, sparsely populated regions in arid to semiarid areas. MT: Mountainous areas. For parameters of the events see Table 17.12.

PLATE 1 Apartment building built on a recent fault through Quaternary alluvial deposits in northern central Tehran. Photographed in 1983.

PLATE 2 Buildings in northern Tehran constructed on the active North Tehran Thrust, as well as on the hanging-wall block of the thrust fault. *Photographed in 2013 (Courtesy of R.Q.M. Tābān).*

1855 *M* 6.9 Ansei Edo (6600 killed); 28 October 1891 *M* 8.0 Mino-Owai (7300 killed); and 1 September 1923 *M* 8.3 Great Kanto (143,000 killed) earthquakes]; Italy [28 December 1908 *M* 7.1 earthquake (>100,000 killed)]; the United States [18 April 1906 *M* 7.9 San Francisco earthquake]; and Chile [the 22 May 1960 *M* 9.5 Valdiva earthquake]. Federal funding of earthquake research in the United States began in 1930 with the expenditure of about US$ 10,000.00 for support of seismology; this had increased to US$ 500,000.00 by

1957. By 1961, the total U.S. federal expenditure for seismological research had risen to almost US$ 30 million. Spurred by the 9 February 1971 M_w 6.6 San Fernando earthquake, Southern California, in 1977, the U.S. Congress established the National Earthquake Hazard Reduction Program with an allocated budget of US$ 55 million (NAS, 2003). During this period and even afterward, and despite the occurrence of numerous destructive earthquakes in Iran and other seismically active developing countries, sufficient steps have not been taken in oil/gas rich Iran to minimize the earthquake risk (Figures 1–3).

In Iran, the 1 September 1962 M_w 7.0 tragic Bu'in earthquake (12,200 killed) resulted in the preparation of the first national code for seismic-resistant building design, which was implemented in 1969 (ISIRI Code No. 519). In 1987, seismic zones map of Iran based on the active fault map of Iran was prepared by M. Berberian, M. Qorashi, A.A. Mo'infar, A.A. Zohurian Izadpanāh, and A. Nāderzādeh (Mo'infar et al., 1987). This map was used for the first time in the second Iranian code for seismic-resistant building design (ISIRI Code No. 2800). The revised building code was officially approved in February 1988. Unfortunately, despite steps taken by earthquake geologists, seismologists, and engineers, no serious effort was made to enforce the code throughout the country, especially in the capital city of Tehran.

The comparison of recent records on the impact of earthquake hazards of similar magnitude in countries of varying levels of development is astonishing (Table 1). During the 17 October 1989 M_w 6.9 Loma Prieta, California, earthquake, only 62 people were killed (Eberhart-Phillips et al., 1994; Vranes and Pielke, 2009), a very low casualty figure for an earthquake of such magnitude. Compare this figure with the 12,200 deaths during the 1 September 1962 M_w 7.0 Bu'in earthquake in Iran. The 22 December 2003 M_w 6.5 San Simeon, southern Santa Lucia Range, central coast of California, killed 2, injured 40, and severely damaged about 40 buildings (McLaren et al., 2008). Four days later, on 26 December 2003 the M_w 6.6 earthquake demolished the desert city of Bam in southeast Iran (population 80,000), killing 26,500 to 43,000 and injuring 30,000 (Berberian, 2005). The difference in casualty numbers and levels of destruction and damage between these two events with almost identical magnitude is directly related to differences in disaster preparedness and risk mitigation as well as accountability and building code practice and enforcement differences between the United States of America and Iran (Table 1).

Strict adherence to building codes during the past few decades in the San Francisco region undoubtedly saved many lives and kept thousands of buildings from collapsing in the 17 October 1989 M_w 6.9 Loma Prieta earthquake; no similar step was undertaken in Iran, Turkey, Caucasus, Turkmenistan, Afghanistan, or Pakistan. In California, building codes and geologic site investigations are mandatory and increasingly accepted; Iran has no comparable public commitment and many houses are built with unreinforced masonry. Despite the frequency of earthquakes in the western United States, the number of deaths due to earthquakes has been very low in the past 50 years as compared

TABLE 1 Comparison of the Post-1900 Earthquake Fatalities in Iran and the United States (Arranged in the Order of Magnitude)

Iran (also See Figure 1 and Table 17.12)				USA[a]			
Date	*Location*	M_w	*Casualty*	*Date*	*Location*	M_w	*Casualty*
				1964.03.28	Prince William Sound, AK	9.2	128
				2010.02.27	[Chile, S. America]	8.8	~500
				1906.04.18	San Francisco, CA	7.9	3000[b]
				1952.07.21	Kern County, CA	7.5	12
1909.01.23	Silākhor	7.4	8,000	1959.08.18	Hebgen Lake/Yellowstone, MT	7.4	28
1978.09.16	Tabas-e Golshan	7.3	20,000	1983.10.28	Borah Peak, ID	(7.3)	2
1990.06.20	Rudbār	7.3	13,000–40,000	1992.06.28	Landers, CA	7.3	3[c]
1997.05.10	Zirkuh	7.2	1568[c]				
1929.05.01	Bāghān	7.1	>3800				
1930.05.06	Salmās	7.1	2514[d]				
1957.07.02	Band-e Pay	7.1	>1500	1949.04.13	Puget Sound, WA	7.1	8
1968.08.31	Dasht-e Bayāz	7.1	10,000	1989.10.18	Santa Cruz County, CA	7.1	63
1979.11.27	Koli	7.1	20[c]				
				2010.01.12	[Port-au-Prince, Haiti]	7.0	>220,000
1962.09.01	Bu'in	7.0	12,200				

1981.07.28	Sirch	7.0	1300[c]				
				1940.05.19	Imperial Valley, CA	6.9	9
				1989.10.17	Loma Prieta, CA	6.9	60
1957.12.13	Fārsinaj	6.8	1200	1918.04.21	San Jacinto, CA	(6.8)	1
1923.09.22	Lālehzār	6.7	290[c]				
1972.04.10	Kārzin	6.7	5,010				
1977.03.21	Khurgu	6.7	152[c]	1994.01.17	Northridge, CA	6.7	60
1979.11.14	Korizān	6.6	171[c]				
1981.06.11	Golbāf	6.6	1400[c]	1971.02.09	San Fernando, CA	6.6	65
2003.12.26	Bam	6.6	31,828-43,200	2003.12.22	San Simeon, CA	6.6	2
1953.02.12	Torud	6.5	>930[c]	1954.12.21	Eureka, CA	(6.5)	1
				1965.04.29	Seattle, WA	(6.5)	7
1911.04.18	Rāvar	6.4	700[c]	1932.06.06	Eureka, CA	(6.4)	1
1990.11.06	E. Furg	6.4	23[c]				
2005.02.22	Dāhuiyeh	6.4	612[c]				
				1915.06.23	Imperial Valley, CA	(6.3)	6
1923.09.17	N. Bojnurd	6.3	157[c]	1925.06.29	Santa Barbara, CA	(6.3)	13
				1933.03.11	Long Beach, CA	(6.2)	115
1968.09.01	Ferdows	6.2	1000	1935.10.19	Helena, MT	(6.2)	2

Continued

TABLE 1 Comparison of the Post-1900 Earthquake Fatalities in Iran and the United States (Arranged in the Order of Magnitude)—Cont'd

Iran (also See Figure 1 and Table 17.12)				USA			
Date	*Location*	M_w	*Casualty*	*Date*	*Location*	M_w	*Casualty*
1941.02.16	Kumirān	6.1	>680[c]				
1999.05.06	Kuhmareh Sorkhi	6.1	26[d]				
2006.03.31	Chālānchulān	6.1	68[e]				
1997.02.28	Shirān	6.0	954	1993.09.21	Klamath Falls, OR	6.0	2
1923.05.25	Kāj Derakht	5.9	2200				
1977.12.19	Dartangal	5.9	665[c]	1987.10.01	Wittier Narrows, CA	5.9	8

[a]*Stover and Coffman (1993); USGS Significant Earthquakes of the World.*
[b]*Hansen and Condon (1989).*
[c]*Sparsely populated remote area.*
[d]*Region evacuated due to a damaging foreshock.*
[e]*Alarmed by foreshocks.*

with losses in the Iranian earthquakes (Figures 1–3). In the twentieth century, earthquakes in North America alone have resulted in more than 4000 deaths (3000 attributed to the 1906 event), while at least 126,000 people have lost their lives during earthquakes in Iran during the same time period.

Table 1 clearly shows that (i) a simple transfer of standards with technical and practical processes from the developed world to a developing country is not easy or practicable; and (ii) the risky system remained intact since 1969 (implementation of the first Iranian code for seismic-resistant building design; ISIRI Code No. 519). The challenge is daunting, especially when the governments in earthquake-prone developing countries (i) do not want to attack the root cause of earthquake destruction; (ii) have not built the capacity for earthquake hazard risk management; and (iii) are not thinking prior to earthquake occurrence, so there is therefore no preparation for the next earthquake disaster. The earthquake hazard with unfathomable destruction and death toll is closely linked with the country's stage of societal development and the accountability of its officials.

Much could have been learned from the multidisciplinary investigations conducted by earthquake scientists and earthquake engineers in Iran in the aftermath of large-magnitude earthquakes. However, the authorities have apparently learned no lessons. The 26 December 2003 M_w 6.6 Bam earthquake and its devastating aftermath may be considered to be a new unprecedented catastrophe; however, in reality, it is the predictable and unfortunate demise of a region hindered by political, socioeconomic, and cultural factors with widespread corruption and no proactive intervention. Tragically, more and more people in hazardous seismically active areas remain vulnerable until the next disaster strikes. Although the Bam earthquake was geologically foreseeable, and the active fault was introduced in 1976 (Berberian, 1976b), there was no effective risk-reducing infrastructure in place during the 27-year period between the introduction of the fault and the strike of the disaster, even 34 years after the implementation of the 1969 Iranian Building Code for seismic-resistant design.

The 1996 GeoHazards International workshop at Ālmāty, Kazakhstan, concluded that half the 6 million people in the capital cities of the five central Asian republics occupy buildings that are extremely vulnerable to collapse during earthquakes, with potential death tolls of up to 135,000 people and at least 500,000 injuries (Geohaz. Int., 1996, 1997). A similar study of Katmandu, Nepal (EERI, 1998), showed that the next major earthquake could kill 40,000 people, seriously injure 100,000, and leave millions homeless. The vulnerability is quite frightening with regards to Tehran, built on the active faults (with a population of ~12–15 million); provincial cities such as Mashhad (~3 million) and Tabriz (~2 million); and others located next to several mapped seismogenic faults with a documented history of several large-magnitude earthquakes (Berberian and Yeats, 1999, 2001, 2014; Berberian, 2005). The losses would have been much more severe if recent earthquakes had been centered in these mega-cities. Hence, the past casualty numbers cannot be used for today or future cases.

Based on the 1996 census (7 million) and available seismotectonics information (mainly based on Berberian et al., 1985), JICA (2000) estimated that if an $M \geq 7.0$ earthquake were to occur along the North Tehran fault (now covered by buildings; Plate 2), more than 130,000 people would be killed and more than 300,000 buildings in Tehran would collapse. Assuming a floating earthquake (a source not previously recognized), the numbers would increase to more than 300,000 deaths and the collapses of more than 510,000 buildings. Considering the present population of Tehran (12–15 million), with buildings constructed over the active faults (Plates 1 and 2), the scenario would be dreadful.

Notwithstanding the importance of studying the major active faults on the Iranian plateau threatening the major urban and rural areas, little has been achieved in this domain and our seismotectonic understanding of them is limited. Despite the visible active fault features on the ground surface, aerial photographs, and satellite imagery, most of the active faults on the Iranian plateau were not mapped or recognized prior to large-magnitude earthquakes. The coseismic surface faults of the 1962 M_w 7.0, 1968 M_w 7.1, 1978 M_w 7.3, and the 1990 M_w 7.3 earthquakes were discovered after the earthquakes had occurred. Amazingly, in the case of the latter, the geological map of the area on scale of 1:100,000 surveyed eight years after the 1990 earthquake (Nazari and Salamati, 1998) does not show the 80-km long coseismic surface fault of the 1990 earthquake. Very few paleoseismicity trench studies have been made across some Iranian faults. Unfortunately, the event dates are not properly constrained and broad error ranges are associated with them. Therefore, we should still rely on macroseismic data of historical and modern earthquakes in the region.

A single 1:100,000 geological map covering a portion of the North Alborz reverse fault (GSI, 2005) shows the dip of this major reverse fault shifting from north-dipping (east of longitude 53°15′E) to the south-dipping fault to the west; this was mapped by two different geologists who could not reach an agreement during the compilation process of their two maps. The fault was shown as a north-dipping reverse fault in Zanchi et al. (2009) (also Zanchi, personal communication, 2 April 2013). The Shāhkuh reverse fault is mapped as a north-dipping fault (GSI, 1989) but is shown as a south-dipping reverse fault in Zanchetta et al. (2009). More work needs to be done on the active faults of the Iranian plateau.

The foreign humanitarian relief and emergency assistance, technical reports, conferences, and workshops that occur in the wake of each devastating earthquake have never been able to break the disaster cycle in the developing countries. A critical element in good governance is missing in almost all these earthquake-prone countries. Retrofitting the critical public structures, exploring risk mitigation, planning, preparedness, and response and recovery are responsibilities of the governments. States and people have paid a high price for the governments' ignorance and neglect of earthquake risk mitigation and emergency practices.

The medium-magnitude Bam earthquake was the most recent devastating example of the failure of disaster risk reduction in a country with rich natural and human resources. In 2008, Iran was the fourth top oil producer in the world, with 4.38 million barrels/day (Figure 2), after Saudi Arabia, Russia and the United States (U.S. Energy Information Administration). Despite these revenues, the country is particularly vulnerable to natural disasters, especially earthquakes. Unlike Afghanistan or Haiti, Iran's exposure to earthquake destruction has not been exacerbated by extreme poverty, but rather by corruption and ignorance of the state in the control mechanisms of governance, and by a lack of accountability, which still permits poor building construction practices.

The first Iranian code for seismic-resistant building design was approved in 1969; it was revised in February 1988. This means that building codes and standards were developed by scientists and engineers 45 years ago. If only 1,000 public buildings were retrofitted through the application of petrodollars every year, we could have improved the performance of 45,000 buildings and infrastructure against earthquakes. Furthermore, seismic hazard maps have been prepared and published since at least 1976 (Berberian 1976a, 1977a, 1981a; Berberian and Mohajer-Ashjai, 1977; Berberian et al., 1983, 1985, 1996, 2000a). Failure to implement building codes, land-use practices, and seismotectonic research findings has resulted in the total destruction of four towns in the Rudbār region of the SW Caspian Sea during the 1990 earthquake and the town of Bam in 2003. The government failed to protect the people's lives and properties and has thereby expanded the earthquake risk in the country by ignorance, irresponsibility, and corruption.

On the Corruption Perception Index (CPI) list prepared by Transparency International, Iran ranked 168 among 180 countries (2009), and 133 among 178 countries (2012); the worst CPI rating was 174 (Transparency Int.). The higher the CPI, the higher the extent of corruption of the state as perceived by indigenous observers. The performance rank clearly indicates that deep roots of corruption have pervaded each and every sector at every level of the state, the society, and the country as whole, which has retrogressed for generations. The high level of corruption accounts for many of the problems that the country faces. It has affected the governance systems overseeing the hazard land use, proper and safe construction standards and approvals, and disaster risk reduction and management. Ignorance and corruption have created one of the greatest man-made disasters in Iran, the destruction of the city of Bam by a medium-magnitude earthquake in 2003.

In 2010, in a measure of a nation's relative peacefulness, Iran was ranked 104 out of 149 countries by the Global Peace Index (GPI, 2010; visionofhumanity.org). Other recent rankings are as follows: 2007: 89; 2008: 89; and 2009: 94. It is evident that the country has not been in a peaceful and stable state, and the situation has been worsening since at least 2007. With unacceptable CPI and GPI rankings, there will not be any monetary or technical support for earthquake risk reduction.

OVERVIEW: TARGETING THE ROOT CAUSES OF THE REGIONAL EARTHQUAKE-FAULT VULNERABILITY

Despite the fact that populations on the Iranian plateau face a high risk from earthquakes, most of the developing countries on and around the Iranian plateau have shown the least progress on priority 4 of the Hyogo Framework of Action to reduce the underlying risk factors. This is partly due to poverty and a lack of resources in some countries (GP-DRR, 2013). It is ironic that a very oil-rich country, with skilled earthquake scientists, engineers, and planners (who prepared the seismic data set and technical standards of the aseismic building codes), has been unable to minimize its earthquake risk despite the advantage of a well-documented seismic history capturing two millennia and well-exposed seismogenic surface faults with active geomorphologic features, which can be extended further more by paleoseismic trench studies. The countries' history shows that a rapidly growing accumulation of urban earthquake risk is occurring and that the states will be unable to protect the people in the event of large-magnitude earthquake disaster. During the recent medium-magnitude earthquakes in the region, almost all communications and lifeline networks failed; governmental help-and-rescue and first-aid operations were chaotic, disorganized, and disastrous; governments were caught by surprise and paralyzed; governments immediately applied for international help-and-rescue teams, food, tents, blankets, medication and donations; people were unaware of the hazards and risks; and the entire incidents were forgotten within a few months without any steps having been taken to minimize the risk in the future. Corruption has doomed any progress in the enforcement of codes for seismic-resistant building design and in retrofitting public structures.

Major active faults on the Iranian plateau have been mapped, their seismic potential has been investigated, and an earthquake history has been compiled since at least 1971. The seismic-resistant building code was prepared by earthquake scientists and engineers, implemented in 1969, and revised in 1988. Unfortunately, the authorities have not shown any interest in seismic-risk minimization, code enforcement, or retrofitting public buildings and infrastructure against large-magnitude earthquakes and are not prepared for the next large-magnitude earthquakes in either urban or rural environments. Clearly, the modern earthquake risks in the region are not within acceptable levels of those of developed countries and regional society is not resilient enough to tolerate the existing rapidly growing earthquake risk. Poor construction practices, antiquated infrastructures, and rapid increases in population concentrated in major cities in recent years are a recipe for an unfathomable casualty rate and destruction during the future large-magnitude earthquakes. Some of the identified factors exposing the region's people to extreme earthquake risks and causing complete destruction and immeasurable death tolls are as follows:

i. Poor risk governance, ineffective policies, lack of accountability of the system, lack of comprehensive earthquake risk assessment and risk-reduction programs, and corrupt institutions from local to national levels (see also Berberian, 2005; Bilham 2009; Ambraseys and Bilham, 2011; Bilham and Gaur, 2013; GP-DRR, 2013; Transparency Int.).

ii. Unsuitable land use (see Berberian et al., 1985).

iii. Poor construction practices; large percentage of seismically unreinforced buildings and public facilities; and vulnerable, inadequate infrastructure.

iv. Lack of stringent building code and inspection, disaster preparedness, and retrofitting programs.

v. Lack of consciousness of earthquake hazard and risk, strategy and management for disaster reduction, state and public education, awareness, and preparedness for seismic hazard.

vi. Delay in emergency search-and-rescue and recovery operations after each earthquake.

vii. Lack of direct communication between scientific, engineering, and architectural communities engaged in earthquake risk minimization and the government.

viii. Lack of national funds, programs, technologies, and opportunities to investigate and understand the nature of earthquakes, earthquake forecasting and alarming, and seismic zoning to minimize earthquake risk.

We should be able to positively impact the reduction of life and socioeconomic losses due to earthquakes in Iran and other developing countries.

Part One

Earthquake Hazard Warning in Oral Traditon and Literature on the Iranian Plateau

It is neither earthquakes, nor buildings that kill the people;
Lack of democracy, accountability, and transparency,
and corruptions in the building industry are the culprits.
Earthquake accountability should be treated as a criminal case.

Throughout the 21st century, earthquakes have produced disastrous destruction and immeasurable death tolls, especially in developing countries. As a result, they have demonstrated the dangers of building unreinforced structures in active fault zones, as well as the undeniable political, social, cultural, and religious challenges involved in improving the situation. In order to better identify and explain seismic hazards on the Iranian Plateau, and in other developing countries prone to earthquakes, we need a profound understanding of past seismicity, active faulting patterns, and the social circumstances in affected areas. Hence, this book utilizes a multidisciplinary social and scientific approach to examine the past with the hope of looking toward a brighter future. We also try to understand how this approach can contribute to our scientific understanding of earthquakes in each social context. This work is, therefore, divided into two parts.

Part One provides the first nationwide and regional survey of earthquakes and earthquake impacts on human communities and social development. In so doing, it focuses on the factors that turn earthquakes into national and regional catastrophes, analyzing why survivors have accepted the status quo since the dawn of civilization, showing no evolution in their thoughts, words, and deeds. The first part of this book also considers the relationship between long-term, time-consuming historical and linguistic research and well-documented social, political, and economic aspects of earthquakes.

Part Two delivers the first comprehensive survey of coseismic active surface faults associated with medium- to large-magnitude earthquakes, as well as seismicity patterns and active faulting on the Iranian Plateau. Therefore, both

Developments in Earth Surface Processes, Vol. 17. http://dx.doi.org/10.1016/B978-0-444-63292-0.00018-1

parts of this book provide a realistic background for seismic hazard evaluation, earthquake risk minimization, and the design of resilient urban and rural structures and infrastructure, as well as the development of resilient societies and cultures.

The person of good intellect says that,
The deeds of good mind must be maintained.
He knows that progressive serenity is the true product of righteousness.

[Gathas, 7, 10]

Chapter 1

Place Names and Linguistic Traces Referring to Prehistoric Earthquakes: Earthquake Hazard Warnings in Oral Traditions

I am close to the beginning of the Earth.

Sohrāb Sepehri (1964)

As one of the strongest forces on Earth and cataclysmically disturbing to human life and property, earthquakes certainly attracted the attention of the inhabitants of the Iranian Plateau in ancient times. Stories and memories of catastrophic earthquakes amid gradually led to legends and myths (Chapter 2) with specific messages and warnings about particular seismic zones.

Linguistic traces and dialectal substrata can be found in the place names of earthquake-prone areas on the Iranian Plateau as well as other active areas. A moderately recent and well-known example outside the Iranian Plateau is the *Temblor* ("Earthquake" in Spanish) *Range* within the California Coast Range along the San Andreas fault, at the southwestern extremity of the San Joaquin Valley, near the Carrizo Plain (Site ID: 254329, 35.19N–119.47W, 3868 ft; USGS/GNIS). The Temblor Range may have been named after the 9 January 1857 *M* 7.9 Fort Tejon earthquake (Sieh, 1978a,b; Grant and Donnellan, 1994; Grant and Sieh, 1994; Fumal et al., 2002; Yeats, 2012), which produced right-lateral offsets on the San Andreas fault along the range (Robert Yeats, personal communication, 18 August, 2013).

The etymology of ancient place names can provide valuable information regarding the origin of major events and history. The toponyms have been used by Iranians for several millennia to explain and warn of geological hazard zones and inform future generations of the risk. In some instances, place names emphasize the seismic sites with recorded historical (pre-1900) earthquakes. In other cases, however, place names refer to areas with no recorded historical seismicity

☆"To view the full reference list for the book, click here."

Developments in Earth Surface Processes, Vol. 17. http://dx.doi.org/10.1016/B978-0-444-63292-0.00001-6

(Berberian, 1991, 1994, 1997a). It was in this part of the world—with its frequent destructive earthquakes and ancient toponyms referring to earthquakes—that an earthquake theory was formulated in ancient days (Chapter 4).

In this section, I present the ancient earthquake hazard warnings that can be found in oral traditions. For most rural areas, local oral tradition and place names provide the only source of information about destructive earthquakes. The etymology of most geographical names analyzed in this study are prolific in connection with the earthquake source areas on the Iranian Plateau. There are no doubt more place names than I have collected in this study. Systematic etymological research in oral traditions coupled with historical, archaeological, and geological studies can provide additional valuable information on the seismicity of remote areas that lack recorded historic seismic data.

Unfortunately, much of the precious oral traditions, folklore, legends, myths, and cultures (Chapters 2–5) have been lost in the modern migrations into towns and cities. The young urban generation has never heard the legends, folklore, or myths. Sadly, a systematic recording and scientific interpretation of these folkloric data have been neglected, and the information may well now disappear.

Since 1971, I have been collecting place names during geological and earthquake field investigations in Iran. Some names in areas that I have not visited are taken from the Joint Operations Graphic (JOG) 1:250,000 topographic maps of the country [NJ38, 39, 40; NI38, 39, 40; NH39, 40, 41; NG39, 40, 41]. Some Persian words have several meanings. For example, *Shekaft* or *Eshkaft* means "cave" and "fracture." *Chāk* means "fissure/fracture" and "thunder," and *Chak* or *Chek* is the sound of dripping water. Entries with question marks denote that the link between the toponomy and earthquake or its associated deformation is not yet certain and requires authentication through further site visits. The Persian *ābād* suffix (lit., "populous," "thriving," "prosperous") is usually added to personal names to form place names. An example is Ashkābād: the place founded or built by King Ashk [Arsacid dynasty; 312 BCE–223 CE]. This is usually mispronounced and misspelled in English and Persian as Eshghābād or Esqābād (the capital city of the Iranian Arsacid dynasty in modern Turkmenistan).

Some of the aforementioned sites can be located in figures found in Chapters 11–16, with the coordinates mentioned in each case (relevant figures are cited for each). For the seismic parameters of earthquakes and seismic sources mentioned in each entry in this chapter, see Tables 14.9–14.11 and 17.10–17.12 in Chapters 14 and 17.

1.1 EARTHQUAKE TERMS USED IN LOCAL DIALECTS ON THE IRANIAN PLATEAU AND THE NEIGHBORING REGIONS

The following terms are still in use by local inhabitants today and referred to in the chronicles and colophons (Table 1.1).

TABLE 1.1 Earthquake Terms Used in the Region Since the Historic Time

Language/ Dialect	Term	Meaning	Notes
Arabic	*Zalzalah*	Earthquake	The Arabic word *zalzala* is used in post-seventh-century Pashtu, Persian, Turkic, and Urdu, when the region was invaded by the Moslem Arabs
	Zalazal	Earthquakes	
	Zalzāl	Earthquake	
	Rajafah	Earthquake	
	Hazzah Arziyyah	Ground shaking	
Armenian	*Yergrā-shārzh*	Earthquake	
	Gednā-shārzh	Earthquake	
	Shārzhum	Shaking, jolt	
	Shārzh	Shake	
	Tsntsum	Shaking	
	Bāduhās Deārn	God's punishment as an earthquake	Grabar (classic) Armenian
Assyrian	*Rodānā*	Earthquake	Modern Assyrian
	Ri-i-bu	Earthquake	Classic Assyrian
Avestan	*Taozhya*	Earthquake	(Haug, 1869)
		Hoar-frost	(Justi, 1864; Spiegel, 1882)
		Avalanches	(Evans, 1880)
		Sorcery	(Darmesteter, 1877, 1879–1887)
			Also see Bahrami and Joneidi (1990)
Baluchi	*Zamin Chand, Bum Chand*	Earthquake	*Zamin*: "Earth"; *Chand*: "shake," "jolt" in Persian
Kurdish/ Kordi: Suraqni	*Bularz*	Earthquake	Shortened form of *Bum Larz*: *Bum*: "Earth"; *Larz*: "shake," "jolt" in Persian
Kurdish/ Kordi: Urāmi	*Bume Lerzeh*	Earthquake	From *Bum Larzeh*: *Bum*: "Earth"; *Larzeh*: "shaking" in Persian

Continued

TABLE 1.1 Earthquake Terms Used in the Region Since the Historic Time—Cont'd

Language/ Dialect	Term	Meaning	Notes
Pahlavi (Middle Persian)	*Bum Chandag [bumchndk']*	Earthquake	From *Bum*: Earth; *Chandag*: "shaking," "jolt" in Pahlavi
Pashto	*Zalzala*	Earthquake	From Arabic *Zalzalah*
Persian	*Buméhén*	Earthquake	*Bumahan* = *Bumahin* [from *Bum*: "Earth, land"; *mahan* = *mathana*: "movement," "shaking"]; *Buméhén* was used by Ferdowsi Tusi (1010) in verse 3376 (ed. Joneidi, 2008, v. 1, p. 231) and by Hakim and Dāvar Vesāl Shirāzi in their poems about the 1853 Shirāz earthquake (Soltāni-Moqadam (2012))
	Bum Larzeh	Earthquake	
	Zamin Larzeh	Earthquake	*Zamin*: "Earth"; *Larzeh*: "shaking" in Persian
Rāji	*Zimin Chandeshæ*	Earthquake	*Zamin*: "Earth"; *Chandesh*: "shaking" in Persian
Sanskrit	bhUkampa	Earthquake	*bhU*: "Earth," "ground," "soil," "land"; *kampa*: "shaking," "tremor," "trembling"
	bhUmicala	Earthquake	
	Kampa	Trembling, tremor, shaking	
Sistāni	*Zami Larza*	Earthquake	*Zami*: "Earth"; *Larzeh*: "shaking" in Persian
Sohi (village near Ābyāneh, Kāshān)	*Bum Chendesh*	Earthquake	*Bum*: "Earth"; *Chendesh* or *Chandesh*: "shaking" in Persian
Tājik	*Zamin Jumbi(sh)*	Earthquake	*Zamin*: "Earth"; *Jonbesh*: "shake," "jolt" in Persian
Tāleshi	*Bumlarz*	Earthquake	*Bum*: "Earth"; *Larz*: "shake," "jolt" in Persian
Tāti	*Bumlarz*	Earthquake	*Bum*: "Earth"; *Larz*: "shake," "jolt" in Persian

TABLE 1.1 Earthquake Terms Used in the Region Since the Historic Time—Cont'd

Language/ Dialect	Term	Meaning	Notes
Turkic	*Deprem*	Earthquake	
	Yer-sarsintisi		
	Kargasha		
	Zelzeleh		From the Arabic *Zalzalah*
Urdu	*Zalzala*	Earthquake	From the Arabic *Zalzalah*
Yazdi Zoroastrian	*Zvin Larz*	Earthquake	From *Zamin*: "Earth"; *Larz*: "shake," "jolt" in Persian

1.2 PLACE NAMES REFERRING TO FREQUENT EARTHQUAKES

The profusion of ancient place names relating to the frequent occurrence of earthquakes in particular zones on the Iranian Plateau reflects the traditional oral method of sending earthquake hazard warnings. These oral traditions can extend the seismic hazard interpretation of the zones addressed.

Bastak (?) [lit., "fastened, knotted, or tied," as knotting votive cloths, or tied by spell and prayer to prevent shaking!]: A town in the Fārs province of southern Irān (27°11′N–54°22′E, +432 m amsl) damaged by the 31 October 1956 M_w 6.6 Gowdeh earthquake.

Bastakābād (?) [as Bastak]: A village in the Barādust District of Urumiyeh.

Bastakān/Bastagān (?) [As Bastak].

Boom-e Hen [lit., "earthquake" in Pahlavi (Middle Persian language); "boom": land, Earth, country, and *Mohan/Massna*: "quake, jolt, shake"]: A village 40 km east of Tehrān (35°45′N–51°51′E), and south of the Moshā fault with recorded earthquakes on 23 February 958, 15 June 1665, 27 March 1830, and 24 November 1966 (Figures 11.2 and 16.11).

Chālkhāmāz (?) [lit., "will not shake," "not shaking" in Turkic; from "Chālkhāmākh": "to shake strongly"]: A village in southern Marāgheh, on the southern foothill of the Sahand volcano (Figure 11.12).

Gaud-e Zelzeleh Zadeh [lit., "Playa Smitten by Earthquake," "Earthquake Shaken Playa" in Persian]: A locality about 12 km south of Kuhbanān, near the Kuhbanān fault (31°22′N–56°22′E and 31°15′N–56°22′E), with recorded earthquakes in ca. twelfth century, ca. fourteenth century, and May 1875 (Figures 16.19 and 16.20).

Kāhtekān [lit., "straw shake" in Persian]: 11.5 km NW of Rostāq and 22 km NW of Furg in the High-Zāgros (28°32′N–55°05′E), located on an active fault with a recorded earthquake of M_w 6.5 on 6 November 1990, to its southeast (Figure 13.14).

Kāl-e Zelzeleh [lit., "Earthquake Creek"]: Northeast of the Ābiz fault (33°56′N–60°15′E) with a recorded earthquake on 10 May 1997 M_w 7.2 along the Ābiz fault (Figures 13.17 and 13.18).

Kharāb Dareh [lit., "Ruined Valley" in Persian]: A common name used in many places.

Kharāb Deh [lit., "Ruined Village" in Persian]: A common name used in many places; for example, a village at Firuzkuh area (35°45′N–52°45′E) near the Firuzkuh and the Moshā faults, with the 20 January 1990 M_w 6.0 Firuzkuh earthquake along the Firuzkuh fault (Figure 16.11). There are numerous ruined or abandoned villages in the epicentral regions of the 1209 and 1389 Neyshābur earthquakes to the east of Neyshābur in the Zebarkhān district (Figures 16.13–16.15) as well as to the southeast of Sabzévār (Berberian et al., 2000a). Some of these villages were not ruined by earthquakes.

Kharābeh [lit., "ruined" in Persian]: A common name used in many places.

Kharāb Shahr (?) [lit., "Ruined City" in Persian]: Tamisheh between Gaz and Kordkui, near the Khazar fault.

Kuh-e Zelzeleh Khiz [lit., "Seismogenic Mountain" in Persian (*Kuh* and *Khiz*) and Arabic (*Zelzeleh*)]: A mountain (2309 m amsl) located 52 km north of Kāshmar and 55 km south southwest of Neyshābur, along the Zelzeleh Khiz (Kadkan) fault (35°41′N–58°28′E), the epicentral region of the 4 May 1940 M_s 6.4 earthquake (Figure 12.7).

Larzaneh [lit., "shaking" in Persian]: A village on the Larzaneh (Lalehband) fault, 3.5 km NW of Ālāsht, Māzandarān (52.5 km SSE of Bābol), in the Alborz Mountains (36°05′N–52°48′E). It is located between the meizoseismal areas of the 1300–1301 [in the east] and the 2 July 1957 [in the west] (Figure 12.10).

Larzini (?) [lit., "shaking" in Persian]: A village in the Margur District of Urumiyeh.

Pālān Tekān Mt. (?) ["saddle shaking"]: A mountain about 3 km south of the Salmās earthquake fault (38°04′N–44°47′E, +2087 m), SW of the Shurgel village (Figure 12.5).

Rud-e Hen (?) [lit., "Quake River" in Persian: "Hen" from Pahlavi *Mohan/Massna*: "quake, jolt, shake"]. A village 42 km east of Tehrān, south of the Moshā fault (35°44′N–51°50′E), with recorded nearby earthquakes on 23 February 958, 15 May 1665, 27 March 1830, 2 October 1930, and 24 November 1955 (Figure 11.2).

Vāyots-Dzor [lit., "Valley of Lament" in Armenian]: The valley of the upper courses of the Mozān [later, Ārpā-Chāi] River was renamed, probably after the 735 destructive earthquake, as the Valley of Lament [Vāyots-Dzor].

The city of Muz [3 km SE of the village Mālishkā of Vāyq region] was destroyed (Stepanos Orbelyan, 13th century; Guidoboni and Traina, 1995). The latter writes that: "… *Everywhere shook, and sounds like human voices rose up from the depths into open air*: *Vāy Dzor*, *Vāy Dzor*" [lit., "Woe Valley, Woe Valley"]. (Guidoboni and Traina, 1995; Babayan, 2006).

Virān Qayeh [lit., "Ruined Rock or Mountain" in Persian (Virān) and Turkic (Qayeh)]: A village in the Qara Āghāj district of southeast Marāgheh (37°03′N–46°53′E).

Zelzeleh village [lit., "earthquake"], Afghanistan, SSW of Ghazni [Ghaznai'n] (33°27′N–68°21′E).

Zelzeleh Kharābeh [lit., "ruined by earthquake" in Arabic (*Zelzeleh*) and Persian (*Kharābeh*)]: A common name used in many places.

Zelzeleh Kuh [lit., "Earthquake Mountain" in Persian (*Kuh*) and Arabic (*Zelzeleh*)]: The correct Arabic word for earthquake is *zalzala* (*zalzal*: "to shake"), which has been modified (Persianized) to *zelzeleh* in modern Persian usage. However, the correct modern Persian term for earthquake is *zamin larzeh*. It is one of the peaks in the Baqerān Mountains, 16 km south of Birjand, and 4.5 km north of Gol village (32°42′N–59°12′E; Figure 11.10). The southern flank and foothill of this mountain is characterized by numerous rock avalanches, and during the 24 November 1987 M_s 4.4 Kāriz Nau, south Birjand, earthquake, rockslides, and avalanches were reported.

Zelzeleh Kuh [lit., "Earthquake Mountain" in Arabic (*Zelzeleh*) and Persian (*Kuh*)]: A mountain in the Khālkumeh Mountain area of the Chahār Dāngeh region, northeast of Mazdeh, south of Kusut in Māzandarān, between Zālemrud/Zaramrud and Tājrud (36°39′N–53°39′E), located in the epicentral region of the 1127 $M_s{\sim}6.8$ Farim and 11 April 1935 M_s 6.2 Kusut earthquakes (Figure 12.10).

Zelzeleh Kuh [lit., "Earthquake Mountain" in Arabic (*Zelzeleh*) and Persian (*Kuh*)]: a N–S mountain range separating the Doroh/Gorāt plain from the Kand Ghinau valley (32°21′N–60°16′E) (Smith, 1876a), located south of Tabas-e Masinān (in Sistān), southeast of Sarbisheh, near the Purang fault (Figure 11.10). Smith (1876a), who visited the area during his Perso-Afghan mission of 1871–1872, wrote that: "*Duruh* [Doroh] *was so destitute of supplies, that though we had intended halting there for two days, we were obliged to push on the next morning to obtain forage for the horses. Our march was but a continuous ascent and descent over the same rocky mountain ranges. For the first fourteen miles the road runs straight over the plain of Duruh in a west-north-west direction, when it reaches the mountain range on the western side, and turning the corner of some low granite hills to the right, arrives at the 'Chah-i-Bermeh', a good well with good water, which is situated in a narrow valley between the granite hills aforesaid and a lofty hill called the 'Kuh-i-Zilzila*

(Hill of Earthquake) [Kuh-e-Zelzeleh; i.e., Earthquake Mountain]. *On the plain were some few tents of shepherds and large flocks; but the grazing here was much more scanty than that obtained in the neighbourhood of the Harut river. We dismounted and breakfast at this well; and remounting at 10:15 A.M. pursued our way along the bed of a watercourse, which ascending gradually for three miles, reaching the summit of the range of the Zilzila hills and the line of the watershed, at an elevation of 500 feet above the level of Duruh plain, or about 4800 feet above the level of the sea. This pass is known as 'Godar-i-Mesum', and from its summit the road gradually descends for three miles through hills, which it clears at the twentieth mile from Duruh, and emerges into the spacious hill-locked valley of Hussainabad, reaching the village at four miles distance from the hill.*"

Zelzeleh Sang [lit., "Earthquake Rock" in Arabic (*Zelzeleh*) and Persian (*Sang*)]: A locality at Gevār about 6 km south of Kuhbanān, south of Asfich, along the Kuhbanān fault (31°18′N–56°16E), with recorded earthquakes in ca. twelfth century, ca. fourteenth century, and May 1875 (Figures 16.19 and 16.20).

1.3 PLACE NAMES REFERRING TO EARTHQUAKE FAULTING AND FRACTURING, FRACTURED OR BROKEN MOUNTAINS, AND CAVES

An amazingly large number of ancient place names define particular localities as having earthquake faulting, consequently providing a warning about the ground deformation effects of large-magnitude earthquakes in the past and mentioning landscape changes (active tectonics). The survival of ancient names may imply the occurrence of multiple events through several millennia. These valuable linguistic traces become significant when they are associated with localities that have had active faulting.

Aznā Chāk Mt. [lit., "Fractured Aznā" Mt.]: along the Kashachāl Fault (36°41′N–50°08′E) in the western High-Alborz Mountains (Figures 11.16 and 13.13; Berberian and Walker, 2010).

Bard-e Shekāft [lit., "Fractured/Faulted Rock" in Persian]: Locality along the Karéhbas fault (29°26′N–52°07′E; Figure 16.2).

Chāk Dareh [lit., "Fractured Valley" in Persian]: A locality south of the Kashachāl fault (36°38′N–50°02′E), western Alborz (Figures 11.6 and 13.13; Berberian and Walker, 2010).

Chāt [lit., "fractured" in Turkic]. A village 60 km WNW of Marāveh Tapeh (37.58′N–55.16′E). The closest recorded earthquake took place on 30 July 1970 (M_w 6.3).

Dagh-e Shekāfteh [lit., "Faulted or Fractured Playa" in Persian]: Playa 2.5 km NW of Esfedeh and 18 km west of Ābiz and the Ābiz fault, 50 km east of Qā'en (33°41′N–59°45′, 1197 m amsl), with recorded earthquakes on 15 February 1549 ($M_s \sim 6.7$), 30 June 1936 ($M_s \sim 6.0$), and 14 November 1979 (M_w 6.6). We can be sure that the playa was fractured during large-magnitude prehistoric earthquakes due to earthquake faulting and/or induced liquefaction and slumping (Berberian et al., 1999) (Figures 13.9,13.17,13.18).

Dareh Eshkāft [lit., "Fractured Valley" in Persian]: Locality in Dālaki region, Borāzjān.

Dochāq/Dochāk (?) [lit., "two fractures" in Persian]: A creek on the way between Ardebil and Meshkinshahr (Figure 13.16).

Eshkaft-e Gāvi [lit., "Cow Cave" or "Cow Fracture" in Persian]: 5 km west of Marvdasht on the north slope of Kuh-e Sabz. At the front of the cave, a shallow ceramic and metal-bearing organic deposit overlies a rocky Pleistocene deposit with a stratigraphic break. The break was apparently due to late Pleistocene–Early Holocene erosion caused by the opening of a hole in the roof of the cave [the cause of the opening is not known yet]. The artifacts collected indicate possible middle or early upper Paleolithic origin (Rosenberg, 1979).

Kamar Kasan Mountain (?) ["Sheared Gorge/Mt./Rock"; "rock cutter" in Turkic. Also possibly a stone quarry]. Located about 42 km NW of Khoi (38°49′E–44°33′E), near the Kamar Kasan fault (NW Khoi; Figure 8.2), with recorded regional earthquakes on 18 April 1843 ($M_s \sim 5.9$), 24 February 1900 (M_s 5.4), 14 March 1970 (M_s 5.2), and 26 May 1977 (M_s 5.4) in the southeast and 24 November 1976 (7.1) in the northwest (Berberian, 1997a).

Kuh-e Aznā Chāk [lit., "Fractured Aznā" Mt.]: A mountain along the Kashachāl Fault (36°41′N–50°08′E; Figures 11.6 and 13.13) in the western High-Alborz Mountains (Berberian and Walker, 2010).

Kuh-e Eshkaft [lit., "Fractured or Cave Mountain" in Persian]: A mountain W. Yazd (31.83°N–54.05°E).

Kuh-e Eshkaft [lit., "Fractured or Cave Mountain" in Persian]: A mountain SW of Kāshān, 18 km SW of Jaushaqān Qāli [destroyed during the 12 May 1844 $M_s \sim 6.4$ earthquake; Figure 16.4], 14 km NW of Meymeh (33°30′N–51°10′E, 2320 m amsl); source of the Mohammadābād River.

Kuh-e Eshkaft [lit., "Fractured or Cave Mountain" in Persian]: A mountain (2550 m amsl) 47 km SW Shahrezā, 1 km south of the Shahrezā-Semirom road, in the lower Semirom region.

Kuh-e Eshkaft [lit., "Fractured or Cave Mountain" in Persian]: A mountain in the Zāgros, south of Borujen (31°30′N–51°09′); the source of the Sulégān River.

Kuh-e Eshkaft Dahāgān/Tāghun [lit., "Dahāgān/Tāghun Fractured or Cave Mountain" in Persian]: A mountain 34 km NW Borujen, NNW Semirom (31°36′N–51°19′E, 2750 m amsl).

Kuh-e Eshkaft Derāz [lit., "The Long Fractured or Cave Mountain" in Persian]: A mountain (2331 m amsl) in the Tashān region, 43 km NW of Behbahān, the source of the Qalāt River.

Kuh-e Eshkāfteh [lit., "Fractured Mountain" in Persian]: West of Birjand (32°51′N–58°49′E; Figure 11.10).
Kuh-e Eshkaft Zard [lit., "The Yellow Fractured or Cave Mountain" in Persian]: A mountain in the Rostam region, 40 km NW of Nurābād (30°28′N–51°26′E, 2224 m amsl); the source of the Kāl Asadi and the Kushk Rivers, immediate west of the Kāzerun fault, 40 km north of Nurābād-e Mamasani.
Kuh-e Eshkasteh [lit., "Broken, Fractured, or Faulted Mountain"]: A mountain (+1520 m amsl) south of Akbarābād Hossein 'Ali village in Khusf region, 35 km southwest of Birjand (Figure 11.10).
Kuh-e Nimeh [lit., "Half Mountain" in Persian]: In Jiroft (Sabzévārān; 28°57′N–57°36′E).
Kuh-e Par? Eshkaft [lit., "The Feather Fractured or Cave Mountain" in Persian]: A mountain (2528 m amsl) 25 km W of Yāsuj.
Kuh-e Sarchāk [lit., "Top Fractured Mountain" in Persian]: A mountain (2205 m amsl) 5 km north northwest of Shekarnāb village, 32 km east southeast of Qazvin, immediately north of the North Qazvin fault, reactivated during the 10 December 1119 $M_s \sim 6.5$ earthquake (Figure 11.6).
Kuh-e Shamshir Borān [lit., "Sword-Cut Mountain" in Persian]: About 35 km southwest of Torud, 2 km northwest of the Torud fault and 6 km west of the Sang-e Sur spring, where a coseismic ground fracture was recorded during the 12 February 1953 Torud earthquake (35°20′N–54°46′E, +1310 m; Figure 12.9).
Kuh-e Shekāf-Reshteh [lit., "String-Fractured Mountain" in Persian]: A mountain (1120 m amsl) southwest of Qal'eh Tal village, 25 km SW of Izeh.
Kuh-e Shekāft [lit., "Fractured Mountain" in Persian]: 30.5 km NE of Shahrezā [Qomsheh] (32°08′N–52°09′E, 1785 m amsl).
Kuh-e Shekaft [lit., "Fractured or Cave Mountain" in Persian]: 60.6 km east southeast of the Kāftar Lake in the High-Zāgros along the Zāgros reverse fault, 26 km southeast of Dehbid (30°29′N–53°26′E, 3182 m), and about 35 km east of the 11 November 1973 M_s 5.5 Qeshlāq earthquake.
Kuh-e Shekaft-e Soleymān [lit., "Solomon's Cave or Solomon's Fractured Mountain" in Persian]: 20.6 km southeast of Izeh (31°39′N–49°57′E, 1689 m amsl).
Kuh-e Shekasteh [lit., "Broken, Fractured, or Faulted Mountain" in Persian]: Located between the Birjand fault [with the 1549 earthquake] in the west and the Nauzād fault [with the 1493 earthquake] in the east, west of Esfezār (32°53′N–59°33′E, +2122 m; Figure 11.10).
Kuh-e Shekasteh [lit., "Broken, Fractured, or Faulted Mountain" in Persian]: (33°37′N–49°53′E).
Kuh-e Shekasteh [lit., "Broken, Fractured, or Faulted Mountain"]: A mountain (+2420 m amsl) 16 km W of Khomain.

Kuh-e Shekasteh [lit., "Broken, Fractured, or Faulted Mountain"]: (33°36′N–58°33′E).

Kuh-e Shekasteh [lit., "Broken, Fractured, or Faulted Mountain"]: (32°42′N–59°33′E).

Kuh-e Shekasteh [lit., "Broken, Fractured, or Faulted Mountain"]: A mountain northwest of Shamsābād village, 53 km W of Taft (31°43′N–53°36′E, 2640 m amsl).

Kuh-e Shekasteh Āb-shaleh [lit., "Āb-shaleh Broken/Fractured/Faulted Mountain" in Persian]: A mountain (1650 m amsl) NW of Sarkavir village in the Daihuk region, 18 km NNW of Nāyband; west of the Nāyband fault. No recorded historical earthquake has been found along the Nāyband fault.

Kuh-e Shekasteh Chāhaki [lit., "Chāhaki Broken/Fractured/Faulted Mountain" in Persian]: 23 km northwest of Ābiz and 13 km southwest of the Ābiz fault (33°46′N–59°43′E, 1403 m amsl) with the 10 May 1997 M_w 7.2 earthquake (Figures 13.9,13.17,13.18).

Kuh-e Shekasteh Chāhuk [lit., "Chāhuk Broken/Fractured/Faulted Mountain" in Persian]: A mountain (1530 m amsl) east of Ja'farābād village, 9 km south of Torbat Haydariyeh, near the Doruneh fault.

Kuh-e Shekasteh-yeh Derisul [lit., "Derisul Broken/Fractured/Faulted Mountain" in Persian]: A mountain (1750 m amsl), 45 km southwest of Sabzévār.

Kuh-e Shekasteh Dochāhi [lit., "Dochāhi (Two-Wells) Broken/Fractured/Faulted Mountain" in Persian]: 26 km northwest of Ābiz and 14 km southwest of the Ābiz fault (33°47′N–59°42′E, 1377 m amsl) with the 10 May 1997 M_w 7.2 earthquake (Figures 13.9,13.17,13.18).

Kuh-e Shekasteh Kalāteh [lit., "Broken/Faulted Fortress Mountain"]: West southwest of Khāf (34°35′N–60°07′E, +1097 m amsl), with a recorded earthquake on 21 October 1336 (Figure 11.9).

Kuh-e Shekasteh Sefid Sang [lit., "White Stone Broken/Fractured/Faulted Mountain" in Persian]: (35°51′N–55°48′E).

Kuh-e Shekasteh-yeh Cheshmeh Bahādār [lit., "Bahādār Spring Broken/Fractured/Faulted Mountain" in Persian]: A mountain (1017 m amsl), 62 km ESE of Biārjomand, 13 km east of Kāl-e Shur-e Khārturān (Qal'eh Moreh), and immediately east of the Kāl-e Shur fault.

Kuh-e Shekasteh-yeh Siāh [lit., "Black Broken/Fractured/Faulted Mountain" in Persian]: (36°03′N–60°10′N).

Kur Cheshmeh Kur [lit., "Blind Spring" in Persian]: (35°43′N–49°46′E), along the Kashachāl fault, northwest of Jirandeh, in western High-Alborz Mountains (Figures 13.12 and 13.13).

Lileh Chāk River [lit., "Fractured Lileh" in Persian]: A locality, crossing the Kashachāl fault (35°46′N–50°04′E; Figure 13.13) in western High-Alborz Mountains (Berberian and Walker, 2010).

Pir Eshkāft ["Old Fracture" or "Old Cave" in Persian]: A locality along the Karéhbas fault in the Zāgros (29°38′N–52°06′E; Figure 16.2).

Qal'eh Shekasteh [lit., "Fractured Castle" in Arabic (Qal'eh: "castle") and Persian (Shekasteh: "fractures, broken, faulted")]: Northeast of Estakhr in the Fārs province (29°54′N–52°54′E), with a recorded earthquake in 1623.
Shekāft [lit., "fracture," "fault" in Persian]: 27 km southeast Dehbid, between Gerdāb-e Bālā and Suriān villages (3333 m amsl).
Shekāft [lit., "fracture," "fault" in Persian]: About 1.5 km east of Karéhbas, along the Karéhbas fault (29°17′N–52°12′E; Figure 16.2), with a recorded earthquake on 8 September 1992.
Shekaft-e Golgol [lit., "Golgol Cave or Fracture" in Persian]: 30 km southwest of Ilam with an Assyrian rock relief (Vanden Berghe, 1973).
Shekasteh [lit., "fractured, faulted, broken," in Persian]: A general term in Persian.
Shekasteh 'Abdoābād [lit., "Fractured/Broken 'Abdoābād" in Persian]: A locality (1470 m amsl) about 66 km northwest of Torbat Haydariyeh.
Shekasteh 'Asāsangi [lit., "Fractured/Broken Wooden Cane" in Persian]: north of Birjand (33°07′N–58°52′E; Figure 11.10).
Shekasteh Bakhtak [lit., "Fractured Nightmare" in Persian]: (31°45′N–58°35′E).
Shekasteh Band-e Mil [lit., "Fractured/Faulted" plus the name of the site (Band-e Mil), in Persian]: A mountain 27 km north of the Jangal fault (34°56′N–59°14′E, 1110 m amsl), with a recorded earthquake on 21 October 1336 (Figure 11.9).
Shekasteh Chāh Cheleh'i [lit., "Fractured/Faulted Chāh Cheleh'i" in Persian]: (31°16′N–58°43′E).
Shekasteh Chāh Goji [lit., "Fractured/Faulted Goji Well" in Persian]: South southwest Torbat Heydariyeh (34°50′N–58°47′E).
Shekasteh Chāh Sad [lit., "Fractured/Faulted" plus the name of the site]: A mountain 15 km northwest of Gonābād (34°29′N–58°33′E, 1320 m amsl). Two earthquakes in 1237–1238 and 1678 destroyed the town of Gonābād (see Chapter 11 and Figure 11.11).
Shekasteh Chāh Tam [lit., "Fractured/Faulted" plus the name of the site (Chāh Tam) in Persian]: 25 km east of Shāh Rakht, along the Chāh Tam fault (33°37′N–60°32′E; Figures 13.17 and 13.18).
Shekasteh Chāhuk [lit., "Fractured/Broken/Faulted Chāhuk" in Persian]: 8 km south southeast of Torbat Haydariyeh (35°11′N–59°14′E, 1496 m amsl), immediately south of the Doruneh fault.
Shekasteh Chāk [lit., "Broken/Faulted Fracture" in Persian]: (35°11′N–59°13′E).
Shekasteh Chek Chek [lit., "Broken/Fractured/Faulted Chek Chek" in Persian]: A mountain, 27 km SW of Torbat Haydariyeh (35.01°N–59.10°E, 1161 m amsl), 20 km north of the Jangal fault, with a recorded earthquake on 21 October 1336 (Figure 11.9).
Shekasteh Chāleh Band [lit., "Broken/Fractured/Faulted Chāleh Band" in Persian]: (35°06′N–59°33′E).

Shekasteh Chilnuk/Chilunak [lit., "Broken/Fractured/Faulted Chilnuk/Chilunak" in Persian]: (33°10′N–58°51′E).
Shekasteh Dāgh [lit., "Broken/Fractured/Faulted Mountain" in Persian (Shekasteh) and Turkic (Dāgh)]: (31°45′N–59°04′E).
Shekasteh Darsiul [lit., "Broken/Fractured/Faulted Darsiul" in Persian]: S. Sabzévār (35°49′N–57°31′E).
Shetasteh Dasht [lit., "Fractured/Faulted Plain" in Persian]: Southwest Khāf (34°30′N–60°04′E; Figure 11.9).
Shekasteh Divār [lit., "Fractures/Faulted/Broken Wall" in Persian]: Located to the west of the 1 April 1962 Chāhak earthquake meizoseismal area (33°03′N–58°29′E, +1632 m; Figure 12.8).
Shekasteh Estuchāh [lit., "Fractured/Faulted Estuchah" in Persian]: Located to the west of the 23 September 1947 Dustābād and to the southwest of the 26 September 1947 earthquakes (33°34′N, 58°27′E, +1365 m; Figure 12.8).
Shekasteh Fāzelmand [lit., "Fractured or Faulted" plus the name of the site (Fāzelmand) in Persian]: 26 km north of the Jangal fault (34°53′N–59°24′E, 1080 m amsl; Figure 11.9), near the Jangal fault, SW of Roshtkhār, with a recorded earthquake on 21 October 1336.
Shekasteh Geleh [Galeh]-band [lit., "Fractured or Faulted" plus the name of the site (Gele-band), in Persian]: 39 km ESE of Torbat Haydariyeh, northwest of Nough and east southeast of Sangān (35°06′N–59°34′E, 1450 m amsl), immediately north of the Doruneh fault (Figure 11.9).
Shekasteh Gesh [lit., "Fractured/Faulted Gesh" in Persian]: East of Salāmi, north of Khāf (34°45′N–60°07′E; Figure 11.9).
Shekasteh Gol-e Ney [lit., "Fractured or Faulted" plus the name of the site (Gol-e Ney: Reed Flower), in Persian]: 32 km north of the Jangal fault (34°58′N–59°06′E, 1220 m amsl), west southwest Sangān, with a recorded earthquake on 21 October 1336 (Figure 11.9).
Shekasteh Gushmir [lit., "Fractured or Faulted" plus the name of the site (Gushmir), in Persian]: Northwest of Dagh-e Mohammadābād, 60 km SW of Qā'en, west of the Chāh Raqsh fault (33°28′N–58°33′E, 1539 m amsl), with recorded earthquakes on 16 February 1941 and 26 September 1947 (Figure 12.8).
Shekasteh Hāji Mir [lit., "Fractured or Faulted" plus the name of the site (Hāji Mir), in Persian]: A mountain, 20 km north of the Jangal fault (34°54′N–59°13′E, 1281 m amsl), W Roshtkhār, with a recorded earthquake on 21 October 1336 (Figure 11.9).
Shekasteh Hesār Sangi [lit., "Fractured/Faulted Hesār Sangi (Stone-Fenced)" in Persian]: (33°07′N–58°52′E, +1857 m); Located in the meizoseismal area of the 1 April 1962 Chāhak earthquake (Figure 12.8).
Shekasteh Hezār [lit., "thousands (numerous) fractures," "highly fractured or faulted" in Persian]: 13 km southeast of Tabas-e Golshan town, along the Tabas fault (33°30′N–57°03′E, 866 m amsl; Figure 13.8). This is a very interesting linguistic trace; local toponyms have been used by the

people in the Tabas-e Golshan area, where there was no record of a large-magnitude earthquake for at least 1000 years prior to the 16 September 1978 M_w 7.3 earthquake. Therefore, it is likely a millennium-old local name given to the hanging-wall of the Tabas earthquake fault; during the 1978 earthquake, everything was shredded by numerous parallel bedding-plane slips with a thrust mechanism (Berberian, 1979a).

Shekasteh Hezārtak [lit., "fractured or faulted" plus the name of the site (Hezārtak), in Persian]: 66 km southeast of Sabzévār, northwest of Kāshmar, west of Ālbulāgh village (35°37N–57°55′E, 1815 m amsl).

Shekasteh Hizu [lit., "Fractured/Faulted Hizu" in Persian]: (31°27′N–58°54′E).

Shekasteh Kālurak [lit., "Fractured or Faulted Creek" in Persian]: 4.4 km west southwest of Kalāteh Korizān, west of Ābgarm [hot/thermal spring], and northwest of Tighāb, near the Ābiz fault (33°52′N–59°44′E, +1850 m amsl; Figures 13.17 and 13.18), located to the north of Dagh-e Shekāfteh (55 km northeast of Qā'en). To the east, this mountain forms the hanging-wall of the Ābiz fault activated during the 15 February 1549, 30 June 1936 M_s 6.0, and 14 November 1979 M_w *6.7* earthquakes; to its west, the 16 January 1979 M_w 6.5 Boznābād earthquake took place.

Shekasteh Kārizi [lit., "Fractured or Faulted Kārizi" in Persian]: north of Tāybād (34°49′N–60°48′E).

Shekasteh Kasuri [lit., "Fractured of Faulted Kasuri" in Persian]: To the west of the 1 April 1962 and to the east of the 16 September 1978 Tabas-e Golshan earthquake meizoseismal areas (33°23′N–58°04′E, +1307 m).

Shekasteh Khun Oshtor [lit., "Fractured or Faulted Khun Shotor (camel's blood)" in Persian]: (31°58′N–59°18′E).

Shekasteh Lisāb [lit., "Fractured or Faulted" plus the name of the site (Lisāb), in Persian]: Southwest of Sabzévār, 130 km WNW of Kāshmar in the Kenār Shahr district (35°21′N–57°04′E, 1220 m amsl).

Shekasteh Malakhi [lit., "Fractured or Faulted" plus the name of the site (Malakhi), in Persian]: 15 km SW of Torbat Haydariyeh, NE of Tājrud, near the Doruneh fault (35°08′N–59°09′E, 1250 m amsl; Figure 11.9).

Shekasteh Maleki [lit., "Fractured/Faulted" plus the name of the site]: A locality along the Chāhak fault (33°15′N–58°51′E, +1497 m); located in the meizoseismal area of the 1 April 1962 earthquake (Figure 12.8).

Shekasteh Mirakhāsh [lit., "Fractured or Faulted" plus the name of the site (Mirakhāsh), in Persian]: West northwest of Birjand, east southeast of Deihuk; to the west of the 1 April 1962 earthquake meizoseismal area (33°05′N–58°20′E, 1544 m amsl; Figure 12.8), a highly faulted and fractured terrain. See Shekasteh Sabz.

Shekasteh Miyān [lit., "Fractured or Faulted" plus the name of the site (Miyān), in Persian]: South of Sabzévār, 93 km NW of Kāshmar (35°29′N–57°30′E, 1370 m amsl), west southwest of the 5 October 1933 Sebeh earthquake meizoseismal area (Figure 12.7).

Shekasteh Miyāndehi [lit., "Fractured/Faulted" plus the name of the site (Miyāndehi) in Persian]: 52 km north of Gonābād, SSW of Miyāndehi village and north of the Kāl-e Shur River, east of Shekasteh Yunesi (34°49′N–58°36′E, 954 m amsl); along the western segment of the Jangal fault (Figure 11.9).
Shekasteh Mohammadābād [lit., "Fractured or Faulted" plus the name of the site (Mohammadābād), in Persian]: In the Mohamadābād Playa, 60 km southwest of Qā'en, with recorded earthquakes on 16 February 1941 (M_w 6.1) and 23 September 1947 (M_w 6.8) (Figure 12.8).
Shekasteh Mohammad Zurā [lit., "Fractured or Faulted" plus the name of the site (Mohammad Zurā), in Persian]: South of Sabzévār and 85 km west northwest of Kāshmar (35°28′N–57°32′E, 1305 m amsl). Southwest of the 5 October 1933 M_s 6.1 earthquake meizoseismal area (Figure 12.7).
Shekasteh Mushkhor [lit., "Fractured or Faulted" plus the name of the site (Mushkhor), in Persian]: South of Sabzévār, 69 km west of Kāshmar (35°15′N–57°44′E, 1050 m amsl), and in the vicinity of the Doruneh fault.
Shekasteh Nilband [lit., "Fractured or Faulted Nilband" in Persian]: Southeast of Torbat Jām and northeast of Tāybād (34°58′N–61°01′E).
Shekasteh Puzbeh [lit., "Fractured or Faulted Puzbeh" in Persian]: (31°48′N–58°58′E).
Shekasteh Qalamsar [lit., "Fractured or Faulted Qalamsar" in Persian]: (34°54′N–57°10′E, 1592 m amsl).
Shekasteh Rish Basteh [lit., "Fractured or Faulted" plus the name of the site (Rish Basteh)]: 19 km south southwest of Torbat-e Jām and northwest of Shahr-e Nau River at Kaj-Āb with right-lateral displacement, in the immediate north of the Torbat-e Jām fault (35°03′N–60°34′E, 1150 m amsl) and near the meizoseismal area of the 24 March 1918 M_w 6.0 earthquake.
Shekasteh Sabz [lit., "Fractured/Faulted Green"]: Highly fractured terrain, west northwest of Birjand, east southeast of Deihuk (33°04′N–58°21′E; Figure 11.10). Also Shekasteh Mirakhash.
Shekasteh Sag Mordeh [lit., "Fractured or Faulted" plus the name of the site (*Sag Mordeh*, or "dead dog"), in Persian]: 11 km northeast of Shāhrakht (33°43′N–60°15′E, 1200 m amsl), near the Ābiz fault, with a recorded earthquake on 30 June 1936 (M_w 6.0) (Figures 13.17 and 13.18).
Shekasteh Sangi [lit., "Fractured or Faulted Sangi" in Persian]: (31°55′N–58°56′E).
Shekasteh Sardāb [lit., "Fractured or Faulted Sardāb" in Persian]: North of Tāybād (34°51′N–60°46′E).
Shekasteh Yunesi [lit., "Fractured/Faulted" plus the name of the site (Yunesi)]: 52 km north northwest of Gonābād, west of Shekasteh Miyāndehi (34°48′N–58°32′E, 905 m amsl); along the western segment of the Jangal fault (Figure 11.9).

Tang-e Chak Chak or Chāk Chāk [lit., "Sliced or Fractured Gorge" in Persian]: At the Qara Āghāj district, southeast of Marāgheh (37.01′N–46.52′E).
Tang-e Chak Chak or Chāk Chāk [lit., "Sliced or Fractured Gorge" in Persian]: West of Furg, between Rostāq and Furq at the High Zāgros near the High-Zāgros fault (Figure 13.14), with a recorded earthquake on 16 November 1990 (M_w 6.5).

1.4 PLACE NAMES REFERRING TO FREQUENT OCCURRENCES OF LANDSLIDES, SLUMPINGS, ROCKFALLS, ROCKSLIDES, AND EARTH FLOWS

The following ancient place names might be related to the secondary effects of earthquakes and signify their hazards.

Ākhmā Qayeh/Qayah (?) [lit., "Fallen Rock," "Rockfall," "Mountain with Rockfall" in Turkic]: A village in northwest of Sahand volvano (Figure 11.12).
Chartāvā Mt. (?) [from *charmākh*, lit., "Springing, Cracking, Splitting Mountain" in Turkic]: Located 26 km NW of Salmās (Figure 12.5), with a recorded earthquake on 6 May 1930 (M_w 7.1).
Chāt Qayeh (?) [lit., "Fractured or Exploded Rock or Mountain" in Turkic]: A village in the Namin District of northeast of Ardebil (Figure 13.16).
Dāsh Ātān [lit., "Rock Slinger," "Rock Thrower" in Turkic]: A village southeast of Bostānābād and Siāh Cheshmeh, along the North Tabriz fault (37°39′N–47°06′E; Figure 11.12).
Dāsh Ātān [lit., "Rock Slinger," "Rock Thrower" in Turkic]: A village 13 km southeast of Marāgheh, south of Sahand volcano (37°19′N–46°22′E; Figure 11.12).
Dushuk Dāgh [lit., "Dropped, Fallen, or Loose Mountain" or "Mountain with Rockfalls" in Turkic]: In northwest Sufiān, near the North Tabriz fault (Figure 11.12).
Jodā Qayeh Mt. (?) [lit., "Separated Rock or Mountain" in Persian (Jodā) and Turkic (Qayeh)]: Located at Māku (39°18′N–44°30′E), near the Shādlu, Kelisā-Kandi fault (Figure 8.2, with a recorded earthquake on 29 April 1968 (M_s 5.5). During the 3 April 1966 earthquake swarm, the large, loosened rock block of Jodā Qayeh, hanging above the buildings of Māku developed a deep fracture, fell, and rolled into the town of Māku (Etelā'āt Newspaper, 4 April 1966, No. 11944).
Kamar Rikhteh [lit., "Fallen Rock or Mountain" in Persian]: 14 km northwest of Kāshmar, north of Sarhauzak village, in the vicinity of Doruneh fault, and northwest of the 25 September 1903 M_w 6.0 *Turshiz* earthquake meizoseismal area (1850 m amsl; Figure 12.7).
Khākriz (?) ["Earth/Soil Fall or Flow" in Persian]: A village in the Germi District of Ardebil and Moghān.
Qārni Yārekh [lit., "Torn Belly" in Turkic]: A mountain, 12 km south of Salmās, along the Salmās earthquake fault (38.09′N–44.43′E; Figure 12.5), with a

recorded earthquake on 6 May 1930 M_w 7.1. This name was given to a large fractured portion of the mountain with a rock-cut Urārtian tomb (Kleiss, 1969; Tchalenko and Berberian, 1974; Berberian and Tchalenko, 1976a).

Yārim Qayeh (?) [lit., "Half Rock or Mountain" in Turkic]: A village in the Iv Oghli District of Khoi, 17 km NE of Khoi (38°39′–45°01′E; Figure 8.2), with recorded earthquakes on 18 April 1843 (M_s~5.9) and 24 February 1900 (M_s 5.4).

Yārim Qayeh (?) [lit., "Half Rock or Mountain" in Turkic]: 18.5 km NNW of Māku (39°27′N–44°25′E), near the Bedāvli, Shādlu, and Kelisā-Kandi faults (Figure 8.2), with a recorded earthquake on 29 April 1968 (M_s 5.5).

1.5 PLACE NAMES REFERRING TO EARTHQUAKE SPRINGS

Since ancient times, we have known that springs appeared or dried up during earthquakes. These spring names refer to earthquakes that may have occurred:

Kur Cheshmeh [lit., "Blind (dried) Spring"]: It is located in the area northwest of Jirandeh (~36°43′N–49°46′E; Figures 13.12 and 13.13), along the Kashachāl fault, where it diverges from the southeastern extremity of the Rudbār fault in the western Alborz. At this location, the Kashachāl fault has a south-facing scarp and shows a series of convincing stream displacements of 150–200 m. It refers to a spring whose water dried up after an earthquake (Berberian and Walker, 2010).

Kur Cheshmeh [lit., "Blind (dried) Spring"]: Located near the western segment of the Ipak fault. During the 1 September 1962 M_w 7.0 earthquake, a new spring appeared (Figure 12.13).

Zelzeleh Au [Āb] [lit., "Earthquake Water (spring)" in Persian]: 26 km southwest of Zarand, along the Jorjāfk fault (30°38′N–56°20′E; Figure 16.19), where a spring flows through a highly fractured and limonitized Jorjāfk fault zone.

Zelzeleh Bulāgh [lit., "Earthquake Spring" in Arabic (Zelzeleh) and Turkic (Bulāgh)]: 22 km east southeast of Salmās (38°10′N–44°53′E; Figure 12.5) and 20 km northeast of the 6 May 1930 (M_w 7.1) earthquake fault (Tchalenko and Berberian, 1974; Berberian and Tchalenko, 1976a).

Zelzeleh Bulāghi [lit., "Earthquake Spring" in Arabic (Zelzeleh) and Turkic (Bulāghi)]: 18 km southwest Khoi (Figure 8.2) and 1 km west of the Qotur fork, north of the railroad (38°09′N–44°52′E), in the epicentral area of the 24 February 1900 M_s 5.4 Khoi earthquake.

1.6 PLACE NAMES REFERRING TO DEFLECTED RIVER COURSE

Kaj-Āb [lit., "Crooked Water"]: The Shahr-e Nau River, flowing northeast, in the Kaj-Āb area, about 23 km south southwest of Torbat Jām (35°04′N–60°34′E, 1049 m amsl). Southeast of Shekasteh ("broken or faulted"), Rish Basteh (35°04′N–60°29′E, +1178 m amsl) shows a sharp bend of 130°, influenced by the NW–SE orientation of the Torbat Jām fault, and then flows along the fault toward the southeast. This sharp-angle disruption to the river course is consistent with the right-lateral motion of the Torbat Jām fault and fold growth of the Neogene beds. An earthquake was recorded on 24 March 1918 (M_s 6.0).

1.7 OVERVIEW

Historical records of earthquakes are heavily biased toward the urban areas located along the major ancient trade routes. In fact, the majority of recorded historical earthquakes represent urban earthquakes: there is little information about the villages, sparsely populated areas, remote locations, mountains, or deserts. On the other hand, local place names and linguistic traces referring to prehistoric earthquakes are interesting tools in filling the gap in historical seismicity and in earthquake hazard warnings in oral traditions.

We find some general references to the seismicity of different regions in the ancient chronicles. Some general historical statements are interesting because they refer to a memory of earthquakes, the dates and macroseismic details of which have been lost. For example, Al-Mas'udi (956) listed areas that he visited that were famous for frequent earthquakes in his time (Berberian, 1997b; Berberian and Yeats, 1999).

The towns of Āmol (along the south shore of the Caspian Sea (Figure 12.10)) and Sirāf (along the Persian Gulf shore in the south), for which there are no pre-943 surviving recorded earthquakes, were known to the local people and Al-Mas'udi (956) as earthquake-prone zones. Al-Mas'udi traveled to the Fārs province in 915 and arrived in Gujārāt in 918. His records on the seismicity of Iran, especially at Sirāf and Āmol, would have been collected from 915 to 918. The earliest known earthquake in the Āmol region is the 1809 ($M_s \sim 6.5$) event; for the Sirāf port (the present Tāheri Port), the earliest known earthquakes are the 17 June 978 ($M_s > 6.0$) and the Spring 1008 ($M_s \sim 6.5$) events. Al-Mas'udi (956) finished his book in 956; he died in the same year, about 22 years before the first recorded Sirāf earthquake of 17 June 978 occurred. This clearly indicates that the recorded seismic history of Iran is not complete. Numerous earthquakes are missing as referred to by the chronicles and ancient place names.

Al-Mas'udi (956; ed. Payandeh, 1986; ed. de Goeje, 1894; ed. Beirut, 1961) reported:

I have visited most of the regions that are famous for their frequent occurrences of great earthquakes; among them are the bilad *[district]* of Sirāf *[the present Tāheri port]* on the Persian Gulf, situated between the mountains and the sea; territory of Saimareh in the Mehrjān Qodzaq[1] at the foot of Mount Kabr *[Kabir Kuh in the Zāgros];*

1. Mehrjān Qodzaq: Mehrgān Kadeh = Saimareh and Dareh Shahr in the western Zāgros, *and of Māsbadzān* district [Mehrgān Qodzaq + Māsbadzan = Ilām in the western Zāgros] of Jebāl district ["Mountain" in Arabic: referring to the Jebāl province/'Erāq-e 'Ajam/Media/Mada, the territory of the ancient Median Tribes].

and the city of Anṭākiyeh *[Antioch]* in the Qonsorin district; and 'Avasem of Shām *[Syria]* at the mountain foot; and velāyat of Qumes[2] which is highly seismic, and the earthquakes are so violent that the springs dry up and gush out in another place. It is a very disturbed city,... And Āmol is another earthquake-prone city which I visited; it is on the foot of the big mountain known as Dabāvand *[Damāvand]*, and it is said that Damāvand *[in Alborz Mts.]* is the highest mountain in the world! Many of the towns of Tabaréstān *[the present Māzandarān]* are subject to earthquakes. And other places as well, besides these. Nonetheless, I have not seen an earthquake larger and longer than this; it was as if a large mass was passing through underground and hitting it and shaking it, and apparently, it was larger than the Earth and was detached from it, and there was a loud roaring noise in the air.

A few years later, Abu-Dulaf (~950; ed. Minorsky, 1955; ed. Tabātabā'i, 1975) wrote:

Between Esfahān and Ahvāz stands the bridge of Idhaj[3] which is one of the wonders of the world; it is built of blocks of stone and spans a dry river (-bed) of great depth. Idhaj is often visited by earthquakes and has many minerals. There grows in Idhaj a kind of alkaline plant (qāqullā)[4], the pressed juice of which is indicated for gout. There also stands an important fire-temple of which the fire continued to burn till the time of Rashid.

Almost 265 years later, Yāqut (1225; ed. Wüstenfeld, 1866–1873) wrote that: "*earthquakes are frequent at Idzaj.*" The oldest recorded earthquake in the region was in 1052; it took place in the time interval between the two visits of Abu-Dulaf and Yāqut.

Mostaufi Qazvini (1340) stated that earthquakes rarely happened at Esfahān; we are now certain of this: "*Earthquakes, or thunderbolts, or rains that cause damage, very seldom occur here*" [Esfahān]. This was later stated by Jean-Baptiste (Sir John) Chardin (1735, 1811), a jeweler and royal merchant to Shāh 'Abbās II Safavid (r. 1732–1736), who added that, on the contrary, *Hircania* [the modern Gorgān province in the northeast] was famous for frequent and furious earthquakes (Chardin, 1811). Later, *Al-Esfahāni* (1891) added that:

Blessed and exalted almighty God has protected the city of Esfahān from windstorms, lightning and earthquakes. He has graced and blessed the inhabitants of this city and has secluded them from these types of calamities; and hence, the city is not located on the earthquake vein; and genuinely earthquakes never occur in the city, except when strong earthquake happening in far distant lands and consequently slightly felt with no damage or destruction in the city. Most people in sleep or walking cannot feel these events.

---

2. Qumes: Qumis, Kumes, Komesh, the modern Dāmghān in northern Iran.
3. Idhaj: Izeh; Māl Amir; lit., "the ruler's property" in the Khuzestān province of southwest Iran.
4. Qāqullā: *Salsola fruticosa, alzkrut, Anglice,* saltwort, or glasswort.

Chapter 2

Earthquake Myths

Do not consider this falsehood or fairy tale.

Ferdowsi Tusi (1010)

In the ancient preliterate societies, unexplainable, unpredictable, and powerful natural disasters, especially earthquakes, played an important role in the oral traditions of the traumatized and surviving populations living in seismic regions. Recurrence of earthquakes and reiteration of the oral traditions among the survivors, together with a rationalization of the disaster, became an effective verbal tool for a primitive earthquake hazard/crisis warning system (addressed in Chapter 1) during numerous earthquake disaster recovery periods in the ancient past.

Unable to depart from their known homeland, these people attempted to transmit their knowledge of its physiography and hazards to the next generations through secular and spiritual oral traditions (Chapter 3). Songs, histories, folktales, and rituals were gradually transformed into complex myths over the millennia.

In this chapter, I review and examine the multifaceted metaphors in the myths and oral traditions that still exist on the Iranian plateau. I try to present the ancient peoples' rationalization of earthquakes, the emotional responses of those who were traumatized, the messages they transmitted to future generations, and steps that they could have taken to prevent or minimize their risks and losses.

Earthquakes were thought to be initiated by storm-gods or "evil-spirited" monsters and demons; during conflicts with good spirits, they dug their way underground and then shook the Earth. As they descended, the Earth trembled and mountains formed. The examples cited here are from ancient religious texts; these were the thoughts held by indigenous peoples prior to the advent of the Zoroastrian Avestan and the Vedic doctrine. I do not categorize them as "religious thoughts"; that category will be addressed later in Chapter 3.

I should mention that most of the Zoroastrian Avestan and the Pahlavi translations into Persian and English are about a century old. There has been

☆"To view the full reference list for the book, click here."

Developments in Earth Surface Processes, Vol. 17. http://dx.doi.org/10.1016/B978-0-444-63292-0.00002-8

relatively little progress in the translation and interpretation of these ancient texts, and the historical veracity of the stories is still a matter of debate. For seismic parameters and locations, see Chapters 11–17 with corresponding figures and tables.

2.1 *CHASHMAG-E DIV* [STORM-MONSTER/DEVIL CHASHMAG] IN PAGAN IRAN (PRE-1200 BCE)

In the Zoroastrian Pahlavi [Middle Persian] texts of the *Bundahishn* [*Bondéhéshn* in Persian; the Zoroastrian "Original/Primary Creation"] and *Denkard* [*Dinkart*; the Zoroastrian "Acts of Religion"], there are brief references to the storm-monster *Chashmag*, who is responsible for earthquakes (ed. West, 1880; Bahār, 1966, 1983, 1990; Berberian, 1991, 1994). A trace of this Iranian pagan belief has been preserved in the Zoroastrian sacred texts. *Chashmag*, referred to as *Div* [*Daeva*/*Dev*/demon] in the texts, might have been an earlier god whose rank was demoted to a demon [*Div*] in the new religion.

In the *Bondéhéshn* (XXVIII/24), *Chashmag-e Div* is described as a storm-monster [*Div*] that generates earthquakes, tornadoes, and typhoons and fights against useful clouds and winds: "Chashmag the demon is who makes earthquakes, and also causes the whirlwind [tornado] which seeks to harm clouds" (ed. West, 1880; ed. Bahār, 1990). The same phrase is mentioned in the Iranian (Greater) *Bondéhéshn*, Chapter XXVII/29 (ed. Anklesaria, 1956). West (1880: XXVIII/24) gave an uncertain and no longer warranted reading of *Kishmak* for "Chashmak," *disastrous (vazandak)* for "earthquake," and *passes over for disturbance* for "to harm clouds."

In Chapter V of the Iranian (Greater) *Bondéhéshn* (ed. Anklesaria, 1956; ed. Bahār, 1990), we read: "7. Other Devs also of innumerable names [Chashmagān Divān] came against the Wind and the Yazads producing rain [Yazads; Avestan Yazata: worthy of being honored or praised]. Their details are long; their motion and contest in retrogression and slowness are manifest in astrology also" (ed. Anklesaria, 1956). There is another reference to Chashmag in Chapter XXI, E, of the same book (ed. Anklesaria, 1956; ed. Bahār, 1990; Berberian, 1991, 1997b; avesta.org; Peterson, 2002): "*As regards earthquake it is manifest that even the same dreadful devs [Cheshmagān Divān] obstruct with sorcery the passage of that wind of life, which is the preserver, whilst moving through the fissures of the mountains, so that it may have no movement therethrough*" (ed. Anklesaria, 1956).

In Book 7 of *Denkard*: *Marvels of Zoroastrianism*, Chapter 2 (Parentage of Zartosht), *Chashmag-e Div* tries to destroy the house of Pourushasp (Zoroaster's father) as well as the village and causes typhoons (ed. West, 1897; Bahār, 1966, 1983, 1990; Berberian, 1991). Accordingly: "*Chashmak, astute in evil, growled thus: 'I will undertake his destruction.' Astute in evil, he rushed away with thrice fifty of the demons who are Karaps* [pagan priests, Avestan karapan] *of Chashmak; and that village was partly uprooted and partly destroyed by him, fellow-workers were ruined, and the number of fellow-eaters of broken victuals, attending the great, was not broken up, among whom was he that had repelled his authority. It is declared that, afterwards, Pourushasp [Zartosht's father] asked again for that Hom from Dukdāub, and he pounded it, and with that cow's milk, into which the nature of the body of Zartosht had come, he here mingled the guardian spirit of Zartosht, and the nature of the body came at once into union with it*" (ed. West, 1897). There is another reference to *Chashmag* in Chapter 4, Book 7 (Marvels of Zoroastrianism) of *Denkard* (ed. West 1897).

The myth of the destruction of the house of Zoroaster's father Pourushasp (Pourush-aspā/Pourushaspā in Avestan/Purushap in Pahlavi; lit., "owner of old horse") by Chashmag-e Div could have originated from a disastrous earthquake. Airyanem-Vaedja, the motherland of Zoroaster [Zarathushtra/Zartosht], is possibly situated in the area north of Khorāsān in the Khārazm Province (south of Lake Khārazm or Ārāl) of the Greater Khorāsān. However, since Doghdu (Pourushasp's wife and Zoroaster's mother) was born and raised in Rhagae (Raga, modern Ray, south of Tehrān), it is realistic to think that they married at Rhagae (Ray in Figure 11.2). Therefore, the earthquake might have occurred in Ray. This could be taken as a record of a major, devastating earthquake in the Ray (or Khārazm?) area in the past ($\sim$1200 BCE); the memory of it was then kept alive in Zoroastrian annals, folktales, and legends. It should be noted that the Khārazm Province and Ray are seismic regions. An earthquake of magnitude $M_s \sim 6.1$ took place in 1208 in Khārazm (Kondorskaya and Shebalin, 1977, 1982; Bune and Gorshkov, 1980), and Ray has been devastated several times in its recorded known history (see Chapter 17; Berberian and Yeats, 2014).

2.2 SUBTERRANEAN WRITHING OF EVIL SPIRIT IN CONFLICT WITH THE EARTH: PROTO-ZOROASTRIAN TRADITION (PRE-1200 BCE)

According to the ancient Iranian myths, earthquakes and chaos during the Evil Spirit's assault on planet Earth ultimately led to the growth of the mountain belts, which then acted as ramparts and shelter against the Evil Spirit [*Ahriman*]. The mountains were thought to be a direct result of the earthquakes caused by the Evil Spirit's assault on the good creation of *Ahurā Mazdā* [in Avestan; *Ohrmazd* in Pahlavi/the Wise-Lord]. This seems to be the earliest known view in which:

(i) Earthquakes are incremental episodes in the "mountain building process (orogeny)";
(ii) Earthquakes accompany abrupt ground surface changes; and
(iii) Earthquakes shape the large crustal structures on planet Earth. Compare the ideas in the following works with the theory of orogeny introduced by the American geologists James Hall (1811–1895) and James Dwight Dana (1813–1895) in 1873.

According to Chapter VI, C of the Iranian (Greater) *Bondéhéshn* (ed. Anklesaria, 1956; ed. Bahār, 1990): "*As the Evil Spirit entered and the Earth trembled* [quaked], *the substance of the mountains was produced in the Earth (on account of the shaking/quaking of the Earth, the mountains were immediately in motion): first, Alburz* [Alborz] *of Divine destiny* [Harāiti Bārez in Zāmyād Yasht], *then the other mountains within the Earth; for as Alburz* [Alborz] *grew up, all the mountains were in motion; for, they have all grown up from the roots of Alburz; at that time, they proceeded from the Earth, like trees, which cause the tendrils to run above and the roots underneath. Their roots were so arranged by connection, passing into one another. And thereafter, [it was not possible] for the Earth to shake from its place. [As] one, says [in the Scripture]: 'The mountain is a great joint of lands'*" (ed. Anklesaria, 1956).

In Chapter VIII of *Bondéhéshn*, regarding the conflict that the evil spirit waged with the Earth (ed. West, 1880), we read that: "*As the evil spirit rushed in, the Earth shook [jonbid] and the substance of mountains was created in the Earth. First Mount Alborz [Harāborzeiti: "High Mountain"] arose; afterwards, the other ranges of mountains of the middle Earth; for as Alborz grew forth all the mountains remained in motion, for they have all grown forth from the root of Alborz. At that time they came up from the Earth, like a tree which has grown up to the clouds and its root to the bottom; and their root passed on that way from one to the other, and they are arranged in mutual connection. Afterwards, about that wonderful shaking out from the Earth, they say that a great mountain is the knot of lands*" (ed. West, 1880).

In Chapter XXXVII/46 of *Dādistān-e Dinik* [the Zoroastrian book of religious opinion and rules] (ed. West, 1882), we find similar consequences of the earthquakes: "*And he smote the ox* [the sole-created or primeval ox, whence all animals are said to have sprung], *he made Goyamard mortal, and he shook the Earth; and the land was shattered, creation became dark, and the demons rushed below, and on all sides, and mounted even to the uppermost third of the sky*" (ed. West, 1880).

In Chapter 7 of the *Selections of Zād Spram* (ed. West, 1897), we also learn of the growing process (mountain building) of the Alborz Mountains as a result of earthquakes. It should be emphasized that ancient Iranians believed that the mighty Alborz Mountain belt made a ring around the ancient known world (Asia and Europe). Hence, they were referring to the Himalayan–Alpine mountain belt and not to the present Alborz Mountains of Iran and Caucasus (Berberian, 1994; see also below). We read that:

And as he (Ahriman) came thirdly to the Earth which arrayed the whole Earth against him – since there was an animation of the Earth through the shattering – Alburz *[Alborz]* grew up, which is the boundary of the Earth, and the other mountains, which are amid the circuit of the Earth, come up 2244 in number. And by them the Earth was bound together and arranged, and on them was the sprouting and growth of plants, wherefrom was the nourishment of cattle, and therefrom was the great advantage of assistance to men. 3. Even so it is declared that before the coming of the destroyer to the creatures, for a thousand years the substance of mountains was created in the Earth especially as antagonism came on the Earth, and settled on it with injury – and it came up over the Earth just like a tree whose branch has grown at the top, and its root at the bottom. The root of the mountains is passed on from one to the other, and is arranged in connection with them, and through it is produced the path and passage of water from below to above, so that the water may flow in it in such manner as blood in the veins, from all parts of the body to the heart, the latent vigor which they possess. 5. And, moreover, in six hundred years, at first, all the mountains apart from Alburz were completed. Alburz was growing during eight hundred years; in two hundred years it grew up to the star station, in two hundred years up to the moon station, two hundred years up to the sun station, and two hundred years up to the sky. After Alburz the Aparsen mountain is the greatest, as it is also called the Avar-royishn ("up-growth") mountain, whose beginning is in Sagastān *[modern Sistān]* and its end unto Pārs *[modern Fārs]* and to Cinistān *[China]*.

ed. West (1897)

The Proto-Zoroastrian/Zoroastrian tradition of the mountains sharing connected roots and being "active and in motion," can be traced in the thoughts of the Iranian poet Maulānā Jalāl U'ddin Rumi [1263–1273; in *Mathnavi-ye Ma'navi*, Book IV:9; ed. Nicholson, 1926]:

And that it *[the mythical Mount Qāf/Alborz]* had become a ring surrounding the (whole) world,
He was amazed at that immense creation (work of God).
He said, 'Thou art the mountain (indeed), what are the others?
For beside thy magnitude they are (but) playthings'.
It [the Mt.] replied, 'Those (other) mountains are my veins,
They are not like unto me in beauty and glory.
I have a hidden vein in every land,
(all) the regions of the world are fastened to my veins.
When God wills an earthquake in any land,
He bids me and I cause the vein to throb.
Then I make to move mightily the vein,
With which the (particular) land is connected.
When He says 'Enough!' my vein rests,
I am (apparently) at rest, but actually I am in rapid motion.
At rest, like the (medicinal) ointment, and very active (efficacious),
At rest, like the intellect, while the speech (impelled) by it is moving.'

Rumi (1263–1273), ed. Nicholson (1926)

About 230 years after *Maulānā Jalāl U'ddin Rumi*, al-Suyuti (1499) mentioned the story of the mother of mountains, the *Qāf Mountain* [the modern Alborz/Elbrus in the Caucasus], and that its roots are connected to all the mountains of the Earth. By the will and command of God, the *Qāf Mountain* shakes its roots and creates earthquakes.

The pre-Zoroastrian [pre-1200 BCE; Boyce, 1989] idea of the storm-demons [Evil Spirit, Ahriman, and the Devil] entering the Earth and setting the Earth atremble, which is addressed in the Zoroastrian Pahlavi Texts, seems comparable to the actions of the Sumerian/Akkadian goddess *Ereshkigal* [ca. the mid-sixth millennium through the early second millennium BCE]. *Ereshkigal* [lit., "Great Lady Under Earth"], the older sister of goddess of fertility, love, sex, and war, Ishtār/Inānnā, was the Sumerian goddess of Irkalla, the land of the dead and underworld. *Ereshkigal* was also considered to be "the originator of earthquakes with her underground cries as roaring of lions" (Weinder, 1939).

2.3 MARUTS, VĀRUNĀ, AND INDRA (IN THE RIG VEDĀ AND THE AVESTĀ; CA. 1500–1200 BCE)

The Rig Vedā [Rg-Veda], one of the four sacred scriptures in Hinduism, composed ca. 1700–1100 BCE, has strong linguistic and cultural similarities with the Iranian Avestā (Boyce, 1989; Oberlies, 1998). In Book V, Hymn LXXXV- Vārunā, we learn that *Vārunā* [the chief lord of natural order; *Varenā* in the Avestā] straightened the mountains' roots, which were loosened by the *Marusts* [the gods of the winds and storms and the companions and friends of Indra], causing earthquakes (ed. Griffith, 1896): "*LXXXV, 4. When Vārunā is fain for milk he moistens the sky, the land, and Earth to her foundation. Then straight the mountains clothe them in the rain-cloud: the Heroes* [the strong Maruts]*, putting forth their vigour, loose them* [loosen the roots of the mountains and make them tremble] . . ." (ed. Griffith, 1896).

A metaphoric reference to an earthquake is also given in Book II, Hymn XI-Indra of the Rig Vedā (ed. Griffith, 1896), where the god Indra's strong thunder trembles Heaven and Earth: "*Indra hath hurled down the magician Vṛtra* [Vritra; the draught demon] *who lay beleaguering the mighty river* [the great cloud that holds the rain]. *Then both the Heaven and Earth trembled in terror at the strong Hero's thunder when he bellowed*" (ed. Griffith, 1896).

In Book II, Hymn XII-Indra of the Rig Vedā (ed. Griffith, 1889) we read about *Indra*'s breath as "moving air" that shakes the mountains. *Indra* reigns over the "intermediate region" [the atmosphere] and conquers with his thunderbolt the demons of drought and darkness. "Moving air" [the wind] was one of the four primary active elements of the ancient world. In ancient India and Iran this "energy" caused earthquakes when trapped in the Earth (see Chapter 4). "*He who, just born, chief God of lofty spirit by power and might became the Gods' protector, Before whose breath through greatness of his valour the two worlds trembled, He, O men, is Indra. He who fixed fast and firm the Earth that staggered, and set at rest the agitated mountains, Who measured out the air's wide middle region and gave the heaven support,*

He, men, is Indra. By whom this universe was made to tremble, who chased away the humbled brood of demons, Who, like a gambler gathering his winnings seized the foe's riches, He, O men, is Indra. Even the Heaven and Earth bow down before him, before his very breath the mountains tremble" (ed. Griffith, 1889).

2.4 SHACKLED GIANT/PAQUA CAUSES EARTHQUAKE AT MOUNT ELBRUZ [ELBRUS, ALBORZ], QĀF VOLCANO, CAUCASUS

According to the ancient Iranian myths, *Azhi Dahāka* [in the Avestan; *Ahi* in the Rig Vedā; *Azhdhāg* in Pahlavi; *Azhdāhā* in Persian; *Zahāk* (Arabicized form of Dahāg)] occupied Iran (Ferdowsi Tusi, 1010). As it is told, *Feraydun* [the legendary *Pishdādiān* Aryan king, hero, and an emblem of victory, justice, and generosity in Persian literature; son of *Āptin,* one of the descendants of the mythical king *Jamshid*], together with *Kāveh Āhangar* ["Kāveh the Blacksmith"], leads a crowd that has assembled in an attack of the tyrannical king, the villainous *Zahāk,* in his palace. *Feraydun* clubs *Zahāk* with his bull-headed mace, takes him to the *Mount Damāvand* volcano in central Alborz Mountains, and leaves him there in chains [*Denkard*, IX.21.8-12]. According to Zoroastrian lore, *Zahāk*'s final end will come at the end of time, when the "promised Savior" appears, and *Zahāk* will break free. But at that future time, *Keresasp* [Garshaspa; Garshāsp, the last *Pishdādiān* Aryan king] will rise again and crush the dragon with his club (*Dādistān-e Dinik* 37/97; ed. Anklesaria, 1911).

As with the Damāvand volcano and the characteristic Zoroastrian/Indo-Iranian legend of a giant atop the highest mountain, the noble *Āzāt Māsis* [*Yazata*; lit., "venerable"], the modern *Ārārāt* volcano, the traditional national symbol of Armenia and the Armenians, was also closely connected with the myths and legends of the *Fiery Devil-Dragon Azhidahāk/Vishāp* (Movses Khorénātsi, 466; ed. Thomson, 1978). Here, among the Armenians and their northern neighbors, arose local versions of the Zoroastrian myth, in which the traditional *Ashy/Azhi Dahāka* yielded his place to native heroes of wickedness, and the traditional mountain was changed from the Damāvand volcano into the Āzāt Māssis [Ārārāt] and the Alborz [Elbruz, Elbrus] in the Caucasus [Qāf].

In Armenian legend, the dreaded figure was *Ārtāvāzd*, the changeling son of King *Ārtākhiāns.* Shortly after his accession to the throne, he went out to hunt wild boars and asses. He disappeared after becoming dizzy and fell, with his horse, down a precipice. People said that he was chained in a cave in *Māssis* with iron fetters, which were constantly gnawed at by two dogs. When the chains were broken, he would come out to rule over the world or to destroy it. They also said that the noise of the blacksmith's hammer on the anvil strengthened those chains. Therefore, even in Christian times, on Sundays and festival days, the Armenian blacksmiths struck their hammers on the anvils a few times, hoping to prevent *Ārtāvāzd* [*Āzhidāhāk, Azhi Dahāka, Azhi; Azhdahā*] from unexpectedly breaking loose upon the world (Ananikian, 1925).

A similar legend exists in characteristic Zoroastrian culture about the *Elbrus* [*Elbruz, Alborz*] volcano in the northern Caucasus Mountains [*Kuh-e Qāf*]. There, the shackled giant chained to the Earth/mountain causes earthquakes, lightning, and thunder (Saga 36 in Colarusso, 2002; Barber and Barber, 2004).

Another legend talks about the great Caucasian sagas, the *Nart* hero *Nasran,* who rode to the Blessed Peak [*Elbruz/Alborz volcano*] to fetch fire from the giant *Paqua*. According to this legend, *Paqua* bound *Nasran*'s body in chains and fastened him to the summit of the Blessed Peak. *Nart Pataraz* then brought back both fire and *Nasran*. On his way to the summit, *Pataraz* fought off one monster after another, who shook the ground as they fell [*As the Soul-Taker toppled over, his groan made the mountains tremble*], until *Paqua* fled, leaving *Pataraz* free to unchain and rescue *Nasran* (Saga 34 in Colarusso, 2002; Barber and Barber, 2004).

2.5 EARTHQUAKES AS A FUNDAMENTAL CAUSE OF MOUNTAIN BUILDING IN THE ZOROASTRIAN IRANIAN CREATION MYTH (~1200 BCE)

As mentioned in the *Bondéhéshn*, according to the Zoroastrian myth, there is a close connection between the "earthquake" and the "growth of the mountain belts." In Chapter LXX of *Dādistān-i Dinik* [Dadestān-e Dinik; lit., "Religious Decisions"] (ed. West, 1882), this connection is mentioned without the presence of evil/*ahriman*:

> The reply is this, that any place where a mountain is not discernible and a river-bed exists it is a fissure (ashkupo); and it is declared as clear that, even before the growth of the mountains, when the Earth was all a plain, by the shaking of the world the whole world became rent (zandako). Even Frāsuyāv *[Afrāsiāb]* of Tur *[Turān]* was especially mighty by causing the construction of channels (vidarg) there where it is mountainous, and also in low-lands, in which there is no mountain, and the shaking in its creation was the formation of great sunken springs and river-beds. And if it has been prepared in, or if it be in a ravine (shikafto) *[fracture]* of, the mountains, the cause, too, of the contraction, thundering, and tearing of a river, if its confinement be in the Earth, is the resistance which it meets in seeking a passage; and as it is a spring of the waters of the Earth, so also it is in the Earth, whose contraction and panting are mighty and full of strength.
>
> Dādistān-i Dinik, ed. West (1882)

2.6 THE SHAKING OF THE SOLID SKY MADE OF STONE (~1200 BCE)

According to the Zoroastrian cosmogony, sky [*asana* or *asan* in Avestan; *asmān* in Persian], the first of the seven creations, was made of hardest stone/rock-crystal; because of its brightness, it was classified as a metal (*Yasnā*, 30.5; *Yashts*, 13.2; *Greater Bondéhéshn*, I.54, Ia.6). We read in *Dādestān-e Dinik* [Religious Decisions], 91.2 (ed. West, 1882) that the sky is also shaken by an evil spirit.

2.7 TREMORS CAUSED BY SROSH'S STRIKE

In Chapter XXVI of the Iranian (Greater) *Bondéhéshn* (ed. Anklesaria, 1956; ed. Bahār, 1990), we read about the metaphoric tremor caused by the mace strike of *Srosh* [the Avestan *Sraosha*, lit., "hearkening"], a *yazata* (a divinity worthy of worship under the Ahurā Mazdā) of obedience, observance, and prayer [the name of the seventeenth day of the month according to the Zoroastrian religious calendar; a spirit being who guards the soul for three days after death; *Sorush* in Persian].

2.8 ATTRIBUTION OF EARTHQUAKES TO THE MOVEMENTS OF ANIMALS HOLDING THE EARTH

Earthquakes were also connected with the movements of animals carrying the planet Earth. Since this is a common belief in other parts of the world, and no particular ancient source is found in ancient Iran, it will be addressed in Chapter 5. The oldest recorded *Earth-Shaking Bull* ["*Bull of Heaven*"] is documented in the *Epic of Gilgamesh*, ca. 2700 BCE [addressed below in this chapter]. Thompson (1937) mentioned that the modern inhabitants of Basra in southern Mesopotamia attribute the earthquake to the movement of the *Buffalo of the Jinn* beneath the surface [*genn, jennies*; supernatural creature in Arabic folklore and later in Islamic teachings (see Surat al-Jinn, the 72nd sura of the Qorān)].

2.9 APOCALYPTIC EARTHQUAKES AND THE IRANIAN IMAGE OF AN ESCHATOLOGICAL LEVELING OF THE MOUNTAINS

Apocalyptic consequences of earthquakes seem to be a common thought in different religions and countries. One of the first references to this image is documented in the ancient Zoroastrian complex cosmogony and addressed briefly in the extant Avestā and the Pahlavi texts [i.e., *Bondéhéshn*, *Menog-e Kherad*, and *Selections of Zād Sparam*]. In this view, the sacred planet Earth was the third good creation of *Ahurā Mazdā/Ohrmazd* [the Wise Lord] after sky and water. As a result of the third insult to the good creation by the evil spirit [*Angra Mainyu/Ahriman*], the evil spirit rushed in, the Earth quaked, and the essence of the mountains was created in the Earth. At the end of time, the active planet Earth becomes flat, with no mountain peaks, valleys, or depressions.

According to Chapter XXXIV of the *Greater (Iranian) Bondéhéshn* (ed. Anklesaria, 1956; ed. Bahār, 1990), at the end of the world, with a leveling of the mountains, the Earth becomes flat: "*This too one says* [it is said in the Avestā], *This Earth will become a plain* [flat], *without height and without bottom; and there will be no hill nor summit, nor dale, nor highland, nor lowland*" (ed. Anklesaria, 1956).

Jāmāsp Nāmag [the Story of Jāmāsp], the Middle Persian book of revelations, adds that the air will be troubled with cold and hot winds blowing, fruits will dry up, and earthquakes will become numerous (Lincoln, 1982). The verse in Lincoln (1982) is not included in the *Pāzand Book of Jāmāspi*: *Prophecies of the Last Millennium* (Modi, 1903).

Although the Avestā and the Pahlavi [Middle Persian] texts make the leveling of the mountain belts on planet Earth the culminating act of the cosmic renovation and use them to metaphorically refer to sociopolitical equality (Lincoln, 1982), the idea and image of leveling the mountains could have come from long observation of rock avalanches, rock falls, and landslides (some triggered by the medium- to large-magnitude earthquakes) as well as erosion and weathering processes continuously acting on the mountains.

The influence of the proto-Iranian/Zoroastrian concept of the leveling of mountains on the sacred active planet Earth, as the culminating act of the cosmic Renovation [*Farāshgard*], can be later seen as an apocalyptic event in the following traditions:

- **(i)** Jewish: Isaiah 2/14, 40/4; Ezekiel 38/20; Habakkuk 3/6; Job 9/5; Psalm 46/2–3; Assumption of Moses 10.4; Apocalypse of Baruch 5.7; Ethiopic Enoch 1.6; Sibylline Oracles 3.777–79, (Winston, 1966; Lincoln, 1982) (see Chapter 3);
- **(ii)** Christian: Corinthians 13/2; Revelation 16/20; Sibylline Oracles 8.236; Lactantius, De ave Phoenice 5–6, (Cumont, 1931; Lincoln, 1982) (see Chapter 3);
- **(iii)** Islamic: Qorān Sura 18.47; 20.105–8; 56.5–6; 69.14; 73.14; 78.20; 8.13, (Bausani, 1959; Duchesne-Guillemin, 1962; Lincoln, 1982); (see Chapter 3) and
- **(iv)** Classical sources: The Oracles of Hystaspes, Book 7 of Lactantius's Divine Institutions (305 AD; Hinnells, 1973); Asclepius 25; Eustatius ad Iliad 16; Seneca, Consolatio Marcus 26.6 (sacred-text.com; Lincoln, 1982).

2.10 EARTHQUAKES IN THE SUMERIAN TEXTS [CA. EIGHTEENTH CENTURY BCE]

The Sumerians were a non-Semitic, non-Indo-European, native people who flourished at the southwest Zāgros Foredeep region from the beginning of the fourth to the end of the third millennium BCE (see Figure 8.1). They were the oldest native people in the region whose written documents in the form of clay tablets (dating largely from approximately 1750 BCE) have been preserved (Kramer, 1944). Being farmers and living in the plain of the Tigris [Arvandrud: i.e., Swift River/Diglat] and Euphrates Rivers (see Figure 8.1) in the Mesopotamia [Miyānrudān], the Sumerians were very alert to the natural forces, such as storms, drought, and floods. Some references to earthquakes, in part possibly metaphoric, can be detected in their writings where earthquakes are attributed to divine activities. Nonetheless, in spite of being away from the Zāgros Mountains foothills, they were fully aware of earthquakes and their destructive forces.

Ea, the Sumerian god of the water and Earth, was responsible for the earthquakes. The belief about *Ea* comes from the Assyrian letter to the king from Balasi, in which he mentions the occurrence of an earthquake (Thompson, 1937): "*Let them perform whatever are the rites for an earthquake; thy gods will cause (it) to pass away; (what) Ea has done, Ea has given release (therefrom). Whoever caused the earthquake, that same provides an incantation for release (therefrom). Was there no earthquake during the times of the fathers or grandfathers of the king? As for me, since I was (then) too small, I did not notice earthquakes*" (Thompson, 1937).

This very old notion later entered into the Judeo–Christian–Islamic beliefs (see Chapter 3).

2.10.1 Ishtār's Revenge and the Earth-Shaking Bull Monster of Heaven in the Epic of Gilgamesh [ca. 2700 BCE]

Gilgamesh was the fifth king of Uruk [Early Dynastic II, First Dynasty of Uruk], ruling ca. 2700 BCE. In the Babylonian epic poem, the *Epic of Gilgamesh*, Tablet VI, we learn that *Ānu* set loose the holy *Bull of Heaven* [the Earth-shaking monster] from out of the sky to terrorize *Gilgamesh*. This created three earthquakes [each time the Bull returned], which swallowed up nine dozen citizens of Uruk. The series of earthquakes shattered the palace of *Gilgamesh* at *Uruk*, killing the beautiful *Shamhat* (Kenneth Sublet, piney-2.com; George, 1999, 2003; Foster, 2001). *Uruk*, which was the ancient city of Sumer and later Babylonia, was located 247 km SE of Baghdad at 31°19′N–45°38′E. During the death of Gilgamesh, the great mountains trembled (Black et al., 1998–2006). The epic shows that the people at Uruk were familiar with earthquakes.

2.10.2 Goddess Inānnā Destroys the Ebih Mountain with Earthquake and Fire at the Rebel Lands [The Western Zāgros Mountains]

Inānnā [the Sumerian goddess of sexual love, fertility, and warfare] destroys the *Ebih Mountain* in the "Rebel Lands/Mountainlands" [in the Western

Zāgros Mountains to the north of Sumer and Akkad (see Figure 8.1)] with a strong "roar" in accordance with her previous "warning" (Kinnier Wilson, 1979). Accordingly, "rocks toppled" from the mountain, and great snakes (oil) spit poison. Kinnier Wilson (1979), quoting George Andrew Reisner (1896), mentioned another verse, where there is a reference to an earthquake and outbreak of fires in the "Rebel Lands."

This seems to be the oldest reference to (i) an earthquake warning possibly due to strong foreshocks; (ii) oil seepage due to strong ground motion and the fracturing of rocks in the Zāgros (the Rebel Lands or Mountainlands); and (iii) possibly catching fire. Kinnier Wilson (1979) speculated that in the ancient Sumerian myth and tradition associated with the goddess *Inānnā*, Mount *Ebih* may be identified with the *Kabir Kuh* [the "Great Mountain" in the modern Iranian Zāgros; for location see Figure 10 in Berberian, 1995].

Kabir Kuh is a 200-km-long anticline in the western Zāgros (32°43′N–48°03′E to 33°39′N–46°13′E) in the area northwest of the city of Dezful (see Figure 10 in Berberian, 1995). This is the location where the massive Saimareh (Saidmarreh) landslip was triggered by the 10,370 ± 120 years BP earthquake (Watson and Wright, 1969; Berberian, 1994). Kinnier Wilson (1979) hypothesized that the earthquake faulted the rock cover above the oilfields, and spontaneous combustion of the escaping gas explains the rest. It should be emphasized that, although Kinnier Wilson's theory is probable, earthquake faulting in the Zāgros basement does not typically penetrate to the surface, cutting the top sedimentary cover. However, the fracturing of oil and natural gas traped in the top sedimentary cover of the Zāgros is probable. Since ancient times, there are two burning mountains in the Zāgros with natural gas fire emerging from the fractures. One is the Tashkuh [lit., "Fire Mt."] located east of Rāmhormoz (31°16′47.6″N–49°48′30.8″E); the other is the Kuh-e Sukhteh [lit. "Burnt Mt."] near Omidiyeh on the southern limb of the Āghājāri anticline (30°49′09″N–49°46′49″E).

Furthermore, in hymns to Inānnā and Nibru for Ishme-Gagan [the fourth king; ca. 1889–1871] in the First Dynasty of Isin [the modern Ishan al-Bahriyat in Iraq], the goddess makes skyquakes and earthquakes (Black et al., 1998–2006). The Sumerian goddess *Inānnā* was associated with the city of Uruk as early as the Uruk Period [ca. 4000–3000 BCE]. Uruk is located on the Euphrates plain [31°19′N–45°38′E], about 200 km to the southwest of the Zāgros Mountains foothills and the Zāgros Foredeep thrust fault (Berberian, 1995).

2.10.3 Sumerian Demon Asag Makes a Rent in the Earth at the Mountainland [the Western Zāgros Mountains] (ca. Eighteenth Century BCE)

In the story of *Ninurtā* [the Sumerian Lord of the Earth/Plough] and *Asag* [the Sumerian monstrous demon; dragon of abyss], we learn that during their battle, *Asag* caused an earthquake and fire in the heart of the Mountainland [the Zāgros] and plain [Mesopotamia], and black poison covered the fields (Thompson, 1937; Kinnier Wilson, 1979 quoting van Dijk, 1962). Kinnier Wilson (1979) interpreted that after the climax of the battle, *Asag* may indeed have emerged as a huge snake of oil that was released by an earthquake and that inundated the valleys of the Mountainlands with its poison. The oil and much of the vegetation appears to have caught fire. He assumed that *Asag* was a configuration of the exposed body of oil in the earthquake area. Although the hypothesis is possible, there is no recorded history of earthquakes in the Zāgros causing an oil seepage that caught fire. Nonetheless, tar rivers are present in the Zāgros, where crude oil flows as small rivers ('Dragon's Blood' near Dehlorān and also in Masjed Solaymān).

Also, at the cry of *Nunurtā*, the Mountainland (the modern Zāgros) and Heaven shook, accompanied by floods and lightning. The mountains were devastated, and the Earth became dark (Kinnier Wilson, 1979). Apparently, on his way back to Nibru, when *Ninurtā* steps into the chariot, heaven and Earth tremble (Kinnier Wilson, 1979). Nibru was an ancient Sumerian city [Akkadian Nibbur; Nippur, in modern Nuffar in Afak al-Qādisya].

In a hymn to *Ninurtā* for *Ur-Ninurtā* [the sixth Sumerian king in the First Dynasty of Isin ca. 1859–1832 BCE], Ninurtā makes heaven and Earth tremble from the east to the west at the Rebel Lands, which causes the great hills to tremble together. This action completely devastates the disobedient populations (Black et al., 1998–2006). This seems to be the oldest reference to earthquakes and motion direction in the Zāgros.

Ninurtā was worshiped in *Nippur*, the sacred city of *Enlil*, located at the modern tell Nuffār in Afak al-Qadaisya, modern Iraq [32°07′N, 45°14′E], about 140 km SW of the Zāgros Mountain foothills and the Zāgros Foredeep thrust fault (Berberian, 1995).

2.10.4 Ishkur's Wrath and Earth Shaking (ca. Eighteenth Century BCE)

Apparently, the Sumerian storm god *Ishkur* [god of rainstorms and vegetation, the great thunderer of heaven and Earth; *Adad* in Akkadian, with a temple in modern Karbalā, Iraq], has the nature of an earthquake. The Sumerians could see a connection between the vibration of thunder (Ishkur) and the trembling of the Earth. It is mentioned that when Adad/Iskur is worth, the Earth trembles. At his wrath, the heavens and Earth quake (Thompson, 1937). During the reign of *Sin-iddinām* (son of Nur-Ādād, the Sumerian ruler, ca. 1785–1778 BCE), Ishkur is praised as the Earth trembles before him (Black et al., 1998–2006). The ca. eighteenth-century BCE wrath of the Sumerian god *Ishkur* has evolved as the "wrath of God" in the Judaic, Christian, and Islamic faiths (see Chapter 3).

2.10.5 The Lament for Sumer and Urim

In a lament for Sumer and *Urim* (the Sumerian name for the ancient city of Ur located in modern Tell el-Mukayyar, Iraq), we read that during a storm that smashed the city, Heaven and Earth trembled and the mountains roared. This was followed by large trees being uprooted, forests destroyed, and orchards stripped of their fruits (Black et al., 1998–2006). Although the reference to an earthquake might have been metaphoric, it is likely that the storm and flooding were accompanied by an earthquake during the summertime. In any event, if there was an earthquake, the roaring of the mountains could be one of the earliest references to the sound of an earthquake.

2.10.6 The Cursing of the City of Agade by Enlil [ca. 2190–2154 BCE]

Agade was the Sumerian city of Akkad, located at the west bank of the Euphrates between Sippar and Kish in modern Iraq. During the reign of *Naram-Suen* [Akkadian king, ca. 2190–2154 BCE], Enlil caused the temples at the city of Agade to shake (Black et al., 1998–2006).

2.10.7 The Death of Ur-Namma [ca. 2112–2095 BCE]

At the death of *Ur-Namma* [Sumerian king, ca. 2112–2095 BCE], Innānā made heaven tremble and the Earth shake (Black et al., 1998–2006).

2.10.8 A Hymn to Utu

Utu, the Sun god and god of justice in Sumerian mythology [*Shamash* in Akkad], was the son of the Moon god *Nānnā* and the goddess *Ningal.* His brother and sister were *Ishkur* and *Inānnā.* A hymn to king Utu mentions that the heavens tremble and the Earth shakes before him (Black et al., 1998–2006; Kasak and Veede, 2001). Although metaphoric, the reference shows that the people were aware of earthquakes in their homeland.

2.11 EARTHQUAKES IN THE BABYLONIAN TEXTS [CA. 1830–1531]

According to the Babylonian mythology [the Evil Demon Series Tablets, in Assyrian and Sumerian], *Ānu* [*Ān* in Sumerian; god of heaven and sky] determines the destinies of the seven gods with evil spirits (Thompson, 1903). In analyzing the features of the rebel seven gods, Kinnier Wilson (1979) mentioned that the fourth god has the nature of an earthquake destroying the Mountainland (i.e., the western Zāgros (Figure 8.1)).

According to Babylonian astrology [2500–670 BCE], an earthquake took place at night in the king's land: "*Omens from Earthquakes. When the Earth quakes in Nisan, the king's land will revolt from him. When the Earth quakes during the night, harm will come to the land, or devastation to the land. From Apla. Last night there was an earthquake. When the Earth quakes in Tebet, the king will sit in the city of his foe. When the Earth quakes in Tebet, the palace of the prince will be smitten and go to ruins. When the Earth quakes in the night, there will be harm to the land (or) devastation to the land. From the Chief Astrologer*" (Thompson, 1904).

2.12 ASSYRIAN BELIEF [934–609 BCE]

In Assyrian beliefs, earthquake, famine, and epidemics were signs of the gods' divine wrath. Special rituals (mass repentance and mourning) were recommended in order to avert them (Wiseman, 1979). This divine wrath and the special rituals later became part of the principle religions. After an earthquake, Muslims must pray.

2.13 EARTHQUAKES IN THE ARMENIAN MYTH: DRAGONS LIVING IN A RAVINE CREATED BY AN EARTHQUAKE AT MOUNT MĀSSIS [ĀRĀRĀT]

Ananikian (1925), who collected the ancient Armenian myths, wrote about dragons living on Mount *Māssis* [*Āzāt* (*Yazata* in Avestan; lit. Venerable, the adorable one), the modern Ārārāt Mountains (see Figure 8.2)]. He mentioned that the volcanic character of the lofty Mount Māssis, with its earthquakes, black smoke, and lurid flames in time of early eruptions, formed the ancient idea that it was associated with *Vishāps* [lit., "with poisonous saliva"; dragons]. He added the epic songs referring to *Ānush*, the wife of the dragon and the mother of the children of the dragon, who lived in the famous ravine in the higher peak of the Mount Māssis.

However, it may be about the children of the dragon, it is incontestable that the dragons themselves were a very real terror for the ancient Armenians. We are told that they lived in a wide ravine left by an earthquake on the side of the higher peak of the Māssis. According to Moses *[Movses Khorenātsi, 466]*, Eznik *[Yeznik Koghbātsi; 5th century]*, and Vāhrām Vārdāpet (quoted by Alishān, 1895), they had houses and palaces on high mountains, in one of which, situated on the Māssis, King Ārtāxiās had enjoyed the dangerous banquet we have mentioned.

Ananikian (1925)

Eghishé [Yeghishé Vārtāpet: 410–475; ed. Boyaijan, 1952] wrote in the fifth century that "*the Vishāps* [dragons] *boasted a gigantic size* [such as volcanic Mount Māssis] *and a terrible voice* [possibly personification of the earthquake sounds and volcanic eruptions]." Eznik [Yeznik Koghbātsi; fifth century] also recorded that "*The lord pulls the vishāp* [dragon] *up through so-called oxen in order to save men from its poisonous breath*" [possible volcanic plume and lava flow].

Ananikian (1925) also recorded a mythological reference to rock falls possibly triggered by volcanic activity and/or earthquakes in ancient Armenia: "*King Erwand* [lit., "serpent" (Ālishān, 1895)] *was proverbially ugly and wicked and possessed an evil eye under the gaze of which rocks crumbled to pieces* [possible personification of the volcanic bombs and/or rock avalanches triggered by earthquakes?]. *These rocks were exposed in the morning to his eyes in order to neutralize their baleful influence during the day. The evil eye is blue. Before it, mountains, even the whole world may flame up (Pshrānk, n.d.)*" (Ananikian, 1925).

A shaking of the ground with or without volcanic activity is another documented observation in ancient Armenia: "*But we have also more definite testimony in early martyrological writing (History of St. Hripsimeans) about dragon worship. The author, after speaking of the cult of fire and water (above quoted) adds: 'And two dragons, devilish and black, had fixed their dwelling in the cave of the rock, to which young virgins and innocent youths were sacrificed. The devils, gladdened by these sacrifices and altars, by the sacred fire and spring, produced a wonderful sight with flashes, shakings, and leapings. And the deep valley (below) was full of venomous snakes and scorpions*'" (Ananikian, 1925).

Armenians living on the active mountain belts with young volcanoes have been at great risk of earthquakes and volcanoes since the ancient pagan days: the memory of this is preserved in their myths. During the 2 July 1840 M_s ~7.4 Ārārāt earthquake, a colossal slide was triggered from above the snow line of Mount Ārārāt (Ambraseys and Melville, 1982; Babayan, 2006). The 7 December 1988 M_s 6.8 earthquake destroyed the three cities of Spitāk, Gumri, and Vānādzor, killing about 25,000 people. Similar events had happened during ancient times, traces of which are preserved in myth.

2.14 OVERVIEW

A glance at the surviving oral traditions and myths of the ancient people (Table 2.1) who lived on the Iranian Plateau and Greater Armenian Highlands reveals their struggle with earthquakes as well as their primitive analyses of this phenomenon. The action and wrath of the evil-spirited storm-monsters were later attributed to the gods and goddesses and, finally, to God in major religions. The people created special rituals to be performed after each event. Astonishingly, the ancient people of the Iranian Plateau hypothesized that the mountain belts were the direct result of earthquakes and that earthquakes were incremental episodes in the process of mountain building. The amazing conclusion—especially as it applies to the growth of the lofty Alborz Mountain belt that encircled their known world with connected roots through a

TABLE 2.1 Regional Earthquake Myths in Chronological Order (See the Text)

Approx. Date	Original Source	Region	Remarks
2700 BCE	Epic of Gilgamesh Tablets	Mesopotamia	The monster Earth-Shaking Bull of Heaven creates an earthquake, damaging the city of Uruk and killing its inhabitants
ca. 1500–1200 BCE	The Vedās	Pagan India	The strong Māruts loosens the root of the mountains and makes them tremble Indra's strong thunder, and his breath as moving air, causes heaven and Earth to tremble
Pre-1200 BCE	Zoroastrian Pahlavi [Middle Persian] Texts (rewritten in the ninth century)	Pagan Iran	Storm-monster Chashmag Div [originally a storm-god, later demoted to a demon] creates earthquakes, tornadoes, and typhoons
Pre-1200 BCE	Zoroastrian Pahlavi [Middle Persian] Texts (rewritten in the ninth century)	Pagan Iran	Ahriman [Evil Spirit] assaults the Earth, causing earthquakes and mountain building (orogeny) Evil spirit shakes the sky Srosh's mace strike causes earthquakes in the east and the west Apocalyptic earthquake levels the mountains
	Ancient legend	Pagan Armenia	Rocks crumble to pieces under the evil eye of King Erwānd
	Possibly Zoroastrian in origin	Pagan Caucasus	The shackled giant chained to the Elbruz [Alborz, Qāf] Mountains in the Caucasus causes earthquakes, lightning, and thunder
Nineteenth century BCE	Sumerian Tablets	Mesopotamia	The goddess Inānnā has the capacity to make the heavens shake and the Earth tremble; Inānnā destroys Ebih Mountain (the western Zāgros)
Nineteenth century BCE	Sumerian Tablets	Mesopotamia	The power to make heaven and Earth tremble was placed in the hands of god Ninurtā

Continued

TABLE 2.1 Regional Earthquake Myths in Chronological Order (See the Text)—Cont'd

Approx. Date	Original Source	Region	Remarks
Eighteenth century BCE	Sumerian Tablets	Mesopotamia	Ea caused earthquakes
Eighteenth century BCE	Sumerian Tablets	Mesopotamia	The goddess Inānnā destroys the Ebih Mountain in the western Zāgros Mountains with earthquakes, avalanches, and fire
Eighteenth century BCE	Sumerian Tablets	Mesopotamia	At the cry of Ninurtā, the god of Nippur city, the mountains shook to and *fro* Ninurtā's chariot creates skyquake and earthquake
Eighteenth century BCE	Sumerian Tablets	Mesopotamia	The wrath of god Iskur causes earthquakes and heavenquakes
Eighteenth century BCE	Sumerian Tablets	Mesopotamia	Heaven and mountains roared and trembled; large trees were uprooted and the forest ripped out
Eighteenth century BCE	Sumerian Tablets	Mesopotamia	During a battle, the Sumerian monstrous demon, Asag created an earthquake and set fire to the western Zāgros Mountains

process of shaking—is what is known today as orogeny (mountain building). In this process, the upward faulting of the mountains is the source of the shaking. The people, living simply in harmony with nature, understood the power of nature and tried to orally communicate their findings to the coming generations. It was in this part of the world where the first recorded accountability building codes were implemented; this was about 3800 years ago during the reign of Hammurabi (1792–1750 BCE). The Code of Hammurabi (tr. King, 1915) has it that:

229. If a builder build a house for someone, and does not construct it properly, and the house which he built fall in and kill its owner, then that builder shall be put to death.

230. If it kill the son of the owner the son of that builder shall be put to death.

231. If it kill a slave of the owner, then he shall pay slave for slave to the owner of the house.

232. If it ruin goods, he shall make compensation for all that has been ruined, and inasmuch as he did not construct properly this house which he built and it fell, he shall re-erect the house from his own means.

233. If a builder build a house for someone, even though he has not yet completed it; if then the walls seem toppling, the builder must make the walls solid from his own means.

Chapter 3

Earthquakes and Religious Thoughts

Where is the hidden source of these mysterious phenomena?

Sohrāb Sepehri (1928–1980)

Since the days of hunter-gatherers and the first Neolithic settlements, when a natural disaster occurred, perhaps first nature, then idols and deities, and finally God were questioned (Murdock, 1934; Guidoboni, 1994; Wright, 2009). In ancient civilizations such as Iran and Armenia that existed on the active Iranian plateau, religious thoughts and beliefs evolved from the proto-Aryan pagan tribal cultures; these beliefs were transmitted orally through generations and were later codified by the Zoroastrian, the Judaic-Christian, and finally the Islamic writings after the Moslem Arab invasion of the plateau in 636–652 CE.

These ancient civilizations developed religious and animistic justifications for earthquakes and their associated destruction, damage, and death tolls. Supernatural powers were invoked to explain the unanswerable questions about earthquakes. Later, earthquakes were thought to be caused by divine manifestations and the wrath of the most gracious, compassionate, and merciful God and a way to test guilt or innocence of the people on Earth. The general thought was that disaster is an admonition from God and, therefore, presupposes an underlying affection for mankind that shows itself in due course as forgiveness. Although the cultural view that such disasters were an "act of fate" psychologically helped the recovery of the survivors, it slowed search-and-rescue, reconstruction, and retrofitting operations and earthquake-risk minimization efforts. Earthquake engineering science and technology in industrialized countries have minimized the impact of earthquakes on societies. Yet, in developing countries, people are placed at great risk from medium- to large-magnitude earthquakes that result in horrific death and destruction, as was recently observed during the 2003 Bam and 2010 Haiti earthquakes.

☆"To view the full reference list for the book, click here."

Developments in Earth Surface Processes, Vol. 17. http://dx.doi.org/10.1016/B978-0-444-63292-0.00003-X

3.1 THE ZOROASTRIAN SCRIPTURES [CA. 1200 BCE]

Except for the issues discussed in Chapter 2, the complete picture of the Zoroastrian religious tradition concerning earthquakes in general and particularly in ancient Iran is not known to us. Earlier I mentioned that during the third insult to the good creation by the evil spirit ("*Ahriman*"), the demons ("*Div*"; the same word found in *divine* and *deity*) rushed in, the Earth quaked, and the essence of the mountains was created in the Earth (*Bondéhéshn* [Zoroastrian]).

According to the Zoroastrian teachings, the Earth and all that is on it—water, fire, soil, and air [the four primeval elements]—is considered to be a sacred manifestation of Ahurā Mazdā [lit., "the Wise Lord"]. Therefore, it is "sacred" and precious; homage should be rendered to every angel to whose care these elements are consigned, and thanks should be offered to the Court of the great Creator. At the end of time, the active planet Earth will become flattened, possibly by earthquakes, and have no mountain peaks or valleys.

3.1.1 The Kāshmar Sacred Cypress-Tree and Earthquake Immunity at Ancient Kāshmar

The famous ancient legend of the sacred Kāshmar cypress tree is pertinent to the seismicity of the ancient city of Kāshmar; the modern Firuzābād village has a large archeological mound and traces of ruined houses to its southwest and southeast (35°07′N–57°56′E, +946 m: Figure 12.7; 49 km west southwest of the modern town of Kāshmar/Turshiz; 15 km south of Bardaskan). According to legend, the power of the vast sacred cypress tree protected the ancient city of Kāshmar from devastating earthquakes while the entire neighboring region suffered from them. Early European travelers who visited Kāshmar from 1783 on did not refer to any earthquakes in the region, and the only recorded seismic event is that of 25 September 1903, when a M_w 6.0 quake took place to the west of Turshiz (modern Kāshmar to the northeast). For seismic parameters of this event, see Table 17.12.

We learn from the Iranian epic literature about two vast and wonderful cypress trees, sacred among the Zoroastrians, which were planted in the ancient Khorāsān Province of eastern Iran. These were located in (1) the ancient city of Kāshmar (Figure 12.7), at the door of the Zoroastrian fire-temple; and (2) Faryumad (Forumad; 36°31′N–56°45′E, +1249 m; 37 km SE of Jājarm). The sacred trees were said to have risen from shoots that Zartosht [Zoroaster] brought from paradise (*behesht*, *minu*). A "spiritual" shoot from paradise with twigs, leaves, and fruits of wisdom and knowledge became a "material" gateway on planet Earth to reach Ahurā Mazdā; hence, the sacred cypress tree and the blessed region became a pilgrimage destination for Zoroastrian believers.

The veneration of trees has been important among the Zoroastrians; the *Amesha Spenta Ameretāt* (Amordād) is lord of all plants and he is worshiped by respecting and praying in front of the plants. It is not for nothing that the center of a Zoroastrian village is marked by a large cypress tree, which is placed either next to or in the courtyard of the fire temple (Fischer, 1991). Herodotus (VII/31) wrote about the Iranian king *Xerxes I the Great Achaemenid* (Khashāyār Shāh; lit., "Ruler of heroes," r. 485–465 BCE), who, one day during his campaign against Greece, was delighted by a wonderful plane-tree and gave it an ornament of gold and assigned one of his Immortals to guard the tree.

Depending upon manuscript variations, the tree was planted either by Master Teacher Zartosht (Zoroaster); King Goshtāsp (Vishtāspa; the Kiyāniān/Kayānid Aryan king of Iran), Zartosht's patron, upon accepting the "Good Religion"; or Minister Jāmāsp, Zartosht's son-in-law and King Goshtāsp's first minister. However, it is clear that the tree was presumed to be of celestial origin, like the fire-censer that Zartosht brought from heaven, and was emblematic of the spreading tree of Zartosht's creed; an image with which Daqiqi and Ferdowsi introduced the whole narrative of the spread of the religion (Bundahishn; Jackson, 1928; 'Adle, 1993; 'Alam, 1993; Berberian, 1994).

Legends in ancient Kāshmar ascribe *earthquake immunity* to a benign influence of the sacred cypress tree, which protected the district from earthquakes. Allegedly, the tree prevented the occurrence of devastating earthquakes in the town of ancient Kāshmar (located near the Doruneh active fault; Figure 12.7) while neighboring regions were frequently devastated by numerous earthquakes. It is not clear if the region has experienced a seismic event since ca. 1200 BCE, however.

However, no such power is claimed for the sacred cypress tree that existed in Faryumad (or other sacred cypress trees in the country; 'Alam, 1993). Facts such as the location of Kāshmar next to the Doruneh active fault; and the survival of old place names, such as "Shekasteh Mushkhor" (lit., "Fractured or Faulted Mushkhor"; Chapter 1), a hilly outcrop of the Neogene deposits along the Doruneh fault (between Bardaskan and Doruneh: 35°15′N–57°44′E; 1050 m amsl; Chapter 1) indicate that the region is seismically active. It is probable that sometime after the cypress tree was cut down in 861 CE, a destructive earthquake took place in the Kāshmar region, for which there is no recorded data. Furthermore, a long interseismic quiescence in Kāshmar prior to 861 probably contributed to the idea of the immunity from earthquakes and gave rise to the myth. Sadly, the sacred old cypress tree of Kāshmar was sacrilegiously cut down by the order of Muslim Caliph al-Mutawakkil 'Abbasid [r. 847–861] in 861 and brought to Baghdād on 9 December 861. The sacred cypress tree at Faryumad was later set on fire by the order of Khārazmshāh in 1142 CE.

3.1.2 Earthquakes During the Advancement of the Turks and the Romans

In Chapter 3 of the Zoroastrian Pahlavi text of *Zand-e Vohuman Yasno* (Bahman Yasht; ed. West, 1880), we read that earthquakes occurred during the invasion the Moslem Arabs, Turks, and Romans. If not metaphoric, the text may refer to an earthquake that took place in Mesopotamia. Nonetheless, it shows that the ancient Iranians were familiar with earthquakes.

3.1.3 Miraculous Opening and Closing of the Earth's Surface

The ancient belief in: (i) opening and closing of the Earth/mountain, enabling certain subjugated people to seek refuge from persecutors; (ii) the opened fracture marker as well as the whole mountain being covered with piles of stones; and (iii) mountain passes opened by the sword strike of a historic/religious hero are documented in most places on the Iranian Plateau. The idea might have evolved through oppressed historic/religious figures who made long-term geomorphological observations of earthquake faulting, fracturing, slumping, numerous landslides, and rock avalanches as well as narrow passes and gorges created either by erosion or by earthquakes and faulting.

Vārtābed Ārākel Tāvrizetzi [Ārākel of Tabriz; 1594–1670], the Armenian celibate priest who was the eyewitness to the 5 February 1641 $M_s \sim 6.8$ Dehkhārqān [modern Āzarshahr, in NW Iran; Figure 11.12] earthquake, wrote about a shepherd and some of his flock being swallowed up by a ground fissure of a possible secondary origin (ed. Bournoutiān, 2006; see also Berberian, 1976a; Berberian and Arshadi, 1976): "They say that a shepherd was driving his sheep in front of him one day, when, in a certain place, during an aftershock, the crust of the Earth cracked and formed an abyss, by which the shepherd and part of his flock were engulfed alive, none of them reappearing. An entire village, built in the fold of a rocky mountain, was covered and disappeared completely, an aftershock having dislodged the solid blocks of rock, which shattered, crumbled and enveloped it with a bed of stones and debris" (tr. Bournoutiān, 2006). Similar observations might have led people in creating folktales.

3.1.4 The Zoroastrian Holy Mountain Sanctuaries, Opening Earth and Water for Protection of the Princesses from Moslem Arab Invaders

The general theme of the escape of members of the Iranian royal family, such as King Yazdgerd III Sassanid [r. 632–642], from the Muslim Arab invaders and their miraculous rescue by Ahurā Mazdā, frequently appears in foundational legends of Zoroastrian sanctuaries intermingled with some pre-Zoroaster goddess of waters in central and southern Iran. In this regard,

Persian toponyms, including "*Bānu*" [woman, lady, princess], "*Dokhtar*" [maiden], "*Khātun*" [lady], and "*Bibi*" [lady, princess, grandmother] usually refer to locations worshiped for the ancient Iranian goddess of waters and fertility, Aredvi Surā Anāhitā (Yasht 13). These toponyms represent the link between the events and legends of the pagan and Zoroastrian Iran with those of Islamic Arab influences (Shahidi, 1954; Sorushian, 1956; Bastani Parizi, 1965; Boyce, 1967; Amir-Mo'ezzi, 2005). In all these legends, the elements of nature [i—Earth: opening and closing the mountains, ground, and caves; ii—water] were used to rescue the innocent nobles in an arid and semi-arid environment. The seeping water from fractures, faults, and/or karst system indicates the link to the ancient Iranian cult of Anāhitā.

The legend of the foundation of the Zoroastrian mountain sanctuaries relates to the fugitive princesses of Yazdgerd III Sāssānid on their escape to Khorāssān during the invasion of the Muslim Arabs in 642. The princesses prayed to Dādār Ahurā Mazdā for help, whereupon the fissures in the mountains opened up and took them in. Similar legend is preserved for the Bibi Shahrbānu shrine in the Ray region of south Tehrān (which is an earthquake-prone region; see Figure 11.2). The details of the tradition for each individual shrine are recorded by Sorushiān (ed. Sotudeh, 1956; also see Hataria, ed. Ankelsaria, 1865; Boyce, 1967; Fischer, 1991).

These mountain shrines are characteristically located at the highly fractured and faulted mountain foots; some of them still have water flowing from the fractures where large, old trees have grown. Examples include the Pir-e Sabz/Chek Chek temple (NW of Yazd), and Naraki, which still have waterfalls today. All these shrines are located on the ancient spots dedicated to the goddess of waters, Aredvi Surā Anāhitā of Yasht 13 (Shahidi, 1954; Sorushiān, 1956; Bastani Parizi, 1965; Boyce, 1967; Amir-Mo'ezzi, 2005). Most of the shrines are dedicated to females who escaped their Muslim Arab persecutors during the Muslim Arab invasion and took refuge in the mountains.

3.1.5 Earth Opening at Kāriyān, Fārs

While visiting Kāriyān [28°08′N–53°31′E, +877 m; 36 km south of Jahrom], Harm [28°10′N–53°29′E, +1.038 m; 6 km NNW of Kāriyān], and Khonj [27°53′N–53°26′E, +736 m; 30 km SSW of Kāriyān], in the Fārs Province of southern Iran in the Zāgros active fold-and-thrust mountain belt with high seismicity rates, Edward Stack (1882) wrote the following legend about the opening of the Earth. Apparently, the most ancient Zoroastrian fire temple, Faranbagh, was located at Kāriyān (Jackson, 1921).

Harm is a large village, with extensive date-groves, and perhaps two hundred houses. It was deserted and in ruins; we could find no quarters there. Karyun *[Kāriyān]* is still larger; it must have had a population of 2000 souls, but we could find only three families in the whole place. Two other forts (besides a modern one) stand in the plain, a mile east of Karyun. One is the Mud Fort (Qala-i-Gili) *[Qal'eh Geli]*, built when Karim Khān *[Zand]* was reigning in Shiraz (1780); it is a square earthwork with a side of 120 yards, and had a tower every twelve yards. The other is the fort of the Fire-well *[Qal'eh-ye Chāh Tashi/Ātashi]*, so called from the discovery of naphtha in a well hard by; it is a tower girt with a wall, on a mound. Forts and well are in ruins now. Karyun stands in the middle of three rocky hills, and these, also, are said to have been fortified. I went up one hill with some men of the village. They stopped at the foot, picked up bones, and said, "These are the bones of men," and proceeded to tell me the following story: Shāh Kāran *[Qaran]* was besieged here by 12,000 Mussulmans *[Muslims]*, when the Arabs first invaded Persia. While they were at their prayers he sallied out. They would not leave their prayers, and he slew them all without resistance. In the Mussulman camp were forty virgins, who thus fell into the hands of Shāh Kāran. These young women, being of virtuous principles, besought deliverance from Heaven, and accordingly the Earth opened and swallowed them all up – except three who fled, with Shāh Kāran and his men after them. One maiden ran across the plain, and up the northern mountains, and was now on the point of capture, when a cave disclosed itself in the mountain-side; she ran in, and was lost. The cave is called the Ghar Bibi *[Ghār-e Bibi; lit., the Lady's Cave]*, or Lady's Cave, to this day, and is well known to have no end. The second maiden fled to the mountains of Khunj *[Khonj]*, far to the south [SSW], and died there of exhaustion. Her shrine, called that of the Bibi Darmānda *[Darmāndeh]*, or Tired-out Lady, is a famous place of prayer for childless wives. The third maiden disappeared in some other mountainside, and water has trickled from the cleft ever since.

Stack (1882)

The region is located at the seismic gap between the meizoseismal areas of the 10 April 1972 M_w 6.7 and 24 April 1960 M_s 5.8 Lār earthquakes (see Figure 6 in Berberian, 1995; Figure 13.4).

3.1.6 Lake Zarivār (Zaribār)

The Kurdish legend has it that the Lake Zarivār [from the Old Iranian word "Zra," lit: "Sea," "Lake"; 35°32′N–46°07′E; 1384 m], adjacent to the Kurdish city of Marivān, occupied the site of the ancient capital of *Daqyānus* [*Decianus; the City of Decius*]. As a punishment for his evil deed, he was swallowed up in an earthquake together with his city, the ruins of which can be seen at the bottom of the lake in calm weather (Rich, 1836; de Bode, 1845; Massé, 1938, Messner, 1954). The lake is located along the Zāgros Main Recent fault, which might have been activated during ancient times and led to the creation of this myth. The only recorded earthquake in the area was in 1310–1311; it destroyed many houses and killed a great number of people in Shahr-e Zur (al-'Umari, 1793; Alsinawi and Ghalib 1975; Ambraseys and Melville, 1982; Berberian, 1994, 1995).

3.2 THE BUDDHA [CA. FIFTH CENTURY BCE]

During the "enlightenment" stage of his life, *Siddhārtha Gautama* [*The Buddhā*; ca. fifth century BCE] encountered the demon *Mārā* [the Buddhist devil], Lord of Desire, and resisted every temptation that the demon could devise. The story goes that the Earth quaked when Mārā, mounted on his elephant, approached the Buddha; Siddhārtha was meditating beneath a Pipal tree [*Bodhi*; "*the Tree of Awakening*"]. He lifted his right hand and reached down and touched the Earth with his finger, calling on the very Earth as his witness; the Earth shuddered. It is also said that the Earth shook when the Buddha died (ORIAS; Grubin, 2010; Carus, 1990; Bodhi, 2000).

3.3 THE OLD TESTAMENT [CA. 450 BCE–200 AD]

There is no clear break between the later religions—Judaism, Christianity, and Islam—in theological terms. The earthquake is simply considered to be a manifestation of divine determination and awe. Any other thoughts are seen as a pagan belief, a denial of the power of Yahweh/Yahu as God, and a heresy. Earthquakes, lightening, thunder, and storms are armaments of a fighting Yahweh. Furthermore, some writers influenced by religious thoughts observe that unusual natural events—an eclipse, comet, or earthquake—occur before sad or disastrous events, such as the death of a prominent political, religious, or national leader; a revolt; an invasion; a massacre; or war. They believe that such calamities were foretold by natural wonders.

The writers of the Old Testament considered earthquakes ["*ra'ash*" in Hebrew] in *Yahweh*'s sovereign control and as his wrath on his chosen Jewish people; they are part of *Yahweh*'s armory for ruling the world in righteousness. Earthquakes, therefore, are considered as divine punishment for the performances in the circus and theater of the heathens, or for their immorality, as well as to remind men of their sins. The following passages are taken from the Old Testament of the Holy Bible, King James Version:

Exodus (Book 2) 19:18—*And Mount Sinai was altogether on a smoke, because the LORD* [Yahweh/Yahu] *descended upon it in fire: and the smoke thereof ascended as the smoke of a furnace, and the whole mount quaked greatly.*

Exodus (2), 20:18—*And all the people saw the thundering, and the lightning, and the noise of the trumpet, and the mountain smoking: and when the people saw it, they removed, and stood afar off.*

Numbers (4) 16:28–34—*28. And Moses said, Hereby ye shall know that the LORD* (Yahweh/Yahu) *hath sent me to do all these works; for I have not done those of mine own mind. 29. If these men die the common death of all men, or if they be visited after the visitation of all men; then the LORD hath not sent me. 30. But if the LORD make a new thing, and the Earth open her mouth, and swallow them up, with all that appertain unto them, and they go down quick into the pit; then ye shall understand that these men have provoked the LORD. 31. And it came to pass, as he had made an end of speaking all these words, that the ground clave asunder that was under them: 32. And the Earth opened her mouth, and swallowed them up, and their houses, and all the men that appertained unto Korah, and all their goods. 33. They, and all that appertained to them, went down alive into the pit, and the Earth closed upon them: and they perished from among the congregation. 34. And all Israel that were round about them fled at the cry of them: for they said, Lest the Earth swallow us up also.*

1 Samuel (9) 14:15—*And there was trembling in the host, in the field, and among all the people: the garrison, and the spoilers, they also trembled, and the Earth quaked: so it was a very great trembling.*

2 Samuel (10) 22:8—*Then the Earth shook and trembled; the foundations of heaven moved and shook, because he was wroth.*

1 Kings (11) 19:11–12—*11. And he said, Go forth, and stand upon the mount before the LORD* [Yahweh/Yahu]. *And, behold, the LORD passed by, and a great and strong wind rent the mountains, and broke in pieces the rocks before the LORD; but the LORD was not in the wind: and after the wind an earthquake; but the LORD was not in the earthquake: 12. And after the earthquake a fire; but the LORD was not in the fire: and after the fire a still small voice.*

Psalms (19) 68:8—*The Earth shook, the heavens also dropped at the presence of God* [Yahweh/Yahu]*: even Sinai itself was moved at the presence of God, the God of Israel.*

Psalms (19) 77:18—*The voice of thy thunder was in the heaven: the lightning lightened the world: the Earth trembled and shook.*

Psalms (19) 104:32—*He looketh on the Earth, and it trembleth: he toucheth the hills, and they smoke.*

Psalms (19) 114:4–7—*4. The mountains skipped like rams, and the little hills like lambs. 5. What ailed thee, O thou sea, that thou fleddest? Thou Jordan, that thou wast driven back? 6. Ye mountains, that ye skipped like rams; and ye little hills, like lambs? 7. Tremble, thou Earth, at the presence of the Lord* [Yahweh/Yahu], *at the presence of the God of Jacob.*

Isaiah (23), 5:25—*Therefore is the anger of the LORD* [Yahweh/Yahu] *kindled against his people, and he hath stretched forth his hand against them, and hath smitten them: and the hills did tremble, and their carcasses were torn in the midst of the streets. For all this his anger is not turned away, but his hand is stretched out still.*

Isaiah (23) 24:18–20—*18. And it shall come to pass, that he who fleeth from the noise of the fear shall fall into the pit; and he that cometh up out of the midst of the pit shall be taken in the snare: for the windows from on high are open, and the foundations of the Earth do shake. 19. The Earth is utterly broken down, the Earth is clean dissolved, the Earth is moved exceedingly. 20. The Earth shall reel to and fro like a drunkard, and shall be removed like a cottage; and the transgression thereof shall be heavy upon it; and it shall fall, and not rise again.*

Jeremiah (24) 10:10—*But the LORD* [Yahweh/Yahu] *is the true God, he is the living God, and an everlasting king: at his wrath the Earth shall tremble, and the nations shall not be able to abide his indignation.*

Ezekiel (26) 38:19—*For in my jealousy and in the fire of my wrath have I spoken, Surely in that day there shall be a great shaking in the land of Israel; 38:20 So that the fishes of the sea, and the fowls of the heaven, and the beasts of the field, and all creeping things that creep upon the Earth, and all the men that are upon the face of the Earth, shall shake at my presence, and the mountains shall be thrown down, and the steep places shall fall, and every wall shall fall to the ground.*

Amos (30) 1:1–2—*1. The words of Amos, who was among the herdmen of Tekoa, which he saw concerning Israel in the days of Uzziah king of Judah, and in the days of Jeroboam the son of Joash king of Israel, two years before the earthquake. 2. And he said, The LORD* [Yahweh/Yahu] will *roar from Zion, and utter his voice from Jerusalem; and the habitations of the shepherds shall mourn, and the top of Carmel shall wither.*

Nahum (34) 1:5—*The mountains quake at him, and the hills melt, and the Earth is burned at his presence, yea, the world, and all that dwell therein.*

Habakkuk (35) 3:6—*He stood, and measured the Earth: he beheld, and drove asunder the nations; and the everlasting mountains were scattered, the perpetual hills did bow: his ways are everlasting.*

Zechariah (38) 14:5—*And ye shall flee to the valley of the mountains; for the valley of the mountains shall reach unto Azal: yea, ye shall flee, like as ye fled from before the earthquake in the days of Uzziah king of Judah: and the LORD* [Yahweh/Yahu] *my God shall come, and all the saints with thee.*

The Hebrew Book of Prophet Habakkuk (3:3–13) develops the picture of storm, lightening, thunder, and earthquake, when Yahweh as God went forth to victory; "with the Light of his arrows" and "the shining of his glittering spears," when "fiery bolts went forth at his feet" (Ward, 1909).

3.4 THE NEW TESTAMENT [CA. 50–150 AD]

In the New Testament, earthquakes are referred to in parallel accounts of the three Gospels of Matthew, Mark, and Luke, wherein, apparently, Jesus refers to earthquakes as a sign of the end of the world, as has been believed by the Zoroastrians. It is considered as a sign of the Lord's presence and will in conveying a reprimand for moral and religious transgression. The following passages are taken from the New Testament of the Holy Bible, King James Version:

Matthews (Book 40) 24:1–3—*1. And Jesus went out, and departed from the temple: and his disciples came to him for to shew him the buildings of the temple. 24:2 And Jesus said unto them, See ye not all these things? Verily*

I say unto you, There shall not be left here one stone upon another, that shall not be thrown down. 3. And as he sat upon the mount of Olives, the disciples came unto him privately, saying, Tell us, when shall these things be? And what shall be the sign of thy coming, and of the end of the world?

Matthews (40) 24:7—*For nation shall rise against nation, and kingdom against kingdom: and there shall be famines, and pestilences, and earthquakes, in diverse places.*

Matthews (40) 27:50–54—*50. Jesus, when he had cried again with a loud voice, yielded up the ghost. 51. And, behold, the veil of the temple was rent in twain from the top to the bottom; and the Earth did quake, and the rocks rent. 52. And the graves were opened; and many bodies of the saints which slept arose. 53. And came out of the graves after his resurrection, and went into the holy city, and appeared unto many. 54. Now when the centurion, and they that were with him, watching Jesus, saw the earthquake, and those things that were done, they feared greatly, saying, Truly this was the Son of God.*

Matthew (40) 28:2—*And, behold, there was a great earthquake: for the angel of the Lord descended from heaven, and came and rolled back the stone from the door, and sat upon it.*

Mark (41) 13:8—*For nation shall rise against nation, and kingdom against kingdom: and there shall be earthquakes in diverse places, and there shall be famines and troubles: these are the beginnings of sorrows.*

Luke (42) 21:10–11—*10. Then said he unto them, Nation shall rise against nation, and kingdom against kingdom. 11. And great earthquakes shall be in diverse places, and famines, and pestilences; and fearful sights and great signs shall there be from heaven.*

Acts (44) 16:26—*And suddenly there was a great earthquake, so that the foundations of the prison were shaken: and immediately all the doors were opened, and every one's bands were loosed.*

Hebrews (58) 12:26—*Whose voice then shook the Earth: but now he hath promised, saying, Yet once more I shake not the Earth only, but also heaven.*

Revelation (66) 16:16–21—*16. And he gathered them together into a place called in the Hebrew tongue Armageddon. 17. And the seventh angel poured out his vial into the air; and there came a great voice out of the temple of heaven, from the throne, saying, It is done. 18. And there were voices, and thunders, and lightnings; and there was a great earthquake, such as was not since men were upon the Earth, so mighty an earthquake, and so great. 19. And the great city was divided into three parts, and the cities of the nations fell: and great Babylon came in remembrance before God, to give unto her the cup of the wine of the fierceness of his wrath. 20. And every island fled away, and the mountains were not found. 21. And there fell upon a great hail out of heaven, every stone about the weight of a talent: and men blasphemed God because of the plague of the hail; for the plague thereof was exceeding great.*

Yeznik Koghbātsi (Yeznik of Kolb) the Armenian Bishop of Bāgrāwānd (fifth century) in his treatise *Refutation of the Sects, A Retelling of Yeznik*

Koghbātso's Apology (Book 1, The Nature of God) wrote that God is the creator of good things, and evil acts with disasters are not act of God:

Some might ask: If God is the creator of good things, then from where are the disharmonies and imbalances which cause earthquakes and floods? From where are the darkness and cold and heat which discomfort man, and from where are the evils which men perpetrate against one another? What causes all this evil, if not a force of evil, which is their source and creator? It is unthinkable that evil acts and events are the work of [p. 20] a good God. For this reason there are people who believe, mistakenly, that along with God there was also Matter (called hiulé from the Greek), from which God created all things and beings. Furthermore, they believe that all evil arises from this shapeless, directionless, imperfect Matter. God saw that it was in need of shaping. Like a skilled architect, He used what He could to shape the good things and threw away the material which was unfit for His creations. They say that the Matter that was thrown away is the source of evil.

Yeznik Koghbātsi, tr. Thomas Samuelian

3.5 ISLAM [CA. 644–656]

In Islam, the period before religion is called the "*Jāhiliyya* [lit., "Ignorance"] *Times*" by the Muslim Arabs; it is viewed as a time not only of paganism but also of barbarism. Therefore, little information is preserved about the earthquake myth in Arabia.

As with Judaic-Christian views and traditions, the Islamic interpretation of earthquakes, which has predominated in Iran since the seventh-century invasion of the Muslim Arabs and commonly found in Muslim accounts since then, is that earthquakes are:

- **(i)** *Allāh*'s Will and Foreordainment; they occur through decree [*qaḍā wa qadar*] of the merciful and the compassionate [*al-rahmān wa al-rahim*] Allāh, and should, therefore, be unquestionably accepted through reliance [*tawakkul*];
- **(ii)** *Allāh*'s admonition to the believers, and a sign of wrath and punishment to miscreants for their sins and deviatory such as adultery, usury, alcohol consumption, and other sins;
- **(iii)** Similar punishments took place during the past;
- **(iv)** People should repent and sin no more;
- **(v)** One of the signs of approaching the Last Judgment Day representing the end of the world [qiyāmat]; and/or
- **(vi)** Astrophysical circumstances, such as planetary constellations or long-tailed stars [comets] and lunar or solar eclipses.

The most famous verse of the *Qorān* dedicated to the earthquake is *Sura* 99, "*Az-Zalzalah*" ["*The Earthquake*"], *Āyāt* 1–8, which states that an earthquake is a sign indicating the end of the world. Similar references are made in Suras

13/31; 16/15; 19/90; 21/30–31; 22/1–2; 31/10; 56/4–10; 67/15–16; 69/13–16; 73/14; and 77/1–10 (Berberian, 1994, 1997b; Qorān).

Sura 2, "Al-Baqarah" [The Cow]/214: "*Or think ye that ye will enter paradise while yet there hath not come unto you the like of (that which came to) those who passed away before you? Affliction and adversity befell them, they were shaken as with earthquake, till the messenger (of Allāh) and those who believed along with him said: When cometh Allah's help? Now surely Allāh's help is nigh.*"

Sura 7, 'Al-A'arāf ["The Heights"]/78: "*So the earthquake seized them, and they became within their home [corpses] fallen prone.*"

Sura 7, 'Al-A'arāf ["The Heights"]/91: "*So the earthquake seized them and they lay (dead), prostrate in their homes.*"

Sura 7, 'Al-A'arāf ["The Heights"]/155: "*And Moses chose from his people seventy men for Our appointment. And when the earthquake seized them, he said, "My Lord, if You had willed, You could have destroyed them before and me [as well]. Would You destroy us for what the foolish among us have done? This is not but Your trial by which You send astray whom You will and guide whom You will. You are our Protector, so forgive us and have mercy upon us; and You are the best of forgivers.*"

Sura 13, '*Ar-Ra'd*' ["*The Thunder*"]/31: "*If there were a Qor'an with which mountains were moved, or the Earth were cloven asunder, or the dead were made to speak, (this would be the one!) But, truly, the command is with Allāh in all things! Do not the Believers know, that, had Allāh (so) willed, He could have guided all mankind (to the right)? But the Unbelievers, never will disaster cease to seize them for their (ill) deeds, or to settle close to their homes, until the promise of Allāh come to pass, for, verily, Allah will not fail in His promise.*"

Sura 14, "Ibrahim" ["Abraham"]/46: "*And they readily schemed their scheming, and their scheming is in the Providence of Allāh, and decidedly their scheming was such that thereby mountains should be removed.*"

Sura 16, "*An-Nahl*" ["*The Bee*"]/15: "*And He has set up on the Earth mountains standing firm, lest it should shake with you; and rivers and roads; that ye may guide yourselves.*"

Sura 19, "*Mayyam*" ["*Mary*"]/90: "*At it the skies are ready to burst, the Earth to split asunder, and the mountains to fall down in utter ruin.*"

Sura 21, "*Al-Anbiya*" ["*The Prophets*"]/30–31: "*30. Do not the Unbelievers see that the heavens and the Earth were joined together (as one unit of creation), before we clove them asunder? We made from water every living thing. Will they not then believe? 31. And We have set on the Earth mountains standing firm, lest it should shake with them, and We have made therein broad highways (between mountains) for them to pass through: that they may receive Guidance.*"

Sura 22, "*Al-Hajj*" ["*The Pilgrimage*"]/1–2: "*1. O mankind! Fear your Lord! For the convulsion of the Hour (of Judgment) will be a thing terrible;*

2. The Day ye shall see it, every mother giving suck shall forget her suckling-babe, and every pregnant female shall drop her load (unformed): thou shalt see mankind as in a drunken riot, yet not drunk: but dreadful will be the Wrath of Allāh."

Sura 29, "Al-'Ankabut" ["The Spider"]/37: "*But they denied him, so the earthquake seized them, and they became within their home [corpses] fallen prone.*"

Sura 31, "*Luqmān*" ["*Luqman*"]/10: "*He created the heavens without any pillars that ye can see; He set on the Earth mountains standing firm, lest it should shake with you; and He scattered through it beasts of all kinds. We send down rain from the sky, and produce on the Earth every kind of noble creature, in pairs.*"

Sura 33, "Al-Ahzāb" ["The Combined Forces"]/11: "*There the believers were tested and shaken with a severe shaking.*"

Sura 52, Al-Toor [Mount Sinai]/9–12: *9. The day will come when the sky will violently thunder. 10. The mountains will be wiped out. 11. Woe on that day to the disbelievers. 12. Who are in their blundering, heedless.*

Sura 56, "*Al-Waqi'a*" ["*The Event*"]/1–10: "*1. When the Earth shall be shaken to its depths, 2. Then will no (soul) entertain falsehood concerning its coming. 3. (Many) will it bring low; (many) will it exalt; 4. When the Earth shall be shaken to its depths, 5. And the mountains shall be crumbled to dirts, 6. Becoming dust scattered abroad, 7. And ye shall be sorted out into three classes. 8. Then (there will be) the Companions of the Right Hand; What will be the Companions of the Right Hand? 9. And the Companions of the Left Hand, what will be the Companions of the Left Hand? 10. And those Foremost (in Faith) will be Foremost (in the Hereafter).*"

Sura 67, "*Al-Mulk*" ["*Dominion*"]/15–16: "*15. It is He Who has made the Earth manageable for you, so traverse ye through its tracts and enjoy of the Sustenance which He furnishes: but unto Him is the Resurrection. 16. Do ye feel secure that He Who is in heaven will not cause you to be swallowed up by the Earth when it shakes (as in an earthquake)?*"

Sura 69, "*Al-Haqqah*" ["*The Reality*"]/13–16: "*13. Then, when one blast is sounded on the Trumpet, 14. And the Earth is moved, and its mountains, and they are crushed to powder at one stroke. 15. On that Day shall the (Great) Event come to pass. 16. And the sky will be rent asunder, for it will that Day be flimsy, And the angels will be on its sides, and eight will, that Day, bear the Throne of thy Allah above them. That Day shall ye be brought to Judgment: not an act of yours that ye hide will be hidden.*"

Sura 73, "*Al-Muzzammil*" ["*Folded in Garments*"]/14: "*One Day the Earth and the mountains will be in violent commotion. And the mountains will be as a heap of sand poured out and flowing down.*"

Sura 77, "*Al-Musalat*" ["*Those Sent Forth*"]/1–13: "*1. By the (Winds) sent forth one after another (to man's profit); 2. Which then blow violently in tempestuous Gusts, 3. And scatter (things) far and wide; 4. Then separate them, one from another, 5. Then spread abroad a Message, 6. Whether of*

Justification or of Warning; 7. Assuredly, what ye are promised must come to pass. 8. Then when the stars become dim; 9. When the heaven is cleft asunder; 10. When the mountains are scattered (to the winds) as dust; 11. And when the apostles are (all) appointed a time (to collect); 12. For what Day are these (portents) deferred? For the Day of Sorting out."

Sura 79, "Al-Nāzi'āt" ["Those Who Drag Forth"]/6: "*On the Day when the (first) commotion commoves.*"

Sura 99, "Az-Zalzalah" ["The Earthquake"]/1–8: 1. "*When the Earth is shaken by a great tremor; 2. When the Earth yields up its burdens; 3. And man shall say: 'What ails it?' 4. On that day the Earth shall tell its story. 5. For that thy Lord will have given her inspiration. 6. On that Day will men proceed in companies sorted out, to be shown the deeds that they (had done). 7. Then shall anyone who has done a minute weight of good, see it! 8. And anyone who has done a minute weight of evil, shall see it.*"

3.5.1 Hadith and Rivāyāt [Post-Eighth Century]

Almost all religions have developed elaborate theories about the end of the world. Their eschatological expectations are based on a series of signs that derive from events predicted in the holy books, narratives, Hadith, and Rivāyāt. *Hadith* [*Sunah*] are religious sayings, actions, or traditions attributed to Muhammad documented by his faithful companions [sahāba]. They provide religious and social guidance and are considered authentic and genuine by his followers. *Rivāyāt* are traditional treatises, reports, and narratives about 14 Innocents [ma'asum] collected and written by the faithful followers [tāba'ain]. There are numerous treatises and books written on the subjects of Hadith and *Rivāyāt* in Arabic and Persian; a review of the entries requires much time and effort. Some of the documented materials are related to apocalyptic signs and events such as earthquake, drought, hunger, and so forth.

3.5.1.1 The Ninth-Century Ahvāz Earthquakes and Remedy for Calming Them

In an article entitled "Some Examples of the Greatness of Imam Javād 'alaihā al-salām" that appeared on the Mouood Web site dated 10 December 2007, we read about the prevention of an earthquake in the city of Ahvāz near the Zāgros foredeep fault in southwest Iran (Berberian, 1995).

Instruction Remedy for Quiescence of Earthquake-The late Shaikh Saduq *[possibly Abu J'afar Muhammad ibn 'Ali Bābuyeh al-Qomi; 917–991]* based on 'Ali ibn Mahzyār Ahvāzi, who was one of the faithful followers of Imam Javād *[the 9th Imām]*, Imām Hādi *[the 10th]*, and Imam Hasan 'Asgari *[the 11th] 'alaihom al-salām, wrote that:*

One day, I wrote a letter to his holiness Abu J'afar, Imām Muhammad Javād alaihā al-salām, as follow: The son of the prophet; Earthquakes happen frequently in the city of Ahvāz and its suburbs. Will his holiness allow us to leave the place and settle in a safe place? And then I sent the letter to his holiness. Imām Alaihā al-salām replied afterwards: Stay at the same place and do no relocate; but fast on Wednesdays, Thursdays, and Fridays. And on Friday perform the Friday Ghosl *[ritual bathing]*; wear clean cloths, and all the families gather in a proper place. And at that place all together perform a fervent prayer, and ask the mighty Allāh to resolve the problems of all. 'Ali ibn Mahzyār, then added that: We all followed the instructions given by Imam Javād alaihā al-salām, and the earthquake diminished and stopped. And after that, all the inhabitants of Ahvaz were safe from the danger of earthquake.

Imām al-Javād, the ninth Imām, was born on 8 April 811 in Medina, Arabia, and was poisoned by his wife in Baghdād on 24 November 835. This documented story indicates that the city of Ahvāz in southwest Iran and its suburbs were seismically active in those days; the letter was sent to the ninth Imām sometimes prior to 24 November 835. Hamzeh Esfahāni (961), Ibn al-Jauzi (1181), Ibn al-Athir (1231), followed by Sani' al-Dauleh (1880–1882) and Shushtari (1952), recorded that in 225 H (12 November 839–20 October 840) many buildings, including the Friday mosque at Ahvāz, collapsed in an earthquake. The mountain overlooking the town was fissured, and many people fled the city to the suburbs and escaped onto ships. The earthquake took place four years after the death of the ninth Imām.

3.5.1.2 Later Conjectures

Later, Moslem historians, scholars, and authors describing earthquake destructions (such as ibn al-Athir, 1231, and al-Suyuti, 1499) were influenced by the apocalyptic descriptions in the *Qorān* verses. Some of the earthquake narratives are directly taken from the Qorānic verses. Earthquakes were described with religious awe and discussed in theological terms.

Similarly, occasional Iranian earthquake descriptions by the Armenian priests, such as Vārtābed Ārākel of Tabriz [1594–1670], include parallel verses from the Bible that state that the calamities are the wrath of God in response to our heavy sins. Historians and religious leaders in previous centuries usually tried to mitigate the fear of the people of their generation by reminding them of more severe historic chastisements that had occurred in the past and advising them to offer more prayers and sacrifices.

3.5.1.3 Five Remaining Apocalyptic Signs

Recently, in a lecture in Qom, Iran, entitled "Five Signs Remain from the 1200 Apocalyptic Signs," Esmā'il Shafi'i Sarvestāni, director of the Mouood Cultural Institute, indicated earthquakes to be an important sign: "*Drought, earthquake, famine, and hunger are signs of the End of the World. In 2009, one billion people worldwide suffered from hungry* [hunger] *and 45 million were affected by [the] AIDS are signs of the End of the World. Every year*

1000 medium [−magnitude] *earthquakes, 18 large* [−magnitude] *earthquakes and one great earthquake shake the Earth*" (mouood.org; 7 November 2009).

3.5.1.4 *Mountain Splitting and Draining of the Dorud Lake [along the Zāgros Main Recent fault]*

The Silākhor Valley, formed by the Zāgros Main Recent fault (Figure 12.1), is a nearly flat depression with layers of light-color clay that has been used for brick manufacturing and agricultural activities. Local inhabitants believe that very long ago, the region was covered by a large lake that was not connected to any sea. Prophet Muhammad cut and split the mountains with his sword, forming Tang-e-Bahrain [Bahrain Pass; now Dorud; Figure 12.1], and the lake water drained to the south toward Dezful and Āb-e Dez (Morgan, 1894; Ambraseys, 1974b; Nabavi, 1985).

We know that Muhammad never traveled to Iran. Furthermore, the Dorud segment of the Zāgros Main Recent fault runs through the Pass, and the fault was reactivated at the same location during the 23 January 1909 M_w 7.4 Silākhor earthquake (Ambraseys, 1974b; Tchalenko and Braud, 1974; Tchalenko et al., 1974b) (see also Chapter 12 and Figure 12.1). It is assumed that the local legend alludes to an earthquake with surface faulting at the same location long before 1909 and prior to the time of Muhammad (570–632).

3.5.1.5 *Shaking of the* Ayvān-e Kasrā, *Extinguishing the Zoroastrian Sacred Fires, and Drying up the* Sāveh Lake *by a Single Earthquake on the Iranian Plateau*

According to early Moslem legends, the following three incidents took place as a result of an earthquake that allegedly occurred on the night of the birth of Muhammad [20 August 570?] during the reign of the Sāssānid Emperor, Khosrow I Anushirvān [r. 531–579 CE]:

- **(i)** The "great lake of Sāveh" in central Iran (34°44′N–50°29′E, +942 m) suddenly dried up;
- **(ii)** The Ayvān (or Tāq)-e Kasrā, the Vault of Khosrow I Anushirvān Sāssānid [the vast ovoid barrel vault at Ctesiphon/Tyspwn/Tisfun; Figure 8.1] at the imperial capital city of the Arsacids (250 BC–224 CE) and the Sāssānids (224–642) [33°05′N–44°34′E, +35 m] was badly shaken and damaged; and
- **(iii)** All the Zoroastrian sacred fires, which were burning for more than 1000 years in numerous fire temples throughout the country, were smothered.

The story is recorded by: *Ya'qubi* (892); Tabari (915); Mojmal al-Tavārikh val-Qesās (1126); Sa'adi (1257); Mostaufi Qazvini (1330); Mostaufi Qazvini (1340); Al-Suyuti (1499); and Sabāhi Bidgoli Kāshāni (d. 1803), who mentioned the triple miracles in one of his poems. Possibly, it was for the first time that Ya'qubi (892) stated that:

And when the messenger of God was born, the devils were denounced and the stars fell down. Then when the Qoraish *[people]* saw that, they became astonished of the fall of the stars, and said: 'this is nothing but the coming of the Qiyāmat *[Resurrection/the Day of Judgment]*.' And the people were struck by an earthquake which reached the whole world to the point that the churches and the synagogues were destroyed, and everything which was worshiped apart from Allāh was rooted up from its place, and the magicians and foretellers became wandered in their work, and their devils were put in chains, and stars appeared that had not seen before, therefore, the Jewish priests were astonished. And the ayvān-e Kasrā was shaken and thirteen of its turrets (pinnacles) fell down and the fire temple of Fars, which was not extinguished for the last three thousand years, was extinguished.

Ya'qubi (892), ed. Mohammad Ebrāhim Āyati (1963), translated from the Persian text.

No recorded earthquake is documented during this time, and it is impossible that the sites of Ctesiphon and Sāveh (580 km apart) together with all the Zoroastrian fire temples throughout the vast country could be destroyed by a single earthquake on a specific night.

3.5.1.6 The 17 June 978 $M_s \gtrsim 6.5$ Sirāf Port Earthquake and Fornication/Usury

Al-Muqaddasi (985) believed that the 17 June 978 $M_s \gtrsim 6.5$ Sirāf Port (modern Tāheri) earthquake occurred because of the corruption of the people of Sirāf.

3.5.1.7 Annual Sacrifice Preventing Earthquakes at Dinévar

From ancient times, the local view in most societies has been that animal sacrifices and related ceremonies are essential to keep the demons, evils, and evil eyes pacified and to deflect their destructive nature. *Al-Ghāzzi* (1782) transmitted the following story that had been told to him.

In Dinévar, there was a shrine to Mumshād Dinévari, with a spring, trees, a mosque, and some private cells, at which two sheep were sacrificed every year. Whenever the sacrifice was forgotten, the village moved. One year, there was a great earthquake and the shrine and all its surroundings were moved bodily to the top of a nearby hill and nothing was altered; but the village was engulfed and no sign of it remained.

al-Ghāzzi (1782), Tāher (1974, 1979), Ambrasye and Melville (1982)

In this religious view, an omission of the annual sacrifice resulted in God's punishment, which engulfed Dinévar, located along the Zāgros Main Recent fault (see Chapter 11 and Figure 11.4). Landslides are common in Dinévar, and the exact date of the above-mentioned event is not known. We know that there were three recorded earthquakes in Dinévar in 912–913, 27 April 1008 ($M_s \sim 7.0$), and 1107–1108 ($M_s \sim 6.5$) (Ambraseys and Melville, 1982; Berberian, 1994) (see also Chapter 11 and Figures 11.4 and 12.11). Out of these three events, two of them were associated with landslides. During the 912–913 event, a mountain near Dinévar named al-Tall [lit., mound, hill, in Arabic] was fractured and collapsed; a large amount of water emerged from beneath it, flooding several villages (al-Qurtubi, 923; Ibn al-Jauzi, 1181; al-Suyuti, 1499; *Al-Ghāzzi*, 1782; Tāher, 1974, 1979; Ambraseys and Melville, 1982; Berberian, 1994). It has been recorded that during the 1008 earthquake, Dinévar sunk into the ground and a large number of people perished (al-Antāki, 1066; ibn al-Jauzi, 1181; ibn al-Athir, 1231; Ambraseys and Melville, 1982). The engulfing, which is mentioned in al-Ghāzzi's story (1782), was only recorded during the 912–913 event (see also Chapter 11 and Figure 11.4).

3.5.1.8 Miraculous Tabriz Earthquake of 8 January 1780 $M_s \sim 7.4$ *Revealing the Lost Tombstone of Noah's Mother in Marand*

Captain Moritz von Kotzebue of the Russian Army, who was in the town of Marand (Figure 11.12) in northwest Iran between May 15 and 17, 1817, narrated a story that had been told to him about a miraculous earthquake that revealed the location of the tombstone of Noah's mother in a mausoleum. He wrote that (Kotzebue, 1819):

> Of Maranda *[Marand town]*, it is likewise asserted by the Armenians, that Noah's immediate descendants settled there, and even that it is the place of his wife's interment. Who could have neglected the sight of such a hallowed ground? Curiosity led us to the spot, and we found that the Moslems had built, on the place where Noah's wife is reported to have been buried, a chapel, with bare walls, which are not so cleanly as the religion of Mahomet *[Mohammad]* prescribes. When the chapel was finished, nobody, however, would undertake to point out the actual spot where the body lay. A miracle solved their doubts. Thirty-eight years ago, during an earthquake *[8 January 1780]*, the ground opened, and two Mollahs (Moslem priests), of whom we saw one in the chapel, together with several inhabitants, witnessed the sudden appearance of a large tomb of stone, which, however, soon vanished in the opening. From that time, true believers have been convinced that Noah's wife lies interred there; although it would seem, that the honour of actual sepulture is a point of issue between her and Noah's mother, as Maranda signifies, in the Armenian language, the 'mother lies here'. This grave, perhaps, contributed to induce the *[Russian]* Ambassador to rest here a day.
>
> Kotzebue (1819)

James J. Morier (1818), who arrived at Marand on 28 May, 1814, referred to Noah's mother's tomb, but did not mention the earthquake. Marand is located near the North Tabriz fault and was destroyed during the October 1786 earthquake (see Chapter 11 and Figure 11.12).

3.5.1.9 The 1836 Earthquake Caused by Passing the Soul

The British traveler G.T. Vigne (1840), who visited Ghazni and Kābul in the summer and early fall of 1836 in his travels to Afghanistan over the Dulaiman Range, recorded the following narrative:

> Several weak shocks of an earthquake took place while I was at Kābul. There are usually about a dozen in the course of the year. Baber describes a very severe one. The Mussulmen *[Muslims]* say that it is owing to the disturbance made by a soul of a great man passing from one place to another.
>
> Vigne (1840)

3.5.1.10 The 5 May 1853 $M_s \sim 6.2$ Shirāz Earthquake and the End of the World

Wills's daughter [in Charles James Wills (1894); the Medical Officer of Britain's Telegraph Department in Iran during 1866–1881], an eyewitness and the only survivor of her family during the 5 May 1853 Shirāz earthquake, gave a stunning and detailed description of the event in *Chapter XII—The Great Earthquake*, meticulously describing her feelings and the people's reaction to a major earthquake.

> ...and that 'Ali Nekki, a fanatical priest *[Mullāh]*, was openly declaring in the great mosque that the end of the world had arrived and as he had great number of followers among poor classes, this had added greatly to the general confusion.
>
> Wills (1894)

3.6 DIVINE WRATH, WILL, WARNING, ORDEAL, AND PUNISHMENT

Since ancient times, people have attributed destructive natural disaster to the wrath or ordeal of gods, without blaming the event on the quality of their dwellings (Chapter 2). They later questioned nature and the gods about the calamity. Although the view of an earthquake as an act of fate is helpful for the psychological recovery of the survivors, the public's ignorance about the hazards of such an event have always retarded the processes of (i) self-help; (ii) search-and-rescue operations; (iii) demand from governments for help; (iv) health services; and (v) construction of earthquake-resistant structures.

Ibn Sīnā [Pursinā, Avicenna; 980–1037] wrote in his treatise 'Kanz al-Masā'il' that:

> In response to the two questions which might be asked regarding earthquake and its genesis: First question is what causes earthquake. The second question is that the philosophers say that whatever exists is good, and evil is a matter of non-existence. How this is possible when earthquake kills a large number of innocent children and people; and why God accepts this and does not prevent its occurrence? In order to answer these

questions, we have to present brief introductions in order to eradicate any doubts from those who do not know.

Now that you learned about earthquake genesis, we will response the second question… Now after many years if suddenly in a corner of the Earth an earthquake occurs and few people and animals die, this cannot be comparable with the numerous goodness and blessings presented by God. Therefore, if almighty God abandons all these goodnesses because of a minute evil, then he should not create human beings, animals, the Earth, and water out of nothing. That is why the philosophers say that abandonment of plentiful good things for a minute evil is a huge mischief.

tr. from the Persian text of Ibn Sinā

Post-Islamic popular perceptions in Iran hold that God punishes people on account of their sins by shaking Earth. Quoting *Khānsāri*'s treatise "*Khulāsat al-Akhbār*" ["*The Perfection of the Narratives*"], Donaldson (1933), wrote that: "*When God desires to send an earthquake as a punishment to any part of Earth, he orders the angel who is the guardian of the mountain to cause the particular stratum that is connected with that part to shake violently*" (tr. Donaldson, 1933).

Al-Majlisi (1668) stated that: "*Whenever 'Ali drew his great sword, Zhu'l-faqār, the mountain of Qāf* [Caucasus; the Zoroastrian mythical Alborz] *trembled.*" For a reference to trembling Mount Qāf, see Chapter 2.

3.6.1 The 855–856 $M_s \sim 7.1$ Ray Earthquake as Wrath of God

In a poem of lament in the form of "qasideh" Poet Qavāmi Rāzi (d. 1164) wrote that the Ray earthquake was a result of the wrath of God because of the immoral deeds and sins of the people. He then recommended charitable deeds and repentance.

3.6.2 The 4 November 1042 $M_s \sim >7.0$ Tabriz Earthquake as God's Annihilation

Qatrān Tabrizi (1009–1072) wrote in his first qasideh poem about the earthquake and praise of Abu Nasr Mamalān that:

God sent annihilation down on the people of Tabriz,
Firmament caused decadence of the affluence of Tabriz

In his second poem, the poet described the power of God in construction and then destruction of everything on the planet. Finally, he stated that the earthquake was the punishment of our immoral deeds and that we are not repentant yet:

This misfortune is fruit of our own wicked acts,
For we did not repent for our unworthy acts.

3.6.3 The 1208–1209 Neyshābur Earthquake as Wrath of God

Mostaufi (1334–1335) stated in his poem that complete destruction of the Neyshābur city was from the wrath of God.

3.6.4 Maulānā Jalāl U'ddin Rumi [1263–1273]

Maulānā Jalāl U'ddin Rumi (1263–1273; in *Mathnavi-ye Ma'navi*, Book IV: 9; ed. Nicholson, 1926), in his poem "the ascent of *Dhu'l-Qarnayn to the peak of the Qāf Mountain,*" described the mythical "*Harā Berezaiti Mountain*" [Alborz] as the "Qāf Mountain" [from "*Kof*" in Pahlavi, meaning mountain]. Furthermore, Rumi clearly mentioned that earthquakes are caused by God's decision:

When God wills an earthquake in any land,
He bids me and I cause the vein to throb.
Then I make to move mightily the vein,
With which the (particular) land is connected.
When He says 'Enough!' my vein rests,
I am (apparently) at rest, but actually I am in rapid motion.

Rumi (1263–1273) and ed. Nicholson (1926)

3.6.5 The 19 October 1336 $M_s \gtrsim 7.0$ Jizd Earthquake

Describing the 19 October 1336 earthquake that destroyed the city of Jizd, Majd al-Din Khāfi (1336) mentioned that: "An earthquake took place in the world as if it was like the appearance of the Day of Resurrection" (see Chapter 11 and Figure 11.9 for this event).

3.6.6 The 30 September 1139 $M_s \sim 7.0$ Ganjeh [Ganja, Gāndzāk, Elizāvetpol, Kirovābād] Earthquake as Wrath of God

Armenian historian Mikhitār Gosh (1130–1213) recorded that the devastating earthquake was the fury of God unleashed upon the world and quoted Job 9:6 and Psalm 103:32 (Guidoboni and Traina, 1995).

3.6.7 The 29 November 1406 $M_s \sim 5.5$ Tātév, Siunik, Armenia Earthquake

A colophon in the Matenādārān, Yerevān, manuscript 9247, fol. 299r, written by Tovmā Mecop'etsi (Zeitouniān, 1991; Karakhanian and Abgaryan, 2004; Guidoboni and Comastri, 2005; Bābāyān, 2006), stated that "such was the quantity of our sins that an earthquake struck the Earth" (Guidoboni and Comastri, 2005).

3.6.8 The 5 February 1641 $M_s\sim6.8$ Dehkhārqān (Āzarshahr) Earthquake Divine Calamity

Mir Baqā' Badakhshi (ca. 1664), in verse 4 of his long 62 verses *mathnavi* poem, described the effect of the earthquake in Tabriz and questioned God about the reason for the divine calamity that had befallen upon Tabriz and its citizens: "*What have we done our Lord that suddenly we were struck by a heavenly calamity?*" (Mir Baqā' Badakhshi, ca. 1664) (see Chapter 11 and Figure 11.12 for this event)

3.6.9 The 14 June 1679 $M_s\sim6.7$ Gārni, Armenia, Earthquake

Eyewitness Zāchāriāh of Kānāker (1627–1699) thought that God's anger fell on the ground during this earthquake: "*June 4, 1679 on the Ārārāt land the God's anger fell on Tuesday at 7 o'clock the ground suddenly roared and after a roar Ārārāt land strongly shook*" (Babayan, 2006; see Figure 8.2).

3.6.10 The 26 April 1721 $M_s\sim7.3$ Shebli Earthquake

In his long poem, *Ordubādi* (eighteenth century) wrote that the earthquake was the result of man's misdeeds against the will of God, greed, a parasitic lifestyle, sins, and rebellion. He added that nothing is left over from the religion of the prophet, and if such a year passes again, there would be no trace of a territory. At the end, he emphasized that nobody can repel the earthquake except us, and that we should be ashamed and scared (see Chapter 11 and Figure 11.12 for this event).

3.6.11 The 7 June 1755 $M_s\sim5.9$ South Kāshān Earthquake

Under the heading of "Some of the Events During the Justice Years of Shāh Tahmāsb Safavid al-Musavi" [r. 1524–1576], Hedāyat (1800–1871) stated that the earthquake occurred as the destiny of the great creator and the influence of the eventful stars (see Figure 16.4).

3.6.12 The 2 June 1824 $M_s\geq6.0$ Kāzerun and 25 June 1824 $M_s\sim6.4$ NW Shirāz Earthquakes

Shirvāni (1780–1837), the strongest nineteenth-century Ne'matollāhi master, indicated that the earthquakes were God's just and due chastisement for the numerous *fatwas* [religious decrees issued by a Muslim leader] issued by Shirāzi mullahs against him (Lewisohn, 1998) (see Figure 16.1).

3.6.13 Heavenly Punishment and Destruction of Shirāz City During the 5 May 1853 $M_s \sim 6.2$ Earthquake

Wills's daughter [in Charles James Wills (1894); the Medical Officer of Britain's Telegraph Department in Iran during 1866–1881], an eyewitness and sole survivor of her family during the event of 5 May 1853, giving a stunning and detailed description of the event in *Chapter XII—The Great Earthquake*, writing: "*It is always said in Persia that the dreadful calamity that destroyed four-fifth of the city of Shirāz was rent as a punishment from heaven for the wickedness of its inhabitants. I do not think that they were more wicked than the rest of the world; but it was my fortune to be present during the events of that awful time, and I know that it was the hand of Providence that preserved me*" (Wills, 1894).

3.6.14 The 19 May 1884 $M_s \gtrsim 6.0$ Central Qeshm Island and 10 January 1897 $M_s \sim 6.4$ Qeshm Town Earthquakes

Poet Mollā Solaymān Khamiri (Mosāfer), the qāzi [judge] of Khamir Port on the Persian Gulf (Nurbakhsh, 1990), composed a *qasideh* poem in 98 duplets about the Qeshm Island earthquake to which he refers as "the first Qeshm earthquake ["zelzeleh-yeh avval"]." In the last two duplets of his poem, Khamiri concluded that: "*Therefore, think that the wrath of God was a punishment.*" In his second poem describing the second earthquake, he reached the same conclusion: "*Once again the Earth shook and damaged the houses... The mountains fell down like a heavy rain because of the wrath of God.*"

3.6.15 The 11 April 1935 M_w 6.6 Kusut Earthquake Divine Punishment

The 1935 earthquake in Sāri and Bābol cities (Figure 12.10) was interpreted as divine punishment for the government's banning of the Muharram religious processions. This led to antigovernment demonstrations that ended in bloodshed in Bābol after the army moved in (Divānbaigui, 1969). Governor General Divānbaigui (1969) wrote that: "*Gradually the pouring rain was becoming more severe and more people gathered in the Sabzeh Maydān* [market] *square. All the people were shouting: 'Yā Hossain'; and they were thinking that the occurrence of the calamity [the earthquake] was the result of the banning the Āshurā's 'chest-striking' processions*" (tr. from the Persian text of Divānbaigui, 1969).

3.6.16 The 1 September 1968 M_s 6.3 Ferdows Earthquake

Almost all the survivors reiterated that the earthquake was the will of Allāh and concluded that everyone has to die one day (Etelā'āt, 1968; Kayhān, 1968; Pace, 1968) (see Chapter 13 and Figure 13.2 regarding this earthquake).

3.6.17 The 10 April 1972 M_w 6.7 Kārzin Earthquake

As in the previous example, almost all the survivors believed that the event was God's will (Etelā'āt, 1972; Kayhān, 1972; Bulloch, 1972) (see Chapter 13 and Figure 13.4 regarding this earthquake).

3.6.18 The 16 September 1978 M_w 7.3 Tabas-e Golshan Earthquake Divine Test

Empress Farah Pahlavi (2004), who with the Shāh visited the devastated town of Tabas-e Golshan immediately after the 1978 devastating earthquake, wrote in her book that: "*Some said the disaster was due to the wrath of God! I spent the day there, trying to help as best I could and allay people's fears*" (Pahlavi, 2004).

During the politically charged and chaotic last days of the Pahlavi dynasty, newly organized religious organizations posted signs in almost all the ruined and damaged villages in the meizoseismal area reminding survivors that the event was a divine test and punishment by Allāh (personal observation during the 40 days of aftershock recording and field work during September–October 1978) (see also Chapter 13 and 13.8 regarding this earthquake).

3.6.19 The 11 June 1981 M_w 6.6 Golbāf Earthquake as God's Ordeal

Excerpts from Āyatollāh Khomeini's message to the earthquake survivors and the people of Iran stated that: "*The great and disastrous earthquake, which caused destruction and death in the Kermān province, has caused grief in everyone. We should know that these unpleasant events are Allāh's ordeals of the tormented people of the earthquake area and the nation. The afflicted people with their patience and tolerance prove that they are firm and are not weak. According to the great Qor'ān, whatever they own or we have are granted by the beneficent Allāh; and we all are depository that soon or late will return the divine consignment to its owner. And for the suffering people of the whole country, it is a great ordeal to see how much we can share with the people overtaken by calamity and how much we will try*" (tr. from Jomhuri Eslāmi, 13 June 1981).

Visiting the area, Prime Minister Mohammad 'Ali Rejā'i declared: "*I am sure that these people are now residing by Allāh; because they deserve being close to the court of Allāh*" (tr. from Jomhuri Eslāmi, 13 June 1981).

Health Minister Dr. Manāfi released the following statement to the nation: "*If human being does not believe in Allāh and in the ideology of Islam, perhaps cannot tolerate. And our recommendation to people is that this is an ordeal performed by Allāh; and God willing, God will grant us patience to be able to pass this ordeal. If the result of this ordeal is positive, God will grant us with the reward*" (tr. from the Persian Etelā'āt, 21 June 1981) (see also Chapter 13 and Figures 13.11 and 14.3 regarding this earthquake).

3.6.20 The 20 June 1990 M_w 7.3 Rudbār Earthquake Divine Test

The state's official line in Tehrān was to describe the 1990 Rudbār earthquake (which killed about 40,000 people) as a divine test; however, other religious leaders called it a punishment from Allāh for a deviation from religious conduct. In a broadcast on the Tehrān state radio, the supreme religious leader Āyatollāh Seyed 'Ali Khāmeneh'i, described the Rudbār disaster as a "divine test," but said Iranians would pass the test if they hurried to assist the victims (IRNA; Katāyon Ghāzi, United Press International, 28 June 1990).

On Thursday, 21 June 1990, Khāmeneh'i issued a condolence letter to Hojat al-Islām Ehsānbakhsh, the Friday Imām of Rasht, the survivors, and the people of Iran, emphasizing that: "... *Human beings are tested by the natural events. We should try to fortify the country against oppression. In fact all these natural events are Allāh's ordeal, and a real and healthy Muslim is the one that when an event occurs, tries to compensate the event with hardworking. Patience towards the bitter events will have a major reward. Those who lost their belovеds should know that the mighty God will reward them with incentives for their sufferings. Of course there is a prerequisite, and that is you should have patience and should not be ungraceful. Sometimes god elevates the promotion stairs by calamity. Sometimes a calamity has a bad appearance, but later on, the occult curtains are pulled aside and progress is achieved*" (Etelā'āt, 1990; IRNA, 21 June 1990; all the Iranian news media).

The reason behind the Rudbār earthquake posed a theological quandary for the religious scholars who hold power in the country's radical Islamic government. President Hojat al-Islām Hāshemi Rafsanjāni, in his Friday sermon (IRNA, 29 June 1930; David Thurber, The Associated Press, 30 June 1990; *The Washington Post*, Saturday, 30 June 1990) stated that: "*An earthquake is a natural phenomenon and of course everything is done by* Allāh. *But that does not mean we should confine ourselves to prayers alone... Neither science nor revolutionary Islam could explain fully what happened in the mountains of northwestern Irān last week. Some may ask why God makes us deserving of such disaster after all these years of great holy war, but it is not like that. The full dimensions of this disaster are unclear to us.*"

A few days after the 1990 earthquake, several mullahs were sent from Tehrān to the meizoseismal area; five days after the earthquake, the odor of dead bodies lingered over the streets of Rudbār and Manjil. As the mullahs handed out copies of a small brochure outlining the rewards of martyrdom and struggle, they promised the survivors great glory for their struggle (Shenon, 1990; for more information about this earthquake, see Chapter 13 and Figure 13.12).

3.6.21 The 28 February 1997 M_w 6.0 Shirān (NW Ardébil) Earthquake

A survivor of the earthquake, who was from Vilādareh village, was asked how it was possible to come to terms with sudden death on such a vast scale. He replied that "*God has sent that and whatever he does, I will accept*" (*The New York Times*, 4 March 1997) (for more information about this earthquake, see Chapter 13 and Figure 13.16).

3.6.22 The 10 May 1997 M_w 7.2 Zirkuh (Qā'enāt) Earthquake Was God's Will

The country's response to this large-magnitude earthquake, when 1568 people killed in a remote desert area, was typically fatalistic. Hossein Rezāpur, who lost his house in the village of Ardékul but not his family, said that it was the will of God, and that "*no matter what we do, we cannot fight what God wants for us*" (Faruqi, 1997) (see also Chapter 13 and Figure 13.18 for this event).

3.6.23 The 26 December 2003 M_w 6.6 Bam Earthquake

When the supreme religious leader Āyatollāh 'Ali Khāmeneh'i visited the completely devastated city of Bam, where between 31,828 and 43,200 people were killed during a moderate-magnitude earthquake, he expressed his sympathy for the earthquake survivors, stating that: "*This disaster reveals God's ordeal; It is in such hardships that we can grow and strengthen our faith*" (IRNA; Persian News Media; kayhannews.com) (see Chapter 14 and Figure 14.6 for more information about this earthquake).

3.6.24 God's Ordeal [12 January 2010]

Dr. Mohebi, an earthquake engineering professor at Ardébil University, stated in a meeting that: "*Natural disasters are one of ordeals of God which could happen anywhere in the world. Even it is possible that a properly designed and constructed seismic resistant building undergo considerable damage*" (NGDIR, 2010). (I will address these statements at the end of this chapter.)

3.6.25 Earthquake as a Sign of Ordeal as Well as Punishment [9 April 2010]

At the end of an article entitled "Why So Much Calamities?," Fātemeh Mohammadi stated that: "*We conclude that any calamity such as earthquake and other difficulties can be considered an ordeal or a punishment. The calamities have different purposes to different people. For example, for the sinners are punishment or notice; whereas for the believers they are considered as ordeal or cleansing. Therefore, it cannot be said that all the earthquake stricken people of a city were punished; or all the sick people and those with difficulties have undergone divine punishment*" (mouood.org, 9 April 2010).

3.6.26 Greater Sin Increases the Earthquake Probability [16 April 2010]

The director of the National Emergency Crises Management quoted a very important message from the president (Ahmadinezhād): "*One of the great religious scholars of Tehrān, Āyatollāh Khoshvaqt, who teaches ethics in Tehrān and a friend of the President, sent me* [the President] *a message that I ask the people to be careful and while avoiding sin to pray to evade disasters*." The director added that a few weeks after the evening prayers in Emām Hassan Mojtabā mosque, Āyatollāh Khoshvaqt spoke about excessive sins resulting in disasters such as earthquakes. He insisted that in order to avert earthquakes, people should avoid sins. This is not a statement from the tenth century, but rather the words of an official in 2010.

In a meeting with the authorities regarding preparedness for a probable earthquake in Tehrān, Āyatollāh Khoshvaqt stressed that: "*an increase of sins results in Allāh's punishment; the scholars and preachers should go among the people and warn them of Allāh's punishment, asking the people not to sin, and pay alms to the poor devout prayers to evade this calamity happening in Tehrān. People also should participate in mass repentance*."

In an interview with rejanews.com, the Āyatollāh insisted that the increased likelihood of earthquakes in Tehrān in the next few months was directly related to the widespread increase in sins in the city. He mentioned a narrative from a Vasā'el al-Shi'iya treatise that describes the link between sins and earthquakes (asriran.com/fa/news/109246, 19 April 2010; rejanews.com).

3.6.27 Allah's Ordeal in Days of Blessing [11 August 2012 M_w 6.1 South Ahar Earthquake]

Hojat al-Islam Mohammad Vā'ez Musavi, the Islamic scholar who is a member of the Assembly of Experts of the Islamic leadership (Majles Khobrégān Rahbari) from Āzarbāijān Province, stated that: "*This ordeal of Allāh* [the August 11, 2012 earthquake that left 258 people dead] *has resulted in strengthening of the national solidarity of people, and once again the grace hand of Allāh will come out of the sleeve. I hope, as in the past, in these days of blessing and beneficence, the authorities, local people and fellow citizens will respond to their human and Islamic responsibilities and accelerate the rebuilding of the ruins and assisting the victims of this event, thereby reducing the suffering of fellow citizens*" (http:/www.tabriz.irna.ir/News/80282722, 18 August 2012).

3.7 GOD'S GRACE, MERCY, AND MIRACLES

Despite believing in the idea of punishment and ordeal, some also see the grace of God during earthquakes. This view is usually expressed when there is less damage, few have perished, some have survived the collapse of structures, and a church or mosque has withstood the earthquake.

3.7.1 The 31 March 1648 $M_s \sim 6.7$, Hāyots Dzor Earthquake

Vārtābed Ārākel Tāvrizetzi [Ārākel of Tabriz; 1594–1670], the Armenian celibate priest and an eyewitness to the 1648 earthquake, recorded that: "*After the earthquake, everyone who had escaped death went to their homes, began to dig the earth, and to pull out people and things from among the debris. Although countless perished, there were many who, through God's grace, survived until they were pulled out from the ruins. Thus, even after ten and five days, they were still pulling out survivors from the debris*" (Bournoutian, 2006).

3.7.2 The 8 January 1780 $M_s \sim 7.4$ Tabriz Earthquake

Mārdiros Khālifā (in Hakobyan, 1951) recorded the following narrative regarding the earthquake: "*On this day, December 17th the day was Friday, there was a tremendous tremor* [āhākin shārzh] *in the city of Tabriz. The whole city collapsed and innumerous people perished and remained under the earth and the eyewitnesses said that the enormous city became like a natural hill and that altogether an appearance of a city was not seen. Also, the surrounding living places and villages Gunāy* [?], *Tāsu* [Tasuj], *Sālmāst* [Salmās], *Koy* [Khoi], *Hormi* [Urumiyeh], *and our city Vān was shaken many times but by the mercy of God no damage was caused*" (Hākobyān, 1951).

3.7.3 The 11 July 1890 $M_s \sim 7.2$ Tāsh Earthquake

The Iran newspaper reported that: "*Continuous raining and repeated earthquakes [mainshock and aftershocks of the 11 July 1890 earthquake in the High Alborz] of these days damaged buildings and public baths of the city [Estārehābād, the modern city of Gorgān]. Nonetheless, as mentioned earlier, thank God nobody was killed or injured; and with the virtue of resorting of people to the Members of the House of the Holy, and blessings of the mourning of Imam Hossein, all the people favored the mercy and grace of God*" (Iran, 16 August 1890, No. 725).

3.7.4 The 17 October 2009 M_l 4.0 SE Tehrān Grace of God Earthquake

Mohammad Rezā Mahmudi, managing director of the Tehrān Province National Disaster Headquarters, Ministry of Interior, announced in a Saturday night meeting regarding the small-magnitude earthquake felt in Tehrān that: "*This is an independent earthquake. It was the grace of God that by this earthquake the trapped energy of the Earth was decreased*" (irna.ir; ngdir.ir, 18 October 2009).

3.7.5 The 20 December 2010 M_w 6.5 South Rigān Earthquake Miracle

Esmā'il Najjār, governor general of Kermān Province, stated in a press conference about the few casualties and little damage from the earthquake that: "*During this dreadful earthquake a miracle occurred which can be attributed to omnipotent divine interaction, and Allāh's favor and kindness. The epicenter of this strong earthquake was located in the Chāh Malek village but no one was killed in that village*" (irna.ir, No. 30138073, 22 December 2010).

Two medium-magnitude earthquakes (M_w 6.5 and 6.2) took place on 20 December 2010 and 27 January 2011 in a remote, very low, and sparsely populated rural desert region of a mostly nomadic and seminomadic population living in tents about 45 km south of Rigān, and 40 km to the south of Chāh Malek (Figure 14.11). The occurrence was mentioned by the governor general, in the southeast Kermān Province, north of Baluchestān Province of southeast Iran.

3.7.6 The 16 April 2013 M_w 7.8 Earthquake Damage Was Averted by the Midnight Prayers

After the occurrence of the large-magnitude, intermediate-depth subducting slab earthquake in a remote and sparsely populated desert area of the Iran–Pakistan border, the governor general of Kermān Province, Hātam Nāru'i, in the emergency headquarters of the province, announced that because of Allāh's grace, the midnight prayers of the believers in the province, and the presence of the pure bodies of the martyrs in the province, the damage, destruction, and casualties of this large earthquake were averted and nobody was killed (Hātam Nāui'i, 2013).

3.8 EARTHQUAKES AND SUPERSTITIONS

Earthquakes are sometimes described as occurring simultaneously with other natural phenomena, such as an eclipse, thunder and lightning, the appearance of a comet, a meteorite fall, a storm, floods, and the death of prominent people.

According to ancient beliefs, an eclipse indicates that God is angry with mankind on account of their sins and sends a dragon to swallow the Sun or the Moon. As such, people make an effort to scare away this monster by climbing upon their housetops or going into their courtyards and making a loud din. It is said that if the vessels they use to make noise are heirlooms, then the sound will be more effective. They also recite prayers as loudly as they can. It is told that once during an eclipse, the prophet Muhammad read the second sura and performed a long prostration ["*ruka*"] in prayer. If there is an eclipse of the Sun during the month of Muharram, grain will be abundant and the cost of living will be low, the ruler will triumph over his enemies, and there will be earthquakes and much sickness (Donaldson, 1933).

In some cases, there have been clear tendencies to associate earthquakes with other events, when clearly such relationships are merely coincidental. The destructive Boznābād (east Iran) earthquake of 16 January 1979 (Figure 16.18) that took place at 13:20 hours local time (09:50 UTC), occurred a few hours before the enforced departure and abdication of the Shāh of Iran, is the best example of this in our day. The Shāh and Empress Farah departed Tehrān Airport on Tuesday 16 January 1979 at 15:08 local time.

The association of earthquakes with other events in superstitious and irrational societies reflects that there are connections between supernatural events; if a calamity such as an earthquake happens prior to an event, the calamity is then considered to be an indicator or a sign. If the calamity follows an event, it is considered to be a punishment. Some examples can be found in ibn al-Jauzi's (1181) interpretation of the few earthquakes that occurred in Baghdād.

3.8.1 Comets, Meteors, the Farghāneh Earthquake, and Death of Caliph

Al-Tabari (850), an Iranian scholar of the ninth century, stated the following in his writing on the celestial signs of coming events:

There appeared in my time also a comet, and the direction of its tail was once to the East and once to the West, and it remained [visible] during a succession of nights. After this a large town in Ferghāna *[Fergana/Farghāneh; 40° 23′ 11″ N-71° 47′ 11″ E]* disappeared *[inkhasafa: sank into ground; swallowed up by an earthquake]* with all its inhabitants; and a host of people rose up against the great king *[i.e. the Caliph]* and he overcame them, but it was not long before he died and God gave the reign to Hārun *[Abu Ja'far Hārun al-Wathiq Bi'llāh; 842-847, grandson of Hārun al-Rashid]*. And I observed in the latter's time one afternoon, when I was at Surra-man-Ra'a *[Sāmarra]*, a star *[meteor]* shooting down from the direction of the south and falling to the north, and it was spread out in the heaven and separated into flashes resembling written letters; and he [the Caliph] died likewise some days later after a violent disease by which he was attacked. Several months before his death there appeared in the heaven an oblong fire remaining from midnight till nearly dawn.

tr. Meyerhof (1931)

Ferghāna [Farghāneh], a town in a large province in Transoxania, in eastern modern Uzbekistan, south of Kazakhstan, is located in a sedimentary basin and was destroyed by an earthquake in 838–839 [40.40°N–71.80°E, (Kondorskaya and Shebalin, 1977, 1982)].

3.8.2 The 4 February 1094 Baghdād Earthquake and Death of the Caliph al-Muqtadāi' 'Abbāsid

The 4 February 1094 earthquake in Baghdād took place shortly before the death of the Caliph al-Muqtadāi' 'Abbāsid [r. 1075–1094] (Ibn al-Jauzi, 1181; Ambraseys and Melville, 1982).

3.8.3 The 3 April 1118 Baghdād Earthquake and Death of the Sultān Ghiyāth ad-Din Mahmud Saljuq

The 3 April 1118 earthquake in Baghdād was later followed by the news of the death of *Sultan Ghiyāth ad-Din Mahmud Saljuq* [r. 1105–1118], who died in Esfahān on April 5, 1118 (Ibn al-Jauzi, 1181; Ambraseys and Melville, 1982).

3.8.4 The 27 February 1130 Baghdād Earthquake and Death of Sultān Mahmud Saljuq

The 27 February 1130 earthquake in Baghdād was followed a year later by the death of *Sultan Mahmud Saljuq* [r. 1118–1131] on September 10, 1131 (Ibn al-Jauzi, 1181; Ambraseys and Melville, 1982).

3.8.5 Holāku Mongol's Invasion of Baghdād [1258]

In discussing the events leading to Holāku (Hulāgu) Mongol's invasion of Baghdād [656/1258], *Rashid al-Din Fazlollāh Hamédāni* (1304) wrote that:

Hesām al-Din the astronomer *advised Hulāgu that it is a bad omen to attack the Caliph family and invade Baghdād. Any king with intention of invading Baghdād and killing the 'Abbāsids will not benefit from his life. He added that in case of Hulāgu's incursion of Baghdād six corruptions will happen: (i) all the horses will die and the troops become sick; (ii) the sun will not rise; (iii) there will not be raining anymore, (iv) strong winds will blow, and (vi) the world will be destroyed by earthquakes. Furthermore, no plants will grow on the Earth and finally the great king [Hulāgu] will die in the same year.*

Nasir al-Din Tusi (1256–1259), in commenting on *Hesām al-Din's* ridiculous predictions, wrote that the system of nature and the cosmos will not be stopped by incidents such as an invasion of Baghdād. Throughout history, massacres such as those carried out by Hulāgo have happened, and those incidents did not have any effect or change on the evolution of the world's orderly system (Berberian, 1997b).

3.8.6 The 28 May 1845 Mashhad Mild Earthquake and Camping on an Ill-Omen Spot

On 28 May 1845, at about 4 h before nightfall, an earthquake was felt two parsangs [ca. 12 km] from Mashhad just at the place where the European traveler General Joseph Pierre Ferrier was preparing to camp for the night. His guide, Hassan, concluded that they were occupying an "ill-omened spot," the tremor was a bad omen, and they had to move the camp location at once (Ferrier, 1857, 1860; Massé, 1938; Messner, 1954).

3.8.7 The 17 January 1895 $M_s \sim 6.8$ Quchān Earthquake During the Absence of Ebrāhim 'Ali ibn Musā al-Rezā

Local inhabitants gave Donaldson (1933) this explanation of the cause of the destructive earthquake in Quchān: Ebrāhim 'Ali ibn Musā al-Rezā, one of the sons of the Imām 'Ali al-Rezā, whose tomb is located in Quchān [Emāmzādeh Ebrāhim in Āstāneh-ye Shahr-e Kohneh], had gone away on a meteor to visit his father, who was buried at Mashhad [the Emām Rezā shrine]. In doing so, he hence left the city of Quchān defenseless against the elements. While his protective presence was temporary removed, the town of Quchān was destroyed. Meteors are said to be heavenly chariots in which the Imams frequently ride. Often, when a meteor is seen in Khorāsān Province, it is declared that Imam Hussain is visiting Imām Rezā, or that Imam Rezā is returning from a visit to Imām Hussain (Donaldson, 1933; Massé, 1938; Messner, 1954).

3.9 OVERVIEW: DOCTRINE OF FATALISM

Earthquakes and their effects on societies were dealt with in different ways from pagan days (Chapter 2) to modern times; however, the basic essentials have not drastically changed. Earthquakes were ascribed to evil spirits, the will of gods, and later to God. God's will was caused either by: (i) people's deviational behavior, such as adultery, usury, lying, alcohol consumption, exposing body parts (including women's hair, etc.); or (ii) the approaching "Last Day" referred to in sura 99 of Qorān. In the first instance, God causes Mount Qāf (substituting the Zoroastrian Alborz surrounding the known Earth), or any other region, and the Bull and the Fish to move and shake the Earth to punish the sinners. Nonetheless, some chronicles found earthquakes to be caused by special planetary constellations and passing comets or eclipses (see also Chapters 2, 4, and 5).

The ancient pagan view of evil spirits, idols, and deities as being responsible for the shaking of the ground was later transferred to Yahwah, God, and Allāh in the monotheistic religions. This belief is still held in the twenty-first century among many nations and officials. We have witnessed authorities, clergymen, the Minister of Health (who is a medical doctor), an earthquake engineering professor (with a Ph.D.), and laymen alike express their beliefs on numerous historical and modern occasions. The main discussion has always been that earthquakes should be considered as a punishment or ordeal. Nobody questions the deep chronic problems associated with earthquake-risk minimization in developing countries, or why the authorities are not prepared for earthquakes and their destructive consequences. Nobody focuses on the enforcement of building codes or a retrofit of the existing public structures. Assessing the historical and modern natural hazards is of paramount importance to mitigate risk in developing countries.

Fatalism is the belief that all events and experiences are predetermined, inevitable, and unalterable and determined by external forces beyond human control. The acceptance of fatalism as central to the dispensationalist tradition by society for millennia has given impetus to the people and their leaders to easily accept the status quo. Fatalism has been rooted in the Iranian mind at least since the Zervanism (Zaehner, 1972), and has been exacerbated since the conversion to Islam in the seventh century and the advent of Shi'aism as the official religion since 1500.

Psychologically, this approach might have helped traumatized survivors of natural disasters in alleviating stress, building up personal resilience, and managing trauma without any other help. However, fundamentally, it has prevented people from becoming aware of the causes of structural failure, fatality, property loss, long-term disaster preparedness, mobilization, prevention, and risk-mitigation measures. Therefore, the tragic results of natural disasters have become a recurring vicious cycle.

Consequently, the disaster of the 2003 M_w 6.6 Bam earthquake, with 40,000 fatalities, is thoroughly analogous with that of the 2010 Port-au-Prince, Haiti; 2004 Sumatra-Andaman, Indonesia; 2002 Nahrin, Afghanistan; 1999 Izmit, Turkey; 1990 Rudbār; 1990 Luzon Island, Philippines; 1998 Spitāk, Armenia; 1978 Tabas-e Golshan; 1976 Moro Gulf, Philippines; 1641 Tabriz; 1271 and 1209 Neyshābur; 1177 and 955 Ray/Tehrān; 856 Komesh/Dāmghān; and 815, 805, 734 Zarang [Old Persian *Zarannka*; Greek *Drangiana*], Sistān earthquakes (see Berberian, 1994, 2005; Tables 17.10 through 17.12 for more information). Hence, the problem is ancient and acutely chronic.

During these long periods and circumstances, the governments and authorities in developing countries have never been held accountable: (i) to the victims of natural disasters; (ii) for controlling the life/safety of citizens and their lifeline infrastructure (especially schools and hospitals; water systems, power supply, roads, etc.); or (iii) for controlling the risks of natural disasters to socioeconomically acceptable levels. The governments (and not the people) in developing countries have long implicitly accepted the risk of losses after each earthquake to the tune of tens of thousands of people's lives and billions of dollars.

Chapter 4

Ancient Earthquake Theories

And I keep in my mind that
I shall not violate the Law of the Earth.

Sohrāb Sepehri (1964)

Early humans, who lived in harmony with the environment and were dependent upon nature, were perhaps the most knowledgeable about nature and may have noticed major changes and disturbances in the order of nature. Due to the destruction of written documents during numerous invasions, no trace has yet been found on the Iranian plateau to reflect the profound effect of earthquakes on the life of the inhabitants. However, having experienced the ground shaking, rockfalls, the rumbling of earthquakes, and coseismic surface ruptures, the inhabitants came to believe that earthquakes were set off by one of the four known primary active elements: air with energy (wind) in the form of trapped underground vapor, which leads to a fracturing of the rigid Earth's surface. This way of thinking has been a deep-seated tradition on the Iranian plateau and extends to modern times. Because the concept of the four active elements is important to ancient earthquake theories, a summary of the development of the idea of the four primordial elements in the old world is organized in chronological order in Table 4.1, which shows that the idea originated on the Iranian plateau before taking hold in classic Greece.

It should be added that Ferdowsi (Shāhnāmeh, 1010), in his "Discourse of the Creation of the World," referred to the ancient Zoroastrian belief of the "Four Elements" and their role in the Creation:

The essence of the four elements were wrought,
And without labour into being.
First when the fire from motion came to flame,
From its heat dryness *[earth]* into being came.
With this at rest, cold *[air]* must itself unfold,
And humid moisture *[water]* issue from that cold.
The mountains were raised, and the rivers first did flow,
And vegetation, too, began to grow.

Rogers (1907) also see Ferdowsi Tusi (1010), ed. Joneydi (2008), vol. 1, pp. 35–50

☆"To view the full reference list for the book, click here."

Developments in Earth Surface Processes, Vol. 17. http://dx.doi.org/10.1016/B978-0-444-63292-0.00004-1

TABLE 4.1 The Chronological Order of Conception of the Four Primeval Elements

No	~Date	Region/Culture	Original Sources	Elements	References
5	450 BCE	Hellenic	Empedocles (490–430 BCE), Tetrasomia	Four roots (Rhizomata) of everything: Enlivening Hera (Earth), Hades (Fire), Shining Zeus (Air), and Nestis, moistening moral springs w/tears (Water). Later, Aristotle added the fifth element, the Aether (see Hinduism)	Empedocles (490–430 BCE); Plato (428–348 BCE); Aristotle (384–322 BCE)
4	Fifth century BCE	Buddhism, India		Earth, Water, Fire, and Air	EB (2009)
3	1500–1200 BCE	Zoroastrianism, Iran	The Avestā: Gāthās (~1500 BCE) The Pahlavi texts (rewritten in the ninth century based on ancient texts)	Four Sacred and Active Elements [Ākhshij]: Earth (Zem, Boom), Water (Ap, Āb), Fire (Ātar; Sun's representative on Earth; later Āzar and Ātash), and Air (Vāyu, Andavāy)/Wind (Vāta)	Boyce (1989)
2	1700–1100 BCE	Hinduism, India	The Vedās: Rigvedā (~1500 BCE)	Earth (Kshiti, Bhumi; Boom in Persian), Water (Jala), Fire (Tejas, Agni), and Air/Wind (Pavan). Void/Aether (Byom, Shunya, Akash)	Boyce (1989), Oberlies (1998)
1	1700–1100 BCE	Babylonian Creation Myth	Enuma Elish Akkadian Cuneiform	Earth, Sea, Sky, and Wind	Smith (1876), Batto (1992), Rochberg (2002)

4.1 AIR MOVEMENT (WIND) THROUGH SUBTERRANEAN FISSURES [PRE-1500 BCE BELIEF]

Our knowledge of the various theories elaborated in the ancient world about the natural causes of earthquakes comes to us through the Iranian, Indian, and Greek myths, philosophies, and cultures. Sporadic allusions to large earthquakes on the Iranian plateau are found in the ancient Indo-Iranian myths, legends, and long-standing oral traditions (see Chapters 1–3). Most western textbooks suggest that in classical Greek times, winds in subterranean caves were believed to be the cause of earthquakes. Aristotle (344–322 BCE) credited the earlier Greek philosopher Anaxagoras (510–428 BCE) as the originator of this idea (Ambraseys and Melville, 1982; Howell, 1986; Guidoboni, 1994) as well as the theory that earthquakes occur when the *ether*, which has a natural tendency to rise, is imprisoned in subterranean cavities and cannot escape because Earth's pores are blocked by rain.

4.1.1 The ca. 1500 BCE Early Iranian and Indian Views

The study of ancient mythological texts such as the *Rig Vedā, Dinkart*, *Yashts,* and *Bondéhéshn* (*Bundahishn*) clearly shows that the Indo-Iranians had early views of the general causes of earthquakes (Berberian, 1991, 1994, 1997b).

4.1.1.1 Bondéhéshn (Bundahishn; the Zoroastrian Primal Creation, Cosmogony)

In the *Iranian* (*Greater*) *Bondéhéshn*, Chapter XXI, E, we read about the theory of air movement through subterranean fissures (ed. Anklesaria, 1956; ed. Bahār, 1990; Berberian, 1997b; Peterson, 2002):

> As regards earthquake it is manifest that even the same dreadful devs *[Cheshmagān Divān]* obstruct with sorcery the passage of that wind of life, which is the preserver, whilst moving through the fissures of the mountains, so that it may have no movement there through. So much Earth, the stability of which is owing to that Wind, is in tremor, and tears asunder. The Wind reaches the base of its passage, and it may either cleave the mountain, or wander over the Earth and the seas, or turn the houses and dwellings upside down. And the place where there is no mountain, the earthquake neither approaches nor comes to sight. For it is a passage within the Earth, which it is not proper to obstruct. When anything obstructs it, the Earth soon splits, and if there is a passage, the earthquake does not appear. For the sinful *[? bazagh, lit., 'damming', in ed. Bahār, 1990]* wind in the world is just as in the body of men. When the sinful wind tarries in their veins for destruction and decrease, and if the wind cannot obtain space for coming and going, the body reckons it a disease, and life seeks an outlet. When it becomes oppressive, the wind of life becomes dormant, and the body dies. And the more the evil admixture, the more oppressive is the sinful wind. So too is it manifest that men become the more sinful. The sinful winds within the Earth, which they also

call dreadful *[Cheshmagān Div, in ed. Bahār, 1990]*, are very oppressive and perpetrate much harm. Their connection too is with these planets [abakhtarān], and they derive greater strength from them.

ed. Anklesaria (1956) and ed. Bahār (1990)

Obviously, this primitive earthquake theory was based on only one of the four primordial elements known in the ancient days as well as the observation that strong ground shaking turns the houses and dwellings upside down. Furthermore, the theory emphasized that earthquakes happen in the mountains, where there is a resistance to fracturing, and that earthquakes cleave and split the mountain belts. This was identified after numerous observations of coseismic earthquake surface faulting at the margins of the mountain belts by the ancient native inhabitants.

Another reference to the wind trapped in subterranean fissures that causes earthquakes is made in Chapter XXXII (ed. Anklesaria, 1956; ed. Bahār, 1990; Peterson, 2002), where we read that:

Just as when the devs *[demons]* obstruct the passages of the wind in the world, earthquake occurs, when the wind of sin *[bazagh, lit., 'damming', i.e., 'damming wind' in ed. Bahār, 1990]* remains in the veins of man, becomes violent, and does not give passage to the wind of life, that spot too catches disease and causes the body to shiver and tremble. 14. There are other growths of sin *[damming winds, in ed. Bahār, 1990]* in the body also, just like other dev-ik growths *[demonic winds, in ed. Bahār, 1990]* that are in the world.

ed. Anklesaria (1956) and ed. Bahār (1990)

In this early and primitive hypothesis, "*Wind,*" as moving "*Air*" [one of the *four primordial elements of the powers of nature,* as the most active and mobile agent among *Earth*, *Fire*, and *Water*], roughly corresponds to our present concept of "elastic energy"; its accumulation in the rocks of the Earth's crust causes coseismic faulting and earthquakes.

The hypothesis was formulated at a time when the belief in the act of demons, devils, and monsters was destroyed by the Zoroastrian monotheism doctrine (in about ca. 1200 BCE; see Boyce, 1989 for a dating of the period of Zoroaster); hence, the ancient scholars were left with the four elements to work with. Today, we think of natural processes as manifestations of energy acting on or through matter. Surprisingly, the ancient Iranians had the same thoughts, though with numerous limitations. They regarded the activities of nature not as indications of supernatural intervention, but as somehow natural and orderly events that could be investigated in light of observation and reason within the framework of the four sacred elements.

The older Iranian legend that was popular prior to this hypothesis attributed earthquakes to a mythological storm-deity or monster called *Chashmag* [*Chashmag-e Div*] (Chapter 2). The important point is that the natural and observable activities of "*Wind,*" the moving air (energy in our present understanding), took the place of the imaginary and therefore inscrutable activities of "Chashmag-e Div," and the science of the ancient native people living on the Iranian Plateau replaced the older superstitions. Having noticed the great earthquakes and surface faulting on the Iranian Plateau (which is highly seismic), the ancient native inhabitants developed the oldest known concept for the cause of earthquakes and coseismic surface faulting: that is, the hypothesis of trapped *winds in subterranean fractures of the Earth.*

4.1.1.2 The Rigvedā and the Avestā

In Book Ten, Hymn CLXVIII of the Rigvedā (ed. Griffith, 1896) we read about the dynamic power of "*Vāyu*" [god of the wind]:

> O the Wind's chariot, O its power and glory! Crashing it goes and hath a voice of thunder. It makes the regions red and touches heaven, and as it moves the dust of Earth is scattered. Along the traces of the Wind they hurry, they come to him as dames to an assembly. Borne on his car with these for his attendants, the God speeds forth, the universe's Monarch. Travelling on the paths of air's mid-region, no single day doth he take rest or slumber. Holy and earliest-born, Friend of the waters, where did he spring and from what region came he? Germ of the world, the Deities' vital spirit, this God moves ever as his will inclines him. His voice is heard, his shape is ever viewless. Let us adore this Wind with our oblation.
>
> ed. Griffith (1896)

After becoming aware of the crashing power of wind, we read in Book Two, Hymn 11 (ed. Griffith, 1896) that the heaven and Earth tremble in a strong wind:

> Indra hath hurled down the magician Vṛtra *[Vritra; the draught demon]* who lay beleaguering the mighty river [the great cloud that holds the rain]. Then both the heaven and Earth trembled in terror at the strong Hero's thunder when he bellowed.
>
> ed. Griffith (1896)

Further on in Book Two, Hymn XII (ed. Griffith, 1896) we read about "*Indra*" [the god who reigns over the intermediate region (atmosphere), and fights against and conquers with his thunderbolt the demons of drought and darkness] whose breath as "moving air" shakes the mountains:

> He who, just born, chief God of lofty spirit by power and might became the Gods' protector, Before whose breath through greatness of his valour the two worlds trembled, He, O men, is Indra. He who fixed fast and firm the Earth that staggered, and set at rest the agitated mountains, who measured out the air's wide middle region and gave the heaven support, He, men, is Indra. Who slew the Dragon, freed the Seven Rivers, and drove the kine forth from the cave of Vala, Begat the fire between two stones [generated lightning between heaven and Earth], the spoiler in warriors' battle, He, O men, is Indra. By whom this universe was made to tremble, who chased away the humbled brood of demons, Who, like a gambler gathering his winnings seized the foe's riches,

He, O men, is Indra ... Even the Heaven and Earth bow down before him, before his very breath the mountains tremble.

ed. Griffith (1896)

It is worth nothing that in the Vedā's contemporary sacred book, the *Avestā: 15, Rām Yasht 9, 45–57* (ed. Darmesteter, 1898), *Vāyu* [Wind] has also been described as a high-energy and crushing element among the four primordial elements:

He goes forwards, backwards, bends backwards, hurling away, hurlying down, destroying, and taking away. The most valiant, the Strongest, the Firmest, crosses over easily, goes along hurling always, crushes at one stroke. Destroys the burrows, spits upon the burrows. Of sharp spear, of the long spear, of the piercing spear; and Greater Vāyu, strong Vāyu, the greatest of great, and the strongest of the strong.

ed. Darmesteter (1898)

Therefore, in both the written Vedic (ca. 1500 BCE; Boyce, 1989) and Avestan cultures (ca. 1200 BCE; Boyce, 1989, 2005), wind [*Vāyu*], "the force of the moving air," was described in the following ways:

- **(i)** The most energetic and dynamic element among the four elements of the sacred nature's power;
- **(ii)** He is the strongest and the greatest and does not take rest or become quiescent or dormant;
- **(iii)** He pushes the waters;
- **(iv)** He crosses easily through everything (since it is the lightest element);
- **(v)** He pierces like a strong and sharp spear in all directions; and
- **(vi)** He has crashing and destroying power, shaking the Earth.

It is with this ca. 1500 BCE Indo-Iranian concept of the element air in motion [Wind] that the theory of air movement through subterranean fissures is described in the *Greater Bondéhéshn* as the cause of the earthquake. Furthermore, we should remember that the storm-demon *Chashmag*, the pagan generator of earthquakes (pre-1200 BCE concept; Chapter 2), was also the creator of tornadoes and typhoons.

4.1.1.3 Denkard

In Book III, Chapter 93 of the Denkard [Acts of Religion], the "wind theory of earthquake" is described as known to the Zoroastrian religion. In this chapter, *Afrāsīāb* [*Frāsiyāv*] is accused of causing destructive earthquakes by imprisoning the wind in the Earth (see below). Denkard was rewritten in the ninth century as a compendium of Zoroastrian belief and customs and included extensive quotes from the lost Avestan texts destroyed during the 330 BCE invasion of Iran by Alexander III of Macedonia.

Be it known that, as air moves about in the body of man, so it always moves about within the Earth for its improvement; and as there is a passage in the body of man for the air to pass out and in which it moves, so is it with the Earth. As, owing to the passage in the body for the movement of air, the body lives, and as, when the air finds no passage in any direction, the body perishes, so, a state of things similar to that of the body happens when the passage in the Earth is stopped up. An injury is caused in the body by irregular ways of behavior productive of change, so when the passage of air is stopped in the Earth by contrary movements, injury is caused, in the direction in which the passage of air, is stopped, owing to no vent being found by it: (injury) *[damage]*, such as was caused by that sorcerer Afrāsiāb the Tur *[of Turān]*, by means of sorcery in rendering large tracts of fertile land unwatered, uninhabited, and desolate. Again, it is made known in the exposition of the good religion that the air within the Earth is always heated by fire and that it rushes up from below, causing injury *[damage]* thereby (to the Earth) and splitting it into two parts *[faulting]*. As in the body of man wind is known to be raised up by fire, so (from a similar cause operating) in the Earth, earthquakes and (other) injuries *[damages]* are caused.

Denkard, ed. Sanjana (1876)

According to the ancient Iranian legends, Afrāsiāb [Frāsiyāv] Turiya [Turānian] was the descendant of Feraidun's son, Tur, who became the Turānian king, hero, demon (div), and archenemy of Iran (Yasnā, 11:7; Yashts, 5:41–43, 9:21–23, 19:56–77; *Bondéhéshn*, 32:6, 13, 33:9; Minug-e Kherad; Denkard; Ferdowsi Tusi, 1010; Yarshater, 1983). Afrāsīāb's devastation of the land by suppressing waters, draining rivers, stopping rain, causing draught and famine (*Bondéhéshn*, 33.5; Ṭabarī, 915, ed. Payandeh; Hamzeh Esfahani, 961, ed. Sho'aār, 1967; Ferdowsi Tusi, 1010), or causing earthquakes [Denkard, III.93] was also expressed through a myth telling of his marriage to the goddess of Earth, Spandārmat, described in Shahrestānhā-yé Irānshahr:

Frāsiyāk asked Spandārmad *[Earth]* as wife and Spandārmad mixed in the Earth. (He) destroyed the city *[of Zarang in Sistān, SE Iran]* and he extinguished the fire *[of Karkoi]*.

ed. Tafazzoli (1989) and ed. Daryāee (2002)

4.1.2 Later Developments

The Greek philosophers elaborated on this idea by trying to substitute other elements such as water and fire as well. A few centuries after the Muslim Arab invasion and destruction of the cultural centers, when all the Iranian, Indian, Greek, and Mesopotamian books were translated into Arabic, the hypothesis was translated from the Greek and other sources into Arabic by Ikhvān al-Safā. The concept was later repeated in different treatises prepared by Karaji (953–1019), Biruni Khārazmi (1025), Ibn Sinā (Pursinā, Avicenna, 980–1037), Bahmanyār ibn-e Marzbān (d. 1066), Nāser Khosrow (1052), Shahmardan ebn-e Abi al-Khair (1095), Esmā'il Esfezāri (1045–1121), Tusi (1167), Sahlān Sāvi (twelfth century), Mas'udi Marvazi (twelfth century), Zakariyā Qazvini (1263), Mostaufi Qazvini (1340), and many others (Berberian, 1991, 1997b).

In Europe, influenced by Aristotle's second book of Meteors, William Shakespeare wrote in *Henry IV*, Part I (ca. 1597), that:

In strange eruptions, oft the teeming Earth
Is with a kind of colic pinched and vexed
By the imprisoning of unruly wind
Within her womb, which, for enlargement striving,
Shakes the old beldam Earth and topples down
Steeples and moss-grown towers. At your birth
Our grandma Earth, having this distemperature,
In passion shook.

Shakespeare (1597)

Due to religious influence, the earthquake is considered a supernatural phenomenon willed by Yahweh, God, and Allāh (Chapter 3).

4.1.3 Fārābi [Al-Fārābi, Alpharabius; 872–950]

Ibn Sīnā (Pursinā, Avicennā; 980–1037), in question No. 2—"Resāleh dar Bayān-e-Hoduth-e-Zelzeleh va Falsafé-yé ān, az Fārābi" ["Epistle in the Occurrence of Earthquake and Its Philosophy" by Fārābi], wrote that:

In the Third and the Fourth Introduction we stated that the 'dry-hot air' *[Arabic 'adkhaneh': plural of 'dokhān', lit., 'smoke' in Arabic]* and the 'wet-hot air' *['abkhareh': plural of 'bokhār', lit., 'Vapor' in Arabic]* are formed in the Earth. Whatever is above the ground was earlier discussed in detail. Whatever appears below the ground is formed from the water vapor. And earthquakes are caused by the 'dry-hot air' *['adkhaneh']*. We also described the formation of water form 'hot-wet air' *['abkhareh']*.

The occurrence of earthquake from 'dry-hot air' *['adkhaneh']* is in this way that since the large amount of 'hot-dry air' which is devoid of water droplets, is light, and is a kind of air, is formed out of its original place. Therefore, it is inclined to escape from an imprisoned location and ascends to reach its natural location. Now if these 'hot-dry air' are located in naturally soft grounds, or areas, which due to large numbers of qanats and deep wells have abundant voids and holes, they can easily escape and would not cause earthquake.

If those 'dry-hot air' are trapped in naturally hard and solid grounds with no voids and holes; grounds which became hard due to severe coldness and freezing the ground surface; or because of extreme heat, humidity is extracted, the passages are blocked and ground becomes hard; the escaping voids are blocked. The air becomes stormy and mobile and shakes the Earth. Sometimes because of its intensity and force, cracks the ground *[earthquake faulting]*, and loud noises *[earthquake sound]* are heard. It has been observed and heard that due to its large amount of force, the dwellings are devastated, deep gorges are formed, and/or huge mountains are derooted and thrown to another place.

It is also possible that the earthquake is caused because deep underground the groundwater created large cavities along its conduits and weakened the ground to the point that large pieces of ground is detached and collapse in the large cavities. The underground trapped air moves and causes earthquakes.

Sometimes in places with excessive winds if opposed winds blow, i.e. winds blow from four sides in a way to block their current paths, it is possible that these opposed winds block the exiting path of the underground 'dry-hot air'. But this assumption is not frequent and the cause of earthquake is the way discussed earlier.

Fārābi in Ibn Sinā; tr. From the Persian text

Fārābi's tenth-century description is similar to that of the Indo-Iranian theory documented from ca. 1500 BCE.

4.1.4 Ikhwān al-Safā [971]

In the compilation work of *Ikhwān al-Safā* (971), the hypothesis of air moving through the subterranean fissures is translated from the Greek text of Aristotle. It mentions that: "*Earthquakes are movements of parts of the Earth where winds/vapors are imprisoned in the bowels of the Earth*" (Halabi, 1981; Berberian, 1997b).

4.1.5 Karaji [953–1019]

Karaji (953–1019, ed. Khadiv-Djam, 1966; Berberian, 1997b; Zāghi, 2007), in the oldest book on hydrogeology, discusses the effects of earthquakes. Karaji was a brilliant hydrogeologist and inventor in the tenth and eleventh centuries; by repeating the ancient earthquake theory and coseismic fracturing, he focused primarily on his observations of changes in the groundwater table and sudden variations in the flow of springs and qanāts after earthquakes.

Earthquake causes springs and sometimes creates new springs or changes the location of existing ones. This is because there are passages in the ground which water moves and appears to the surface as a spring. There are subterranean channels in the ground, through which water flows, issuing from the surface spring. The soil that surrounds the channels is hard. An earthquake may be formed by the explosion of a compressed steam under the ground. Spring water gushes out of these pores in the soil and openings that are closed to the center of the Earth. Then water will run out of one of these openings, closing the first passage. Sometimes the stores steam of confined ground water causes a crack in the Earth, making an opening to the surface, thus forming a new spring. This phenomenon has often been reported... What I have said applies only to regions where the ground consists of different kinds of soil *[the structure is heterogeneous]*. If the ground is uniformly hard or soft [homogeneous] the down flow of spring and qanāt water occurs very rarely. Based on what was said sometimes there are different variations in the flow capacity of qanāts during earthquakes. A further observation connected with earthquakes is that just as there are moving and static waters underground, there is also moving and static vapor. When this vapor is compressed it cracks the Earth and escapes. This mechanism also causes earthquakes.

Karaji, ed. Khadiv-Djam (1966), Berberian (1997b), Zāghi (2007)

4.1.6 Biruni Khārazmi [1030]

Biruni Khārazmi (1030; Sachau, 1910; ed. Soduqi-Sahā, 1983; Berberian (1997b)) related the earthquakes to crustal faulting and briefly mentioned landslips:

The calamities which sometimes hit the Earth from up and down have different quantities and qualities. In most cases the Earth has experienced unimaginable severe disasters which were not predictable or treatable. Some disasters such as storms, or earthquakes which are associated with crustal faulting, inundation by water, burring rocks or ashes, thunder and lightning, or rock avalanches, etc., all cause damage and destruction. Hence, a vast area loses its population and after a period of passing the calamity, the area becomes alive again show signs of life.

Biruni (1030), ed. Sachau (1910), ed. Soduqi-Sahā (1983)

In the same treatise (1030), Biruni Khārazmi added that:

The disasters which from time to time befall the Earth, both from above and from below, differ in quality and quantity. Frequently it has experienced one so incommensurable in quality or in quantity, or in both together, that there was no remedy against it, and that no flight or caution was of any avail. The catastrophe comes on like a deluge or an earthquake, bringing destruction either by the breaking in of the surface, or by drowning with water which breaks forth, or by burning with hot stones and ashes that are thrown out, by thunderstorms, by landslips, and typhoons; further, by contagious and other diseases, by pestilence, and more of the like. Thereby a large region is stripped of its inhabitants; but when after a while, after the disaster and its consequences have passed away, the country begins to recover and to show new signs of life, then different people flock there together like wild animals, who formerly were dwelling in hiding-holes and on the tops of the mountains. They become civilized by assisting each other against common foes, wild beasts or men, and furthering each other in the hope for a life in safety and joy. Thus they increase to great numbers; but then ambition, circling round them with the wings of wrath and envy, begins to disturb the serene bliss of their life.

Biruni (1030), ed. Sachau (1910), ed. Soduqi-Sahā (1983)

4.1.7 Avicennā [Pursinā, Ibn-e-Sinā; 980–1037]

Avicennā (1020) in Part II, Section V of his peripatetic (ed. Foroughi, 1937, 1940) considered the cause of earthquakes to be the vapor which, having formed under the influence of the heavens, had been locked up inside the

Earth without any means of escape except through eruption and violence (also see: Nasr, 1963, 1980, 1993; al-Rawi, 2002). He also noticed topographic changes of the crust of the Earth, or as he put it, the upheaval of the Earth crust during violent earthquakes. It is probable to assume that Avicenna, who made amazing geological observations and interpretations, witnessed upward displacement of the Earth's crust due to earthquake faulting (reverse faulting or blind thrusting) in the eleventh century.

The formation of heights *[mountains]* is brought about by [i] an essential cause, and [ii] an accidental cause. The essential cause [is concerned] when, as in many violent earthquakes, the wind, which produces the earthquakes, raises a part of the ground, and a height is suddenly formed *[uplift]*.

Ibn Sinā, tr. by al-Rawi (2002), Nasr (1963, 1980, 1993)

In the same treatise, in which he developed the concept of uniformitarianism and law of superposition in geology, *Avicennā* described the earthquake as phenomena associated with orogeny (mountain building). In discussing the formation of mountains, he explained:

Either they *[mountains]* are the effects of upheavals of the crust of the Earth, such as might occur during a violent earthquake, or they are the effect of water, which, cutting itself a new route, has denuded the valleys, the strata being of different kinds, some soft, some hard... It would require a long period of time for all such changes to be accomplished, during which the mountains themselves might be somewhat diminished in size.

Ibn Sinā, tr. by al-Rawi (2002), Nasr (1963, 1980, 1993)

4.1.8 Bahmanyār ibn Marzbān [d. 1066]

Bahmanyār ibn-e Marzbān (ca. 1066), a student of Avicennā, narrated his master's views of earthquakes (Nurāni and Dānesh-Pazhuh, 1983; Berberian, 1997b):

Earthquake is movements occurring in deeper parts of the Earth causing shaking of whatever is on or below the crust. The earthquake is caused by wind, vapor, or smoke. There are little earthquakes in summer, and more earthquakes occur due to the lunar eclipse.

tr. from the Persian text

Although his observation about the time of the earthquake is incorrect, he brought up the preliminary notion of lunar effects on the planet Earth and its earthquakes.

4.1.9 Nāser Khosrow [1069]

Nāser Khosrow Qobādiyāni (1069, ed. Corbain and Mo'in, 1953; Berberian, 1997b) repeated the ancient earthquake theory:

If asked what is earthquake? We say that there are pores/hollow areas in the Earth where the vapors accumulate, and since the amount of vapor exceeds the pore capacity, the vapor shakes the Earth and fractures it, and then the vapor is released. The whole Earth will not shake but only the places will shake that have entrapped vapors.

Nāser Khosrow (1069), tr. from the Persian text

4.1.10 Shahmardān ibn Abi al-Khair [1095]

Shahmardān ibn Abi al-Khair (1095; ed. Jahānpur, 1983; Berberian, 1997b) repeated the ancient earthquake theory, but added an observation about the ground continuing to shake (possibly due to the foreshocks, the mainshock, and aftershocks), emphasizing the fracturing of the crust and referring to sulfurous gases emerging from fractures in mountains (possibly active volcanoes). In general, he was amazed at the power of the trapped air/gases.

Regarding earthquakes: whenever the vapors are in the open space or in the porous Earth, they create wind. But when the vapors are formed under the ground and have no access to open space and atmosphere, due to their heat, they ascend and since the Earth is rigid and does not let the vapors escape, the accumulated vapors shake the Earth until crack the Earth through which they can escape and the motion can then stop. If the strength of the vapor equals the strength of the Earth, the shaking and earthquake last several days. Then if the vapor prevails, the Earth fractures. If the vapor cools down and loses its strength, it calms down and cannot fracture the crust. If the heat of the vapor exceeds or additional vapor is added and gets stronger, the Earth fractures and the vapor emerge from it. If the quantity of vapor is large and its flow is continuous, there will be continuous air or wind emerging from the fracture; such as the sulfurous areas in the mountains where there is a continuous air/wind flow out of it. And the flow is sometimes so strong that if we toss a stone in the hole, the stone will be tossed out of the hole and land of the ground somewhere. If the vapor is composed of smoke emerging from a well, and we throw some flammable material in, it will be ignited and burn, and because of the excess heat, smoke is ascending.

4.1.11 Esmā'il Esfézāri [1045–1121]

Esmā'il Esfézāri (1045–1021; Berberian, 1997b) repeated *Shahmardān ibn Abi al-Khair*'s statement regarding earthquakes. Since no reference is cited, it is not clear if *Esfézāri* was referring to *Shahmardān's* book or if both authors benefitted from a single unknown earlier source.

4.1.12 al-Ghazāli [1058–1111]

Abu Hāmed Muhammad al-Ghazāli (Alghazel; 1058–1111) in Problem XVII: Refutation of their belief in the impossibility of a departure from the natural course of events, of his treatise (ed. Halabi, 1982; ed. Sāhib Ahmad

Kamāli) brought a religious aspect to the natural phenomena; he discussed the possibility of rain, thunder, and earthquakes as caused by the power of the prophet's soul and related all the activities to Allāh.

4.1.13 Ahmad Tusi [1167]

Tusi (1167; Berberian, 1997b) wrote a small section on earthquakes by linking the ancient theory to the will of God. He appears to be the second scholar after al Ghazāli to do so. He also mentioned that the earthquakes caused changes in the seawater level along the shore lines.

Earthquake Chapter—Know that earthquakes are willed by God. It is caused by excessive pressure of vapor entrapped in the bowels of the Earth, when the ground is rigid the Earth is shaken, the crust is fractured and the vapor is released. It is similar to a boiling pot with a tight lid; if the boiling prevails the pot explodes and the vapor escapes. Earthquakes can cause disturbances along the shorelines, and the properties close to sea shake and impacted by the water and waves, I have heard from people of Jilān *[Arabicized for the Persian 'Gilān' province of the SW Caspian Sea]* that when the Kabudān *[Blue]* Sea is disturbed by waves, our cities shake. A person told me that when the Kabudān was agitated by high tides, the city of Ardébil *[38°.14′N-48°.17′E, +1,346 m amsl]* was shaken; and the distance of Ardébit to the Kabudān Sea is 12 farsang, and God knows *[12 × 6 = 72 km]*.

Kabudān Sea is the former name of Lake Urumiyeh; however, it is located about 240 km to the west of Ardébil, with the Caspian Sea located 55 km to the east of Ardébil. The author seems to be confusing the names of the lakes. The people were probably referring to the 23 February 958 $M_s \sim 7.1$ earthquake in the Ruyān region, which might have caused some changes to level of the Caspian Sea.

Ahmad Tusi seems to be one of the early scholars who referred to changes in the course of rivers during earthquakes. Of course, his view that most earthquakes occurred in areas with no mountains or wells is incorrect. However, his last reference seems to mention earthquake-triggered landslides that uprooted the trees:

In our days, in the year 561 *[7 November 1165-27 October 1166]* an earthquake took place in Kuhestān *[lit., Mountains, possibly referring to the Jebāl, ancient Media/Mād province of central Iran]* and lasted for seven days. It did not cause any damage in the Jebāl province, but the news reached us that Az Zangān city was devastated *[editor: or, news reached from Rangān that city was devastated]*, and a mountain fell down where it dammed the river and the region lost its water supply. The water course was changed and destroyed another area. Most earthquakes occur in areas with no mountains and no wells. Therefore, if all the subterranean passages are blocked, the moving vapor fractures the crust. In the Kuhestān *[lit., mountainous area]* there are hundred thousands of mountains and hundred thousand wells and kahrizes *[qanāts]*.

In the year 562 *[1166-67]* an unbelievable earthquake took place alongside of Arvand *[Alvand or Arvandrud?]* where there were trees. All the trees were cut and their roots were in the air and stood upside down, and the Earth was fractured.

tr. from the Persian text

There is an error in the Persian texts regarding "Az Zangān" [or "Rangān," in the editorial note]. This should not be confused with the Iranian city of Zangān/Zanjān. Abrishami (1990) erroneously stated that: "*Without any doubt, the earthquake took place in the Zanjān region.*" The phrase "*Az Zangan*" should be read as "*Arzangān city*," which is Erzincā/Ezuncān, or the modern Erzinjān in Asia Minor (39°44′N–39°28′E) that was destroyed by the 1166–1168 earthquake (Ambraseys, 2009). Apparently, the typographic error was caused by the addition of a "dot" on the Persian letter "R" which transformed it to the letter "Z" in Persian; therefore, the phrase "*the city of Arzangān*" was changed to "*the city az* [from]-*Zangān*," which has no meaning or significance. Furthermore, no earthquake has been reported in the city of Zanjān (36°40′E–48°29′E), which is situated about 720 km to the SE of Arzanjān. Further, it should be emphasized that no earthquake was reported at Alavand (Hamédān) or Arvandrud (Ambraseys and Melville, 1982; Ambraseys et al., 1994; Berberian, 1994), and mention of "Arvand" by Ahmad Tusi should refer to a site damaged by the 1166–1168 Erzinjān (Erzincān) earthquake.

4.1.14 Sahlān Sāvi [Twelfth Century CE]

Sahlān Sāvi (*Sāvoji*, twelfth century), a contemporary of *Soltān Sanjar Saljuq* [1118–1157] discussed earthquakes in his treatise. He also described the ancient theory and stated that if the faulting is in a city, the city will be devastated. This seems to be one of the early references of active faults destroying a city. He also referred to the rockfalls from mountains and the deformation of the Earth's crust along the fault lines.

If those vapors formed underground surface are not cooled down by cool air and become cold water, and the heat does not decrease, and the crust is hard and lacks pores; then when the hot vapors ascend and move, due to the rigidity of the ground cannot escape. Obviously, its power exceeds and cracks the crust; and if this faulting is in a city, that city will be devastated. There are plenty of underground voids; when the power of vapors increase to fracture the Earth's crust, all the lands near those ground fractures including the mountains collapse and fall into the underground openings. Most earthquakes are created by smoky vapor formed underneath the ground surface. Obviously the subterranean pores, such as cavern or underground rooms, have weak ceilings, the ceilings will collapse and the air contained in the pore will tremble. It then will ascend and shakes the surface of the Earth which is above it. It is possible an earthquake takes place in another area and a piece of mountain collapses and hits the ground, the ground in that region of several farsangs *[1 farsang = 6 km]* will shake.

tr. from the Persian text

4.1.15 Sohrévardi [1186]

Shahāb al-Din Sohrévardi (1186, ed. S.J. Sohrévardi, 1982), discussed various geological phenomena [minerals, springs, earthquakes], noting the importance of light in the changes he observed. According to the doctrine of Sohrévardi—he used Zoroastrian Gnostics in almost all of his writings—the earthquakes were caused by entrapped subterranean water vapors in the bosom of the Earth; vapor is caused by heating water, and heat is light; therefore, the principal cause of all these changes was "light."

4.1.16 Mas'udi Marvazi [Second Half of the Twelfth Century CE]

Mas'udi Marvazi (twelfth century; Berberian, 1997b) accepted the theory of crustal faulting in the mountainous areas with subterranean noises. He mentioned that earthquakes seldom occur in salty, sandy, and unconsolidated soils. He also referred to surface faulting and the sounds of subterranean earthquakes.

> When wet or dry vapor or both are born in the Earth, and the Earth is rigid and its pores are clogged, the vapor cannot escape and is imprisoned. And since its quantity as well as temperature increases, the heat forces it to ascent to the Earth surface; and since the Earth is rigid and its pores are closed, the Earth shakes. If its power is so that it could rupture the crust, the vapor can pass through and the Earth becomes tranquil. If the coldness of the Earth decreases the temperature of vapor, it will stay there and the earthquake will diminish. If they cannot overcome each other, as much as there is resistance, there will be earthquakes. Most earthquakes occur in the mountains. No earthquakes take place in the salty, sandy and unconsolidated lands; however, if it happens it should be very rare. This is because these lands have large pores and openings *[it is porous]* and vapors cannot be imprisoned in them. It is possible that the vast amount of the subterranean entrapped vapor moves around and causes subterranean noises. There are also plenty of examples that when earthquake is diminished and a portion of the Earth is faulted, water flows out of the fractures, and in some cases fire discharges.

4.1.17 Maulānā Jalāl U'ddin Rumi [1273]

In Book IV: 9 of his masterpiece, *Maulānā Jalāl U'ddin Rumi* (1273) rejected the idea of earthquakes being caused by entrapped vapors in the Earth and stated that this idea was believed by ignorant people. He emphasized that earthquakes were caused by God's will and decision: "*In the opinion of him whose intelligence does not perceive this, Earthquakes are caused by terrestrial vapours.*"

4.1.18 Zakariyā Qazvini [1263]

In Chapter 6 of his book, *Zakariyā Qazvini* (1263) considered the vapors fracturing the Earth and causing damage and destruction.

4.1.19 Afzal al-Din Mohammad Narāqi Kāshāni [Thirteenth Century AD]

Narāqi Kāshāni (thirteenth century, ed. Minavi, 1987) repeated the ancient view of earthquakes after a discussion of the four elements and meteorology.

4.1.20 Ayub Denisari [1272]

Denisari (1272) wrote that "*...When smoke is imprisoned in the Earth it creates earthquake.*"

4.1.21 Ansāri Dameshqi [1300]

Ansāri Dameshqi (1300), in his compilation of the existing knowledge of his time, described the old theory and added that: "*Earthquakes mostly occur in cities located in the mountainous areas; it is usually very strong and dangerous; it can split the mountains, swallow the rivers, and break walls and forts. If the ground is hard and lacks any cavity, the vapors shake the Earth to escape. If the shaking of the vapors is near the surface, it will fracture the crust and escapes out.*"

4.1.22 Mostaufi Qazvini [1340]

Mostaufi Qazvini (1340; Berberian, 1997b), while describing the Tabriz earthquakes, wrote about the effects that the wells and qanāts had in releasing the pressure of the compressed subterranean air in the city.

> Now this same, up to the present date, during the three hundred years that have elapsed since his prediction, has proved to be perfectly true; for though the city has been many times visited by earthquakes these have caused no great ruin; and the reason would appear to be that there being now numerous underground water-channels carried through the subsoil *[qanāts]*, also many wells dug down into the same, the Earth's vent-holes are so opened, that its puissant vapours no longer are dangerously compressed, and therefore, violent earthquakes do not occur.
>
> Mostaufi Qazvini (1340) and ed. Le Strange (1919)

Despite Mostaufi's belief in the releasing effect of the qanāts and wells in the city of Tabriz, devastating earthquakes took place in Tabriz in 1550–1551, 1641, and 1780 (see Figure 11.12).

4.1.23 al-Suyuti's Rejection of the Theory [1499]

Jalāl al-Din al-Suyuti (1499; ed. Sa'adāni, 1971) rejected the theory of the entrapped air in subterranean fissures. Influenced by Islamic theology, al-Suyuti emphasized that earthquakes occurred solely by Allāh's decree [qaḍā wa qadar] and were therefore not the result of natural causes such as entrapped vapors, as the [*abkhirah*] philosophers had erroneously believed.

4.1.24 Kayhān International [19 August 1970]

As discussed above, most ancient scholars cited the lack of damaging earthquakes to the wells and qanāts located under the cities or villages, which released the powerful vapors below the ground before they could build up. What is surprising is to find approval of this idea in an article in the Kayhān International Newspaper (10 August 1970) attributed to the director of the Institute of Geophysics, University of Tehran, who allegedly stated: "*The two million cesspools underneath Tehran city act as buffers against earthquakes!*" (Kayhān International Newspaper, 19 August 1970). This solidifies the theory of the element air/wind, at least from the time of the ancient sacred Vedā and the Avestā in the second millennium BCE to modern times.

4.2 MORE RECENT UNCONVENTIONAL THOUGHTS

4.2.1 Yellow Orpiment Causes Earthquakes [1807]

Auguste de Bontemps (1807), who experienced the 11 July 1807 Tasuj earthquake while residing in Tabriz (Figure 11.12), wrote a bizarre idea about earthquake occurrence: "*I was shown yellow orpiment* [zarnikh-e zard] *several times, which the same as arsenic sulphur* [arsenic sulfide; orpiment, As_2S_3], *which were extracted from deep Earth. It is possible that the earthquakes and underground explosions are created by decaying of this matter. Yellow orpiment is used for removing body hair in the public baths*" (Bontemps, 1807).

4.2.2 Salt Plugs or Folding Forces [1922]

Richardson (1922), who experienced three earthquakes during January 1922 [the first week of January, 1922, and 28 January 1922] at the Qeshm and Hengām Islands in the Persian Gulf, wrote that: "*It should be interesting to know if the earthquakes are connected with saline intrusion* [salt plugs]. *A more feasible explanation, however, is that they represent crustal re-adjustment consequent on the more violent folding forces in operation during Bakhtiāri and post-Bakhtiāri times*" (Richardson, 1922). Sayyās (1987), in his strange article entitled "*Is the core of the planet Earth void?*," wrote that: "*the occurrence of earthquake and volcanism is due to the penetration of the sea water and salt in the depth, and their decomposition into O and H!*" Paradoxically, the abstract of his paper was accepted and published in the conference proceedings by the Geological Survey of Iran in 1987 in Tehran.

4.2.3 Underground Explosions Caused the 5 October 1948 M_w 7.2 Ashkābād Earthquake

The Etelā'āt (7 October 1948) wrote that the large-magnitude 1948 Ashkābād earthquake (see Figure 16.3) was caused by underground explosions that were recorded by meteorological and seismographic stations in Europe. They acknowledged that those explosions occurred in Central Asia and Siberia and caused tremors in the North Pole and parts of the northeast Iran.

4.3 OVERVIEW

An examination of the surviving ancient Indo-Iranian texts from 1500 to 1200 BCE (Tables 4.1 and 4.2) clearly demonstrates an early awareness of the people who lived on the seismically active Iranian–Indian Plateau. They offered one of the oldest generic causes of earthquakes using an energy source: that is, the hypothesis of winds in the locked subterranean fractures of the Earth. They selected one of the most active agents of the four known primordial elements (the moving air/wind), which by applying extra accumulated pressure to the rocks, resulted in faulting in the Earth's crust.

The hypothesis was based on analyzing complex phenomena with limited or almost no modern scientific knowledge except for accurate observations; they were in search of an energy source for the shaking (among the four known primordial elements) that caused the strong shaking of the ground and rumbling noise following intense fracturing and faulting of solid rocks. This early theory was favored by the majority of the later Iranian, Indian, and Greek scholars (Table 4.2).

As discussed, the ancient Indo-Iranians suggested that:

i. Earthquakes were caused by the accumulation of moving air (wind, air, vapor, and/or gases) in subterranean voids and fissures;
ii. The increasing pressure of the trapped and moving air underneath the active Earth's crust, which cannot condense at such depths or escape through the voids, fractures, wells, or qanāts to the surface, shakes the ground and causes surface fracturing;
iii. Earthquakes are movements occurring in deeper parts of the Earth;
iv. If fractures and faults are in the cities in the mountainous areas, they can devastate the cities;
v. Earthquakes were associated with fracturing, faulting, landslide, rockfalls, deflection of the rivers' courses, topographic changes (upheavals of Earth's crust), changes in the groundwater table, sudden variations in water flow in springs and qanāts, and changes in the seawater level along the shorelines;
vi. Earthquakes are not predictable or treatable;
vii. Lunar influences could also cause earthquakes.

The hypothesis was rejected later, in about the eleventh century, when scholars stated that earthquakes are the will of Allāh. Apparently, following the 1 November 1755 Lisbon earthquake, John Michell wrote for the first time that earthquakes are waves set up by the shifting masses of rocks miles below the surface (Guidoboni, 1994, 1998; Ben-Menahem, 1995).

TABLE 4.2 Evolution of the Theory of the Element Air in Motion/Wind as the Cause of Earthquakes

Period	Culture/Person	Element/Theory Causing Earthquakes	Source
Pre-1500 BCE	Pre-Vedā/Gāthā Pagan Period	Four elements	Smith (1876a), Batto (1992), Rochberg (2002)
ca. 1500 BCE	Rigvedā	Four elements	Boyce (1989), Griffith (1896)
ca. 1200 BCE	The Avestā	Subterranean winds causing earthquakes	Boyce (1989, 2005), Berberian (1997b)
624–545 [ca. 585] BCE	Thales of Miletus	Water/humidity on which the Earth floats like a vast ship	Seneca, 86 (6.6.1), Guidoboni (1994, 1998)
550–330 BCE	The Achaemenian Empire		
585–528 [ca. 550] BCE	Anaximenes of Miletus	Air causing earthquakes	Guidoboni (1994, 1998)
499–401 BCE	PERSIAN–GREEK WARS [499: Ephesus; 494: Keys of Cyprus and Marsya; 494: Lade; 490: Marathon; 480: Salamus and Thermopylae; 479: Plataea and Mycale; 466: Euymedon River; and 401: Cunaxa]		
440 BCE	Herodotus, 484–414 BCE [*Histories*, 440 BCE] reported that the Iranians venerate and make sacrifices to the Sun and the Moon and the Earth, to Fire, to Water, and to the Winds		
499–428 BCE	Anaxagoras	Aether causing earthquakes	Aristotle; Guidoboni (1994, 1998), Berberian (1977b)
Late fifth century BCE	Democritus of Abdera	Movement of subterranean water	Seneca, 86, Guidoboni (1994, 1998), Berberian (1997b)
Fifth century BCE	Archelaus	Wind causing Earthquakes	Guidoboni (1994, 1998)
336–328 BCE	Callisthenes of Olynthus (ca. 360–320 BCE), Greek historian and friend of Theophrastus (see below), was appointed to attend the Persian campaign of Alexander III of Macedon (336–330 BCE) on the recommendation of his uncle and tutor, Aristotle (see below) to bring scientific material (Brown, 1949; E.B.)		

Continued

TABLE 4.2 Evolution of the Theory of the Element Air in Motion/Wind as the Cause of Earthquakes—Cont'd

Period	Culture/Person	Element/Theory Causing Earthquakes	Source
336–330 BCE	Invasion of Alexander III of Macedon: Devastation of the Persian resources, lives, properties, institutions, and infrastructure		
384–322 [ca. 340] BCE	Aristotle	Dry exhalation (pneuma) trapped inside the Earth shakes the Earth as it is trying to escape. Earthquakes usually occur at night and midday and in spring and autumn	Aristotle, 350 BCE (Meteorologica, 1.2, 2.7.365a), Guidoboni (1994, 1998), Berberian (1997b)
373/370–287 BCE	Theophrastus	Collapse of subterranean caves by pneuma, causing earthquakes	Guidoboni (1994, 1998)
341–270 BCE	Epicurus	All the elements can cause earthquakes	Guidoboni (1994, 1998)
328–270/278 BCE	Strato of Lampscacus	Dynamic relationship between pneuma and heat and cold	Guidoboni (1994, 1998)
4 BCE–65 CE [ca. 45 CE]	Lucius Annaeus Seneca	Summary of the views on earthquakes in the Greco-Roman tradition, caused by the three elements of water, air, or fire inside the fourth element, the earth	Guidoboni (1994, 1998), Ben-Menahem (1995)
132 CE	Chang Heng	The earliest seismoscope	Ben-Menahem (1995)
224–642 CE	The Sasanian Empire		
260 AD	The Gundishapur Academy Establishment in Khuzestān, SW Iran (during the reign of Shāpur I Sassanid; 241–272 CE)		
636–661 CE	Muslim Arab Invasion		
971 CE	Ikhwān al-Safā Treatise	Winds entrapped in subterranean Earth	ed. Halabi (1981), Berberian (1997b)
973–1048 CE	Biruni Khārazmi		Tahqiq Mal al-Hend; Berberian (1997b)
980–1037	Avicennā [Pursinā]		Berberian (1997b)

d. 1066	Bahmanyār ibn Marzbān	Wind, vapor, smoke, lunar eclipse can cause earthquakes. He is a follower of Avicenna	Berberian (1997b)
1003–1088	Nāser Khosrow	Entrapped subterranean vapors causing earthquakes and fracturing	Jāme' al-Hekmatin, Berberian (1997b)
953–1019	Karaji	Entrapment of subterranean air, coseismic fracturing, & variations in groundwater flow	Karaji (953–1019), ed. Khadiv-Djam (1966), Berberian (1997b), Zāghi (2007)
1095	Shahmardān ibn Abi al-Khair	Entrapment of ascending hot subterranean vapor, causing earthquake and fracturing	Nozhat Nāmeh 'Alā'i, Berberian (1997b)
1045–1150	Esmā'il Esfézāri	Entrapment of ascending hot subterranean vapor, causing earthquake and fracturing	Resāleh Āthār 'Alavi, Berberian (1997b)
1161	Ahmad Tusi	Entrapment of ascending hot subterranean vapor, causing earthquake and fracturing	'Ajā'ib al-Makhluqāt, Berberian (1997b)
Twelfth century	Sahlān al-Sāvi	Entrapment of ascending hot subterranean vapor, causing earthquake and fracturing	Al-Resāla al-Sanjariya fi al-Kā'enāt al-Onsoriya, Berberian (1997b)
1186	Sohrévardi	Light causing heat+water=water vapor; entrapped water vapor, therefore, light, causing earthquake	Hekmat al-Ishrāq
Second half of the thirteenth century	Mas'ud Marvazi	Entrapment of ascending hot or cold subterranean vapor, causing earthquake and fracturing	Resāleh Āthār 'Alavi; Berberian (1997b)
1205–1283 [1263]	Zakariyā Qazvini	Entrapment of subterranean vapor and smoke, causing earthquake and fracturing	'Ajā'ib al-Makhluqāt wa Gharā'ib al-Maujudāt, Berberian (1997b)
Thirteenth century AD	Afzal al-Din Mohammad Narāqi Kāshāni	Entrapment of subterranean smoke, causing earthquake and faulting	Mosanafāt (ed. Minavi, 1987)
1272	Amin ad-Din Ayub Denisari	Entrapment of subterranean smoke, causing earthquakes	Navāder al-Tabādor le-Tohfata al-Bahādor (ed. Modares Razavi, 1971)

Continued

TABLE 4.2 Evolution of the Theory of the Element Air in Motion/Wind as the Cause of Earthquakes—Cont'd

Period	Culture/Person	Element/Theory Causing Earthquakes	Source
1340	Mostaufi Qazvini	The qanāts in the city of Tabriz acted as a conduit to release the pressure on the Earth caused by the accumulation of vapors and gases, causing harmless aftershocks of the 1042 earthquake	Nozhat al-Qolub, Berberian (1997b)
1601–1680 [1664]	Athanasius Kircher	Motion of fire inside a system of channels within the Earth working against the rocks that block the motion, causing earthquakes	Guidoboni (1994, 1998), Ben-Menahem (1995)
1696	Johann Ahan	The air trapped inside the Earth is mixed with flammable material, causing earthquakes	Ben-Menahem (1995)
1638–1712 and 1645–1715 [1701]	Martin Lister and Nicolas Lemery	The internal fire producing earthquakes and volcanoes was produced by chemical means through a mixture of iron, sulfur, and salt	Guidoboni (1994, 1998), Ben-Menahem (1995)
1707–1788 [1749]	Comte de Buffon	Explosion of inflammable materials, such as the fermentation of pyrites, causing earthquakes, produces a quantity of heated air in subterranean chambers that could escape horizontally for a great distance through underground tunnels and caves. This would explain why earthquakes are felt over a long distance	Guidoboni (1994, 1998), Ben-Menahem (1995)
1761	John Michell [following the Lisbon earthquake of 1 November 1755]	Earthquakes are waves set up by the shifting masses of rocks miles below the surface	Guidoboni (1994, 1998), Ben-Menahem (1995)

Chapter 5

Earthquake Folklore and Legends

In the form of every tale,
There is for those of insight a portion of meaning.

Khāqāni Shervāni (1121–1190)

Most likely, there are proverbs regarding earthquakes on the Iranian plateau that have been repeated throughout generations. Although I have gone to some lengths to collect the most representative folklore and legends, I have not been able to find many proverbs. However, they deserve respect and should be systematically collected. I hope the following fragments can provide the basis for further critical analyses.

5.1 PLANET EARTH ON THE HORN OF A BULL

This myth appears to be common in the region. The cow represents the element Earth in myths, folklore, and legends. We present some recorded instances below.

5.1.1 The Bull Myth in Iran

Traditionally, older generations, especially in the rural areas, have believed in a regionally widespread notion that earthquakes can be attributed to the position of the Earth on the horns of a bull that rests on a fish in the sea. When the bull is tired, or when there is too much injustice and sin in the world, the bull becomes impatient and shifts the Earth from one horn to the other, resulting in earthquakes. Some believe that earthquakes occur when the Earth falls directly onto the bull's horn (Massé, 1938; Messner, 1954; Donaldson, 1933).

Valle (1677), the Italian traveler who visited Iran from January 1617 to January 1623, wrote a letter from Esfahān dated 18 December 1617 (Bull, 1989), stating that the people believed that the fabric of the world was sustained on the horns of a cow or an ox; when the cow shakes herself, an earthquake occurs.

☆"To view the full reference list for the book, click here."

Developments in Earth Surface Processes, Vol. 17. http://dx.doi.org/10.1016/B978-0-444-63292-0.00005-3

Rev. Wilson (1895), experiencing the 3 May 1883 earthquake in Āzarbāijān, recorded that: "*One Persian explanation of the earthquake was that the ox which, with twenty-one horns, supports the Earth was engaged and was tossing his horns*" (Wilson, 1895).

5.1.2 Reference of the Bull in Khayyām and Khāqāni

There is a reference to the bull in a quatrain [Rubā'i; verse of four lines] attributed to poet 'Omar Khayyām Neyshāburi [1048–1122; ed. Edward Fitzgerald]:

There is a bull in the sky and its name is Pleiades,
Another bull lies hidden under the Earth.
Open your wisdom eyes and see intellectually,
Between these two bulls, behold a handful of donkeys.

The Iranian poet Khāqāni Shervāni, 1121-1190 also referred to this ancient folklore in one of his poem: *"The heavenly-sphere lion lays down his outfit on the cow of the Earth."*

5.1.3 The Parallel Jewish Moroccan Bull Myth

In the Moroccan myth, recorded orally by the Israel Folklore Archives (IFA 4396), God sets the world on one horn of a giant bull. However, when the people sin, the world becomes heavier and the bull tosses the planet Earth from one horn to the other, causing earthquakes and other catastrophes (Schwartz, 2004).

5.1.4 The Parallel Caucasian Bull Myth

In Caucasus and Kyrgyzstan mythology, a giant bull carrying the Earth on its horn produces earthquakes (Tributsch, 1982).

5.1.5 The Nauruz [Nowruz; The Iranian New Year] and the Bull

According to an ancient Iranian legend, the Earth's axis turns on one horn of a giant bull. Once a year, on the vernal equinox, the bull tosses its burden from one horn to the other so deftly that the shift can be observed only by watching the delicate movements of a highly sensitive egg on a polished, slippery surface. The traditional Iranian Nauruz ritual is setting a (table) cloth ["sofreh"] with seven iconic dishes corresponding to the seven creations and the seven holy immortals protecting them. A bowl of clear water [representing the sky], containing an egg or an orange [representing the planet Earth] and either a leaf of a rose or a green leaf, and a mirror [representing sincerity] are placed on a table. A plain hard-boiled egg is placed on the middle of the mirror. It is believed that at the exact moment of the New Year, when the Sun enters into the new sign of the zodiac (the spring equinox; first day of the Iranian new year), the bull tosses the Earth from one horn to the other horn. Consequently, the egg rolls on the mirror, the orange flips over in the bowl of water, and the floating leaf moves (Donaldson, 1933; Massé, 1938; Messner, 1954; iranicaonline.org/nowruz).

5.1.6 The Parallel Indian Sikhs Bull Myth

In panaas 3 of the *Sri Guru Granth Sahib Ji*, the holy book of the Sikhs, we find a reference to the popular mythical bull, which is patiently holding the planet Earth on its horn: "*The mythical bull is Dhārmā, the son of compassion; this is what patiently holds the Earth in its place. One who understands this becomes truthful. What a great load there is on the bull!*" [www.sikhism.us/sikh-youth/8180-the-mythical-bull.html].

5.2 SATURDAY EARTHQUAKES IN THE MANICHEAN SOGDIAN BELIEF

The seven weekdays of the Manichean Sogdians are related to the planets [the name of a planet followed by "time" (jmnw)]. Saturday [kyw'n (Kewān, Persian Kayvān; planet Saturn) jmnw (zhmnu; time)] in the Manichean Sogdians nomenclature means "the day of Saturn" (Mackenzie, 1964; Gharib, 2009). Henning (1945) reported a Manichean text (written in Western Middle Iranian; i.e., Parthian and Middle Persian) as follows: "*When there is an earthquake on a Saturday during the day, illness and sickness will attack*" (Henning, 1945).

5.3 EARTHQUAKE-INDUCED ROCK AVALANCHES AND THE FAITH OF THE INFIDELS

The activity of the Dorud fault has created numerous earthquake-induced rock avalanches (see Figure 12.1 for the location). Local residents and shepherds believe that the avalanches fell upon a town because it pagan infidels did not accept Islam. A large stone in the shape of a column, approximately 6 m tall and made of Jurassic dolomitic limestone, was left behind; locals believe it to be the petrified leader of the town (Nabavi, 1985). A similar folk story is told about the "Khāk Malgeh" stone ruins of Poshtkuh [2 km south of Daryāb; about 9 km from the junction of the Khorramābād–Borujerd–Dorud junctions]. The local shepherds believe that "*The stones and dirt fell down on the homes and heads of the infidels*" (tr. from Nabavi, 1985).

Lieut-Colonel H.A. Sawyer (1894), who explored the Lake Gahar region in 1889 (see Figure 12.1 for the location), found the ruins of an ancient Armenian stone village [the infidel village mentioned by the local inhabitants]. It had been deserted on account of frequent earthquakes that triggered large rockfalls in the upper reaches of the Dareh Luku, approximately 30 km SE of Lake Gahar, 10 km NW of Arjanak, along the Dorud segment of the Zāgros Main Recent fault. The fault was activated during the 23 January 1909 M_w 7.4 Silākhor earthquake (Figure 12.1).

5.4 EARTH MOVEMENT LEGENDS IN SISTĀN AND BALUCHESTĀN

George Passman Tate, a survey officer who explored the remote northwest frontier of British India in 1886–1905 in an expedition during the sojourn of the Boundary Commission under Sir Arthur McMahon, became enamored of the land and its peoples. Over the course of 20 years, he charted desolate places; breakfasted with tribal chiefs; and sat at campfires, drinking in the local legends. Based on the local traditions he heard from Sistāni tribesmen, Tate (1910–1912) told of legends of Hāmun-e Sistān as well as Earth movements in southern Sistān, where volcanic activities have also occurred.

> There is no doubt the reason of the fable that the ancient lake of Seistān *[Sistān]* was filled by genii acting according to the orders of the celebrated Pir Khizr *[Pir Khezr[1]]:* as this task occupied but half a day, the country was called Nimroz *[Nimruz]*. All the legends of saints after whom the highest summits of mountains are named in the country to the south of Seistān are associated with Earth movement. Ramadthān Ghaibi *[Invisible Ramazān]* and Pulchota *(Mikh-e Rostam Mt.[2])* disappeared in chasms that opened in the Earth's surface and closed after those mythical personages had been received within. Pir Sultān vanished into the Koh-i Sultān *[Kuh-e Soltān volcano]* in an outburst of flame. Sheikh Husen *[Shaikh Hossain]* was also swallowed by the Earth opening; and this is supposed to have taken on the peak now called after him. These ancient legends become significant when we find them associated with localities which are either close to or actually show traces of having been centres of volcanic activity.
>
> Tate (1910–1912)

5.5 BURSTING OF THE HAFTVĀD'S WORM AT THE BAM CITADEL AND SHAKING THE WHOLE REGION

The Bam Citadel [Arg-e Bam (see Figure 14.6 for the location)] may have been destroyed during the reign of King Ardeshir Bābakān Sāsānid (r. 224–241), the founder of the Sāsānian dynasty. Since then, there have been at least 10 phases

1. Khezr: al-Khaḍir; lit. "The Green (Saint)" in Arabic. The green prophet or saint, Elijah, was an encounter with Moses. See Qorān, sura al-Kahaf 18:65–82. Possibly Sorush (Srush) of the Zoroastrians.
2. Mikh-e-Rostam Mountain: Lit., "The Nail of Rostam"; a single conic mountain east of Hāmun-e-Lorā in modern Pakistani Baluchestān at 29° 10′N–65°05′E.

of documented reconstruction at the citadel (Berberian, 2005). The oldest recorded partially standing structure in the Bam Citadel was the 1751 reconstructed mosque (Gaube, 1979). Much of the citadel dates from the late eighteenth and nineteenth centuries. The citadel and the city of Bam, located by the Bam fault (see Figure 14.6), was completely destroyed during the 26 September 2003 M_w 6.6 earthquake (Berberian, 2005).

Local legend has it that the Bam Citadel was the place where the gigantic "Haftvād Worm" was housed in the old times,[3] or *Haftānbokht* [in the Pahlavi text of *Kārnāmak Artakhshir Pāpakān;* 272–273] during the reign of the *Pārthian* governor of Kermān and Bam. The oldest reference to this legendary gigantic Worm is found in the *Kārnāmak Artakhshir Pāpakān* (Anklesaria, 1935; Mashkur, 1950; avesta.org; Markus-Takeshita, 2001; Shahbāzi, 2003a; Shāki, 2002). According to local legend, the Bam Citadel [*Arg-e Bam, Qal'eh Haftvād*, or *Qal'eh Bahman*] was the capital of *Haftvād* of Kermān. It is said that *Haftvād* resided in Kerman's *Qal'eh Dokhtar,* near *Qal'eh Ardeshir,* built on the mountainous outskirts of the city of Kermān. These toponyms might be remnants of the Haftvād's legend. However, according to a second folk etymology, the Worm burst open with noise at the Bam Citadel and shook the region (Sanjana, 1896; Shahbāzi, 2003a; iranicaonline.org).

In accordance with the *Kārnāmak,* the fortress was destroyed by the order of *Ardeshir Bābakān Sāsānid* [r. 224–241], and not by bursting of the Worm. *Ferdowsi Tusi* in *Shāhnāmeh* (1010), first "symbolically" noted that the entire region was shaken by a very loud rumble that had burst from the Worm's throat when two youths accompanying *Ardeshir* poured molten metal (lead and bronze) into the pool where the Worm lay. *Tabari* (915), followed by Bal'ami (d. 996), does not mention the story of the Worm; however, he addressed Ardeshir's campaign in Kermān and the area to the south (Levy, 1967; Pāyandeh, 1975; Bosworth, 1999) (Table 5.1).

It seems that at that time, *Ardeshir Bābakān Sāssānid* [r. 224–239], by conquering Kermān and Bam, killed the *Kerm-e Haftvād* at the Bam Citadel ["*the worm burst with a big bang noise which rocked the area*"], completely destroyed the citadel and killed most of its inhabitants, put an end to the rule of *Haftvād*, built the new village of Kolālān or Kojārān [possibly the old quarter of Kurzān in western Bam], and brought the "*seven fires of Bahrām*" to the new village. The entire episode of the Haftvād Worm rests on the rationalization of historical events of an unknown nature; the legendary element could be a mixed metaphoric reference to a "destructive earthquake" or even a "conquering battle" by King Ardeshir against the ancient city of Bam and its Pārthian governor, Haftvād (iranicaonline.org; Berberian, 2005).

---

3. Haftvād Worm: Kerm-e Haftvād'. This may refer to the first silkworm industry developed in Bam. The city was famous for its silk products, and the "Haftvād worm" brought prosperity to the Haftvād family and the region. Also, see Haftvād in Shāhnāmeh of Ferdowsi (1010).

TABLE 5.1 Summary of the Haftvād/Haftānbokht Worm Legend and Destruction of the Bam Citadel in 224 AD

Author	Book	Cause of Destruction of the Bam Citadel	Cited Section
Mostaufi Qazvini	Nozhat al-Qolub [1340]	–	Chapter XIII, Bam
Anonymous writer from Asadābād, Hamédān	Mojmal al-Tavārikh val Qesas [520/1126]	Summary of the legend based on the *Shāhnāmeh*	Bahār (1318/1939, p. 60)
Ferdowsi Tusi	Shāhnāmeh [1010]	A rumble burst forth from its [worm's] throat so heavy that it shook the pool and the whole region about it ["*Tarāki bar āmad ze holqum-e-oui; Ke larzān shod ān kondeh-o-boom-e oui*"]	Ashkānian/744
Bal'ami [d. 996]	Tārikh Bal'ami	The Persian text of Tabari	ed. Bahār (pp. 820, 879)
Tabari [839–923]	Tārikh al-Rasul wal-Moluk (Tabari) [915]	Ardeshir defeated Balāsh, the last Ārsācid ruler at Kermān, then marched to the coastland and cut the ruler in half with his sword	ed. Bosworth, (1999) V, 10; ed. Pāyandeh (1975)
Anonymous	Kārnāmak Ardeshir Pābakān [272–273]	Ardeshir Bābakaān commanded the destruction of the Fortress	Chapter 8/16

5.6 OVERVIEW

A preliminary survey of both old and recent Iranian folklore and legends shows that earthquakes and their destructive effects have deep roots in the people's daily beliefs and stories.

Chapter 6

Earthquakes in Epic Literature

O brother, a story contains meaning as a measure contains grain;
A wise man is concerned only with the grain of meaning,
not with the measure.

Maulavai Rumi (1207–1273)

Sporadic symbolic allusions to ground shaking and earthquakes are found in the Iranian national epics, indicating that there is a powerful earthquake psyche in the minds of the country's inhabitants.

6.1 THE SHĀHNĀMEH (THE EPIC OF THE KINGS) OF FERDOWSI TUSI (1010)

There are a few metaphoric references to possible earthquakes in the national Iranian epic literature of the *Shāhnāmeh* written by master poet Ferdowsi Tusi (1010) (see Rogers, 1907; Zimern, 1926; Levy, 1967; Poure Davoud, 1968; Curtis, 1993; Tafazzoli, 1996; Joneidi, 2008; iranicaonline.org, for the following entries). Ferdowsi was born and lived at Tus in the Khorāsān province of northeastern Iran, which is an earthquake-prone region.

6.1.1 Creation of the World

After the world was created and the four primordial elements were shaped, the heavens were shaken together and the mountains and sea were formed [Discourse on the Creation of the World in the *Shāhnāmeh*]. This view was definitely influenced by the ancient Zoroastrian beliefs addressed in Chapter 2.

6.1.2 Zahāk's Dream and the Trembling of the Palace of Horned Columns

When Feraidun was born, villainous Zahāk dreamed that he saw a youth who was slender like a cypress tree. The youth came toward Zahāk, carrying a cow-head mace, and, with it, struck Zahāk to the ground. The tyrant was

☆"To view the full reference list for the book, click here."

Developments in Earth Surface Processes, Vol. 17. http://dx.doi.org/10.1016/B978-0-444-63292-0.00006-5

disturbed by the portentous dream; he awoke with such loud exclamations that "the palace of a hundred columns trembled at the sound" (Zahāk, 3/50, in Ferdowsi Tusi, 1010).

6.1.3 Sām's Battle with Demons of Mazāndarān and Gorgsārān

In a letter to King Manuchehr, *Sām* [father of Zāl] describes that "our army lost their confidence; I, therefore, shouted with a roaring voice that the Earth shook. The fighters, who had lost their confidence and were retreating, began to fight" (Manuchehr, 14/917, in Ferdowsi Tusi, 1010).

6.1.4 Sām's Battle with the Monster of the Kashafrud [Kashaf River]

The monster of Kashaf River [*Azhdehā of Kashafrud*] emerges from the river resembling a huge mountain, his mouth reminiscent of a black tree with hot breath. *Sām* [father of Zāl] metaphorically mentions that the crust of the Earth shook from the cry of the monster (Manuchehr, 15/1034, in Ferdowsi Tusi, 1010). The modern Kashafrud (River) is located near Mashhad in the Khorāsān province of northeastern Iran near the active Kashafrud fault (see Figure 16.4 for the location). Kashafrud is close to the hometown of Ferdowsi Tusi.

6.1.5 Trial of Zāl by the Mobeds [Earthquake with Wind]

One of the last challenging questions asked by one of the mobeds during their trial of Zāl after he was summoned by the king contained the following story, in which Zāl had to solve the metaphors. The wise men from the citadel of a mountainous country erected buildings on a plain whose tops reached to the moon. Suddenly, there was an earthquake, and the city completely disappeared. The wind blew with the earthquake and raised grief and lamentation in the world (Manucher, 3376 and 3401, in Ferdowsi Tusi, 1010, v. 1, ed. Joneydi, 2008; Zaehner, 1972).

6.1.6 Rostam's Fight with the Magic Azhdehā [Dragon]

During Rostam's pass through the *Seven Labors' [Haft Khān]*, Rostam, with the help of his horse *Rakhsh*, at the third Khān fought with a "magic azhdehā [dragon]" and defeated him. From the cry of the magic dragon, the whole land became fractured (Pādeshāhi-yeh Kay Kāvus va raftan-e oo bé Māzandarān, 7/269, Ferdowsi Tusi, 1010, ed. Joneydi, 2008).

6.1.7 Rostam's Fight to Save King Kay-Kāvus

In *Rostam*'s journey to save the life of sovereign *Kay-Kāvus*, who was captured by the Divs [demons] of Māzandarān, Rostam passes through the *Seven Labors [Haft Khān]*. With the help of Olād, Rostam had to reach the *Alborz Mountain* at the Iran–India border to find *Kay Qobād* and bring him to Iran. In his descriptive instructions, *Olād* metaphorically tells Rostam that the mountain trembles like a willow tree in fear of the chieftain of demons, the White Demon [Div-e Sepid] (Pādeshāhi Kay Kāvus va raftan-e oo be Māzandarān, 9/490 afzudeh, in Ferdowsi Tusi, 1010, ed. Joneydi, 2008; see also references cited in 6.1). It is interesting to note that the continuation of the present Alborz and Binālud Mountains into northern India was still called the Alborz Mountain, a memory of the Zoroastrian Alborz encircling the planet (i.e., the present active Alpine–Himalayan belt).

6.1.8 Rostam's Fling of the Akvān-e Div's Stone into China

In the story of *Bizhan and Manizheh* of *Shāhnāmeh* (Khaleghi-Motlagh, Akvān-e-Div, iranicaonline.org), Afrāsiāb throws Bizhan into the pit and covers him with the heavy stone of Akvān-e-Div [*Akvān/Akoman, the demon*], which Akvān had hurled from the sea to the forests of China. Rostam removes the stone of Akvān-e Div, casting it again into the forests of China. The surface of the Earth was shaken after the stone hit the ground (Dāstān-e Bizhan va Manizheh, 1086, in Ferdowsi Tusi, 1010).

6.1.9 The Iranian's Hardship During the Time of King Kay-Khosrau

The next symbolic case mentioned in the *Shahnāmeh* is the experience of the Iranian army prior to its suffering from snow and perishing through hardship during the time of Kay-Khosrau, possibly in the area of the Āzargoshnasb fire sanctuary in Kordestān, west of Iran. The Iranian army clamored while the sun lost its path; children and women rushed to the streets and the bazaar; the whole Earth was shaken, and the lords were astonished. The king told the Iranians that they should learn lessons from this event (Ferdowsi Tusi, 1010). The area was stuck by the 4 July 1880 Takht-e Soleymān earthquake.

6.1.10 Watchman Roaring Like Earthquake

During the war between Dārāb and the army of Rome, and the retreat of the Romans, a metaphoric reference to an earthquake is made in Shāhnāmeh as "the clamor of the watchman roared like an earthquake" (verse 28699, v. 3, p. 606, ed. Joneydi, 2008).

6.1.11 Bursting of the Haftvad's Worm at the Bam Citadel and Shaking the Whole Region

Addressed earlier in Chapter 5.

6.2 THE FRACTURED AND SHAKING MOUNTAIN

In his heroic epic "Garshāsp-nāmeh" [the Epic of Garshāsp], Asadi Tusi (d. 1072) describes the metaphoric Ezhdehā [Dragon] and mentions that it lived in a fracture, or cave, on top of the "fractured mountain" [Shekāvand Kuh]. Afterward, when the hero approaches the valley, a "shaking mountain" [Kuh-e Jonbān] clearly appeared.

6.3 THE DAREDEVILS OF SĀSSOUN [SĀSSOUNTSI TĀVIT]

As with the Shāhnāmeh of Ferdowsi Tusi (1010), there are some sporadic metaphoric references to earthquakes in the popular national Armenian folk epic, *Daredevils of Sāssoun* [Sāssountsi Tāvit]. This is an amalgamation of folktales from 1000 years ago that were collected in 1893 by Gāregin Servāntstiān, the bishop of the Armenian Apostolic Church (Surmeliān, 1966; Hāroutyunian, 1997). In 1939, Armenians the world over celebrated the 1000th anniversary of the Sāssoun saga.

Nikonov (1991) pointed out that the epic of the Daredevils of Sāssoun (Sāssountsi Tāvit version by Grigoryan and Grigoryan, 1977) can be considered as a source of information about earthquakes in ancient Greater Armenia in the Asia Minor. Guidoboni and Traina (1995) argued that, as may be deduced from the earliest references to the Daredevils of Sāssoun in the accounts of Portuguese travelers [*T. Tenreiro*' and "*Mestre Affonso*"] in the sixteenth century when they visited Bidlis [Bitlis], Khlāt [Ākhlāt (Ahlat; western Lake Vān)], and Sāssoun (Gulbenkian and Berberian, 1971; Hāroutyunian, 1997), the epic of Sāssoun, while belonging to the ancient times, in fact contains references to many recent earthquakes. "*Itynerario*" of "*Mestre Affonso*," compiled after a journey to Armenia between 1565 and 1566, mentions the town of *Akhlāt* situated on the shore of Lake Vān and recalls the legend of David of Sāssoun. Describing the city of Khlāt [Ākhlāt], both Affonso and Tenreiro maintain that it was destroyed. The foundation of the fortress of Ākhlāt and nearby buildings with cupolas were traditionally ascribed to David (Gulbenkian and Berberian, 1971; Hāroutyunian, 1997).

According to this legend, as it was recounted to the Portuguese traveler Antonio Tenreiro, who was in Armenia between 1523 and 1559, David of Sāssoun destroyed the town and then rebuilt it with characteristic domed buildings [*kubbet; Gonbad*]. Apparently, as Guidoboni and Traina (1995) mentioned, the *kubbet* of Akhlāt date back no earlier than the late thirteenth century; the destruction of the town, which the epic attributed to the giant

David, the destroyer of mountains and cities, can instead be attributed to the earthquake of 3 October 1275 (Guidoboni and Traina, 1995).

6.3.1 The Fight Between Bālthāsār and Sānāsār and the Dragon

A dragon did not permit the spring water to flow from the top of a mountain down to the Green City. Twin brothers *Bālthāsār* and *Sānāsār* had carried two huge round olive-pressing stones to the top of the dragon mountain, when suddenly the sky clouded, hiding the sun, and the mountains reverberated with an awful roar. They saw a monstrous serpent (Surmeliān, 1966). This might be a reference to a volcanic eruption with related earthquakes.

6.3.2 The Dispute and Fight Between the Twins Bālthāsār and Sānāsār

Twin brothers *Bālthāsār* and *Sānāsār* decided to fight each other after Sānāsār found out that the daughter of the king of the mythical Copper City had greeted Bālthāsār twice in her letter but Sānāsār only once. During their fight, the rocks echoed with their shouts and the Earth shook under their feet (Surmeliān, 1966).

6.3.3 The Fight Between Meherr [Mehr] and the Sultān of Egypt over Paying Taxes

During the fight between the Armenian hero and the agents of the Caliphs of Baghdād [representing the struggling and suffering of the Christian Armenian peasants who were taxed to death by the occupiers], the mountains shook and crumpled in an earthquake; the world shook from the dreadful clamor of this combat (Surmeliān, 1966).

6.3.4 Young David's Anger After Losing His Herd

After the young David drove his herd to a mountain meadow, he took a nap. When he woke up and noticed that his herd was gone, he ran. The ground shook under his feet, and the rocks echoed loudly with the thunder of his steel boots (Surmeliān, 1966).

6.3.5 Meherr [Mehr] Junior's Search of the Rock of Vān, Raven's Rock

During his search for the Rock of Vān, Raven's Rock, Meherr Junior rode up into the mountains of Ākhlāt and arrived at the summit of Mount Sipān [volcano] north of the Lake Vān in Greater Armenia. He drew his sword, cut the iron and gold chains, and freed himself and his horse from the clutches of this prince. The Earth shook and gave way under the feet of his horse (Surmeliān, 1966).

6.4 THE EPIC OF AMIR HAMZA SĀHEB QARĀN

In the *Adventures of Amir Hamza* [*Dāstān-e Amir Hamzeh Sāheb Qarān* (lit., "Lord of the Auspicious/Felicitous Conjunction")] (Pritchett, 1991, 2004; Farooqi, 2000, 2008), Hamza [Hamzeh], a character who assumes the name and identity of Mohammed's uncle in certain instances, undertakes to spread the Faith. He falls in love with Mehrnegār, the daughter of the Iranian king Nushirvān [Khosrow I Anushirvān Sāssānid; r. 531–579], and ventures from the Isle of Ceylon to the fabulous realms of the Qāf [Caucasus] Mountain (the Alborz), the region inhabited by Pari (spirits who have been denied paradise), Parizāds, Devs, Jinns, and other nonhuman species, under the dominion of king Shāhpal, son of Shāhrokh and the ruler of the Qāf Mountain. As foretold, at the Qāf Mountain, Hamza is betrothed to Āsmān Pari and marries Mehrnegār (Pritchett, 1991, 2004; Farooqi, 2000, 2008).

Throughout the story, Amir Hamza was called "*the earthquake of Qāf*" [the earthquake of the Caucasus/Alborz Mountain]. The reason for this epithet among others is not clear; it may have been because he was such a great fighter and spent a lot of time in the realm of the Qāf Mountain, though not entirely successfully (Frances W. Pritchett, personal communication, 27 March 2009).

It is also possible that because the growth of the mythical Qāf Mountain [the Caucasus, or the mythical Alborz in the Zoroastrian Avestā and the Pahlavi Texts, i.e., Bondéhéshn (Bundahishn), where the mythical Alborz Mountain is around the world] was traditionally associated with earthquakes [see also Chapter 3, *Maulānā Jalāl U'ddin Rumi's* poem regarding the ascent of *Dhu'l-Qarnayn* to the peak of the Qāf Mountain]. It is not clear why Shāhpal, the king of the Qāf Mountain, was not called so. Nonetheless, it is interesting that in this epic literature, the "*earthquake*" is attributed to the "*Qāf Mountain*" (the mythical Alborz addressed in Chapter 2).

6.5 OVERVIEW

Since primary and secondary epics are inherently and culturally concerned with a serious subject such as heroic deeds or events, importing earthquakes might be considered very important. In reviewing the national epic literature of the people living on the Iranian Plateau, we see that the authors were fully aware of the power of earthquakes; we think they might have experienced a major earthquake during their lifetime, especially when we find out that they were living in earthquake-prone regions close to active faults. For example, Master Ferdowsi Tusi (creator of Shāhnāmeh, 1010) lived in Tus, near an active fault and in between the two active mountain belts of Kopeh Dāgh in the northeast and Binālud in the southwest, with numerous documented historical earthquakes (see Chapters 16 and 17; Figures 16.14, 16.15). Moreover, most of the locations mentioned in the epic literature are earthquake-prone regions.

Furthermore, we can trace some of the elements addressed in the epic literature back to the ancient myths (Chapter 2), religious thoughts (Chapter 3), and folklores and legends (Chapters 4 and 5). In the case of the Amir Hamza, the mythical Zoroastrian Alborz Mountains are definitely replaced by the Qāf (Caucasus) Mountains, which are the northwestern continuation of the active Alborz Mountains. The highest summit of the former is still named Elbrus [i.e., the Alborz].

Chapter 7

Earthquake Poems and Chronogrammatic Verses

Three earthquakes occurred at three times,
In five hundred and a bit when the city turned into a plain;
The second event was in six hundred sixty six,
And the third earthquake took place in eight hundred and eight.
Anonymous poet quoted in Hākem Neyshāburi (998, tr. lit. from the Persian edition of Kadkani, 1996)

The Iranian plateau is highly seismic, with abundant historical documentation of earthquakes; however, more earthquakes have certainly occurred than have been reported in written historical source materials. The scarcity of books (this was before the invention of writing and printing), the frequent invasions and destruction of libraries, and book-burning ceremonies have resulted in inconsistent earthquake coverage. Large amounts of macroseismic data, including oral sources on historical earthquakes, may remain, scattered and lost, in ancient manuscripts, archives, gazetteers, and so forth.

Apparently, a book devoted exclusively to earthquakes, called *Kitāb al-Zalāzil* (the Book of Earthquakes), by al-Hāfiz ibn 'Asākir (d. 1176) has been lost (Baird-Smith, 1842; Sprenger, 1843; Ambraseys, 1961). Presumably, this was the *Kitāb al-Indhar bi-Huduth al-Zalāzil* mentioned by Yāqut (1226), quoted by al-Suyuti (1499). It is probable, however, that Ibn 'Asākir's treatment of the subject was cursory, or al-Suyuti might have quoted from it more frequently (Ambraseys, 1961). al-Suyuti (1499) also mentioned the work of Abu-Ja'far Muhammad ibn-Musāal-Khārazmi on earthquakes, which is also lost.

Poetry has been an important art in the Persian culture and the main vehicle for communication and transmission of personal, local, and national political and cultural ideas. Systematic study of chronogrammatic verses in contemporary poems and odes gives dates of events along with some important macroseismic information. Furthermore, some poems give additional information about the accompanying aftershocks, hardships, droughts, famines, and epidemics. Others

☆"To view the full reference list for the book, click here."

Developments in Earth Surface Processes, Vol. 17. http://dx.doi.org/10.1016/B978-0-444-63292-0.00007-7

convey the power of God through construction and destruction, ask forgiveness, and remind people about the end of the world and the judgment day.

The earliest literary works preserved are the sacred scriptures of the Zoroastrian religion in the Avestā (Chapters 2–4). The Avestā is a most difficult book, and in it are imbedded some 17 poems (the actual words of Zoroaster) known as Gāthās dated ca. 1200 BCE (Boyce, 1989). The narrative in the Avestā is epical. Evidence exists of a great deal of other writing in the epic form, though little has survived, either in the old language or in Middle Persian (Pahlavi); however, a lengthy epic tradition existed when Ferdowsi Tusi (1010) sat down to incorporate the entire legendary past of Greater Iran into one monumental poem (*Shāhnāmeh; the Epic of the Kings*), which is discussed briefly in Chapter 6.

The modern Persian language alphabets date from the 636–652 CE Moslem Arab invasion and conquest. The first recorded Persian poems are a few verses from the early ninth century. Apparently, 'Abbās Marvi composed a poem in honor of Hārun al-Rashid's son, Al-Ma'mun, in 809 CE. The Arab geographer, Ibn Khurdādbih (846), cited a Persian verse by Abu Taqi al-'Abbās ibn Tarkhān. The Tārikh-e Sistān' (ed. Bahār, 1935) quoted a Persian verse written by Abu Wasif who was asked by Ya'qub the Saffārid [r. 868–900] to translate some complimentary Arabic lines.

Occasionally, Persian poets described or eulogized the earthquake affected areas of their hometowns. Seldom, the contemporary poets provide chronogrammatic verses that give the year of the event in the last line. Usually, the Arabic *Abjad* numerals are used to yield the date of an event [a decimal numeral system in which the 28 letters of the Arabic alphabets are assigned numerical values]. Some contemporary poets, like Qatrān Tabrizi (1009–1072), have used poems to illustrate their own description of disasters. As we see in this chapter, this type of poetry can be used as an important primary macroseismic source for data about historical earthquakes and locations of active faults.

For seismic parameters of the discussed earthquakes and their locations, see Chapters 11–17 with the corresponding figures and tables.

7.1 QAVĀMI RĀZI, THE 855–856 RAY EARTHQUAKE

Poet *Badr al-Din Qavāmi Rāzi* (d. before 1164) in a lamenting poem in the form of *qasideh*[1] wrote that the ancient city of Ray (southern suburb of modern Tehrān; Figure 11.2) was destroyed [first verse] and 350,000 people perished [second verse] in 241 H [22 May 855–11 May 856]. Although Rāzi did not mention where the highly exaggerated death toll number was concentrated

1. 'Qasida' in Arabic, 'chekāmeh' in Persian: a form of poetry which typically runs more than 50 lines, and sometimes more than 100.

or his sources, the exaggerated figure is somehow close to some reported figures of the 22 December 856 Komesh earthquake [Kumis, Qumis; the modern city of Dāmghān], located about 270 km to the east northeast of Ray. Nonetheless, the death toll is highly overstated and covers both events of 855–856 and indicate the large magnitudes [$M > 7.0$] of the two separate events. The poet insisted on the idea that the earthquake was a calamity imposed by God in response to immoral activities and sins of the local people. He then recommended charitable deeds and repentance.

7.2 DĀVUD IBN TAHMĀN AL-BAIHAQI, THE 22 DECEMBER 856 KOMESH EARTHQUAKE

Ibn Funduq (Fandoq; d. 1169) quoted a *qasideh*[1] poem in Arabic by the contemporary poet *Dāvud Ibn Tahmān al-Baihaqi* regarding the 22 December 865 earthquake at Komesh [Kumis, Qumis Qumes; the modern city of Dāmghān; Figure 11.1], loosely translated as: "*News reached from the land of Baihaq* [modern Sabzévār] *that a long-lasting earthquakes took place during the night and all the houses and shops were destroyed; causing numerous people and beast of burden losses*" (Ibn Funduq quoting Dāvud Ibn Tahmān al-Baihaqi; ed. Bahmanyār, 1938).

In his third quoted qasideh [which is quoted as "from a learned person of Baihaq"], *Ibn Funduq* repeated the same poem as quoted from the contemporary poet *Dāvud ibn Tahmān* in connection with the Qumesh earthquake of 22 December 856. The difference between these quoted qasidehs is mentioning "Qumes" in the older quoted verse [for the 856 earthquake] and "Baihaq" in the later version [for the 1052 Baihaq (modern Sabzévār) earthquake]. It is not clear if *Dāvud ibn Tahmān* and/or *Ibn Funduq* have used the same qasideh for both earthquakes, or it is a later alteration (Ambraseys and Melville, 1982; Berberian et al., 1996).

7.3 QATRĀN TABRIZI, THE 4 NOVEMBER 1042 TABRIZ EARTHQUAKE

Abu Mansur Qatrān's (1009–1072) Persian collection of poems [*Divān*] is composed of 3000–10,000 couplets [a pair of lines of verses, usually consisting of two lines that rhyme and have the same meter]. His two "qasidehs"[1] on the 4 November 1042 Tabriz earthquake (Figure 11.12) have been much praised and are regarded as a true masterpiece (ed. Nakhjavāni, 1954; Rypka, 1968). Apparently, his first qasideh was composed, while Qatrān was in Tabriz and witnessed the earthquake, and the second one was composed when he left the ruined city of Tabriz and stayed in one of the cities of Arān (Kasravi, 1929; Zokā', 1989; Mansuri, 1991). The following is a summary of Qatrān's findings in his two very long poems.

(i) *Qatrān's First poetic qasideh Ode on the 1042 Tabriz Earthquake and in Praise of his patron Abu Nasr Mamalān*:

In his first qasideh poem, Qatrān recorded the following effects of the earthquake:

- The topography of the ground was changed: He wrote that the highland became low and the lowland raised high, whereas gravels piled up into mountains and mountains were reduced to gravels.
- The Earth components were deformed: He noted that the ground surface was fractured, the trees were bent or toppled, water evaporated [change in the water flow of the rivers, springs and ground water], and mountains slid (rock falls, landslides).
- High buildings and tall trees all collapsed resulting in the traces of trees and building debris.

(ii) *Qatrān's Second poetic qasideh Ode on the 1042 Tabriz Earthquake and in Praise of his patron, Abu Mansur Mamalān and his son:*

In his second qasideh poem on the 4 November 1042 Tabriz earthquake and an Ode to the Amir Abu Nasr Mamalān [Abu Mansur Vahsudān; Ravādiān family prince] and his son, Qatrān recorded additional interesting issues:

- The poet started with describing the power of God in construction and then destruction of everything on the planet.
- Qatrān then referred to Tabriz as a distant location. This may imply that he left the ruined city after the earthquake.
- The 1042 earthquake took place in the 200-year-old city of Tabriz. This indicates that he, as well as the people of Tabriz, was aware of the 858 earthquake which took place 190-lunar year prior to the 1042 event.
- The earthquake was of high magnitude and within an hour all the people died and the whole city was demolished.
- Those who died were saved from misfortune, whereas the survivors were in grief, desperate and suffering from hunger, thirst, and famine. They sold themselves and their souls for a piece of bread. Finally, nobody escaped loosing children or siblings.
- The devastating earthquake took place during the night.
- Such calamity was of high magnitude and has not been seen in the world.
- The earthquake was the punishment of immoral deeds of the people and they are not repentant yet.

7.4 IBN FUNDUQ (D. 1169), THE 2 MAY 1052 BAIHAQ (SABZÉVĀR) EARTHQUAKE

Ibn Funduq (d. 1169), after meeting several old men remembering the details of the 1052 earthquake, wrote that the year was remembered by the people as "*the Year of Earthquake*" ["*Sāl-e Zelzeleh*"]. He mentioned three poems in Arabic describing the event. The composer of the first poem is not acknowledged. The second poem is by Nasr ibn Ya'qub (Baihaqi, 1170; Ambraseys and Melville, 1982; Berberian et al., 1996). His third poem, which is quoted as "*from some learned Baihaqis,*" is the same as quoted from the contemporary poet Dāvud ibn Tahmān al-Baihaqi in connection with the Komesh (Qumes in Arabic, modern Dāmghān) earthquake of 22 December 856 (discussed earlier). The difference between these quoted qasidehs is mentioning "Qumes" in the older quoted verse [for the 856 earthquake], and "Baihaq" in the later version [for the 1052 Baihaq earthquake].

Ibn Funduq clearly mentioned that the poet Dāvud ibn Tahmān al-Baihaqi was contemporary to the 856 Komesh earthquake, whereas when referring to the identical poem for the 1052 Baihaq earthquake, he only stated that a learned person from Baihaq has written the qasideh. We know that the 856 earthquake took place during the night (mentioned in the poem), but the time of the 1052 earthquake is not known. It is probable that the second version referring to Baihaq is copying of the Qumes (Komesh) poem and changing Qumes to Baihaq.

7.5 NEZĀMI GANJAVI [1141–1209], THE 30 SEPTEMBER 1139 GANJEH [GANJA, GĀNDZĀK, JANZA, ELIZAVETPOL, KIROVĀBAĀD] EARTHQUAKE

The Persian poet *Nezāmi Ganjavi* (1191) described the effect of a terrible earthquake by which the city of Ganjeh was destroyed on the eve of Saturday (Minorsky, 1951; Guidoboni and Traina, 1995). Storey et al. (2004) mentioned that in all probability the reference is made to the "Rabi" I, 590/February or March 1194, which must have been in very fresh memory. The poem was in fact dedicated to the Atābak A'aḍam Nosrat al-Din Abu Bakr ibn Mohammad, who died in 1210, but the final verses glorify a third prince, who is not easily identifiable. Minorsky (1951) argued that this part of the poem had been rewritten, but that in any case the original version referred to the Atābak Abu Bakr and his son Bishkin, who can be identified with "Beshken the courageous" of the *Georgian Chronicle* (Guidoboni and Traina, 1995).

Nezāmi was born about 2 years after the earthquake; therefore, he did not experience the event. After praising prince Biskhin, Nezāmi described the terrible earthquake "*on the eve of Saturday.*" The poem ends with the restoration of the city of Ganjeh by the prince (Nezāmi Ganjavi, 1191; ed. Vahid Dastgerdi, 1963). Summary of Nezāmi Ganjavi's long poem is as follows:

- The earthquake ruptured the sky and the cities disappeared.
- The mountains and plains trembled severely and the planet was covered by dust.
- Like the sky, the Earth became restless.

- The chain of the firmament separated, and the joints of the Earth broke.
- The Earth was so compressed from its dejection that the mountains became fractured or faulted.

7.6 MAULĀNĀ JALĀL U'DDIN BALKHI (RUMI) [1207–1273], EARTHQUAKES AT THE QĀF [ALBORZ] MOUNTAIN

Maulānā Jalāl U'ddin Rumi (1273; Book IV:9; ed. Nicholson, 1926) composed a poem about the ascent of *Dhu'l-Qarnayn* [lit., "Two-Horned one," or "Lord of Two Epochs"] to the peak of the *Qāf Mountain* on his eastern quest for the water of life. There he meets the angel *Esrāfil*[2] waiting to blow the trumpet of the Judgment Day. According to Qorān, the first blow will destroy everything (69:13, 14), while the second blow will bring all human beings back to life (35:51).

Based on Sura 18 [Al-Kahf (The Cave)]/83–98 in Qorān mentioning *Dhu'l-Qarnayn,* Ibn al-Kathir and others supposed that *Dhu'l-Qarnayn* represents "Alexander III of Macedonia" (Briant, 1982; Shahbāzi, 2003b). However, others suggested that *Dhu'l-Qarnayn* represents "Cyrus the Great Achaemenid": 550–530 BCE (Shahbāzi, 2003b). The *Qāf Mountain* is the same as the mythical Avestan *Harā Berezaiti* [lit., the "*Lofty Watchpost,*" the mythical mountain in Yasht (19.1) and the Greater Bondéhéshn (IX.1)] and *Harburz,* and *Alborz* in the contemporary Persian language. The Qāf/Alborz Mountain has been considered in the Iranian myth as: (i) the mother of all the mountains circling the planet Earth; (ii) its roots were connected to the roots of all the mountains; and (iii) its growth has been associated with earthquakes and mountain building (Chapter 2).

Rumi's Book IV poem entitled: "*Ascent of Dhu'l-Qarnayn to the Qāf Mountain*" (section 137) talks about the mountain and its earthquake:

How Dhu'l-Qarnayn went to Mount Qāf and made petition, saying:
'O Mount Qāf, tell me the majesty of the Attributes of God';
and how Mount Qāf said that the descriptions of His Majesty is ineffable,
since (all) perceptions vanish before it;
and how Dhu'l-Qarnayn made humble supplication, saying:
'Tell of His works that thou hast in mind and of which it is more easy for thee to speak':
"Dhu'l-Qarnayn went towards Mount Qāf,
He saw it was (made) of pure emerald.
And that it had become a ring surrounding the (whole) world,
He was amazed at that immense creation (work of God).

---

2. 'Esrāfil: Raphael; angel responsible for signaling the coming of Judgment Day by blowing a horn ('sur') and sending out a Blast of Truth.

He said, 'Thou art the mountain (indeed), what are the others?
For beside thy magnitude they are (but) playthings'.
It [the Mt.] replied, 'Those (other) mountains are my veins,
They are not like unto me in beauty and glory.
I have a hidden vein in every land,
(all) the regions of the world are fastened to my veins.
When God wills an earthquake in any land,
He bids me and I cause the vein to throb.
Then I make to move mightily the vein,
With which the (particular) land is connected.
When He says 'Enough!' my vein rests,
I am (apparently) at rest, but actually I am in rapid motion.
At rest, like the (medicinal) ointment, and very active (efficacious),
At rest, like the intellect, while the speech (impelled) by it is moving.
In the opinion of him whose intelligence does not perceive this,
Earthquakes are caused by terrestrial vapours.

Rumi (1273) and ed. Nicholson (1926).

Rumi undoubtedly describes the mythical "*Harā Berezaiti Mountain*" under the name of the "Qāf Mountain." Furthermore, he clearly mentions that the earthquakes are caused by God's decision. Rumi evidently rejects the idea of earthquakes being caused by the entrapped vapor in the Earth, and states that the idea is believed by the ignorant people.

It is interesting to note that as discussed in Chapter 6, throughout the story of the *Adventures of Amir Hamza* (Pritchett, 1991, 2004; Farooqi, 2000, 2008), Amir Hamza has been called "*the earthquake of Qāf*" [Mountain].

7.7 HAMDOLLĀH MOSTAUFI, THE 1208–1209 AND 1270 NEYSHĀBUR EARTHQUAKES

Mostaufi (1334–1335) made a serenade that:

- The first event completely destroyed the city of Neyshābur in 605 H (16 July 1208–6 July 1209; Figure 16.14), where only a mosque stood among the ruins. The name of the mosque is not given; however, based on Jovaini (1260), it should be the Mani'ee mosque of Neyshābur, built by Mani' al-Makhzuni, who died in 1070–1071 at Marvrud (Mo'jam al-Boldān in Dehkhodā).
- Buildings collapsed in the abyss.
- The villages and the fields were destroyed. This may indicate that the qanat water supply system was severely damaged.
- The earthquake took place in the morning and the day became dark as night.
- The survivors left the ruined city and stayed in the field;
- Later, they build a new city and settled in Shādyākh (Figure 16.13).

- About 64 years later the 1208–1209 earthquake, the second earthquake in 1270 ruined Shādyākh and completely destroyed all the structures built after the 1209 event.
- Black water emerged from the ruins.
- The survivors built a new city on another spot (‘Abdollāh Quchāni and Maysam Labbāf Khāniki kindly provided the original manuscript of Mostaufi, 1334–1344; October 16, 2013).

7.8 MAULĀNĀ TĀJ AL-DIN PURBAHĀ JĀMI, THE 7 OCTOBER 1270 NEYSHĀBUR EARTHQUAKE

The poet *Maulānā Tāj al-Din Purbahā Jāmi* (d. 1299) was a contemporary of the Mongol Il-Khān Abāqā [r. 1265–1282; son of Hulāgu Mongol], and was closely associated with the family of great statesmen (Jovaini, 1260; Minorsky, 1956). In his poetic qasideh ode on the 1270 Neyshābur earthquake (Figure 16.14), *Purbahā Jāmi* described the damage to different structures [minaret, mosque, library, school] and then gave the date of the rebuilding of the city as early May 1721. The poem is quoted by Hāfiz Abru (1414); Fasih Khāfi (1442); Sani’ al-Dauleh (1883–1886); and Gerāyeli (1978).

7.9 MAJD AL-DIN KHĀFI, THE 19 OCTOBER 1336 AND THE 21 OCTOBER 1336 JIZD-ZUZAN EARTHQUAKES

Majd al-Din Khāfi (1336), a contemporary local writer, while describing the 1336 earthquake(s), gave two distinct dates in two poems in the same text. In the first poem he gave the date of 19 October 1336, followed by a short prose and another poem with the date of Monday 21 October 1336. The order of the dates, consistent with the flow of the prose, indicates occurrence of two large-magnitude earthquakes ruining a large area and killing about 20,000 people. He described the strong ground motions as if the Earth was off-centered, and the shaking of the world as that of the Day of Resurrection (Khāfi, 1336, ed. Farokh, 1346/1967). Although the poet might have mistaken the dates or the mistakes were introduced by the later copiers, it is probable that there were two earthquakes occurring on two different faults (see Chapter 11 and Figure 11.9). A summary of Khafi’s writing is summarized below:

The 19 October 1336 Earthquake

- Nobody remembers such an intense earthquake at Jizd (Figure 11.9) for the last thousand years.
- The earthquake took place at night in the world as if it was the Day of Resurrections.
- At early dawn, the shock was so strong as if the Earth or the firmament became off-centered.

- Earth was mixed with the air and the horizon became inclined.
- Within a few minutes about 20,000 people perished under debris.
- The ruler Malek Ghiyath al-Din Firuz, rushed to and fro in his palace repeating "it is the Last Day," when suddenly the palace was overturned on top of the ruler.

The 21 October 1336 Earthquake

- The earthquake took place during night in the Khāf region (Figure 11.9).
- Many people perished.
- All the houses, palaces, and mosques were ruined.

7.10 ANONYMOUS POET IN HĀFEZ ABRU, THE 10 FEBRUARY 1364 HARĀT EARTHQUAKE

Hāfez Abru (1414) quoted a poem by an anonymous poet (a dear person) about the 1364 earthquake at Harāt, when the Sun [al-Shams] was in Pisces [borj al-Hut].

7.11 MAULĀNĀ LOTFOLLĀH NEYSHĀBURI, THE 23 NOVEMBER 1405 NEYSHĀBUR EARTHQUAKE

Fasih Khāfi (1442) quoted a quatrain by a contemporary poet *Maulānā Lotfollāh Neyshāburi* [died in 1413 in Esparis/Esfaris, the modern Qadamgāh] regarding the 1405 earthquake (Figure 16.14) (Melville, 1980):

A city in which obligatory prayers were performed voluntarily;
In the concourse of whose people there assembled caravaneers;
Was thrown down when her land was shaken by an earthquake;
In a moment or two, the highest places became the lowest.

tr. Melville (1980)

Lotfollāh Neyshāburi's qasideh, quoted by Fasih Khāfi (1442), Hāfiz Abru (1414), Sani' al-Dauleh (1883-1886), and Gerayeli (1978), has a chronogrammatic verse at the last line ["*Dam-e-ruz-e-Qiyāmat*"] giving the year of the event in 808 H [1405]. He also mentioned that more than 30,000 people were killed during the earthquake (see Chapter 16 and Figure 16.14).

7.12 ANONYMOUS POET IN KHALIFA NEYSHĀBURI, THE 1145, 1270, AND 23 NOVEMBER 1405 NEYSHĀBUR EARTHQUAKES

Khalifeh Neyshāburi (see Melville, 1980; Ambraseys and Melville, 1982; Hākem Neyshāburi, 998; ed. Kadkani, 1996) quoted the following quatrain, giving the dates of three earthquakes at Neyshābur (Figure 16.14):

Three earthquakes^ occurred at three times,
In five hundred and a bit* when the city turned into a plain;
The second event was in six hundred sixty six**,
And the third earthquake took place in eight hundred and eight***.

^: The poet misses the 605 H [1209 CE] earthquake taking place between 540 [1145] and 808 [1405].
*: Referring to the 540 [1145] earthquake.
**: 666 (sic), possibly referring to the 669 [19 Safar 669/7 October 1270] earthquake.
***: Referring to the 30 Jumāda 808 [23 November 1405] earthquake.

The same chronogrammatic quatrain, with some variations, is given by *Daulatshāh Samarqandi* (1487) and *Shirvāni* (1837). *Daulatshāh Samarqandi* (1487) quoted an anonymous poet, referring to him with adjective "*'azizi*" [In Arabic, lit. "dear," "a dear friend," "honored person"]. In the same book, *Daulatshāh Samarqandi* used the same adjective several times [pp. 166, 171, 179, 189]. Melville (1980) and Ambraseys and Melville (1982) erroneously considered the word "'azizi" as the name of a poet *'Azizi*, and added that the correct identity of the poet *'Azizi* is not certain (see Chapter 16 and Figure 16.14).

7.13 'ABDI BAYG SHIRĀZI, THE 1567 QARACHEH DĀGH [ARASBĀRĀN] EARTHQUAKE

Zayn al-'Abidin 'Ali 'Abdi Bayg Shirāzi (1515–1580) in *Takmilāt al-Akhbār* (in Tarbiyat, 1935; Storey, 1927; Kholāsat al-Tavārikh, Ed. Eshrāqi, 2004) gave a chronogram in the last verse dating the 974 H [1567] earthquake at the castle/prison of Qahqaheh in the Arasbārān region of the northwest Iran.

7.14 MIR BAQĀ' BADAKHSHI, THE 5 FEBRUARY 1641 DEHKHĀRQĀN (ĀZARSHAHR) EARTHQUAKE

A long 62 verses *mathnavi* of *Mir Baqā Badakhshi* (ca. 1664; also in Zokā', 1989), describes the event, and the chronogrammatic verse at the third line before the end ["*Ghami bar dāman-e-guiti raqam zad*"] gives the date of the earthquake [1050 H; referring to the 1641 earthquake (Figure 11.12)]. According to Nakhjavāni (1964), Daulatābādi (1964), and Zokā' (1989), the chronogram yields the date 1060 H/1650 ["ghami" = 1050 + "*i*" (from the end of "quiti") = 10; which gives the total of 1060 H/1650]. This misinterpretation of the chronogrammatic verse resulted in a new entry in the Tabriz earthquake lists in the year 1060 H/1650 CE by *Hāj Hossein Āqā-yeh Nakhjavāni* (1964), Kasravi (1948), Daulatābādi (1964), Berberian and Arshadi (1976), Melville (1981), Ambraseys and Melville (1982), and Zokā' (1989).

The reason for adding "*i*" = 10 borrowed from the end of the third word after "ghami = 1050" is not warranted; the chronogram can be read as "ghami," which yields the correct date of 1050 H/1641. Therefore, Badakhshi's poem refers to the 1641 event and not an earthquake in 1650. No record of the 1650 devastating earthquake is documented in the literature. Tavernier (1681) who was in Tabriz in 1655 mentioned a destructive earthquake in 1651 (sic., 1641) destroying Shām-e Ghāzān monument; we know that the monument was destroyed during the 1641 earthquake (Vārtābed Ārākel Tavrizetsi, Ārākel of Tabriz; 1594–1670; ed. Bournoutian, 2006).

Badakhshi's *mathnavi* describes the following catastrophic effects of the 1641 earthquake (Figure 11.12):

- *Severe Earthquake Shaking*: The poet mentioned that the Earth shook so severely like a sea; the minarets, houses, and walls fell down; people were pulled out from underneath of the collapsed debris; the mountains collapsed; the ground surface fractured; and finally the aftershocks continued for days and nights.
- *Ground Fracturing*: The Earth fractured in all directions.
- *Mass Slope Movements*: Mountains separated from each other and pulverized like dust. In the Sorkhāb Mountain [north of Tabriz] a hill of pulverized cinnabar [shanjarf; HgS] appeared.
- *Continuous Shaking* (aftershocks): Earth was shaking for days and nights like the heart beats of a lover (see also Chapters 11 and 16 and Figures 11.12 and 16.7).

7.15 ANONYMOUS POET, THE 30 JULY 1673 MASHHAD EARTHQUAKE

A poem in the Goharshād mosque in Mashhad narrates that because of the Earth shaking (Figure 16.15) the congregational mosque was ruined (Berberian et al., 2000a).

7.16 TABIB ORDUBĀDI, THE 26 APRIL 1721 SHEBLI EARTHQUAKE

Two chronogrammatic verses in a long poem by *Tabib Ordubādi* (eighteenth century) unmistakably yield the date of the earthquake on 28 Jumada II 1133 H [26 April 1721] (verse 8) and again 1133 [1721] (last verse). Despite this, Melville (1981) and Ambraseys and Melville (1982) wrote that the poem dated the event in "1139 AH (August 1726–August 1727)," which is not the case in the Persian poem.

In his poem, *Ordubādi* recorded that the 1721 earthquake flattened the city of Tabriz (Figure 11.12) and described a few issues:

- *Mysteriously Informed*: The voice of an invisible person informed the calamity will strike, flattening the city of Tabriz.
- *Complete Destruction*: Suddenly from top and bottom an earthquake destroyed the world. The sky became dark from the dust. All the buildings, public baths, schools, shops, and the whole bazaar were flattened; No wall remained standing.
- *High Casualty and Injuries*: Thousands of people were killed under debris; those who survived had broken feet, hands, waists, and fractured skulls.
- *The Earthquake Was the Punishment of Our Deeds*: Ordubādi clearly mentioned that the earthquake was the result of people's wrong deeds against the will of God, greed, parasitic lifestyle, sins, and rebellion. He added that nothing is left from the religion of the prophet and if a year passes such, you will not see any trace of a territory. At the end, he emphasized that nobody can repel the earthquake except you, and we should be ashamed and scared (see also Chapters 11 and 16, Figures 11.12 and 16.7).

7.17 THREE POETS, THE 15 DECEMBER 1778 KĀSHĀN EARTHQUAKE

Contemporary poets Hātef Esfahāni, and Āzar Shāmlu narrated in verses the misery caused by the 1778 earthquake (Figure 16.4) to the people of Kāshān (Ghaffāri Kāshāni, 1796; Zarrābi, 1870; Nakhjavāni, 1964). Sabāhi Bidgoli's poems are about the event itself; rebuilding the bazaar of Kāshān with chronogrammatic verses indicating 1195 H (1781) by the order of 'Abdol Razāq Khān Kāshi, the governor of Kāshān; finishing the reconstruction of the city in 1198 H (1783); and the destruction of his own house and the death of his wife and his children.

The contemporary poet, *Hātef Esfahāni* (1783) in qasideh poems No. 5 and 38 writes about the 1778 earthquake and the subsequent restorations. His poems describe: (i) the earthquake took place during midnight and flattened the buildings; (ii) repair of his own house in 1780; (iii) constructions ordered by *'Abdol Razāq Khān-e-Kāshi*, the governor of Kāshān in 1782; and (iv) restoration of the Kāshān congregational mosque, bazaar, and the city wall in 1782. The chronogram of the last verse of qasideh poem No. 38 ("*shod masjed Kāshān tāzeh*") yields the date 1196 H (1782) for the restoration of the Kāshān congregational mosque.

The second contemporary poet *Sabāhi Bidgoli Kāshāni* (d. 1803) composed poems with verses on the 1778 Kāshān earthquake (Nakhjavāni, 1964; Narāqi, 1966). The poet describes that: (i) the houses and bazaar were destroyed; (ii) rebuilding the bazaar of Kāshān with chronogrammatic verses indicating 1195 H (1781) by the order of *'Abdol Razāq Khān-e Kāshi*, the governor of

Kāshān; (iii) finishing the reconstruction of the city in 1783; and (iv) the destruction of his own house and the death of his wife and his children.

The chronogrammatic verses provided by poet at the last line of another poem ["*zé 'Afv Dāvar āfāq Kāshān did ārāmi*"] yield the year of 1198 H (1783), which is the finishing date of construction of the city. However, Nakhjavāni (1964) considered it as the date of a destructive earthquake in Kāshān, and enters an erroneous earthquake date for the city.

In his quatrain [*robā'i*] Āzar (1722–1780) in Ghaffāri Kāshāni (1796), who lost his relatives in the 1178 earthquake, wrote that a great earthquake destroyed life on the Earth (see also Figure 16.4 for the location of the destroyed site).

7.18 THREE POETS, THE 8 JANUARY 1780 TABRIZ EARTHQUAKE

Contemporary poets Zonuzi, Hādi Hamédāni [Nesbat], and Bābā-ye Hamédāni (in Zokā', 1989) serenaded verses about the 1780 Tabriz earthquake (Figure 11.12).

Zonuzi (1801) wrote two chronogrammatic proses yielding the date of 1194 H (1780) for the Tabriz earthquakes; the first chronogram is attributed as the "*this (recent) earthquake*" [the mainshock] and the second attributed as the "*first earthquake*" [the foreshock]:

(i) "*In zelzeleh az zelzeleh-ye avval shaded-tar bud*" [=1194 AH for the mainshock of 1 Muharram 1194/8 January 1780]; and
(ii) "*Bali zelzeleh sakht bud*" [=1193 AH, foreshock on 30 Dhu'l-Hijja 1193/7 January 1780].

Zonuzi also added a quatrain including a chronogram in the last verse as "*Az zelzeleh shod Kharāb*" [=1194 AH/1780]. *Nāder Mirzā* (1883–1885) stated that the last chronogram ["*Az zelzeleh shod Kharāb*"] is composed by the writer of *Riāz al-Janat* [i.e., Zonuzi]; however, Zonuzi himself wrote that the chronogrammatic quatrain is composed by somebody else. The quatrain is also mentioned with some modification by Daulatabadi (1964) and Nakhjavāni (1964).

Tabātabā'i Tabrizi (1875) quoted a poem by the contemporary poet, *Hāji Hādi Hamedāni (Nesbat)*, and stated that it gives the reconstruction date of the Tabriz city wall and its gates in the last chronogram verse as "*Zé nau sad-e-sekandar gashteh ābād*" yielding 1782 (Nakhjavāni, 1964). In fact the chronogram yields 1194/1780. This indicates that: (i) the reconstruction was conducted in the same year of the earthquake; and (ii) the Abjad date was miscalculated in *Tārikh Owlād al-Athār* by the later inscribers (Daulatābādi, 1964; Nakhjavāni, 1964; Zokā', 1989).

Tabātabā'i Tabrizi (1875) also quoted a similar poem by a contemporary poet *Hāji Āqā Bābā-ye Hamedāni* with a chronogram "*Hesār-e sekandar thāni*" yielding 1196 H/1782 (Nakhjavāni, 1964). As with the above case, the chronogram yields the year 1194 H/AD 1780 (Daulatābādi, 1964; Nakhjavāni, 1964; Zokā', 1989). This indicates that: (i) the reconstruction was conducted in the same year as the earthquake; and (ii) the Abjad date was miscalculated in *Tārikh Owlād al-Athār* by the later inscribers. The inscription, which is carved on a block of marble [3.5 × 0.37 × 0.30 (thick) m], is stored in the Tabriz Museum (Daulatabadi, 1964; Zokā', 1989).

Irvāni (Qodsi, 1696–1739) has two poems regarding the earthquake and the chronogrammatic verses of both of them yield 1193 H/1780. The first poem with chronogram "*Afsus afsus zé ahl va nās-e-Tabriz*" [=1193 H] is attributed to Mirzā Moslem Irvāni (Qodsi) by Daulatabadi (1964) and Nakhjavāni (1964). However, Sayyed Yunesi (in Fehrest Ketābhāyeh Khatti Ketābkhāneh Melli Tabriz) cataloged the poem under the name of 'Abbāsqoli ibn Muhammad Khān Thāni [Qodsi] (Zokā', 1989). In this poem, the poet speaks of the month of Muharram and mentions that the ground surface was fractured (the 10th verse of the poem).

The second poem of *Qodsi* (Daulatābādi, 1964; Nakhjavāni, 1964) starts with the date of the earthquake as "*Friday night, the last day of Dhu'l-Hijja*" ["*Shab-e Shanbeh kān salkh-e Zihajjeh bud*"], that is, Friday 29 Dhu'l-Hijja 1193 H, which is the last day of the 1193 lunar year. Qodsi mentions that the "*mostaufiān*" [state accountants] estimated 100,000 young and old people were killed during this "*rajafah*" [earthquake]; which is highly overestimated. He then describes the strong earthquake destruction and estimates 100,000 people killed. The chronogram of the last line "*Qasr kharāb*" yield 1193 H/1780.

7.19 VESĀL SHIRĀZI (1782–1845), THE 25 JUNE 1824 NW SHIRĀZ EARTHQUAKE

The effect of this earthquake is briefly described in a poem by Vesāl Shirāzi (1782–1845).

7.20 MIRZĀ AHMAD ADIB SHAIBĀNI, THE 12 MAY 1844 KĀMU EARTHQUAKE

Adib Shaibāni (in Narāqi, 1966; Zarrābi, 1870) in a long odd verse laments the casualties caused by the earthquake on a Saturday afternoon between the 20th and 30th Rabi'-II, 1260 AH, in the area southwest of Kāshān (see Figure 16.4 for the location of the event).

7.21 THE JUNE 1851 QUCHĀN-MA'DAN EARTHQUAKE

Shākeri (2002) based on Sani' al-Dauleh (1883–1886) and Tarāyeq al-Haqāyeq of Ma'sum 'Alishāh (v.III) referred to a chronogram, "*Quchān*

Kharāb Shod" [1267 H], which yields the year of the earthquake in the Quchān and the Neyshābur Turquoise Mine (see Figure 16.17 for the location of the event).

7.22 VESĀL SHIRĀZI'S SONS, THE 5 MAY 1853 SHIRĀZ EARTHQUAKE

Contemporary poets Vaqār (1817–1846), Hakim (1819–1857), Dāvar (1822–1865), Farhang (1828–1891), and Tohid (1830–1869) Vesāl Shirāzi, sons of poet Vesāl Shirāzi, have poems describing the effects of the earthquake on their city Shirāz (Soltāni-Moqadam, 2012). *Vaqār* in a poem in the form of *Qasideh* provided chronogrammatic verse at the last line ["*Ah kharab-ast*"], giving the year of the event in 1269 H/AD 1853 (Nakhjavāni, 1964).

Hakim and Dāvar in their poems mentioned that the ground was fractured. Dāvar in his poems stated that: (i) the center of the Earth is disturbed; (ii) the shock was so severe that teeth and eyes were tossed out of skulls; and (iii) everything collapsed with one or two shocks and 13,000 people were killed.

7.23 AFSAR KERMĀNI, THE 17 JANUARY 1864 CHATRUD, KERMĀN EARTHQUAKE

Afsar Kermāni (ed. Afsari, 1977) in a poem describes the earthquake to Mohammad Esmā'il Khān Vakil al-Molk, the governor of Kermān [1860–1869], who was out of the town during the earthquake (Figure 16.10). He states that: "*This non-Muslim earthquake! occurred when you left this territory. It destroyed the property and disturbed the peasants during night. The earthquake destroyed my strong house which was built like the Alexander Dam and flattened it like the Lut city* [referring to the Kavir-e Lut desert]. *The effect of the earthquake to the people of Kermān was worse than the invasion of Changiz Khān* [the Mongol]" (translated from the Persian poem).

7.24 HĀJI 'ALIREZĀ SAYYĀH, THE 23 DECEMBER 1871 QUCHĀN EARTHQUAKE

Shākeri (2002) narrated a chronogrammatic verse "*Quchān-e-jadid kharāb shod*" which yields the date of 1288 H/1871 (see Figure 16.17 for the location of the event).

7.25 MOLLĀ SOLAYMĀN KHAMIRI, THE 19 MAY 1884 QESHM ISLAND EARTHQUAKE

Poet Khamiri (Mosāfer), the qāzi [judge] of Khamir Port on the Persian Gulf (Nurbakhsh, 1990), composed a qasideh in 98 duplets about the Qeshm Island earthquake which he attributed as "the first Qeshm earthquake" ["zelzeleh-yeh avval"]. The chronogrammatic verse ["*ghāreq*"] yields the date 1301 H/1884. The poet became blind in his childhood while he was living in Bandar Khamir on the Persian Gulf.

Khamiri serenaded that: "*On the early night of 24 Rajab [1301] when some people were asleep, some walking, and others sitting, an earthquake appeared in the sky. About 31 villages of the Island were demolished. These were: Ramchāh, Giyādān, Heler* (?), *Zirāng, Dargāhān, Kārévān, Ramkān, Band Hāj 'Ali, Kusheh, Sarāhān* [now Sarāvān], *Jijijiyān, Turiyān, Biriyān* [today a garden is left of the ruined village], *Gol* [Gel?], *Khargu, Kermun, Dayrestān, Suzā, Masan, Khāledin, Zainabi, Tomsinti, Māfun* [was located near Kārévān], *Bākui* [Kui, kuhee], *Pay Posht, Kahurān Siyāh, Lāft, Guri, and Khurkharān* [now Dehkhodā]. *The earthquake lasted for six months days and nights, and the people were in hardship and pain. The news has it that 124 people were killed* (tr. From the Persian poem).

7.26 GARAKĀNI, THE 11 JULY 1890 TĀSH EARTHQUAKE

Garakāni has created a chronogram of "*Tāsh biyā bebin zir-o-zebar ze-zelzeleh*" providing the date of 1307 H/1890 for the earthquake (Figure 11.15) (recorded by Berberian and Arshadi in 1977 at Tāsh).

7.27 ANONYMOUS POET, THE 17 NOVEMBER 1893 SOUTH QUCHĀN EARTHQUAKE

Riyāzi Heravi (quoted in Shākeri, 1967, 1981, 2001) narrates that in 1311 H/1893 the city of Quchān (Figure 16.17) became upside down.

7.28 HĀJ MOHAMMAD TAQI BAQĀ'I, THE 17 JANUARY 1895 QUCHĀN EARTHQUAKE

The chronogrammatic verse narrated by the poet at the last line of one of his poems ["*Bi gharq*"] yields the year of 1312 H/1895 (Figure 16.17) (Nakhjavāni, 1964).

7.29 MOLLĀ SOLAYMĀN KHAMIRI, THE 10 JANUARY 1897 QESHM EARTHQUAKE

Poet Mollā Solaymān Khamiri (Mosāfer), the qāzi [judge] of Khamir Port on the Persian Gulf (Nurbakhsh, 1990), composed a qasideh attributed as "the second Qeshm earthquake" ["Zelzeleh-yeh Thāni"]. The poem sounds as a letter written and sent on Tuesday 18 Dhu'l Qa'da 1314 [20 April 1897] to his relatives at Lāft village of the Island describing the event. The earthquake took place on 10 January 1897/1314 6 Sha'ban 1314, whereas Kabābi (ed. Eqtedāri, 1963) wrote that "*the event took place Monday night of 14 Sha'ban 1316/1898.*" It should be mentioned that Monday 12 Sha'ban 1316 falls on 26 December 1898.

Khamiri narrated that: "*Once again the Earth shook and damaged many houses. The mountains fell down like a heavy rain because of the wrath of God. Two people from the city* [of Qeshm] *were killed in the mountain. The soul and body of a woman was injured and her father died.*"

7.30 SA'METE (?) BORUJERDI AND ANONYMOUS POET, THE 23 JANUARY 1909 SILĀKHOR EARTHQUAKE

Contemporary poet, Sa'mete Borujerdi (1909) narrated the effect of the earthquake and seems to be the first person who attributed the earthquake by the name of the district (Ambraseys and Moinfar, 1973; Tchalenko and Braud, 1974). Ambraseys and Moinfar (1973) referred to two short poems composed by natives of Borujerd and Kermānshāh, lamenting the victims of the Silākhor earthquake. Unfortunately, I have not been yet able to consult the reference while living in the United States (see Figure 12.1 for the location of the event).

7.31 TWO POETS, THE 6 MAY 1930 SALMĀS EARTHQUAKE

Mirzā 'Ali 'Asgar Urumiyeh'i (Mohit) in three chronogrammatic verses of: (i) "*Gardid zé yek zelzeleh Salmās kharāb,*" (ii) "*Kharāb shod Salmās,*" and (iii) "*Hova sha'i 'Azim*" (the latter in Arabic) gives the date of the earthquake in 1348 H/1930 (Nakhjavāni, 1964; Zokā', 1989).

Kāzem Rajavi, another contemporaneous poet from Salmās (Figure 12.5), narrated the earthquake effect in a *qasideh* (Zokā', 1989). The last chronogram verse yields 1348 H/1930.

7.32 TEBYĀN AL-MALEK (REZĀ'I), THE 3 AUGUST 1936 TABRIZ EARTHQUAKE

Mirzā Tebyān al-Malek (Rezā'i) in his chronogrammatic verse of "*Che sakht zelzeleh bā howl chon padid āmad*" yields the year of the earthquake in 1315 S/1936. Nakhjavāni (1964) and Zokā' (1989) dated the event as Tuesday 13 Mordād 1315 S which should be 12 Amordād.

7.33 RECENT EVENTS

Numerous poets including Mehdi Soheili (1968 Dasht-e Bayāz earthquake), Mehdi Borhāni, Nāhid Kabiri, Manuchehr Sagart, Simin Behbahāni (1990 Rudbār), Simin Behbahāni and Akhtar Sā'és (1997 Zirkuh), Orkideh Behruzān (2003 Dāhuiyeh), 'Ali Esfahāni, Vidā Farhudi and Simin Behbahāni (2003 Bam), and many more narrated odes lamenting the disastrous earthquakes in Iran.

7.34 OVERVIEW

Numerous contemporaneous poets by expressing their sentiments have narrated odes lamenting disasters at least since the ninth century. Valuable macroseismic data can be extracted from their poems covering a wide variety of issues such as time and day of earthquake, location, size of earthquake and type of destruction or damage, ground fracturing and faulting, rock fall and landslide, casualty numbers, reconstruction dates, and much more. Poet Ali Esfahāni (residing in Canada), in his poem entitled "*Keep on Saying My God is Great,*" questions the validity of the concept of "*God-the Compassionate*" and his wrath and ordeal rooted in the religious psyche of the Iranians and the authorities. His poem, which was serenaded after the 2003 disastrous Bam earthquake (Chapter 14 and Figure 14.6), ends by narrating: "*Would you, for the sake of your kind God, define kindness for me!*"

Chapter 8

Earthquake-Related Inscriptions

. . .The dust of the Earth will not discolor it
Nor will the passage of time efface its pattern. . .

Farrokhi Sistāni (d. 1038)

Numerous inscriptions on historical monuments of the Iranian plateau—in particular, mosques, churches, palaces, and clay tablets—serve as important sources of information about the construction of, damage to, and repairs to those monuments (Kaphadarian, 1946, 1975; Hanaway, 1977; Bivar and Yārshāter, 1978). There are many inscriptions related to repairs; only a fraction refer to work that was required because of earthquakes. Inscriptions and royal orders (Farmān, in Arabic) explicitly mentioning destruction or extensive repairs following earthquakes and epigraphic material referring to remission of tribute to the public exchequer following an earthquake provide valuable and indisputable evidence about the locations and, quite often, the size of earthquakes. Inscriptions on various buildings in the city of Mashhad testify to the widespread damage caused by the 30 July 1673, earthquake (Berberian et al., 2000a). An inscription on a stone plaque written in the old Armenian language (*Grābār*) found on the northern wall inside the Saint Tāddeus Armenian Monastery, located in the area south of Māku (ancient Armenian Her) in northwestern Iran, is the only reference to the damage caused by and the reconstruction made necessary by the 1319 earthquake in this area (see Berberian and Yeats, 1999).

Methodical recording and photographic documentation of epigraphical inscriptions relating to earthquakes on the Iranian plateau and the renovation of structures that were damaged and/or destroyed have not yet been conducted. Furthermore, some inscriptions have been stolen or damaged. A systematic study of inscriptions on historical buildings, dates of reconstruction and repair, causes of damage, and general archaeoseismicity investigations (Chapter 10) should yield further information about seismic activity.

The following is an attempt to salvage some of these valuable data. For parameters of the earthquakes and their sources discussed in this chapter,

☆"To view the full reference list for the book, click here."

Developments in Earth Surface Processes, Vol. 17. http://dx.doi.org/10.1016/B978-0-444-63292-0.00008-9

see Tables 14.9–14.11 and 17.10–17.12 in Chapters 14 and 17, respectively. Locations can be found in the referenced figures in the cited chapters.

8.1 CA. 1263 BCE NINEVEH-ĀSSUR EARTHQUAKE

Soon after Shalmaneser I came to the throne (1263–1234 BCE), his country suffered greatly from an earthquake, which knocked down Ishtar's temple in the city of Nineveh (Ninveh; 36°22′N–43°09′E) (Figure 8.1) and (perhaps) Āssur's temple in the city of Āssur (Āshur; 35°27′N–43°′15′E). Fire broke out in the latter building and destroyed it completely. These disasters did not deter the young monarch. Rather, they appear to stimulate him to set out on a career of conquest, to secure treasure and slaves to reconstruct the temples without delay. Details of the earthquake were inscribed in Assyrian on a clay tablet discovered at Nineveh, Assyria (McKenzie, 1915; Mallowan, 1966; Thompson and Hamilton, 1932; Thompson, 1937; Ambraseys and Melville, 1982; Ambraseys, 1988a, 2009).

> *[5]* When the Temple of Ishtar, the lady of Nineveh, my lady, fell in ruins [which] aforetime Shamshi-Ādād *[6]* had built (and) after him Āshur-uballit, my father, restored: that temple [was ruined] throughout by an earthquake *[7]* ... its weaknesses I bonded and (its) fallen part from its foundation to [its roof I built: the stone tablet] and stone cylinders of Ashur-uballit. [I restored] *[8]* anew to their places....
>
> Thompson and Hamilton (1932)

Ishtār was the goddess of fertility, love, sex, and war in the Babylonian pantheon. She was the Assyrian and Babylonian counterpart to the Sumerian Inānnā, and her temple at Nineveh ensured the city's importance as a national religious center.

Data are insufficient to give a tentative epicentral location. The city of Nineveh (36°22′N–43°′09′E) is located near the modern city of Mosul, in the Zāgros Foredeep to the south of the Zāgros Mountain Front master basement fault. Āssur (35°27′N–43°′15′E), the modern Qal'āt Sherqāt, is located approximately 115 km to the south of Nineveh and just to the north of the Zāgros Foredeep master basement fault (Berberian, 1995). Nonetheless, it seems there was serious damage to Nineveh (Figure 8.1).

In his article, "Tracing Assyrian Scholarship," Stephan Maul of the University of Heidelberg referred to an Assyrian clay tablet describing a ritual that is supposed to avert disaster caused by an earthquake. This indicates that the danger of earthquakes was quite familiar to the ancient Assyrians (Mual, 1997). Data reveal the first recorded knowledge of the seismic activity of the Zāgros Mountain belt.

8.2 CA. 1170 BCE [1187–1150 BCE] NINEVEH EARTHQUAKE

Nearly a century after the previous earthquake at Nineveh (Figure 8.1), which took place during the reign of Shalmaneser I (see above), another earthquake caused damage and destruction at the Ishtār temple during the reign of

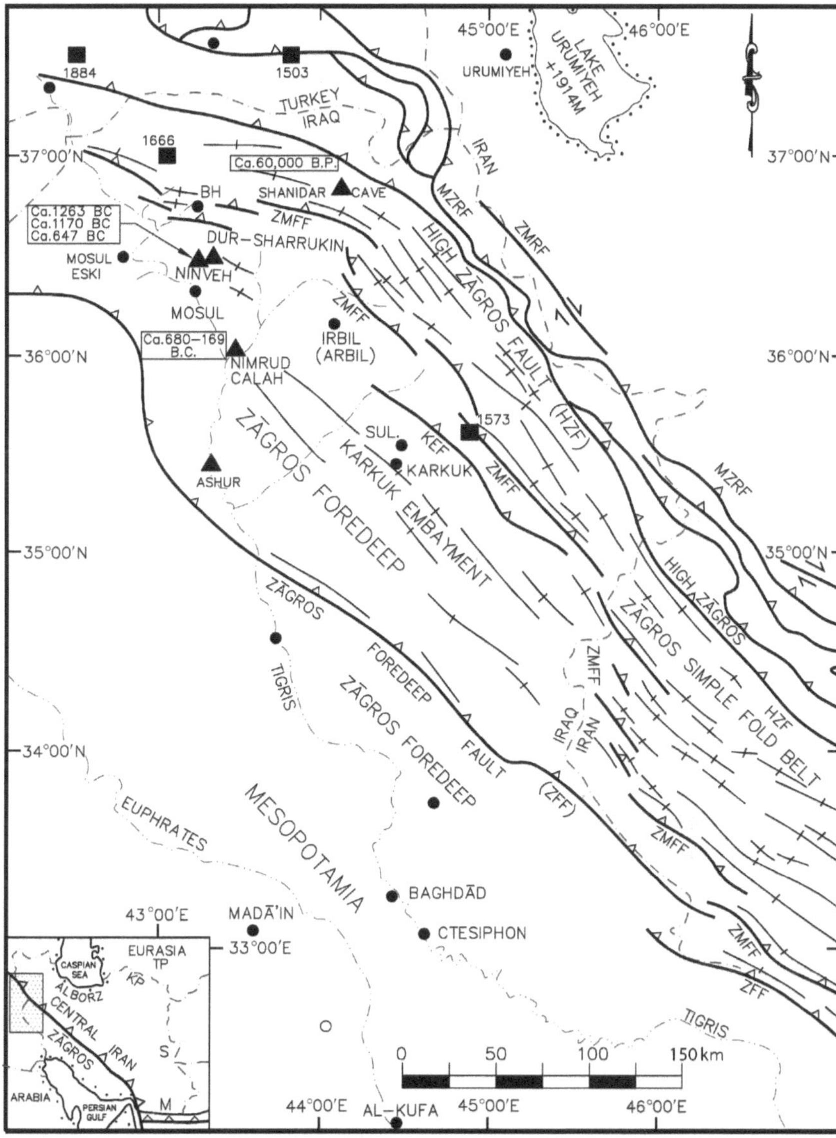

FIGURE 8.1 Historical seismicity map of the northwestern Zāgros fold-and-thrust mountain belt in southwest Iran and northeast Iraq. Reverse faults are shown with teeth on hanging-wall side. Thin lines with short bars indicate fold axes. Archeological sites are indicated with filled triangles; the dates of prehistoric earthquakes are shown in boxes. Approximate locations of the pre-1900 earthquakes (Ambraseys and Melville, 1982; Ambraseys, 2009; Berberian, 1994) are marked by filled squares. For sources, see the text. See also Tables 17.10–17.12 for the seismic parameters of the earthquakes. Inset lower left, Kp: Kopeh Dāgh, M: Makrān, S: Sistān, TP: Turān Plate (Eurasia). *Modified from Figures 1 and 10 in Berberian, 1995.*

Āshur-dan [ca. 1187–1150 BCE]. The two destructive earthquakes at Nineveh were inscribed on a clay tablet by the order of Ashur-rish-ishi [ca. 1120 BCE] (Thompson, 1937; Ambraseys and Melville, 1982; Ambraseys, 1988a,b, 2009).

When the namiru of the Great Gate of the Lions' Heads of the Great Court of *[9]* Ishātār [of Nineveh], my lady, which aforetime in the time of Shalmaneser, the king of Assyria, was ruined in an earthquake, (and) [Shalmaneser], a king who preceded me and restored its ruins; *[10]* a second time it had been shaken in an earthquake in the [time of Ashur-d] an, [the king] of Assyria, who begot my father, and those namiri had been unsettled and had fallen in ruin; for fifteen tipki from the coping to the 'beam of the house' I pulled down, *[11]* I took (it) down. I raised (it) to fifty tipki, increasing it by thirty-five tipki more than before; with iaeri of stone I surrounded them.

Thompson (1937)

The second event seems to be the one that caused damages at Calah [Nimrud] (Mallowan, 1966). Nimrud [Kalhu, Calah, Kalakh; 36°22′N–43°09′E, +224 m] is located about 32 km to the southeast of Nineveh, in the Zāgros Foredeep, and between the Zāgros Foredeep fault in the south and the Zāgros Mountain Front fault in the north (Berberian, 1995). The approximate time interval between the two events of ca. 1263 BCE and ca. 1170 BCE is a century (Figure 8.1).

One of the Assyrian clay tablets found above the sun-dried brick foundation of the temple of Ishtār (not far from the great slab of Āshunasirpāl) during the 1930 excavation at Nineveh was a letter to the king at the royal court describing an earthquake and its consequences. The letter was written by a scribe from one of the outlying towns (Thompson, 1937):

Unto the king, my lord, thy servant Nabu-shuma-ukin, the A.BA. Greeting unto the king, my lord; may Nabu and Marduk be gracious to the king, my lord. On the 21st of Elul *[Elulu, lit., 'Harvest' in Akkadian]* an earthquake took place; all the back part of the town is down; all the wall at the back of the town is preserved, (except) thirty and a half cubits there from being strewn and fallen on the near-side of the town. All the temple is do[wn], (but) I am glad (to say that) the gods of the king are all safe. The casing (?) of the window of the temple . . . Of the portico, of one cubit (?) of . . . has fallen. One of the . . . therefrom of this temple has fallen. As for the namiru of the Great Gate nearby, of the Great Gate. . . have fallen. (Of) Bit. . .; (Of) the keep (guard-house) of the town, one or two houses therefrom have fallen. Let the Chief-[Architect(?)] come and inspect.

Thompson (1937)

8.3 THE 680–669 BCE NIMRUD EARTHQUAKE

In two separate letters [tablets] from Balasi, a priest of Nimrud (Figure 8.1), to Esrahaddon [r. 680–669 BC], two earthquakes were recorded as having been felt at Calah/Nimrud (Waterman, 1929–1931; Ambraseys, 1988a,b, 2009).

The significance, that is, of the earthquake, is this: since it has taken place, let them perform whatever the rites are for an earthquake. Your gods will give prosperity. Did Eā do it, Eā will surely release (from it). Whoever caused the earthquake, that (god) also has provided a releasing incantation. In the times of the fathers and ancestors of the king no earthquake took place (and) I, myself, because I am without understanding, did not perceive the earthquake.

Waterman (1929–1931, No. 355).

Now for a whole day there was an earthquake. When the Earth quakes for a whole day, the prince of the land will be taken away.

Waterman (1929–1931, No. 344)

Thompson (1937) translated a portion of the aforementioned text: "*Let them perform whatever are the rites for an earthquake*; *thy gods will cause* (*it*) *to pass away*; (*what) Eā has done, Eā has given release (therefrom). Whoever caused the earthquake, that same provides an incantation for release (therefrom). Was there no earthquake during the times of the fathers or grandfathers of the king? As for me, since I was (then) too small, I did not notice earthquakes*" (Thompson, 1937; Waterman, 1929–1931).

8.4 CA. 647 BCE DUR SHĀRRUKIN [KHORSĀBĀD] EARTHQUAKE

An earthquake on Ādār 9 that was strongly felt at the city of Dur Shārrukin (modern Khorsābād; Figure 8.1), is mentioned in a letter sent to Āshurbanipāl [?; 669–627 BCE] (Waterman, 1929–1931; Thompson, 1937; Ambraseys, 1988a,b, 2009):

To the king my lord, your servant Kisir-Āshur. May it be well with the king my lord. From the city of Nikiā I have come to the city of Dur Shārrukin, (and) they inform me that an earthquake took place in the city of Dur Shārrukin on the ninth day of the month Ādār. The king my lord will at once speak saying, 'Is there any structure whatsoever (left) in the fortress?' It is well with the mausoleum, the temple towers, the palace, the fortress (and) with the houses of the entire city. May the heart of the king my lord be of very good cheer.

Waterman (1929–1931)

Dur-Sharrukin [Fortress of Sargon; the Assyrian capital city of Sargon II (722–705); modern Khorsābād; 21 km NNE of Mosul: 36°30′N–43°13′E, +310 m (Figure 8.1)] is located in the Zāgros Foredeep, and between the Zāgros Foredeep fault in the south and the Zāgros Mountain Front fault in the north (Berberian, 1995). After 523 years, following the two earthquakes of ca. 1263 BCE and ca. 1170 BCE in Nineveh, a third earthquake was recorded in this region (Figure 8.1).

8.5 THE 17 MARCH 1016–16 MARCH 1017 SHIRĀVĀKĀN, ĀNI EARTHQUAKE

An inscription in the Armenian cathedral of Holy Cross (Surb Khāch) at Hāgbāt in Shirākāvān [40°38′N–43°44′E; near the ancient Armenian capital city of Āni], refers to the restoration of the cathedral after an earthquake (Hakobyan, 1951; Khachaturian, 1974; Guidoboni and Traina, 1995; Guidoboni, 1997; Babayan, 2006; Ambraseys, 2009):

> In 465 *[Armenian year; 17 March 1016-16 March 1017]* I, Honovār, son of Mukān, by grace of God have reconstructed the holy cathedral which was ruined by a violent earthquake.
>
> Hakobyan (1951) and Guidoboni and Traina (1995)

8.6 THE 1052 ARRAJĀN EARTHQUAKE

An inscription on the Shustar congregational mosque (32°02′N–48°51′E), located 207 km to the northwest of Arrajān (30°65′N–50°32′E), refers to repairs carried out in 1053, 1284, 1526–1527, 1761–1762, and 1798–1799 (Meshkāti, 1970). The first reference (1053) could be to renovations completed after the 1052 earthquake, a strong earthquake with a large area of damage (Ambraseys and Melville, 1982; Berberian, 1994). Arrajān is located at the footwall of the Zagros Mountain Front master fault (see Figure 8 in Berberian, 1995). Three other earthquakes are known to have occurred in this same region in 1085, 1988, and 1991.

8.7 THE 1301 FARIM [PARIM] EARTHQUAKE

An inscription above the entrance of the mausoleum of Bāyazid Bastāmi at Bastām (36°29′N–55°00′E) dated the first half of Shawwal 700/June 1301 (Meshkāti, 1970; Ambraseys and Melville, 1982) refers to repairs made to the mausoleum. Bastām is located about 160 km from the 1301 earthquake meizoseismal area (Figure 12.10). However, the connection has not yet been proved.

8.8 THE 1319 ST. THĀDDEUS EARTHQUAKE

A stone plaque written in the old Armenian language [“*Grābār*”], found on the northern wall inside the Surp Thādei Vānk Armenian Monastery [St. Thāddeus; Plates 8.1 and 8.2] located in the area south of Māku [ancient Her in the Greater Armenia] in northwestern modern Iran (Figure 8.2), refers to the destruction of the monastery and its rebuilding by the Abbot Zākāriā in 1329 (Berberian and Yeats, 1999). This is the only preserved contemporary artifact regarding the earthquake (Plate 8.2). A few years ago, treasure hunters tried to destroy the stone plaque.

PLATE 8.1 The twelfth-century Armenian Monastery of Surp Thādei Vānk [St. Thāddeus] in northwest Iran, south-southwest of Māku [ancient "Her" in Greater Armenia], which was destroyed by the 1319 earthquake (Berberian, 1977h, 1997a). The present church consists of two wings. The older wing, of dark stone, was constructed in 1329 (see inscription in Plate 8.2) after the 1319 earthquake destruction; the newer portion was added during the 1820s. See Figure 8.2 for the location. *(Photographed by John Tchalenko, 1973, during our study of the 1930 earthquake coseismic surface fault).*

PLATE 8.2 A stone plaque, written in the old Armenian language ["Grābār"], on the northern wall inside the Surp Thādei Vānk Armenian Monastery [St. Thāddeus; see Figure 8.2 and Plate 8.1] located in northwest Iran, south-southwest of Māku [ancient "Her" in Greater Armenia]. This stone inscription depicts the destruction from the 1319 earthquake and the reconstruction of the Monastery in 1329 (see the text for translation). This invaluable inscription is the only surviving contemporary document to provide any information about the 1319 earthquake. See Figure 8.2 for the location.

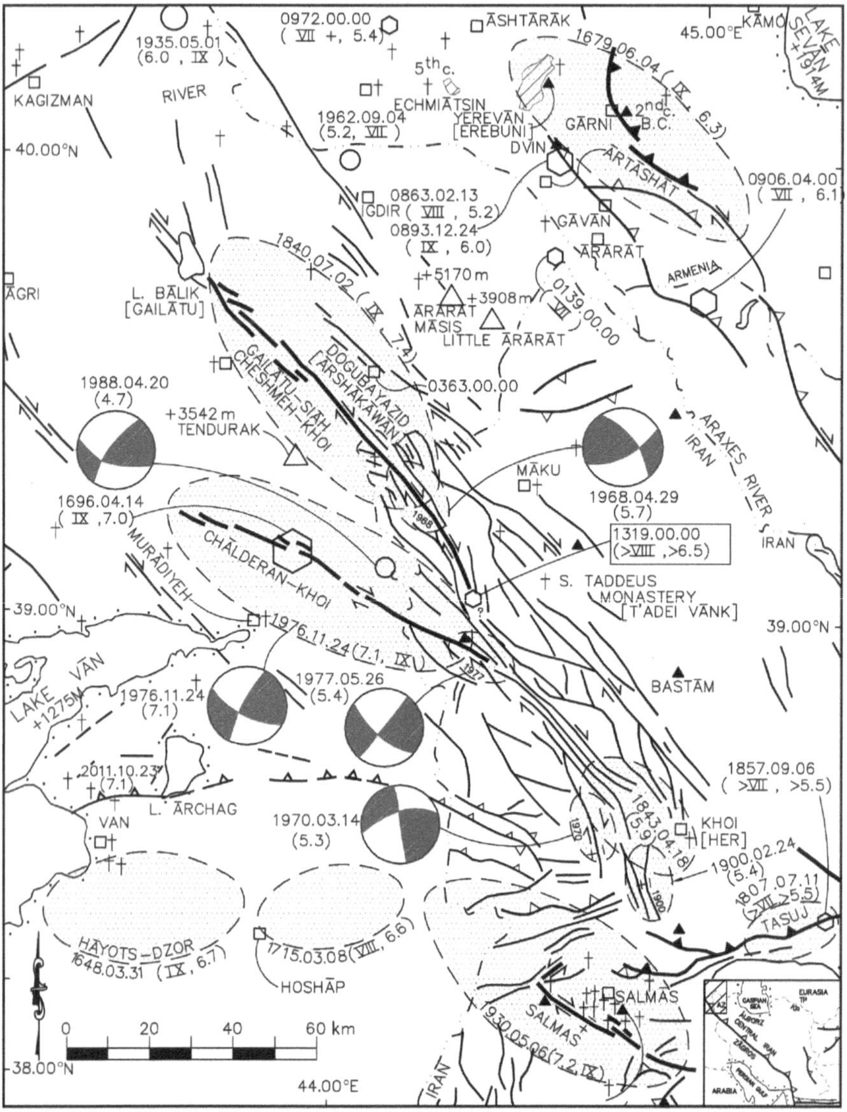

FIGURE 8.2 Seismicity and active fault map of the "West Āzarbāijān right-lateral Shear Zone" (WASZ) in northwest Iran. Reverse faults are shown with teeth on hanging-wall. Strike-slip faults are shown with arrows. Faults without teeth or arrows: sense of slip unknown. Faults have been modified after Adamia et al. (1974), Berberian (1981a, 1983d, 1988, 1997a), Berberian and Arshadi (1976), Berberian and Qorashi (1987), Cisternas et al. (1989), Westaway (1990), Philip et al. (1992), Saroglu et al. (1992), Berberian et al. (1992), Karakhanian (1993), Avanessian (1994), NSSP (1995), and this study. Meizoseismal areas of the earthquakes (shown by ellipses) and macroseismic epicenters modified from Berberian (1981a, 1983d, 1984, 1988, 1997a), Ambraseys and Melville (1982), Berberian and Qorashi (1987), Ambraseys (1988a,b), Ambraseys and Adams (1989), Ambraseys and Finkel (1987, 1995), Nikonov (1991), Trifonov et al. (1994),

This glorious and heavenly temple *[the Saint Thaddeus Armenian Monastery]* was built with solid foundation on the grave of the Apostle Thaddeus which collapsed by an earthquake *[in 1319]* because of our numerous sins. I, Der Zākār *[Bishop Zākāriā]*, the son of Mr. Manuel as the humble servant of the Apostle Thaddeus with the help of God started to repair it; now that times are harder than before when there was persecution against Christians and many churches were destroyed. This is why we established this memorial so that our parents, cousins, and our successor priest Dirātzoo are remembered before great God and bloodless lamb of God; in addition, our genuine brother Bedros and Sārkis who in accordance with their capacity contributed with their belongings. It was finished in 778 *[Armenian year/1329 CE]*

Abbot Zākāriā (1329) (Plate 8.2)

In the continuation of the chronicle of Samuel of Āni [Sāmvel Ānètsi; 1113–1166], written by his followers, two different dates are given for this event: 1308 and 1319, with 65 to 75 people killed in the monastery (Guidoboni and Comastre, 2005). The St. Thāddeus Armenian monastery (Figure 8.2) is located in the vicinity of the Siyah Cheshmeh-Khoi right-lateral strike-slip fault (Berberian, 1997a; Berberian and Yeats, 1999), along the Western Āzarbāijān Shear Zone (see Chapter 13 and Figure 13.6).

8.9 THE 8 MAY 1336 KUSHEH, CENTRAL QESHM ISLAND EARTHQUAKE

A wooden inscription in Arabic, now in the National Museum of Iran in Tehrān (Plate 8.3), tells of the 26 Ramaḍān, 736 H [8 May 1336] earthquake that caused the destruction of a mosque in the Kusheh village on Qeshm island in the Persian Gulf (Quchāni, 1988). As with the 1319 earthquake, this inscription is the only surviving contemporary document to offer information about the earthquake.

The mosque was originally built in 858–859; Sadid al-Saltaneh (Kabābi, 1963) saw the inscription in the mosque by the mausoleum of Borkh al-Asvad in the Kusheh village prior to its transferral to the museum in Tehrān. Quchāni (1988) published both the Arabic and Persian translation

FIGURE 8.2 and Balassanian et al. (1995). The approximate epicenter of the 1319 earthquake that destroyed the St. Taddeus Armenian monastery (marked by a cross) is indicated with a hexagon (see Plates 8.1, 8.2, and 8.5). Major historical Armenian churches are marked by a cross. Triangle: Historical monument. Hexagon: Pre-1900 earthquake. Queried where uncertain. Fault plane solutions: 29 April 1968 M_s 5.5 Gol (McKenzie, 1972); 14 March 1970 M_s 5.2 Badalān (Jackson and McKenzie, 1984); 24 November 1976 M_w 7.0 Chālderān (Jackson and McKenzie, 1984); 26 May 1977 M_s 5.7 Mokhur (Jackson and McKenzie, 1984); 20 April 1988 M_w 5.5 Chālderān (best-double-couple HRVD CMT solution). Inset bottom right: Map of Iran showing the boundary with Arabian plate (line with teeth). AZ, Āzarbāijān; KP, Kopeh Dāgh; TP, Turān Plate. *Modified from Berberian (1997a) and Berberian and Yeats (1999).*

PLATE 8.3 A wooden inscription written in Arabic relates the 26 Ramaḍān, 736 Hijra [5 May 1336] Kusheh earthquake that destroyed the Kusheh village in central Qeshm Island in the Persian Gulf (Quchāni, 1988). As with the 1319 earthquake, this irreplaceable inscription is the only surviving contemporary document that provides any information about the earthquake. See the text for translation.

of the wooden inscription, along with five photographs of it (Plate 8.3). The following is the English translation.

This mosque was rebuilt in 244 Hijra *[858-9]* ... *(missing)*... The building was intact until 26 Ramaḍān 736 *[8 May 1336]* when the sun inclined from the mid-sky towards the west, the foundations of the Earth trembled and a strong earthquake occurred and destroyed the building of this honorable mosque... *(missing)*...; the son of the late king Kordānshāh, son of Solghor, son of Mahmmud... *(missing)*... rebuilt the mosque which was finished in Rajab 737 Hijra *[February 1337]*.

Sadid al-Saltaneh (Kabābi, 1963) stated that the current mosque was built in 1307 H [1889–1890] and the mosque was destroyed during the 1303 H/May 1885 [sic., 19 May 1884] earthquake. Rā'in (1971) wrote that the old Borj [sic., Borkh] mosque was repaired [renovated or reconstructed] in 244 H. (838) [sic., 858] on a Zoroastrian fire temple.

8.10 THE 1598 MASHHAD EARTHQUAKE

Godard (1941) reported an inscription on a tile (kāshi) referring to the repair of the cupola of the Goharshād mosque in 1008 H [1598]. Apparently, the tile is kept in the repository of the sanctuary (Ambraseys and Melville, 1982; Berberian et al., 2000a).

8.11 THE 20 APRIL 1608 ALAMUTRUD EARTHQUAKE

Ambraseys and Melville (1982) stated that the restoration and repair work carried out on the Damāvand congregational mosque as recorded by an inscription dated 1024 H/1615 (Meshkāti, 1970) may be associated with the 1608 earthquake. Damāvand is located about 180 km to the southeast of the meizoseismal area of the 1608 earthquake (Berberian and Walker, 2010; for the locations, see Figure 11.6). The Damāvand congregational mosque, built in 1322, has been repaired several times (Meshkāti, 1970; Shaibāni, 1986, 1990) in 1416, 1521, 1615, 1621, 1670 (possibly after the 1665 earthquake), and 1831 (after the 1830 earthquake). The old mosque was demolished in 1958 to be replaced by a more up-to-date structure. At present, only the lower part of the two massive columns and the Saljuk minaret [ca. 1000–1218] remain (Matheson, 1972; Shaibāni, 1986).

8.12 THE 1614 MEGRELI, WESTERN GEORGIA EARTHQUAKE

On the authority of Brosset (1849), Ambraseys and Melville (1982) reported three inscriptions on icons stating that the Armenian church of the Virgin Mary at Tchaishi was destroyed by an earthquake and rebuilt by the faithful people.

8.13 THE 5 FEBRUARY 1641 DEHKHĀRQĀN [ĀZARSHAHR] EARTHQUAKE

The congregational mosque of Osku, which was destroyed during the 1641 earthquake, was rebuilt two years later. An Arabic inscription (Plate 8.4) mentions a reconstruction by Hāj Safi-Qoli that was completed within two years in 1052/1642 (Zokā', 1989). For the location of the site, see Chapter 11 and Figure 11.12.

PLATE 8.4 An Arabic stone inscription in the Osku [old Dehkhārqān] congregational mosque indicating reconstruction of the mosque in 1052 Hijra [1642 CE] after its destruction by the 5 May 1641 Dehkhārqān earthquake. The inscription was created by Hāji Safi-Qoli. For the location, see Figures 11.7 and 11.12. *Photograph by Ma'ruf, in Zokā' (1989).*

8.14 THE 1665 DAMĀVAND EARTHQUAKE

Sani' al-Dauleh (1800–1802) mentioned an inscription, uncovered by a painter in 1879, on a beam on the wall to the left of Mehrāb of the Damāvand congregational mosque. It referred to earthquake damage (without mentioning the date of the event) as well as restoration efforts carried out in 1081 [11 May 1670–30 April 1671], during the reign of Shāh Solaymān Safavid [r. 1666–1694] (Ambraseys and Melville, 1982). Meshkāti (1970) reported five inscriptions in the Damāvand congregational mosque that referred to restorations in 1416, 1521, 1615, 1621, 1670 (possibly after the 1665 earthquake), and 1831 (after the 1830 earthquake) (Shaibāni, 1986, 1990). Damāvand is located by the Moshā fault (for the location, see Figure 11.2).

8.15 THE 30 JULY 1673 SHĀNDIZ EARTHQUAKE

Pickett (1984) reported four inscriptions by Mohammad Rezā Emāmi on tiles in the Imam Rezā shrine referring to repairs and restorations carried out after the 1084 H/1673 earthquake. The repairs were made in 1675 (Sykes, 1911; Pope 1965; Berberian et al., 2000a).

8.16 THE 1675–1678 (?) WINTER GONĀBĀD EARTHQUAKE

According to local tradition at Gonābād (Tābandeh, 1969), an earthquake on the cold winter night of 1089 [1678] devastated the town of Gonābād as well as all the villages around it. Only one child and the mosque [possibly the Gonābād congregational mosque, built in 1212] survived (Tābandeh, 1969; Ambraseys and Melville, 1982). The exact date of the event has not yet been established. The Gonābād or Bidokht faults (with fresh displacements visible on aerial photograph nos. 29305 and 29307, Worldwide Aerial Surveys, Inc., project 158, scale 1:55,000) could be possible sources of the 1675 and/or 1238 earthquakes (Berberian and Yeats, 2001; see Chapter 11 and Figure 11.11).

No evidence of these two earthquakes is reported from the town of Qā'en. An inscription in the Qā'en congregational mosque (Figure 11.8) records restoration work in 1086 H [18 March 1675–5 March 1676], with no reference to any earthquakes (Meshkāti, 1970; Nāderi, 1980). If the restoration of the Qā'en congregational mosque [located 83 km to the southeast of Gonābād] was due to the 1675 Winter (?) event, the earthquake should have occurred sometime between 1086/1675 and 1087/1678.

It should be noted that the twentieth-century earthquake cluster in the Qā'en/Dasht-e Bayāz area, which consisted of 13 earthquakes of $M > 6.0$

in 61 years [1936–1997; see Chapter 16 and Figure 16.8], did not cause destructive intensities at the Qā'en congregational mosque of 1368 (Berberian and Yeats, 1999, 2001).

8.17 THE 15 DECEMBER 1778 KĀSHĀN EARTHQUAKE

An inscription above the mehrāb and the plaster work on the inside of the dome of the Kāshān congregational mosque (Figure 16.4) refers to repairs in 1780, 1782, and 1793 (Meshkāti, 1970). The first repair [1780] seems to be due to the 1778 earthquake (see Figure 16.4). The 1780 post-earthquake repair inscription inside the shabestān is written by *Ibn Muhammad Hossein Muhammad Kāzem Esfahāni* and mentions the work of master *Bāqer Esfahāni* (Meshkāti, 1970).

The 1782 post-earthquake reconstruction inscription (Meshkāti, 1970) is a poem written by *Sayyed Ahmad Hātef Esfahāni* (d. 1794) that gives a chronogrammatic verse of "*Shod masjed-e-Kāshān tāzeh,*" which yields 1196 H/1782 (see also Chapter 7).

The 1793 maintenance inscription, which was conducted by the order of *'Abdol Razāq Khān-e Kāshi*, the governor of Kāshān (Meshkāti, 1970), reads: "*After the earthquake destruction, and repairs conducted by 'Abdol Razāq Khān. Written by 'Abdolkarim, 1207* [1793]."

There are two additional inscriptions dated 1780 regarding reconstruction of the Kāshān congregational mosque (see Chapter 7).

8.18 THE 8 JANUARY 1780 TABRIZ EARTHQUAKE

Tabātabā'i Tabrizi (1875) quoted an inscribed poem by contemporary poet *Hāji Āqā Bābā-ye Hamedāni* with a chronogram "*Hesār-sekandar thāni*" yielding 1194 H/1780 (Nakhjavāni, 1964). This indicates that the reconstruction was carried out in the same year as the earthquake. The Abjad numeral date was miscalculated in Tabātabā'i Tabrizi (1875) by later inscribers. The inscription, apparently originally carved on a marble plaque on the Dāl-o Zāl mosque (Kārang, 1958), is stored in the Tabriz Museum (Daulatābādi, 1964; Zokā', 1989). For the location and seismic fault of the area, see Figure 11.12.

8.19 THE 1805–1806 MĀZANDARĀN EARTHQUAKE

Rabino (1928) wrote about an inscription in the Bārforush [modern Bābol; Figure 12.10] mosque referring to the 1805–1806 earthquake destruction and reconstruction of the structure in 1220 H [1 April 1805–20 March 1806]. "*Another inscription informs us that the mosque was destroyed by an earthquake during the reign of Fath 'Ali Shāh* [Qājār; r. 1794–1834] *and that it was rebuilt by Mir Muhammad Husayn in 1220* [1805–1806]" (Rabino, 1928).

8.20 THE 16 DECEMBER 1808 TĀLÉQĀN EARTHQUAKE

The last line of an inscribed poem located on a blue clay tile in Emāmzādeh Ja'far [built in 1549] at Pishvā, southwest of Tehran [35°18″N–51°43′E, +942 m; see Figure 11.2], refers to a repair date with Abjad numerals in 1227 H/1812 during the reign of Fath 'Alishāh Qājār (Meshkāti, 1970). Ambraseys and Melville (1982) mentioned that the repairs at the Pishvā mausoleum may be associated with the 1808 earthquake in Taléqān. The distance between Pishvā and Tāléqān is about 130 km.

8.21 THE 1809 ĀMOL EARTHQUAKE

An inscription on the door of the 'Abdolhaq mausoleum in Zirāb-e Savādkuh village refers to damage from the 1809 earthquake (see Figure 12.10): "*In the name of God, the Mercy-giving, and the Merciful! In the year 1224* [H.] *by God's will, the creator and the builder of day and night, and the provider of light and fire, a severe earthquake and a major accident occurred in this district. It almost destroyed all the old/ancient buildings including this holly, illuminated and pure place. With the grace of Karbalā'i Abutāleb and Darvish Esmā'il Hossein this great building was built that all the people benefit from the pilgrimage of this great person*" (tr. from the Persian text of Sotudeh, 1970).

8.22 THE 1812 JOLFĀ [JUGHĀ] EARTHQUAKE

Without citing any reference, Ambraseys and Melville (1982) under the 1812 earthquake [An earthquake caused extensive rockfalls in Julfā (Ouseley, 1823)], wrote that: "*An inscription in the church of St. Tāddeus dated 1229/1813 refers to repairs of the structure.*" I have not been able to verify the source of the referred inscription and the nature of the 1813 repairs of the St. Thāddeus Armenian monastery (Plate 8.5). For the location of Jolfā and the monastery, see Figures 8.2 and 13.6.

Oskiān (1942), Kleiss (1971), Ter-Grigoriān (1971), and Uluhogiān (1980) do not mention any repairs and/or inscriptions dated 1813. The historical St. Thāddeus Armenian monastery is located about 97 km to the west northwest of Jolfā, while another historical Armenian monastery, St. Stepānos (Plate 8.6; Figure 13.6), is located about 15 km WNW of Jolfā. I consulted Hakhnazariān (1980) and Hofrichter (1980) about the latter and was unable to find any evidence of inscription and/or repair in 1813. In a response to my request, Harmut Hofrichter (Universität Kaiserslautern; letter dated 2 July 1993) wrote

PLATE 8.5 A close-up view of the twelfth-century Armenian Monastery of Surp Thādei Vānk [St. Thāddeus] in northwest Iran, south-southwest of Māku [ancient "Her" in Greater Armenia]. See Figure 8.2 for the location, as well as Plates 8.1 and 8.2. *Courtesy of Shojādel (2008).*

PLATE 8.6 The ca. eleventh- to sixteenth-century Armenian Monastery of Surp Stepānos Nākhāvekā [St. Stephen the Protomartyr] (Figure 13.6), located about 15 km WNW of Jolfā, to the south of the Araxes (Aras) River. The monastery was damaged during several earthquakes. It is regarded as one of the masterpieces of Armenian architecture. *(Photographed by John Tchalenko, 1973, during our coseismic surface rupture mapping of the 1930 earthquake).*

that: "*I do not know any inscription dating 1813 and referring to repairs of buildings neither in the case of St. Tāddeus nor in that of St. Stepānos.*"

Sir William Ouseley (1823), who felt a slight earthquake in Tabriz on 23 June 1812 while residing in Major D'Arcy's house, and was in Jolfā on

6 July 1812, did not mention the date of the earthquake that caused landslides in Jolfā. He wrote that: "*Beyond this point the view represents, connecting the rock, several steep and lofty masses of rock had lately fallen, during earthquakes, and indeed the whole country, for many leagues about Julfā, bespeaks some ancient and most tremendous convulsion of nature, which seems to have torn the hills into uncommon forms, leaving their outlines broken and irregular.*"

Ouseley's (1823) description of the 23 June 1812 event at Tabriz is as follows: "*On the twenty-third (of June), a little before two o'clock, the thermometer being of 66, a slight shock was felt in most parts of the city but not, (as many persons declared) in all; a high and sudden wind immediately preceded it. I happened to be in Major D'Arcy's house, writing at a table, which was perceptibly, although momentarily, shaken; but I should scarcely have supposed that the tremor proceeded from an earthquake, had not a servant hastily entered the room and cautioned me against the zelzeleh* [earthquake]*; whilst several Persians, the Russians and others, ran out into an open court, the safest place on such occasions*" (Ouseley, 1823).

8.23 THE 27 MARCH 1830 LAVĀSĀNĀT EARTHQUAKE

An inscription in the Qom congregational mosque [built in 1135; Meshkāti, 1970] refers to repairs carried out in 1246 H/1830 and 1248/1832 (Meshkāti, 1970) after the 1830 earthquake. Ambraseys and Melville (1982) only mentioned the 1832 repairs. The city of Qom [34°38′N–50°52′E, +926 m] is located about 128 km to the south-southwest of the city of Tehrān (which was damaged during the earthquake) and about 153 km to the south-southwest of the center of the meizoseismal area located on the Moshā fault (Berberian and Yeats, 1999, 2001; for the location, see Figure 11.2).

8.24 THE 12 MAY 1844 KĀMU EARTHQUAKE

An inscription on the upper frieze of the south ayvān of the Esfahān Shāh mosque gives the year 1261 H [1845] as the time when the superficial repairs of the 1844 earthquake damages were carried out, during the reign of Mohammad Shāh Qājār [r. 1834–1848] (Godard, 1937; Burgess, 1942; Zander, 1972). Ambraseys (1979) and Ambraseys and Melville (1982) erroneously gave masjed Jāme' [congregational mosque] instead of masjed Shāh (Soltāni mosque).

The work of the humble Shojā', son of the late ostād *[master]* Kāzem, mason from Esfahān. On the southern ayvān's upper band one can see the written date of 1261 (1845). It represents the repairs that followed an earthquake during the reign of Mohammad Shāh Qājār. The earthquake dislocated *[affected/fractured/partially destroyed]* this part *[southern ayvān]* of the building. Bricks were broken and large fractures appeared. Both minarets were completely detached *[loosened, sheared, torn, disconnected]* from the main body of the building. Mohammad Shāh only limited himself to replacing the covering kashis *[ceramic tiles]* that had been broken and have placed at the top of the building; this inscription dating to 1261 H *[1845]* that gives him credit *[glory]* for a repair that was never carried out.

Godard (1937, tr. from the French text)

An inscribed Kāshi (tile) installed above the entrance to the *'Āliqāpu* grand palace [lit., "Sublime Gate"] in Esfahān indicates that repairs were carried out in 1274/1857 (Galdieri, 1979, 1983).

The city of Esfahān is located about 105 km to the south southeast of the 12 May 1844 $M_s \sim 6.4$ Kāmu (southwest Kāshān) earthquake epicenter (see Figure 16.4).

8.25 THE JUNE 1851 SARVÉLĀYAT EARTHQUAKE

Inscriptions on the Bālāsar Seminary (madreseh) in Mashhad records repairs carried out in 1680 [during the reign of Shāh Solaymān Safavid], 1854–1855, and 1937 (Meshkāti, 1970). The 1854–1855 repairs might have been carried out as a result of the May 1851 $M_s \sim 6.9$ Sarvelāyat (southeast Quchān) earthquake (Ambraseys and Melville, 1982). However, there is no recorded evidence for this speculation. The city of Mashhad is located about 130 km to the southeast of the epicenter of the 1851 earthquake (see Figures 16.17 and 16.15 for the locations).

During the same earthquake, the dome of the Quchān mausoleum of Emāmzādeh Soltān Ebrāhim and the walls of the haram [harem] collapsed (see Chapter 16 and Figure 16.17). The structure was repaired in 1286 [1869–1870] (?) and a poem by Hāji 'Alirezā [Sayyāh] was inscribed on the tiles [kāshi] (Shākeri, 1967).

8.26 THE 21 DECEMBER 1862 SHIRĀZ EARTHQUAKE

A poem dated 1206 [1888–1889] carved on the mausoleum of Emāmzādeh Abu Eshāq (d. 820–821) at the Sabz Pushān Mountain, west of Dehak [28 km SSE of Shirāz], refers to the reconstruction of the mausoleum. Although it does not mention the cause of the damage/destruction, it is probable that the reconstruction was necessary after the 1862 earthquake.

8.27 THE 23 MARCH 1879 SE BOZQUSH EARTHQUAKE

An inscription on the stone above the eastern gate of the Dāsh masjed [lit., "stone mosque"] in Tark village indicates that the rebuilding of the mosque was completed in 1282 H?/1865? (Berberian, 1976c). The year 1282 [1865] seems to be a typographic error, and the original inscription should be reviewed (see Figure 11.14 for the location).

8.28 OVERVIEW

A review of the surviving earthquake-related inscriptions preserved in historical monuments and stored in museums reveal important data about the prehistoric and historical seismicity of the area. In the case of the ca. 1263 BCE, ca. 1170 BCE, ca. 680–669 BCE, ca. 647 BCE, CE 1319, and 1336 earthquakes, the inscriptions are the only surviving contemporary documents that relay information about the earthquakes. The engravings also reveal the first ever recorded earthquake in the Zāgros fold-and-thrust mountain belt that occurred in ca. 1263 BCE as documented in the Assyrian tablet. This evidence has pushed the lower boundary of the recorded historical Zāgros earthquakes from the seventh century CE to 1263 BCE in a fold-and-thrust mountain belt where paleoseismic trench studies cannot help because the majority of seismic faults are blind reverse faults. The inscriptions show information about the severity of the shocks, the damages to palaces, and the duration of earthquake sequences. They also mention that the price of land dropped after earthquakes and that the people performed earthquake rituals requesting prosperity from the gods. These rituals are still being performed in most societies, especially in the Moslem world.

Part Two

Dynamic Phenomena Associated with Earthquakes on the Iranian Plateau

In Part One, we review the effects of destructive earthquakes on humans and their cultures, and we show that, despite centuries of experience with earthquakes and modern developments in science and engineering, people on the Iranian plateau continue to experience unfathomable loss of life and the destruction of buildings and infrastructure built in active fault zones. In Part Two, we present original information regarding the long-term seismicity and active faulting associated with medium- to large-magnitude earthquakes since ca. 280 BCE. Seismologists must review information from such a long period because short-period teleseismic data cannot alone guide a study of the frequency of earthquakes in the past and the future. In order to assess earthquake hazards, a much longer period of seismic and active fault data is needed if one wishes to find out when and where earthquakes happened in the past.

We review the basic principles of active tectonics, earthquake seismology, earthquake geology, and geomorphology, with an emphasis on field evidence and macroseismic information provided by historical coseismic surface faulting and the reassessment of seismic parameters for the last twenty centuries. The region covered by this book is an exceptional orogen that can help us develop a long-term perspective on seismicity and active faulting patterns. We can then apply this perspective to other earthquake-prone areas, and the described accounts of long-term earthquake history and coseismic surface faulting can contribute to our scientific understanding of this phenomenon, with implications for engineering. As a result, this study is important for the mitigation of earthquake risk throughout the world.

Developments in Earth Surface Processes, Vol. 17. http://dx.doi.org/10.1016/B978-0-444-63292-0.00019-3

Chapter 9

Active Tectonics and Geologic Setting of the Iranian Plateau

If the Earth opens up her mystery,
Will reveal her beginning and end.

Shāhnāmeh, Ferdowsi Tusi (1010); Story of Nushzād with Kasrā, 38635, v. 4, p. 482, ed. Joneidi (2008)

The Iranian plateau is an extensive active crustal deformation and seismic activity zone located between the stable Arabian and Eurasian plates. Its present high elevation, active deformation, and seismicity with complex interactions of active thrusts and strike-slip faults (Figure 9.1) are caused by the driving convergence forces of the plates. The plateau is characterized by different tectonic units with inherited structures organized in diverse directions that have undergone long and complicated plate tectonic evolution since the Late Proterozoic Era. The plateau consists of a composite system of collision-oblique transpressive fold-and-thrust mountain belts with active reverse and strike-slip faulting, range-and-basin terrains, active subduction zones, recent volcanic activity, variable crust thicknesses and rigidity, and relatively stable aseismic blocks of different dimensions with low topographic relief and nearly flat areas. The plateau is generally covered by deserts (Stocklin, 1968a, 1974, 1977; Berberian and Berberian, 1981; Berberian and King, 1981a,b; Berberian et al., 1982; Berberian, 1983a,b,c, 1984a,b, 1989).

9.1 TECTONIC CONTEXT: STRUCTURAL PROVINCES OF THE IRANIAN PLATEAU

The present political boundaries of Iran cover an area of approximately 1,648,000 km^2 in the center of the Iranian plateau with various tectonic and topographic features of active mountain belts resulting from a juvenile continental collision with intense seismic activity, various suture zones, and quasi-rigid blocks. Several major structural provinces found on the Iranian

☆"To view the full reference list for the book, click here."

Developments in Earth Surface Processes, Vol. 17. http://dx.doi.org/10.1016/B978-0-444-63292-0.00009-0

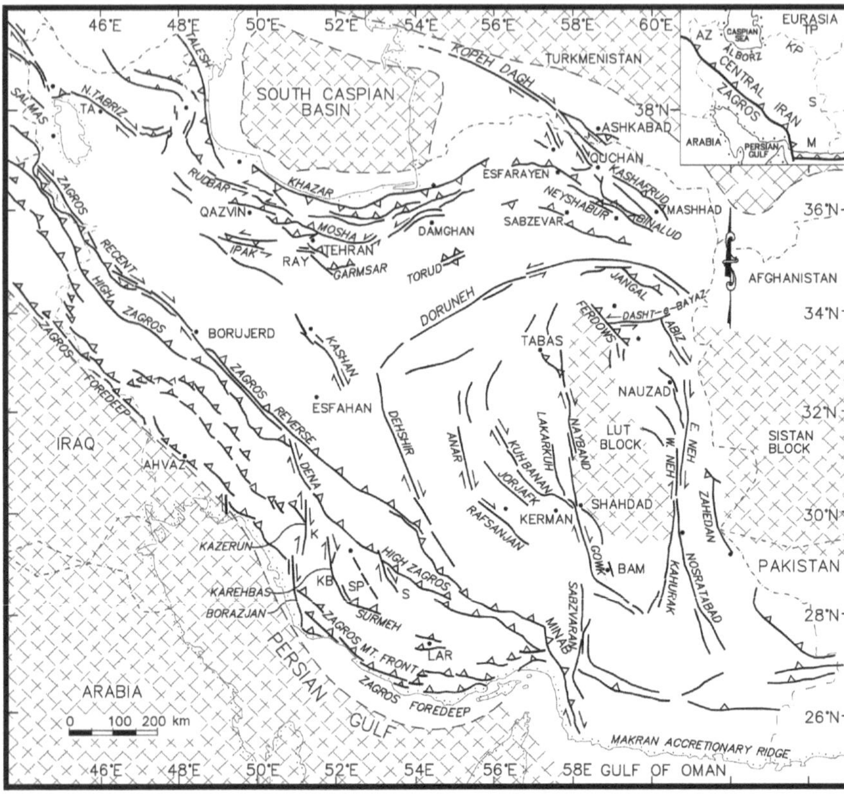

FIGURE 9.1 Major active faults of Iran (names in italics). Reverse faults shown with teeth on hanging-wall. Strike-slip faults shown with arrows. Faults without teeth or arrows: sense of slip unknown. Inset top right: Map of Iran showing boundary with Arabian plate (line with teeth). AZ, Āzarbāijān; KP, Kopeh Dāgh; M, Makrān; S, Sistān suture zone. TP, Turān Plate. Rigid blocks are cross-hatched. *Modified from Berberian (1976e, 1981a, 1995, 1997a, 2005) and Berberian and Yeats (1999, 2001).*

plateau are characterized by a varied structural history, style, and evolution (Figure 9.2). The crusts of diverse structural provinces, ranging in age from the Late Neoproterozoic Era to the Tertiary Period, have undergone numerous collisional orogenies, with the final orogeny having occurred during the Late Neogene Era subsequent to the complete closure of the Neo-Tethyan Ocean during the Early Tertiary Period.

9.1.1 The Zāgros Fold-and-Thrust Mountain Belt (The High-Zāgros with Neo-Tethys Ophiolite Nappes, Zāgros Simple Fold, and Zāgros Foredeep)

The Zāgros is a more than 1500-km-long NW–SE trending mountain belt of extensive crustal deformation and intense seismic activity with pronounced

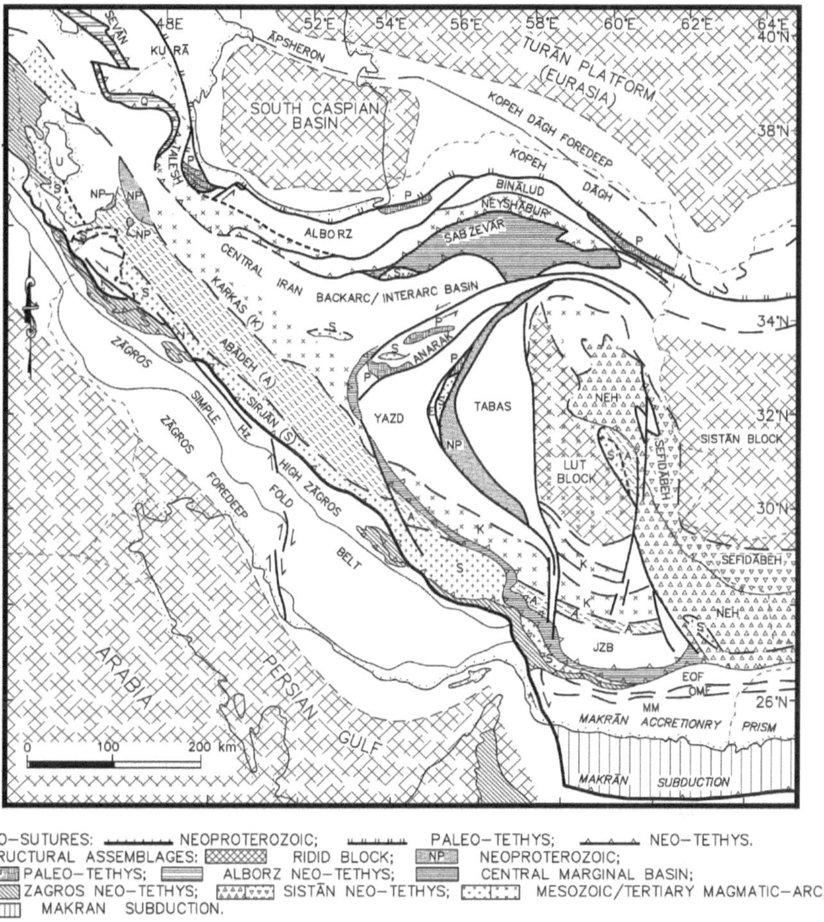

FIGURE 9.2 Simplified tectonic units of the Iranian Plateau with the main subdivisions and geosutures. A, Ābādeh; E, Middle Eocene Chāpedoni complex; EOF, Eocene–Oligocene flysch (Makrān); JZB, Jāz Muriān Basin; K, Karkas; MM, Miocene molasse (Makrān); OMF, Oligocene–Miocene flysch (Makrān); NP, Neoproterozoic; P, Paleo-Tethys nappe; Q, Qaradāgh (Arasbārān); S, Sirjān; U, Lake Urumiyeh. *Modified from Berberian and King (1981a), Berberian and Berberian (1981), Berberian et al. (1982), Berberian (1983b), and other data cited in the text.*

topography (Figures 9.1 and 9.2). It evolved from the closure of the Neo-Tethys by a complex northeast subduction (present-day geographic directions) underneath the Sirjān belt of the southwestern Central Iranian active continental margin (Stocklin, 1968a; Berberian and King, 1981a,b; Berberian et al., 1982; Berberian, 1983a,b, 1989, 1995). The belt consists of 6–14 km of sedimentary rocks covering the Precambrian metamorphic basement with a thick sequence of Late Upper Vendian–Lower Cambrian Hormoz Salt deposits acting as a décollement zone (James and Wynd, 1965; Stocklin, 1968a,b; Berberian, 1976a, 1995; Berberian and King, 1981a,b).

9.1.2 The Sirjān Variscan-Cimmerian Belt, Mesozoic Magmatic Arc, and Tertiary Forearc

The belt with little topographic relief is a narrow (~200 km) but continuous NW–SE trending continental crust longer than 1600 km in southwest Central Iran, which has SW-verging thrust stacks and nappes. It lies between the Zāgros fold-and-thrust mountain belt in the south and southwest along a number of strongly dismembered and mylonitized ophiolite assemblages and the Karkas Tertiary magmatic arc (Central Iran) in the northeast and can be traced from northwest Iran through the Makrān accretionary wedge in the southeast (Figure 9.2). The belt is composed of numerous Paleozoic and Mesozoic Barrovian-type (including glaucophane-schist) and Abukuma-type metamorphosed volcano–sedimentary (magmatic arc) complexes with obducted Paleozoic and Mesozoic ophiolites along the collision suture lines (Berberian, 1976d, 1977b, 1983b, 1989; Berberian and Alavi-Tehrani, 1977a,b; Berberian and King, 1981a,b; Ghasemi et al., 2002).

9.1.3 The Ābādeh Mesozoic Back-Arc Basin

The Mesozoic narrow back-arc basin is 75 km wide and located between the Sirjān (in the southwest) and Karkas (in the northeast) belts (Figure 9.2).

9.1.4 The Karkas Tertiary Magmatic Arc

The belt (formerly known as Urmia-Dukhtar [sic.]), which is a NW–SE trending magmatic arc with pronounced topographic features, is ~50 km wide and 1800 km long, parallel to the Zāgros Neo-Tethyan suture zone. It lies to the northeast of the Sirjān belt and runs parallel to the Sirjān and the Zāgros belts in the southwest of Central Iran (Figure 9.2). It is composed of extensive and almost continuous Eocene to Miocene calcalkaline Andean-type magmatic activity formed during the subduction of the Zāgros Neo-Tethys oceanic lithosphere under the Sirjān belt of the Central Iranian active continental margin (Berberian and Berberian, 1981; Berberian and King, 1981a; Berberian et al., 1982; Berberian, 1983a,b, 1989).

9.1.5 The Alborz–Tālesh Mountain Belt

The Alborz Mountains of northern Iran form a high arc of mountains with pronounced topography around the southern coast of the Caspian Sea from the Tālesh in the west to their junction with the Kopeh Dāgh in the east (Figure 9.2). It forms part of the northern margin of the Arabian–Eurasian collision zone joining the Lesser Caucasus of Armenia in the northwest to the Paropamisus Mountains of the northern Afghanistan in the east. The structural trend of the Alborz changes from N110°E in the west to N80°E in the east with a marked hinge at about 52.50°E. The range is higher than the Kopeh Dāgh and central Iran, with many summits in the range of 3600–4800 m amsl, culminating in the Quaternary Volcano of Damāvand (5671 m amsl) in the center of the belt. With its restricted width of 60–120 km, the Alborz is extremely steep, with flanks that abruptly join the plains along major thrust faults on both sides.

To the west, the Alborz joins the Tālesh Mountains, a N–S range along the western coast of the Caspian Sea (Figure 9.2). The Tālesh is narrower than the Alborz, with a typical width of <50 km. The range is dominated by N–S folds and thrusts, which swing smoothly into an E–W trend at the northern end of the range where it meets the flat plains of the Kurā Basin (Berberian and King, 1981a; Berberian, 1983a, 1997a). The range contains a thick Paleogene volcaniclastic sequence. To the east, the Alborz joins the Binālud Mountains south of the Kopeh Dāgh (Figure 9.2).

9.1.6 The Kopeh Dāgh Fold-and-Thrust Mountain Belt (Mesozoic Basin) with the Central Kopeh Dāgh Shear Zone

The Kopeh Dāgh forms a NW–SE range of mountains separating the Turān shield (Turkmenistan; part of the stable Eurasia) in the north from the Alborz, Binālud, and northern Central Iran in the south. Structurally, the rocks of the Kopeh Dāgh are distinct from those of Alborz, Central Iran, and Zāgros and belong to the Turān shield (Figures 9.1 and 9.2). The tectonostratigraphic units of the Kopeh Dāgh and the northern Alborz closely represent the Paleo-Tethyan Ocean, as the Alborz became sutured to the Turān shield of Eurasia. The belt is up to 3000 m in elevation, rising 2000 m above the Turkmen plains.

Shortly after the collision of the northern Alborz with Laurasia and elimination of the Paleo-Tethys, the Kopeh Dāgh became the foreland area at the southern Turān platform and underwent extensional deformation and fault-controlled subsidence along the Paleo-Tethyan suture zone. The northeast flank of the Kopeh Dāgh (the Āqdarband inlier) is underlain by the Hercynian basement of the Turān shield. The Turkmenistan borehole data (Lyberis and Manby, 1999) show that the Mesozoic rocks are underlain by Devonian to Early Carboniferous basalts, andesites, and tuffs alternating with coal layers. These rocks are succeeded by 50–850 m of a Triassic tectonic mélange composed of sandstones, chlorite schist, black shale, andesite, and tuffs with a lack of Middle Carboniferous-Permian rocks.

During the Late Triassic–Early Jurassic, the Kopeh Dāgh-Alborz landmass was unconformably covered by the Kashafrud/Shemshak foreland molasse deposits. The basal conglomerate contains detrital rock fragments of volcanics, granite, metamorphics, and Triassic red sandstone and tuffs (Madani, 1977; Berberian and King, 1981a). After a Bathonian lacuna, a Callovian-Late Jurassic carbonate platform adjacent to a deeper marine basinal carbonate environment developed on the northern margin of the Kopeh Dāgh

(Afshar-Harb, 1979; Lasemi, 1995). The Kopeh Dāgh Basin then became shallower, and emergence took place over the area, producing the Early Cretaceous continental red deposits. Hence, the Jurassic–Cretaceous boundary on the Kopeh Dāgh margin is marked by an erosional unconformity possibly corresponding to global marine regression (Afshar-Harb, 1979; Berberian and King, 1981a).

9.1.7 The Central Iranian Range-and-Basin Blocks

Central Iran is a complex mosaic of several microcontinental blocks with little topographic relief and different structural histories once separated by minor marginal ocean basins. It is a triangular-shaped region of the interior Iranian Plateau and is bounded by the marginal fold-and-thrust belts of the Zāgros and Makrān in the south, the Alborz and Kopeh Dāgh in the north, and the Sistān Accretionary wedge in the east (Figure 9.2). The area is dominated by narrow mountain ranges and intermontane depressions. The mountain ranges and the lowlands of Central Iran are separated by deep-seated reverse and strike-slip faults (Figures 9.1 and 9.2).

Some of the Central Iranian blocks contain the Late Neoproterozoic basement (Stocklin, 1968a; Takin, 1972; Berberian and King, 1981a,b; Ramezani and Tucker, 2003; Hassanzadeh et al., 2008; Alirezaei and Hassanzadeh, 2012), and the Gondwanian Paleozoic platform cover sedimentary environment similar to those of the Zāgros, Arabia, and Alborz (Berberian and King, 1981a). Upper Vendian–Lower Cambrian Hormoz Salt (Rāvar/Desu Series in the Kermān area) is exposed in the southern Tabas block (Figure 9.2) and evaporites with volcanic/volcaniclastics and cherty dolomites also developed both in the Tabas and the Yazd Blocks (Berberian and King, 1981a).

Central Iran is composed of the Qom Back-Arc/Interarc Basin, Neyshābur magmatic arc, Anārak Variscan-Cimmerian zone, Yazd Mesozoic Basin, Chāpedoni Mid-Eocene Arc, Bon-e Shuru Late Neoproterozoic magmatic arc, Tabas Mesozoic Basin, Lut Block, and Dehsalm Block (Figure 9.2).

9.1.8 The Sistān Accretionary Prism and Suture Zone

The accretionary prism with little internal topographic relief forms a narrow N–S belt in eastern Iran. Unlike the Alborz and Zāgros Mountains, which broaden and rise in summit elevations toward their central areas, the Sistān Mountains in east Iran are narrowest and lowest in the center of the belt (Figure 9.2). It is a highly deformed forearc basin recording the destruction of an arm of the Neo-Tethys during Senonian–Paleocene times (~89–55 Ma) and consequent collision of the Afghan Hirmand (Hilmand) (in the east) and Lut (in the west) cratonic blocks.

The Sistān accretionary complex is composed of two accretionary prisms: (i) the Late Cretaceous (Senonian) to Eocene (~89–34 Ma) Neh and Ratuk complexes (ii) separated by sediments and volcanics of the pre-Maastrichtian (>71 Ma) Sefidābeh forearc basin. The basin developed as a result of Early Late (?) Cretaceous rifting between the once-connected Lut and Afghan cratonic blocks, but was relatively short-lived, eventually closing as the two blocks converged again in a NE–SW direction (in the present geographic orientation). This convergence was responsible for considerable folding and faulting in pre-Neogene times and for emergence of the area during the Late Eocene to Oligocene (Freund, 1970; Berberian, 1981b, 1983b, 1989; Berberian and King, 1981a; Camp and Griffis, 1982; Tirrul et al., 1983; Berberian et al., 2000a; Brocker et al., 2013).

9.1.9 The Makrān Accretionary Wedge and Subduction Zone

The Makrān E–W accretionary wedge covers the 900-km-long coastal ranges of Iran and Pakistan, with 125–200 km width formed above the Makrān active subduction zone (Figure 9.2). Subduction was probably started during the Paleocene, leading to an initiation of accretion during the Eocene and the modern accretionary prism since the Late Miocene. It is composed of approximately 6000 m of sediments above a low-angle ($\sim 5°$) subduction zone, with the deformation front approximately along the 3000-m depth contour in the north Oman Sea (Jacob and Quittmeyer, 1979; Berberian and King, 1981a; White, 1982; White and Louden, 1982; Palvis and Bruhn, 1983; Harms et al., 1984; McCall, 1985, 1997, 2003; Platt et al., 1985, 1988; Laane and Chen, 1989; Byrne et al., 1992; McCall and Kidd, 1992; Carbon, 1996; Ravaut et al., 1997; Kopp et al., 2000; Hosseini-barzi and Talbot, 2003; Smith et al., 2013).

The Makrān accretionary wedge is composed of the Inner Makrān Ocean Ophiolite Zone (Jurassic to Paleocene), the Sirjān microcontinent sliver (Paleozoic metamorphic basement with Mesozoic Carbonate Forearc), the MakrānTrench Zone Ophiolitic Mélange (Mesozoic–Early Paleogene subduction zone), the Eocene–Oligocene Flysch Zone, the Oligocene–Miocene Flysch Zone, the Miocene Neritic zone, and the Miocene–Pliocene Neritic–Molassic Sediments (McCall, 1985, 1997; McCall and Kidd, 1992).

9.2 TECTONIC EVOLUTION OF THE IRANIAN PLATEAU

The broad crust of the Iranian Plateau, stretching from the Persian Gulf to the Caspian Sea, covers the complete orogen of extensive crustal deformation, complex tectonic history, and topographic contrasts between the converging Arabian and Eurasian plates (Figure 9.2). It is an agglomeration of diverse continental fragments with inherited structural trends that separated from the Gondwanian Passive margin, traveled thousands of kilometers, and accreted to the margins of Laurasia/Eurasia during different collisional orogenies since the Late Neoproterozoic Era. Paleo- and active deformations and topography are distributed irregularly over a very broad area.

Pronounced deformation and topographic contrast are mainly concentrated along the northern and southern marginal mountain belts (the Tālesh, Alborz, Binālud, Kopeh Dāgh, and Zāgros) surrounding quasi-aseismic blocks with low topographic relief. It also includes the Karkas Tertiary magmatic arc with high topography in between. Less internal relief (the Sirjān Mesozoic magmatic arc, Ābādeh Mesozoic back-arc basin, Qom Tertiary back-arc basin), and low relief areas such as the Central Kavir and the Lut desert are located in Central Iran (Stocklin,1968a, 1974, 1977; Berberian and King, 1981a; Berberian et al., 1982; Berberian, 1983a,b, 1989) (Figure 9.2).

More data are required to unravel the complex tectonic history of the Iranian Plateau, however, we try to simplify the events. Tectonic evolution of the Iranian mountain belts may be divided into two main phases, namely Pre-Tethyan (Late Neoproterozoic to Permian) and Tethyan (Permian to Present). The Pre-Tethyan evolution was governed by: (i) the late Neoproterozoic–earliest Paleozoic orogeny; (ii) the Variscan/Caledonian; and (iii) Hercynian movements. The Tethyan evolution took place during two partly overlapping events in time: (i) the Paleo-Tethyan (Permian to Jurassic) and (ii) the Neo-Tethyan (Jurassic to the Present) (Berberian, 1983b, 1989).

The Late Neoproterozoic–earliest Paleozoic orogeny was documented in a few stable areas on the Iranian Plateau (Stocklin, 1968a,b; Berberian and King, 1981a,b; Ramezani and Tucker, 2003; Hajialioghli et al., 2007a,b, 2011; Hassanzadeh et al., 2008; Horton et al., 2008; Moazzen and Oberhänsli, 2008; Saki et al., 2008a,b, 2010, 2012; Karimpour et al., 2009, 2010, 2011; Moazzen et al., 2009, 2013; Saki, 2010, 2011).

The Late Neoproterozoic movements were followed by a widespread rifting phase over the entire area. During the Paleozoic era, the whole region was a relatively stable continental platform with epicontinental shelf deposits that lacked major magmatism (except for rift volcanism) or folding and metamorphism, excluding some Variscan wedges trapped along the Paleo-Tethyan suture zone in the northern Alborz, Anārak, and the Sirjān belts (Stocklin, 1974; Berberian and King, 1981a,b; Berberian, 1989; Zanchetta et al., 2009; Zanchi et al., 2009).

The Paleozoic quiet platform condition was disturbed by the widespread late Emsian–Eifellan (regression during the Variscan/Caledonian movements?) and the Late Carboniferous (regression during the Hercynian movements?) hiatuses. Some Variscan core tectonic slices were trapped in the Tālesh Mountains of the northwest Alborz (Zanchetta et al., 2009, 2013), Nakhlak-Anārak arc-trench belt (Bagheri and Stampfli, 2008; Zanchi et al., 2009), and the Sirjān belt (Ghasemi et al., 2002).

The Paleozoic extension with Cambrian to Permian volcanism (Berberian and Berberian, 1981; Berberian and King, 1981a) and Early Carboniferous Mishu mafic–ultramafic MORB magmatism (Saccani et al., 2013) led to the Iranian continental breakup and the birth of the Paleo-Tethys ocean governed by the main north-dipping (present geographic orientation) subduction zone of the Paleo-Tethyan oceanic crust beneath southern Laurasia during the Permian–Triassic interval (Davies et al., 1972; Stocklin, 1974; Berberian and King, 1981a; Berberian, 1983a,b, 1989; Horton et al., 2008). Two segments of the Paleo-Tethyan suture belt exist in Iran (Figure 9.2): (i) the Tālesh, Gorgān, and Mashhad nappes in the north (Berberian et al., 1973; Berberian, 1983a; Zanchetta et al., 2009, 2013; Zanchi et al., 2009); and (ii) Nakhlak-Anārak arc-trench belt of Central Iran (Bagheri and Stampfli, 2008; Balini et al., 2009; Zanchi et al., 2009; Kargaranbafghi et al., 2012).

The entire imbricate packages of the Paleo-Tethys assemblages (Tālesh, Gorgān, and Mashhad) are thrust southward onto the rocks of the Alborz and northern Central Iranian platform (ranging from Cambrian to Middle Triassic). The entire deformed ensemble, together with the Paleo-Tethyan foreland and hinterland, are in turn unconformably overlain by the Rhaetic–Liassic Shemshak Formation (Berberian, 1983a,b, 1989; Wilmsen et al., 2009; Zanchetta et al., 2009, 2013; Zanchi et al., 2009). Therefore, the Paleo-Tethys must have closed along the Tālesh–Khazar (South Caspian)–Mashhad line sometime during the Triassic to Liassic interval (Figure 9.2).

The northward motion of the Central Iranian continental fragment(s) was responsible for: (i) the main north-dipping subduction zone of Paleo-Tethys beneath Laurasia; (ii) the closure of the Paleo-Tethys along the Tālesh, Khazar, Mashhad line in northern Alborz; (iii) the collision of the Nakhlak-Anārak arc-trench system with the Yazd block; and (iv) the apparently simultaneous opening of the Neo-Tethys in the south, with at least two branches: (i) the South Alborz Neo-Tethys (Sevān-Qaradāgh/Arasbārān-south Alborz-Sabzévār) in the north and (ii) the Zāgros Neo-Tethys in the south (Figure 9.2).

During this time, Central Iran internally disintegrated and gave birth to the intra-Central Iranian narrow marginal oceans. The initial Neo-Tethyan rifting commenced possibly during the Permian and is indicated by the switch from the generally quiet platform detritics to open marine sedimentation and facies–thickness changes across normal faults. The Neo-Tethyan Ocean opening in the Iranian foreland took place largely during the Late Triassic–Early Jurassic period (Berberian and King, 1981a; Berberian, 1983a,b, 1989).

While the southern Alborz underwent weak-to-moderate Paleozoic–Triassic deformation with conformable Jurassic sediments covering the older rocks, the northern Alborz showed: (i) intense compressional deformation resulting from collisional orogeny with metamorphism and crystallization, complex nappe structures and dominant southward vergence and retrochariage; (ii) magmatism, Triassic dykes, and ophiolites; (iii) Pb–Zn mineralization; and (iv) pronounced angular unconformity with postorogenic Norian foreland molasse deposits of the Shemshak and Kashafrud Formations (Figure 9.2).

Throughout Jurassic and Cretaceous extension dominated the entire area. The Kopeh Dāgh marginal basin (without ophiolite) was developed during medial Jurassic to Cretaceous (Berberian and King, 1981a; Berberian, 1983a,b, 1989). The South Alborz Neo-Tethys, possibly a marginal basin

(Sevān-Qaradāgh/Arasbārān-South Alborz-Sabzévār marginal Neo-Tethyan oceanic arm) in the north (Berberian et al., 1981; Berberian 1983b, 1989), separated the Central Iranian platform from the Alborz–Binālud (southern Eurasia); whereas the Zāgros ocean in the south separated the former from the Gondwanian foreland (the present Zāgros belt). The central marginal Neo-Tethys in Central and east Iran also formed during this process.

A major Neo-Cimmerian (Jurassic–Cretaceous) movement strongly affected most of the Central Iran. After this movement, the Early Cretaceous Biābānak flysch (slightly metamorphosed; GSI, 1972b) deposited along the eastern margin of the Yazd block and the western margin of the Middle Eocene Chāpedoni metamorphic complex, Posht Bādām Triassic arc, and Bon-e Shuru Neoproterozoic magmatic arc (Ramezani and Tucker, 2003; Figure 9.2).

The South Alborz Neo-Tethys Ocean was later eliminated during the Late Cretaceous–Early Paleocene contraction movements mainly by the north–northeast-dipping subduction beneath the Caucasus–Binālud ranges with the formation of the Sabzévār–Neyshābur–Mashhad Andean-type magmatic arc (Berberian et al., 1973, 1981; Stocklin, 1974, 1977; Berberian and King, 1981a; Berberian, 1983a,b, 1989; Guest et al., 2006a,b).

Subduction of the Zāgros ocean beneath Central Iran formed the Jurassic–Cretaceous Alvand–Bazmān Andean-type magmatic arc along the Sirjān belt and the Nā'in-Bāft-inner Makrān marginal basin (with ophiolites) during Jurassic to Paleocene (Figure 9.2). During the late Cretaceous–Paleocene contraction, widespread ophiolite nappes were emplaced onto the: (i) Sevān-Qaradāgh/Arasbārān-Sabzévār, (ii) Nā'in-Bāft-Inner Makrān, (iii) southern Makrān, (iv) Sirjān, and (v) High-Zāgros-Oman belts.

While the northern Alborz underwent weak-to-moderate Mesozoic deformation with subhorizontal Jurassic and Cretaceous deposits without any Jurassic or Cretaceous unconformities and lacked widespread Paleocene-basal Eocene conglomerates, the southern Alborz showed: (i) intense, deformed Jurassic and Cretaceous deposits, with Lower–Upper Jurassic and Lower–Upper Cretaceous unconformities showing south-verging folds with north-dipping thrusts; (ii) Campanian limestone unconformably overlies Cenomanian–Turonian conglomerate; (iii) rapid facies changes in the Cretaceous deposits; (iv) whole deformed and uplifted sequences of the southern Alborz unconformably covered by the Paleocene–Lower Eocene basal conglomerate of the Fājān and Ziārat Formations; and (v) with magnetic-anomaly patches along the southern Alborz (Figure 9.2).

The Middle Eocene collision of the Lut and the Afghan blocks in eastern Central Iran resulted in widespread ophiolite emplacement, west-verging imbrications, and folding of the East Iranian flysch along the Sistān suture zone with the Senonian–Eocene Sefidābeh forearc basin and the Neh and Ratuk complexes with ophiolites (Tirrul et al., 1983). The Middle Eocene collision along the Sistān Suture seems to be contemporaneous with the Middle Eocene Chāpedoni metamorphic complex farther to the west (Ramezani and Tucker, 2003).

During the Paleogene, synchronous Andean-arc and back-arc spreading stages were developed above the subduction zone of the Zāgros Neo-Tethys ocean in Central Iran. This is characterized by calcalkaline magmatism along the Kakas belt (Berberian et al., 1982) and alkali basalt volcanism in northwest and northern Iran. Apparently, the intraplate block faulting and predominantly alkali basaltic volcanism mainly developed in northern Central Iran and the northwestern Alborz (Berberian and King, 1981a; Berberian et al., 1982; Berberian, 1983a,b, 1989; Ballato et al., 2011; Verdel et al., 2011), leading to the formation of two marginal basin complexes (back-arc spreading stage above the Zāgros subduction zone): (i) the South Caspian marginal basin in the north with oceanic crust; and (ii) the Oligocene–Miocene Lower Red-Qom Formation back-arc/inter-arc basin in northern Central Iran without oceanic crust (Berberian and Yasini, 1983; Berberian. 1989).

Deposition of the Middle–Late Oligocene to Early Miocene reefal limestone of the Āsmāri Formation in the Zāgros (James and Wynd, 1965) and the Qom Formation in northern Central Iran (Bozorgnia, 1966) indicates that the Iranian Plateau was close to sea level during the Burdigalian stage about 15 Ma ago. The diachronous deformation of the High Zāgros apparently started during the Messinian (Late Miocene, about 8 Ma; Homke et al., 2004) creating the present compressional tectonic regime of the Iranian mountain belts.

Based on published paleomagnetic and geologic data since 13 Ma, Austermann and Iaffaldano (2013) stated that the data indicate an approximately 30% slowdown of the Arabia–Eurasia convergence from 5 Ma to present day due to orogenic uplift along the Iranian Plateau. Continued east–northeast convergence of the continental blocks resulted in thickening and shortening of continental crust by active folding, reverse and strike-slip faulting, elevation of the Iranian fold-thrust mountain belts, subduction of the Makrān, and the present seismicity of the Plateau.

9.3 THE NEOGENE REGIONAL CHANGE IN KINEMATICS

Comparison of the direction of maximum horizontal shortening deduced from the Neogene axes of folding with the horizontal component of the compressional axes deduced from focal mechanism solutions in the 1970s (Berberian, 1976e,f), and more recently with GPS horizontal velocity data (Nilforoushan et al., 2003; Vernant et al., 2004), clearly show a post-Neogene regional change in kinematics (Figures 9.3 and 9.4). Furthermore, studies of a complex tectono-sedimentary evolution support the idea of major changes in

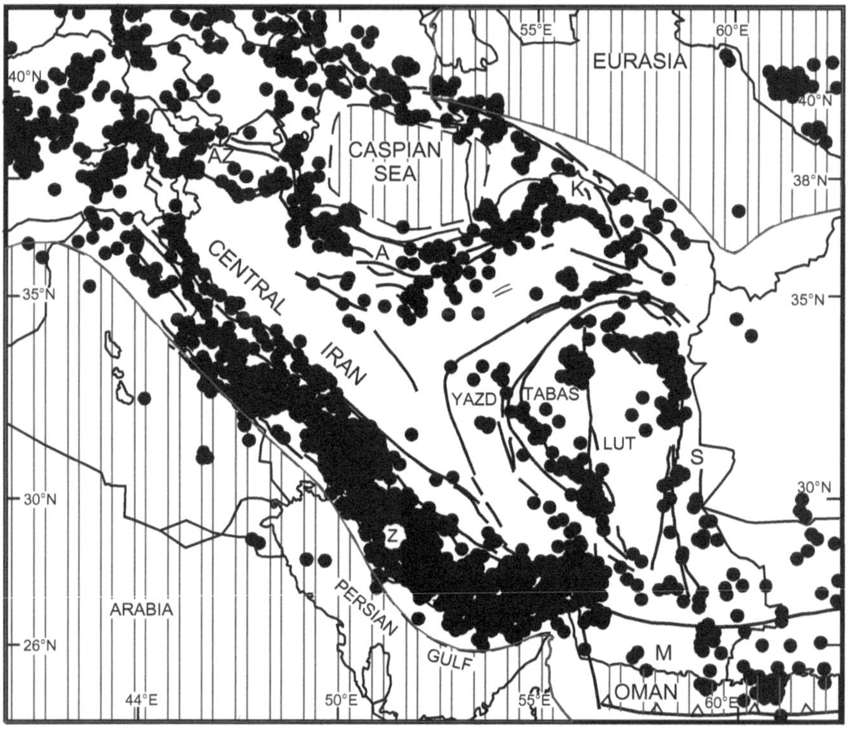

FIGURE 9.3 1964–1998 epicenter map of Iran (Engdahl et al., 1998). Most seismicity and active deformation occur in the marginal mountain belts of the Zāgros (Z: in the southwest and south), Alborz (A: in the north), Kopeh Dāgh (K: in the northeast), Makrān (M: in the southeast), and Sistān (S: in the east).

the kinematic regime (Axen et al., 2001; Jackson et al., 2002; Allen et al., 2003; Guest et al., 2006a,b; Ritz et al., 2006; Zanchi et al., 2006; Ballato et al., 2008, 2011, 2013; Landgraf et al., 2009; Solaymani Azad et al., 2011).

This major change might have resulted from oblique compression due to the combination of the Arabia–Eurasia north–northeast convergence and northwest motion of the rigid South Caspian Basin (Jackson et al., 2002). The Neogene northeast shortening direction (Berberian, 1976e,f; see Figure 9.4) was changed to a nearly north–northeast direction, a more orogen-perpendicular orientation (oblique shortening and transpressional deformation). Ritz et al. (2006) suggested that the beginning of the northwestward motion of the South Caspian Basin to Eurasia and/or its clockwise rotation took place in the Pleistocene. This has caused modern-day oblique strain partitioning into longitudinal subparallel thrusts and left-lateral strike-slip motion on separate fault systems in the Alborz (Berberian et al., 1992; Jackson et al., 2002; Berberian and Walker, 2010).

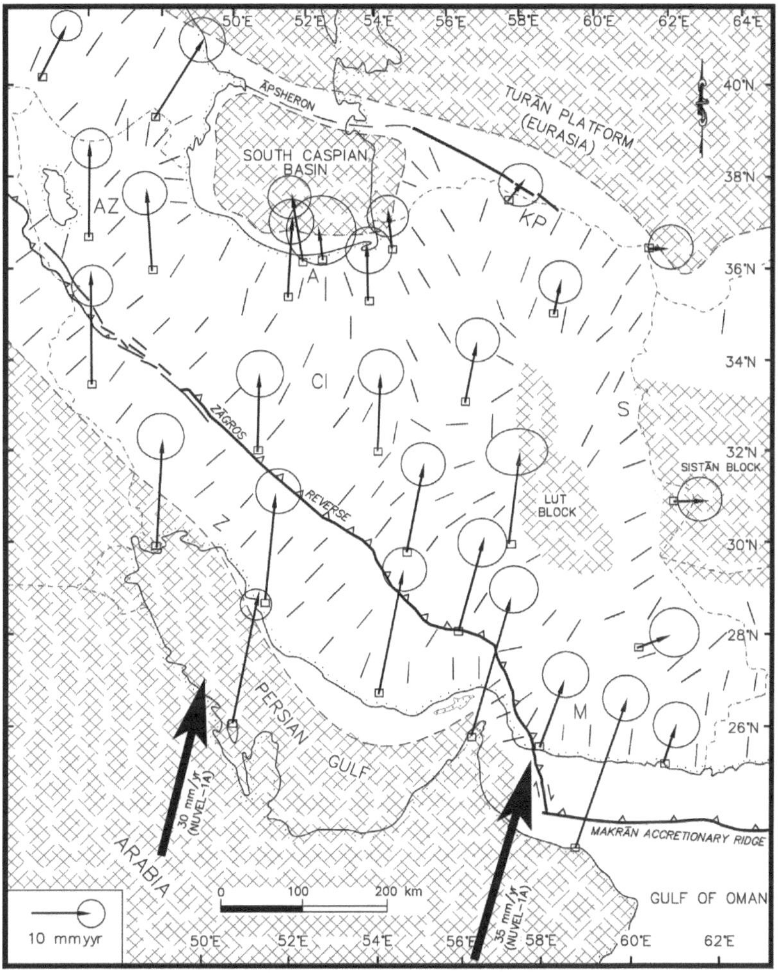

FIGURE 9.4 GPS horizontal velocities and their 95% confidence ellipse in a Eurasian-fixed reference frame for the period 1999–2001 (Nilforoushan et al., 2003; Vernant et al., 2004) on a background of post-Neogene maximum shortening direction (short bars) deduced from Berberian (1976e). The GPS velocities decrease to zero in both north of Kopeh Dāgh in the northeast and east of Sistān in the east, suggesting that the Arabia–Eurasia convergence is accommodated within the political boundaries of the modern Iran. The figure also shows Neogene regional change in kinematics and how the NNE motion of Arabia relative to Eurasia is absorbed in the Iranian Plateau. A, Alborz; AZ, Āzarbāijān; CI, Central Iran; KP, Kopeh Dāgh; M, Makrān; S, Sistān; Z, Zāgros.

9.4 ACTIVE TECTONICS

Active deformation of the Iranian Plateau, which involves intracontinental shortening almost everywhere on the Plateau with subduction of the Oman oceanic lithosphere along the Makrān subduction zone, is spread in a 1000-km wide zone from the Persian Gulf and Makrān in the south to the rigid South Caspian Basin in the north (Figures 9.1, 9.3 and 9.4). The active deformation, which is the result of the continued convergence between the Arabian and Eurasian plates, is mainly accommodated by: (i) the major fold-and-thrust mountain belts of the Zāgros, Tālesh-Alborz, and Kopeh Dāgh (Figures 9.1 and 9.2) with pronounced topography; (ii) along strike-slip and thrust faults of the Central Iran surrounding relatively aseismic blocks of the Lut and Afghan Helmand blocks with less topographic relief; and (iii) along the Makrān subduction zone of southeast Iran and Pakistan (Figures 9.3 and 9.4).

The major patterns of active faulting in Iran (Figure 9.1), which will be discussed in the following chapters, are as follows:

- **(i)** Longitudinal reverse faulting in the Zāgros (in the south and southwest), Tālesh (west of the Caspian Sea), Alborz (south of the Caspian Sea), Binālud and Kopeh Dāgh (NE), Kuhbanān Mountains (SE), Sistān suture zone (E), and Makrān acrretionary prism (SE).
- **(ii)** NW–SE reverse faulting in Central Iran.
- **(iii)** Transverse right-lateral strike-slip faulting in central Zāgros.
- **(iv)** Range-parallel right-lateral strike-slip faulting along the northwestern boundary of the Zāgros and Central Iran (the Zāgros Main Recent fault (ZMRF)).
- **(v)** NW–SE Central Kopeh Dāgh right-lateral Shear Zone (Central Kopeh Dāgh Shear Zone (CKDSZ); in the NE).
- **(vi)** NW–SE Western Āzarbāijān right-lateral Shear Zone (WASZ; in the NW).
- **(vii)** N–S Afriz Shear Zone south of the Ferdows fault (E).
- **(viii)** N–S right-lateral strike-slip faulting along the western (Nāyband, Gowk, Sabzvārān) and eastern (Neh, Kahurak) margins of the aseismic Lut block.
- **(ix)** NNW–SSE right-lateral strike-slip faulting in Central Iran (Kāshān, Dehshir, Rafsanjān, Jorjāft, Kuhbanān).
- **(x)** E–W left-lateral strike-slip faulting north of the Lut block (Dasht-e Bayāz, Doruneh).
- **(xi)** Longitudinal left-lateral strike-slip faulting in the Alborz (Rudbār, Āstāneh, Dāmghān).
- **(xii)** Active Makrān subduction zone (SE).

9.4.1 The Present-Day GPS-Based Deformation

The recent GPS data convergence rate in the Iranian Plateau suggests a converging rate at about 25 mm/year at a longitude of 56°E in a NNE-SSW direction (Nilforoushan et al., 2003; Vernant et al., 2004; Figure 9.4). This convergence rate is about 10 mm/year lower than the earlier global plate tectonic models (the NUVEL-1A) using a combination of Afro-Eurasia and

Arabia–Africa motions based on global seafloor spreading and earthquake fault-slip vectors (DeMets et al., 1990, 1994; Jackson, 1992; Jestin et al., 1994; Chu and Gordon, 1998; Sella et al., 2002; Kreemer et al., 2003; McClusky et al., 2003). It should be noted that the available Iranian GPS data are sparse (28 sites in 1999–2000 in an area of 1,648,000 km^2) and cannot be used to determine the precise deformation style throughout the Plateau or along individual faults. However, the GPS measurements coupled with earthquake data and geological fault-slip measurements have advanced our understanding of the active deformation on the Plateau (Djamour et al., 2010, 2011; Mousavi et al., 2013).

The GPS, seismicity distribution, and active geomorphology data show that the active compression and shortening in the Iranian Plateau is accommodated by distributed active folding and complex faulting in high fold-and-thrust mountain belts of the Zāgros (in the SW), Alborz (north), Kopeh Dāgh (NE), Central and East Iranian ranges and basins, and the Makrān accretionary wedge in the southeast (Figure 9.4). Active tectonics and seismicity appear to be concentrated along the boundaries of relatively aseismic stable blocks in Central Iran and the South Caspian Basin (Figures 9.3 and 9.4). The GPS velocities of points relative to Eurasia decreases to zero in the surrounding regions of the Iranian Plateau to the north (the Turān plate/Turkmenistan) and the east (the Farāh/Helmand block of Afghanistan and Pakistan). These regions appear to be quasi-aseismic with low topography and are possibly deforming at much lower rates at the present day (Nilforoushan et al., 2003).

The present-day GPS-based deformation of Iran (Figure 9.4) shows that approximately 10 mm/year shortening is occurring in the Zāgros fold-and-thrust mountain belt of southwestern Iran; the rest is partly distributed in the Alborz and the Kopeh Dāgh (~8 mm/year) and east Iran. The GPS-derived velocities decrease to zero at the northern and eastern borders of Iran (Nilforoushan et al., 2003; Vernant et al., 2004; Djamour et al., 2010, 2011; Mousavi et al., 2013).

9.4.1.1 The Zāgros Fold-and-Thrust Mountain Belt

Nearly half of the GPS-derived convergence rate between the Arabian plate and the Central Iran (10 ± 4 mm/year) is accommodated in the Zāgros by north–south crustal shortening and strike-slip faults oblique to the strike of the belt (Figure 9.4). The main shortening direction ranges from N7°E in the southeast to N3°W in the northwest. The convergence rate is decreasing from 9 ± 2 mm/year in the southeastern Zāgros to 7 ± 2 mm/year near the longitude of Kāzerun in central Zāgros and to 4.5 ± 2 mm/year in the northwest (Tatar et al., 2002; Nilforoushan et al., 2003; Vernant et al., 2004). Slip vector directions derived from the focal mechanism data of the Zāgros thrust earthquakes show an angle of 35–40° to the east of the GPS-derived vectors, suggesting partitioning of strain between reverse and strike-slip faults in the Zāgros (Jackson, 1992; Berberian, 1995; Maggi et al., 2000a,b; Talebian and Jackson, 2002).

The Zāgros Main Recent fault (Tchalenko and Braud, 1974; Tchalenko et al., 1974b; Berberian, 1976a, 1995) is an active right-lateral strike-slip fault of about 640 km length, which more or less follows the NW–SE trend of the Main Zāgros reverse fault (the Neo-Tethyan geosuture) separating the Central Iranian range-and-basin province to the northeast from the Zāgros fold-and-thrust belt to the southwest (Figure 9.1). It marks an abrupt cut-off of intense seismic activity of the Zāgros (Figure 9.3) adjacent to the less seismic Sirjān belt of the southwestern Central Iran (Berberian, 1995; Maggi et al., 2000a,b; Talebian and Jackson, 2002). A right-lateral offset of a geological marker bed along the Nahāvand and Dorud segments of the Zāgros Main Recent Fault was reported by Gidon et al. (1974). Later, Talebian and Jackson (2002) reported a right-lateral offset of 50–70 km and estimated a horizontal slip rate of 10–17 mm/year. By using GPS data, Vernant et al. (2004), arrived at slip rate of 3 ± 2 mm/year. Walpersdorf et al. (2006) suggested a horizontal slip rate of 4–6 mm/year with 3–6 mm/year perpendicular shortening. Later, based on his three-dimensional mechanical modeling using GPS data, Nankali (2011) arrived at 2.3 mm/year horizontal slip rate. Finally, Alipoor et al. (2012) suggested a slip rate of 1.6–3.2 mm/year horizontal slip rate based on 16-km displacement along the fault.

9.4.1.2 The Transition Zone Between the Zāgros and the Makrān

The GPS measurements in this complex transition zone (Figure 9.4), with a mixture of shortening and N–S right-lateral strike-slip faulting, indicates N–S convergence between Oman and Central Iran of about 11 ± 2 mm/year (Nilforoushan et al., 2003; Vernant et al., 2004; Peyret et al., 2009).

9.4.1.3 The Makrān Subduction Zone

Most of the GPS-derived 30 mm/year shortening produced by Arabian–Eurasian convergence is accommodated in the Makrān subduction zone. GPS data shows velocities of 25 ± 2 mm/year near the Strait of Hormoz (Figure 9.4). The shortening rates for the Gulf of Oman range from between 11 ± 2 mm/year near Jāsk port in the west Makrān (in the transition between the Zāgros and Makrān), and $19.5 + 2$ mm/year in the central Iranian Makrān coast, where the present-day seismic activity (Figure 9.3) is quite low (Nilforoushan et al., 2003; Vernant et al., 2004) and most of the seismic activity is related to the bending of the subducting slab at intermediate depths (Laane and Chen, 1989; Byrne et al., 1992; Peyret et al., 2009; Barnhart et al., 2014).

9.4.1.4 The Central Iran

Derivations from coherent behavior of Central Iran, with internal deformation less than 2 mm/year, are smaller than $\sim 10\%$ of the overall Arabia–Eurasia convergence, indicating that Central Iranian block can be regarded as composed of quasi-rigid blocks (Sirjān, Karkas, Lut; Figures 9.3 and 9.4) moving together and transmitting motion to the north (Nilforoushan et al., 2003; Vernant et al., 2004).

9.4.1.5 Eastern Iran

The GPS data near the eastern Iranian border, east of longitude 61°E, show that the displacement rate is low (Figure 9.4), indicating that little deformation is taking place in this area near the Afghan block (Nilforoushan et al., 2003; Vernant et al., 2004), and the seismicity (Figure 9.3) abruptly decreases (Jackson and McKenzie, 1984). The GPS-derived velocity from the Zābol area in eastern Iran with respect to the Central Iran is 16 ± 2 mm/year of N–S right-lateral shear between Central Iran and Afghan Hirmand (Hilmand) block (Figure 9.4), and the deformation is mostly accommodated by N–S right-lateral strike-slip faulting (Figure 9.1) along the western (8 mm/year) and eastern (8 mm/year) sides of the Lut stable aseismic block (Nilforoushan et al., 2003; Vernant et al., 2004). These faults with active geomorphology have shown seismic activity in the past (Freund, 1970; Mohajer-Ashjai et al., 1975; Berberian and Yeats, 1999; Berberian et al., 1999, 2000a,b, 2001; Walker and Jackson, 2002; Berberian, 2005). To the north of the Lut block, the shearing is taken up by both E–W left-lateral and N–S right-lateral strike-slip faulting (Figure 9.1) with clockwise block rotation (Freund, 1970; Berberian and Yeats, 1999, 2001; Walker and Jackson, 2004; Berberian, 2005; see also Chapter 16).

9.4.1.6 The Alborz and the South Caspian Basin

The Alborz Mountains in northern Iran accommodate shortening between the Central Iran and the South Caspian Basin (Figure 9.4). Nearly, half of the 20-mm/year GPS-derived N–S convergence between Arabia and Central Iran is accommodated in the Alborz Mountains. The N–S shortening in central Alborz is 8 ± 2 mm/year (Nilforoushan et al., 2003; Vernant et al., 2004). The shortening rate absorbed by the Alborz and the South Caspian Basin is 14 ± 2 mm/year (Vernant et al., 2004; Jackson et al., 2002).

The GPS data (Djamour et al., 2010) indicated 2.0 ± 1 mm/year of left-lateral displacement along the Āstāneh fault, west of Dāmghān, where Hollingsworth et al. (2010) suggested 1.7–2.5 mm/year slip.

More recent GPS data suggest that the South Caspian Basin is moving up to ~7 mm/year relative to Eurasia at an azimuth of 317°N, constraining a rigid block rotation around an Euler pole that is farther away than previously thought. This maximum relative motion between the South Caspian Basin and its surroundings result in ~7 mm/year of right-lateral strike-slip motion along the Ashkābād fault and 4–6.5 mm/year of left-lateral strike-slip motion within the eastern Alborz Mountains (Mousavi et al., 2013). The GPS block modeling data shows 4.4–6.5 mm/year left-lateral displacement and 0.5–2.0 mm/year of shortening across the eastern Alborz at a latitude of 55°E (Mousavi et al., 2013).

9.4.1.7 The Kopeh Dāgh

The Kopeh Dāgh fold-and-thrust belt of northeast Iran (Figure 9.4) accommodates the N–S deformation between the Turān plate (in the north) and the Lut block and the northern Central Iran (in the south; Figure 9.2) at the northern limit of the Arabian–Eurasian collision zone. Short-term GPS data (Nilforoushan et al., 2003; Vernant et al., 2004; Masson et al., 2007) indicate a convergence rate of 6.0 ± 2 mm/year (range-parallel right-lateral shear of 2–4 mm/year) for the Kopeh Dāgh, which is close to the N–S geological convergence slip rate of 8.0 ± 2 mm/year (Shabanian et al., 2009a).

Apparently, the western Kopeh Dāgh together with the eastern Alborz is accommodating the westward extrusion of the South Caspian Basin, with the right-lateral Ashkābād fault showing ~5 mm/year of along-strike motion and up to 6.7 mm/year at the longitude of Marāveh Tappeh in the west (37°54′N–55°′E), reducing to 4.3 mm/year at the longitude of Shirvān (37°23′N–57°55′E) in the east (Mousavi et al., 2013).

Recent GPS data (Mousavi et al., 2013) show that ~7 mm/year of range-parallel extension with a trend of N35°W in the central Kopeh Dāgh is accommodated in the Central Kopeh Dāgh Shear Zone along the Bāghān and Quchān strike-slip fault zone with counterclockwise vertical axis rotation of the fault blocks (see Chapter 16). This allows the faults to accommodate extension between the eastern Kopeh Dāgh and the south Caspian Basin (Hollingsworth et al., 2006, 2010; Mousavi et al., 2013).

Mousavi et al. (2013) estimated 3.5–5.4 mm/year of cumulative right-lateral slip along the Bāghān and Quchān faults. At this rate, the 45 km of geologically recorded right-lateral displacement (Hollingsworth et al., 2006; Shabanian et al., 2009a) would accumulate in 8.3–15 Ma. The long-term slip-rates of 2.8 ± 1.0 mm/year and 4.3 ± 0.6 mm/year for the Bāghān and Quchān faults, respectively (Shabanian et al., 2009a,b), with bedrock displacement by 10 and 15 km, might have gotten started about 4 Ma (Mousavi et al., 2013; see Chapter 16).

9.4.2 Seismicity

The Iranian Plateau is a unique place with a well-documented historical earthquake record of at least 2000 years (Berberian, 1976a,e,g–i, 1977a–i, 1981a, 1994, 1995, 1997a, 2005; Ambraseys and Melville, 1982; Berberian and Yeats, 1999, 2001, 2014). The earthquakes that are associated with crustal shortening, with distributed deformation along reverse and strike-slip faults, have resulted from continental collision between Arabia and Eurasia and their continued convergence. Seismic activity occurs in structural zones in different tectonic blocks with inherited older structural trends.

Analysis of the historical seismicity (cited above) and the post-1900 instrumental seismic data (Jackson and McKenzie, 1984, 1988; Engdahl et al., 2006) shows that the intracontinental deformation and seismicity is concentrated along major mountain belts surrounding quasi-aseismic blocks of the Central Iran and the South Caspian basin (Figure 9.3). Waveform modeling analysis shows that except for the Makrān subduction zone, the average thickness of the seismogenic layer is about 15 km and most intracontinental earthquakes in the Iranian Plateau occur in the upper crust (Jackson and McKenzie, 1988; Maggi et al., 2000a,b; Tātār, 2001; Jackson et al., 2002; Hatzfeld et al., 2003). The seismic history also shows periods of relative enhanced activity along active faults with coseismic surface ruptures separated by periods of relative quiescence of several centuries to millennia. We will return to this issue in the coming chapters.

9.4.2.1 The Zāgros Fold-and-Thrust Mountain Belt

The Zāgros shows intense seismic activity spread across the belt with low- to medium-magnitude earthquakes (Figure 9.3; Table 17.10–17.12). No large-magnitude earthquake ($>M_w$ 7.0) has ever been recorded along the Zāgros belt, except possibly the event in about 9000 BCE that induce the colossal Saimareh (Saidmarreh) landslip along the northeastern flank of the Kabirkuh anticline (Berberian, 1994). Earthquakes in the Zāgros belt along the southwest of the Plateau are confined to the upper continental crust with depths less than 20 km, with no evidence for a seismically active subduction slab dipping NE beneath Central Iran (Maggi et al., 2000a,b; Jackson et al., 2002; Talebian and Jackson, 2004; Tatar et al., 2004a; Engdahl et al., 2006). The majority of seismicity in the Zāgros occurs along the late Neoproterozoic basement longitudinal reverse and transverse right-lateral strike-slip master faults, decoupled from the top sedimentary cover folds along the Upper Vendian–Lower Cambrian Hormoz Salt decollement, with smaller events occurring throughout the belt (Berberian, 1976a, 1981a, 1995; Berberian and Navai, 1977, 1978; Berberian et al., 1977; Berberian and Papastamatiou, 1978; Maggi et al., 2000a,b; Talebian and Jackson, 2004).

Berberian (1995) introduced the basement master blind faults (both north-dipping longitudinal basement reverse and transverse strike-slip faults), some (such as the High-Zāgros, Zāgros Mountain Front, Surmeh, and Zāgros Foredeep reverse faults as well as the Kāzerun, Karébas, Sabzpushān, and Sarvestān strike-slip faults) cutting through the top sedimentary cover, controlling major earthquakes, surface folding, and large-scale topographic steps, which may or may not necessarily break the surface during earthquakes (Figure 9.1). In this case, certain folds are formed by a mixture of basement fault propagation and fault-bending above north-dipping reverse faults nucleating within the pre-Hormoz Salt basement.

9.4.2.2 *The Transition Zone Between the Zāgros and the Makrān*

This complex region in the southeast of the Zāgros (Figures 9.1 and 9.3) is a transition zone from continent–continent collision of the Zāgros in the west to the Makrān subduction zone in the east (Molinaro et al., 2004; Regard et al., 2004, 2005, 2010; Yamini-Fard et al., 2007). In this transition zone between the Zāgros and the Makrān, seismicity is usually occurring up to 30–45 km in the crust, which is consistent with low-angle underthrusting of the Arabian basement underneath the southeast Central Iran. Waveform modeling shows a northward increasing of earthquake depth from typically 8 km near the coast to about 30 km (with a low-angle thrust mechanism) in the area about 50 km north of the Neo-Tethyan geosuture, the Main Zagros Thrust fault (Maggi et al., 2000a,b; Talebian and Jackson, 2004; Bayer et al., 2006; Engdahl et al., 2006; Peyret et al., 2009).

9.4.2.3 *The Makrān Oceanic Subduction Zone*

Along the Makrān of southeast Iran, the Arabian (Oman) seafloor is being subducted beneath the Makrān accretionary wedge of the southeastern Central Iran (Figures 9.1–9.4). The area is characterized by shallow northward-dipping subduction-related mantle seismicity and low-level upper crust earthquakes. The subduction-related earthquakes occur to depths of at least 150 km within a northward-dipping subduction slab. Most of the Makrān subduction zone deeper events show down-dip T-axes, indicating slab-related events. The low-level upper crust events in Makrān have a median depth of about 25 km (Jackson and McKenzie, 1984; Quittmeyer and Kafka, 1984; Laane and Chen, 1989; Byrne et al., 1992; Maggi et al., 2000a,b; Engdahl et al., 2006; Smith et al., 2013; Barnhart et al., 2014).

9.4.2.4 *The Central Iran Range-and-Basin*

Central Iran (Figures 9.1 and 9.3) is characterized by large-magnitude reverse, right-lateral, and left-lateral strike-slip earthquakes surrounding quasi-rigid blocks (see Chapter 16 for further details).

9.4.2.5 *Eastern Iran*

Seismicity in Eastern Iran surrounds the relatively stable aseismic blocks such as the Lut in the west and the Afghan block in the east (Figures 9.1 and 9.3), with N–S right-lateral, E–W left-lateral, and NW–SE reverse faulting with numerous large-magnitude earthquakes (Berberian and Yeats, 1999, 2001; Walker and Jackson, 2004; Berberian, 2005). Waveform modeling along the Gowk strike-slip faulting to the west of the Lut block showed earthquakes at two different depths. The 1981 M_w 7.0 Sirch earthquake produced a coseismic surface rupture with moderate offsets (10–40 cm) and a centroid depth of 18 km. However, the 1998 M_w 6.6 Fandoqā earthquake with a centroid depth of 5 km produced coseismic rupture with offsets up to 3 m (Berberian et al., 2001). See Chapter 16 for further discussion.

9.4.2.6 The Alborz Mountains

Active geomorphological indicators, earthquake slip vectors, and GPS data indicate that the Alborz Mountains are spatially partitioned into components of shortening and left-lateral strike-slip (Berberian et al., 1992; Jackson et al., 2002; Nilforoushan et al., 2003; Berberian and Walker, 2010; Djamour et al., 2010; Mousavi et al., 2013). Seismic activity in the Alborz Mountain belt occurs in the upper crust (Jackson et al., 2002; Engdahl et al., 2006) with large-magnitude earthquakes (Berberian, 1983a, 1994).

9.4.2.7 The Tālesh Mountains

The Tālesh Mountains in the western South Caspian Basin are tectonically distinct and different from the Alborz Mountains (Figures 9.2 and 9.3). Seismicity in the Tālesh Mountains is characterized by infrequent events in the lower crust indicating westward underthrusting of the South Caspian Basin underneath of the Tālesh Mountain belt (Berberian, 1983a; Jackson et al., 2002; Engdahl et al., 2006).

9.4.2.8 The Kopeh Dāgh Fold-and-Thrust Mountain Belt

Unlike the Zāgros, the Kopeh Dāgh (Figures 9.1 and 9.3) shows less frequent seismicity associated with large-magnitude earthquakes along longitudinal reverse and right-lateral strike-slip faults obliquely cutting the range in the Central Kopeh Dāgh Shear Zone (see Chapter 16 for further details).

9.4.2.9 The South Caspian Basin

The South Caspian Basin (Figures 9.1 and 9.3) seems to be an aseismic rigid block surrounded by active mountain belts of Tālesh (in the west), Alborz (in the south), and Kopeh Dāgh (in the east) with crustal seismicity (Berberian, 1983a; Priestley et al., 1994, 2001; Jackson et al., 2002).

Chapter 10

Archeoseismicity

Discovering of several skeletons at the depth
crushed by falling debris may indicate occurrence of an earthquake.
Skeletal remains of a mother protecting her two children in her arms,
show that the woman was protecting her children with her body during a danger;
and they all died at the same position.
Apparently, the man was in squatting position when he was grounded on his left side.
The left foot is dismembered; the right foot is folded;
and the rest of the smashed bones found near the folded foot.

Ghirshman (1938), Sialk South III_5 Period, ca. 3800 BCE

Ancient settlements in the arid and semiarid tectonically active regions were generally clustered along seismogenic faults at the foot of the mountain belts or in the depressions formed along the strike-slip faults. Although active faults and folds provided a fault-controlled, secure water supply, arable lands, and passageways through the mountains, it is clear that ancient settlements located near active faults were affected by medium- to large-magnitude earthquakes. Ancient earthquakes left their marks on a number of monuments and archeological strata (McGuire et al., 2000; Meghraoui et al., 2003; Galadini et al., 2006; Marco, 2008; Niemi, 2008; Pérez-Lopez et al., 2009; Reicherer et al., 2009; Silva et al., 2011). Like modern seismometers and accelerometers, these "archeoseismic indicators" or "archeo-seismometers" record information about the frequency, intensity, and nature of ancient earthquakes.

Archeological cultural layers and ancient monuments that exhibit on-fault and off-fault abrupt seismogenic disturbance, deformation, distortion of stone blocks, fracturing, offsetting, differential settlement, structural damage, and destruction provide valuable data about ancient earthquakes, faulting episodes, and their approximate seismic parameters. Such data can expand our knowledge of the temporal and spatial distribution of medium- to large-magnitude earthquakes, their recurrence periods, and long-term seismic hazard assessment along the adjacent active faults. It should also reveal how the structures responded to strong ground motion during fault movements and how people reacted—technologically and socially—to earthquake damage and destruction

☆"To view the full reference list for the book, click here."

Developments in Earth Surface Processes, Vol. 17. http://dx.doi.org/10.1016/B978-0-444-63292-0.00010-7

by making structural renovations to construct resilient buildings or confronting the decline, abandonment, and relocation of their communities.

Historically, large-magnitude earthquakes resulting in the collapse of a community have provoked significant social, political, and economic reactions in one of two forms: (i) *in situ* (epidemics; social unrest; rioting; cultural changes; a premature decline in the local economy, demography, and political importance; a slight shift in settlement location; resilience; and construction of more sustainable buildings), and (ii) *ex situ* (site abandonment, migration, and a relocation of settlement) (see references cited in the previous page).

It should be reiterated that neither archeoseismic indicators nor the paleoseismology of the active faults in Iran have been studied to the extent that they have been in many other regions. We must therefore rely on the limited archeological soundings, excavation reports, and historic photographic documentation captured during excavations by archeologists (see Figures 3-12 in Berberian et al., 2012; Figures 10-19 in Berberian et al., 2014; and Plate 12.15 in Chapter 12 on this study). There are approximately 45,000 archeological sites on the Iranian Plateau (CHTHO, 2011.03.07) that can help to expand the rich historical seismic dataset (see Tables II.1-II.5 in Berberian, 1994).

Despite the numerous archeological sites in close proximity to active faults on the Iranian Plateau (Berberian, 1994; Berberian and Yeats, 2001; Berberian et al., 2012, 2014), limited archeoseismic data have been extracted thus far from older archeological studies (Table 10.1). Moreover, evidence exists for site abandonment (Wilkinson, 1986) and shifts in settlement locations after some large-magnitude earthquakes as well as postseismic structural innovations that have enabled the construction of structures able to withstand strong ground motion (Berberian et al., 2012, 2014).

Table 10.1 presents the current archeoseismic knowledge of the Iranian Plateau. Clearly, the list is not complete and more work is needed to complete the dataset. Nonetheless, it clearly shows that large-magnitude earthquakes on the Iranian plateau have devastated ancient settlements for at least 7000 years. Despite the long sequences of occupation at sites situated close to active strike-slip faults, such as Tol-e Sepid (~6000 years; Berberian et al., 2014; 15 km north of Nurābād, see Figure 16.1) and Sialk (~8000 years; Berberian et al., 2012; see Figure 16.4), only a few archeoseismic events have been detected at these sites thus far (two events at the former and four at the latter; see Table 10.1). Unexpectedly, no obvious archeoseismic indicators (earthquake-related damage) were clearly detected during the 8000-year settlement record of Tol-e Nurābād (Figure 16.1) located close to Tol-e Sepid along the Kāzerun fault (Berberian et al., 2014). These shortcomings clearly indicate that more attention should be given to archeoseismic study during excavations by experienced multidisciplinary teams.

Furthermore, important research questions about the reported archeoseismic events (Table 10.1) remain unanswered. These include the magnitude and intensity of the earthquakes at each site; the extent of meizoseismal areas;

TABLE 10.1 Summary of the Discovered Archeoseismicity Events in Iran and Neighboring Areas

Approx. Date	Location [00°00′N–00°00′E]	Archeo. Period/Phase	Earthquake Effect	$\sim I$	$\sim M_S$	Active Fault (Mech.)	References
1220–1221: Invasion of the Mongol Hordes							
1145	Vineyard Tapeh, Shādyākh, Neyshābur [36°10′–58°49′, +1183 m]	–	Neyshābur Earthquake: Collapsed walls and ceilings; two skeletons under fallen debris	>VIII	>6.0	Binālud?, Neyshābur? (R)	Wilkinson (1975, 1986), Berberian and Yeats (1999)
1153: Invasion of the Ghuzz (Oghuz) Turk Nomads							
Twelfth century	Bishāpur [29°46′39.72″–51°34′16.58″, +877 m]	–	Bishāpur Earthquake: Collapse of the Islamic seminary bldg., three mosques, and toppling of Shāpur's statue	>VIII	>6.0	Kāzerun (RLSS)	Moqaddasi (985); Sarfarāz (1987, pers. com., 1994), Berberian (1994), Berberian et al. (2014)
1066 May	Qā'en, E. Iran [33°43′–59°11′, +1447 m]. Southern suburb of the city	–	Qā'en Earthquake: Complete destruction of the great old Qā'en congregational mosque (at Shāhzādeh Hossein mound)	>VIII^{+}	>6.5	Pāvak? Boznābād (RLSS)? Āvash (LLSS)	Naderi (1989), Berberian and Yeats (2001), Sorush (2007), Labbāf Khāniki (2013)
713–762 AD	Bishāpur [29°46′39.72″–51°34′16.58″, +877 m]	–	Bishāpur Earthquake: Destruction of the Triple Ayvān bldgs. and SW Unfinished Columned Hall				Ghirshman (1956, 1971), Berberian et al. (2014)

Continued

TABLE 10.1 Summary of the Discovered Archeoseismicity Events in Iran and Neighboring Areas—Cont'd

Approx. Date	Location [00°00′N–00°00′E]	Archeo. Period/ Phase	Earthquake Effect	$\sim I$	$\sim M_S$	Active Fault (Mech.)	References
636–662: Invasion of the Moslem Arabs							
531–590 AD	Bishāpur [29°46′39.72″–51°34′16.58″, +877 m]	–	Bishāpur Earthquake: Collapse of the Anāhitā Temple and the Triple Ayvāns	>VIII+	6.9	Kāzerun (RLSS)	Sarfarāz (pers. com., 1994), Berberian (1994, 1997c, 1998) Berberian et al. (2014)
Post-second half of the third century	Mil-e Ezhdehā [30°06′–51°27′, +927 m]	Early Sassanid	Collapse of top of tower, deep vertical fractures, shifted stone blocks	?	?		Berberian et al. (2014)
293–303 AD	Bishāpur [29°46′39.72″–51°34′16.58″, +877 m]	–	Bishāpur Earthquake: Collapse of the high cupola of Shāpur's Hall of Audience	>VIII	>6.0	Kāzerun (RLSS)	Ghirshman (1956), Sarfaraz (1987, pers. com., 1994), Berberian (1994), Berberian et al. (2014)
274–578[a] AD	Kangāvar [34°30′–47°57′, +1488 m]	–	Kangāvar Earthquake: Destruction, fire, and reconstruction of the Anāhitā temple	>VIII	>6.8	Zāgros Main Recent (RLSS)	Kāmbakhsh-Fard (1994), Berberian (1994), Āzarnoush (2009), Berberian and Yeats (2001)
10 BC–10 AD	Mithrādātkart (Nesā), Kopeh Dāgh [37°57′–58°12′, +350 m]	–	Mithrādātkart Earthquake: complete destruction of Mithrādātkart city	IX	7.1	Main Kopeh Dāgh (RLSS)	Gorshkov (1947a,b), Golinsky (1977), Kondorskaya and Shebalin (1977, 1982), Berberian (1994), Berberian and Yeats (2001)

100 BC	Qal'eh Kali, Fahliān [30°13′27.03″–51°26′54.74″, +837.5 m]	IV	? Collapse of Phase II structures	?	?	Kāzerun (RLSS)	Potts et al. (2009), Berberian et al. (2014)
336–330 BC: Invasion of Alexander III of Macedonia							
400–200 BC	Qal'eh Kali, Fahliān [30°13′27.03″–51°26′54.74″, +837.5 m]	II	Qal'eh Kali Earthquake: Collapse of the Achaemenid structures, differential settlement	>VIII	>6.5	Kāzerun (RLSS)	Potts et al. (2009), Berberian et al. (2014)
Seventh century BC	Sialk, Kāshān [33°58′07″–51°24′15″, +967.94 m]	Sialk VI	Sialk VI Earthquake: Collapse of Iron Age III exterior wall, two long fractures cutting the mound	?	?	Kāshān (RLSS)	Helwing (2005), Berberian et al. (2012)
1000–800 BC	Mārlik, W. Alborz [36°51′–49°29′, +293 m]	–	Mārlik Earthquake: Disturbed objects; dented and crumpled metal vessels; overturned gold vessel; broken, displaced, and rotated stone slabs beneath buried bodies	?	?	Rudbār (LLSS)	Negahbān (1964, 1984, 1990, 1996), Berberian et al. (1992), Berberian (1994), Berberian and Yeats (2001), Berberian and Walker (2010)
1200–900 BC Warm/Dry Climate (Kay and Johnson, 1981)							
1650–1600 BC	Godin [34°31′06.29″–48°04′06.52″, +1485 m], Giyān [34°10′53.34″–48°14′37.83″, +1563 m], NW Zāgros	Godin III:2	Godin Earthquake: Complete destruction and abandonment of Godin and Giyān sites	>VIII	>7.0	Zāgros Main Recent (RLSS)	Young (1968, 1969), Young and Levine (1974), Berberian (1994), Berberian and Yeats (2001)

Continued

TABLE 10.1 Summary of the Discovered Archeoseismicity Events in Iran and Neighboring Areas—Cont'd

Approx. Date	Location [00°00′N–00°00′E]	Archeo. Period/ Phase	Earthquake Effect	$\sim I$	$\sim M_S$	Active Fault (Mech.)	References
2000–1500 BC	Sagzābād, NW Bu'in, [34°48′58.96″–49°57′06.47″, +1248 m]	Sagzābād 9 Trench B	Sagzābād Earthquake: Complete destruction of Sagzābād w/complete but crushed skeletons of animals under debris	>VIII	>7.0	Ipak (LLR)	Negahbān (1971, 1973, 1974a,b, 1976, 1977), Berberian et al. (1983), Berberian (1994), Berberian and Yeats (2001)
2000 BC	Āk Tapeh, Kopeh Dāgh [38°14′–58°42′, +97 m]	–	Āk Tapeh Earthquake: Complete destruction of Āk Tapeh structures	IX	>7.0	Gyaurs (R)	Gorshkov (1947a,b), Golinsky (1974), Kondorskaya and Shebalin (1977, 1982), Bune and Gorshkov (1980), Berberian (1994), Berberian and Yeats (2001)
2200–1700 BC (ca. 4.2 ka) Drought/cooling event (Weiss et al., 1993; Weiss, 2000; Booth et al., 2005; Stevens et al., 2006; Staubwasser and Weiss, 2006) [possible loss of archeoseismic data]							
3030 BC	Tol-e Sepid, Fahliān [30°15′08.83″–51°29′03.76″, +854 m]	18/17 Bānesh/ Kaftari	Tol-e Sepid earthquake: Fracture w/normal displacement; destruction, abandonment	~IX	~7.3	Kāzerun (RLLS)	Petrie et al. (2007, 2009), Berberian et al. (2014)
Post-3200 BC	Sialk, Kāshān [33°58′07″–51°24′15″, +967.94 m]	Post-Sialk IV_2	Sialk Earthquake: Long fracture cutting all the excavated Sialk IV_2 structures	?	?	Kāshān (RLSS)	Berberian et al. (2012)

3600–3000 BC (ca. 5.2 ka) Drought/cooling event (Weiss, 2000; Stevens et al., 2006) [possible loss of archeoseismic data]							
3800 BC	Sialk, Kāshān [33°58′07″–51°24′15″, +967.94 m]	Sialk III_5/III_6	Sialk III_5 Earthquake: Complete destruction of Sialk III_5, several skeletons covered by debris; fractures cutting walls and floors	>VIII^+	>6.5	Kāshān (RLSS)	Berberian et al. (2012)
Early 5th Mill. BC	Qara Tapeh, near Qomrud [34°43′90″–51°03′89″, +850 m]		?	?	?	Alborz, Kushk-e Nosrat, Sarājeh	Kāboli (2000)
3850–680 BC	Tol-e Sepid, Fahliān [30°15′08.83″–51°29′03.76″, +854 m]	23/22 Lapui	Tol-e Sepid Earthquake: Fracture w/reverse displacement	~IX	~7.3	Kāzerun (RLLS)	Berberian et al. (2014)
ca. 4300 BC (ca. 6.3 ka) Drought (Singh et al., 1990; Fleitmann et al., 2003, 2007; Lézine et al., 2007; Djamāli et al., 2010) [possible loss of archeoseismic data]							
5000 BC	Sialk, Kāshān [33°58′26″–51°24′27″, +952.02 m]	Sialk I_5/II_1	Sialk I_5 Earthquake: Collapsed walls covered by sand storm deposits; abandonment and settlement shift	?	?	Kāshān (RLSS)	Berberian et al. (2012)

Continued

TABLE 10.1 Summary of the Discovered Archeoseismicity Events in Iran and Neighboring Areas—Cont'd

Approx. Date	Location [00°00′N–00°00′E]	Archeo. Period/Phase	Earthquake Effect	$\sim I$	$\sim M_S$	Active Fault (Mech.)	References
6500–6000 BC (ca. 8.2 ka) Drought/Cooling Event (Weiss, 2000; Rohling and Pälike, 2005; Weninger et al., 2006; Staubwasser and Weiss, 2006; Fleitmann et al., 2007; Ebbesen et al., 2008; Gronenborn, 2009; Akkermanns et al., 2010) [possible loss of Archeoseismic data]							
7400 BC	Hotu Cave, northern Alborz [36.69–53.50°]	Mesolithic	Cave ceiling collapse(?)	?	?	Khazar (R)	Coon (1952, 1957), McBurney (1968), Berberian (1994)
10,000 BC	Zard-e Sāhel, Kāshān [33°55′41.62″–51°23′51.88″, +1031 m]	Neilithic	Zard-e Sāhel Earthquake(?): Drying up of the paleo-travertine springs	?	?	Kāshān (RLSS)	Berberian et al. (2012)
Ca. 60,000 BP	Shānidar Cave, Barādust Mts., Zāgros [36°50′5.30″–44°13′8.56″, +768 m]	Neanderthal	Five major rockfalls from the cave's ceiling collapse crushed Neanderthals I–III and V (N. I, III, and V: ^{14}C dated >45,000 BP; N. II, IV, VI–IX: 65,000–100,000 BP)	?	?	Zāgros Mountain Front (R)	Solecki (1955, 1963, 1964, 1971, 1977, 1997), Berberian (1994), Bartsiokas (2003)

Note: Please note that further research is required to constrain the dates and in some cases the events.
$\sim I$, approximate equivalent intensity of earthquake; Mech, mechanism of earthquake faulting (R, reverse fault; RLSS, right-lateral strike-slip fault; LLR, left-lateral oblique reverse fault; LLSS, left-lateral strike-slip fault); $\sim M_s$, approximate equivalent surface-wave magnitude.
[a]*The earlier date of AD 224–549 (Berberian and Yeats, 2001) was bracketed on data by Kāmbakhsh-Fard (1994). The present dates are bracketed from new data by Āzarnoush (2009).*
Organized in chronological order; modified after Berberian (1994), Berberian and Yeats (2001), and Berberian et al. (2012, 2014).

the periodicity of the large-magnitude earthquakes on a given fault; and the human, technological, and architectural responses to ancient large-magnitude earthquakes.

The information presented in Table 10.1 (which should be observed cautiously) indicates that the existing archeological data, although not aimed at earthquake archeology, provides indicators of a few large-magnitude, hitherto unknown earthquakes. Paleoseismic fault-trenching studies coupled with detailed archeoseismic investigation and radiometric dating of active alluvial/fluvial deposits should constrain earthquake chronology, their source parameters, and periodicity.

Chapter 11

Pre-1900 Coseismic Surface Faulting

For two farsangs [12 km] between Nauzād and Mask villages the ground was fissured to such a depth that the bottom of the fissures was invisible.
Historian Mo'in al-Din Esfezāri (1494), in the first known description of a coseismic surface fault rupture associated with an earthquake by a contemporary observer in Irān

Earthquake activity along the mountain belts on the Iranian plateau is related to an ongoing collision between Arabia and Eurasia. The collision has produced the active Iranian mountain belts; its active folding and faulting is associated with numerous catastrophic earthquakes that have been documented over time and have led to archeological and mythological investigations (Chapters 2–10). The primary effects of earthquakes are coseismic faulting and strong ground shaking (Chapters 11–14). Active buried (blind) reverse faults, which do not extend to the surface of the Earth, can cause active flexural-slip folding and flexural-slip faulting over the blind faults (Chapter 15).

Since the Neolithic age, people in semiarid to arid regions of Central Iran—unaware of earthquake fault and rockfall hazards—have settled on arable lands: (i) at the feet of mountains [some created by active mountain-front reverse faults]; (ii) at foothills [at the feet of folds above active blind reverse faults]; and (iii) along the strike-slip faults, where they could make use of surface water (springs, rivers) and groundwater (*qanāts*; underground aqueducts). Since ancient days, the major trade routes have run along these active fault features, mainly exposed along the northern and the southern foothills of the Alborz Mountains [in the north and northeast], the areas of the North Tabriz fault in the northwest, the Kāshān fault in Central Iran, and the Kuhbanān fault in the southeast, and many more have provided water, arable land, and life (Figure 9.1). Periodic movements along these active faults during large-magnitude earthquakes have caused massive destruction and loss of life, devastating the unfortunate and unprepared communities, disrupting water supplies and agricultural infrastructures, and causing the decline of numerous settlements.

☆"To view the full reference list for the book, click here."

Developments in Earth Surface Processes, Vol. 17. http://dx.doi.org/10.1016/B978-0-444-63292-0.00011-9

Detailed studies of coseismic surface faulting associated with different earthquake mechanisms in diverse structural provinces (Chapter 9 and Figure 9.2) are vital in understanding the coseismic dynamics of the ground rupturing associated with each earthquake. The results of such investigations can be synthesized into generic models of earthquakes and rupture propagation and used in seismic hazard evaluation studies of the strong ground motion accompanying earthquakes. Furthermore, the lessons learned can be applied to the other active mountain belts throughout the world and help our understanding of the active tectonics of planet Earth.

From an active tectonics point of view, the Iranian plateau is a diffuse, deforming continental plate-boundary zone of intracontinental active strike-slip and reverse faults as well as an active subduction zone. It experiences very strong and destructive earthquakes similar to southern California, New Zealand, Venezuela, Argentina, the Andes, the Atlas, and other active parts of the world (Figures 9.1, 9.3, 9.4; Chapters 16 and 17; Tables 14.10–14.11 and 17.10–17.12). However, unlike California, in Iran, we can draw on 2000–6000 years of historical and archeological data to work out coseismic surface ruptures and earthquake histories and study their interactive patterns (Chapter 16). Despite the existing rich earthquake database, which has some limitations on the nature of the recorded macroseismic data (Berberian, 1976h, 1977b–e, 1981a, 1994, 1995, 1997a, 2005; Ambraseys and Melville, 1982; Guidoboni and Traina, 1995), assigning most pre-1970 events to specific active fault sources in a relatively complicated arrangement of densely populated active faults (Chapter 16) has always been a challenge. I address this issue in the following chapters.

In most reports about the historical earthquakes in southwest Asia, few attempts are made to determine the causative fault for each event; the assigned magnitudes may be overestimated and far exceed the capability of the faults (Kondorskaya and Shebalin, 1977, 1982; Melville, 1980, 1981; Ambraseys and Melville, 1982; Ambraseys, 1988a; Nikonov, 1991; Guidoboni and Traina, 1995; Babayan, 2006). Bearing in mind that it is very difficult to assign historical events to the surface faults with certainty, little has been done to associate individual events with a specific active fault or specific fault segment (Berberian, 1976b,c,e,j,k, 1977a, 1981a, 1994, 1995, 1997a, 2005; Berberian and Yeats, 1999, 2001; Berberian and Walker, 2010). A twentieth-century teleseismic database has not been able to resolve a great many location and foci errors (Ambraseys, 1978a; Berberian, 1979b) (Chapters 12–14).

Contemporary and near-contemporary notes from numerous unpublished chronicles, or very little-known published sources, in Persian, Armenian, Assyrian, and Arabic reveal important coseismic surface faulting information. For example, unpublished documents in Armenian and Assyrian recount details of great importance about the little-known Salmās earthquake of 6 May 1930, M_w 7.1 (Chapter 12); the causative Salmās fault rupture at the surface; and the resultant ground deformation (Tchalenko and Berberian, 1974;

Berberian and Tchalenko, 1976a). Scores of similar case histories of possible coseismic surface faulting have been found and verified in the field.

Despite the many historical earthquakes told of in Persian, Armenian, and Arabic chronicles, the historical seismicity of the Iranian plateau is not uniformly documented throughout the past few millennia; these chronicles have many shortcomings. Toward the end of this period, Khanikoff, the imperial Russian consul-general at Tabriz and an amateur scientist and orientalist, studied the small 4 October 1856 earthquake that triggered the Cacciatore-type seismoscope in the Tabriz observatory (in Abich, 1857, 1858). He prepared the first isoseismal line for an earthquake in Iran based on his findings. He also recoded the time, date, and intensity of earthquakes felt in Tabriz in 1843, 1844, 1848–1849, and 1851–1854; some of the earthquakes originated in Khoi (located about 124 km to the northwest of Tabriz). Khanikoff (1861) also mentioned earthquakes felt in the Neyshābur turquoise mine area. During the same period, the world seismicity map of Mallet (1858) showed active regions of the Iranian plateau, with the highest historical seismicity band passing through northern Iran. Since 1961, Nicholas Ambraseys and his students have studied the historical seismicity of the region, which has resulted in the analyses of many earthquakes, especially in Iran.

In this chapter, I present and analyze the historical accounts of the dynamic phenomena associated with earthquakes on the Iranian plateau. I have combined geology, geophysics, remote sensing, a knowledge of ancient Persian history, and archeology in my analysis of the historical seismic data and coseismic surface ruptures along active faults. The study highlights an acute earthquake-faulting hazard associated with the known historical seismicity and coseismic faulting on the Iranian plateau. The following examples are taken from original sources in which localized damage data suggest the activation of particular surface faults. Some of these texts clearly refer to coseismic surface faulting and ground deformations, while others refer to secondary ground deformation, such as landslips and lurching. Although I have tried to separate coseismic faulting from nontectonic deformation, some cases still remain controversial; there are not enough data about the extent of damage or the affected region covering a large area with several active faults. I do not discuss these cases here.

The approximate surface-wave magnitude estimates were taken from Ambraseys and Melville (1982), Ambraseys and Adams (1989), Brommer and Ambraseys (1989), Berberian (1994, 1997a, 2005), Ambraseys and Finkel (1995), Ambraseys (1997), Berberian and Yeats (1999, 2001), Ambraseys and Bilham (2003a,b), and Ambraseys (2009). These magnitudes were derived from macroseismic information embedded in written accounts calibrated against M_s values. Therefore, they present poorly constrained estimates. Some overexaggerated magnitudes have been adjusted in this study. Intensity (MMI) estimates are based on historical accounts and a study of damaged monuments. Active fault data, unless otherwise specified, are the result of the author's 43 years of field work. Tables 14.9–14.11 present summaries of the pre- and post-1900 coseismic surface fault ruptures; Tables 17.10–17.12 show full seismic parameters. Some of the proposed pre-1900 meizoseismal areas shown in the figures of this chapter are not constrained, or are only partially constrained, and should be observed advisedly.

11.1 THE BREAKUP OF KOMESH MOUNTAIN DURING THE "RELIGION WAR" [CA. 1200 BCE]

During the war between *Vishtāsb*[1] and *Arjāsb*,[2] an earthquake split Komesh (Kumes, Qumis) Mountain and apparently triggered huge landslides and rock avalanches. The event was probably considered a miracle and was recorded as a "beneficial earthquake." This is one of the oldest references to the surface deformation (coseismic rupture or unstable slope failure) associated with a large earthquake (Berberian, 1994, 1997b).

In Chapter IX of the *Iranian (Greater) Bondéhéshn*[3] (Bundahishn; ed. Anklesaria, 1956; ed. Bahār, 1990; Berberian, 1994, 1997b), the event is addressed as follows: "The Damāvand Mountain, that on which Bevarasp is fettered, is from the same *Patashkhārgar*[4]." Even Mount Komesh was connected with Georgia [*sic* or Gorgān. In ed. Bahār, 1990, "Even Mount

1. Vishtāsb. Vishtāspa/Kai-Vishtāsp/Kavi Vishtāspa or Goshtāsb: lit., "owner of a thin horse"; the King of Iran from the Kiyāniān (Kayānid) dynasty during the time of Zartosht/Zarathustra, who accepted the Mazdā-worshipping religion; the last king of the old history derived from Avestā (see Christensen, 1931).

2. Arjāsb. Arjāsp/Aregad-aspa/Arejat-aspa: lit., "owner of a valuable horse"; the King of the Turanian Turkic Hiyun/Hvyaona/Khiyon/Huns; Hephthalites/Chionites (see Bailey, 1930–1932, 1972; Cereti, 1996).

3. Bondéhéshn (Bundahishn). From "Bun-dahishn" in Pahlavi meaning "Original/Primal Creation," "Cosmogony"; a religious text of the Zoroastrians, written in the Pahlavi language [Middle Persian; the language of Sassanian Iran—224–642 CE]. The book in its present state appears to be a collection of fragments relating to the cosmogony, mythology, and legendary history taught by Mazdāyasnian tradition. Bundahishn is preserved in two versions, the "Indian" and the "Greater Iranian," the latter being the finer work. It has 36 chapters. The first seven deal with cosmogony proper; they tell of the initial conflict of Urmazd [Ahurā Mazdā; Ormazd, Hormoz; the Avestan name for the chief deity of Zoroastrianism and the source of light and embodiment of good] and Ahriman [the Avestan Angra Maiynu; the destructive one; evil] and the creation of the world, first in a "Minu or Minavi" [spiritual or embryonic] state and then in its "material" state. Chapters 8 through 32 are mainly a description of the universe. Chapter 14 tells the story of the origin of man: the semen of Primeval Man [the Avestan Gayo-mareta, Giyumars], whom Ahriman had killed, fell onto the Earth and generated a rhubarb plant [rivās], from which the first human couple sprang. Chapter 28 expounds the doctrine of man as a microcosm. Chapters 33 through 36 deal with the history of the world, from creation to resurrection (see also West, 1880; Christensen, 1931; Anklesaria, 1956; iranicaonline.org; Hinnells, 1973; Bahar, 1983, 1990; EB, 1985; Behzadi, 1989; Berberian, 1994, 1997b).

4. Patashkhārgar/Padishkhārgar: lit., "dish-shaped Mountain"; referring to the "arcuate form" of the Alborz Mountains. From "Patishkhār/Padishkhār": lit., "dish, bowl, plate"; and "gar": mountain.

Komesh and Gorgān are connected to it (Patashkhārgar)]; as they call Mount Komesh the mountain which 'Had come to Help', that by which Vishtāsp [Goshtāsp] had defeated Arjāsp, the mountain which was broken there from that mountain in the midst of the battlefield [the Miyāndasht Mountain was broken/separated from that (Komesh) Mountain]. They say, 'In the war of Religion, when defeat was with the Iranians, it [the Miyān Dasht Mountain] fell down in the midst of the battlefield out of these mountains from which it broke, and the Iranians were saved thereby. They named it 'Come to help'. . .' " (ed. Anklesaria, 1956, with corrections from the Persian translation of the Pahlavi language in ed. Bahār, 1990).

Almost similar phrases are given in *Bondéhéshn,* XII/32–33 (ed. West, 1880; Christensen, 1931; Poure Dāvoud, 1977a): "*From the same Padashkhārgar unto Mount Kumis, which they call Mount Madofryād* [matanfaryād: lit., Come-to-help]—*that in which Vishtāsp routed Arjāsp–is Mount Miyān-i-dasht* ["Miyān Dasht": lit., "mid-plain"], *and was broken off from that mountain here. They say, in the war of the religion, when there was confusion among the Iranians it broke off from the mountain, and slid down into the middle of the plain; the Iranians were saved by it, and it was called 'come-to-help' by them.*" (ed. West, 1880).

Joneidi (2008) stated that *matan Faryāt [come-to-help] is* a misreading of *Mithrā Faryāt [Mehr Faryāt; yāri resāni Mehr; Help from Mithrā]*. He added that the Iranians were in a defensive position in this war.

The places addressed in different texts [such as *Zand-e Vohuman [Bahman] Yasht* (ed. West, 1897; ed. Peterson, 2002) and *Yadègār-e Zarirān* (*Memoirs of Zarir*; Horne, 1917; Weimer, 2002)] may indicate that the conflict between Vistāsp and Arjāsp occurred in different places during several wars, the last of which apparently happened near Komesh Mountain (modern Dāmghān), which is addressed in *Bondéhéshn* (Figure 11.1). The event occurred during the time of Zoroaster (Zarathushtra; Zartosht), that is, ca. 1500–1200 BCE (Boyce, 1989).

I have not been able to locate "*Mount Miyāndasht*" that *Bondéhéshn* refers to. The modern village of *Miyāndasht* [36°25′N–56°03′E, +1242 m] in the medieval Komesh province is located about 160 km east–northeast of Komesh [modern city of Dāmghān, where the devastating 856 earthquake occurred], and 40 km east of Meyāmay [36°24′N–55°39′E, +1082 m], on the route between Meyāmay and Sabzévār (Figure 11.1). The distance between Miyāndasht and the Meyāmay active fault is 11 km (Figure 11.1). No paleo-landslide and/or paleo-rock avalanche has been recorded in the Miyāndasht area. Except for the "*Religion War*" case addressed above, there is no account of an earthquake along the Meyāmay active fault in the surviving recorded history of the region. Therefore, the event may have taken place near the modern city of Dāmghān (old Komesh), possibly along the Dāmghān fault (Figure 11.1).

The city of Komesh (modern Dāmghān) was devastated during the 22 December 856 earthquake, which took place along the Dāmghān/Āstāneh fault system (Figure 11.1; see the description of the 865 event below).

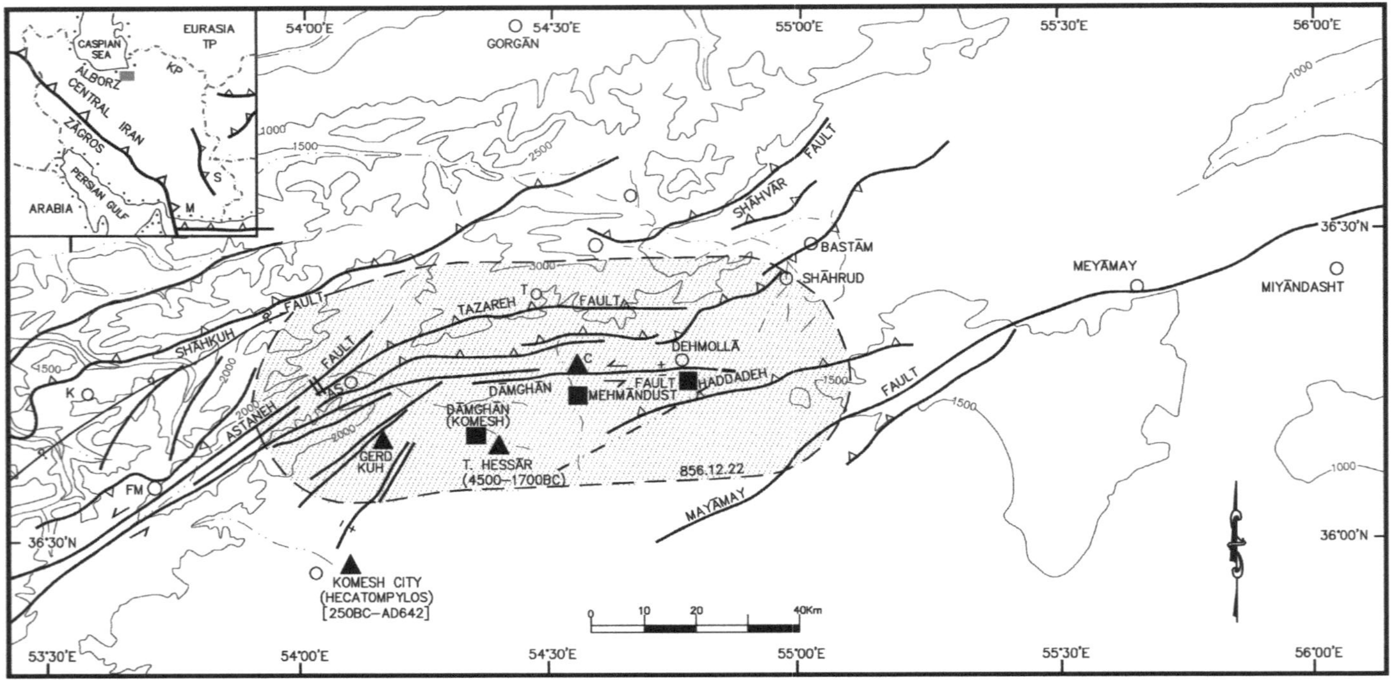

FIGURE 11.1 Historical seismicity and fault map of the Miyāndasht-Komesh (modern Dāmghān) area. Throughout the book: reverse faults shown with teeth on hanging-wall. Strike-slip faults shown with arrows. Faults without teeth or arrows: sense of slip unknown. + and –: sense of vertical motion. Dashed where uncertain. Fault names are also added. Faults drawn from 1:55,000 scale aerial photographs (Worldwide Aerial Surveys, Inc., 1956), satellite imagery (Landsat-TM), Berberian (1976e), and numerous field investigations and author's internal reports of the GSI since 1971. Approximate partially constrained meizoseismal area (stippled, shaded with dashed-line ellipse showing possible outer edge of intensity VIII on MMI) of the 856 earthquake is shown by a dashed-line ellipse. Where the date of the earthquake is shown, it is given by year.month.day. Triangle: archeological site with approximate date. Filled squares: documented sites destroyed by the 856 earthquakes. Other villages are shown by blank circles. AS, Āstaneh;FM, Fulād Mahaleh; K, Kiyāsar. Topographic contour lines in meters. Major river beds in broken lines and dots. Inset top left: location of the site on the Iranian plateau between Arabian and Eurasian plates (KP, Kopeh Dāgh; M, Makrān; S, Sistān; TP, Turān Plate). *Modified from Berberian et al. (1996). References as cited in the text.* Some of the proposed pre-1900 meizoseismal areas shown in the figures of this chapter are not constrained, or are only partially constrained, and should be observed advisedly.

al-Suyuti (1499) wrote that during the 856 Komesh earthquake, mountains were shattered and the ground opened up in many places so extensively that it could easily have swallowed a man. It is likely that *Bondéhéshn* is referring to a ca. 1200 BCE earthquake that took place in the Dāmghān area, much before the 856 Komesh earthquake. If the case is established by paleoseismic trench study, it may indicate an interseismic interval of about 2000 years.

11.2 CA. 280 BCE GROUND DEFORMATION AT THE ANCIENT CITY OF RHAGAE [OLD RAY, SOUTHERN TEHRĀN]

The ancient city of *Rhagae* [Ragae; modern Ray in the southern suburb of the modern capital city Tehrān] has been destroyed or damaged several times by large-magnitude earthquakes in ca. 280 BCE (312–280 BCE), 855–856, 864; 958 (?), 1177 (?), and 1830 (Ambraseys, 1974a; Ambraseys and Melville, 1982; Berberian et al., 1985; Berberian, 1994; Berberian and Yeats, 1999, 2014). Some historic large-magnitude events should have been associated with coseismic surface ruptures along the southern Alborz Mountains and the northern edge of Central Iran (see Chapter 16).

Duris of Samos, a near-contemporary source (Strabo, Geography: I.3.19; 64 BCE–23 CE), described the ca. 280 BCE earthquake at Rhagae, including the "*Caspian Gates*," modern "*Tang-e Sardarreh*" at 35°17′N–52°08′E, +966 m (Anderson, 1928; Hansman, 1968; Standish, 1970; Ambraseys, 1974a; Bosworth, 1983), where the ground ruptured, diverting the flow of the river: "*Duris wrote that Rhagae in Media* [Māda, Mād] *has received its name because the Earth about the Caspian Gates had been 'rent'* [the root of the classic Greek verb here used is '*rhag*'?] *by earthquakes to such an extent that numerous cities and villages were destroyed, and the rivers underwent changes of various kinds*" (Strabo, Geography, I.3.19).

Whereas *Poseidonius of Apamenia* (History, second century BCE; Strabo, Geography, XI.9.1) mentioned that 2000 villages and many cities were destroyed and implied that Rhagae [Ray, 35°36′N–51°26′E, +1076 m, south of modern Tehrān] itself was involved in ground deformation: "*As for the Parthian country, it is not large; at any rate, it paid its tribute along with the Hyrcanians in the Persian times, and also after this, when for a long time the Macedonians held the mastery. And, in addition to its smallness, it is thickly wooded and mountainous, and also poverty stricken, so that on this account the kings send their own throngs through it in great haste, since the country is unable to support them even for a short time. At present, however, it has increased in extent. Parts of the Parthian country are Comisene and Chorene, and, one may almost say, the whole region that extends as far as the Caspian Gates and Rhagae and the Tapyri, which formerly belonged to Media. And in the neighborhood of Rhagae are the cities Apameia and Heracleia. The distance from the Caspian Gates to Rhagae is five hundred stadia* [55 miles/88.5 km; the actual distance is about 44.7 miles/72 km], *as*

Apollodorus says, and to Hecatompylus, the royal seat of the Parthians, one thousand two hundred and sixty. Rhagae is said to have got its name from the earthquakes that took place in that country, by which numerous cities and two thousand villages, as Poseidonius says, were destroyed" (Strabo, Geography, XI.9.1; tr. Hamilton and Falconer, 1854).

Although in classical Greek, *Rhag* or *Ragh* means to "break asunder," or "be rent, of the Earth in an earthquake" (Liddall, 1961; Ambraseys, 1974a), it is implausible that a Greek name was given to an ancient Iranian city with the Avestan name of *Ragha* and the Old Persian name of *Raga* [in the Avestā; lit., "Bright," "Lightened"], because it was rent by earthquakes. Rhagae is also mentioned in several ancient texts, such as the Jewish book of Tobit; Vendidād, 1.16; *Yasnā* 19.18; and the sixth century BCE Bisutun [Behestān, Baghestān] inscription of Dārius I Achaemenid the Great in Kermānshāh (Dārayavaush; r. 522–486 BCE). Hence, its ancient name is much older than the Macedonian invasion of Iran in 334–330 BCE.

Apollodorus of Artemita (in Strabo, XI.8.6) stated that Rhagae lays 500 stadia [92.5 km (S. Attic or Ptolemy) to 96.1 km (S. Olympic)] to the south of the Caspian Gates (Strabo, Geography, XI.8.6). The city of Ray lies exactly 73 km to the northwest of the "Caspian Gates" (heading 119°) [if the latter were known as the modern Tang-e Sardarreh; lit., "the Sardarreh Pass/defile"] and not 92 km to the south of it as mentioned by Apollodorus of Artemita. Tang-e Sardarreh is located on the hanging-wall of the Garmsār thrust at the southeastern end of the Pārchin fault (Berberian et al., 1985; Berberian and Yeats, 2001) (Figure 11.2). Records of this event by the Greeks, the survival of the information over a long period, and widespread destruction with topographic changes suggest that the earthquake was a large-magnitude [$M_s > 7.0$] event with the surface rupturing possibly along the Pārchin fault (Figure 11.2). Paleoseismic trench study is needed to prove this link.

The Pārchin fault is an approximately 70-km-long fault with NW–SE strike and is located to the southeast of the city of Ray and northwest of the Garmsār thrust (Figure 11.2). It clearly cuts the Quaternary alluvial deposits of the plain in the area east of the Jājrud Valley mouth (aerial photograph Nos. 4827 and 4828, Worldwide Aerial Surveys, Inc., Project 158, scale 1:55,000; see photograph 4.29, p. 103 in Berberian et al., 1985), not far from the controversial Pārchin military nuclear complex (35.52°N–51.77°E). Regarding the historical seismicity and pattern of coseismic surface faulting in the Ray-Tehrān area, see Chapter 16, Tables 14.9–14.11, and 17.10–17.12.

11.2.1 Paleoseismicity

The second event introduced by Nazari et al. (2009a) from a trench across the Tālėqān fault [36°07′24.75″N–51°12′48.28″E] was dated between 3470 and 1540 calibrated years BP (unconstrained). The authors speculated that

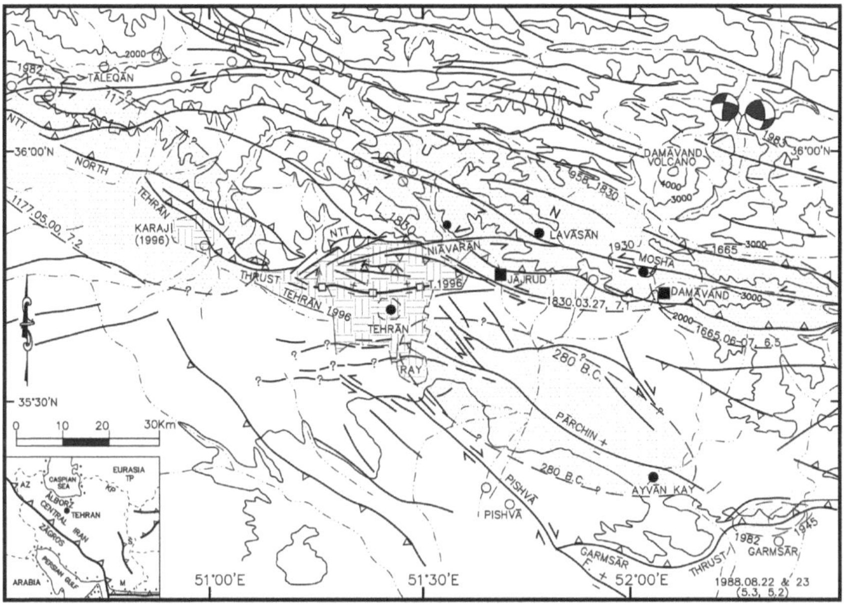

FIGURE 11.2 Historical seismicity and fault map of the Tehran–Ray area (crosshatched with 1996 city limit—T.1996) in the southern central Alborz Mountains, north of Central Iran. The 1996 limit of the city of Karaj is also shown to the west of Tehran. Fault symbols defined in Figure 11.1. Short line with + and – (vertical displacement) and small blank squares in Tehran: groundwater anomaly/cascade in the aquifer. The nature of the three escarpments in the Ray area of south Tehran (North Ray, South Ray, and Kahrizak) is not established. Meizoseismal areas of large-magnitude earthquakes (approximate outer edge of intensity VIII) with their dates (year.month.day) and approximate surface-wave magnitudes are shown by ellipses with broken lines; queried where uncertain and not properly constrained. Fault plane solutions of the 25 and 26 March 1983 Bāijān earthquakes (M_s 5.5 and 5.4, respectively), NNE of the Quaternary Damāvand Volcano: best-double-couple Harvard CMT solutions. Circles: villages and towns. Squares: two sites destroyed by the 1830 earthquake. *Modified from Berberian et al. (1983, 1985), Berberian and Yeats (1999, 2001, 2014), and Berberian (2005).*

their second event can be associated with the 312–280 BCE earthquake and estimated a maximum coseismic offset of between 2.26 and 5.10 m. It should be noted that the 312–280 BCE event destroyed the ancient city of Ray and villages to its east; its meizoseismal area is located about 110 km southeast of the Tāléqān fault (Figure 11.2).

11.3 THE 22 DECEMBER 856 M_S ~7.2+ KOMESH EARTHQUAKE GROUND DEFORMATION

The sole contemporary document about this event mentions a strong earthquake of long duration that occurred at night in the land of Komesh [Kumes, Arabicized Qumis; modern city of Dāmghān]: all the houses were destroyed, killing many people (Tahman al-Baihaqi, 856). About 47 years after the earthquake, *ibn Rusteh Esfahāni* (903) visited the area and wrote that, from Komesh city to Haddādeh village (near the modern Dehmollā), the houses and caravanserais destroyed by the earthquake were still visible (Figure 11.3).

Later chronicles combined data from at least three different earthquakes (including the 855 at Ray and 856–857 in Neyshābur; both far from the Komesh epicentral area) into a single megaearthquake and stated that about 200,000 people [a highly exaggerated and unrealistic figure] were killed in the area and 45,096 in the city of Dāmghān [also exaggerated] (Ya'qubi, 892; Tabari, 915). About 325 years after the earthquake, *Ibn al-Jauzi* (1181) wrote that mountains fractured, some fractures on other mountains closed up, and loud sounds were heard from the ground as well as the air. Almost 643 years after the event, *al-Suyuti* (1499) stated that the mountains and the ground were fissured so extensively that one could enter into them (see Berberian et al., 1996 for details). It was about 670 years after the earthquake when Khāndmir (1521) added that: "*half of Dāmghān and 1/3 of town of Bastām* (the latter town mentioned for the first time in the chronicles) *were*

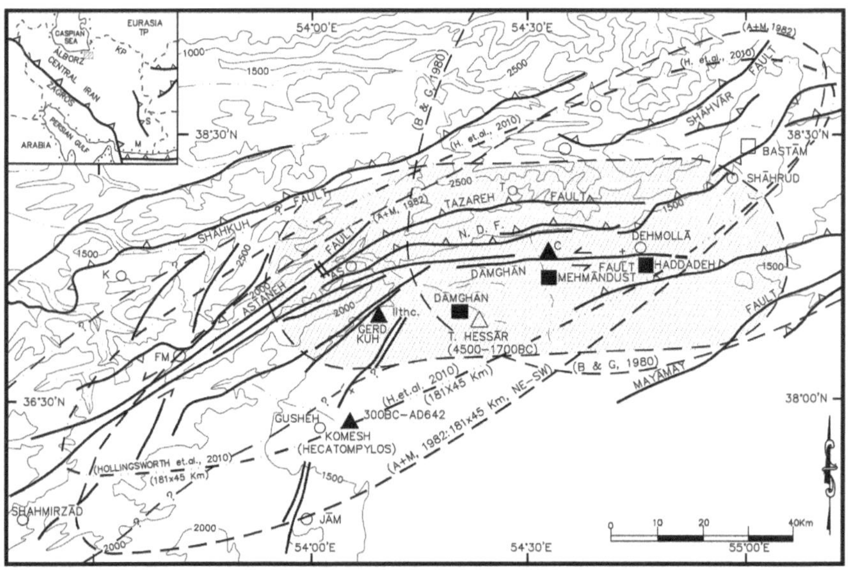

FIGURE 11.3 Proposed meizoseismal area of the 22 December 856 Komesh (modern Dāmghān) earthquakes in the southern edge of the eastern central Alborz Mountains. Symbols as in Figure 11.1. N.D.F, North Dāmghān fault. The overexaggerated and speculative meizoseismal areas proposed by Ambraseys and Melville, 1982 (A+M, 1982), Hollingsworth et al., 2010 (H. et.al., 2010), and Bune and Gorshkov, 1980 (B+G, 1980) crossing several active fault features are also inserted by broken line ellipses. Short parallel lines (=) across the Āstāneh fault (NW of Dāmghān): location of paleoseismic trench (Hollingsworth et al., 2010). Triangles: Archeological sites with dates. C, destroyed caravanserai during the 856 earthquake. AS, Āstāneh; FM, Fulād Mahalleh; K, Kiyāsar; T, Tazareh. See also Plates 11.1 and 11.2. *Modified from Berberian et al. (1996).*

destroyed"; he also stated that Ray, Jorjān, Neyshābur, and Esfahān were hit by the same earthquake! (combining the damage from at least three known earthquakes in diverse areas in 855, 856, and 856–857).

Ambraseys and Melville (1982), followed by Hollingsworth et al. (2010) and Mousavi et al. (2013), assigned an exaggerated magnitude of M_s ~7.9 to this earthquake, listed 200,000 casualties, and drew an exaggerated and unconstrained meizoseismal area defined as the area within the highest isoseismal of ~X (MMI) with an elongated narrow ellipse of about 180 × 48 km, along a NE–SW direction (N58°E) (Figure 11.3). Ambraseys and Melville (1982) also assigned a 100-km-long surface faulting to this earthquake and added that the actual fault length is probably longer than shown in their Table 3.1, p. 110; however, no fault map was provided. Berberian (1976b,c, 1981a, 1994), Berberian and Qorashi (1985), and Berberian et al. (1996) introduced the Āstāneh–Dāmghān fault system as the major active fault in the area with late Quaternary evidence of deformation (Figure 11.3; Plate 11.1).

The damage data discussed and scrutinized in Berberian and Yeats (2014) and in this study do not support the maximum destruction region covering an area of 181 × 45 km with several active faults and cutting the structural trend of the area with M_s ~7.9 earthquake, as presented by Ambraseys and Melville (1982), or a maximum destruction area 150 km long presented by Hollingsworth et al. (2010), and are not supported by the existing data

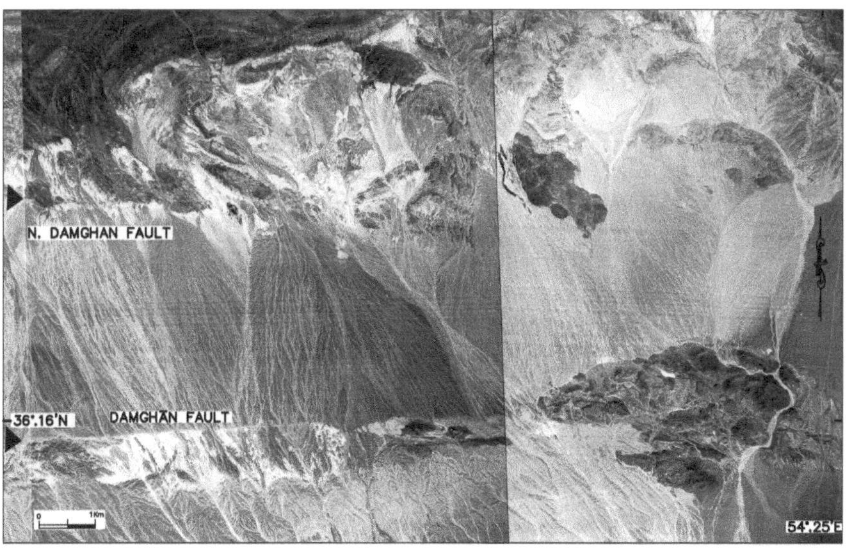

PLATE 11.1 Aerial photograph of the Dāmghān and North Dāmghān faults in the area north of the city of Dāmghān (old Komesh). See Figures 11.1 and 11.3 for the location and the historical seismicity. Worldwide Aerial Surveys, Inc. (1956), Project 158. Cartographic Department of the Imperial Army of Iran, Joint Operations Graphics, USA, WWS (Berberian, 1976b).

(Figure 11.3). Hence, the statement of Hollingsworth et al. (2010) that "*the 856 AD Qumis earthquakes is the most devastating to have affected Iran in recorded history, and the fifth most destructive globally*" cannot be warranted.

Based on the empirical relations between surface fault rupture and earthquake moment magnitude (Wells and Coppersmith, 1994; though we should not place much weigh on), the 65-km-long E–W left-lateral strike-slip Dāmghān fault (from the fault bend in the west to its eastern tip; Figure 11.3; Plate 11.1) is capable of generating an M_w ~7.2 earthquake with ~2.4 m left-lateral displacement (Berberian and Yeats, 2014). The extant historical monuments at Dāmghān (former Komesh) show that no major earthquake has occurred in the area since the 856 event. Nonetheless, two earthquakes took place in the area in 1102 and 1852 (Ambraseys and Melville, 1982; Berberian, 1994; Berberian et al., 1996) for which little macroseismic information is available. The ISC and USCGS epicenter of the 10 November 1967 *mb* 5.0, which was felt in Gorgān (SE of the Caspian Sea) is located near the Āstāneh fault (Figure 11.3; Plate 11.2). As mentioned earlier, the ca. 1200 BCE Komesh earthquake might have taken place in the same area. Properly designed paleoseismic trenches along the Dāmghān and Āstāneh faults are required to resolve the outstanding issue of the 856 and ca. 1200 BCE events.

11.3.1 Paleoseismicity

Without offering supporting data, Hollingsworth et al. (2010) assumed that the 856 earthquake took place along the Āstāneh fault (32 km west of the destroyed city of Dāmghān). Furthermore, they did not rule out reactivation of either the Dāmghān or North Dāmghān faults during the 856 earthquake (Figure 11.3; Plate 11.1). Hollingsworth et al. (2010) likewise assumed that the destruction of the Dāmghān region during the 856 earthquake might also have resulted from multiple slip events or clusters of earthquakes along the Āstāneh fault (Figure 11.1; Plate 11.2) over a short period of time. This cannot be confirmed: almost all the chronicles refer to the occurrence of a single earthquake on 22 December 856 in Dāmghān.

A paleoseismology study of a single trench within a sharp bend across the Āstāneh fault (Figure 11.3) indicated three events since 11.6 ka with very large error bars: (i) 9600–4600 BCE, (ii) 4600–600 BCE, and (iii) 600 BCE–1300 CE, with an estimated average repeat time of 3.7 ± 4.2 kyr (Hollingsworth et al., 2010). The trench is located at the fault bend, and the ages of faulting are poorly constrained. The authors concluded that their last event (600 BCE–1300 CE) is consistent with the 856 earthquake. However, it should be noted that during the time interval of 600 BCE–1300 CE, a less well-documented earthquake took place with some effects at the Gerdkuh fortress (Dezh-e Gonbadān; 36°09′N–54°09′E, +1578 m), 18 km west of Dāmghān in 1102 (Ambraseys and Melville, 1982) (Figure 11.3). However, no record of damage to Dāmghān was reported. Hollingsworth et al.'s (2010) trench on the

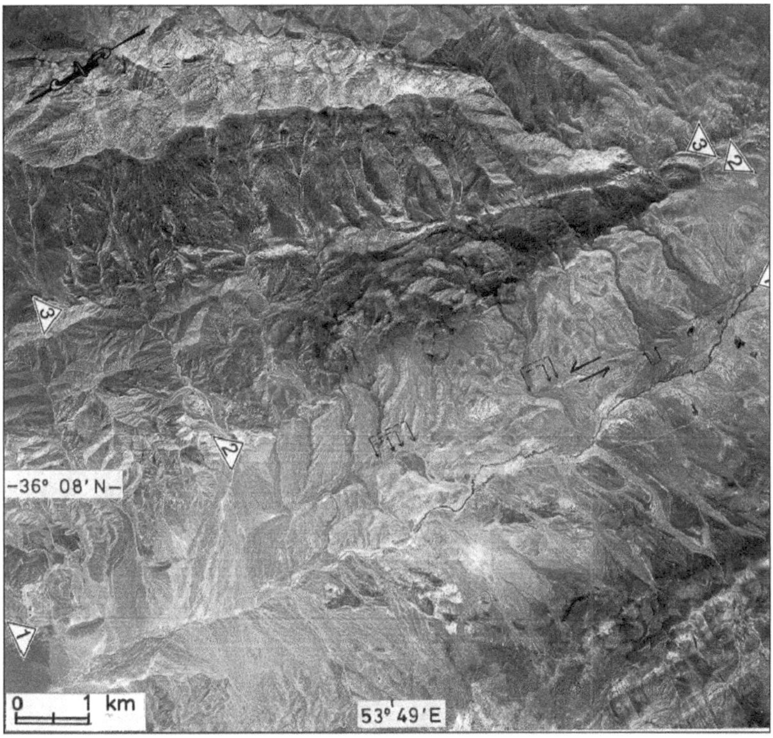

PLATE 11.2 Aerial photograph of the Āstāneh left-lateral strike-slip fault (triangle number 1) in the area west-northwest of the city of Dāmghān (old Komesh). See Figures 11.1 and 11.3 for the location and the historical seismicity of the area. Worldwide Aerial Surveys, Inc. (1956), Project 158. Cartographic Department of the Imperial Army of Iran, Joint Operations Graphics, USA, WWS.

Āstāneh fault may give evidence for a rupture consistent with the 856 earthquake, but the displacement in the trench offers no support. They also suggested that the Dāmghān and North Dāmghān faults participated in the 856 earthquake, but offer no evidence for this interpretation (Figure 11.3; Plates 11.1 and 11.2).

11.4 THE 913 DINÉVAR EARTHQUAKE GROUND DEFORMATION

'Arib ibn Sa'ad al-Qurtubi (932), Ibn al-Jauzi (1181), and al-Suyuti (1499) stated that, Al-Tall [lit., "hill" in Arabic], a mountain near Dinévar fissured and collapsed and water emerged from beneath it that flooded a number of villages. It is likely that the apparent landslide and flooding were triggered by an earthquake with surface faulting along the Zāgros Main Recent fault. Dinévar [Dināvar; 34°37′N–47°28′E, +1471 m] is located on the Zāgros Main Recent fault scarp on a triangular-shape landslide at the valley leading from the northeast (Figure 11.4; Plate 11.3).

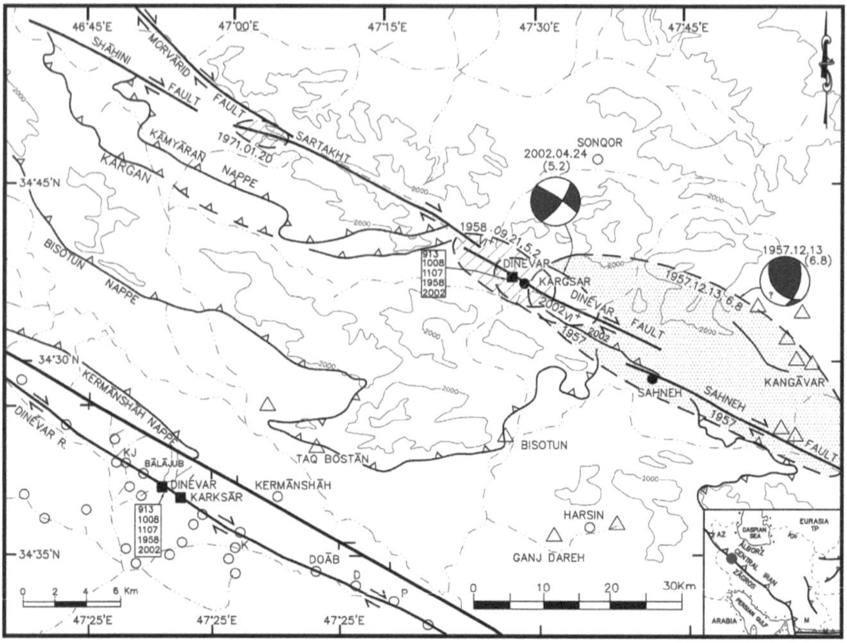

FIGURE 11.4 Historical seismicity along the Sahneh, Dinévar, Sartakht, and Morvārid segments of the Zāgros Main Recent right-lateral strike-slip fault in western Iran. Symbols as in Figure 11.1. The 1957 and the 2002 fault plane solutions: McKenzie (1972), and best-double-couple Harvard CMT solutions, respectively. Inset lower left: Dinévar fault segment. DB, Dastjerd Bālā; K, Karaj; KJ, Kalehjub. See also Plate 11.3. *Modified from Berberian and Yeats (2001).*

In the Dinévar-Nahāvand region (Figure 11.4), the Zāgros Main Recent fault sustained surface rupture in two 20th-century medium-magnitude earthquakes: the Fārsinaj earthquake on 13 December 1957, M_w 6.8 (Dinévar underwent an intensity of VII), and the Firuzābād earthquake of sequence of 14 and 16 August 1958 (M_w 5.7 and 6.8, respectively). The 1957 earthquake was accompanied by surface rupture on the Sahneh and Dinévar strands, and the 1958 earthquake resulted in surface rupture on the Nahāvand strand and a portion of the Garrin strand of the Zāgros Main Recent fault. The 21 September 1958 M_s 5.2 Kargsār earthquake (see Figure 12.12 in Chapter 12) was associated with ground rupture along the Zāgros Main Recent fault (see Chapters 12 and 16 for discussion; Ambraseys et al., 1973; Ambraseys and Moinfar, 1974; Tchalenko and Braud, 1974; Berberian, 1995). The Godin [Gowdin] and Kangāvar archeological sites with archeoseismic indicators (Berberian and Yeats, 2001; see also Chapters 10 and 12) lie within the intensity VII (MMI) isoseismal line of the 1957 earthquake, and those sites and the Giyān mound lie within the intensity VII (MMI) isoseismal line of the 1958 earthquake (see Chapter 12, and Figures 12.11 and 12.12).

PLATE 11.3 Aerial photograph of the Dinévar (D)-Karksār (K) area along the Zagros Main Recent fault (marked by NW-SE trending black arrows in the middle left and lower right, and ZMRF). For location, see Figure 11.4. Worldwide Aerial Surveys, Inc. (1956), Project 158. Cartographic Department of the Imperial Army of Iran, Joint Operations Graphics, USA, WWS. The scale bar is in the lower left (0-1-2 km). *Modified from Tchalenko and Braud (1974).*

Recorded data show that the Dinévar region experienced earthquakes in 913, 27 April 1008 (M_s~7.0), September 1107 (M_s~6.5), 13 December 1957 (M_w 6.8), 21 September 1958 (M_s 5.2), and 24 December 2002 (*mb* 4.6) (Figure 11.4). Regarding the pattern of coseismic earthquake surface rupture along the Zāgros Main Recent fault and the historical seismicity of the area, see Chapter 16.

11.5 THE 6 AUGUST–3 SEPTEMBER 943 M_S ~7.4 EARTHQUAKE-INDUCED GROUND DEFORMATION OF THE SAMALQĀN [ASHKHĀNEH] VALLEY

Abu Dulaf (950) witnessed the catastrophic ground subsidence of about 600 ft. (182.8 m, a highly exaggerated figure) and the pouring of water into the sunken ground from each direction in the Samalqān Valley (modern Āshkhāneh) caused by the 943 earthquake. Almost a century later, the near-contemporary Gardizi (1048) mentioned that in Dhu'l-Hijja 331

[6 August–4 September 943], a catastrophic earthquake in the districts of Nessā destroyed many villages, killing more than 5000 people. This statement was later followed by Ibn al-Athir (1231) and al-Suyuti (1499). Ambraseys and Melville (1982) assigned a surface-wave magnitude of M_s 7.6 for this event, which seems to be an overestimate.

Abu Dulaf (ca. 950) was in Neyshābur when he heard of 30 villages that had sunk into the ground in the Samalqān or Shoqān valley. He returned to see the situation and wrote that:

Thence I travelled in the open country (mafāza) of Khwārazm *[Khārazm]* and saw here many traces of Arab and non-Arab kings. Here there are plenty of trees and woods. Snow does not fall here but it rains continuously and hardly ever stops. (This open country) adjoins the districts of Nisabur [*Neyshābur*] and a district known as AS.S. QĀN (al-Shoqān, Suqān?) [*Esesqān/Samalqān; the present Āshkhāneh*]. Some years (ago) over thirty of its villages sank (into the ground). A hurricane passed over [Suqān?], and lifting red sand from fractures of the ground and carried it in the air over the distance of 150 (farsakhs?) [*900 km*] and deposited the sands in the Nisabur and Tus districts. This is what I witnessed and what I learnt myself because I crossed this district when it was extremely flourishing and had numerous gardens and running (takharruq) streams. But hardly had I settled in Nisabur when the news reached me that (Suqān?) had sunk and I went back to look at it. I saw that it had sunk into the Earth to (a depth) of some 100 man-heights or more and waters were flooding it coming from (every) sides.

Abu Dulaf, ca. 950; ed. Minorsky, 1955; ed. Tabātabā'i, 1975; corrected from the original Arabic and Persian versions.

The editors of the Arabic (Muhammad Zaki Khalil, 1953; in Minorski, ed. 1955), the English (ed. Minorski, 1955), and the Persian (Tabātabā'i, 1975) versions of Abu Dulaf's text of ca. 950 added in parenthesis the "Shuqān," after the original name of "Esesqān" given by Abu Dulaf himself. Esesqān is the corrupt form of Samalqān [or Esbinqān; the modern Āshkhāneh (Māneh and Samalqān)] located on the southern embankment of the Atrak River, to the north of the Ālā Dāgh Mountains of the eastern Alborz (Figure 11.5). The Samalqān valley is a prosperous, green valley with numerous water resources, arable fields, and villages. In contrast, the Shoqān Valley, located to the south of the Ālā Dāgh Mountain, is an arid region with few villages, limited fields, and scarce water resources (Figure 11.5). Therefore, Abu Dulaf was definitely referring to the Samalqān [Esesqān in Abu Dulaf's original Arabic text] valley and not Shoqān, as stated by the three editors of his original manuscript.

Regarding the source of the earthquake, it should be noted that Samalqān/Ashkhāneh is located at the foot of the Ashkhāneh thrust fault, which does not show impressive Quaternary evidence of movement, whereas the Shoqān Valley [25 km south of Samalqān/Ashkhāneh] is located at the foot of the Shoqān (Robāt-e-Qarabil) fault, showing young Quaternary evidence of movement with large vertical throw of the plain and numerous young

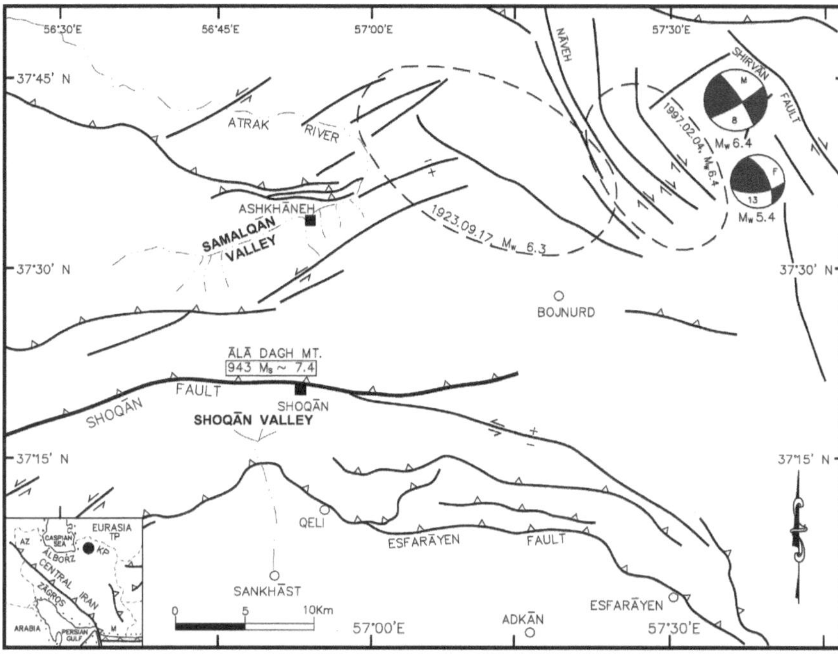

FIGURE 11.5 Location of the 943 earthquake, southwest of the Central Kopeh Dāgh Shear Zone (CKDSZ). Symbols as in Figure 11.1. Focal mechanisms of the 1997 Nāveh earthquakes in the Central Kopeh Dāgh Shear Zone (CKDSZ) constrained by body wave modeling with centroid depths: Hollingsworth et al. (2006, 2007). F, Foreshock; M, Mainshock. The Shoqān fault might have some left-lateral motion (the fault has not been studied).

triangular facets (Figure 11.5). It is, therefore, highly probable that an earthquake on the Shoqān fault [dipping north toward the Samalqān/Ashkhāneh valley (located on the hanging-wall block of the fault)] triggered ground failure. The ground was already saturated by heavy rains, as described by Abu Dulaf (950) when he passed the area. Hence, the slopes were already unstable prior to the earthquake. The earthquake had significant effects on the streams, possibly disrupting and damming the flow of water in the valley, and overwhelmed more than 30 flourishing villages in the Atrak Valley (Figure 11.5).

11.6 THE 23 FEBRUARY 958 M_S ~7.1 RUYĀN EARTHQUAKE GROUND DEFORMATION

A number of individual earthquakes [at least the two events of the 23 February 958 Ruyān in the Alborz (Figure 11.2), and the 958 April Holvān (Sar-e Pol-e Zahāb in the Zāgros)] are confused and combined in historic accounts (contemporary Ibn al-‘Amid, in Biruni Khārazmi, 1025; Ibn Miskawayh, d. 421/1030; ibn al-Jauzi, 1181; ibn al-Athir, 1231; al-Dhahabi, 1315; Bar Hebraeus, 1286; al-Suyuti, 1499; Ibn al-‘Imad, d. 1678). Despite this setback, it is clear that the

Ruyān earthquake (Figures 11.2 and 11.6) was associated with widespread landslide/rock avalanche; it is very likely that it was also associated with extensive coseismic surface faulting (Ambraseys, 1974a; Ambraseys and Melville, 1982; Berberian, 1994; Berberian et al., 1985; Berberian and Yeats, 1999, 2001) along the Moshā fault (Berberian and Yeats, 2014).

Biruni Khārazmi (1025) quoted Abu al-Fazl ibn al-'Amid's account of the effects of the 958 Ruyān earthquake. Ibn al-'Amid was a writer learned in several branches of scholarship who became prime minister for the Buyid ruler Rokn al-Dauleh [Rukn al-Dawla, r. 935–976]. The contemporary author reported: "*Ibn al-'Amid related in his book, Fi Binā' al-Mudun* [On the Construction of Cities], *that an earthquake took place in Ruyān, not long ago, which made two mountains collide and tumble down, and that the debris of the collision blocked the course of the rivers which ran between them, and that the waters of the rivers receded and formed a lake. This is what usually happens when the water has no outlet, like the Dead Sea which is formed by*

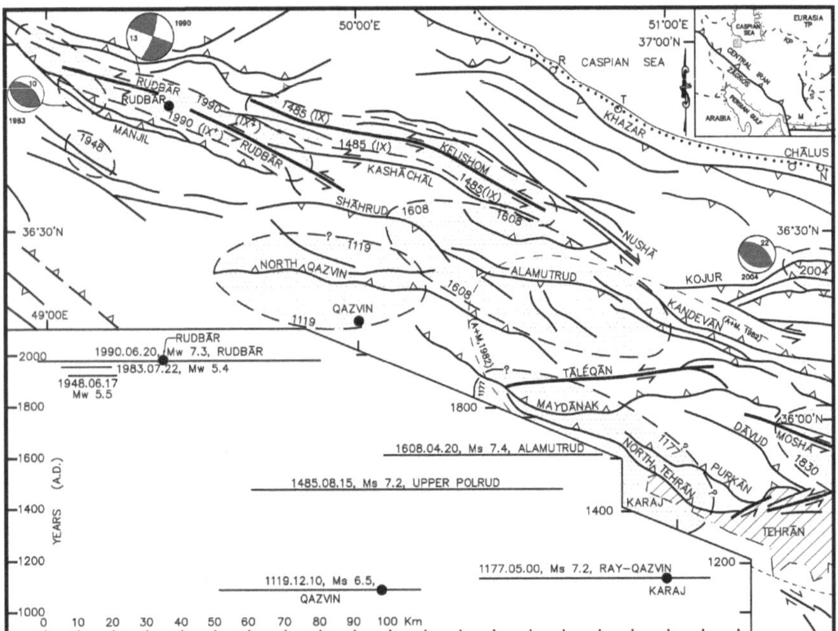

FIGURE 11.6 Meizoseismal areas of the large-magnitude historical earthquakes since 958 (top) and time–space diagram of the earthquakes (bottom) of the area northwest of the capital city of Tehran (hatched; lower right), along the southern margin of the Alborz Mountains. Queried where uncertain. Symbols as in Figure 11.1. The highly exaggerated meizoseismal area of the 958 earthquake by Ambraseys and Melville (1982) is marked by (A+M, 1982). The 958 earthquake took place in the meizoseismal area of the 1830 event to the SE (lower right corner of the figure; see Figure 11.2 and Berberian and Yeats, 2014). Focal mechanisms of the earthquakes constrained by body wave modeling with centroid depths: 1983 and 1990 (Berberian et al., 1992; Berberian and Walker, 2010); 2004 (Tatar et al., 2007). *Modified from Berberian et al. (1983, 1985), Berberian and Yeats (1999, 2001, 2014), Berberian (2005), Berberian and Walker (2010), and Berberian (2005).*

the water of the Jordan River" (Biruni Khārazmi, 1025; ed. 'Ali, 1967, quoting contemporary writer ibn al-'Amid) (Berberian and Yeats, 2014).

Large landslides along rivers are known in the Mobārakābād area (along the Moshā fault: 35°46′N–51°58′E) and at Galukān (south of Ushān along the Lashgarak River, south of the Moshā fault; 35°51′N–51°32′E), both are located farther east of Ruyān and are not to be mistaken for the earthquake mentioned by Ibn al-'Amid. Landslides also developed farther west in the Tāléqān area to the west of Ruyān (Berberian et al., 1985; Berberian, 1994; Figures 11.2 and 11.6).

In Persian, *Ruyān* [or *Rudān*] means a mountainous area with numerous rivers [Rud = River]. According to Ibn al-Faqih (903), Istakhri (951), Ibn Hawqal (978), and Yāqut (1225), the Ruyān and the Ray Mountains are connected, and the way to Ruyān is from Ray (Figure 11.2). The Ray Mountains correspond with the crest of the present Tochāl Mountains and the southern Qasrān [Arabicized "Kuhsārān"; i.e., Mountains] north of Tehrān (in the south), and the Ruyān Mountains (in the north) covering the Lavāsānāt, Inner Qasrān [Kuhsārān], and Rudbār-e Qasrān mountainous districts (Sotudeh, 1970; Karimān, 1977). Therefore, the location of Ruyān given by Le Strange (1905) as in the area southwest of Kalār (Kalārdasht; SW Chālus: 36°30′N–51°09′E, and the Valasht Lake at 36°32′N–51°17′E) is not justified (Berberian and Yeats, 2014). The old city of Ruyān (later called Kojur: 36°23′N–51°43′E, +1495 m) was located farther to the north (Rabino, 1928; Stark, 1934). This suggests that the earthquake might have been associated with the Moshā fault to the east of the Tāléqān fault (Berberian et al., 1983; Berberian and Yeats, 2001, 2014; see the latter for the revised location of the meizoseismal area). Properly designed paleoseismic trench study is needed to find the source fault of this event (Berberian and Yeats, 2014).

Two-hundred twenty-three years after the earthquake, Ibn al-Jauzi (1181), in describing the 958 earthquake at Holvān and Baghdād, wrote about two events at Ray [23 February 958] and Holvān [April 958], and added Tālèqān for the first time. Tālèqān was not part of Ray or Ruyān (Figure 11.6). He wrote:

> In the year 346H *[5 April 957-24 March 958]*, in the region of al-Ray and its dependencies, occurred a dreadful earthquake; the calamity extended as far as Khalwan *[Holvān; referring to the April 958 Holvān earthquake]*, the greater part of which was engulfed and the Earth vomited up the bones of the dead, and waters burst forth. A mountain in al-Ray was cleft asunder, and a village with its inhabitants was suspended between heaven and Earth during half a day, then it was swallowed up gradually. The Earth was rent in a mighty chasm and fetid waters came forth with volumes of smoke. Also at Tāléqān, the ground was rent and in places sunk into the Earth, and there escaped of its inhabitants only about 30 persons, and 150 of the villages of al-Ray were swallowed up. *(Ibn al-Jauzi, 1181; Ambraseys, 1968, 1974a, Nabavi, 1972; Berberian et al., 1985; Berberian and Yeats, 2014).*

It is not clear if Ibn al-Jauzi (1181) is describing two different events at Ruyān/Ray and at Tāléqān. He definitely amalgamated two different earthquakes of Ruyān and Holvān.

This statement was later reported by al-Suyuti (1499), about 541 years after the earthquake. For the first time, al-Suyuti (1499) reported two events and then added Tālèqān. It is not clear if there was a break in the aftershock activity, or if the Tālèqān event took place sometime after the one at Ray (Ruyān): "*At Shahr-e Ray and about that town; it lasted forty days, then it discontinued for some time, but it again returned. The earthquake extended to Tāléqān in particular. At Shahr-e Ray 150 settlements sunk into the ground, and a mountain sunk, and an enormous chasm opened from which water and smoke gushed out*" (Ambraseys, 1961, 1968, 1974a; Nabavi, 1972; Berberian et al., 1985; Berberian and Yeats, 2014).

Ambraseys and Melville (1982) presented a meizoseismal area in the shape of an elongated ellipse that was 150 km long and 60 km wide covering an area from Alamut in the northwest to Palangvāz (confluence of the Jājrud and the Damāvand Rivers) in the southeast (Figures 11.2 and 11.6). The historic data do not support such a vast meizoseismal area. Therefore, the magnitude M_s 7.7 of this event presented by Ambraseys and Melville (1982) seems to be exaggerated. The unconstrained meizoseismal area given by Berberian and Yeats (2001) was mainly based on (i) Ibn al-Jauzi [secondary source, 223 years after the event] and (ii) the location of the 1830 meizoseismal area to its east. We cannot constrain the alleged damage area in Tāléqān (Figure 11.6; for the revised location of the 958 earthquake see Figure 11.2 and Berberian and Yeats, 2014). Regarding the pattern of coseismic surface rupture and historical seismicity of the area, see Chapter 16.

11.6.1 Paleoseismicity

Nazari et al. (2009a) stated that the first event observed in their trench study across the Tālèqān fault (Figure 11.6) could coincide with the 958 and/or the 1177 earthquakes [the distance between the trench across the Tālèqān fault and Ray is about 65 km]. The authors showed two to three (?) unconstrained events with estimated magnitudes M_w >7.0 during the past 5300 years with large error bars and a maximum estimated recurrence period of 3760 years. Their estimated minimum horizontal and vertical slip rates are 0.6–1.6 and ~0.5 mm/year, respectively.

The youngest event described by Nazari et al. (2009a) is considered to be younger than 80 CE by an interpretation of colluvial units without evidence for a clear faulting episode. Hence, no displacement could be measured for this event (if any). Nazari et al. (2009a) speculated that their colluvial unit 20 could correspond to a deposit following strong historical earthquakes of 855–856, 958, 1177, 1665, or 1830, which could be associated with the Tāléqān fault. The 855–856 earthquake was felt strongly at the city of Ray, 70–80 km southeast of the Tāléqān fault. The 1177 earthquake, which damaged the cities of Ray and Qazvin, could also be attributed to the western segment of the North Tehrān Thrust (Figure 11.6). The 1665 event took place along the Moshā fault, about 145 km southeast of the Tāléqān fault. Finally, the meizoseismal areas of the 958 and 1830 events were located about 40 km northeast of Tehrān and 80 km southeast of the Tāléqān fault (Figures 11.2 and 11.6; also discussed later; see Berberian and Yeats, 2014 for the proper location of the 958 Ruyān earthquake north of Ray in the meizoseismal area of the 1830 Lavāsānāt earthquake). A properly designed paleoseismic trench study with constrained dates should resolve this issue.

11.7 THE 27 APRIL 1008 M_S ~7.0 DINÉVAR EARTHQUAKE GROUND DEFORMATION

Al-Antāki (1066) wrote that Dinévar sunk into the ground and was completely destroyed. Over 16,000 people were killed, with even more perishing in landslides (Figure 11.4). About two centuries after the earthquake, Bar Hebraeus (1286) stated that:

> And in this year *[398H/17 September 1007–4 September 1008]* there was great abundance. And a kor of wheat was sold for twelve dinars. And a great earthquake took place in the city of Dinawar *[Dinèvar]*, and it destroyed very many buildings; sixteen thousand people were brought out from under the dust and buried, besides those which the Earth swallowed up when it was rent open. And, moreover, a violent black wind blew in Tagrith and Dakukah, and it destroyed many houses, and tore up very many palm-trees and olive-trees by the roots; and great ships were sunk in the Sea of Persia *[the Persian Gulf]*.
>
> Bar Hebraeus (1286)

As mentioned earlier (regarding the 913 earthquake) Dinévar [Dināvar; 34°37′N–47°28′E, +1471 m] is located on the Zāgros Main Recent fault scarp (Figure 11.4) and on a triangular-shape landslide at the valley leading from the northeast. The hills to the north are composed of Pleistocene unconsolidated and poorly cemented marly conglomerate and sandstone (Figure 11.4). Regarding the pattern of coseismic surface rupture along the Zāgros Main Recent fault and historical seismicity of the area, see Chapter 16.

11.8 THE 4 NOVEMBER 1042 M_S ~7.0$^+$ TABRIZ EARTHQUAKE GROUND DEFORMATION

The 1042 Tabriz earthquake, with a reported casualty list of 40,000 (Figure 11.7), was associated with widespread landslide/rock avalanches and very likely associated with extensive coseismic surface faulting along the North Tabriz fault (Berberian and Arshadi, 1976; Berberian, 1981a, 1994, 1997a; Berberian and Yeats, 1999).

Contemporary poet *Abu Mansur Qatrān Tabrizi* (1009–1072), who was residing in the city of Tabriz during the 1042 earthquake, wrote in his first

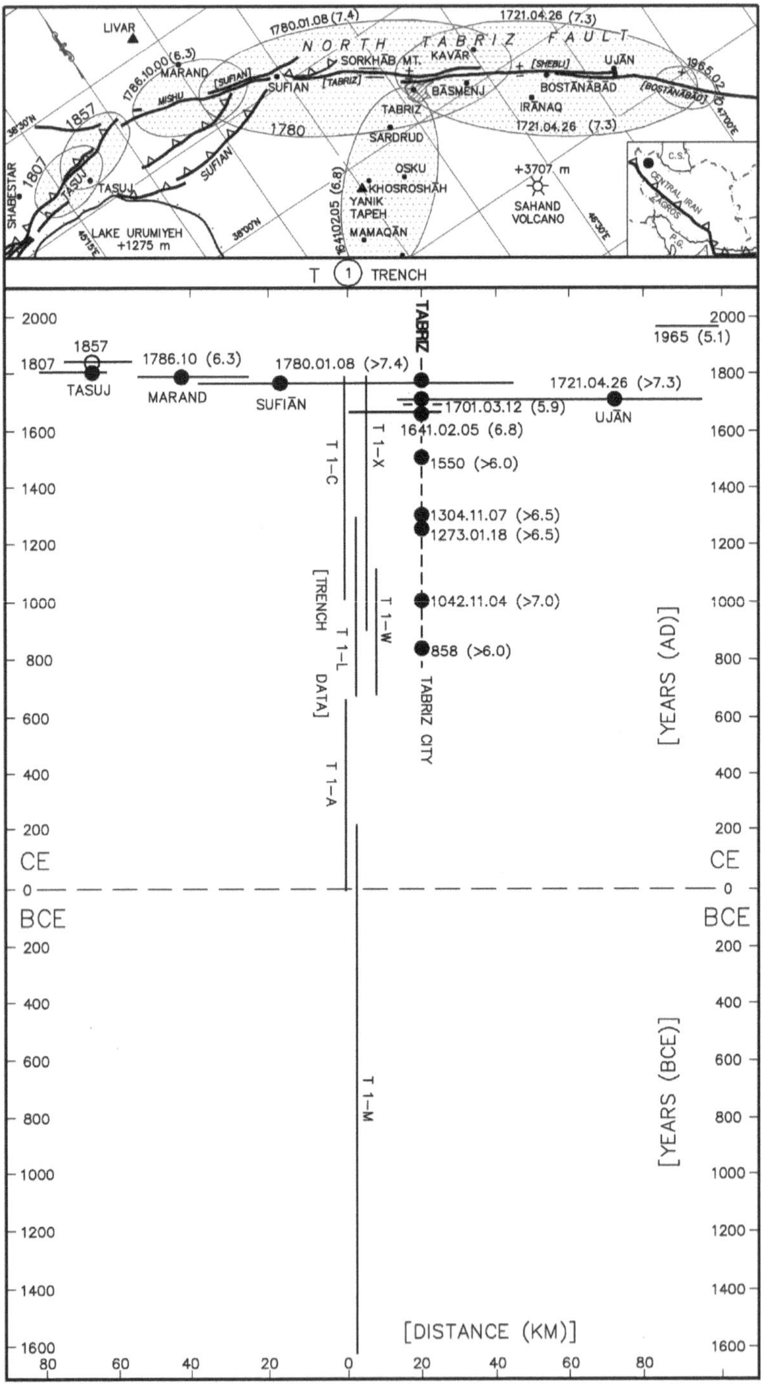
NORTH TABRIZ FAULT
LIVAR
1780.01.08 (7.4)
1721.04.26 (7.3)
1786.10.00 (6.3)
MARAND
SORKHAB MT.
KAVAR
UJAN
SUFIAN
TABRIZ
BASMENJ
BOSTANABAD
IRANAQ
1721.04.26 (7.3)
1965.02
1857
1807
1780
SARDRUD
OSKU
KHOSROSHAH
YANIK TAPEH
MAMAQAN
1641.02.05 (6.8)
+3707 m
SAHAND VOLCANO
SHABESTAR
TASUJ
LAKE URUMIYEH
+1275 m
C.S.
CENTRAL IRAN
ZAGROS
P.G.
T 1 TRENCH
1857
1807
TASUJ
1786.10 (6.3)
MARAND
1780.01.08 (>7.4)
SUFIAN
1721.04.26 (>7.3)
UJAN
1965 (5.1)
1701.03.12 (5.9)
1641.02.05 (6.8)
1550 (>6.0)
1304.11.07 (>6.5)
1273.01.18 (>6.5)
1042.11.04 (>7.0)
858 (>6.0)
TABRIZ
TABRIZ CITY
T 1–C
T 1–X
T 1–L
T 1–W
T 1–A
T 1–M
[TRENCH DATA]
[YEARS (AD)]
[YEARS (BCE)]
CE
BCE
[DISTANCE (KM)]

qasideh poem that the topography of the ground changed, the ground fractured, trees bent, and mountains slid down during the earthquake (Chapter 7).

The city of Tabriz has experienced great earthquakes in 858, 4 November 1042, 18 January 1273, 7 November 1304, 1550, 5 February 1641, 12 March 1717, 26 April 1721, and 8 January 1780 (Figure 11.7; Berberian and Yeats, 1999). For patterns of coseismic surface rupture along the North Tabriz fault and historical clustered seismicity of the area, see Chapter 16.

11.8.1 Paleoseismicity

Limited paleoseismological trenches in separate areas across the fault indicate at least three strong earthquakes since 33.5 ka. The first trench was dug in the area to the northwest of the city (Hessami et al., 2003) in the meizoseismal area of the 1780 earthquake; the second was located to the southeast of the city (Solaymani Azad, 2009; Solaymani Azad et al., 2009) in the epicentral area of the 1721 earthquake (Figure 11.7).

Based on the lack of properly dated materials, poorly constrained ages of the events, and a lack of correlation between the two nearby trenches [100 m apart; 20 km NW of Tabriz], Hessami et al. (2003) defined four surface faulting paleo-earthquakes during the past 36.0 ka with unconstrained dates along the northwestern segment of the North Tabriz fault:

1780 CE [1190–1780]: The date is based on speculation that the youngest event observed in the trench was the 1780 documented earthquake.
910 ± 250 CE [660–1320].
320 ± 320 CE [1–640].
700 ± 920 BCE [1620 BCE–220 CE] (Figure 11.7).

With no radiometric dating of the displaced alluvial deposits, Hessami et al. (2003) reported a 4 ± 0.5 m horizontal slip/event, 3.1–6.4 mm/year minimum horizontal slip rate along the fault, and 0.5–0.8 mm/year vertical slip rates along the North Tabriz fault; these cannot be justified due to the speculative dates of the displaced alluvial deposits. With a high level of uncertainty, average recurrence intervals of 350–1430 years with a mean recurrence interval of 821 ± 176 years were also estimated (Hessami et al., 2003). This is a much larger value than 250 years (considering the entire set of recorded earthquakes

FIGURE 11.7 Earthquake chronology of the North Tabriz fault. Meizoseismal area of the well-documented, large-magnitude, historical earthquakes (top) and space-time diagram of historic earthquakes (since 858; both modified from Berberian and Yeats, 1999), and preliminary unconstrained paleoseismic evidence for surface faulting-rupture earthquakes (bottom; based on Hessami et al., 2003). Location of the trench 1 (two nearby trenches) in the area southeast of Khājeh Marjān is shown in between the two portions of the figure. Vertical bars indicate unconstrained ages. Symbols as in Figures 11.1 and 11.6. Distances are along the strike with respect to the city of Tabriz. Earthquake source does not seem to be periodic but seems clustered. See also Plates 11.6 and 11.7.

in Figure 11.7; Berberian and Yeats, 1999) and 700 years (considering the strongest events; see Berberian and Yeats, 1999; Figure 11.7). Hessami et al. (2003) concluded that the northwestern segment of the North Tabriz fault does not appear to present a major seismic potential for the near future, perhaps because the last clustered earthquake (1780) took place to the northwest of the fault (Figure 11.7).

The second paleoseismological study in the southeast of the city suggests that three earthquakes have occurred during the past 3300 years (Solaymani Azad, 2009; Solaymani Azad et al., 2009), implying a longer recurrence interval for earthquakes along the North Tabriz fault than those calculated from the historical seismic data (see Berberian and Yeats, 1999; Figures 11.7). See Chapter 16 for information about clustered earthquakes in the region and a pattern of coseismic rupture along the North Tabriz fault.

11.9 THE 1066 MAY M_S ≳6.5 QĀ'EN EARTHQUAKE GROUND RUPTURE

Based on a document sent to the capital city of Baghdād, Ibn al-Jauzi (1181) wrote that: "*In Jumada II 458* [30 April-28 May 1066] *there was an earthquake in Khorāsān which caused the mountains to split; it cleft hills, overturned towns together with their inhabitants, and it leveled them to the ground in such a way that but few people escaped; most buildings lay in ruins, and it is impossible to ascertain the number of those who perished*" (Ibn al-Jauzi, 1181; Ambraseys, 1974a). Similar accounts were reported later by Ibn al-Athir (1231), al-Dhahabi (1315), and al-Suyuti (1499). The "*splitting of mountains*" and "*cleaved hills*" may refer to coseismic surface faulting and/or mass slope slides in the meizoseismal area of the earthquake.

No precise indications exist as to where the shocks occurred, and later accounts tend to confuse the issue (Ambraseys, 1974a; Ambraseys and Melville, 1977, 1982; Berberian, 1994; Berberian and Yeats, 2001). However, archeological trenching beneath the foundation of the present Timurid congregational mosque of Qā'en [built in 1368; Figure 11.8] revealed collapsed walls of the earlier Seljuq structure [1000–1218] over broken and upside-down Seljuq pottery containing watermelon seeds and a few apricot pits, possibly as the result of an earthquake during the Seljuq dynasty (Nāderi, 1980; Boqrāt Nāderi, personal communication, 1991).

Archeological excavations at the Shāzdeh Hossein mound in the southern suburbs of the modern city of Qā'en (about 310 m to the SSE of the present mosque; Figure 11.8) revealed ruins of the old congregational mosque of Qā'en built upon ruins of an early seventh-century simple mosque, which itself was built on a Sassanid building foundation [224–642] with Pahlavi inscriptions (Sorush, 2007; Labbāf Khāniki, 2013). The excavations revealed that this was the famous congregational mosque visited and described by Istakhri (951), Jayhāni (976), Ibn Hawqal (978), the anonymous writer of

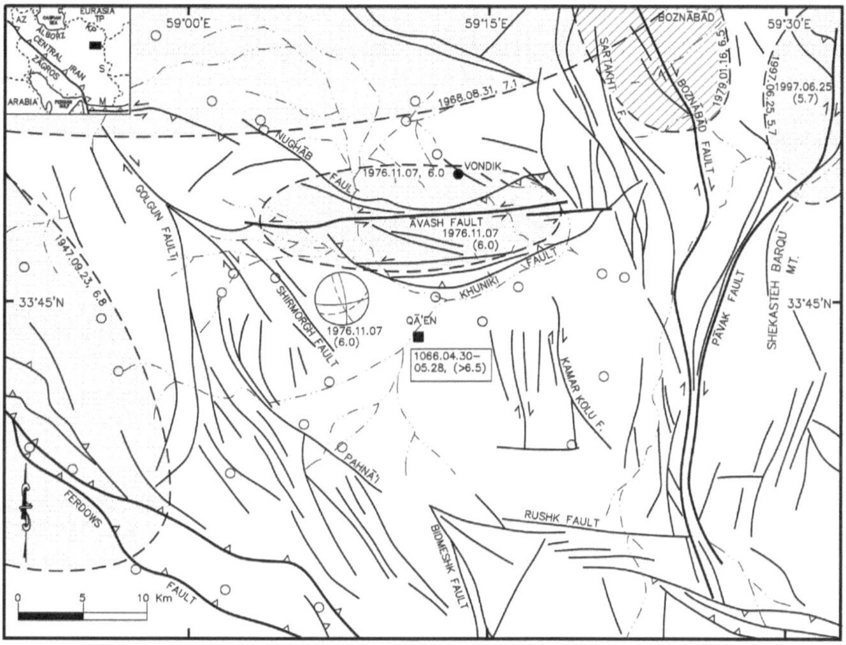

FIGURE 11.8 Location of the city of Qā'en (filled square) destroyed by the 1066 earthquake in the center of the interacting active faults of the area. Symbols as in Figure 11.1. Focal mechanisms of the 1976 Vondik earthquake with two subevents constrained by body wave modeling: Baker, 1993 (see Chapters 8 and 16). *Modified from Berberian and Yeats (1999, 2001).*

Hodud al-'Alam (982), Moqaddasi (985), and Nāsser Khosrow (1052). When Nāsser Khosrow (1052) visited the Qā'en old congregational mosque, it was known as the "*Great Mosque*"; he described it as "*the largest arch to be seen in all Khorāsān province.*" Qā'en, as the capital city of ancient Qohestān province in southern Khorāsān, was the largest city in the province and merited a large congregational mosque.

Information such as (i) the ruined nature of the old congregational mosque (Labbāf Khāniki, 2013), (ii) the fact that later writers such as Mostaufi Qazvini (1340) did not mention the great congregational mosque of Qā'en, (iii) the last reference to the intact mosque was recorded by Nāsser Khosrow (1052) (14 years before the 1066 earthquake), and (iv) the present location of the ruined mosque in the southern suburb of the modern city (congregational mosques are built in the city center) indicate that the great congregational mosque of Qā'en was completely destroyed by the 1066 earthquake and the survivors rebuilt a smaller town (the modern Qā'en) to the north of the larger ruined city (Figure 11.8).

The present Timurid congregational mosque of Qā'en (Wilber, 1955; Tābandeh, 1969; Nāderi, 1980; Golombek and Wilber, 1988) was built in 1368 on the site of a Seljuq [ca. 1000–1218] structure that also apparently collapsed during an earthquake (Nāderi, 1980). The mosque, which was restored in 1393, 1675, and 1847 (buttresses added), experienced intensities of VI (MMI) during the 31 August 1968 (M_w 7.1) Dasht-e Bayāz earthquake, after which the buttresses were replaced by a concrete reinforcement (Figure 11.8). The mosque was again damaged (intensity VII) during the Vondik, north Qā'en, earthquake of 7 November 1976 (M_w 6.0), and experienced an intensity of V (MMI) during the 10 May 1997 (M_w 7.2) Zirkuh earthquake (Berberian and Yeats, 1999; Berberian et al., 1999).

Although the destruction, which has been documented by archeological evidence and written accounts, is clearly related to an earthquake in Qā'en, it is not certain which of the many active faults near Qā'en was the source of the 1066 earthquake (Figure 11.8). The site is located 30 km south of the Dasht-e Bayāz Fault, 60 km west of the Ābiz Fault, 9 km south of the Āvash fault, and 20 km west of the Pāvak/Boznābād fault (Berberian and Yeats, 1999, 2001; Figure 11.8).

The complete destruction of the congregational mosque of Qā'en suggests that the earthquake source was close to the city; we thus reject the Dasht-e Bayāz and Ābiz faults as possible sources. The Āvash left-lateral strike-slip fault (Berberian and Yeats, 1999, 2001), in the mountains approximately 9 km north of Qā'en, the Pāvak/Boznābād right-lateral strike-slip fault, located 20 km east of Qā'en, and the southeastern segment of the Ferdows reverse fault, located 26 km southwest of Qā'en, are possible candidate sources for the May 1066 event (Berberian and Yeats, 1999, 2001). Furthermore, shorter faults are also located to the west (Shirmorgh fault) and east of Qā'en (Kamar Koḷu fault set; Figure 11.8). The Āvash Fault [~30 km long, with the 7 November 1976, M_w 6.0 Vondik earthquake], seems to be short for such a large-magnitude earthquake. In this case, the Pāvak/Boznābād fault system [with the 16 January 1979 M_w 6.5 Boznābād earthquake], and 'Kuh-e Shekasteh Barqu' [lit., the "Barqu Broken/Faulted Mountain"] on its eastern side (Figure 11.8), being 6 km closer to Qā'en than the Ferdows Fault, is a more suitable candidate for the 1066 earthquake (Berberian and Yeats, 1999, 2001). The Ferdows congregational mosque (1200) and the Soltān Mohammad 'Abed mausoleum of Kākhk (1553) postdate the 1066 event; the effect of this earthquake on these towns is, therefore, not known.

The pattern of coseismic surface rupture and clustered historical seismicity of the area is discussed in Chapter 16.

11.10 THE 1237–1238 AND WINTER 1678 M_S >6.5 GONĀBĀD EARTHQUAKES

Apparently, localized destruction with excessive casualties during the two earthquakes in 1237–8 and 1678 ruined the town of Gonābād (Tābandeh, 1969). The two E–W trending Gonābād and Bidokht faults to the immediate south of Gonābād with fresh surface displacements (aerial photograph Nos. 29305–29307, Worldwide Aerial Surveys, Inc., Project 158, scale 1:55,000) seem to be the responsible faults for these events (Berberian and Yeats,

2001). See also the 1678 event discussed later in this chapter and Figure 11.11 for earthquake location and the adjacent faults.

11.11 THE 19 OCTOBER 1336 (AND 21 OCTOBER 1336?) M_S ≥7.0 JIZD/ZUZAN EARTHQUAKE(S?) GROUND DEFORMATION

The contemporary Majd al-Din Khāfi (1342) mentioned the dates of two earthquakes in two poems: (i) 12 Rabi' I, 737 [19 October 1336] and (ii) Monday 14 Rabi' I, 737 [21 October 1336]. Majd al-Din Khāfi (1342) added that the city of Jizd was devastated and 20,000 were killed. Furthermore, he wrote that the leader, Malek Ghyāth al-Din Firuz, was killed when his palace collapsed on him. He did not mention the location of the palace; however, it is assumed to have been in the city of Zuzan (Figure 11.9). About

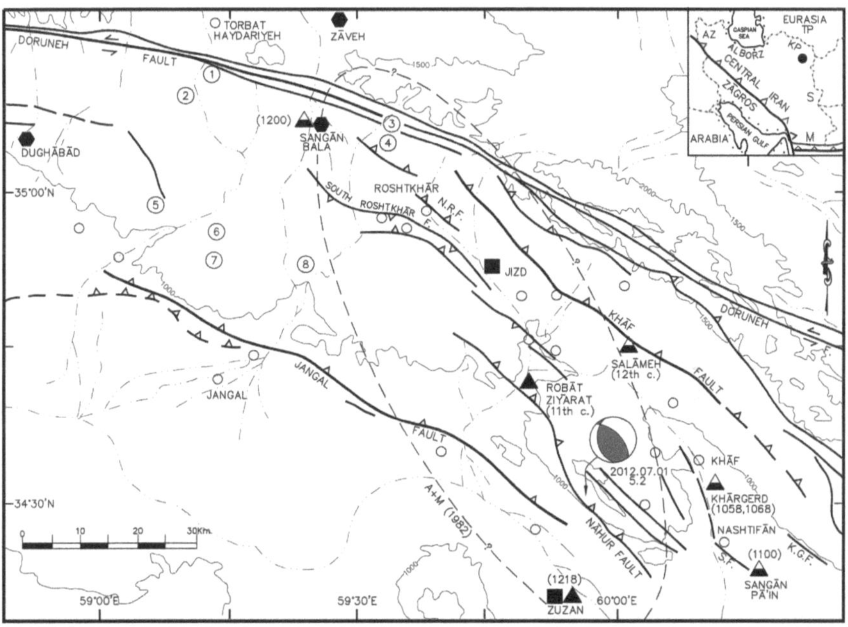

FIGURE 11.9 Epicentral area of the 1336 Jizd and Zuzan towns earthquake(s?) in eastern Iran. Symbols as in Figure 11.1. N.R.F, North Roshtkhār fault; S.F, Sangān fault. Filled squares: sites destroyed or heavily damaged. Filled triangles: destroyed or damaged archeological sites. Half-filled triangles: destroyed or damaged archeological sites with unknown cause. Filled hexagons: cholera epidemic sites following the earthquake, from Sangān to Zāveh (old center of Torbat Haydariyeh) and Dughābād; 11,000 reported killed. Dashed-line ellipse with A+M (1982): speculative and unconstrained meizoseismal area of the 1336 earthquakes from Ambraseys and Melville (1982), crossing various fault lines. Ancient toponyms of oral earthquake hazard warning: 1—Shekasteh (broken/faulted/fractured) Chāhuk. 2—Shekasteh (broken/faulted/fractured) Malaki. 3—Shekasteh (broken/faulted/fractured) Galehband. 4—Shekasteh (broken/faulted/fractured) Gushmir. 5—Shekasteh (broken/faulted/fractured) Golnay. 6—Shekasteh (broken/faulted/fractured) Band-e Mil. 7—Shekasteh (broken/faulted/fractured) Hājimir. 8—Shekasteh (broken/faulted/fractured) Fāzelmand. *Modified from Berberian and Yeats (1999, 2001) and Berberian (2005).*

a century later, historian Fasihi Khāfi (1442) gave just one date, the year of the earthquake in 737H [1336 CE], and stated that 30,000 people were killed in the boroughs of Zuzan and Jizd. The distance between the two ruined cities of Zuzan (in the south) and Jizd (in the north) is about 60 km (Figure 11.9).

Ambraseys (1975), Ambraseys and Melville (1977, 1982), and Ambraseys and Jackson (1998) introduced a 100-km-long coseismic surface fault rupture associated with the 21 October 1336 event with an azimuth of N155°E. However, none of the authors provided a fault map associated with this earthquake. In order to cover both destroyed cities of Jizd [in the north] and Zuzan [60 km to the south], Ambraseys and Melville (1982) presented a meizoseismal area as an elongated ellipse of 115 km long, striking N155°E. All the active faults in the area have a strike of N130°E; there is a 25° difference between the long axis of the unconstrained meizoseismal area and the trend of the active fault lines in the region (Figure 11.9).

Study of the aerial photographs, satellite imagery, and field survey of the region show at least six active surface faults cutting the recent alluvial deposits. These are (Figure 11.9):

The Khāf fault: ~80 km long (Berberian, 1981a) [aerial photograph Nos. 28859, 28860, 26908, 28780, 28759, 28859, and 28762, Worldwide Aerial Surveys, Inc., Project 158, scale 1:55,000].
The Jangal thrust: ~90 km long (Berberian and Yeats, 1999, 2001) [aerial photograph Nos. 26920, 26963, and 29436].
The South Roshtkhār fault: ~60 km long [aerial photograph Nos. 28779 and 28759].
The North Roshtkhār fault: [aerial photograph No. 28780].
The Kāl-e-Guryāb fault: [aerial photograph No. 29443].
The Sangān-e-Pā'in fault (see Figure 11.9).

The city of Jizd, now a mound south of the Akbarābād village [34° 52′N–59°43′E, +1058 m], is located in the Khāf valley, about 62 km north northwest of Zuzan [34°21′N–59°52′E, +825 m] and 50 km northwest of Khāf [Khāfud/Rui; 34°34′N–60°08′E, +971 m]. Jizd city is situated adjacent to the south Roshtkhār fault and about 7 km to the west of the Khāf fault (Figure 11.9). Since 1979, the ancient town of Jizd [now a mound of dirt] has become a destination for illegal excavators, looters, and heritage thieves.

As with the city of Jizd, the city of Zuzan (Figure 11.9) was completely destroyed, including the palace of Ghiyath al-Din Firuz, the Lord of Zuzan (he died under the collapsed debris, as mentioned above). The only remaining structure was the badly damaged (with collapsed parts of walls and the main vault) Hanafite mosque and School/College [madrasa] of Zuzan built in 1218 (Godard, 1949; Wilber, 1955; Meshkāti, 1970; Blair, 1985; Khosravi, 1987;

'Adle, 1988). The remaining structure shows long and deep fractures, especially in the south ayvān (portico) of the school (see Figures 1 and 9 in Blair, 1985), and the west ayvān of the mosque (Figure 3 in 'Adle, 1988). The school and the mosque, as well as the palace of the Malek of Zuzan, were destroyed during the 1336 earthquake.

Mostaufi (1340, ed. Le Strange, 1919) wrote that "*The chief towns of its* [Khāf district] *dependency are Salāmeh, Sanjan* [Sangān], *and Zawzan* [Zuzan]*; and in this last Malik* [Malek, prince, governor] *Zawzani* [Zuzani] *built a mighty palace.*" Although Mostaufi wrote after the earthquake, the written accounts of Khāfi (1342), who stated that the palace was destroyed during the earthquake, clearly indicate that Mostaufi's information is based on a second-hand account of the pre-1336 earthquake destruction: "*Malek Ghiyāth al-Din Firuz, who, according to one of his intimates, rushed back and forth, when the earthquake struck, between the center of his palace [miyān-e kushk] and the veranda [Soffeh], and was saying the Day of Resurrection has arrived [qiyāmat āmad]. Suddenly the whole place was overturned on top of him and the king disappeared*" (Khāfi, 1342, ed. Farrokh, 1966).

Zuzan [34°21′N–59°52′E, +825 m] is located about 25 km to the southwest of the Khāf valley, about 45 km to the southwest of the Khāf fault and 35 km south of the South Roshtkhār fault. At these distances, a complete destruction of the city and its properly built mosque and school (Figure 11.9) is un realistic. The complete destruction of Zuzan and its nearby villages [including Robāt Ziyārat] may indicate that the Jangal active thrust fault (Berberian and Yeats, 1999, 2001) was possibly reactivated; Zuzan is located about 12 km south of the Jangal fault. This may indicate that there were probably two earthquakes, on October 19 and 21, as mentioned by Majd al-Din Khāfi (1342).

Local contemporary sources [Khatib Abolfazl Zuzani, in Fasih Khāfi (1442), both natives of the Khāf town], wrote that: "*In the year 737H* [1336 CE] *an earthquake took place in the Khāf district known as the Jizd earthquake, and Khatib Abolfazl Zuzani says that 30,000 people from the town of Zuzan to the town of Jizd were killed; and as a result of this earthquake, a cholera epidemic broke out in the Zāveh* [district] *which accounted for 11,000 death from Sanjān* [Sangān Bālā] *to Dughābād*" (Figure 11.9). This historic information shows that heavy destruction and disruption of the water supply system as well as a large number of dead bodies caused the epidemic cholera outbreak in the northern part of the epicentral area (Figure 11.9). If the contemporary native of Khāf, Majd-e Khāfi, is correct, it is likely that the first event [19 October 1336] took place along the Khāf fault, destroying the town of Jizd, and the second event [21 October 1336] was triggered on the Jangal thrust with destruction of the town of Zuzan (Figure 11.9). Further paleoseismic research may resolve the ambiguities.

11.11.1 Additional Archeoseismic Evidence

The Sangān-e Pā'in congregational mosque, thought to have been built in the late 1100s (Khosravi, 1987), is located about 22 km south of Khāf [34°23′N–60°15′E, +937 m] (Figure 11.9). It had two porticoes: the one standing opposite to the qebleh portico to the east was destroyed by the earthquake, and a Shabestān [nocturnal place of rest] was built in its place. The remains of the collapsed portico roof can be seen on the wall of the Shabestān. The columns of the qebleh portico [with an arched and ribbed roof] have been tilted so that the distance between the columns at the base is shorter than that under the ceiling (Khosravi, 1987). The Gonbad mosque of Sangān-e Pā'in was apparently built in 1140–1141 (Khosravi, 1987). The reported date is before the earthquake occurred; it is not clear whether it was rebuilt after the event.

The Khargerd [Khurgerd; lit., "the Sun City"; 34°32′N–60°10′E, +1031 m] congregational mosque, Nezāmiyeh Madrasa, and Madrasa Ghiyāthiyeh are located about 34 km northeast of Zuzan and 5 km southeast of Khāf (Figure 11.9). The effect of the 1336 earthquake on the 1058 Khargerd congregational mosque (Meshkāti, 1970) is not known. The 1068 Nezāmiyeh Madrasa (Herzfeld, 1935; Godard, 1965) is in ruins, possibly due to the 1336 earthquake, and only the south ayvān (portico) remained. The 1442–1446 four-porched Madrasa Ghiyāthiyeh (Wilber, 1955; Meshkāti, 1970; Golombek and Wilber, 1988), though built after the earthquake, shows large vertical fractures [see the figures on pages 101 and 102 in Daneshdoust, 1980].

The Khāfrud congregational mosque (1502–1503; Golombek and Wilber, 1988) is located 34 km northeast of Zuzan [34°34′N–60°08′E, +971 m] (Figure 11.9). The destroyed Robāt Ziyārat mosque, built in the eleventh century (Labbāf Khāniki, 1988), is located about 38 km north of Zuzan and 6 km south of Fadak [34°42′N–59°49′E, +1026 m], on the eastern bank of the Kāl-e Shur [Rud-e Fadak] River bed (Figure 11.9). Only the ruined walls, partially inscribed with large Kufic bands, are preserved. The mosque was definitely severely damaged during the 1336 earthquake.

The twelfth-century structure of the Kushk-e Salāmeh [Salāmeh Palace], which was ruined, possibly by the 1336 earthquake, and covered by piles of sand, was later rebuilt (Khosravi, 1987). The two domed roofs of the Sangān-e Bālā "Shāh Sanjān mausoleum" of early 1200 (Khosravi, 1987) are collapsed. The time and cause of the destruction is not known, however, so it could have happened during the 1336 earthquake.

A review of the historical and archeological data indicates that the two towns of Jizd and Zuzan as well as the eleventh-century mosque at Robāt Ziyārat were ruined (Figure 11.9). The archaeological monuments at Sangān Pā'in [1100 AD], Khārgerd [1058, 1068], Salāmeh [twelfth century], and Sangān Bālā [1200] were destroyed or damaged, but the cause of their destruction has not yet been documented. The congregational mosque of Khāf was built in 1502, and we have no data of the pre-1336 earthquake structure in Khāf.

11.11.2 Oral Earthquake Hazard Warning

The various hills on the hanging-wall block of the Jangal active thrust fault (faulted anticlines in the Neogene molasse; Berberian and Yeats, 1999, 2001) have local toponyms of *Shekasteh* (i.e., "broken" or "faulted"). Shekasteh Golney, Shekasteh Band-e-Mil, Shekasteh Haji Mir, and Shekasteh Fāzelmand indicate important oral earthquake hazard warnings from ancient times (see Chapter 1 for details and Figure 11.9 for the locations).

No major earthquake has been recorded in the Jizd and Zuzan area since the 1336 event(s?). On 1 July 2012, an earthquake of M_w 5.2 took place about 17 km north northeast of Zuzan on the hanging-wall block of the Nāhur thrust (to the SE of the Jangal thrust; a member of the Jangal thrust) with a NW–SE trending thrust CMT solution (HRVD) (Figure 11.9). The earthquake damaged several houses in Zuzan and in the villages of Neqāb, Baqsāni, Nashtinān, Qāsemābād, and Robāt Kāl Jangi.

11.12 THE 10 JANUARY 1493 M_S ~7.0 NAUZĀD EARTHQUAKE FAULTING

The account of *Esfezāri* (1493), who was a contemporary bureaucrat in the *divān* [chamber of law court; council of state] of Soltān Hussain Bāyqāra during the 1493 earthquake, seems to be the first known description by a contemporary observer in Iran of a coseismic surface rupture accompanying an earthquake (Berberian and Yeats, 1999). The earthquake took place along the Nauzād active reverse fault (Berberian, 1976c, 1981a, 1994) (Figure 11.10). Esfezāri (1493) wrote that:

> What happened in Qohestān[5] was an earthquake which took place before the midday prayers on Friday 21 Rabi' I, 898 [*10 January 1493*], affecting three villages of Qohestān which were accused of their intrepidity, bad belief, and following Hasan-e Sabbāh [*the Assassin*], God's curse upon him. In the villages of Nauzād and Mask of the Boluk [*rural district*] of Mo'menābād, as well as Darmiyān, the houses were leveled with the ground causing much destruction. Most of the inhabitants of Nauzād were perished and Mask suffered heavily with large death toll. There was no casualty in Darmiyān. It is reported that for two farsangs [~*12 km*] between Nauzād and Mask villages the ground was fissured to such a depth that the bottom of the fissures was invisible.
>
> Esfezāri (1493). Tate (1910–1912) almost repeated the information given by Esfèzāri (1493).

The distance between Nauzād and Mask is about 8 km, and both villages are located on the Nauzād reverse fault (Berberian, 1976c, 1981a, 1994;

5. Qohestān (Kārizèstān, Qanātestān). Qohestān is the Arabicized form of Persian "Kahestān," lit., the place flourished by Kah, water (i.e., Kahriz, Kāriz, or Qanāt[6]). It was the name of the ancient province of southern Khorāsān.

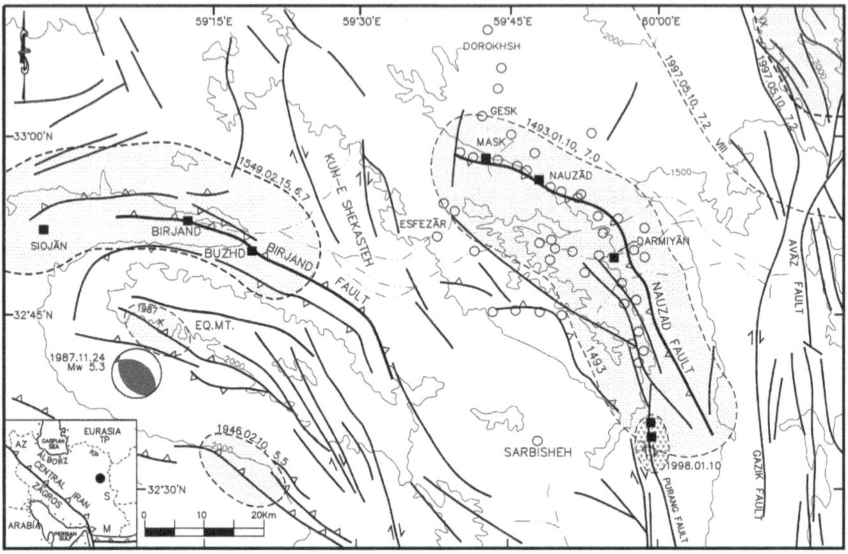

FIGURE 11.10 Epicentral areas of historical earthquakes in the Nauzād and Birjand area in eastern Iran. The 1493 Nauzād earthquake was associated with coseismic surface rupture documented by a writer of the time. The southern extension of the 1493 meizoseismal area is not constrained. Symbols as in Figure 11.1. Filled squares: villages destroyed by the earthquakes. 1987 focal mechanism: best-double-couple Harvard CMT solutions. EQ.MT. in the meizoseismal area of the 1987 earthquake in the west: toponym "Earthquake Mountain" (*Kuh-e Zelzeleh* in Persian), an ancient oral warning. See also Plates 11.4 and 11.5. *Modified from Berberian and Yeats (1999, 2001) and Berberian et al. (1999).*

Berberian and Yeats, 1999, 2001). Darmiyān is located 2.5 km southwest of the Nauzād fault in the mountain valley, and 17 km SE of village of Nauzād (Figure 11.10; Plate 11.4).

The Nauzād reverse fault (Berberian, 1976c,e, 1981a, 1994, 1997a; Berberian and Yeats, 1999, 2001) is an arcuate fault striking approximately N110°E in the northwest [from Esaik to Tashmān for about 26 km] and N150°E in the southeast [from Tashmān to the west of Pāhvāz for about 50 km; Figure 11.10]. Along the Nauzād reverse fault, the Paleocene flysch of the Neh Complex is thrust from the southwest onto the Quaternary alluvial deposit fans of the northeast. Several villages [including Esaik, Sargaz, Mask, Nimdeh, Nimrāh, Khunikbāz, Nauzād, Chek, Takht baig, Tajvij, Ja'farābād, Tasvand, Furg, and many more] are located along the fault line and have access to spring water (Figure 11.10; Plates 11.4 and 11.5).

Ambraseys and Melville (1982) presented an approximately 12-km-long fault line striking N145°E, bifurcating from the Nauzād fault at Nimrāh village toward the northwest, passing the immediate west of Gask and Furkhās villages. They described this fault line as: "*At Nim-i deh a series of faint surface traces of terraces facing northeast were found traversing the valley*

PLATE 11.4 Eroded scarp of the Nauzād fault ruptured during the 1493 earthquake (Figure 11.10). Looking to the west-southwest with Mask village. For location, see Figure 11.10. *Photographed in 1972 (courtesy of John Tchalenko).*

PLATE 11.5 Close-up view of the eroded Nauzād fault (near the dirt road) ruptured during the 1493 earthquake northwest of Nauzād (Figure 11.10). *Photographed in 1972 (courtesy of John Tchalenko).*

alluvium heading northwesterly. They can be followed to the southwest of Gask and Malikabad and they are thought to be a clue to the continuation of the Nauzad fault in a northwesterly direction under the valley alluvium." The inferred fault line cannot be substantiated in the field nor on the aerial photographs, satellite imagery, and geological maps of the area (GSI, 1972a, 1991, 1992a). The authors showed a NE–SW trending short fault line along a tributary to the north of Furg village, which cannot be validated either (see figure 3.9 in the cited source). Ambraseys and Jackson (1998) and Ambraseys et al. (2002) reported a 30-km-long coseismic surface rupture; though probable, its length cannot be substantiated by the existing historic data.

About 56 years after the 1493 Nauzād earthquake, the seismicity migrated cross-fault to the west and an earthquake of M_s ~6.7 took place on 15 February 1549 along the Birjand reverse fault (Figure 11.10). No major earthquake has been recorded in the Nauzād area since the 1493 event. The 10 April 1998 M_w 5.8 earthquake took place about 45 km to the south southeast of Nauzād, possibly along the Purang fault (Figure 11.10).

11.12.1 Oral Earthquake Hazard Warning

Although no previous earthquake was documented along the Nauzād fault in the historical annals, ancient place names such as Kuh-e Shekasteh [lit., "Broken, Fractured, or Faulted Mountain" in Persian] along the fault might refer to earthquakes in this area. This could be considered as an oral earthquake hazard warning that has survived throughout millennia (Chapter 1).

11.13 THE 6 JULY 1505 M_S ~7.3 PAGHMĀN, NW KĀBOL EARTHQUAKE FAULTING

The Mughal ruler *Bābur* ["The Tiger"; Zahir-ud-din Muhammad; 1483–1530] in *Bāburnama* [Memoirs of Bābur] recorded the effects of the 1505 M_s ~7.3 earthquake, including the extensive surface rupture along the base of the Paghmān mountain range extending from Isterghah [Istarghij, 34°54′N, 69°04′E; a town north of Istalif] to Maidān [the Paghmān fault, northern section of the Chaman fault]. The interesting coseismic ground deformation recorded by Bābur (Beveridge, 1922) is as follows:

> (c. An earthquake): At that time there was a great earthquake such that most of the ramparts of forts and the walls of gardens fell down; houses were leveled to the ground in towns and villages and many persons lay dead beneath them. Every house fell in Paghmān village, and 70 to 80 strong heads-of-houses lay dead under their walls. Between Pagh-mān and Beg-tut a piece of ground, a good stone-throw wide [the throw of a catapult] may-be, slid down as far as an arrow's-flight; where it had slid springs appeared. On the road between Istarghach and Maidān the ground was so broken up for 6 to 8 yighāch [*Farsang; 36-48 miles/57.9-77.2 km*] that in some places it rose as high as an elephant, in others sank as deep; here and there people were sucked in. When the Earth quaked, dust rose from the tops of the mountains. Nuru'l-lah the tambourchi [drummer] had been playing before me; he had two instruments with him and at the moment of the quake had both in his hands; so out of his own control was he that the two knocked against each other. Jahāngir Mirzā was in the porch of an upper-room at a house built by Aulugh Beg Mirzā in Tipa'; when the Earth quaked, he let himself

down and was not hurt, but the roof fell on some-one with him in that upper-room, presumably one of his own circle; that this person was not hurt in the least must have been solely through God's mercy. In Tipa' most of the houses were leveled to the ground. The Earth quaked 33 times on the first day, and for a month afterwards used to quake two or three times in the 24 hours. The begs and soldiers having been ordered to repair the breaches made in the towers and ramparts of the fort [Kabul], everything was made good again in 20 days or a month by their industry and energy.

Babur, 1483–1530; tr. Beveridge, 1922.

11.14 THE 1550 $M_S \geq 6.0$ TABRIZ EARTHQUAKE GROUND DEFORMATION

About 295 years after the 1550 earthquake, al-'Umari (1793; in Melville, 1981; Ambraseys and Melville, 1982) wrote that: "*The ground moved in perceptible waves and caused some fissuring in the mountains; a mountain near the city split into four pieces and a great deal of smoke* [dust] *poured out, blottering out the horizon*" (al-'Umari, fol. 190r, in Melville, 1981; Ambraseys and Melville, 1982). It is probable that the earthquake was associated with surface faulting along the North Tabriz fault, but it was definitely associated with mass slope sliding, causing dust (Figure 11.7). See Chapter 16 for a pattern of coseismic faulting and clustered historical seismicity of the Tabriz area and along the North Tabriz fault.

11.15 THE 5 FEBRUARY 1641 $M_S \sim 6.8$ DEHKHĀRQĀN [MODERN ĀZARSHAHR] EARTHQUAKE GROUND DEFORMATION

Vārtābed Ārākel Tāvrizetsi [Ārākel of Tabriz (1662), the celibate priest in the Armenian Church living in the city of Tabriz], was an eyewitness to the 5 February 1641 earthquake (Berberian, 1976a, 1984a,b; Berberian and Arshadi, 1976; ed. Bournoutian, 2006). His reference to the "exposure of the inner Earth," "sudden opening of the Earth's crust," and "appearance of a deep chasm" might refer to surface faulting and/or lurching. This earthquake took place to the south of the North Tabriz fault and was not associated with slip on that fault (Figures 11.7 and 11.12).

On Friday, February 5, in the year 1090 of the Armenian calendar [*1641 CE*], there occurred a sudden, unexpected, and terrible earthquake in the city of Tāvriz [*Tabriz*] and its surrounding districts. The tremors were so strong that the entire crust of the Earth shook like a leaf that flutters in a hand or in a bowl of water and vibrates in different directions. The ground rumbled terribly; it thundered from the depths of the Earth. Then the earthquake began and it demolished, destroyed, and ruined all the city structures. Many of the magnificent and marvelous buildings, spacious and tall, which stood in the city-known to all who have seen it-from ancient times and [left to us] from kings, fell down and turned into debris in the twinkling of an eye. During the earthquake a famed building known as Ustāshāgart [*'Ostād-Shāgerd', lit., 'Master-Apprentice' in Persian; both Khanlarian, 1982, and Bournoutian, 2006, assumed this building as a caravanserai; while in fact, it was a famous mosque built in 1340-1*] also collapsed and covered many camels and other beasts that had found shelter there from the cold winter snow. The domes and minarets of mosques, which raised to dizzying heights in the sky, fell down from on high, collapsed, crashed down to Earth and disappeared.

The glorious, and immense, magnificent tall building, known to all, and called Shamghāzān [*Shanb Ghāzān; lit., 'Tomb' (in Mongolian) of Ghāzān Khān Ilkhānid (r. 1295-1304), built in 1298-1303 in Ghāzāniyeh town, west of the outskirts of the city of Tabriz*], located not far from the city, on the banks of the Shur River [*'Aji Chāi', lit., 'Bitter River', in Turkic; 'Talkeh Rud' in Persian*], was also ruined. All the roofs collapsed and only the walls remained standing. The walls were also shaken from their foundation, cracks appeared on all four corners, and the walls crumbled on four sides. All of it remains in ruins to this day. As to the houses, palaces, [*public*] baths, and putkay, that is the carvanserais, who can list how many of them were destroyed to their foundation? Many people, animals, cattle, equipment, chattel, depots, and goods remained buried under ground. People dug the earth and stones for twenty or thirty days looking for them and, with difficulty, managed to pull out the corpses and belongings.

Although the earthquake caused great loss to the city of Tabriz, it caused greater losses-destruction, toppling, death of people and cattle-in the surrounding districts, for example in the district of Khosrvshāh [*Khosrowshāh*] and Usku [*Osku*] and the village of Dukhurkān [*Dehkhārqān; modern Āzarshahr*]. The strong and terrible shocks exposed the inner Earth, and black waters, resembling streams, gushed out and the waters flowed in their own channels. Three days later, there was another strong earthquake; the above-mentioned streams disappeared, but others appeared. Small streams also appeared in various places. When the water accumulated, it reminded one of the former black waters. As long as the earthquake continued, there were streams of black water; the moment the earthquake stopped, they disappeared as well.

They say that, one day, a shepherd had taken a large flock of sheep to pasture, when the earthquake began. The Earth's crust suddenly opened, a deep chasm appeared; the shepherd and part of his flock fell into it, were covered by Earth, and disappeared. The same happened to some village, constructed under a rocky cliff. When the earthquake happened, the solid rocks of the cliff separated [from the mountain] and fell down [on the village]. It rushed down like a drift of sand, stone, and earth. It completely covered the village and totally annihilated it.

The earthquake occurred during winter time, in the season of severe cold weather. The winter was very snowy. The Earth shook not once, twice, or three times and then stopped; but [the aftershocks] lasted six months. During the first two months, the tremors were frequent; they occurred throughout the day, in daytime and during the night. They occurred five or six times, sometimes more, sometimes less. This happened every day. After two months the tremors occurred every day, for twenty or thirty days. For this reason, everyone was alarmed, terror-stricken, and panicked from fear of the earthquake [aftershocks], which would start without warning. Although many ran out of their homes and were saved from the collapsing structures and death, there were many others who remained under the debris and perished from suffocation. That winter, people could not bear to live in their homes; they all went out and lived in cabins and tents-some by the doors of the house, some in vineyard and orchards, and others in similar appropriate places until the wrath of God was pacified. For it became clear to all the inhabitants of the land and the universe that this was not an ordinary earthquake, but it was an obvious retribution, punishment, and wrath of God, which had descended upon them, as it had on the people of Nineveh [Jonah 1.2].

Vārtābed Ārākel Tāvrizetsi; tr. Bournoutian, 2006, with permission.

The causative fault of the 1641 earthquake along the southwestern foothill of the Sahand volcano is not known (Figures 11.7 and 11.12). The meizoseismal area of this earthquake has a long axis trending NE–SW, nearly perpendicular to the North Tabriz fault (Berberian and Yeats, 1999). The meizoseismal area is poorly constrained to the southeast due to lack of villages on the northwestern steep slopes of the Sahand volcano.

The 1641 earthquake, which took place off the North Tabriz fault (Figures 11.7 and 11.12), might have occurred on a line along the Talkhehrud fault: (i) a NNE–SSW-tending cross blind thrust (similar to the 1968 Dasht-e Bayāz and Ferdows sequence; Berberian, 1981a; Berberian and Yeats, 1999); (ii) a conjugate strike-slip fault to the North Tabriz fault (similar to the 1986 North Palm Spring and the 1992 Joshua Tree, Mojave Desert, California conjugate strike-slip earthquakes; Yeats, 2012; the South Rigān earthquakes of 20 December 2010 and 27 January 2011; Walker et al., 2013a); or (iii) a normal fault responsible for the formation of the Tabriz plain as a pull-apart basin covered by the Sahand volcanic lava and ash. In any case, the 1641 earthquake might have increased the Coulomb failure stress along the adjacent North Tabriz fault, triggering the clustered sequence of 1721–1786 earthquakes along the fault (Figure 11.7; see also Chapter 16 for discussion).

It is interesting to note that the earliest known Iranian isoseismal map for the 4 October 1856 small-magnitude earthquake felt in Tabriz (which did not cause damage but triggered a Cacciatore-type seismometer in Tabriz) drawn by Khanikoff (in Abich, 1858) has an elongated shape trending ENE–WSW (N70°E; the seismometer showed the shock had a direction of E23°16′N; followed by a second shock in the W31°12′S; Khanikoff in Abich, 1858). Moradi et al. (2011) showed a very short ENE–WSW-tending normal fault to the southwest of Tabriz; however, neither their recorded seismicity nor the focal mechanism solutions correlate with the introduced fault and the nature of a pull-apart-basin southwest of Tabriz.

An exceptional statistic, unlike any data gathered after a historic earthquake on the Iranian plateau, was collected by Mohammad Nasir after he visited the meizoseismal area upon the order of Rostam Khān Sepahsālār [commander-in-chief] Beglarbegi [governor] of Āzarbāijān (Zokā', 1989; Yahyā Zokā's letter dated 18 November 1991). The report was then delivered to the king, Shāh Safi Safavid [r. 1629–1642] who died 15 months after the earthquake on 12 May 1642 (Table 11.1). For historical earthquake clusters and patterns in the Tabriz area and coseismic surface faulting along the North Tabriz fault, see Chapter 16.

11.16 THE 1678 GONĀBĀD EARTHQUAKE

According to Tābandeh (1969), a destructive earthquake in the Khorāsān province destroyed many villages. The town of Gonābād was completely ruined and the casualties were excessive. Only the old Gonābād congregational mosque, built in about 1212 (Meshkāti, 1970), withstood the shock. Only one person survived. The town was resettled by the survivors from outlying villages. This was the second recorded earthquake in Gonābād since the 1238 event (Figure 11.11).

Labbāf Khāniki (1997), quoting Nabipur, a native of Gonābād, stated that about 3302 m south of the mouth of the Gonābād Qasabeh Qanāt,[6] in the Gāvkamus area [located in the eastern part of the town of Gonābād at Ku-ye Shahri or Qasabeh-ye Shahr], the qanāt was dammed by an earthquake that had taken place about 400 years before [the 1678 earthquake]. The qanāt line was then reconstructed by Mirzā 'Alinaqi Riyābi form 3302 m to the qanāt mouth in the town.

Although qanāts usually collapse by the strong ground shaking during earthquakes, the referred point of damming of qanāt water is located where the qanāt line crosses the Gonābād–Bidokht active fault system showing Quaternary vertical displacement with some indications of left-lateral slip motion and young fault scarp (Figure 11.11). The northern fault block covered by the Recent alluvium is downthrown, and the southern block is raised and shows numerous drainage incision cutting the Pliocene (?)-Early Quaternary light-colored clay deposits (Figure 11.11). This may indicate that the fault system with active geomorphological features was possibly reactivated during the 1678 Gonābād

6. Qanāt (Kahriz, Kāriz). Ancient Iranian irrigation technology in semiarid and arid regions of Central and East Iran. It is composed of several well-like shafts (with intervals ranging from 15 to 100 m; with average 50–70 m in the Khorāsān province), connected by gentle sloping tunnels of several hundred to thousand kilometers long, directing the shallow aquifer of slightly higher elevations in the alluvial deposits, especially at the footwall of active reverse faults. In some cases, the mother shaft taps the elevated water table of the tip of reverse hanging-wall block, where the aquifer is sealed by the active fault gauge. The oldest discovered qanāt system in Iran dates back 4,000 years in an archaeological site at Semnān (Mehryar and Kabiri, 1986). The abandoned and displaced qanāt lines in the Dasht-e Bayāz and Gonābād area record several generations of older qanāts destroyed by the active fault (discussed later). In both aforementioned cases, the ancient local inhabitants were aware of the earthquake effects on the water supply system.

TABLE 11.1 Contemporary Statistics of the 5 February 1641 Dehkhārqān Earthquake by Mohammad Nasir

No. of People Killed in the City [Tabriz] and the District				Houses Destroyed [Excluding Walls, Mosques, Mausoleums/Religious Buildings, Shops, and Caravanserais]				Houses Damaged	Prince Navāb Khāni's Donation [For Purchasing the Shroud (Kafan) for Burying The Dead Bodies of the Poor]
				City	*District*			*City*	
Men	*Women*	*Boys and Girls*	*Total*	*[Tabriz]*	*Dehkhārqān*	*Dezajrud [Dizaj]*	*Vidhar [Osku], Sarsarā, Gāvkān, & Sardrud*	*[Tabriz]*	
5984	5144	1475	12,613	2064	2495	3779	2251	2031	130 Tumān of Tabriz

For locations, see Figure 11.12.

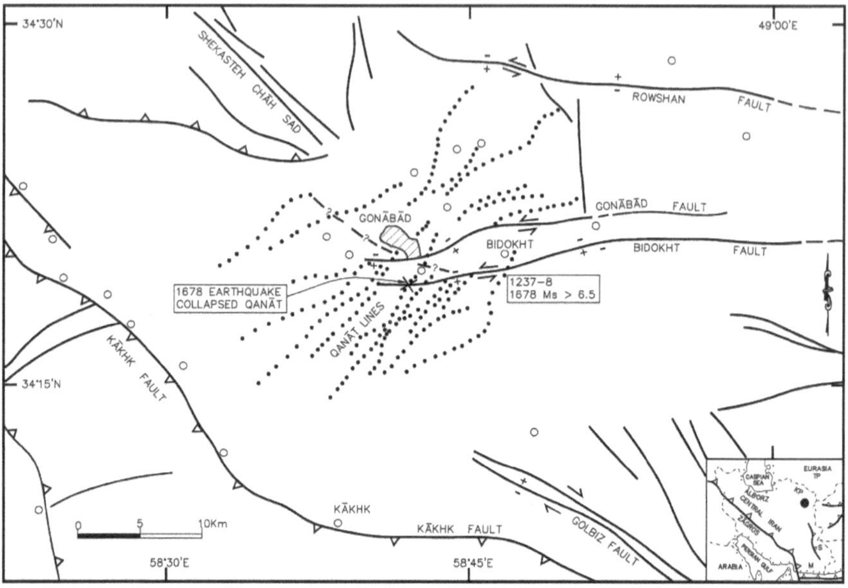

FIGURE 11.11 Epicentral region of the historical earthquakes at Gonābād in eastern Iran. Dotted lines show numerous qanāt lines. The collapsed qanāt during the 1678 earthquake is marked by an X along the Bidokht fault. Symbols as in Figure 11.1.

earthquake, damming the major qanāt line of Gonābād supplying water to the town. The fault has a nearly E–W curved orientation of more than 30 km (aerial photograph Nos. 29305–29307, Worldwide Aerial Surveys, Inc., Project 158, scale 1:55,000). For the historic seismicity, exceptional earthquake clusters, and coseismic surface rupture pattern in the area, see Chapter 16.

11.16.1 Oral Earthquake Hazard Warning

The local toponym of the "Shekasteh (broken, faulted) Chāh Sad Mountain" about 15 km to the northwest of Gonābād indicate an important oral earthquake hazard warnings from ancient times (see Chapter 1 for details and Figure 11.11 for the location).

11.17 THE 26 APRIL 1721 M_S ~7.3 SHEBLI [SE TABRIZ] EARTHQUAKE FAULTING

The earthquake reportedly killed 40,000 people and destroyed the city of Tabriz along with most of its historical monuments. Sir Harford Jones Brydges (1834), who visited the area in the summer of 1809 [88 years after the 1721 earthquake], wrote his observation of at least a 35-km-long coseismic surface rupture along the North Tabriz fault (Figure 11.12):

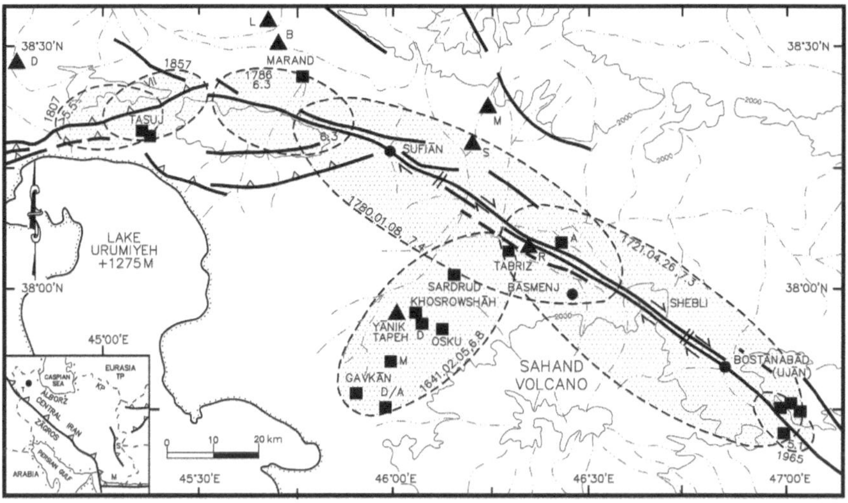

FIGURE 11.12 Meizoseismal area of the historical large-magnitude earthquakes of the Tabriz region northwest of Iran. The 1721 and 1780 earthquakes were associated with coseismic surface ruptures. Symbols as in Figure 11.1. Filled squares: sites destroyed by earthquakes. A, Ārpādarasi; D/A, Dehkhārqān (modern Āzarshahr); M, Mamaqān. Filled triangles: Archeological sites. B, Bāruj; D, Dizaj; M, Mujumbār S. Hripsimeh (seventeenth century) and S. Anerevuyt (1810) Armenian churches; L, Livār; R, Rushdiyeh mound; S, Sohrol (Sorhul) S. Hovānes (1840) Armenian Church (Alpago-Novello, 1988). *Modified from Berberian (1997a) and Berberian and Yeats (1999).*

Between the camp *['Chemen Oujān'; 'Chaman-e Ujān', lit., 'the Ujān Meadows']* and Bosmeech *[Bāsmenj]*, we passed over ground which some years before had been rent by a succession of earthquakes in the most extraordinary manner, and on the left hand of the road *[west]*, I was shown a mountain, riven at that time from top to bottom. This dreadful calamity took place in the year 1724 *[*sic*; 1721]*, and cost Tauris or Tabreze *[Tabriz]* the lives of 100,000 of its inhabitants.

Brydges, 1834.

Brydges's observation was carried out (i) 88 years after the 6 April 1721, (ii) 29 years after the 8 January 1780, and (iii) 23 years after the October 1786 earthquakes. The 1786 earthquake meizoseismal area was in the Marand area to the northwest (Figure 11.12) and as such was not part of his observation. The 1780 coseismic surface rupture was reported mainly at the foot of the Sorkhāb Mountains in the north of the city and northwest of the Shebli pass (Figure 11.12; discussed below). Furthermore, Brydges was told that the surface fractures occurred during the 1724 [*sic*, 1721; with three years error] earthquake. His observation and the information provided by the local people 88 years after the earthquake clearly shows that this large-magnitude earthquake was associated with a surface faulting that extended for at least 35 km from Bāsmenj [20 km SE of Tabriz] to Ujān [55 km SE of Tabriz], through the Shebli high-pass [36 km SE of Tabriz] along the North Tabriz fault (Figure 11.12) (Berberian and Arshadi, 1976; Berberian, 1981a, 1994, 1997a; Berberian and Yeats, 1999). For historical earthquake clusters and the pattern of coseismic surface faulting of the Tabriz area, see Chapter 16.

11.18 THE 8 JANUARY 1780 M_S ~7.4 TABRIZ EARTHQUAKE FAULTING

The city of Tabriz, together with 400 villages, was destroyed; an exaggerated 200,000 casualties were reported. The city's monuments were destroyed or severely damaged (Plate 11.6). At least six valuable accounts were recorded (see below) regarding the coseismic surface fault rupture along the North Tabriz

PLATE 11.6 Outer portal of the Tabriz Blue mosque (masjed-e Kabud/Jahānshāh; built in 1465) severely damaged by the 1780 earthquake. Note the massive adobe brick walls, clad with kiln tiles, disintegrated by the strong ground shaking of the 1780 Tabriz earthquake. Note the deep fractures developed in very thick walls of this splendid fifteenth-century structure. The well-constructed portal did not collapse, whereas the modern buildings in Iran have been leveled to the ground during recent earthquakes. *Photographed in 1906 by Sven Hedin (published with permission from the Sven Hedin Foundation, Stockholm, http:/svenhedinfoundation.org).*

fault associated with the 1780 earthquake (Figure 11.12), which destroyed the area from southeast Tabriz to Sufiān and beyond to the northwest (Berberian and Arshadi, 1976; Berberian, 1981a, 1994; Berberian and Yeats, 1999).

i. The observation of the coseismic surface faulting along the North Tabriz fault as well as a large-scale landslide by *Mirzā Mohammad Rafi' Tabrizi*, quoted by *Donboli* (1813), is very interesting: "*The Sorkhāb Mountain* [to the north of Tabriz city] *was split apart and a very deep fissure was formed. Its width was two zar'* [~2 meters] *and its length about two farsang* [~12 km] *oriented from the south to the east* [leading southeasterly]. *About two farsang* [~12 km] *east of Tabriz, a portion of meadowland, requiring more than 20 mann* [~60 kg] *grain seeds for sowing, was displaced from its original location for about a quarter of a farsang* [~1.5 km]. *The* [displaced] *meadowland is still visible and is well known* [by everybody]" (Donboli, 1813).

ii. *al-'Umari* (1793) likely describing the vertical throw along the North Tabriz fault in the northern part of the city during the 8 January 1780 Tabriz earthquake, wrote that "*the ground* [at Tabriz] *settled down four meters.*"

iii. Twenty-one years after the earthquake, *Mirzā Mohammad Hassan Zonuzi* (1801) recorded that "*Destruction occurred within a radius of 12 farsangs* [~72 km] *from Tabriz city; ...large* [very long] *fissures developed in most mountains, and long fissures extended for approximately 7 farsangs* [~42 km] *along the road to Shebli* [36 km SE of Tabriz] *joined together*" (Zonuzi, 1801). The locations addressed by Zonuzi (1801) are found along the North Tabriz active fault line (Figure 11.12).

iv. *Colonel Gaspar Drouville* (1825), who served in the Iranian army in 1812–1813, wrote a sentence similar to that of Donboli (1813): "*The Sorkhāb Mountain* [north of the city of Tabriz] *was splitted* [faulted] *by deep fissures of 2 zars* [~2 meters] *wide and 2 farsangs* [~12 km] *long trending southeast... Apparently, all these difficulties were not enough for its* [Tabriz's] *destructions that few years later* [1780], *a disastrous earthquake leveled it with the ground and more than 90,000 people were buried below its ruins... A strange phenomenon took place during the Tabriz earthquake. When the Earth was moving severely from the east to the west, suddenly in the northwest of Tabriz between the city and the mountain, a grey color rock mass of 2 miles* [3.3 km] *long, 100 m wide [50 tavaz], and 10 m height was thrown up from the Earth. The mentioned rock mass was composed of a mixture of sulphur and sand and still is preserved in grey color. This clearly contrasted with the red color of the mountain* [Kuh-e-Sorkhāb; lit., the Red Colored Mountain, in Persian] *and the green grasses grown aground the area*" (Colonel Gaspar Drouville, 1825).

This is a very interesting observation by Drouville (1825). Visiting the area 32 years after the earthquake, he describes a fault plane 2 miles long and 10 m height (possibly exaggerated or not a true displacement along the fault) in a nearly E–W direction in the area to the northwest of the city between the city and the mountain; that is, along the North Tabriz fault at the foot of the Sorkhāb Mountain (Figure 11.12) (Berberian and Arshadi, 1976; Berberian and Yeats, 1999).

v. Captain Moritz von Kotzebue of the Russian Army, who was in the town of Marand in northwest Iran (Figure 11.12) between 15 May and 17 May 1817, recounted a vague story told to him about ground fracturing during the 1780 earthquake at Marand (Kotzebue, 1819): "*Of Maranda* [Marand town], *Thirty-eight years ago, during an earthquake* [1780], *the ground opened, and two Mollahs (Moslem priests), of whom we saw one in the chapel, together with several inhabitants, witnessed the sudden appearance of a large tomb of stone, which, however, soon vanished in the opening*" (Kotzebue, 1819). Although von Kotzebue could not see the fracture linked to a myth revealing the tombstone of Noah's mother, it is likely that the northwestern segment of the North Tabriz fault near the town of Marand was reactivated during this earthquake (Figure 11.12). Nonetheless, the North Tabriz fault passes to the south of Marand and the location cited by Kotzebue could have been in the city.

vi. Finally, *Mohammad Rezā Tabātabā'i Tabrizi* (1875), about 95 years after the 1780 earthquake, recorded that: "*The ground of the city of Tabriz and its suburbs, especially in the northern district of the city, frightening fractures and dreadful faults have been seen by some people. If a large stone was thrown into the fractures, no impact noise in such dreadful abyss could be heard by anybody.*"

The above accounts unequivocally indicate:

i. Coseismic surface faulting trended NW–SE along the North Tabriz fault (Figure 11.12);

ii. A succession of deep surface ruptures was observed between the city and the Sorkhāb Mountain in the north and to the northwest of the city of Tabriz;

iii. A minimum surface fault length of 42 km, fault width ranging between 2 and 100 m, and fault scarp/throw ranging between 4 and 10 (?) m were observed;

iv. The gray-colored fault scarp, which was thrown up from the ground during the earthquake, contrasted with the background red color of the Sorkhāb Mountain (Plates 11.7 and 11.8);

v. The strong ground motion directed from the east to the west, possibly indicating rupturing propagated from the SE to the NW along the North Tabriz fault (Figure 11.12); and

vi. This was a large-magnitude earthquake (Plate 11.6).

It should be emphasized that the only person who referred to the extension of the succession of coseismic surface ruptures to the southeast of the city of

PLATE 11.7 The North Tabriz fault scarp northeast of the Tabriz city airport. The Miocene red molasses (left, north) is raised above the recent alluvial deposits (right, south). *Photographed in 1975. For location, see Figures 11.7 and 11.12.*

Tabriz for ~42 km toward the Shebli high pass [36 km SE of Tabriz] was Zonuzi (1801); the four other observers of that time reported the faulting in the north and the northwestern part of the city. Therefore, it is likely that Zonuzi confused his observation and/or reporting of the ground ruptures of the 1721 southeast Tabriz earthquake with those of the 1780 event (Figure 11.12). For historical earthquake clusters and the pattern of coseismic surface faulting of the Tabriz area, see Chapter 16.

11.19 THE 1824 M_S ~>6.5 HARĀZ EARTHQUAKE DUBIOUS FAULTING

Bell (1840), who visited the area in spring of 1837, wrote that "*Between Kuhrud and Bul Qalam* [along the Harāz River valley near Bāyjān, the location of the 25 and 26 March 1983 earthquakes (Figure 11.2)] *there is some*

PLATE 11.8 Close-up view of the North Tabriz fault scarp northeast of the Tabriz city airport. The Miocene red molasses (left, north) is raised above the recent alluvial deposits (right, south). *Photographed in 1975. For location, see Figures 11.7 and 11.12.*

evidence that in this locality the shock was associated with permanent ground deformations. The piers of a masonry bridge [the Hājji 'Ali Moshā'i bridge across the valley along the old Harāz road] *built on solid rock and destroyed by the earthquake seemed as if they could never have been intended to support the same arch, so different was their parallel... and the opposite sides of the ravine had no doubt suffered displacement by a tremendous earthquake, which occurred about eleven or twelve years before* [i.e. 1825]" (Bell, 1840).

This account is quoted as evidence of ground deformation of a possible tectonic origin by Ambraseys and Melville (1982), who wrote that the 1824 earthquake took place not far from the Manjil earthquake of 10 June 1990 [the latter event took place 268 km to the northwest of the location mentioned at the Harāz River]. Ambraseys and Melville (1982; table 3.1, p. 110) and Ambraseys and Jackson (1998; table 1, p. 392) considered this event associated with surface faulting without referring to the regional fault system or presenting a fault map. Nonetheless, it should be mentioned that the Harāz gorge and its tributary valleys have very steep unstable slopes and are prone to numerous mass movements, landslides, and rock avalanches triggered by earthquakes and even by winter snowfall. Rock avalanches and landslides killed the majority of the people during the 25 March 1983 M_w 5.5 earthquake in the same region (Figure 11.2). The ground deformation recounted by Bell (1840) definitely refers to landslips and rock avalanche displacing the piers of the bridge built on the slopes, not to coseismic surface faulting at that location.

11.20 THE 27 MARCH 1830 M_S ~7.1 LAVĀSĀNĀT EARTHQUAKE FAULTING

Historical accounts reveal that during the 1830 earthquake, the town of Damāvand and its mosque were heavily damaged and about 500 people were killed (Figure 11.2). The caravanserai at Jājrud was ruined—possibly damaged by the mainshock and collapsed by the 6 April aftershock. Old houses in Tehran collapsed, the remaining structures were damaged, and 30 people were killed in the capital city of Tehran (Conolly, 1838; Watson, 1866; Ambraseys, 1974a; Berberian et al., 1985). The limited macroseismic data indicate that the Moshā fault near the junction with the North Tehran fault (west of Damāvand; Figure 11.2) was reactivated (Berberian et al., 1985; Berberian and Yeats, 1999). No report of coseismic ground rupture is documented; however, geomorphic evidence and paleoseismic trench study reveal ground rupture possibly associated with this earthquake.

Optically stimulated luminescence (OSL) dating of a sample collected in a trench across the Moshā fault west of Āb-e ‘Ali ski resort [alluvial unit 4 cut by faulting, in trench 2 at site 4; 35°46′N–51°59′E] yielded an age of ~1120 OSL years (Solaymani Azad et al., 2011). The date may correspond with two known earthquakes of 1665 and 1830 along the Moshā fault. However, the macroseismic data indicate that the faulting might have been related to the 1830 event (Figure 11.2).

11.21 THE 1838 M_S ~7.0 NOSRATĀBĀD EARTHQUAKE FAULTING

Drawing on local tradition, which allegedly appeared in *Tārikh-e Dorrāni* [*The History of Dorrāni*], Ambraseys and Melville (1982) reported that in the area between Shushaki, Nosratābād, and Gurgaz to the south, the ground opened up, and from Haidarābād to Qal’eh Gorg, the ground was transformed into a mountain and blocked the narrow defiles (Figure 11.13). No other reference has been found on this issue; I have not been able to corroborate this in any history book. Iraj Afshār Sistāni (personal communication, 1989, Tehrān) thought that this could be the Tārikh-e Dorrāni by Ahmad Khān Dorrāni of

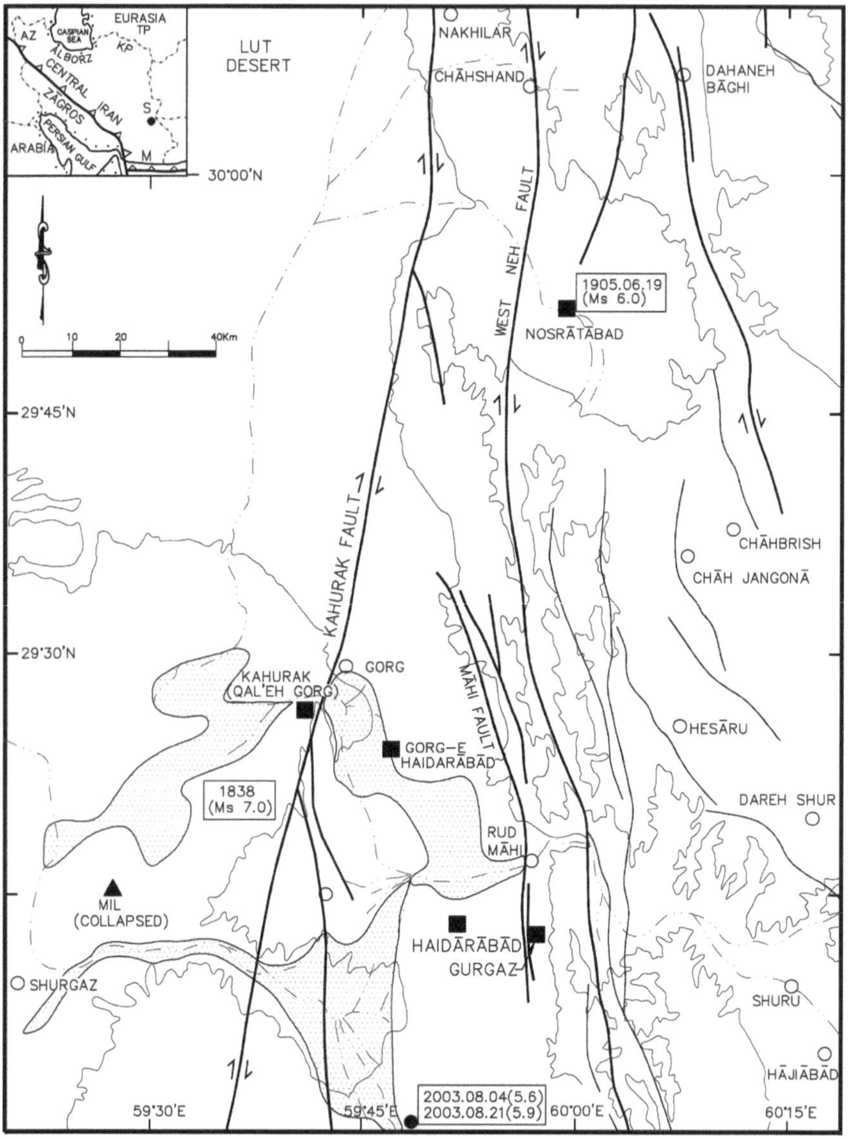

FIGURE 11.13 Epicentral areas of the 1838 and 1905 earthquakes southeast of the Lut desert. Destroyed sites are shown by filled squares. Triangle shows the location of the brick signal-tower (Mil-e Nāderi) ruined by the 1838 earthquake. Symbols as in Figure 11.1. *Modified from Berberian and Yeats (1999).*

Kābol. There is also a reference to Tārikh Sultāni Durrāni (Muhammad Khān ibn Musa Khan Durrāni, 1881), which I have not been able to consult.

Apparently, the twelfth-century brick tower called Mil-e Nāderi, located about 15 km to the northeast of Shureh-gaz, was destroyed by the earthquake

(Figure 11.13). Smith (1876a) visited the area during his Perso-Afghan mission of 1871–1872 and wrote that: "*The other tower is eleven miles beyond Shor-Gez* [Shureh-gaz; 29°09′N-59°20′E, +435 m], *but is now almost completely in ruins, having been destroyed by an earthquake about twenty-five years ago* [?; 1871-25 = 1846?]. *It is said to have been twice the height of the former one* [i.e., the extant Mil-e Nāderi]. *From appearance, the conclusion is warranted that Nader Shah* [Afshārid, r. 1736-1739] *merely repaired these towers, which seems to be of a date far antecedent to this epoch*" (Smith, C.B.E., 1876).

Nosratābād [29°51′N–59°58′E, +1119 m], the former Nasirābād, was referred to as Nasratābād [*sic*] by Tate (1910–1912) and Ambraseys and Melville (1982). Based on the aforementioned tradition, Ambraseys and Melville (1982) drew a 70-km long straight dashed fault line from the area south of Shushaki (in the NNW) to the area east of Gurgaz (in the SSE). Later, Ambraseys and Jackson (1998) and Ambraseys et al. (2002) claimed a 70-km-long coseismic surface faulting with an azimuth of N170°E. No straight fault line extends from Shushaki to Gurgaz, or from Kahurak to Haidarābād, as shown in figure 3.19 in Ambraseys and Melville (1982). The Kahurak fault has a NNE–SSW trend at this locality, and the West Neh fault, to the south of Nosratābād, is nearly N–S extending from the area west of Nosratābād to the east of Gurgaz (Figure 11.13). The same surface fault length with the same azimuth is given by Ambraseys and Jackson (1998), which cannot be justified.

The existing legendary data do not allow the causative fault among the Kahurak, West Neh, and Māhi Quaternary faults (Figure 11.13) to be pinpointed. Based on (i) the recency of the fault rupture trace in the playa; (ii) rivers being blocked by mountains (right-lateral displacement along the fault); and (iii) the destruction of the historic watch tower of Mil-e Nāderi located 18 km to the west of the Kahurak fault, and Haidarābād village about 22 km to the east of the fault, it is likely that the Kahurak fault was reactivated during this earthquake (Figure 11.13). The southernmost extremity of this fault was reactivated during the 20 December 2010 M_w 6.5 South Rigān earthquake (see Chapter 14).

The only earthquake known in the Nosratābād area is the 19 June 1905 M_s 6.0 (Figure 11.13); during this event, the Nosratābād fort [built in about 1855] together with its qanat and a fort were apparently destroyed (Ambraseys and Melville, 1982). Epicenters of the 4 August 2003 (5.6) and 21 August 2003 earthquakes are located near the eastern branch of the Kahurak fault (Figure 11.13). No macroseismic data are available for these earthquakes from this remote arid desert area.

11.22 THE 5 MAY 1853 M_S ~6.2 SHIRĀZ EARTHQUAKE

The earthquake destroyed the city of Shirāz and killed about 9000 people. Contemporary poets Hakim (1819–1857) and Dāvar (1822–1865), sons of the poet Vesāl Shirāzi, stated in their poems that the ground was fractured during the 1853 devastating earthquake at Shirāz (Soltāni-Moqadam, 2012). Unfortunately, no additional data have survived and the nature of the aforementioned coseismic ground fractures is not known.

11.23 THE 1877 M_S ~5.6 ĀB-E GARM (SIRCH) EARTHQUAKE

Houtum Schindler (1881), followed by Ambraseys et al. (1979), reported that the 1877 earthquake ruined the villages of Āb-e Garm, Sirch, Hasanābād, Deh Qoli, and Hashtādān. Ground cracks appeared in the mountains near Āb-e Garm, and the hot springs dried up.

It seems that the earthquake was associated with surface faulting along the Gowk fault (see Figure 13.11). The surface faulting along the Gowk fault with an earthquake of M_s ~5.6 is likely. The 20 November 1989 M_w 5.8 South Golbāf earthquake was associated with 11- and 8-km long surface ruptures along the west and the east side of the valley formed by the Gowk fault (Figure 13.10). The 18 November 1998 M_w 5.3 Chahār Farsakh earthquake (Figures 14.3 and 14.4) was also associated with at least 4 km of surface faulting along the Gowk fault (Berberian et al., 1984, 2001; Berberian and Qorashi, 1994). For additional information about historical earthquake clustering and the coseismic rupture pattern of the Gowk fault, see Chapters 13, 14 and 16.

11.24 THE 22 MARCH 1879 M_S ~6.7 SE BOZQUSH EARTHQUAKE

The earthquake ruined many villages north of Miyāneh and killed more than 2000 people at the southern and eastern foothills of the Bozqush Mountains in northwest Iran (Figure 11.14). After finding a 2-km-long section exposure of a recent fault in the area about 1 km north of Sārighamish, located in the meizoseismal area of the earthquake (Plate 11.9), Berberian (1976c) reported that the fault could have been associated with the 1879 earthquake. The fault was described as a high-angle reverse fault with an approximate strike of N170°E and dip of 75°SW. The southwestern block, which is composed of silicified allunite-bearing breccia of Miocene age, is thrust over the northeaster Quaternary alluvial deposits with a 3-m-thick gauge zone (Plate 11.9; see also figures 5–7, pp. 150 and 151 in Berberian, 1976c). The 1976 observed short section could be a part of a nearly N–S Germirud fault with a length of about 60 km (Berberian, 1988, 1997a). The 1879 event might have occurred on either the Germirud fault or the eastern section of the nearly E–W to NE–SW South Bozqush fault system, which also shows active escarpments (Figure 11.14). The NE–SW (N30°E) elongated meizoseismal area of the 1879 earthquake proposed by Ambraseys and Melville (1982) follows neither the South Bozqush nor the Germirud faults (Figure 11.14). An angle of 140° exists between the long axes of meizoseismal area and the reported azimuth of the faulting reported by Ambraseys and Melville (1982) (cf. Table 3.1, p. 100

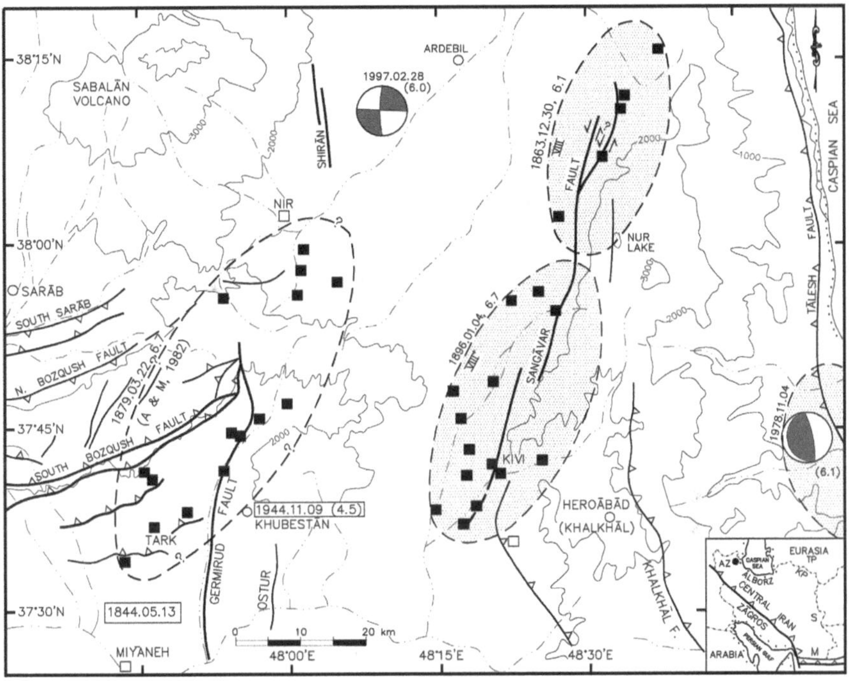

FIGURE 11.14 Meizoseismal areas of the historical earthquakes in the Bozqush–Germirud and Khalkhāl area of northwest Iran. Symbols as in Figure 11.1. The meizoseismal area of the 1879 SE Bozqush earthquake (Ambraseys and Melville, 1982) is not constrained. Focal mechanisms of the earthquakes constrained by body wave modeling: 1978 (Berberian, 1983a; Jackson et al., 2002); 1997 (Jackson et al., 2002). See also Plate 11.9. *Modified from Berberian (1997a) and Berberian and Yeats (1999).*

and figure 3.24, p. 66). The case requires further field investigation and paleoseismic trench study along both faults.

The seismic sequence in this region started with the 30 December 1863 M_s ~6.1 earthquake along the northern segment of the Sangāvar fault. Sixteen years later, seismicity migrated cross-fault to the southwest with the 1879 M_s ~6.7 SE Bozqush earthquake. Seventeen years later, the seismicity migrated to the east with the 4 January 1896 M_s ~6.7 earthquake along the southern segment of the Sangāvar fault (Figure 11.14). The 9 November 1944 *mb* 4.5 Khubestān earthquake along the Germirud fault caused slight damage in a few villages located along the fault (Berberian, 1988).

11.25 THE 11 JULY 1890 M_S ~7.2 TĀSH EARTHQUAKE

The earthquake destroyed many mountainous villages and was associated with large landslides (Figure 11.15). Sieberg (1932) wrote that the 1890 earthquake at Tāsh in the eastern Alborz was associated with ground fractures and

PLATE 11.9 Recent faulting along the Germirud River. The west southwestern block of Miocene rocks (right) is thrust over the east northeastern Quaternary alluvial deposits (left). *Photographed in 1975. For location, see Figure 11.14.*

rockfalls. This was followed by a report by Ambraseys (1975) of a 5-km-long ground deformation of tectonic origin associated with this event with an azimuth of N60°E. Ambraseys and Melville (1982) refuted coseismic ground deformation of tectonic origin suggested by Ambraseys (1975). However, Ambraseys and Jackson (1998) later reported a 10-km-long uncertain and spurious surface faulting associated with the 1890 earthquake. No evidence of coseismic surface faulting has yet been established for the 1890 earthquake.

Landslides and rockfalls are recorded in the Tāsh and Shāhkuh area; however, the Shāhkuh reverse fault in the immediate north of Shāhkuh Bālā village looks very sharp on the satellite imagery and in aerial photographs (Figure 11.15). The fault line has not been studied in detail. The fault is mapped as a north-dipping fault (GSI, 1989), whereas it is shown as a south-dipping reverse fault in Zanchi et al. (2009).

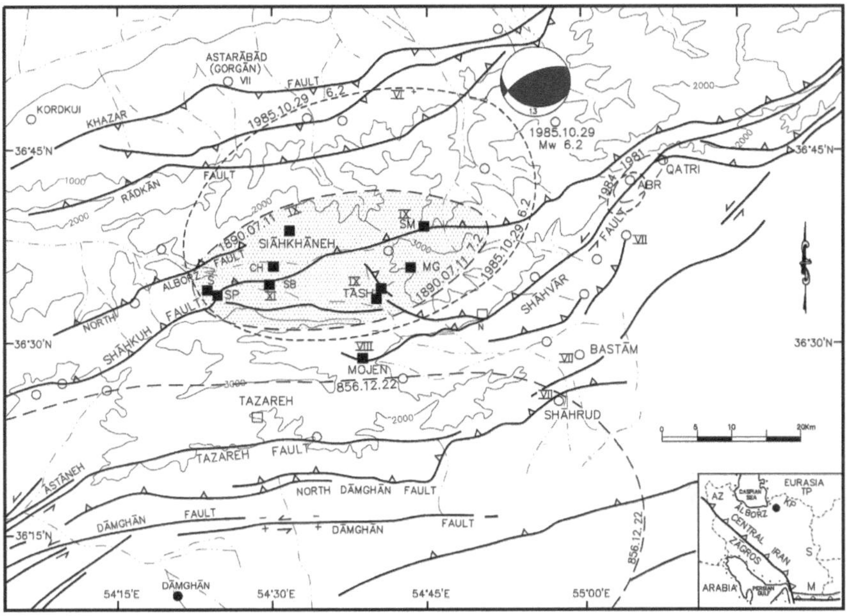

FIGURE 11.15 Fault map and meizoseismal areas of historical earthquakes of the eastern Alborz Mountains, southeast of the Caspian Sea. Symbols as in Figure 11.1. The dip of the Shāhkuh fault is uncertain (see the text). Filled squares show destroyed villages by the 1890 Tāsh earthquake. CH, Chahār Bāgh; MG, Marta' Ganju; S, Shirkuh; SB, Shāhkuh Bālā; SM, Siāhmarz Kuh; SP, Shāhkuh Pā'in. Intensities IX, VIII, and VII (MMI) are those of the 1890 earthquakes. Focal mechanisms of the 1985 earthquake constrained by body wave modeling with centroid depth (Jackson et al., 2002). *Modified from Berberian et al. (1996).*

Tāsh is located in a valley to the north of the Shāhvār fault and the footwall of the Shāhkuh reverse fault, both striking NE–SW (Berberian and Qorashi, 1988; Figure 11.15). Meizoseismal area of the 1890 earthquake presented by Ambraseys and Melville (1982) is elongated along a N60°E line cutting the structural trends of the Alborz Mountains and does not follow the major regional active faults of the area. The northeastern part of their proposed meizoseismal area covers the northeastern section of the Shāhkuh fault, whereas the southwestern end covers the southwestern tip of the Shāhvār fault (the faults were not shown on their figure 3.26). The existing macroseismic data are spread along the Shāhkuh fault (Figure 11.15). The region is also located on the hanging-wall of the Khazar reverse fault at the northern foothill of the Alborz Mountains (Berberian, 1981a, 1983a; Berberian and Qorashi, 1988).

On 29 October 1985, the Nomal earthquake of M_w 6.2 took place nearly in the meizoseismal area of the 1890 Tāsh earthquake (Figure 11.15); however, the slight damage was distributed over a larger area. Body-waveform modeling of the 1985 earthquake indicated thrusting with centroid depth of 13 km (Priestley et al., 1994). This event may have taken place along the Khazar fault (Figure 11.15) and did not cause severe damage.

The Shāhvār fault, with sharp Quaternary deformation [aerial photograph Nos. 3950 and 4037, Worldwide Aerial Surveys, Inc., Project 158, scale 1:55,000], is located to the southeast of the 1890 earthquake meizoseismal area. Although the fault was associated with the 9 August 1981 (M_s 4.9) and the 11 May 1984 (M_s 4.5) Qatri and Abr earthquakes (Berberian and Qorashi, 1988), it may not have been activated during the 1890 earthquake (Figure 11.15). Mousavi et al. (2013) claimed that the 1890 earthquake likely occurred on the Abr fault segment. Following Hollingsworth et al. (2010), Mousavi et al. (2013) used the misleading term of the "Shahrud earthquake" for the 1890 event, which took place 35 km to the northwest of Shāhrud in the Alborz Mountains. Javidfakhr et al. (2011) stated a left-lateral horizontal offsets of about 420 ± 50 and 400 ± 50 m for the Abr and Khij faults, respectively (the Shāhvār and Khoshyailāq faults of Berberian and Qorashi, 1988). Quartz-rich samples, dated from two different fan surfaces using an *in situ*-produced ^{10}Be cosmogenic dating method, indicated minimum exposure ages for the abandonment of the alluvial fan surfaces of 115 ± 14 ^{10}Be kyr along the Abr Fault (Shāhvār fault in Berberian and Qorashi, 1988 and this study; Figure 11.15) and of 230 ± 16 ^{10}Be kyr along the Khij Fault, implying slip rates of about 3–4 and 1–3 mm/year, respectively (Javidfakhr et al., 2011).

11.26 THE 4 JANUARY 1896 M_S ~6.7 SANGĀVAR EARTHQUAKE

The earthquake killed more than 1400 people and destroyed many villages in northwest Iran. The *Irān Newspaper* (8 February 1896) reported that two villagers were swallowed up by a ground fissure during the 1896 earthquake: "*The ground ruptured and took two people in up to the neck while they were walking from one village to another, and then the rupture closed and killed both. After two days a man passing the area noticed two heads sticking out of the ground. The bodies were then dug up and buried in another place. A dry mountain was ruptured and water started to flow out*" (*Irān Newspaper*, No. 874, Saturday 8 February 1896, Tehrān; Berberian, 1976c).

It is likely that the villagers were killed by lurching, rock avalanche, or landslides. However, the meizoseismal area of the earthquake is located along the southern segment of the Sangāvar fault, which might have experienced coseismic surface rupture (Figure 11.14). Along the northern segment of the Sangāvar fault, the 1896 earthquake was preceded by the 1863 M_s ~6.1 earthquake showing a southward migration of seismicity along the fault (Figure 11.14). The 10 March 1969 earthquake caused slight damage (MMI V) at Kivi along the Sangāvar fault.

11.27 OVERVIEW: PRE-INSTRUMENTAL PERIOD COSEISMIC SURFACE RUPTURES

The purpose of identifying and characterizing seismically active faults and detecting crustal deformation is to evaluate seismic hazard and minimize earthquake risk for urban and rural areas such as the Iranian plateau, California, Venezuela, New Zealand, and other vulnerable parts of the world where the concentration of population and material properties are high and future earthquakes could be catastrophic. It should be noted that almost all the urban and rural settlements in the semiarid and arid areas on the Iranian plateau are located along the active faults. The research is also important for active tectonics studies and future paleoseismic trench investigations.

Analyzing the scattered, surviving pre-1900 chronicles of the Iranian plateau, with uncertainties in their meizoseismal areas, showed limited references to coseismic ground deformations (both of primary tectonic surface faulting and of secondary origin not directly due to faulting). However, it is difficult without properly designed paleoseismic trench studies to confidently assign causative faults to the historic earthquakes. Nonetheless, by utilizing the valuable chronicles, knowledge of active faulting on the Iranian plateau, fault-zone geomorphic analysis and fresh topographic expression of fault offset, remote sensing tools, and geophysical investigation, we were able to correlate some surface ruptures with large-magnitude historical earthquakes where enough macroseismic data lead to a linear zone of damage distribution. In some cases, the wealth of historical data permitted the recognition of the minimum rupture lengths of the pre-instrumental period, large-magnitude earthquakes (see Figures 11.1–11.15 and Tables 14.9–14.11). Great care was taken to avoid errors in earthquake dating, the possibility of more than one earthquake occurring in a particular region, and defining the correct causative faults in areas with high concentration of active faults (see Chapter 16).

The earliest known reference to coseismic ground deformation (surface rupture and/or rock avalanche) was documented in the ancient Zoroastrian Pahlavi texts regarding the breaking of the Komesh Mountain (modern Dāmghān) during an earthquake ca. 1200 BCE (Figure 11.1). As expected, a single paleoseismic trench study conducted outside the meizoseismal area of the earthquake across the Āstāneh fault could not provide evidence to establish the pre-historic (the 1200 BCE) and historic (856) earthquakes, their coseismic faulting, and/or establish the magnitudes of described events with a large range of dating errors (Figures 11.1 and 11.3).

Ground deformation associated with the ca. 280 BCE earthquake near the ancient large city of Rhagae (modern Ray in the southern suburb of the capital city of Tehrān) is described by a source who wrote at that time that during the earthquake, the ground ruptured and diverted the flow of the river (Figure 11.2). This is the first recorded stream bed diversion associated with an earthquake-faulting event on the Iranian plateau.

In the case of the 1336 earthquake, the contemporary data and the pattern and dimension of the damaged area show possible evidence for two large-magnitude earthquakes occurring along two nearby faults within a 3-day interval (Figure 11.9).

The oldest coseismic surface faulting inspection and description was conducted by Mo'in al-Din Mohammad Zamchi Esfezāri [the contemporary historian, calligrapher, and clerk to the court of Sultān Hossein Mirzā Bāyqarā Timurid (r. 1469–1505)] describing a minimum 12-km long earthquake faulting associated with the 10 January 1493 M_s ~7.0 Nauzād earthquake, which took place to the northeast of his birth place, Esfézār (located 17 km to the southwest of the coseismic surface faulting of the earthquake; Figure 11.10). Of course we do not expect a historian in 1493 to identify the total length of the coseismic surface rupture in a remote and sparsely populated desert area of eastern Iran.

Contemporary observers of the 1780 earthquake in Tabriz described a 2-m wide fault zone along 12- to 42-km-long fault with at least a 4-m vertical scarp in a NW–SE direction, coinciding with the orientation of the North Tabriz fault (Figures 11.7 and 11.12). Local inhabitants showed a European traveler an unusual 10 m-high (?) escarpment created by the same event.

The study also revealed cross-strike triggering of earthquakes and migration of seismicity during the 1493 and 1549 earthquakes along the Nauzād and Birjand faults (Figure 11.10), and the 1863, 1879, and 1869 earthquakes along the Sangāvar and Germirud or South Bozqush faults (Figure 11.14). It also showed along-strike migration of seismicity along a single fault with clustered earthquakes during the 1721, 1780, and 1786 earthquakes along the North Tabriz fault (Figure 11.12), and the 1863 and 1896 earthquakes along the Sangāvar fault (Figure 11.14). It documented earthquakes on conjugate faults destroying the city of Tabriz in 1641 and 1721 (Figure 11.12). The study also showed occurrence of the 958 and 1830 earthquakes along the central segment of the Moshā fault, north of Tehran, with a 876-year interval (Figure 11.12).

Clearly, earthquakes and their causative coseismic surface ruptures are subject to different interpretations. A controversy always exists over choosing the correct causative fault, or whether surface deformation is the result of primary tectonic faulting or a secondary adjustment due to strong ground shaking (unstable slope failure, slumping of the ground, liquefaction, and qanāt collapse). The cases described here as well as the accompanying (Figures 11.1–11.15) are a work in progress and subject to modifications and corrections by additional research, especially by paleoseismic trench investigation in the future.

Chapter 12

1900–1963 Coseismic Surface Faulting

Some disasters such as earthquakes
are associated with crustal faulting.

Biruni Khārazmi (1030)

It is well known that the instrumental coverage of earthquakes during this period was inadequate until the early 1960s, when the Worldwide Standardized Seismographic Stations (WWSSN) and United System of Seismic Observations (ESSN) were established. Seismicity in the early instrumental period (1900–1963) is still poorly understood, even for basic parameters such as earthquake locations, magnitudes, and focal depths. In some cases, this is the result of limitations in the distribution of seismogenic stations, response characteristics, timing of the early seismometers, and uncertainties in the crustal velocity model used. This chapter deals with the pre-WWSSN period coseismic surface ruptures.

Before 1961, the only two analog seismic stations in Iran, in Tehran (TEH; with Kashi-Afshār) and Shiraz (SHI; with P. Stahl)—which had begun operations in 1958 and 1959, respectively, through the Institute of Geophysics, University of Tehran (established in 1957 in two small rooms)—did not publish amplitude-period data. Hence, unlike Japan and California, where dense networks of seismic stations exist, teleseismic data in Iran cannot be used to locate causative faults of earthquakes, especially from 1900 to 1963 (Berberian, 1979c).

Unfortunately, because of a lack of earthquake geologists and seismologists on the Iranian plateau for most of this period, much of the macroseismic data of the early earthquakes in this period are lost. The first incomplete epicenter map of Iran during this period was prepared by De Ballore (1906), who showed few epicenters in major cities with no earthquake faults. Later, perhaps for the first time, Sieberg (1932) prepared a seismic zoning map of Iran with apparent zones of destruction, damage, and no damage, with large-, medium-, and low-magnitude earthquakes and a few speculated curved fault lines.

☆"To view the full reference list for the book, click here"

Developments in Earth Surface Processes, Vol. 17. http://dx.doi.org/10.1016/B978-0-444-63292-0.00012-0

The news of the 23 January 1909, M_w 7.4 Silākhor earthquakes, which killed more than 8000 people, reached Tehrān through consular couriers and appeared in Irān's Persian newspaper in late April 1909 (Irān, 61, 1327S/1909 CE). It took almost two weeks by mail for the news to reach the Russian consulate in Kermānshāh from Borujerd (a distance of 158 km). N. Nikolski, the Russian consul in Kermānshāh, sent a reconnaissance team, including the Russian consular agent Asadollāh-Mirzā, Sergeant Kurkin, and two cossaks, to find the location of the event. The group reached the epicentral area on 15 February 1909 (Nikolski, 1909a–d; Sablin, 1909; Ambraseys and Moinfar, 1973; Ambraseys, 1974b). Under the Anglo-Russian agreement of 31 August 1907, the country was occupied and divided into British (in the south) and Russian (in the north) spheres of influence.

Abdalian was the first Iranian geologist who studied the 1953 Torud earthquake (Abdalian, 1953). In January 1958, Francesco Peronaci (Geophysical Institute, Rome, 1958a,b, 1959, 1971) visited the meizoseismal area of the 14 and 16 August 1958 Firuzābād earthquake sequence and prepared four reports. Hagiwara and Naito (1959) also visited the southeastern part of the meizoseismal area and published an incomplete intensity distribution map. Toward the end of this period, Abdalian (1963a) prepared a preliminary seismotectonic map of Iran.

Comparison of the Iranian earthquake epicenters based on *P*-wave arrival time at teleseismic distances with well-determined restricted damage zones of moderate-magnitude earthquakes clearly indicated large errors, some as large as 500 km, in epicentral and focal depth determinations, especially for early period earthquakes (Ambraseys, 1978a; Berberian, 1979c). This large-magnitude location error is caused by poor station geometry around the Iranian plateau and mistakes in the assumed Earth velocity model and arrival times, among other factors.

The mean error in the instrumental and relocated epicenters of the medium- to large-magnitude Iranian earthquakes range from 500 km (in 1903) to 200 km (1918), 40–100 km (1940), and 30 km (1963), which is not acceptable for detailed seismotectonic and seismic risk studies (see Berberian, 1979c for discussion). Because of considerable location errors in the instrumental and relocated epicenters and a lack of adequate micro-earthquake surveys, it is impossible to locate the activity of most faults in the Middle East from the data supplied by national and international seismic agencies (Berberian 1976c,e, 1979c, 1981a, 1983c).

The reported focal depths of Iranian earthquakes based on arrival times by international agencies for this period suffered a greater magnitude of error than those of the epicenter locations. Most of the depths were fixed at an arbitrary depth of 33 km. Many Iranian relocated earthquakes were reported at depths greater than 90 km (i.e., Nowroozi, 1976) and have quoted errors of depth location that suggest that they could not have been shallower than 30 km. However, events such as 19 February 1924 (h 115 km), 13 April 1935 (140 km), 11 May 1945 (149 km), 30 June 1948 (114 km), 2 July

1957 (262 km), 4 April 1961 (230 km), 2 September 1962 (105 km), and 4 September 1968 (92 km) caused such localized intense destruction and damage that it appears that depth estimates must also be in error (Berberian, 1979c). This problem was later resolved by the synthetic *P* and *SH* waveform analysis of the earthquakes.

In order to reduce error, some teleseismic data of the Iranian earthquakes have been relocated (Nowroozi, 1971; Nabavi, 1972; Engdahl et al., 1998, 2006, among others). However, uncritical use of the data, especially for the moderate-magnitude Iranian earthquakes with shorter source dimensions, can lead to incorrect causative faults for the earthquakes.

First motion fault-plane solutions based on the observed polarities of *P*-wave onset of the short- and long-period instruments of the early days were also unconstrained and unreliable. Early solutions were provided by Shirokova (1962), Canitez and Ucer (1967), Canitez (1969), and Chandra (1981, 1984). The 1953 Torud earthquake focal mechanism, one of the early mechanisms, does not match the active structures of the region (discussed later).

Most of the earthquakes that occurred from1900 to 1963 were not studied immediately afterward and were not tied to individual faults. Therefore, the principal aim of this chapter is to document the causative surface faults of the 1900–1963 earthquakes and present the best catalog of the seismicity (Chapter 17) and active faults of the plateau (Chapter 14). This study results in a better understanding of the seismotectonics and presents a coherent and consistent seismogenic picture of active deformation of this vast area.

12.1 THE 23 JANUARY 1909, M_W 7.4 SILĀKHOR EARTHQUAKE

The 23 January 1909 Silākhor earthquake killed more than 8000 people, destroyed more than 65 villages (out of about 130 villages in the fertile Silākhor Valley), demolished the town of Bahrain (modern Dorud), and slightly damaged the city of Borujerd. The damage and destruction in the mountains to the southeast (with no roads) is unknown. The Russian consul in Kermanshāh, Nikolski (1909a–e, in Ambraseys, 1974b), observed landslides and linear cracks in the ground in many places past Zargarān village, near the Āb-e Dez River (~20 km NW of Bahrain, the modern city of Dorud; Figure 12.1). He reported that the linear cracks did not exceed 3 arsin (2.1 m/84 in.) in depth; their width varied between several versok (1 vershok = 4.44 cm) and 4 arsin (2.8 m/112 in.) and the length of some cracks extended to several versts (1 verst/versta = 1.0668 km/3500 ft.) (Nikolski, 1909a–e, in Ambraseys, 1974a,b). Nikolski, (1909a–e) clearly referred to coseismic faulting along the Dorud segment of the Zāgros Main Recent fault (Figure 12.1) and its fresh slickensides.

Sablin (1909, in Ambraseys, 1974b) in an appendix to the Russian Mission to Iran, added that at Khosrowābād village, located about 2 km to the northeast of the Zāgros Main Recent fault [~13 km northwest of Bahrain, modern

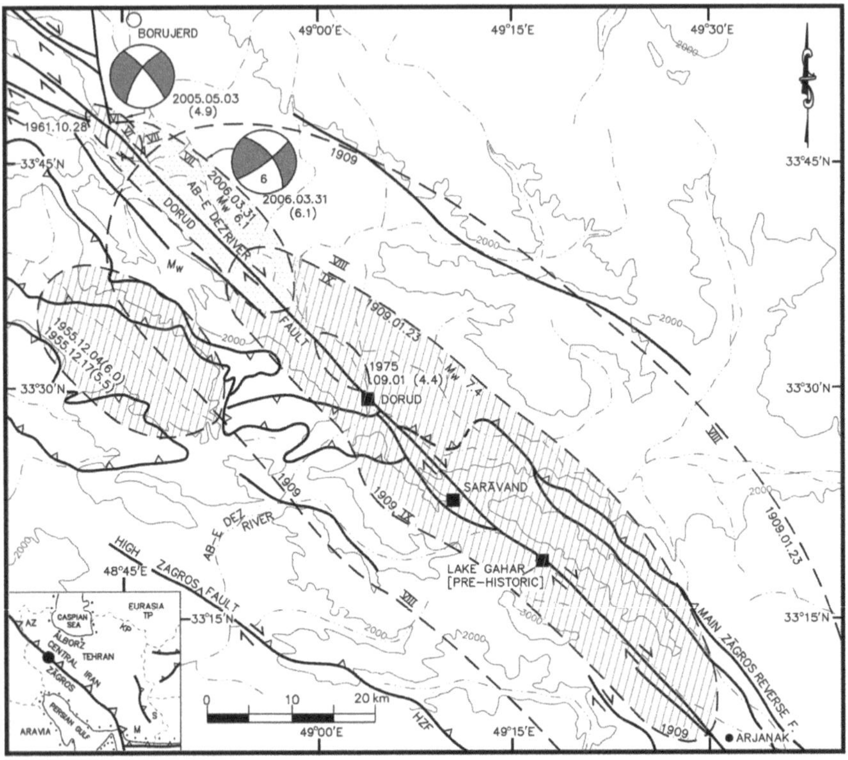

FIGURE 12.1 Meizoseismal area of the 23 January 1909 M_w Silākhor earthquakes along the Dorud fault segment of the Zāgros Main Recent fault. Symbols as in Figure 11.1. The city of Dorud (previously Bahrain) and the location of the Lake Gahar, formed during a prehistoric, earthquake-triggered, large rock avalanche, are shown by filled squares. Focal mechanisms of the 2006 M_w 6.1 Chālānchulān earthquake constrained by body wave modeling with centroid depth (Peyret et al., 2008) and 2005 South Borujerd earthquakes (best-double-couple HRVD CMT solution) have also been added (the 1961 and 2005 earthquakes took place in the same area). *Modified from Berberian (1995).*

city of Dorud], the ground opened up in many places, tearing houses apart and swallowing walls, and water came out of the ground, flooding the ruins. He added that the Cheshmeh-Āb village was turned upside down when the Earth split in two from east to west (SE to NW) and a river now flows into a chasm (Sablin, 1909, in Ambraseys, 1974a,b). During the 1971 site visit (Tchalenko and Braud, 1974), some survivors remembered a large linear crack into which the Āb-e Dez River near Gushehpol disappeared for two days.

The 1909 earthquake was associated with coseismic surface rupture along the Dorud segment of the Zāgros Main Recent fault; only 45 km was visited in 1971 (Ambraseys and Moinfar, 1973; Tchalenko and Braud, 1974; Tchalenko et al., 1974b). Due to high topography, rugged and uninhabited terrain, and a lack of access to roads, the continuation of the surface deformation to the southeast of Sarāvand was not visited. The fault segment that was seen was from Kalāngāneh/Kalāngonā in the northwest (33°38′N–48°53′E, +1591 m) to Sarāband/Sarāvand in the southeast (33°22′N–49°09′E, +2128 m; Figure 12.1). The size of the meizoseismal area, the magnitude of the earthquake, and intensity of ground deformation further to the southeast of Sarāvand [Lake Gahar (33°18′N–49°17′E, +2518 m) and Arjanak area (33°06′N–49°34′E, +2282)] suggest that coseismic rupture extended to the southern portion of the meizoseismal area, increasing the length of the coseismic surface rupture to a minimum of 70 km (Figure 12.1).

The fault scarp is located in the flat alluvial part of the Silākhor Valley, except between Lāvān and Kalāngonā [~5 km] where the surface ruptures separate the plain from the alluvial fans originating in the mountains to the southwest. The northeastern side of the fault scarp was always downthrown by up to 1 meter (visited 62 years after the earthquake in 1971). The fault trace cuts a meander of the Āb-e Dez River near Gushehpol (33°33′N–48°59′E, +1573 m) at the approximate location where the inhabitants in 1971 had indicated that the river had disappeared into a crack for 2 days. Southeast of Dorud (Figure 12.1), the alluvial scarp is continued by a marked fault in the bedrock of the Shotorān Kuh and was reactivated during the earthquake at Dar Āstāneh and as far as south as Sarāvand (Tchalenko and Braud, 1974).

The Zāgros Main Recent fault is a NW–SE trending right-lateral strike-slip active fault separating the Zāgros fold-and-thrust belt (in the SW) from the Central Iran range-and-basin (in the NE) in the northwest of the Zāgros (Figure 12.1). It follows the geological suture zone of the Neo-Tethys and the Main Zāgros reverse fault, where the crust between Zāgros and Central Iran is weak (Berberian and King, 1981a; Berberian et al., 1982; Berberian, 1989). The fault is a zone of localized crustal deformation and has been associated with historic and instrumental period earthquakes (Tchalenko and Braud, 1974; Tchalenko et al., 1974b; Berberian, 1995; Berberian and Yeats, 2001).

Long-term slip rates derived from geological observations in the Zāgros area were proposed by Berberian and King (1981b), Berberian (1995), Talebian and Jackson (2002), Bachmanov et al. (2004), Vernant and Chery (2006), Walpersdorf et al. (2006), Yamini-Fard et al. (2006), Peyret et al. (2008), and Authemayou et al. (2009). Assuming a Pliocene (3–5 Ma) initiation of the Zāgros Main Recent fault and a cumulative lateral slip of 50 km along that fault, Talebian and Jackson (2002) stated a long-term slip rate of 10–17 mm/year. Based on the offset of a river valley incised into a surface of likely postglacial age (?), Bachmanov et al. (2004) gave an estimate of 10 mm/year. Authemayou et al. (2006), based on the Quaternary geological offsets, suggested a slip rate of 5–7 mm/year. Later, Authemayou et al. (2009) evaluated the total geological displacement rates along the fault at 3.5–12.5 mm/year [with adding the Fārsān (2.1–5.9 mm/year) and Ardal (1.5–6.5 mm/year) segments].

The GPS data are not in agreement with these long-term estimates. The maximum slip rate along the Zāgros Main Recent fault deduced from GPS measurements would be at 4.0 ± 2.5 mm/year if the fault achieves complete partitioning of the 4.7 ± 2 mm/year of north-trending shortening across the western Zāgros (Nilforoushan et al., 2003; Vernant et al., 2004). The instantaneous slip rate along the fault deduced from GPS measurements (Vernant et al., 2004) would be of 5.0 ± 1.5 mm/year if the fault is considered to achieve total partitioning of the 7 ± 2 mm/year of N–S shortening measured around 50°E. Based on local GPS network surveys, Walpersdorf et al. (2006) suggested 4–6 mm/year of cumulated right-lateral strike-slip across the whole northwestern Zāgros. By adopting a 3D viscoelastic finite element model for simulating lithospheric deformation with faults, Nankali (2011) predicted a maximum 2.3 mm/year right-lateral slip rate of the Zāgros Main Recent fault on its central segment.

A color map prepared by the Seismotectonic Department of the Geological Survey of Iran as Figure 13 (ngdir.ir, 2006) shows 57 km of coseismic surface faulting associated with the 1909 Silākhor earthquakes at a distance of 6–12 km to the northeast of the Dorud fault segment of the Zāgros Main Recent fault. The assumed fault, which is drawn along the northeastern edge of the Silākhor plain in a region with no active fault features, cannot be substantiated.

For the pattern of coseismic surface rupture and historical seismicity along the Zāgros Main Recent fault, see Chapter 16.

12.2 THE 18 APRIL 1911 M_S 6.4 RĀVAR EARTHQUAKE

The earthquake destroyed the town of Rāvar and the nearby sparsely populated villages in a desert region of the Kermān province; about 700 people were reported dead. There is a controversy regarding the coseismic surface faulting associated with this earthquake and its location. Ambraseys and Melville (1982) believed that the earthquake was associated with faulting west of Abdirjān, extending for a few kilometers in a south–southwesterly direction. On their meizoseismal map (figure 3.33, p. 74), Ambraseys and Melville (1982) showed a minimum of 15 km coseismic surface faulting with a general azimuth of N155°E. However, the location of villages, rivers, and coordinates on this map suffer some unknown errors. If their figure is scaled and overlain on a 1:250,000 topographic map based on the coordinates (31°N and 57°E), the location of the Rāvar town falls 10 km to the east of its correct spot. If the overlay is fixed with locations of Rāvar, Rayhān, Hauz-e Panj, and Esma'ilābād, then the longitude 57°E falls about 8 km to the west of its correct position. In the first case, the highlighted NNW–SSE earthquake fault cuts the N–S Lakarkuh fault trend, which is incorrect. In the second case, the highlighted earthquake fault cuts the E–W thrust faults, which is also incorrect (Figure 12.2).

These two fitting trials and study of the aerial photographs do not support the highlighted earthquake fault of figure 3.33 in Ambraseys and Melville (1982). Later, Ambraseys and Jackson (1998) reported a 15-km-long spurious surface faulting striking N155°E with 50 cm of vertical displacement. The fault trend and reported displacement cannot be verified. The meizoseismal area shown in Berberian (2005) is also not constrained.

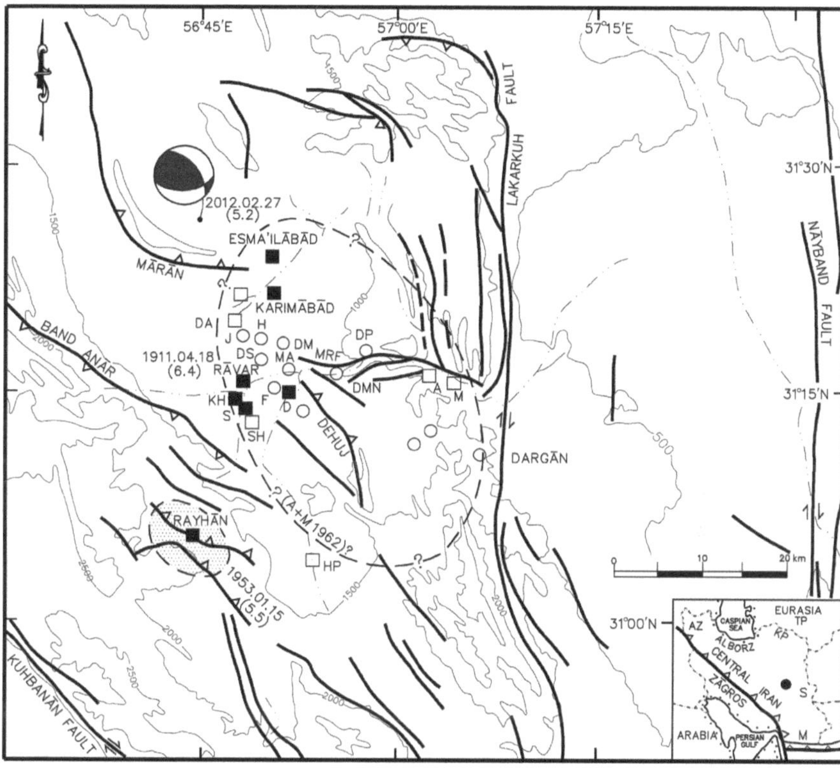

FIGURE 12.2 Epicentral area of the 18 April 1911 M_s 6.4 Rāvar earthquake in eastern Iran. Symbols as in Figure 11.1. Filled squares: documented collapsed sites. A+M, 1982: the unconstrained and uncertain meizoseismal area of the 1911 earthquake. Fault-plane solution of the 2012 earthquake (best-double-couple HRVD CMT solution). A, Ābdarjān; D, Dehuj; AD, Deh 'Ali; DM, Deh Mudi; DMN, Deh Miyān; DP, Deh Pā'in; DS, Dehshir; F, Fathābād; H, Hematābād; HP, Hauz Panj caravanserai; J, Jalālābād; K, Karimābād; KH, Khairābād; M, Maki; MA, Mehdiābād; MRF, Miyānrud fault; S, Shahrābād; SH, Sharifābād; Y,Yusefābād. *Modified from Berberian (2005).*

On 7 February 2012, an earthquake of M_w 5.2 took place about 20 km north northwest of Rāvar with NW–SE trending reverse fault CMT solution (HRVD) on the hanging wall of the Mārān fault (Figure 12.2). It is not clear which of the Mārān, Band Anār, Mahirud (MRF in Figure 12.2), and Lakarkuh faults was responsible for the occurrence of the 1911 earthquake.

12.3 THE 22 SEPTEMBER 1923 M_S 6.7 LĀLEHZĀR EARTQHUAKE

The earthquake destroyed several villages and killed about 300 people in the remote mountainous area of the Kermān province of southeast Iran. The mainshock, together with its two aftershocks of 23 September 1923 (M_s 5.4) and

18 January 1924 (M_s 5.4), took place in the Lālehzār area, immediately south of the Rafsanjān fault (Berberian, 1976b,c,e; Berberian et al., 1984; Berberian and Qorashi, 1989a). The meizoseismal area given by Ambraseys and Melville (1982, figure 3.36, p. 77) covers the E–W Lālehzār fault (the fault is not shown in their figure) for which there is no evidence of Quaternary movement (Figure 12.3). During the field work of September 1984, Berberian and Qorashi (1989a) mapped the Chamanrang thrust fault with recent active fault features (visible both on the aerial photographs and the ground) to the southwestern foothills of the Ākhorak Mountains (Figure 12.3). Asghar Zol'alā, an old man from Qal'eh Asgar village, mentioned that the shock with underground noise came from the Ākhorak Mountain in the north (Berberian and Qorashi, 1984, 1989a). The damaged villages are mostly concentrated along the Chamanrang thrust and on its footwall (Figure 12.3).

To the northwest of the Chamanrang thrust, at about 29°43′N–56°32′E along the Rafsanjān fault, Fattahi et al. (2011) reported ~1.5 km long fresh fractures with right-lateral offset of ~3.0 m and vertical displacement of ~0.5 m and speculated that the ground fractures resulted from the 1923 earthquake. Based on the 3.0 m displacement, Fattahi et al. (2011) estimated a

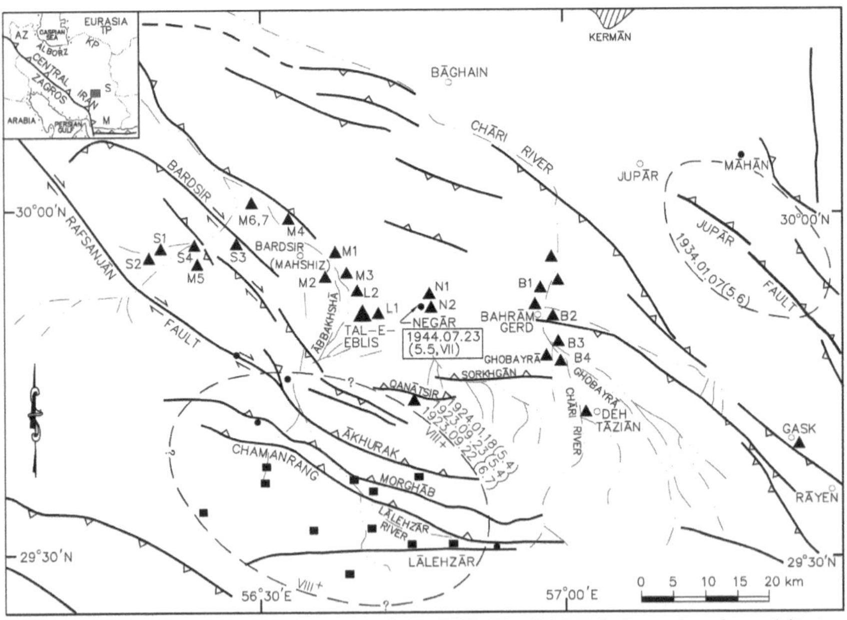

FIGURE 12.3 Epicentral and fault map of the 1923 M_w 6.7 Lālehzār earthquake and its two aftershocks in the area south-southwest of the city of Kermān in southeast Iran. Symbols as in Figure 11.1. Documented destroyed sites are shown by filled squares. The 1944 earthquake damaged the historical village of Negār, killing a few people. The archeological sites of Tal-e Eblis (Iblis Mound: 5500–3500 BCE) and Ghobayrā (3rd Millennium BCE) together with B, M, and S sites have also been added (Caldwell 1967; Bivar, 2000). *Modified from Berberian and Qorashi (1984, 1989a) and Berberian (2005).*

magnitude of 7.0–7.5 and surface rupture length of 65–125 km for the event. The estimated surface rupture parameters do not match the 1923 M_s 6.7 earthquake. Paleoseismic trench study and archeoseismic investigation of numerous archeological sites such as Tal-e Elbis (5500–3500 BCE; Caldwell, 1967) and the Ghobayrā (3rd millennium BCE; Bivar, 2000) mounds will help explain the seismic history and surface rupture of the area (Figure 12.3).

12.4 THE 1 MAY 1929 M_S 7.3 BĀGHĀN EARTHQUAKE

This large-magnitude earthquake killed more than 3800 people (the exact number is unknown) and injured more than 1100 people in 300 villages in the Central Kopeh Dāgh Shear Zone (CKDSZ), northeast of Iran (Figure 12.4).

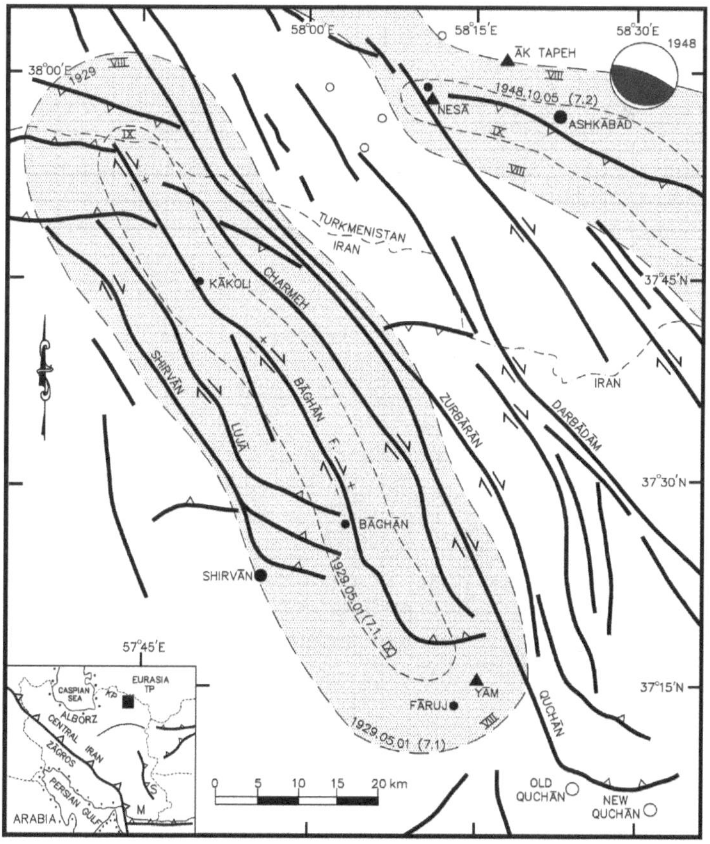

FIGURE 12.4 Meizoseismal area of the 1 May 1929 M_w 7.1 Bāghān earthquake along the Bāghān right-lateral strike-slip fault in the Central Kopeh Dāgh Shear Zone (CKDSZ) of northeast Iran. Symbols as in Figure 11.1. Fault-plane solution of the 1948 Ashkābād earthquake (McKenzie, 1972). *Modified from Tchalenko (1975), Afshar-Harb (1979), GSI (1986a), Berberian and Yeats (2001), Hollingsworth et al. (2008), Shabanian et al. (2009a).*

Wilson (1930), based on the *Daily Telegraph* of 6 May 1929, wrote that: "*a cleft three yards wide* [2.7 m] *was opened between the towns* [villages] *of Khaki* [Kāki/Kākoli] *and Bagham* [Bāghān], *to the east of the Tehran-Isfahan* [sic, Esfarāyen] *road, the cleft extending to a distance of 18 miles* [28.9 km]. *The towns of Kuchan* [Quchān] *and Rubat* [Robāt] *also suffered severely, huge fissures in the ground being opened up, one being, according to a report in the Times of 9 May 1929, 24 miles* [38.9 km] *long and 9 feet* [2.7 m] *wide. The total casualties were subsequently given in an official report from a government inspector at Kuchan* [Quchān] *as 3253 persons killed, 1121 injured, 83 villages destroyed, and 6542 cattle killed*" (Wilson, 1930). The distance between Kākoli and Bāghān, where earthquake faulting along the Bāghān fault was reported, is 36 km (Figure 12.4).

M. Pashinskiy, chief of the Khairābād Meteorological Station near the Russian border (in Popov, 1940), wrote that there was a horrifying crash from the talus in the mountains and landslides and fissures in the surrounding areas. Nazarevsky (1932) reported some earthquake fissures at the area immediately northeast of the Germāb [Garmāb] village in southern Russia (present Turkmenistan), along the northeastern embankment of the Mergen Ulya River running NW–SE parallel to the river, indicating sliding along the northern embankment of the river course. He also added that all buildings at Garmāb were destroyed and the Sakiz Āb River to the south of Garmāb was waterlogged.

The earthquake produced coseismic ground ruptures about 70 km in length with a vertical motion of about 2 m on the northeast side of the Bāghān fault (Figure 12.4). No fresh strike-slip motion was documented in 1974 when the area was visited (Tchalenko et al., 1974b; Tchalenko, 1975; Ambraseys and Melville, 1982).

The Bāghān right-lateral strike-slip fault is a member of the "CKDSZ" of closely spaced NW–SE trending parallel right-lateral strike-slip faults cutting and displacing the Kopeh Dāgh fold-and-thrust belt (Figure 12.4). The CKDSZ resembles the Eastern California Shear Zone (Sauber et al., 1986, 1994). A total right-lateral offset of about 10 km can be measured between the displaced Upper Cretaceous Ābderāz Formation outcrops at about 37°35′N–58°00E (Afshar-Harb, 1979; Hollingsworth et al., 2006). The total length of the fault is 80 km and is capable of right-lateral slip of about 3.0 m (Wells and Coppersmith, 1994). Hollingsworth et al. (2008) and Shabanian et al. (2009a) suggested a right-lateral slip-rate of 1.0 and 2.8 mm/year, respectively.

12.5 THE 15 JULY 1929 M_S 6.0 ANDIKĀ EARTHQUAKE

Ambraseys (1975) reported >1-km-long ground deformation, striking N150°E, associated with this earthquake. A similar statement is reported by Ambraseys and Jackson (1998) as uncertain surface faulting. No evidence of coseismic surface faulting has yet been established associated with this medium-magnitude earthquake in the Zāgros. The meizoseismal area of the earthquake is located on the foot-wall of the Zāgros Mountain Front Reverse fault (Berberian, 1995).

12.6 THE 6 MAY 1930 M_W 7.1 SALMĀS EARTHQUAKE

Preceded by an alarming foreshock of M_s 5.4 (10:03 local time on 6 May), which forced the people to stay outdoors at night, the mainshock (01:34 local time on 7 May) destroyed two cities of Dilmān (Salmās) and Kohneh Shar, killing about 2514 people in the Salmās Plain and the surrounding mountains of the northwest Iran (Figure 12.5; Plates 12.1–12.4). The alarming M_s 5.4

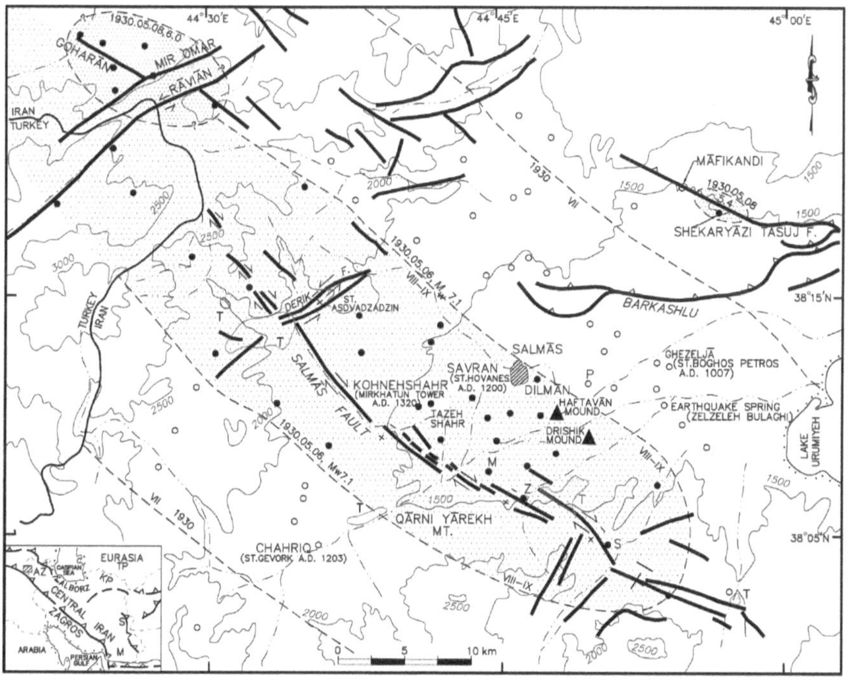

FIGURE 12.5 Meizoseismal area and fault map of the May 6 1930 M_w 7.1 Salmās earthquake and its two strong aftershocks of 8 May 1930 in the northeast and northwest of the mainshock in the northwest of Iran. Symbols as in Figure 11.1. Triangle: archeological mound. Crosses: Historical Armenian churches. To the east of the city of Salmās, Dilmān was the old city destroyed by the 1930 earthquakes. Meizoseismal area of the Shekaryāzi aftershock in the northeast is well constrained. Despite the documented destroyed sites (filled circles), the meizoseismal area of the northwestern aftershock is not constrained and could be elongated along the Rāvian and/or Mir 'Omar faults (see the text). M, Malham; P, Pāyājuk (the birthplace of the Armenian novelist and poet Rāffi/ Hākob Melik-Hākobiān: 1835–1888); S, Shurgel; T, Recent travertine deposit; Z, Zaviehjuk. See also Plates 12.1–12.7. *Modified from Tchalenko and Berberian (1974), Berberian and Tchalenko (1976a), Berberian (1981a, 1997a).*

PLATE 12.1 Historic photograph of the 1930 earthquake ruined city of Dilmān (Salāms; rebuilt as Shāpur). The buildings were adobe and brick structures with wooden beams used in the roof. *Courtesy Yahyā Zokā' from an unknown photographer.*

PLATE 12.2 Historic photograph of the 1930 destruction at Dilmān (Salāms) showing the type of structures. Only the load-bearing masonry pillars and thick walls are standing. Wooden beams of the collapsed ceilings are scattered in the debris. For the location, see Figure 12.5. *Courtesy Yahyā Zokā' from an unknown photographer.*

foreshock of 6 May 1930 (07:03 UTC) caused localized destruction and casualties in a small region centered at 38.15°N–44.75°E, south of the city of Salmās with possible short ground rupture.

Several eyewitnesses recorded the coseismic surface faulting and associated ground deformation during the 1930 Salmās mainshock. A Reuters dispatch from Tehran on 12 May 1930 (*The Times*, 13 May 1930; Tchalenko and Berberian, 1974; Berberian and Tchalenko, 1976a) wrote that the emergency vehicles were unable to approach Salmās, as new rifts had been formed.

PLATE 12.3 The 1930 ruined mosque at Dilmān (Salāms). Only sections of the portal arches and thick load-bearing masonry pillars are standing. For the location, see Figure 12.5. *Courtesy Yahyā Zokā' from an unknown photographer.*

PLATE 12.4 Close-up view of Plate 12.3, showing details of the structure. For the location see Figure 12.5. *Courtesy Yahyā Zokā' from an unknown photographer.*

The Tabriz Newspaper (10 May 1930) quoted a cable dated 7 May 1930 that at Dir [sic., Derik] cracks appeared in the mountains (Tchalenko and Berberian, 1974; Berberian and Tchalenko, 1976a). If this date is correct (?, reported on May 10), then this report becomes very important—in the mountains NW of Derik, the Derik fault was reactivated with surface ruptures. This indicates that the Derik fault (a short fault conjugate to the Salmās fault) moved prior to the 8 May M_s 6.2 aftershock, possibly during the mainshock, which ruptured along the Salmās fault (Figure 12.5).

The same newspaper (24 May 1930) described rockfalls and cracks in the mountains west of Kohneh Shahr. The cracks started from Boghāz-e Zarindarreh and continued to the Kuh-e Shomāl. The cracks after Chahriq crossed the Aliguli hills and then turned southward to Qārni Yārekh (Tchalenko and Berberian, 1974; Berberian and Tchalenko, 1976a). The fractured section (from Zarrindarreh to Qārni Yārekh) covers a section of the Salmās coseismic surface fault of more than 10 km (Figure 12.5).

Abel Zayia (1930a,b), a Persian Lazarist of the French Mission in Rezāiyeh [later Urumiyeh], was the eyewitness of the Salmās earthquake mainshock. Abel Zayia's (1930b) observation, made the morning after the earthquake (Tchalenko and Berberian, 1974; Berberian and Tchalenko, 1976a), stated that at the entrance to the Salmās Plain, there was an E–W [NW–SE] oriented fissure, 3 km long, following the base of the mountain; the southern part of the fissure being about 2 m higher than the northern part. This indicated a partial lowering of the plain along the Salmās coseismic surface fault (Figure 12.5; Plates 12.5 and 12.6).

Abel Zayia (1930a) then gave information about the observed minimum fault length and the amount of vertical displacement: "*I began to descend in*

PLATE 12.5 The 6 May1930 M_w 7.1 Salmās earthquake, right-lateral, strike-slip fault rupture at the eastern part of the old Malham Armenian cemetery, down-throwing the northeastern part (left side) of once nearly horizontal cemetery, looking southeast. The fault scarp with 4 m throw runs from the right-hand side of the photograph to the background. The line of tombstones delineating the far edge of the cemetery shows a narrow graben developed at the foot of the coseismic fault scarp. The oldest Armenian stone masonry church in Malham, S. Zorāvār, dates back to 1641; the cemetery should have been constructed on nearly flat ground at least during that period. For the location, see Figure 12.5. *Photographed in 1973 by John Tchalenko. Courtesy of photograph pool of Tchalenko and Berberian (1974) and Berberian and Tchalenko (1976a).*

PLATE 12.6 The Zaviehjuk segment of the 6 May1930 M_w 7.1 coseismic surface rupture of the Samlās fault. Aerial view of the fault trace (between arrows) passing through Zaviehjuk (Z) and the Armenian Malham cemetery (M). There are several topographical steps north of, and parallel to, the Salmās fault scarp, the most obvious one marking for a short distance the limit between mountain foot alluvial fans and cultivated fields of the southern edge of the Salāms plain northwest of Zaviehjuk. North is at the top of the aerial photograph. For the location, see Figure 12.5. National Cartographic Center of the Imperial Government of Iran, 1:20,000 aerial photograph series (Tchalenko and Berberian, 1974; Berberian and Tchalenko, 1976a).

the Salmās Plain, that I saw the ground opened up to the left; on the Āghiān and Zivadjouk side, on the slope of the mountain, a deep opening with a length of 4 km. The water, which previously came out 6 meters lower down, is now coming out above the road." Abel Zayia's description of the surface faulting in the named location is a perfect match with the Salmās fault line (Figure 12.5; Plates 12.5 and 12.6).

Franssen (1930), of the Lazarist Mission and writing from Rezāiyeh [now Urumiyeh] on 26 May 1930 (Tchalenko and Berberian, 1974; Berberian and Tchalenko, 1976a), also gave data on coseismic surface rupture length and its vertical motion: "*I traveled all the areas situated between Keuhnachahar* [Kohneh Shahr] *and Patavour, these being localities situated close to the supposed center of the earthquake. Everywhere the ground is crevassed. According to Mr. Abel* [Zaya], *there is an opening, which is two meters deep and about twenty kilometers long. I saw a whole field of barley lowered by at least one meter*" (Franssen, 26 May 1930).

Franssen's account from Rezāiyeh on 8 June 1930 (Tchalenko and Berberian, 1974; Berberian and Tchalenko, 1976a) added that the fault resulting from this earthquake starts at Ārāwoul, in the direction of the lake [Urumiyeh], therefore from west to east; it measures a length of 20 km. The depth of the fault [vertical drop] is variable, but at certain localities it is more than 2 m deep with a width of 3 m.

Brunk (1930) wrote that on several places cracks were formed on the surface. The opening was measured to be 40 m long, 10–15 cm wide, and 15–20 cm deep.

Two still visible spectacular fault-breaks were associated with the 1930 earthquake: (i) the Salmās Fault, with a NW–SE strike at the southern edge of the Salmās Plain (Figure 12.5; Plates 12.5 and 12.6) and (ii) the Derik Fault, conjugate to the former in a NE–SW direction near the mountain village of Derik (Plate 12.7; Figure 12.5). Maximum displacements on these faults were: (i) for the Salrnās Fault, 4 m right-lateral [at 38°07′08″N–44°45′27″, +1430.1 m] and over 5 m vertical (NE downthrown) and (ii) for the Derik Fault, left-lateral (?) by an unknown amount, and about 1 m vertical (NE downthrown?). The nearly close components of normal [5 m, in many places] and right-lateral strike-slip [4 m] displacements along the Salmās fault may imply an ENE–WSW slip vector for this event (Figure 12.5). The combination of the right-lateral movement along the Salmās fault and left-lateral

PLATE 12.7 The 1930 coseismic Derik fault north of the destroyed old Derik village. Vertical displacement at this locality is about 1 m. For the location, see Figure 12.5. *Photographed by John Tchalenko in 1973. Courtesy of photograph pool of Tchalenko and Berberian (1974) and Berberian and Tchalenko (1976a).*

motion along the Derik fault has resulted in downthrowing of the Salmās plain (Tchalenko et al., 1974b; Tchalenko and Berberian, 1974; Berberian and Tchalenko, 1976a; Berberian, 1981a, 1997a).

The Salmās Fault is the longer of the two faults and produced the larger displacements. It is formed by four separate *en echelon* segments referred to here as: (i) the Ākhiān segment, (ii) the Mahlam cemetery segment (Plates 12.5 and 12.6), (iii) the Kohneh Shahr segment, and (iv) the Dowshivān segment (Figure 12.5). We were able to trace only 43 km of the surface faulting along the Salmās fault, 43 years after the earthquake during 1973 field work (Tchalenko and Berberian, 1974; Tchalenko et al., 1974b; Berberian and Tchalenko, 1976a). The fault is a member of the Western Āzarbāijān Shear Zone (WASZ); sub-parallel NW–SE striking right-lateral strike-slip faulting of northwestern Iran (Berberian, 1997a), showing strain partitioning and clockwise rotation of blocks (Jackson, 1992; Copley and Jackson, 2006).

Two aftershocks of 8 May 1930 took place off the Salmās fault: (i) the Shekaryāzi aftershock at 15:05 UTC (M 5.4) occurred along the Tasuj reverse fault located about 28 km to the northeast of the Salmās fault (20 km NE of the Salmās city) and (ii) the Goharān (South Qotur) aftershock at 15:35 UTC (M_s 6.2) took place about 22 km northwest of the Salmās–Derik fault junction (38 km northwest of the Salmās city), either on: (a) the NW–SE trending shorter en-echelon right-lateral Goharān faults (?, parallel to the Salmās fault), or (b) the NE–SW-trending longer Mir ‘Omar and Raviān left-lateral strike-slip faults (parallel to the Derik fault; about 22 km to the NW) (Figure 12.5). The area has not been visited since the earthquake; however, the longer length of the NE–SW trending Mir ‘Omar and Raviān faults together with the magnitude of this aftershock (M_s 6.2) may indicate that this set of faults were reactivated, destroying several villages on 8 May 1930 that had not been destroyed by the mainshock of 6 May 1930 (Figure 12.5).

An earthquake of M_s 3.7/mb 4.2 (ISC), with a centroid moment tensor (Zur_RMT) of right-lateral strike-slip faulting striking N111°E, took place on 24 January 2004 in the Salmās area. Unfortunately, the epicenters calculated by NEIC, THR, and ISC have large location errors and do not match each other.

12.7 THE 28 NOVEMBER 1933 M_W 6.2 NW BEHĀBĀD EARTHQUAKE

The earthquake destroyed several remote, small desert villages in the playa northwest of Behābād in the Yazd province, killing an unknown number of people. Ambraseys (1975) reported a >12-km-long coseismic ground deformation striking N150°E with 100 cm vertical displacement. The fault length is later reduced to 5 km with 50 cm of vertical displacement in Ambraseys and Jackson (1998). According to local information, the earthquake caused ground fractures in bedrock in the area about 6 km to the southwest of

'Aliābād-e Mollā 'Alirezā (Ambraseys et al., 1979). This location falls near the Kuhbanān fault, showing a clear recent fault trace with right-lateral strike-slip as well as vertical displacements (northeastern block downthrown) along the northwestern segment of the Kuhbanān right-lateral strike-slip fault. The fault limits the southwestern edge of the Behābād playa called Darreh Boland-e Behābād (Figure 12.6).

See Chapter 16 for historical seismicity and coseismic surface rupture pattern along the Kuhbanān fault.

12.8 THE 30 JUNE 1936 M_S 6.0 ĀBIZ EARTHQUAKE

The earthquake of 30 June 1936, which caused considerable damage in the region between Bamrud and Ābiz, killing 10, is the first known earthquake that happened along the Ābiz fault (Berberian et al., 1999; see Figures 13.9 and 13.11 in Chapter 13). The earthquake is still remembered by some inhabitants of Kalāt-e Naudeh and Firuzābād, who confirmed that damage extended to Estend and that cracks in the ground extended from Naudeh to Hajiābād (a distance of about 8 km length). They confirmed that the surface ruptures were particularly large where they crossed the dry bed of Āb-e Bamrud (Bamrud River). The indications are that this earthquake ruptured a segment of the Ābiz right-lateral strike-slip fault over a distance of perhaps as much as 12 km that also moved during the 10 May 1997 M_w 7.2 Zirkuh earthquake (Berberian, 1981a; Berberian and Yeats 1999, 2001; Berberian et al., 1999, 2001). See Chapter 16 for historical seismicity and coseismic surface rupture pattern along the Ābiz fault.

12.9 THE 4 MAY 1940 M_W 6.4 ESTĀYESH EARTHQUAKE

The earthquake destroyed several small and semidesert remote villages west northwest of Kadkan, killing an unknown number of people. Khosravi (1987), quoting 'Abdolhamid Maulavi's article published in a magazine (*Majalleh Farhang Khorāsān*, volume 3, no. 8), wrote that "*About 25 years ago in one of the hills near Kadkan village, which looked natural, a fracture developed by an earthquake and an entrance with a pre-Islamic golden decorated coffin became visible. Hansan 'Ali Afsari, brother of prince Mohammad Hāshem Mirzā Afsar, the previous parliament member, confiscated the coffin and it is not known where he took it.*" Meizoseismal area of this earthquake (Figure 12.7) is located along the Zelzeleh Khiz (lit., "Earthquake-Prone"; Kadkan) active fault (Berberian and Qorashi, 1989a).

12.9.1 Oral Earthquake Hazard Warnings

Although no pre-1900 earthquake data have been preserved in the chronicles, ancient toponyms preserved in oral earthquake hazard warnings, such as "Earthquake-Prone Mountain" (Kuh-e Zelzelhkhiz, just to the WSW of the

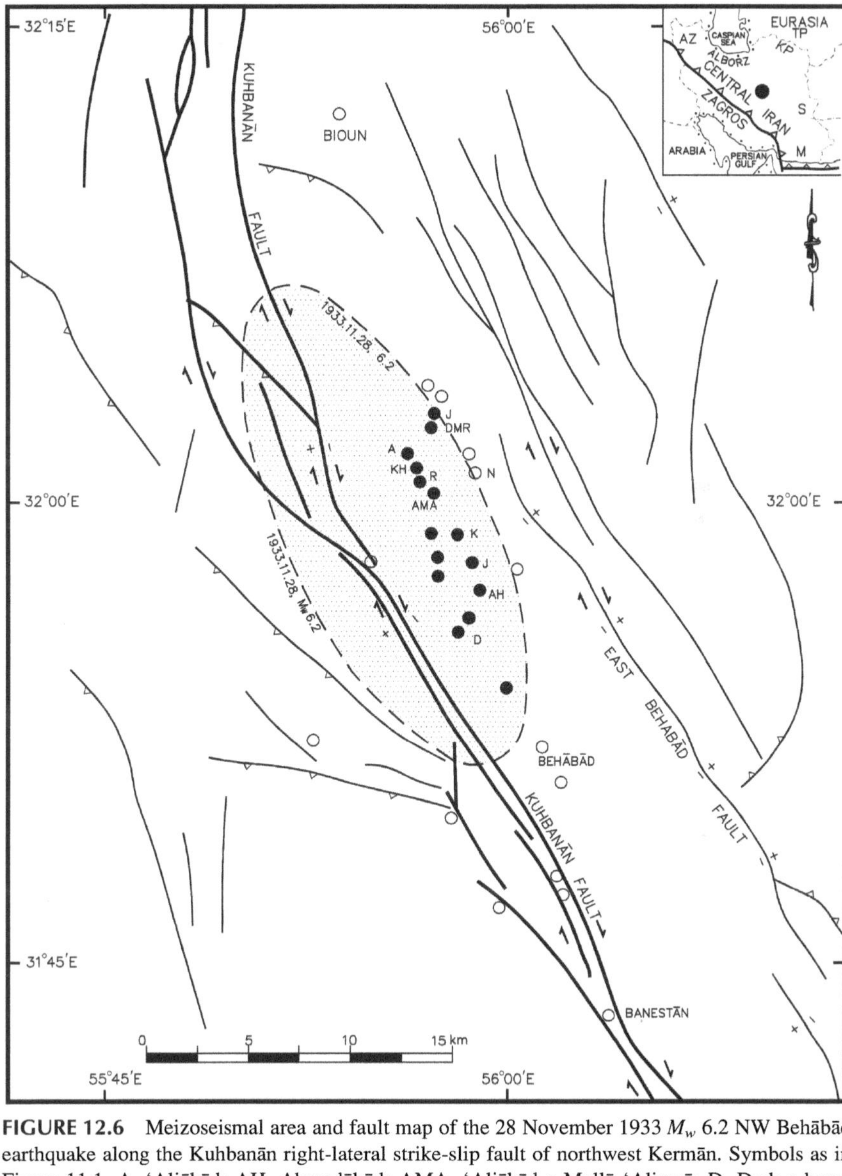

FIGURE 12.6 Meizoseismal area and fault map of the 28 November 1933 M_w 6.2 NW Behābād earthquake along the Kuhbanān right-lateral strike-slip fault of northwest Kermān. Symbols as in Figure 11.1. A, 'Aliābād; AH, Ahmadābād; AMA, 'Aliābād-e Mollā 'Alirezā; D, Darb-e hauz; DMR, Deh Mohammad Rafi'; J, Janatābād; K, Karimābād; KH, Khairābād; N, Nosratābād; R, Rahimābād.

1940 destroyed village of Estāyesh), "Thousands (Numerous) Fractures" ("Highly Fractured or Faulted"; Shekasteh Hezār), and "Burnt City" (Shahr-e Sukhteh) show several location names indicating a region with earthquake history (Figure 12.7). A ruined Safavid (1500–1722) Fort is also located in

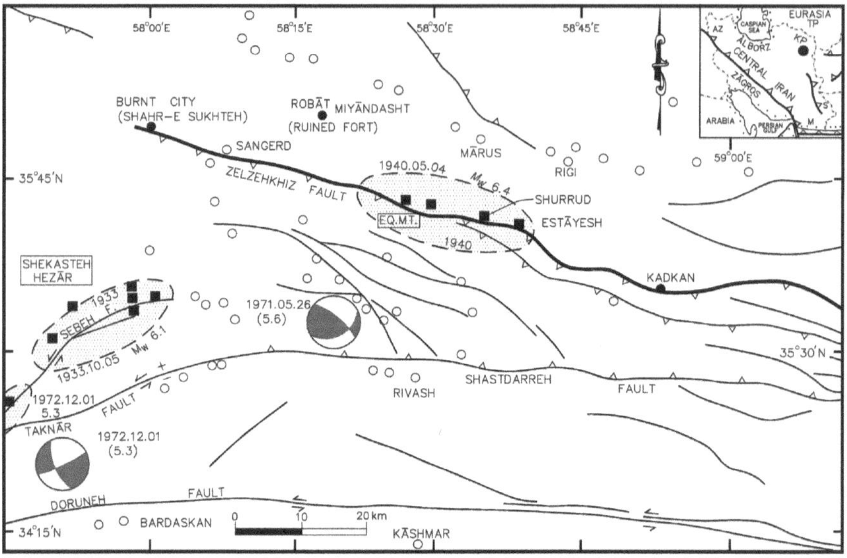

FIGURE 12.7 Meizoseismal area and fault map of the 4 May 1940 M_w 6.4 Estāyesh earthquake along the Zelzelehkhiz fault in the Kadkan area of northeast Iran, north of the Doruneh fault. Symbols as in Figure 11.1. EQ. MT (in the epicentral area of the 1940 earthquake): "Earthquake Mountain" (*Kuh-e Zelzelehkhiz* in Persian/Arabic). In the West, *Shekasteh Hezār* means "Highly Fractured/Faulted Mountain" (see the text and Chapter 1 for an interpretation). Fault-plane solutions: Jackson and McKenzie (1984).

Robāt-e Miyāndasht to the northwest of the 1940 meizoseismal area (Figure 12.7); the cause of its ruined status is not clear.

12.10 THE 16 FEBRUARY 1941 M_S 6.1 KUMIRĀN EARTHQUAKE

The earthquake destroyed several sparsely inhabited small desert villages, killing 680 of 920 inhabitants (74% of the population; Table 12.1). Local inhabitants noted that the 1941 earthquake was associated with more than 12 km of surface faulting with a vertical offset of about 50 cm (downthrown to the west) in the area south of Qominjān in the bedrock. Weathered en echelon cracks, observed in 1978, suggest a right-lateral strike-slip faulting (Ambraseys and Melville, 1977, 1982) along the N–S striking Chāhak fault (Berberian and Yeats, 1999, 2001; Figure 12.8; Plates 12.8 and 12.9).

The Chāhak fault is an almost N–S westward-convex arcuate fault in a N–S right-lateral shear-zone south of the Ferdows reverse fault in eastern Iran (Berberian, 1976a; Berberian and Yeats, 1999, 2001). The fault starts from Kumirān Mountain in the north and then forms the contact between the plain in the east and the bedrock in the west with a clear Holocene scarp (see aerial photograph No. 28384, Worldwide Aerial Surveys, Inc., Project 158,

TABLE 12.1 Statistics of the 16 February 1941 M_s 6.1 Kumirān Earthquake (See also Figure 12.8)

Village/Town	Population at the Time of the Earthquake	No. of People Killed	No. of People Injured	*I* (MMI)	Destruction	Damage	Minor Damage
Ā̄friz 33°27′N–59°00′, +1425 m	?	–	+	VII	4 houses	Fort and Emāmzādeh Zaid ibn Musā	
'Aliābād	?	–	+	VII		x	x
Bestāq	?	–	–	VI			x
Chāhak 33°16′N–58°54′E, +1382 m	?	+	+	VIII	x		
Dusābād	?	–	–	VI			x
Khur				VI^+			
Kumirān	?	+	+	VIII	x		
Mohammadābād (old) 33°32′N–58°47′E, +1336 m	920	680	+	VIII	x	Old caravanserai	
Musaviyeh 33°17′N–58°54′E, +1382 m	?	–	+	VII	5	x	
Nosratābād 33°26′N–58°58′E, +1393 m	?	–	–	VII		x	
Nurah	?			VIII	x		
Qaisar	?	–	+	VII		x	

Continued

TABLE 12.1 Statistics of the 16 February 1941 M_s 6.1 Kumirān Earthquake (See also Figure 12.8)—Cont'd

Village/Town	Population at the Time of the Earthquake	No. of People Killed	No. of People Injured	*I* (MMI)	Destruction	Damage	Minor Damage
Qarān	?	+	+	VIII	x		
Qominjān 33°33′N–58°52′E, +1366 m	?	+	+	VIII	x	x	
Sarāyān	?	–	–	VI^+			x
Shavangān	?	–	–	VI			x
Tājkuh	?	–	–	VI	–	–	x
Tighdar	?	–	–	VII	5 houses	x	
TOTAL		~700					

+, unknown number of people; –, no casualties; x, sites damaged or destroyed.
Modified after Nabavi (1972), Ambraseys et al. (1972), Ambraseys and Melville (1977, 1982), my field visit and an interview with 'Ali Jom'eh Rudbāri of Band-e Nau village.

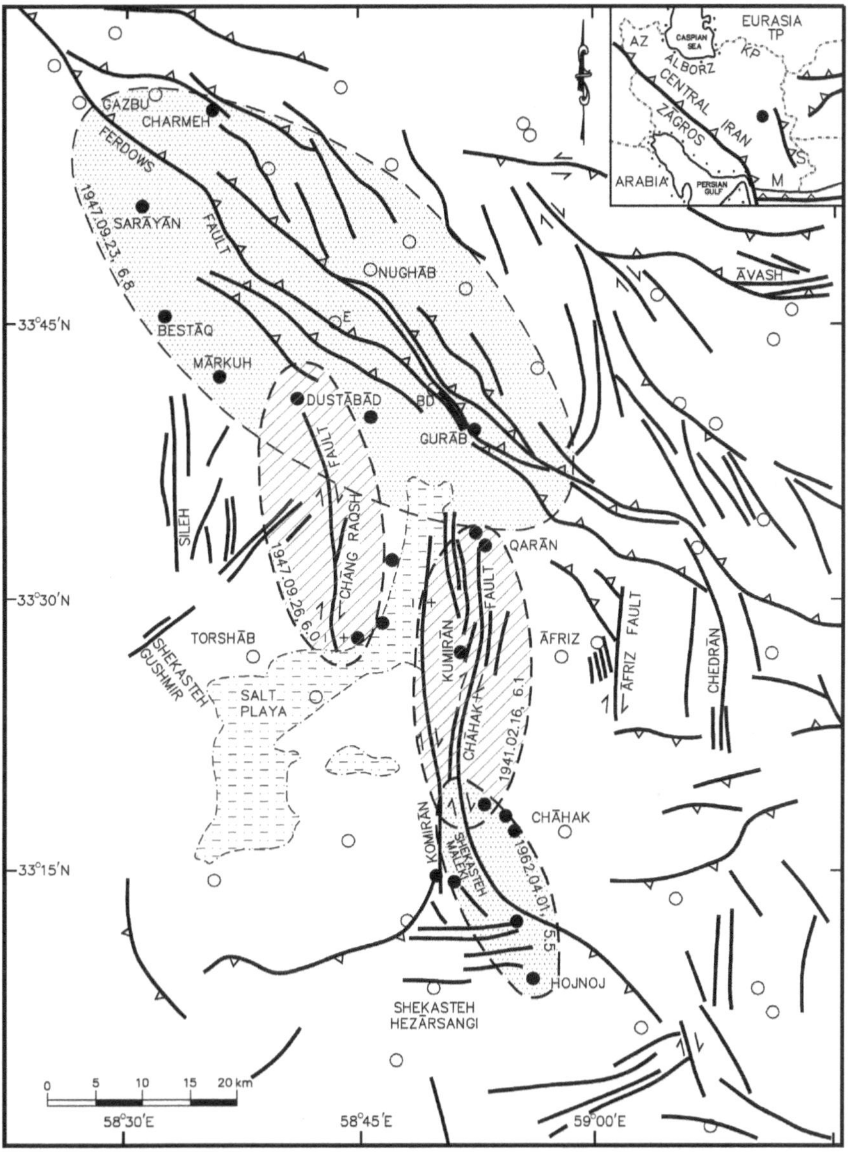

FIGURE 12.8 Epicentral and fault map of the 1941, 1947, and 1962 earthquakes in the Āfriz Shear Zone of the Qohestān province of eastern Iran. Symbols as in Figure 11.1. The previously amalgamated meizoseismal areas of the 23 September 1947 M_w 6.8 Charmeh and 26 September 1947 M_w 6.0 Dustābād earthquakes are separated in this study. *Modified from Berberian and Yeats (1999, 2001) and Berberian et al. (1999).*

PLATE 12.8 Coseismic surface rupture of the 16 February 1941 M_w 6.1 Kumirān earthquake along the Chāhak fault in eastern Iran. For the location, see Figure 12.8. Looking northeast. *Photographed in 1978.*

PLATE 12.9 Coseismic surface rupture of the 16 February 1941 M_w 6.1 Kumirān earthquake along the Chāhak fault in eastern Iran. For the location, see Figure 12.8. Looking north. *Photographed in 1978.*

scale ~ 1:55,000) for a length of at least 40 km (Plates 12.8 and 12.9). The southern segment of the fault (east of Shekasteh Maleki) was reactivated during the 1 April 1962 (M_s 5.5) Chāhak earthquake, showing a southward migration of seismicity along the fault; no surface rupture was reported by the inhabitants for this event located in a remote desert area (Figure 12.8).

In the northeast corner of the Lut Block, there are several parallel, left-stepping N–S right-lateral strike-slip faults with recent active fault features (the Āfriz Shear Zone; Figure 12.8; Berberian and Yeats, 1999, 2001). These faults are located in proximity and to the southwest of the NW–SE Ferdows thrust. To the north, the N–S right-lateral strike-slip faults die out as they approach the Ferdows thrust. The right-lateral, N–S trending faults of the "Āfriz Shear Zone," some of which are associated with the 1941, 1947, and 1962 earthquakes, are as follows (from the SE to the NW; Figure 12.8; Berberian and Yeats, 1999, 2001):

The Shedrān fault: 22 km long (in the SE).
The Āfriz Fault: 15 km long.
The Chāhak Fault: westward-convex, 35 km long; ruptured during the 16 February 1941, M_s 6.1, Kumirān, and 1962 M_s 5.5 Chāhak earthquakes.
The Kumirān Fault: 35 km long.
The Chang Raqsh Fault: 25 km long; ruptured during the 26 September 1947, M_s 6.0, Dustābād earthquake.
The Sileh Fault: 18 km long (in the NW).

Two of these six N–S right-lateral strike-slip faults were associated with the post-1900 earthquakes. The other faults might have been ruptured during historical earthquakes for which we have no data from the desert area. The NW–SE Ferdows thrust (Figure 12.8) was associated with the two earthquakes of the 23 September 1947 (M_w 6.8) Charmeh along the southeast, and the 1 September 1968 (M_w 6.3) Ferdows along the northwest segment (discussed in Chapter 13). See Chapter 16 for a unique coseismic surface rupture pattern and historical seismicity in the Qohestān province of eastern Iran.

12.10.1 Oral Earthquake Hazard Warnings

Although no pre-1900 earthquake data have been preserved in the chronicles, ancient geographical place names serving as a kind of oral earthquake hazard warning show several location names that point to a region with earthquake history (Figure 12.8):

Shekasteh Gushmir [lit., "Fractured or Faulted" +the name of the site (Gushmir), in Persian].
Shekasteh Mohammadābād ["Fractured or Faulted" +the name of the site (Mohammadābād), in Persian].
Shekasteh Maleki ["Fractured/Faulted" +the name of the site].
Shekasteh Hessarsangi ["Fractured/Faulted Stone Wall"] (see Chapter 1 for details).

12.11 THE 27 JULY 1946 M_S 5.5 PENJÉWIN EARTQHUAKE FAULTING

Apparently, the earthquakes of 27 July 1946 and 17 August 1946 partly destroyed Penjéwin (35°37′N–45°56′E) and killed six people (Rothé, 1948; Tchalenko, 1977; Ambraseys and Melville, 1982). Based on an unpublished report by Hitchen (1946), Ambraseys (1978b), and Ambraseys and Jackson (1998) reported 2-km-long surface faulting associated with this earthquake with an azimuth of 45°. No fault map or other information is presented. The earthquake took place in the High-Zāgros Mountain near the Iranian border of Iraq, northwest of Marivān. My correspondence with some Iraqi geologists failed to obtain a copy of Hitchen (1946).

12.12 THE 23 SEPTEMBER 1947 M_W 6.8 CHARMEH AND 26 SEPTEMBER 1947 M_W 6.0 DUSTĀBĀD EARTQHUAKES

The earthquakes ruined several remote, small desert villages, where about 400 people were reported dead. Ambraseys and Melville (1982), based on data collected in the field with the help of the inhabitants, showed a single earthquake damage zone by a NNW–SSE (N164°E) ellipse of 65 × 28 km with a N–S coseismic faulting of 20 km long at its southern end (figure 3.48, p. 91 in Ambraseys and Melville, 1982).

Local inhabitants in 1978 reported at least two sets of ground fractures:

i. A N–S faulting of at least 20 km long with right-lateral strike-slip displacement of about 100 cm and vertical drop of 80 cm at Qermez Mountain [4 km southeast of Dustābād; west side downthrown]; and
ii. A NW–SE fracturing of at least 6-km long between Bādāmak [33°41′N–58°49′E, +1620 m] and Gurāb [33°38′N–58°51′E, +1565 m] villages.

The first reported coseismic surface rupture is along the north–south right-lateral Chang Raqsh strike-slip fault (Figure 12.8), with recent active morphology clearly visible on the aerial photographs (Nos. 27864 and 28382, Worldwide Aerial Surveys, Inc., Project 158, scale 1:55,000). It is a N–S right-lateral strike-slip fault cutting through the Qermez and Chang Raqsh Mountains south southeast of Dustābād and north of Band-e Nau (Berberian and Yeats, 1999, 2001). It is a member of the local N–S "Āfriz Shear Zone" of the eastern Iran (Figure 12.8).

The second NW–SE trending coseismic surface rupture is located in the folded and uplifted Neogene molasse sediments near the frontal tip of the southeastern segment of the Ferdows thrust, about 15 km to the east of Dustābād (thick line between BD and Gurāb along the Ferdows thrust in Figure 12.8). This may indicate either bedding-plane slip with thrust mechanism (flexural-slip faulting) above the frontal tip of the Ferdows thrust uplifting the northeast hanging-wall block, or rupturing of the Ferdows thrust (Figure 12.8).

The very short surface ruptures on two diverse faults with two different mechanisms, the size and shape of the damage zone in Ambraseys and

Melville (1982, figure 3.48, p. 91), and the M_w 6.8 have become an unresolved puzzle. Later, Walker et al. (2011), in their calibrated relocation attempt, could not relocate the 23 September 1947 M_w 6.8 epicenter near the N–S Chang Raqsh fault in the southern part of the damage zone.

Both Ambraseys and Melville (1982) and Walker et al. (2011) did not consider the second earthquake, M_w 6.0, 25 September 1947, which took place 3 days after the first event. The two events were considered by Berberian and Yeats (1999, 2001) but their meizoseismal areas were not constrained, and are corrected here. Newspaper reports (Irān, 2–8 October 1947) and additional site visits confirmed the destruction of small villages in a sparsely populated desert area from Charmeh (eight killed, 30 injured) and Sarāyān (two killed, six injured) in the north northwest to Dustābād (200 killed, 100 injured) and beyond in the south southeast (Tables 12.2 and 12.3). Damage extended further northwest to Ferdows (northwest of Charmeh in Figure 12.8). A new scenario can be offered, considering that (i) the short N–S Chang Raqsh strike-slip fault is not capable of creating a M_w 6.8 earthquake; (ii) the damaged villages along the NW–SE trending Ferdows reverse fault in the NW and along the Chang Raqsh strike-slip fault in the southeast; and (iii) there were two different sets of surface rupturing (Figure 12.8).

The first event (23 September 1947, M_w 6.8) took place along the NW–SE trending Ferdows thrust, destroying villages along the fault. Earthquake destruction and damage extended to areas such as Bādāmak, Gurāb, Estakhr [33°44′N–58°43′E, +1621 m], Sarāyān [33°51–58°31E, +1445 m], Qal'eh Qolāa' [33°51′N–58°38′E, +1907 m], and Bestāq [33°45′N–58°33′E, +1358 m], all along the Ferdows thrust and far from the 26 September 1947 surface faulting, indicating activation of the southeastern section of the Ferdows thrust during the 23 September 1947 M_w 6.8 Charmeh earthquake (Figure 12.8). Later, the northwestern section of the Ferdows thrust became reactivated during the 1 September 1968 (M_s 6.2) and the 4 September 1968 (M_s 5.5) Ferdows thrust earthquakes (Berberian, 1981a; Berberian and Yeats, 1999, 2001). I discuss this issue in Chapter 13.

The 23 September 1947 M_w 6.8 Charmeh earthquake was followed by the 26 September 1947 M_w 6.0 Dustābād earthquake, which destroyed villages along the N–S trending Chang Raqsh fault. The Dustābād village, located on the northern tip of the Chang Raqsh fault and to the southwest of the Ferdows thrust, suffered complete destruction with the most casualties (Figure 12.8).

From 1941 to 1968, there seems a temporal clustering of earthquakes and loading of adjacent faults in the region characterized by N–S "Āfriz Shear Zone" and NW–SE thrust faults with strain partitioning. Earthquake interaction and triggering took place between the Chāhak right-lateral strike-slip fault (1941, 1962), the Ferdows thrust (1947 and 1968), and the Chang Raqsh (1947) right-lateral strike-slip fault in the region bordering the northeastern Lut Block, west of the northwestern Sistān Block (Figure 12.8).

TABLE 12.2 Statistics of the 23 September 1947 M_w 6.8 Charmeh Earthquake (See also Figure 12.8)

Village/Town	Population at the Time of the Earthquake	No. of People Killed	No. of People Injured	*I* (MMI)	Destruction	Damage
Bādāmak 33°41′N–58°49′E, +1620 m	?	?	?	VIII^+	x	
Barāz	?	?	?	VIII		x
Bestāq 33°45′N–58°33′E, +1358 m	?	?	?	VIII^+	x	
Charmeh 33°56′N–58°35′E, +1877 m	2000	8	30	VIII^+	x	
Dustābād 33°40′N–58°40′E, +1428 m	500	170	58	VIII^+	Damaged by both events	
Estakhr 33°44′N–58°43′E, +1621 m	?	?	?	VIII^+	x	
Gurāb 33°38′N–58°51′E, +1565 m	?	?	?	VIII^+	x	
Karch 33°37′N–58°54′E, +1697 m	?	?	?	VII		x
Khunik	?	?	?	VII		x
Mārkuh				VIII^+	x	
Naughāb	?			VII^+		

Qarān	?	?	?	VII	Damaged by both events and 1941	
Qolāa' Castle 33°51′N–58°38′E, +1907 m	–	–	–	VII+		x
Qominjān 33°33′N–58°52′E, +1366 m	?	?	?	VII	Damaged by both events and 1941	
Sarāyān 33°51–58°31E, +1445 m	?	2	6	VII+	x	
Tajan	?	?	?	VII		x
Tighāb 33°39′N–58°45′E, +1417 m	?	3?	?		Damaged by both events	
Tighdar	?	?	?	VII		x
Zu stone masonry Dam 33°56′N–58°31′E, +1744 m	–	–	–	VII+		x
Total		Several hundreds				

+, unknown number of people; –, no casualties; x, sites damaged or destroyed.
Modified after Iran Newspaper, 2 & 5 October 1947, Bozorgnia (1966), Nabavi (1972), Ambraseys and Melville (1977, 1982), my field visit and interview with 'Ali Jom'eh Rudbāri of Band-e Nau village.

TABLE 12.3 Statistics of the 26 September 1947 M_s 6.0 Dustābād Earthquake (See also Figure 12.8)

Village/Town	Population at the Time of the Earthquake	No. of People Killed	No. of People Injured	*I* (MMI)	Destruction	Damage
Ban-e-Nau	?				x	
Chāh Tāleb	?			VII$^+$	x	x
Dustābād 33°40′N–58°40′E, +1428 m	?	?	?	VII$^+$	Already damaged by the 09.23 event	
Hauz-e-Shāh Ghiyāth	?					
Hesāriān? [Dehsārun?, Dehsārān?]	?	2	7		x	
Mārkuh	?	+	+	VII$^+$	Damaged by both events	
Mināb?	?				x	
Mohammadābād (new) 33°32′N–58°47′E, +1336 m	?	+	+	VII$^+$	x	
Qal'eh Kohneh	?	+	+	VII$^+$	x	
Tighāb 33°39′N–58°45′E, +1417 m	?	3 (?)	+	VII$^+$	Damaged by both events	
Torshāb	?					
TOTAL		?				

+, unknown number of people; –, no casualties; x, sites damaged or destroyed.
Modified after Iran Newspaper, 2, 5, 7, & 8 October 1947, Bozorgniā (1966), Nabavi (1972), Ambraseys et al. (1972), Ambraseys and Melville (1977, 1982), my field visit.

The northern section of the Chāhak right-lateral strike-slip fault ruptured during the 16 February 1941, M_s 6.1, Kumirān earthquake. Six years later, during the NW propagating of seismicity, the southern section of the Ferdows thrust ruptured during the 23 September 1947 M_w 6.8 Charmeh earthquake. This was followed 3 days later by the 26 September 1947 M_w 6.0 Dustābād earthquake along the N–S Chang Raqsh strike-slip fault (the fate of the

6 October 1947 M_w 5.6 aftershock is not known). Fifteen years later, the southern segment of the Chāhak strike-slip fault reactivated during the 1 April 1962, M_s 5.5, Chāhak earthquake (Figure 12.8). The seismicity migrated northward when the Dasht-e Bayāz and the northwestern section of the Ferdows faults were reactivated during the 1968 earthquake sequence (Berberian and Yeats, 1999, 2001). See Chapter 16 for pattern of coseismic surface ruptures and historical seismicity of the region.

A similar pattern was addressed earlier: The 6 May 1930 M_w 7.1 Salmās, 8 May 1930 (15:05 UTC) *mb* 5.4 Shekaryāzi, and 8 May 1930 (15:35 UTC) South Qotur earthquakes occurred along separate strike-slip and reverse faults (Figure 12.5).

Because of the 23 September 1947 M_s 6.8 Charmeh earthquake, which had occurred 3 days before the Dustābād event, people were on guard and mostly living outdoors. Nonetheless, the macroseismic data of the three earthquakes of 16 February 1941 M_s 6.1 Kumirān, 23 September 1947 M_w 6.8 Charmeh, and 26 September 1947 M_s 6.0 Dustābād earthquakes are merged and in some cases difficult to separate (see Figure 12.8).

12.13 THE 19 JANUARY 1950 M_S 5.5 DEHNAU 'ASSALUYEH EARTHQUAKE

James and Ghashghaie (1960) wrote that E. K. Cullingham—who reported on the Southern Fārs earthquake of January 1950—described an earthquake that was strikingly similar to the one that occurred in Lār [24 April 1960]. The damage was restricted to narrow, well-defined areas. The shock was of brief duration and transmitted more readily along the WNW–ESE structural strike than at right angles to it. Cullingham suggested that the earthquake of 1950 may have been caused by a fault associated with the offset plunge end of the Kuh-e 'Assalu anticlinal complex. The earthquake, located along the Zāgros Mountain Front reverse fault, was not associated with coseismic surface rupturing (Berberian and Tchalenko, 1976b; Berberian, 1995).

12.14 THE 12 FEBRUARY 1953 M_S 6.5 TORUD EARTHQUAKE

This earthquake took place at the edge of the Great Salt Kavir of Central Iran, destroying eight small desert villages, damaging three others, and killing about 930 people, mainly in the remote desert village of Torud (Table 12.4; Plate 12.10). Field observations (Abdalian, 1953; Berberian, 1976c) indicate that the Torud earthquake was associated with surface deformation along the ENE-trending Morghāb-Bidestān fold zone (Figure 12.9). The kinematics of the surface ruptures were controversial, and both left-lateral and thrust focal mechanisms have been suggested (Shirokova, 1962; Gansser, 1969; Ambraseys and Moinfar, 1977a; Jackson and McKenzie, 1984; Walker and Jackson, 2004; Eshraghi and Jalali, 2006).

TABLE 12.4 Statistics of the 12 February 1953 M_w 6.5 Torud Earthquake (See also Figure 12.9)

Village/Town	Population at the Time of the Earthquake	No. of People Killed	No. of People Injured	No. of Livestock Perished	*I* (MMI)	Destruction	Damage	Minor Damage
Anārak	?	–	–	–	IV			
Behshahr	?	–	–	–	IV			
Bidestān	330	18	15	?	VIII	85 out of 100 houses	x	
Chāh Jām							6 houses	
Chāh Musā	?				VII			
Damāvand	?	–	–	–	IV			
Dāmghān	8900 (in 1956)	–	–	–	V			
Dashtgerd	?	–	–	–	V			
Dizak	?	–	–	–	VI			x
Emāmzādeh Pir Mardān [Nur 'Alā]	?				VIII			
Emāmzādeh Shāh Oliyā	?	–			VII			
Goleki	?	–	–	?	VII			x
Gonbad Kāvus	?	–	–	–	IV			
Gorgān	?	–	–	–	IV			
Hossainābād	?	–		?	VIII	6 houses		

Hossainiān 35°13′N–54°33′E, +913 m	40 houses	4	?	?	VIII	A few	x	
Jājarm	?	–	–	–	IV			
Kalārsham	?				VII			
Kāshmar	?	–	–	–	IV			
Khur	?	–	–	–	VI			x
Mahābiyeh	?				VII			
Mehdiābād 35°16′N–54°43′E, +918 m	15 families	2	?	?	VIII	x		
Miyāndasht	?	–	–	–	V			
Moallémān 35°13′N–54°34′E, +874 m	?	–	?	?	VII		x	
Mozafarābād	?				VII			
Razeh	?	–		?	VII^{+}	10 out of 20 houses		
Reshm 35°16′N–54°29′E, +1290	?			?	VII			x
Sālārān	?				VII			
Sadfi	?			?		Partially		
Sangsar	?	–	–	–	V			
Satveh 35°16′N–54°41′E, +926 m	500	8	10	?	VIII	50%		

Continued

TABLE 12.4 Statistics of the 12 February 1953 M_w 6.5 Torud Earthquake (See also Figure 12.9)—Cont'd

Village/Town	Population at the Time of the Earthquake	No. of People Killed	No. of People Injured	No. of Livestock Perished	*I* (MMI)	Destruction	Damage	Minor Damage
Semnān	?	–	–	–	V			
Shāhrud	?	–	–	–	V			
Shesh	?				VII			
Shisheh	?				VII			
Tabas	?	–	–	–	IV			
Tehrān		–	–	–	IV			
Torbat Haydariyeh	?	–	–	–	IV			
Torud 35°25′N–55°01′E, +809 m	2100	720	160	?	VIII	x		
Total		930[a]				1800 houses		

+, unknown number of people; –, no casualties; x, sites destroyed of damaged.

[a]*On 22 January 1954, the Red Lion and Sun Society reported that 1200 people had been killed. The number was later changed to 530 in its 26 February 1954 report.*

Modified after Press Reports; Abdalian (1953), Parham (1971), Berberian (1976c), Ambraseys and Moinfar (1977a).

PLATE 12.10 The 1953 earthquake ruins at Torud, photographed in 2013. *Courtesy of Lars Larson, The Sven Hedin Project, svenhedin.com.*

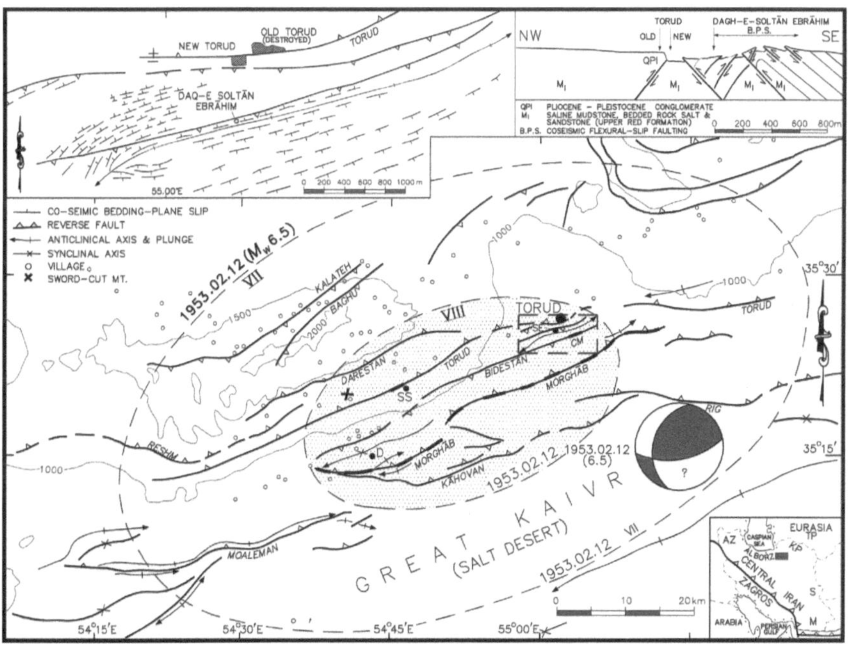

FIGURE 12.9 Epicentral and fault map of the 12 February 1953 M_w 6.5 Torud earthquake in north Central Iran along the northern margin of the Great Kavir salt desert. Symbols as in Figure 11.1. Short-period, fault-plane solution with unconstrained nodal planes: Shirokova (1962, 1967). SS, Sang-e Sur with ground deformation. The Sword-Cut Mountain is marked by an X. Inset top left: Flexural-slip fault map (bedding-plane slip with thrust mechanism: short lines with bars along both flanks of the double-plunged anticline) of the meizoseismal area. The Torud site was relocated after the 1953 earthquake. Inset top right: Simplified cross-section across the anticline shown in the inset top left. See also Plates 12.11 and 12.12. *Isoseismal lines modified after Abdalian (1953).*

The most intense ground disturbance, referred to specifically only by Abdalian (1953), occurred about 1 km southwest of Torud at the small farm of Soltān Ebrahim. Here, Abdalian (1953) described earthquake fractures following the strike of the Neogene molasse beds over several kilometers, with a principal fracture 15 m wide showing a south-facing scarp 20–140 cm high, and other parallel fractures. Surface ruptures were also reported from the area north of Sang-e Sur (SS in Figure 12.9) and south of Dahaneh, SW of Torud (Figure 12.9; Plates 12.11 and 12.12).

Ambraseys (1975) reported >8k-long surface faulting striking N65°E with 160 cm vertical displacement. Gansser (1969) believed that the earthquake was caused by the fault located at the edge of the Torud Quaternary terrace, between the terrace and the alluvial plain [the Torud fault; Figure 12.9]. Although Ambraseys and Moinfar (1977a) concluded that there was no conclusive evidence that the earthquake was associated with surface faulting, Ambraseys and Jackson (1998) later reported 8-km-long coseismic thrust faulting with 140 cm of vertical motion associated with this earthquake. They did not provide a map or photograph showing the location of the reported fault.

PLATE 12.11 Coseismic flexural-slip faulting (bedding plane slip with thrust mechanism) developed during the 12 February 1953 M_w 6.5 Torud earthquake by folding above the Morghāb blind thrust at the Soltān Ebrāhim playa 1 km southwest of Torud (for the location, see Figure 12.9, top left inset). The Neogene molasse beds dip north northwest. Looking west southwest. *Photograph courtesy of John Tchalenko in Berberian (1976b).*

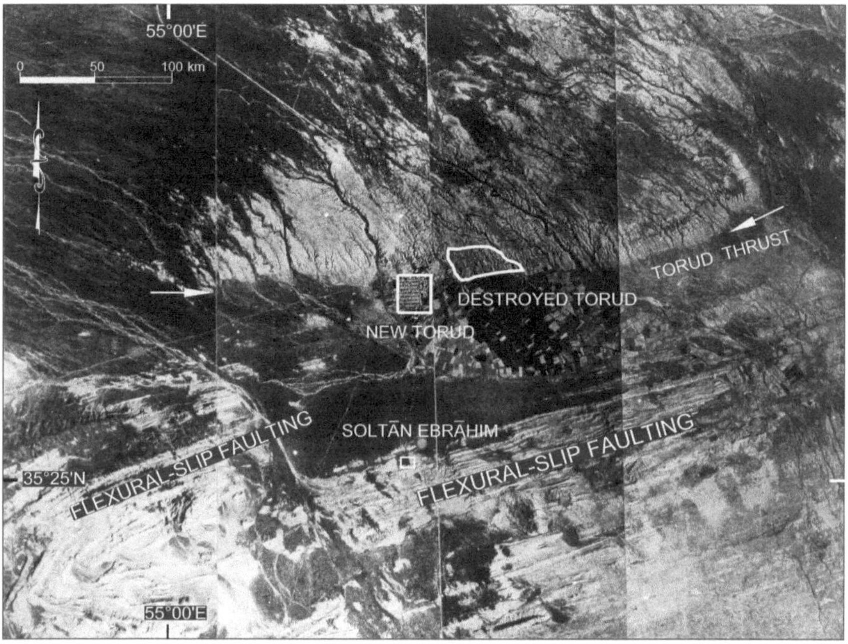

PLATE 12.12 Aerial view of the meizoseismal area of the 12 February 1953 M_w 6.5 Torud earthquake. For the location, see Figure 12.9. National Cartographic Center of the Imperial Government of Iran, 1:20,000 aerial photograph series. *(modified from Berberian, 1976b)*

In fact, the surface ruptures reported were intensive coseismic "bedding-plane slip with thrust mechanism"' developed during flexural-slip folding of the Neogene molasse deposits south of Torud (Plates 12.11 and 12.12; Figure 12.9) (Berberian, 1976c, 1979a). The case is similar to the bedding plane slips with thrust mechanism (flexural-slip faulting) associated with coseismic fold growth above thrust faults documented during the 16 September 1978 M_w 7.3 Tabas-e Golshan (Berberian, 1979a) and the 24 February 1994 M_w 6.2 Sefidābeh earthquakes (Berberian et al., 2000b). Walker (2003) suggested that the bedding-plane slips at Torud were developed by fold growth above a blind thrust fault.

Based on the information provided by local residents, Ambraseys and Moinfar (1977a) reported a short coseismic ground fracture at Sang-e Sur spring [at the southern foot of the Dushākh Mountain; 35°20′N–54°46′, +1167 m; "SS" in Figure 12.9], located a few 100 m northwest of the Torud–Satveh road, 26 km west southwest of Torud. The ground fracture is located at the contact of the Quaternary alluvium (to the SE) and the Eocene volcanics (to the NW). Apparently, the southeastern alluvium was downthrown. Although Ambraseys and Moinfar (1977a) considered this surface fracture as slumping of the ground where the brackish spring water had dried up during the earthquake, the reported 1953 fracture at San-e Sur is located along the NE–SW trending

Torud fault, along which the northwestern Eocene volcanic and metamorphic rocks are thrust over the southeastern playa of Torud-Reshm (Figure 12.9).

Four short-period, fault-plane solutions with a thrust mechanism were published for this earthquake by Shirokova (1962, 1967), Sobuti (1963), Canitez and Ucer (1967), and Wicken and Hodgson (1967). One of the planes of the fault-plane solution given by Shirokova (1962) shows a strike direction that corresponds with the regional structures, but indicates a steep south southeast-dipping reverse fault; the southern block being upthrust with a small right-lateral horizontal component over the northern block. The slip vector then strikes N135°E and the horizontal component of the compressional axis lays N40°W. The west northwest dipping nodal plane cuts the regional structural trends. The nodal planes in this mechanism do not match the regional structure. Jackson and McKenzie (1988) assumed the following fault parameters for the seismic fault of this earthquake: strike: 120°, dip: 45°, rake: 90°, and M_o: 1.0×10^{26} dyne cm.

The Torud region has been subjected to earthquakes in 1915, 22 July 1927 (M_s 6.3, Dasht-e Kavir), 11 April 1928 (Satveh), 6 April 1939 (Satveh), and 27 August 2010 (M_w 5.8, Kuhzar). Among these recorded earthquakes, the 1953 event was the largest that directly hit Torud. The 27 August 2010 M_w 5.8 earthquake took place at Kuhzar area (35.48°N–54.49°E), northwest of Torud: 4 people were killed and about 700 houses destroyed.

12.14.1 Oral Earthquake Hazard Warning

About 6 km to the west northwest of San-e Sur, the western tip of the Daushākh Mountain is called "Shamshir Borān" [lit., "Sword-Cut Mountain"]. This name may come from an oral hazard warning tradition of earthquake faulting/fracturing in the area (marked by "X" in Figure 12.9).

12.15 THE 2 JULY 1957 M_W 7.1 BAND-E PAY EARTHQUAKE

This earthquake, which took place in the mountainous region of the High Alborz, totally destroyed more than 120 villages and caused more than 1500 casualties (Figure 12.10). Vrolyk (1957), the chief of the central civil protection service at Algiers who was sent on a mission to the disaster area, followed by Rothé (1969), reported that near the reinforced concrete building hotel in the village of Āb-e Garm [36.10°N, 52.30°E] a fault (*faille* in Vrolyk, 1957) developed in the ground and went through the building. In fact, photographs of the hotel taken by Savage (1957) suggest that the fracture in question was due to slumping of incipient sliding triggered by the earthquake (Tchalenko, 1974).

Tchalenko (1974) mentioned that some fault movements seem, however, to have taken place along the Amirrud Fault zone at Mangol in the Harāz Valley. He added that there were also unconfirmed reports of large ground fissures, several kilometers long, in the mountains north of Sangchāl, one of the

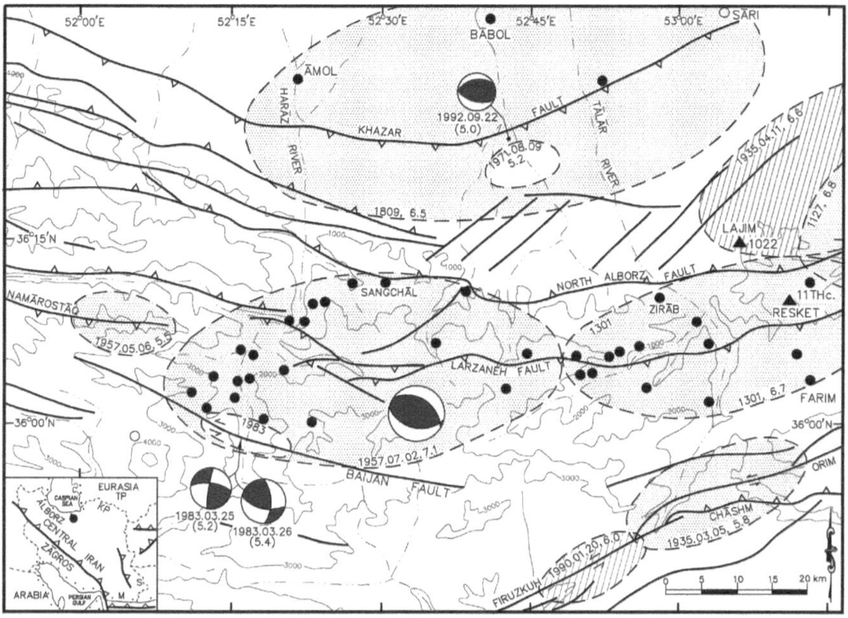

FIGURE 12.10 Meizoseismal areas and fault map of the east Central Alborz. Symbols as in Figure 11.1. Fault-plane solutions of the 1983 and 1992 earthquakes (best-double-couple HRVD CMT solution); 1957 (Shirokova, 1962; McKenzie, 1972) with unconstrained nodal planes.

most severely damaged villages of the earthquake. It bears mentioning that the reported Amirrud minor fault by Tchalenko (1974) is located outside the northern edge of the meizoseismal area. This minor fault cannot be assumed as the causative fault. The reported fracture north of the Sangchāl village, located at the northern edge of the meizoseismal area, was a landslide and not a fault (Figure 12.10).

Ambraseys (1975) reported >3-km-long surface faulting with an azimuth of N120°E. Ambraseys and Jackson (1998) later reported the same parameters under the spurious surface-faulting category. No evidence of coseismic surface faulting associated with this earthquake has yet been established. The reason is that parts of the meizoseismal area are extremely difficult to access, and the situation was aggravated by numerous rock avalanches and landslides after the earthquake that blocked the main access mountainous routes. Therefore, the macroseismic data are incomplete and, for some regions, missing entirely (Tchalenko, 1974).

Short-period, fault-plane solutions for this earthquake have been published by Shirokova (1962, 1967), Sobuti (1963), Wicken and Hodgson (1967), Canitez and Ucer (1967), and McKenzie (1972). Since the observations of polarities were obtained from short-period records, the solutions are less reliable than the long-period ones. Neither nodal plane of McKenzie's (1972) solution is well controlled (Figure 12.10). Two fault-plane solutions

(Shirokova, 1962; McKenzie, 1972) show a thrust mechanism. McKenzie's solution shows a fault plane with a strike: 256°, dip: 44°, rake: 259°, slip vector: 30°. Jackson and McKenzie (1988) assumed the following fault parameters for the seismic fault of this earthquake: strike: 120°, dip: 45°, rake: 90°, and M_o: 2.7×10^{26} dyne cm. Seismic moment of this earthquake obtained from surface waves and an equation from Ekstrom and Dziewonski (1988) is given by Pacheco and Sykes (1992) as 0.44×10^{20} Nm.

The meizoseismal area of the earthquake (Hagiwara and Naito, 1959; Gansser, 1969; Tchalenko, 1974; Ambraseys and Melville, 1982; Persian press reports) covers the Lazaneh reverse fault dipping south (Berberian and Qorashi, 1989b). The eastern section of this fault was associated with the 1300–1301 Farim (Parim) destructive earthquake (Figure 12.10).

12.16 THE 13 DECEMBER 1957 M_W 6.8 FĀRSINAJ EARTHQUAKE

This earthquake killed more than 1130 people, injured about 900, and destroyed or severely damaged more than 210 villages located in the area between the Zāgros and Central Iran. The mainshock, which was followed by the 1958 events to the southeast of its meizoseismal area (Figure 12.11), seems to be part of the sequence that began during the 23 January 1909, M_w 7.4, Silākhor earthquake along the Zāgros Main Recent fault (Figure 12.1). The 1957 earthquake revealed an unusual meizoseismal area and fault-plane solution.

The mainshock and its numerous strong damaging aftershocks enlarged the meizoseismal region in a NW–SE as well as NE directions. Consequently, a highly asymmetric distribution of the earthquake intensity was recorded (Figure 12.11). Both intensity and damage distribution decrease with distance to the southwest of the NW–SE axis almost 20 times faster than to the northeast. Vibrational energy radiated away to the northeast of the Zāgros Main Recent fault line far more efficiently than to the southwest. Villages to the southwest of the fault mainly built on the bedrock, especially southwest of the Sahneh–Bisutun road, suffered little or no damage. An attenuation coefficient of 0.5% to the northeast [destruction extended to Fārsinaj (23 km to the NE of the Zāgros Main Recent fault), and the Gerāreh (55 km to the NE of the same fault) areas], and 2.5% to the southwest [where destruction became unequal and decreased rapidly] was estimated (Ambraseys et al., 1973; Tchalenko and Braud, 1974). The Zāgros Main Recent right-lateral strike-slip fault dips to the northeast (Figure 12.11).

Based on information received from local inhabitants in spring of 1973, Ambraseys et al. (1973) introduced six sets of coseismic ground fractures; in contrast, Tchalenko and Braud (1974) reported only two sets of fractures (NW Sahneh and at the Sarāb-e Bidsorkh spring) both along the Sahneh segment of the Zāgros Main Recent fault. The six sets of ground ruptures reported by the villagers (in Ambraseys et al., 1973; Tchalenko and Braud, 1974) are as follows:

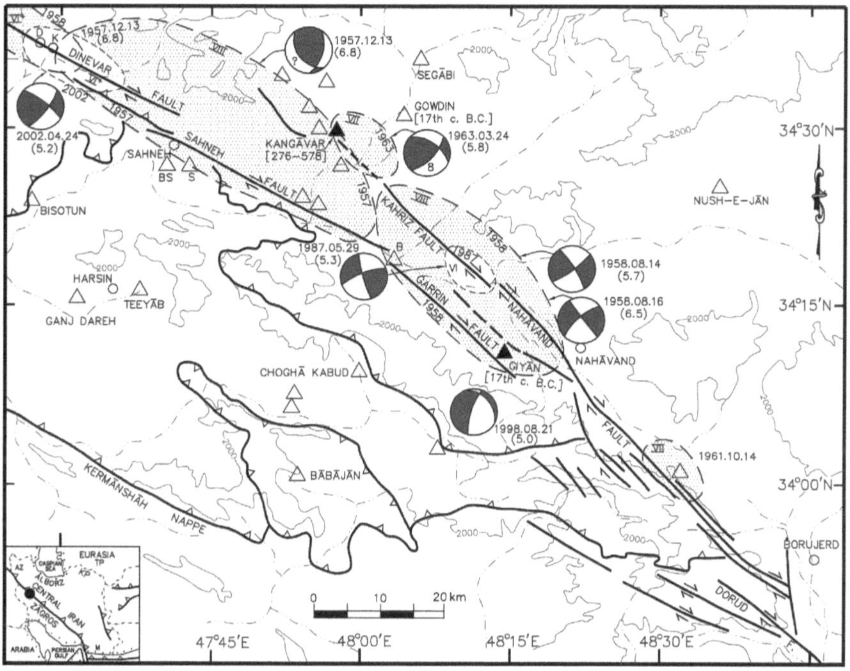

FIGURE 12.11 Meizoseismal areas along the central segments of the Zāgros Main Recent fault in western Iran. Symbols as in Figure 11.1. The amalgamated meizoseismal areas of the 1958 three-event earthquake are shown as a single ellipse. Fault-plane solutions: 1957 and 1958 based on the short-period polarities (Shirokova, 1962; McKenzie, 1972; Canitez, 1969); 1963 (Ni and Barazangi, 1986); 1987, 1998, and 2002 (best-double-couple HRVD CMT solution). B, Barafraq; BS, Bozorg-e Sahneh; D, Dinévar; K, Karksār. *Modified after Berberian (1995) and Berberian and Yeats (2001).*

1. *Kargsār–Karaj Fracture*: A NW–SE trending fracture, along the Zāgros Main Recent fault, with the northeastern block (mountain side) downthrown by 1 m reported by some villagers.
2. *NW Sahneh Fracture*: A significant NW–SE ground fracture about 3.2 km from Sahneh, on the new road to Fārsinaj and 0.5 km to the east of that point.
3. *Qasemābād–Tushmālān Fracture*: A NW–SE fracture with the ground broken up into steps. Possibly associated with the occurrence of rockfalls and slides.
4. *Kangarshāh Oliā–Kārgolān Fracture*: A NW–SE trending fracture associated with extensive upheaval of the ground with landslides and rockfalls.
5. *Soltān Tāher–Chomāq Tappeh Fracture*: A NW–SE fracture with ground settlements that disrupted footpaths for a few kilometers from Dehāsiāb to Kuh-e Nakuchāl.
6. *SW of Kuh-e Soflāleh (SW of Fārsinaj) Fracture*: A NW–SE ground rupture of unknown origin located to the NE of fracture No. 5, to the SW of Kuh-e Soflāleh.

Short-period, first-motion, fault-plane solutions (Shirokova, 1967; McKenzie, 1972) show that the earthquake apparently involved nearly N–S oriented thrust faulting with a small lateral component, which does not match the regional fault pattern (Figure 12.11). The solution by McKenzie (1972) gives two nodal planes of (i) N136°E dipping 50°SW with approximately E–W slip vector, and (ii) N06°E dipping 52°ESE. The solution given by Shirokova (1962) is similar to the corresponding plane striking N125°E, dipping SW and the slip vector oriented at N70°E. The solutions are not reliable and no N–S fault exists in the area (Figure 12.11).

It is difficult to resolve the structural complexity of this event for these reasons: (i) the 1957 earthquake coseismic surface ruptures were not mapped immediately after the earthquake; (ii) the meizoseismal area was covered and complicated by numerous sets of nappe structures; (iii) the short-period, first-motion, fault-plane solutions of the early days was less reliable; and (iv) there was a nonlinear meizoseismal area and expansion of damage further to the northeast, typical of the trend of aftershocks of the Zāgros Main Recent fault.

The NW–SE extension of the meizoseismal area coincides with the Sahneh and Dinévar segments of the Zāgros Main Recent fault (Figure 12.11). However, the highly asymmetric damage distribution toward the northeast may: (i) indicate a deeper event with more complicated movement on buried faults, as happened during the 27 July 1981 (M_w 7.0, with centroid depth of 18 km) Sirch earthquake along and beneath the Golbāf strike-slip fault (Berberian et al., 2001; discussed in Chapter 13); or (ii) relate to the NE-dipping Zāgros Main Recent fault plane (Figure 12.11). As with the 1981 Sirch earthquake case, the main ruptures in the 1957 earthquake might have occurred on different, deeper parts of the Zāgros Main Recent fault system, producing only a minor reactivation of the shallower faults at the surface (see Chapters 13 and 16 for discussion).

12.17 THE 14 AND 16 AUGUST 1958 (M_W 5.5, 5.7, 6.6) FIRUZĀBĀD EARTHQUAKE SEQUENCE FAULTING

During this earthquake sequence, about 170 villages were destroyed or damaged, more than 130 people were killed, and more than 200 people were injured in the Nahāvand area. After the 13 December 1957 Fārsinaj earthquake and its aftershocks (discussed above; Figure 12.11), the 16 August 1958 M_w 6.6 earthquake, preceded by two strong and damaging shocks of 14 August 1958 [11:27 UTC, M_s 5.7; and 15:26 UTC, M_s 5.5] took place on the Zāgros Main Recent fault to the southeast of the 1957 earthquake (Figure 12.12). The timing and overlap of the meizoseismal areas of the 1957 and 1958 earthquakes indicate temporal clustering and loading of the adjacent fault segments, with the earthquake sequence propagating to the southeast. The 1957 and the 1958 events were part of the earthquake sequence that began on 23 January 1909 M_w 7.4 Silākhor earthquake (Figure 12.1) along the Zāgros Main Recent fault (Berberian and Yeats, 2001).

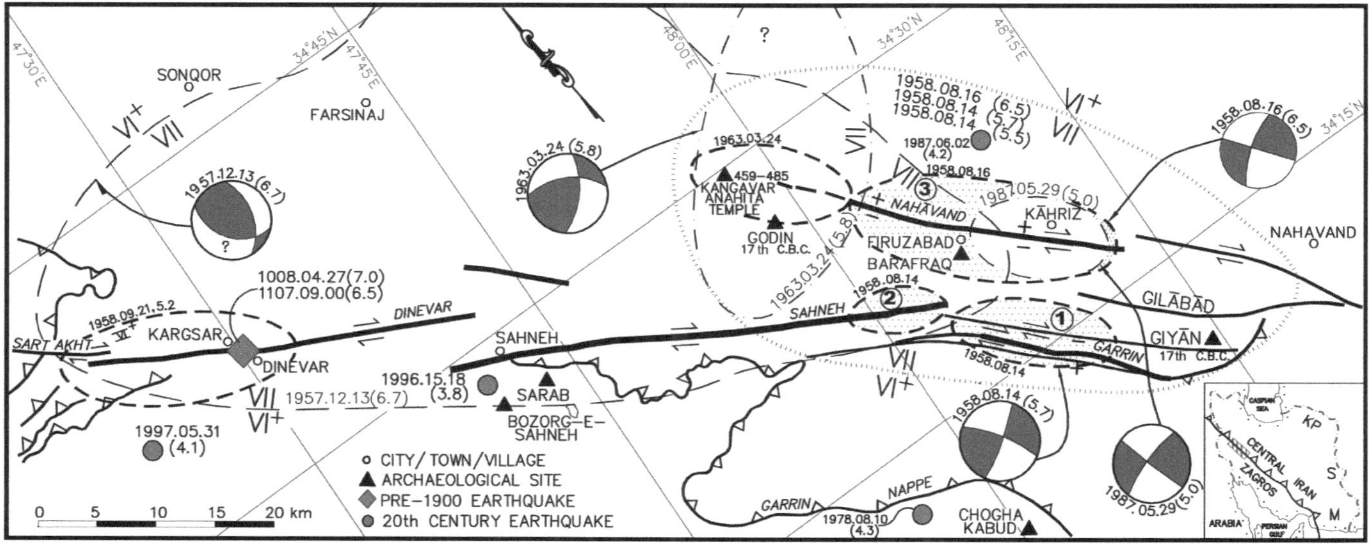

FIGURE 12.12 Separated meizoseismal areas and causative faults with coseismic surface ruptures of 1: the 14 August 1958 (11:27 GMT) M_w 5.7 Givaki; 2: 14 August 1958 (15:26 GMT) M_w 5.5 Kalādeh; and 3: 16 August 1958 M_w 6.5 Firuzābād earthquakes along different segments of the Zāgros Main Recent right-lateral strike-slip fault in western Iran. Symbols as in Figure 11.1. Fault plain solutions as in Figure 12.11. *Modified from Berberian (1995) and Berberian and Yeats (2001).*

Unfortunately, the meizoseismal areas of the three 1958 earthquakes, as well as the ground deformations associated with them, were not documented immediately afterward. Furthermore, teleseismic data with large location errors cannot be used to separate the areas of the three 1958 events. All the reports address the accumulated impacts of the damaging events as foreshocks and the mainshock (Montandon, 1957; Hagiwara and Naito, 1959; Rothé, 1969; Nabavi, 1972; Ambraseys and Moinfar, 1974; Tchalenko and Braud, 1974), as shown in Figure 12.11.

Based on information provided by local villagers in 1971 (Tchalenko and Braud, 1974) and 1973 (Ambraseys and Moinfar, 1974), both authors reported three sets of surface faulting, as follows (two sets by the former and three sets by the latter):

1. *SE of Larehkuh [Cheshmeh Māhi to Givaki] in the SW*: This is a NW–SE trending ground fracture of 15-km long with maximum vertical displacement of 2 m (southeastern side downthrown). The southeastern part of the reported faulting coincides with the Garrin fault segment, and the northwestern part of the reported fault is along the southeastern part of the Sahneh fault segment of the Zāgros Main Recent fault (Figure 12.12).
2. *NE of Larehkuh [Firuzābād; Barreh Farākh to Kāriz and Leylān] in the NE:* On either side of Firuzābād, a 20-km-long NW–SE trending surface rupture with southwestern block downthrown by a few tens of centimeters was developed. This surface rupture coincides with the Nahāvand segment of the Zāgros Main Recent fault (Figure 12.12).
3. *SW of larehkuh [Dehkohneh–Kalādeh] in the SW:* A NW–SE ground rupture about 8 km long with the northeast block downthrown. The ground rupture is located along the southeastern part of the Sahneh segment of the Main Recent fault (Figure 12.12).

It seems that the Nahāvand-Kāhriz, Garrin, and southern tip of the Sahneh segments of the Zāgros Main Recent fault were reactivated during the 1958 three-earthquake sequence (Figure 12.12).

12.17.1 Coseismic Noncontemporaneous Ruptures of Parallel Fault Segments

As described above, both Tchalenko and Braud (1974) and Ambraseys and Moinfar (1974)—on the authority of the local residents and their own site visits—reported three sets of coseismic surface ruptures along the two parallel segments of the Zāgros Main Recent fault (8 km apart): the Garrin and the Sahneh segments in the southwest and the Nahāvand segment in the northeast (Figure 12.12). Ironically, the intensity map of the area (Ambraseys and Moinfar, 1974) shows that: (i) the area of total destruction with loss of life (VIII^{+}) is bifurcated in the northwest along the two aforementioned fault segments; and (ii) the meizoseismal area is enlarged in the NE–SW direction. Ambraseys and Moinfar (1974) reported that within the meizoseismal area, damage was most serious along the fault-zone in the region between Leylās and Barfarāgh [the Nahāvand segment, NE of Larehkuh] as well as in the Gāmāsiāb valley in the region of Rudbāri and Rāzini [the Garrin and the Sahneh segments, SE of the Larehkuh] (Figure 12.12).

There is no doubt that both separated fault segments on either side of the Gāmāsiāb River, which are about 8 km apart from each other, were reactivated during the three 1958 events. Due to the large magnitude of the epicentral error, teleseismic data do not shed any light on the case. Considering the two strong and damaging shocks of 14 August 1958 [11:27 UTC, M_s 5.7; and 15:26 UTC, M_s 5.5], and the 16 August 1958 [M_w 6.6] earthquake, the following scenario of the 1958 earthquake sequence coseismic surface rupture observed in this study (Figure 12.12) is likely:

1. *The 14 August 1958 M_s 5.7 (11:27 UTC) Givaki Earthquake*: This event may have taken place along the Garrin fault segment with ~15 km of coseismic surface rupture. The southwestern bifurcated total destruction intensity zone covers the surface faulting of this event with a possible intensity of ~VII$^+$ (MMI). The August 16 earthquake destroyed the already damaged buildings. This event took place about 8 months after the 13 December 1957 M_w 6.8 earthquake to the southeast (Figures 12.11 and 12.12).
2. *The 14 August 1958 M_s 5.5 (15:26 UTC) Kalādeh Earthquake*: Probably occurred along the Sahneh fault segment with ~8 km surface rupture and intensity of ~VII (MMI) (Figure 12.12).
3. *The 16 August 1958 M_w 6.6 Firuzābād Earthquake*: The earthquake ruptured the longer northeastern Nahāvand segment with >20 km of surface rupture of the Zāgros Main Recent fault. The bifurcated, total destruction intensity zone (VIII$^+$) in the NE covers the surface faulting of this event, extending the damage zone in the northwest and southeast directions (Figure 12.12).

Shirokova (1962, 1967) and Canitez (1969) prepared short-period, fault-plane solutions for the 14 August 1958 M_s 5.7 earthquake that shows a right-lateral strike-slip motion on a fault trending northwest. The short-period, fault-plane solution of the 14 August 1958 M_s 5.5 event by Canitez (1969) shows a right-lateral strike-slip motion on a fault-plane trending NW. Short-period, fault-plane solutions are also available for the 16 August 1958 earthquake by Shirokova (1962) and Canitez and Ucer (1967), and Canitez (1969). Shirokova's solution shows a right -lateral strike-slip solution; whereas Canitez's shows a left-lateral strike-slip solution on a fault trending NW. Shirokova's solution was wrongly plotted in McKenzie (1972) and Jackson and Fitch (1979). It is not possible to assess the reliability of these short-period solutions. Jackson and McKenzie (1988) assumed the following fault parameters for the seismic fault of August 16 earthquake: strike: 130°, dip: 90°, rake: 180°, and M_0 1.4×1026 dyne cm.

See Chapter 16 for coseismic surface rupture pattern and historical seismicity along the Zāgros Main Recent fault.

12.18 THE 21 SEPTEMBER 1958 M_S 5.2 KARGSĀR EARTQHUAKE

The earthquake destroyed seven villages, killed 16, and injured about 60 people in Kargsār. Based on information received from local residents in spring of 1973, Ambraseys et al. (1973) and Ambraseys and Moinfar (1974) reported a ground fracture in the Kargsār–Karaj area (Figure 12.11). The Karaj villagers reported a NW–SE-trending fracture with the northeastern block (mountain side) downthrown by 1.0 m. However, villagers from Kargsār mentioned that the southwestern block settled and in places the ground opened up over short distances after the 21 September 1958 earthquake (one of the aftershocks of the 16 September 1958 earthquake). The latter report is more likely correct. The reported ground deformation is in the proximity of the Zāgros Main Recent fault (Figure 12.11).

At least three pre-1900 earthquakes took place in 913, 1008, and 1107 in the Karksār–Dinévar area along the Zagros Main Recent fault (Figure 11.4). For further discussion about the seismic history of the Zagros Main Recent fault, see Chapter 16.

12.19 THE 24 APRIL 1960 M_S 5.8 LĀR EARTHQUAKE

James and Ghashghaie (1960) reported that small cracks were developed during the 1960 earthquake. On a map of the Lār area, the authors drew a 2.5-km fault line labeled as "ground fracture" in the alluvial deposits that connects to Kuh-e Qemez (the Red Mountain). The reported ground fracture with NNE–SSW strike is located between the town of Lār in the south southwest and the Kuh-e Qermez in the north northeast, with the east–southeast block downthrown. The nature of this fracture is not known; we only know that this medium-magnitude earthquake in the Zāgros was not associated with coseismic surface faulting (Berberian, 1976a, 1995).

12.20 THE 1 SEPTEMBER 1962 M_W 7.0 BU'IN EARTHQUAKE

This large-magnitude earthquake destroyed 91 villages, killing 12,225 and seriously injuring 2800 people (fatalities represented 11.6% of the total population of 142,029 in the epicentral area of 5387 km^2) in the area south of the city of Qazvin. Over 21,300 houses in 300 villages were damaged beyond repair or partially destroyed, 180 of them with loss of life (Ambraseys, 1962, 1963; Berberian et al., 1983).

The earthquake was associated with surface ruptures of unknown length along the Ipak fault with left-lateral, oblique-reverse displacements

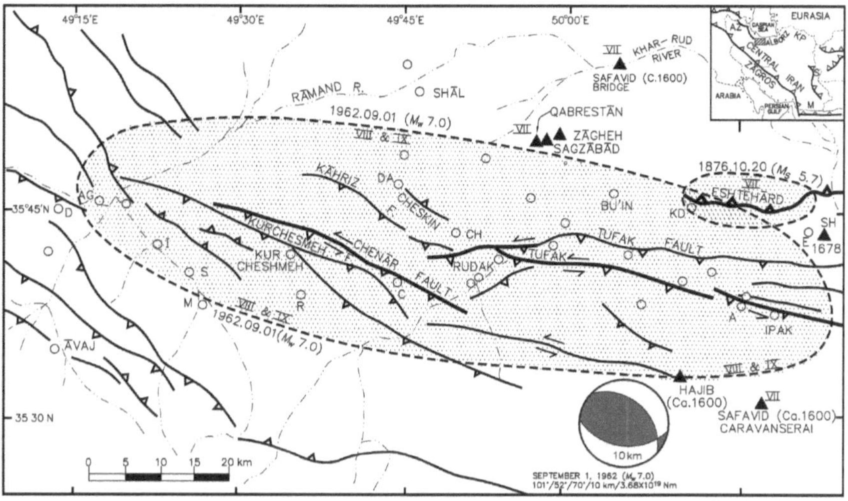

FIGURE 12.13 Meizoseismal area of the 1 September 1962 M_w 7.0 Bu'in earthquake along the Ipak fault with coseismic surface rupture, and the 20 October 1876 $M_s \sim 5.7$ Kalehdareh/Kolādareh (KD) earthquake along the Eshtehārd thrust. Symbols as in Figure 11.1. Focal mechanism determined by waveform modeling with source parameters (Priestley et al., 1994). A, Āhangarān; AG, Āb Garm; C, Chenār; CH, Cheshkin; D, Dakhrājin; DA, Dānesfahān; E, Eshtehārd; I, Ilderjin; R, Razak; S, Sagmasābād; SH, Shāh Solaymān shrine (built in 1678). See also Plates 12.13 and 12.14. *Modified from Berberian et al. (1983) and Berberian and Yeats (2001).*

(Figure 12.13; Plates 12.13 and 12.14). Bedding-plane slips with a thrust mechanism (flexural-slip faulting), landslides, and reactivation of old thrusts or beddings (in the Eocene volcanic, Cretaceous carbonate rocks, or between the Cretaceous carbonates and the Eocene volcanics; due to gravity in Chenār, Kuh-e Miyānband, and Rostamābād) were misreported as coseismic surface ruptures by almost all the authors in the early 1960s.

Discontinuous ruptures deduced from a few and widely spaced observations with unconfirmed total length of 55–103 (?) km, with vertical displacements of 40–76 cm, accompanied by left-lateral horizontal slip of 15–50 cm were reported by different authors. Bearing in mind that the western and central sections of the surface rupture were not visited; the highest displacements were recorded along the eastern segment of the coseismic surface rupture during the widely spaced observation traverses.

Inconsistent, discontinuous coseismic surface fault maps (with short segment ruptures in a few visited sections) were provided by Saraby and Foroughi (1962), Ambraseys (1963, 1965), Mohajer and Pierce (1963), Abdalian (1963b), Omote et al., 1965), and Ambraseys and Melville (1982). Except for a few locations, the central and western parts of the meizoseismal area were not visited after the earthquake because of the mountainous terrain and few or no access roads (Berberian et al., 1983; Berberian and Yeats, 2001). Ambraseys and Melville (1982) and Ambraseys and Jackson (1998) reduced the originally reported

PLATE 12.13 Eroded coseismic surface rupture of the 1 September 1962 M_w 7.0 Bu'in earthquake 365 m north of Āhangarān, south of Qazvin, west of Tehran. See Figure 12.13 for the location. Looking east. *Photographed in 1975.*

PLATE 12.14 Eroded coseismic surface rupture of the 1 September 1962 M_w 7.0 Bu'in earthquake north of Āhangarān, south of Qazvin, west of Tehran. See Figure 12.13 for the location. Looking southeast. *Photographed in 1982.*

coseismic fault length from 103 (Ambraseys, 1963, 1965) to 85 km with slightly increased vertical and horizontal displacements of 80 and 60 cm, respectively.

Inconsistency occurs in three discontinuous fault maps prepared by Ambraseys (1963 and 1965) and Ambraseys and Melville (1982). Ambraseys (1963, 1965) published two coseismic surface fault maps of 103-km in length that differs from each other. Ambraseys (1965) stated that the difference between the 1963 and 1965 fault maps was due to the fact that the triangulation control system used in the 1963 fault map had been improved somewhat in 1965. The third fault map (Ambraseys and Melville, 1982) differs from those previously published. These three different and discontinuous coseismic surface rupture maps only cover the eastern section of the epicentral area with short and discontinuous fault segments, whereas the western section is very unclear in nature (Berberian et al., 1985; Berberian and Yeats, 2001). Ambraseys (1963, 1965), followed by Ambraseys and Melville (1982), and Berberian and Yeats (2001) inaccurately reported coseismic north- and south-dipping reverse faults, whereas in fact the seismogenic fault dips south–southwest (Figure 12.13).

The earthquake was also associated with intense bedding-plane slip with a thrust mechanism in the heavily fissured area (6 km long × 2 km wide) of the Rudak-Tufak area in the Neogene and Pliocene molasse deposits (Figure 12.13), which was not identified by the original investigators. This indicated flexural-slip folding above the blind Tufak thrust to the north of the Ipak fault (Figure 12.13).

Zāre' (2002), by presenting a chain-link pattern of an imaginary interweaving fault map, proclaimed that a "seismogenic knot" was responsible for the 1962 earthquake. He incorrectly stated that the surface ruptures of the earthquake occurred in between the important conjunctions of the NW–SE lineaments and the E–W faults.

Schmidt et al. (2011) prepared active faults of the epicentral region (the coseismic Ipak and the Cheskin blind thrust) with a slight modification of the original maps (Ambraseys, 1963; Berberian and Yeats, 2001) to adhere to topographic/geologic features. Their mountain border, south-dipping thrusts were not activated during the 1962 earthquake. Furthermore, the length of the Cheskin north-dipping blind thrust (at the southwestern edge of the Cheskin anticline) is grossly exaggerated to a 17-km long straight line with its southeastern forked branch, totaling to a 28-km long fault line.

The WNW–ESE-trending Ipak reverse fault with slight left-lateral slip consists of multiple segments connected in a complex network (Figure 12.13; Plates 12.13 and 12.14). At least four main segments dipping south–southwest are recognized. The two eastern segments (Ipak and Rudak) strike N115°E and are 12 and 26 km in length, respectively (Figure 12.13). The North Rudak segment, which strikes nearly E–W, is about 13 km long and is right stepped with reference to the Rudak segment. The fourth segment (Chenār), which is located in the mountains to the west, is about 30 km long and strikes N115°E (Figure 12.13). The total length of the four segments is about 70 km.

12.20.1 Seismology

The exact locations of the two strongest aftershocks of September 04 (*mb* 5.6) and October 13 (M_s 5.7), which added to the destruction and damage, are not known. Considering the large location error (Berberian, 1979c; Ambraseys, 1978a), most of the teleseismically recorded aftershocks were located to the west of the mainshock epicenter. A few houses collapsed at Eshtéhārd (in the east; "E" in Figure 12.13) during the aftershock of 3 November 1964 (*mb* 5.1). The epicenter of this event was located to the northeast of the 1962 earthquake, in the area north of the north-dipping Eshtéhārd thrust (Figure 12.13), which was activated during the 20 October 1876 M_s 5.7 Kalleh Darreh earthquake (Berberian et al., 1983; Berberian, 1994). For westward propagation of seismicity in the Ipak area, see Chapters 14 and 16.

A fault-plane solution of the main shock determined from *P*-wave first motion (Petrescu and Purcaru, 1964; Wu and Ben-Menahem, 1965; McKenzie, 1972) showed mainly thrust faulting. *P* & *SH* waveform modeling demonstrated a long source time function of 24 s. At a rupture velocity of 3 km/s, this can account for 72 km of faulting. The estimated seismic moment of 3.68×10^{19} Nm (M_w 7.0) can account for 1.4 m of displacement on a 72-km-long fault, extending to a centroid depth of 10 km with a dip of 52° toward SSW (striking N101°E) with predominantly reverse motion and a smaller left-lateral component. The misfit slip vector was N42°E (Priestley et al., 1994; Figure 12.13; Table 12.5).

12.20.2 Oral Earthquake Hazard Warning

Although no previous earthquake is documented in the area (except for the 1876 event along the Eshtéhārd thrust in the northeast; Figure 12.13), ancient place names such as *Kur Cheshmeh* ["Blind [dried up] Spring"] along the western segment of the coseismic fault may indicate a historical event preserved in the ancient oral earthquake hazard warnings (see Chapter 1 for details).

TABLE 12.5 ***P*** **and** ***SH*** **Body-Waveform Inversion Source Parameters of the 1 September 1962 Bu'in Earthquake (Priestley et al., 1994) (See also Figure 12.13)**

Earthquake	Time (UTC)	Coordinate (N°–E°)[a]	Strike (°)	Dip (°)	Rake (°)	Slip Vector Azimuth	Centroid Depth (km)	M_o (Nm) ($\times 10^{19}$)	M_w
1962.09.01	19:20	35.55–49.83	101	52	70	41	10	3.68	7.0

[a]*Epicenter from Engdahl et al. (2006).*

12.20.3 Long Term Seismicity and Archeoseismicity

Structural analysis of the joints developed in the Eocene volcanic rocks at the eastern section of the Ipak fault zone defined four tension joint sets. The persistently developed youngest of these (J4) indicated a post-Neogene local maximum principal stress axis acting at N56°E (variable from N28°E to N86°E), which gives a left-lateral movement for the geological fault (Berberian, 1976l).

Bachmanov et al. (2004) reported cumulative vertical offset of presumed Lower to Middle Pleistocene deposits (exact age unknown) of 2–3 m along a fault 8.5 km to the southwest of Dānesfahān (1 km west of Surāvejin village; not on the 1962 surface coseismic rupture of the Ipak fault). The authors stated an accumulative left-lateral slip of 85–90 m from the "oldest generation of alluvial fans" near the village of Ipak. They also reported a left-lateral offset of 25 and 30 m of Late Pleistocene terraces (exact age unknown) on the southern slopes of the Jaushālu Ridge across the eastern segment of the Ipak fault (exact location unknown).

Archaeological sites of Zāgheh (7170–6300 years BP), Qabrestān (6215–4950 years BP), and Sagzābād (4050–2350 years BP) (Fāzeli et al., 2005; Schmidt and Fāzeli, 2007; Fāzeli Nashli et al., 2011; Quigley et al., 2011; Schmidt et al., 2011) underwent intensity VII (MMI) during the 1962 earthquake (Figure 12.13). The Kharaqān monuments (1067 and 1093; Stern, 1966; Stronach and Cuyler Young, 1966; Meshkāti, 1970), which were damaged (VIII) during the 2002 event to the west, indicate a regional seismic quiescence of about 935 years for the western part of the 1962 meizoseismal area (see Chapter 14).

The archeological investigation of "Trial Trench B" at "Level 9" (late third millennium BCE) at the eastern side of the Sagzābād mound revealed many complete but crushed skeletons of domesticated animals, lying side by side under collapsed walls, perhaps having perished in a stable during an earthquake in about 2000–1500 BCE (Negahban, 1971, 1973, 1974a,b, 1976, 1977; Berberian and Yeats, 2001). "Trench A," opened at the southeast corner of the mound, produced 36 organized strata indicating settlements from the late 4th millennium BCE to the middle of the 1st millennium BCE of the Median or Achaemenid culture (Negahban, 1973, 1977, 2006). A clear, unnoticed fracture can be seen in the lower one-third of "Trench A," covered by an unfractured strata indicating occurrence of an earthquake at this level (Plate 12.15). The exact date of this event is not known. Numerous correspondences with archeologists in Tehran (after the death of Ezatollāh Negahbān in Philadelphia) did not resolve the issue.

Excavation near the Sagzābād mound also showed displacement of a charcoal layer (not dated) and rotation of blocks in alluvial deposits (see plates 8.1 and 8.2 on p. 103 in Berberian et al., 1983). More recent excavations (Talā'i, 1998) revealed: (i) a Late Bronze Age (ca. 2000–1600 BCE) square brick column collapsed toward the northwest (in "Trench 700"); (ii) the tilting of a

PLATE 12.15 Trench A (late fourth to first millennium BCE) at the southeast corner of the Sagzābād mound (see Figure 12.13 for the location). A clear fracture can be seen in the lower one-third of the "Trench A," covered by an unfractured strata possibly indicating the occurrence of an earthquake at this level. The two fractures are highlighted (shown by arrows) and the unfractured stratum is highlighted by a white dashed line over the figure on page 476 in Negahban (2006).

Late Bronze Age wall toward the southwest; and (iii) the NW–SE fracturing of an area approximately 2000 and 3000 m^2, splitting a pottery vessel embedded in the floor into two pieces [from "Trench NXX" to "Trench OXX" in the Late Bronze age layers (ca. 2000–1600 BCE) to the Early Iron Age (ca. 1450 BCE)] (H. Talā'i, personal communication, 25 April 1999, in Berberian and Yeats, 2001). If these deformations were caused by an earthquake, it occurred in ca. 2000–1500 BCE.

Schmidt et al. (2011) and Quigley et al. (2011), without presenting paleoseismicity data, stated that simultaneous motion along the Cheskin and the Ipak faults created M_w 7.1 earthquakes from ca. 8830 to 2150 years BP with a return period of about 500 to 1000 years. Archeological investigation of the remarkably long settlement records in the area from 7000 BP to 2300 BP (with a 900-year gap between the Qabrestān and the Sagzābād occupation: 4950–4050 BP) does not reveal several large-magnitude earthquakes with

>1000-year recurrence intervals proposed by Schmidt et al. (2011). However, it should be emphasized that no archeoseismic investigation was carried out during the excavations.

It should be emphasized that based on the surface morphology, the surface length of the Cheskin anticline is about 7.5 km (Figure 12.13), and a blind thrust of about 10 km is incapable of releasing elastic energy to create a large-magnitude earthquake. Furthermore, there is no evidence of simultaneous rupture of the Cheskin and Ipak faults in the past. It did not happen during the 1962 M_w 7.0 earthquake, and no teleseismic activity has yet been recorded along the Cheskin blind thrust since the 1962 earthquake.

12.21 OVERVIEW: 1900–1963 COSEISMIC SURFACE RUPTURES

In modern times, the inspection by a qualified researcher of coseismic surface deformation immediately following an earthquake began in 1953 with Dr. Setrāk Ābdālian; he investigated the 1953 M_w 6.5 Torud earthquake during national political turmoil and the 1962 M_w 7.0 earthquake (a year prior to his death). The study of large-magnitude earthquakes began in 1962 with missions sent by UNESCO at the request of the Imperial Government of Iran for the purpose of post-disaster reconstruction.

The number of the known coseismic surface faulting events during the period of 1900 to 1963 is very low: the sites were not visited by trained earthquake geologists after each event, and large errors in the teleseismic locations for the Iranian Plateau meant that the causative faults could not be located. Any events with inconclusive evidence of surface faulting were deleted from this study. Therefore, Tables 14.9–14.11 represent an incomplete set of data of a work in progress.

Out of about 120 earthquakes of $M>5.5$ that occurred between 1900 through 1963 on the Iranian Plateau, we have been able to assign very few to their coseismic surface ruptures (Table 14.9). We hope that future investigations will increase this number. Nonetheless, it should be mentioned that during this period, some earthquakes took place along the Zāgros Mountains and the Makrān subduction zone and its accretionary prism. Except for the Zāgros master longitudinal reverse faults (the High-Zāgros, Zāgros Mountain Front, and Zāgros Foredeep faults) and the transverse strike-slip faults (Kāzerun, Karébas, Sbazpushān, and Sarvestān faults) cutting through both basement and top sedimentary cover (Berberian, 1995), earthquakes in the Zāgros Mountains, even the largest ones (the 1972 M_w 6.7 Kārzin, and 1977 M_w 6.7 Khurgu) failed to produce surface faulting due to decoupling of the basement (where most earthquakes occur) and the top sedimentary cover along the Upper Vendian–Lower Cambrian Hormoz Salt and other tectonically incompetent decollement beds (Berberian, 1976a, 1977a, 1995; Berberian and Papastamatiou, 1978).

Chapter 13

1964–1997 Coseismic Surface Faulting

When the house is washed away by flood,
or destroyed by earthquake,
you will search for a broken half door under rubble
for carrying the rotten body of your beloved one.

Jalāl Āl-Ahmad (1964)

Although coverage of earthquakes around the world was considerably improved by the establishment of the Worldwide Standardized Seismographic Stations (WWSSN) and United System of Seismic Observations (ESSN) in the early 1960s, only two Iranian seismographic stations were operating in Tehran (TEH) and Shiraz (SHI) in 1962, with a secondary local station at the Sefidrud Dam (SEE and WRI). These stations were incapable of locating earthquakes and calculating seismic parameters, and published no amplitude-period data. Three more were added in 1964 and 1965: the Tabriz (TAB), Mashhad (MSH), and Kermānshāh (KER) seismic stations.

The WWSSN's first full year of operation was 1964; the number of worldwide earthquakes reported by USCGS and ISC considerably increased that year. As mentioned in Chapter 12, the mean error in the instrumental and relocated epicenters of the medium- to large-magnitude Iranian earthquakes in 1962 was about 30 km, which is not acceptable for detailed seismotectonic and seismic risk studies (see Ambraseys, 1978a; Berberian, 1979c for discussion). This magnitude of epicentral location error has led to an association of earthquakes with incorrect faults.

Furthermore, teleseismically recorded earthquakes in Iran prior to about 2000 cannot be correlated to a particular fault. Routine teleseismic locations were usually in error because of (i) inadequate station coverage, (ii) systematic and random reading errors (poor timing), and (iii) bias because of differences between the real Earth and the Earth velocity model used in the location as a spherically symmetric Earth model. Reappraisal and refinement of epicentral

☆"To view the full reference list for the book, click here"

Developments in Earth Surface Processes, Vol. 17. http://dx.doi.org/10.1016/B978-0-444-63292-0.00013-2

locations of the early Iranian earthquake instrumental data is impractical because there is no reliable input data. Therefore, many of the existing regional and global parametric instrumental catalogs are not reliable.

As discussed in Chapter 12, the focal depth calculations based on the arrival times for the Iranian earthquakes reported by international agencies or early relocation efforts suffered greater errors. Focal depths of about 100 km beneath the Zagros Mountains of southern Iran led to the misleading assumption of active subduction of the Arabian plate underneath Central Iran.

The quality of the first-motion fault-plane solutions based on the observed polarities of *P*-wave onsets on the long-period instruments improved during this period (Petrescu and Purcaru, 1964; Wu and Ben-Menahem, 1965; Niazi, 1969; McKenzie, 1972; North, 1972, 1973; Nowroozi, 1972; Dewey and Grantz, 1973; Berberian, 1979a, 1982, 1983a; Berberian et al., 1979a,b; Jackson and McKenzie, 1984; Ni and Barazangi, 1986; and many more).

The source parameters of the Iranian earthquakes were improved by the advent of synthetic seismogram techniques during the 1970s by the Harvard and USGS centroid moment tensor (CMT; with inherent limitations) for earthquakes $>M_w$ 5.5; this database has been available since 1977. Later, *P* and *SH* body waveform modeling constrained the centroid seismic parameters of the earthquakes, with ±4 km in depth, ±5° in fault dip, ±10° in fault strike and rake, and a proper estimate of seismic moment. As a result, Maggi et al. (2000a,b) and others showed that, unlike the Tien Shan, Himalayas, and East African Rift, the seismicity of the Iranian plateau is confined to a 20-km thick, strong seismogenic layer of the upper crust.

The majority of the earthquakes of this period (1964–1997) occurred before the development of geodetic techniques for imaging earthquake ground deformation. However, field studies, aerial photographs, satellite imagery, and seismological investigations have helped develop a broad understanding of the coseismic faulting associated with earthquakes on the Iranian plateau. Satellite imagery (ERTS-A, Landsat) was used for the first time for active fault and earthquake studies in Iran in 1976 (Berberian, 1976a, 1977a).

I started the systematic field and literary study of modern and historical earthquakes in 1971 by establishing the first Department of Tectonics and Seismotectonics in Iran at the Geological Survey of Iran. Six important volumes dedicated to seismotectonics knowledge of the Iranian plateau (with comprehensive texts and colored maps) as well as seismic hazards in urban areas (such as Tehran and Qazvin) were published for the first time in Iran (Berberian, 1976a, 1977a, 1983c; Berberian et al., 1983, 1985, 1996).

The 1976 volume included the first seismotectonics map of Iran (1:2,500,000), an epicenter map of Iran (1900–1976; 1:5,000,000), a generalized fault map (1:5,000,000), areas of destructive earthquakes (fourth century BCE–1976 CE; 1:5,000,000), macroseismic epicenters of destructive and damaging earthquakes (1900–1976), documented earthquake faults, Quaternary faults, and pre-Quaternary faults in Iran (Berberian, 1976a–l).

The 1977 volume included the macroseismic epicenters of the Iranian earthquakes; a historical seismicity (pre-1900) map of Iran (1:5,000,000); maximum intensity, isoseismal, and intensity zone maps (fourth century BCE–1977); an intensity zone map (fourth century BCE–1900 CE; 1:5,000,000), maximum intensity of earthquakes (1900–1977; 1:5,000,000); an isoseismal map (1900–1977; 1:5,000,000); an intensity zone map (1900–1977; 1:5,000,000); a seismic zones map; and a list of historical and twentieth-century earthquakes in Iran (Berberian, 1977a–k).

Field work of the 11 June 1981, M_w 6.6 Golbāf and 28 July 1981, M_w 7.0 Sirch earthquakes as well as the mapping of active faults and earthquake-fault hazard studies of the Greater Tehran and Qazvin quadrangles (each quadrangle covering an area of 100–150 km) were conducted during Iran's dark chapter: the devastating Iran–Iraq War (22 September 1980–20 August 1988), when gasoline for cars (including vehicles for field work) and food items were scarce and rationed, and almost all parts of the country were under the threat of bombardment. If studies had not been carried out in those days, under very difficult and unusual conditions, the data on the surface ruptures of the two earthquakes would have been lost, and most of the faults in the capital city of Tehran would have been covered by buildings. The majority of the active faults cutting the recent alluvial deposits can no longer be seen in Tehran.

In 1994, on behalf of the Geological Survey of Iran, I invited my friend James Jackson of the University of Cambridge to participate in a joint research project to study the active tectonics of the Iranian plateau as well as train young Iranian and British geologists and research students in the field. By studying both satellite imagery and field observation, this cooperation resulted in a better understanding of the seismicity, active tectonics, and surface geomorphological features associated with active faulting of the Iranian plateau and provided detailed insights into active fault behavior and recent crustal deformation in Iran (Berberian et al., 1999, 2000b, 2001; Jackson et al., 2002; and many more—see the References section of this work).

13.1 THE 31 AUGUST 1968 M_W 7.1 DASHT-E BAYĀZ EARTHQUAKE

The Dasht-e Bayāz [lit., "White Plain"] earthquake of 31 August 1968 killed about 10,000 people in a remote, arid area of eastern Iran, left 70,000 people homeless; and destroyed or damaged about 160 villages with a total population of 112,100. Out of 23,000 houses in the area, about 9600 (41%) were completely destroyed. Based on postearthquake aerial photographs taken by the National Cartographic Center of the Imperial Government of Iran (scale ~1:7500), Ambraseys and Tchalenko (1969), Eftekhar-nezhad et al. (1968), and Tchalenko and Ambraseys (1970) mapped the coseismic surface ruptures formed along the western segment of the Dasht-e Bayāz fault that crossed the Nimboluk plain [lit., "the plain of rural half-district"]. Tchalenko and

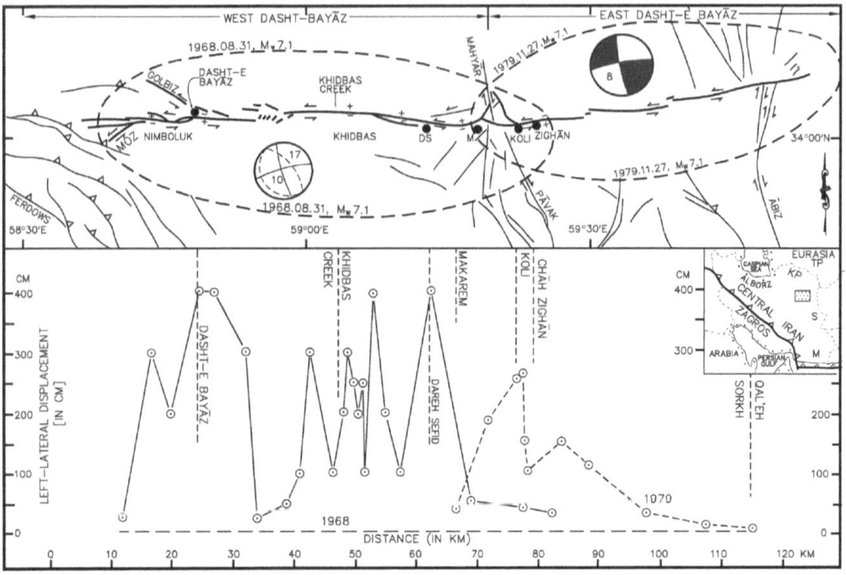

FIGURE 13.1 Top: Meizoseismal areas of the 31 August 1968 M_w 7.1 Dasht-e Bayāz and 27 November 1979 M_w 7.1 Koli earthquakes along the western and eastern segment of the left-lateral strike-slip Dasht-e Bayāz fault. Fault and other symbols as in Figure 11.1. DS, Darreh Sefid; M, Makārem; Moz, Mozdābād. 1968 and 1979 focal mechanisms constrained by body-wave modeling with centroid depths (Walker et al., 2004, 2011). Bottom: Measured amounts of left-lateral displacement along the Dasht-e Bayāz fault in cm; 1968 (Ambraseys and Tchalenko, 1969, 1970; Tchalenko and Berberian, 1975); 1979 (Nowroozi and Mohajer Ashjai, 1980/1981). Observed points at which offsets were measured have been joined by lines and do not imply that the offsets varied between those points in the jagged manner indicated by the profiles. Names of the villages added for reference. See also Plates 13.1–13.6.

Berberian (1975) mapped the Khidbas [lit., "sufficient unripe ear of corn"] section to the east, where the fault passes through bedrock (Figure 13.1; Plates 13.1–13.3). The earthquake is erroneously dated as 30 August in Walker et al. (2011).

The earthquake was associated with at least three sets of coseismic surface fault breaks: (i) the 70-km-long E–W-trending fault break (the western segment of the Dasht-e Bayāz fault), (ii) the NW–SE-trending fault branch (the Golbiz fault), and (iii) the NE–SW trending fault branch (the Mozdābād fault; Moz in Figure 13.1).

13.1.1 The E–W Trending West Dasht-e Bayāz Surface Fault Break

This earthquake was associated with a set of coseismic left-lateral strike-slip surface fractures over a length of 70 km caused by the reactivation of the

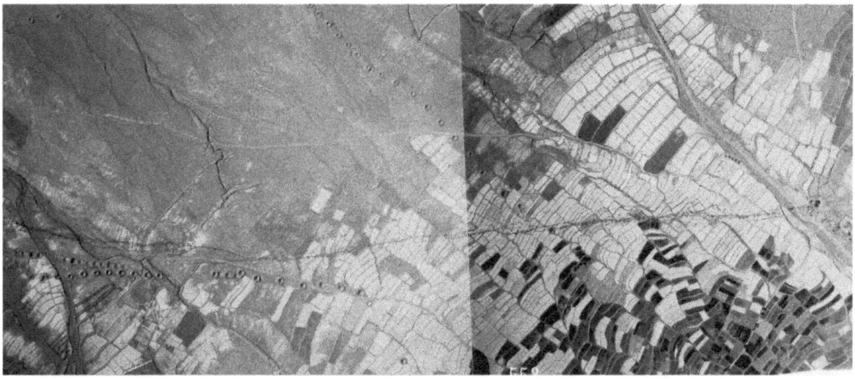

PLATE 13.1 Aerial photographs of the area southeast of Dasht-e Bayāz village, taken immediately after the 31 September 1968 M_w 7.1 Dasht-e Bayāz earthquake along the western segment of the Dasht-e Bayāz left-lateral strike-slip fault in eastern Iran. See also Figure 13.1. *Photographs (Nos. 560-558) courtesy of the National Cartographic Center of the Imperial Government of Iran, October 1968. The average distance between two circular access shafts of qanāts is about 50 m.*

PLATE 13.2 The 31 September 1968 M_w 7.1 coseismic surface rupture of the western segment of the Dasht-e Bayāz left-lateral strike-slip fault in alluvial deposits in the area west of the Khidbas creek. Looking west, photographed in 1972. See Figure 13.1 for the location. *Courtesy of John Tchalenko (1972).*

western segment of Dasht-e Bayāz fault (Figure 13.1). The trace of the earthquake fault (Plates 13.1, 13.2, and 13.4) extends from the western Kuh-e Siāh Range (58°35′E), across the Nimboluk plain, to the Kuh-e Meykay Mountains, and east of Chāh Zighān (34°01′N–59°24′E, +1.251 m) in the east (Eftekhar-nezhad et al., 1968; Ambraseys and Tchalenko, 1969; Tchalenko and Ambraseys, 1970). The Khidbas section to the east, where the fault passes through the Mesozoic bedrock (Plates 13.3 and 13.6), starts where the trace enters the mountains and extends about 18 km eastward (Tchalenko and

PLATE 13.3 The 31 September 1968 M_w 7.1 Dasht-e Bayāz earthquake fault in bedrock about 2.2 km east of the Khidbas creek (see Figure 13.1 for the location). The coordinates of the center of the photograph is added. *Courtesy of the National Cartographic Center of the Imperial Government of Iran, October 1968.*

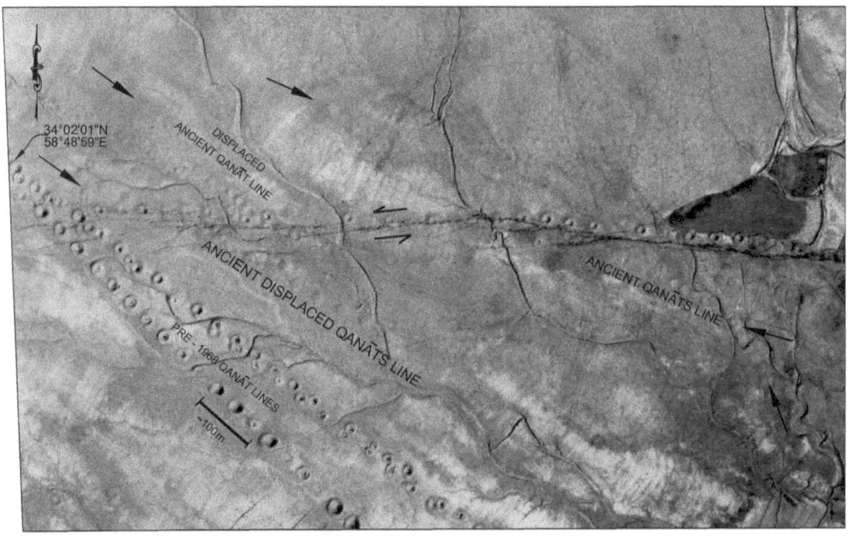

PLATE 13.4 Aerial photograph of the 31 September 1968 M_w 7.1 Dasht-e Bayāz earthquake surface rupture in the area north-northwest of Miām and east of Dasht-e Bayāz village (see Figure 13.1 for the location). Left-lateral offset of the stream beds can be seen. Two lines of NW–SE-trending younger qanāts (traditional ancient Iranian underground aqueduct system; indicated by their circular access shafts) are located to the left (west). A set of older abandoned NW–SE trending qanāt lines can be seen to the northeast and north of the younger qanāt lines. Additional old, abandoned east–west trending qanāt lines are also present in the vicinity of the Dasht-e Bayāz fault trend. The older qanāts were damaged and displaced by historical earthquakes along the fault (dates unknown). *Courtesy of the National Cartographic Center of the Imperial Government of Iran, October 1968.*

Berberian, 1975). Beyond this section, the fault continues for another 30 km and dies out in the Rud-e Shur Desert. The maximum components of displacement that accompanied the 1968 earthquake were measured in the Nimboluk plain and amounted to 4.5-m left-lateral and 2.5-m vertical (Figure 13.1). All horizontal displacements were left-lateral, but vertical displacements varied along the trace. The longest sections, however, and the entire Khidbas section in particular, showed a relative lowering of the southern block (the Nimboluk Plain) (Figure 13.1).

Coseismic surface deformation along straight segments of the Dasht-e Bayāz fault in alluvial deposits of the Nimboluk plain was made up of a systematic array of Riedel shears interspersed with compressional ridges (Tchalenko and Ambraseys, 1970). The width of the surface rupture zone ranged from 2 m to about 100 m but broadened in the vicinity of major dilational strep-overs (at the SW of Dasht-e Bayāz and the west of Khidbas creek) with cross-strike dimensions greater than 1 km (Figure 13.1; Tchalenko and Ambraseys, 1970). Where the near-linear surface rupture trace was exposed in the Mesozoic bedrock (the Khidbas section), field mapping revealed that most of the strike-slip displacement was restricted to a shear zone less than 40 cm wide (Tchalenko and Berberian, 1975). Similar contrasts between surface faulting in alluvium and bedrock were also observed during the 26 June 1992, M_w 7.3, Mojave Desert, southern California earthquake (Sieh et al., 1993).

Horizontal displacement along the coseismic surface rupture decreases in the area between the eastern Nimboluk plain section and the western Khidbas section, and a gap of about 4 km becomes apparent between the two sections. Furthermore, the general alignment of the Khidbas fault trace is displaced about 1 km to the north with respect to the Nimboluk trace (Figure 13.1). Measured vertical and horizontal earthquake displacement decreases consistently as the gap is approached from the east or the west, suggesting a genuine discontinuity in the displacement field. A remarkable feature about this gap is that the step from one trace to the other is to the left; that is, it is opposite to the one taken when proceeding from one Riedel shear to the next in a left-lateral, en-echelon set of fractures. This "anti-Riedel" configuration is therefore contrary to the one characterizing a natural evolution of a wrench fault (Tchalenko and Berberian, 1975).

Two step-overs (dilational jogs) with cross-strike dimensions of about 1 km are particularly prominent within the broad infrastructure of the Dasht-e Bayāz left-lateral strike-slip fault. These features developed in (i) the gap area between the Bedrock (Khidbas, in the east) and the alluvium (Nimboluk, in the west) segments and (ii) the area to the SW of Dasht-e Bayāz (Figure 13.1). Abrupt changes in the measured slip occur in the vicinity of these features. An area of intense sandblows immediately adjacent to one of the jogs is also noteworthy (Tchalenko and Ambraseys, 1970; Tchalenko and Berberian, 1975; Sibson, 1986).

The Dasht-e Bayāz left-lateral strike-slip fault consists of a 70-km-long west segment that ruptured during the 31 August 1968 M_w 7.1 Dasht-e Bayāz earthquake and a ~55-km long east segment that ruptured 11 years later during the 27 November 1979 M_w 7.1 Koli earthquake (Figure 13.1). The 1979 earthquake and two smaller events on the N–S Ābiz fault to the east (see Figure 13.18) occurred during a period of civil disturbance and were not studied soon after the earthquakes in as much detail as the 1968 earthquake. The two segments of the Dasht-e Bayāz fault are separated by the north–south-trending, Mahyār right-lateral, strike-slip fault. The intersection is marked by structural complexity, including a local change of strike, a zone of splays of the Dasht-e Bayāz fault, and a right-step of about 1 km on the Mahyār fault (Figure 13.1). The 1968 earthquake was part of an unusual 13 large- to medium-magnitude earthquake cluster during a short period of 61 years from 1936 to 1997 on complex fault systems (Berberian and Yeats, 1999, 2001; Berberian et al., 1999). See also Chapter 16 for coseismic surface rupture pattern and historic seismicity of the area.

Analysis of the post-Pliocene tension fractures, the Pliocene conjugate shears, and the pre-Pliocene tension joints showed that the compression direction remained constant and ranged between 47° and 55° during these three periods. In this respect, the 1968 earthquake deformation was similar to the early deformations; it is probable that the fault has been subject to left-lateral movement throughout its recent history (Tchalenko and Berberian, 1975).

13.1.2 NW–SE Trending Golbiz Fault Branch

This fault begins about 1 km east of the Gonābād-Qā'en road, crosses the road about 900 m to the north of the Dasht-e Bayāz fault, and continues toward Maysur mountain to the NW trending N100°E (Figure 13.1). Fresh surface faulting was about 20 km long with 30–50 cm left-lateral and 30 cm vertical displacements [northeastern block downthrown] (Ambraseys and Tchalenko, 1969; Eftekhar-nezhad et al., 1968).

13.1.3 NE–SW Trending Mozdābād Fault Branch

The Mozdābād fault is about a 6 km long, NE–SW-trending fault branching off the Dasht-e Bayāz fault in the area southwest of Kehtak (Figure 13.1). Ambraseys and Tchalenko (1969) reported reactivation of about 1.5 km along the fault, with small left-lateral displacement and no apparent vertical deformation in the area southwest of Kehtak, which coincides with the northeastern section of the Mozdābād fault (MOZ in Figure 13.1). They also reported a 1–5 m wide zone of deformation in the area about 3 km north of Marghzār (Eftekhar-nezhad et al., 1968), which covers the southwestern section of the Mozdābād fault.

13.1.4 Seismology

Long-period, first-motion polarities provide a well-constrained nearly E–W left-lateral strike-slip faulting of the early part of the rupture (Niazi, 1969; McKenzie, 1972) in keeping with the mapped coseismic surface rupture. The complexity of the body phases, low value of the rupture speed as studied from the analysis of the surface-wave spectra, reported long duration of shaking, and complicated pattern of slickensides produced by the coseismic surface faulting indicate a multiple source for the mainshock (Niazi, 1969). The *P* waveforms of the Dasht-e Bayāz mainshock are complicated, and many of the *SH* waves were off the scale on the old WWSSN long-period instruments. The shape of the *P*-waveform pulses in the east and the west directions are quite similar, but relatively compressed in the east, which suggests that the fault rupture began near the western end and propagated eastward in an irregular shape, at times decelerating to pause for several seconds (Walker et al., 2004, 2011). The secondary faults observed in the field were probably produced during the initial 5 s of this process (Niazi, 1969; Berberian et al., 1999; Walker et al., 2004, 2011).

A free inversion of the long-period waveforms produces: (i) the mainshock with an initial nearly E–W, left-lateral, strike-slip solution (M_w 7.10) with a release of 90% of the moment and (ii) a separate subevent with a NW–SE reverse mechanism about 8 s after the initial onset (M_w 6.44) with a release of 10% of the moment (Walker et al., 2004; Table 13.1; Figure 13.1). The initial event is compatible with both the first-motion polarities and the coseismic surface faulting. The inversion results gave a source time–function duration of ca. 18 s, which gives a fault length of ca. 60 km, with typical rupture velocity of ca. 3.5 km/s. Because of an extra pulse in the *P*-wave onsets of the southern stations, the inversion is led to a depth of about 17 km. A centroid depth of 17 km would double the seismogenic thickness and halve the average slip.

TABLE 13.1 ***P* and *SH* Body-Waveform Inversion Source Parameters of the 31 August 1968 M_w 7.1 Dasht-e Bayāz Earthquakes and its aftershock (Baker, 1993; Walker et al., 2004, 2011)**

Earthquake	Time (UTC)	Coordinate (N°–E°)[a]	Strike (°)	Dip (°)	Rake (°)	Centroid Depth (km)	M_0 (Nm) ($\times 10^{18}$)	M_w
1968.08.31 (S1)	10:47	34.04–58.95	254	84	5	17	5.478	7.10
1968.08.31 (S2)			320	70	90	10	4.573	6.44
1968.09.11	19:17	33.97–59.52	78	90	16	06		5.6

See also Figure 13.1.
[a] *Epicenters are from Engdahl et al. (2006).*

It is likely that the double pulse comes from the second subevent, which is matched with a relatively small event (M_w 6.4) 24 km east of the mainshock, but neither this, nor its orientation are well resolved (Walker et al., 2004).

The strongest aftershock took place on 11 September 1968. Body waveform modeling indicated left-lateral strike-slip motion with M_w 5.6/strike 078°/dip 90°/rake 16°/06 km (Baker, 1993).

13.1.5 Previous Earthquakes

About 3 km to the east-southeast of Dasht-e Bayāz (1.5 km to the northwest of Miām), the 1968 east–west surface rupture with 240–450 cm left-lateral displacement has cut and displaced the pre-1968, constructed double qanāt lines trending northwest (exact date of construction unknown). The 1:7500 scale aerial photographs taken by the National Cartographic Center of the Imperial Government of Iran immediately after the earthquake (Plate 13.4) in the same area show traces of (i) an ancient abandoned and eroded qanāt line displaced left-laterally by the Dasht-e Bayāz fault at 34°02′01″N–58°48′59″E (Plate 13.4) and (ii) other set of qanāt lines that follow the fault line (directly north of the fault line and to the east of the abovementioned location). Numerous old abandoned qanāt lines also all indicate that older qanāt lines were damaged by unrecorded historical earthquakes along the western segment of the Dasht-e Bayāz fault much before the 1968 earthquake (Ambraseys and Tchalenko, 1969; Ambraseys et al., 1969). The previous earthquakes definitely disrupted the active water supply system, and after each event, the villagers tried to construct new qanāt lines (Plate 13.4).

Further to the east along the same fault, a heavily eroded ruin of a square-shaped fort (the date and cause of destruction are unknown) lies to the immediate south of the 1968 surface rupture (3.8 km to the ENE of Miām and 0.5 km SW of Chāh Khondri) at 34°02′08″N–58°53′17″E (Plate 13.5). At this location, about 2 m of left-lateral and 30 cm vertical motion were measured after the earthquake (Ambraseys and Tchalenko, 1969; Ambraseys et al., 1969). Paleoseismic trench study should reveal the pre-1968 earthquake history of the area.

About 400-m to 4-km cumulative geological offset of unknown age is recorded in the bedrock section of the western Dasht-e Bayāz fault, whereas cumulative Holocene (?) left-lateral displacement of 18–28 m is documented at the Khidbas creek (34°02′09″N–59°01′33″E; Plate 13.6) in the same area (Tchalenko and Berberian, 1975). A left-lateral offset of 10 m is also measured across an old qanāt line (Ambraseys and Tchalenko, 1969). By applying the 4000-year-old qanāt system discovered in Semnān (Mehryar and Kabiri, 1986) to the ancient Dasht-e Bayāz qanāt system, a "minimum" slip-rate of 2.5 mm/year was determined by Berberian and Yeats (1999). The 250-cm left-lateral slip measured during the 1968 earthquake in this section (Tchalenko and Berberian, 1975) could have accumulated in about

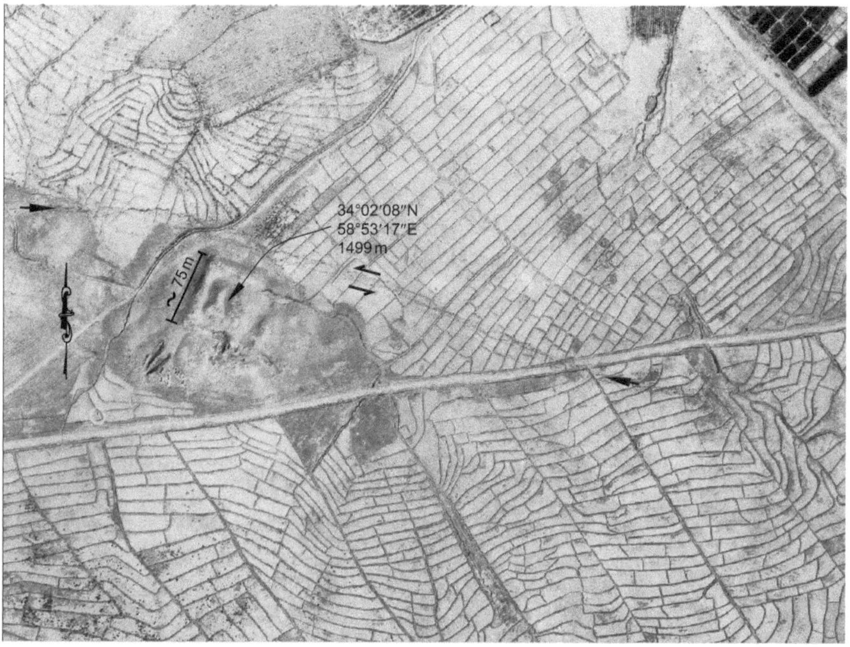

PLATE 13.5 Aerial photograph of a destroyed historical fort and houses located to the immediate south of the 31 September 1968 M_w 7.1 Dasht-e Bayāz earthquake surface rupture, about 500 m southwest of Chāh Khondri (coordinates and elevation are added). See Figure 13.1 for the location. *Courtesy of the National Cartographic Center of the Imperial Government of Iran, October 1968.*

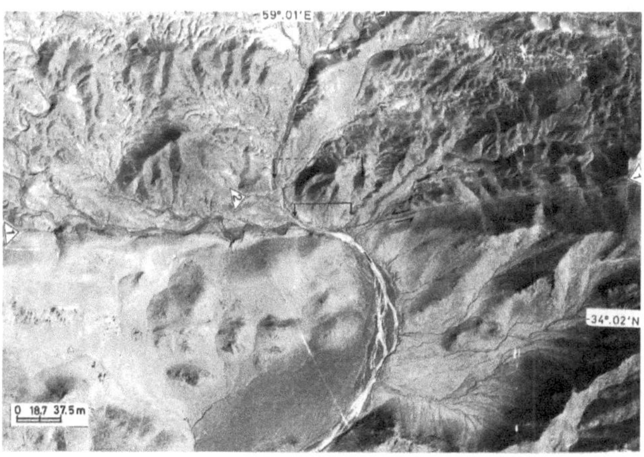

PLATE 13.6 Aerial photograph of the 31 September 1968 M_w 7.1 Dasht-e Bayāz earthquake surface rupture showing cumulative left-lateral displacement of the Khidbas creek (34°02′–59°01′E, +1553 m). See Figure 13.1 for the location. *Courtesy of the National Cartographic Center of the Imperial Government of Iran, October 1968.*

1000 years (at a rate of 2.5 mm/year). Of course, the age of the qanāt system at Dasht-e Bayāz is not known and the system might be younger than the oldest qanāt discovered by archeologists in the Semnān area north of the Central Iranian Great Salt Desert (Dasht-e Kavir).

13.2 THE 1 AND 4 SEPTEMBER 1968 (M_W 6.3 AND M_W 5.5) FERDOWS EARTHQUAKES

The 31 August 1968 M_w 7.1 Dasht-e Bayāz earthquake on the western segment of the Dasht-e Bayāz left-lateral strike-slip fault was followed 20 h later by the 1 September 1968 M_w 6.3 earthquake, and 3 days later by the second event of M_w 5.5, with reverse fault focal mechanism in the Ferdows town region, about 70 km west of Dasht-e Bayāz (Figure 13.2). The first Ferdows earthquake almost totally destroyed the town of Ferdows (killing 750 out of 11,000 people) and ruined a number of villages that had been only slightly

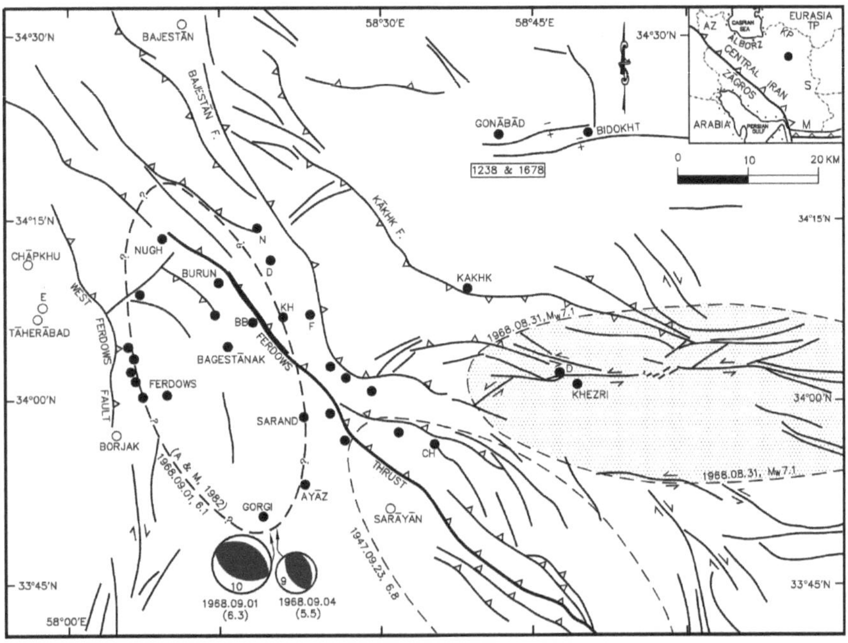

FIGURE 13.2 Epicentral area of the 1 and 4 September 1968 Ferdows earthquakes (M_w 6.3 and 5.5, respectively) to the west of the Dasht-e Bayāz earthquake meizoseismal area. Fault and other symbols as in Figure 11.1. The elongated ellipse (A+M, 1982) covering the town of Ferdows is an unconstrained meizoseismal area of both Ferdows events; some villages located to the east of Ferdows were considered destroyed by the Dasht-e Bayāz earthquake (see the text for discussion on this subject and the causative thrust fault of the two events). Focal mechanisms constrained by body-wave modeling with centroid depths (Walker et al., 2003, 2004). BB, Bāghestānak Bālā; CH, Charmeh; D (NE Ferdows), Degushā; D (in the east), Dasht-e Bayāz; E, Ebrahimābād; F, Fathābād; KH, Khartut; N, Nārestānak.

affected by the 31 August earthquake. The earthquake caused great damage at the extreme west part of the Dasht-e Bayāz meizoseismal area. Most studies of the Dasht-e-Bayāz earthquake failed to separate the effects of the two consecutive events. The Ferdows earthquakes could be interpreted as a separate event, just as the 28 June 1992 M_w 6.5 (15:05 UTC) Big Bear, California, left-lateral strike-slip earthquake is considered a separate event from the right-lateral strike-slip 28 June 1992 M_w 7.3 (11:57 UTC) Landers earthquake to the east (Hauksson et al., 1993; Sieh et al., 1993).

Based on the prominent active expression in the late Quaternary topography and cutting of the terraces and the western limbs of the Neogene asymmetric folds to the western end of the Dasht-e Bayāz strike-slip fault (NE of Ferdows in the Bāghestān area; Figure 13.2), together with the reported coseismic surface deformations and site visit of the Ferdows thrust active fault features, Berberian (1979a, 1981a), Berberian et al. (1999), and Berberian and Yeats (1999, 2001) suggested that the two earthquakes of 1 and 4 September occurred on the NW–SE-trending Ferdows thrust fault (Figure 13.2). The epicenters of the both Ferdows earthquakes (with thrust mechanism) fall onto the footwall of the Ferdows thrust, but the Iranian teleseismic location errors of up to 30 km for the large-magnitude earthquakes for this period (Ambraseys, 1978a; Berberian, 1979c) were thought to be the reason. The idea was later followed by Walker et al. (2003) when they visited the site.

The Ferdows thrust fault (Berberian, 1981a) cuts the western limb as well as parts of the eastern limb of the Neogene molasse anticline (aerial photograph Nos. 28071-3, 27961, 27017-8, 27203-4, and 26040, Worldwide Aerial Surveys, Inc., Project 158, on scale of 1:~55,000). The molasse deposits are folded and thrust above the Ferdows plain. The fault has a sharp trace in the area between Bāghestānak-e Bālā and Burun (thicker fault line in Figure 13.2). The Ferdows thrust fault was not checked immediately after the 1 and 4 September 1968 events; however, Ambraseys (1975) reported a 20-km-long fault striking N125°E with 10 cm of vertical displacement. Ambraseys and Melville (1982) reported a 4-km-long surface rupture trending NW–SE and dipping NE, near Eshratābād, between Khartut and Rahmatābād. The fractures were along the strike of the Neogene molasse deposits and could also indicate bedding-plane slip with a thrust mechanism due to coseismic flexural-slip folding above the Ferdows thrust fault (Figure 13.2).

Calibrated relocations of the two Ferdows medium-magnitude earthquakes, along with its small-magnitude aftershocks (*mb* 4.5–4.9) by Walker et al. (2011), show that the epicenters occur about 8–12 km to the west of the Ferdows thrust into the footwall of the fault. Despite the absence of a west-facing, active scarp of the mountains to the west of the Ferdows town and the lack of active deformation, Walker et al. (2011) inferred that the fault to the west of Ferdows (with no active fault feature) was responsible for the Ferdows earthquakes (see the West Ferdows fault in Figure 13.2). Although this is possible, the lack of site visits immediately after the earthquake makes both alternatives plausible. Nonetheless, it should be noted that the West Ferdows fault (in two separate segments of 11 and 18 km with a 7-km gap in between, and no active fault features at the surface), stops at the latitude of Ferdows (~34°N), whereas the damage zone continues 20 km farther to the southeast (see figure 3.54, p. 103 in Ambraseys and Melville, 1982, and Figure 13.2 in this study). Furthermore, villages located on the footwall of the West Ferdows fault, such as Chāpkhu, Ebrahimābād (both 6.5 km to the west of the fault), Tāherābād (9 km to the west of the fault), and Borjak (9 km SW of Ferdows; Figure 13.2) were not damaged by the 1 September 1968 earthquake (Ferdows is located 9 km to the east of the West Ferdows fault). See Chapter 16 for coseismic rupture pattern and historical seismicity of the area.

13.2.1 Seismology

Long-period, first-motion polarities provide a well-constrained NW–SE orientation with the thrust fault-plane dipping to the NE of the early part of the rupture (Niazi, 1969; McKenzie, 1972). *P* and *SH* body-waveform inversion for the Ferdows earthquakes (Walker et al., 2003, 2004) indicated NW–SE thrust motion dipping about 35° to the NE (Table 13.2).

13.3 THE 14 FEBRUARY 1971 M_W 5.7 SEROKHI EARTHQUAKE

During this earthquake, the adobe houses and walls were destroyed in Serokhi and Kohneh Kalāteh (VII). Damage and destruction were greater in Serokhi (VII$^+$). About 30% of the houses were destroyed and 50% fissured in Jilān (Figure 13.3). An old woman was killed in Hunestān, where most houses fissured. A wall in Kubān and part of the fort wall in Asgharābād collapsed. A house collapsed in Rayābād, and houses fissured in Sharifābād and Bakrān. Altogether, 50 people were injured, many livestock were killed, and the shock was felt in Meyāmey, Shāhrud, and Sabzévār (Figure 13.3).

TABLE 13.2 ***P*** **and** ***SH*** **Body-Waveform Inversion Source Parameters of the Two 1968 Ferdows Earthquakes (Walker et al., 2003, 2004)**

Earthquake	Time (UTC)	Coordinate (N°–E°)[a]	Strike (°)	Dip (°)	Rake (°)	Slip Vector Azimuth	Centroid Depth (km)	M_o (Nm) (×10^{18})	M_w
1968.09.01	07:27	34.04–58.20	307	37	95	32	10	2.959	6.25
1968.09.04	23:24	34.03–58.31	340	34	100	58	09	0.2163	5.49

See also Figure 13.2.
[a]*Epicenters are from Engdahl et al. (2006).*

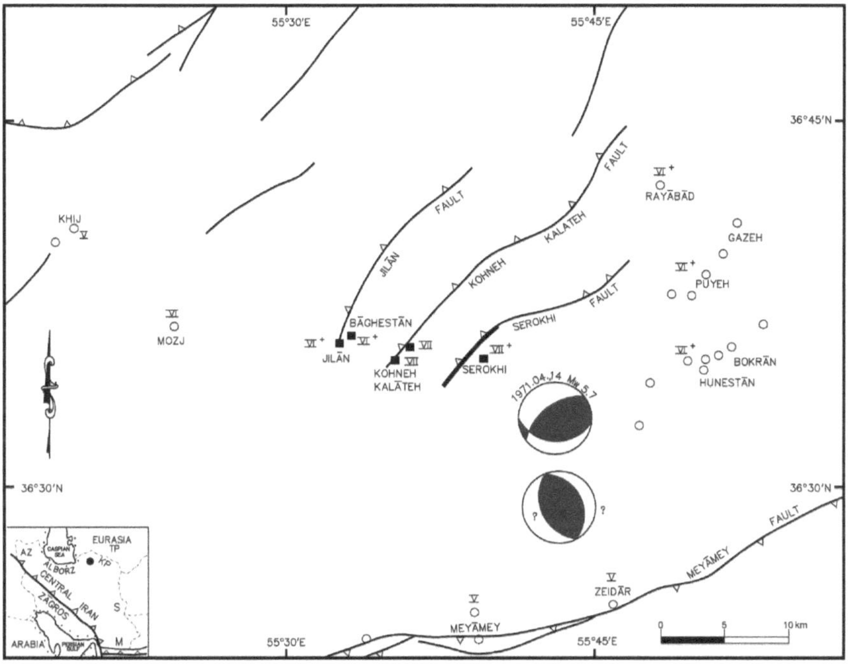

FIGURE 13.3 Epicentral area of the 14 February 1971 M_w 5.7 Serokhi earthquake in the area southeast of the Caspian Sea in the southern Alborz. Fault and other symbols as in Figure 11.1. Thick fault line shows the coseismic surface rupture length along the Serokhi thrust. Top fault-plane solution (Jackson and McKenzie, 1984); bottom solution is unconstrained P-waveform inversion (Priestley et al., 1994).

A field survey in 1981 (Berberian unpublished) revealed a fresh earthquake thrust fault with a strike of N60°E dipping northwest, cutting the Quaternary alluvial deposits in the area northwest of Serokhi. The total fresh fault length was measured at 2 km and the fault cut the qanāt line of Kohneh Kalāteh (Figure 13.3). The three parallel faults of Serokhi, Kohneh Kalāteh, and Jilān show recent active morphology in the field (aerial photograph Nos. 2046-2048; Worldwide Aerial Surveys, Inc., Project 158, scale 1:55,000) and satellite imagery.

13.3.1 Seismology

Long-period, first-motion polarities provide a well-constrained NE-trending thrust fault (Jackson and McKenzie, 1984) compatible with the observed surface rupture along the Serokhi fault (Figure 13.3). However, P-waveform inversion (Priestley et al., 1994) indicated NW-trending thrust faulting, perpendicular to the general structural trends of the region. The P-waveform inversion constrained the centroid depth to 11 ± 5 km and seismic moment

(4.046×10^{17}; Priestley et al., 1994), but due to the lack of *SH* seismograms other parameters, especially the strike and slip vector azimuths, are not constrained (Figure 13.3).

13.4 THE 10 APRIL 1972 M_W 6.7 KĀRZIN EARTHQUAKE

The earthquake, which was preceded by a number of alarming foreshocks, killed 5010 and injured 1710 people in 50 villages; this was about 20% of the population of the Kārzin Borough of the Fārs province in the central Zāgros (Figure 13.4). Haghipour et al. (1972) introduced two WNW–ESE-trending, very long surface faults by dotted lines and considered them as a "*fault, probably reactivated.*" Their northern line in the Qir plain has a N120°E trend with a 32-km length; whereas the southern one in the Dasht-e Shur/Afzār plain has a strike of N110°E and a length of 71 km. They introduced the small central sections of these two speculated faults as lines of

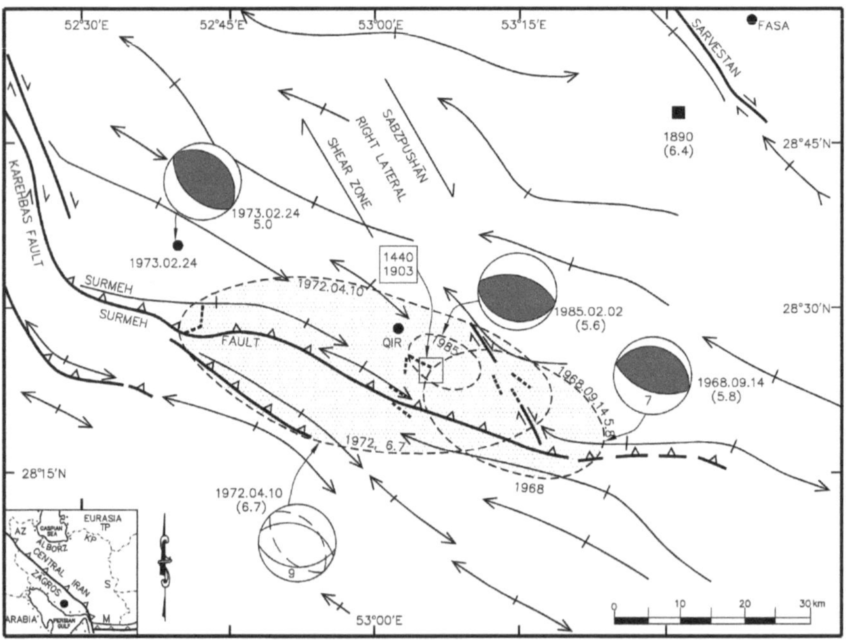

FIGURE 13.4 Meizoseismal areas of earthquakes in the Surmeh reverse fault area (at the southern end of the Karéhbas right-lateral fault) and to the south of the Sabzpushān shear zone of central Fars in the Zāgros. Fault and other symbols as in Figure 11.1. Focal mechanism of the 10 April 1972 M_w 6.7 Kārzin and 14 September 1968 M_w 5.8 Tang-e Ru'in earthquakes constrained by body-wave modeling with centroid depths (Ni and Barazangi, 1986; Baker et al., 1993, respectively). Fault-plane solutions of the 1973 and 1985 earthquakes (best-double-couple HRVD CMT solution). See also Plate 13.7. *Modified from Berberian (1995).*

"*extreme lurching, tension cracks, and few pressure ridges.*" No evidence of such coseismic surface faulting was observed.

Sobouti et al. (1972) introduced a 15-km-long "*fracture and trace of fault*" in an ~N136°E direction, which could not be traced in the field. Ambraseys (1975) reported a 20-km-long fault break with an azimuth of N120°E. Later, Ambraseys and Jackson (1998) changed the statement to a 20-km long spurious surface faulting striking N120°E with 5 cm horizontal and 25 cm vertical displacements.

The discontinuous surface fractures mapped by Ambraseys et al. (1972) and Ambraseys and Melville (1982) can be categorized in two groups: (i) the easternmost NNW-striking, en-echelon fractures with right-lateral displacement and (ii) the rest of the fractures with mostly NW-trending ground ruptures (with two cases of almost N–S fractures) in the southeast Qir and to the west of the meizoseismal area (Figure 13.4; Plate 13.7). The first group of the fractures is located in the Sabzpushān right-lateral strike-slip shear zone and may indicate that the shear zone was reactivated. The second group of fractures developed on the folds above the Surmeh reverse fault, indicating flexural-slip faulting on the Surmeh anticline (Figure 13.4; Plate 13.7). Almost all the surface deformations took place above the Surmeh reverse

PLATE 13.7 NW–SE-trending, en-echelon surface fractures associated with the 10 April 1972 M_w 6.7 Kārzin earthquake, indicating right-lateral strike-slip motion in the area north of Tang-e Ru'in in the Sabzpushān right-lateral shear zone (see Figure 13.4 for the location). *Courtesy of John Tchalenko (1972).*

fault, the only reverse fault in the central Zāgros along which the Paleozoic sediments have surfaced (Berberian, 1995).

The Kārzin earthquake possibly took place along the Surmeh reverse fault at depth (Figure 13.4), which did not propagate to the surface (Berberian, 1995). Known earthquakes in the area were recorded in 1440 (M_s~6.9), 14 November 1903 (VII at Qir), 14 September 1968 (M_w 5.8), 24 February 1973 (M_s 4.8), and 2 February 1985 (M_w 5.6) (Berberian, 1995). The pre-1900 earthquakes left no historical monuments in the Kārzin borough of the cultural rich Fārs province to be damaged by the 1972 Kārzin earthquake.

13.4.1 Seismology

The 1972 earthquake focal mechanism showed pure thrusting along ~E–W plane (North, 1972, 1973). Synthetic waveform modeling of this earthquake was calculated for a double event, with the second subevent 13.7 s occurring after the first with a different strike (Table 13.3; Baker et al., 1993). Although both subevents show a thrust mechanism, the second subevent faulting is almost parallel to the Sabzpushān shear zone and parallel to the first set of ground ruptures documented along the same zone.

13.5 THE 2 JULY 1972 M_W 5.3 MISHĀN EARTHQUAKE

Berberian and Tchalenko (1976b) reported a 10-km-long surface rupture with Holocene cumulative throw of 4 m, reactivated by the 1972 Mishān earthquake in the Zāgros. The observed fresh surface fractures strike N110°E with a dip of 84–90°N (Figure 13.5). Despite its thrust focal mechanism (Jackson and McKenzie, 1984; Ni and Barazangi, 1986), the Mishān surface fractures showed normal characteristics such as fresh open tension fractures and parallel open fissures (Berberian and Tchalenko, 1976b; Plates 13.8 and 13.9; Figure 13.5). This was followed by Ambraseys (1975) who reported a 10-km long surface break with a normal mechanism and 400 cm vertical displacement.

TABLE 13.3 ***P*** **and** ***SH*** **Body-Waveform Inversion Source Parameters of the 10 April 1972** M_W **6.7 Kārzin Earthquake (Baker et al., 1993)**

Earthquake	Time (UTC)	Coordinate (N°–E°)[a]	Strike (°)	Dip (°)	Rake (°)	Centroid Depth (km)	M_o (Nm) ($\times 10^{18}$)
1972.04.10	02:06	28.41–52.78	288	49	99	9	8.445
			322	40	98	10	3.709

See also Figure 13.4.
[a]*Epicenters are from Engdahl et al. (2006).*

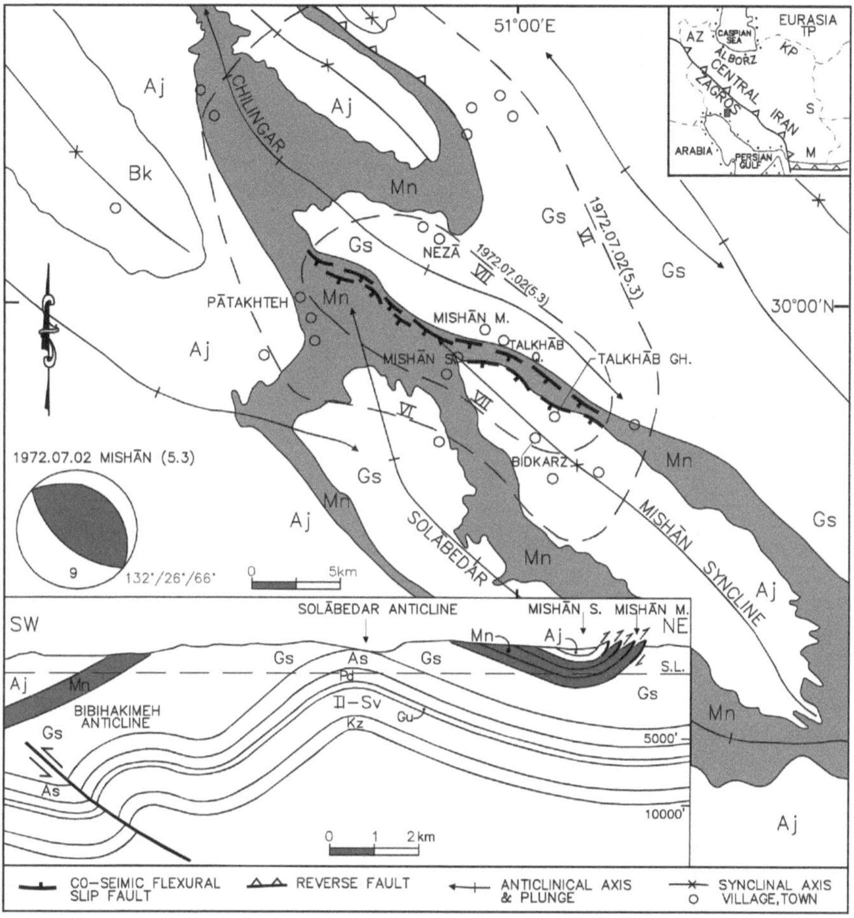

FIGURE 13.5 Meizoseismal are of the 2 July 1972 M_w 5.3 Mishān earthquake and reactivation of the flexural-slip faults at the surface. Bottom: Simplified geological cross section. Geological Formations: AJ, Āghā Jāri (Mio-Pliocene); As, Āsmāri (Oligocene); Gs, Gachsārān (Miocene); Gu, Gurpi (Upper Cretaceous); Il-Sv, Ilām-Sarvak (Upper Cretaceous); Kz, Kazhdomi (Lower Cretaceous); Mn, Mishān (Miocene); Pd, Pābdeh (Eocene). Fault-plane solution constrained by body-wave modeling with centroid depth (Ni and Barazangi, 1986). See also Plates 13.8 and 13.9. *Modified from NIOC (1975).*

Ambraseys and Melville (1982), Shāhpasandzādeh and Zāre' (1996), and Ambraseys and Jackson (1998) denied the reactivation of the surface fractures during the 1972 Mishān earthquake. Aerial *photographs* taken prior to the 1972 earthquake (figure 13 in Berberian and Tchalenko, 1976b, with an erroneous caption; see Plate 13.8) show the existence of the escarpment prior to the 1972 event. The surface reactivation of the existing escarpments with fresh fractures was confirmed by (i) local inhabitants at the time and

PLATE 13.8 Aerial photograph of the 2 July 1972 M_w 5.2 Mishān earthquake area taken prior to the earthquake showing possible flexural slip faulting at the surface (marked by arrows). See Figure 13.5 for the location. *Courtesy of the National Cartographic Center of the Imperial Government of Iran, and Berberian and Tchalenko (1976b).*

PLATE 13.9 The reactivated scarp of the 2 July 1972 M_w 5.3 Mishān earthquake showing cumulative vertical displacement of 3–4 m with fresh open fractures at the foot of the scarp near Mishān-e Soflā (see Figure 13.5 for the location). Looking west. *Courtesy of Berberian and Tchalenko (1976b).*

(ii) fresh reactivation features observed and photographed in 1974 (2 years after the earthquake; figures 8 and 9 in Berberian and Tchalenko, 1976b).

Two trenches were excavated across the Mishān surface scarps by Bachmanov et al. (2004) where four thermoluminescence ages with three seismic events associated with vertical offsets ranging from 0.5 to 1.0 m were obtained. Despite some reverse stratification date results, the authors stated that the penultimate event took place after about 4000 years BP.

It is clear that the surface deformation reported by Berberian and Tchalenko (1976b) was a secondary structure developed on the top sedimentary cover by possible flexural-slip folding in the tectonically incompetent Mishān Formation (Early–Middle Miocene marls) near the contact with the incompetent Gachsārān Formation (Early Miocene evaporate) along the northeastern limb of the Mishān syncline (Figure 13.5; Plates 13.8 and 13.9).

13.6 THE 7 NOVEMBER 1976 M_W 6.0 VONDIK EARTHQUAKE

This earthquake killed 16 people and injured 260 in the area north of Qā'en and affected the flow of water in the qanāts around Vondik (or Vandik) (Figure 11.8). The qanāt line of Khunik Bālā collapsed. Moazami Goudarzi and Ghaderi-Tafreshi (1977) reported inconclusive ground deformation of 9 km in length with an azimuth of N140°E. This was repeated by Ambraseys and Melville (1982). The authors neither gave a fault map nor showed a photograph of the ground rupture, and its location is not known. Furthermore, the reported azimuth cuts the general structures of the meizoseismal area of the earthquake.

About 36 h after the mainshock, aftershocks were recorded by a portable array of seismographs. Niazi (1977) described the distribution of aftershocks in a zone of nearly 50 km length with a N115°E direction. The focal depths of the located aftershocks are generally confirmed to the upper 16 km of the crust. Niazi (1977) concluded that one of the nodal planes derived from the polarity of the teleseismic *P*-wave was consistent with the trend of the aftershock activity and defined a steep fault plane with a substantial left-lateral component of motion.

The fault-plane solution (Jackson and McKenzie, 1984) shows E–W, left-lateral, and N–S, right-lateral, strike-slip nodal planes, with the following parameters for the seismic fault of the earthquake: strike: N–S, dip: 90°, rake: 80°, and M_o: 0.7 × 1026 dyne-cm (Figure 11.8). The authors suggested that the earthquake involved right-lateral, strike-slip faulting on the north–south nodal plane in the fault-plane solution.

Based on the damage distribution and active fault features, Berberian et al. (1999) and Berberian and Yeats (1999, 2001) attributed the earthquake to the E–W-trending Āvash fault with late Quaternary, left-lateral stream displacement running in the area north of Qā'en (Figure 11.8). Calibrated relocation of the mainshock and two of its aftershocks (9 November 1976 *mb* 5.1

and 19 March 1977 *mb* 4.7; Walker et al., 2011) supported slip on the Āvash fault. For the pattern of coseismic rupture of a unique cluster and historic seismicity of the area, see Chapter 16.

13.6.1 Seismology

P and *SH* waveform modeling of the event shows two subevents; the first nearly purely strike-slip and close to that obtained by Jackson and McKenzie (1984) and the second smaller subevent with the N–S-striking nodal plane rotated 10° clockwise from the first (Table 13.4; Figure 11.8).

13.7 THE 24 NOVEMBER 1976 M_W 7.0 CHĀLDERĀN EARTHQUAKE

This earthquake, which took place in eastern Turkey near the Iranian border and along the Western Āzarbāijān Shear Zone (WASZ), destroyed much rubble masonry and many adobe houses in both Turkey and Iran. It was associated with a roughly 50-km-long, almost purely right-lateral faulting with a WNW–ESE strike. The maximum horizontal offset near its northwest end amounted to as much as 350 cm, decreasing to 230 cm in the middle near Chālderān, and to 10 cm at its southeast end (Figure 13.6). Vertical movements were small, with an overall slight depression of the southwest block (Toksoz et al., 1977, 1978). Despite destruction of several villages (VIII) on the Iranian side, no field investigation was carried out after the event to see if the coseismic surface rupture continued into Iran along the WASZ. The 26 May 1977 M_w 5.7 Mokhur aftershock destroyed several villages and killed some people in the Mokhur region in Iran. This case was also not investigated (Figure 13.6).

TABLE 13.4 *P* and *SH* Body-Waveform Inversion Source Parameters of the 7 November 1976 M_w 6.0 Vondik Earthquake (Baker, 1993)

Earthquake	Time (UTC)	Coordinate (N°–E°)[a]	Strike (°)	Dip (°)	Rake (°)	Centroid Depth (km)	M_o (Nm) ($\times 10^{18}$)	M_w
1976.11.07	04:00	33.82–59.18	84	79	12	8	1.115	6.0
			67	52	353	10	0.3048	

See also Figure 11.8.
[a]*Epicenter is from Engdahl et al. (2006).*

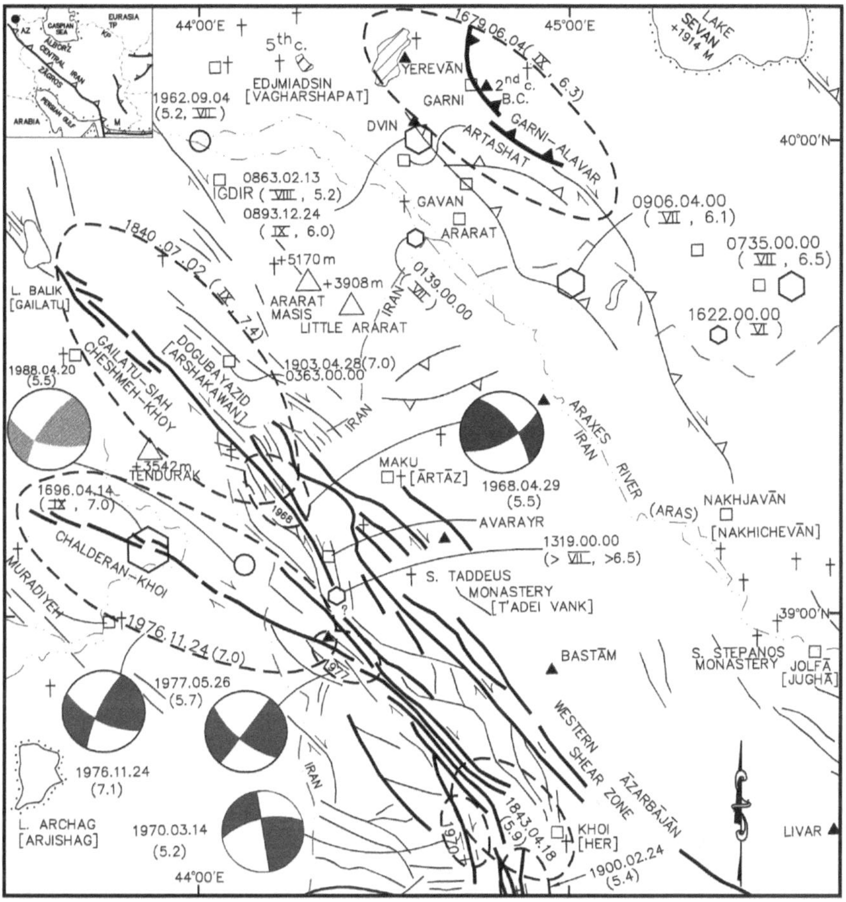

FIGURE 13.6 Meizoseismal areas of the earthquakes and their causative faults along the Western Āzarbāijān Shear Zone (WASZ) of northwest Iran. Fault and other symbols as in Figure 11.1. Fault-plane solutions: 29 April 1968 M_s 5.5 Gol (McKenzie, 1972). 14 March 1970 M_s 5.2 Badalān, 24 November 1976 M_w 7.0 Chālderān, and 26 May 1977 M_w 5.7 Mokhur (Jackson and McKenzie, 1984). 20 April 1988 M_w 5.5 Chalderān (HRVD CMT). *Modified from Berberian (1997) and Berberian and Yeats (1999).*

13.8 THE 21 MARCH 1977 M_W 6.7 KHURGU EARTHQUAKE

The earthquake killed 152 people, injured 556, and caused destruction over an area of 550 km^2. The shock damaged 1500 houses beyond repair and killed 1.3% of livestock in the area north of Bandar ‘Abbās port in southeast Iran. Sobouti et al. (1978) reported minor cracks (in a direction of N75°W) and undulating of the ground in the area where houses and buildings were seriously damaged. On their map, the authors (figure 3 in Sobouti et al., 1978) showed about 26 km of continuous surface “fractured area” in the center of the highest intensity zone marked VIII–XI (MMI). A second "fracture area" of about 7 km long was also shown to the western part of the highest intensity zone, to the south of the one mentioned above.

Despite the location of the meizoseismal area of the earthquake along the master basement fault of the Zāgros Mountain Front (Berberian, 1995), detailed field study (Berberian and Papastamatiou, 1978) and visits to the surface faults of the area did not show any coseismic surface rupture associated with this event.

13.9 THE 6 APRIL 1977 M_W 5.9 NĀGHĀN EARTHQUAKE

This earthquake took place in the mountainous area of Chāhār Mahāl Bakhtiāri in the High Zāgros, south of Shahr-e Kord. It killed about 366 and injured about 200 in the region of Nāghān, Joqdān, and Ardal, which covers an area of about 700 km^2. It damaged 2100 houses beyond repair and killed 0.7% of livestock in the area. Eight schools collapsed; 37 were damaged.

A field investigation (Berberian and Navā'i, 1977, 1978) did not show coseismic surface faulting. Nonetheless, in hindsight, a 1-km-long surface fracture striking N150°E in the area northwest of Ardal near the bedrock was noticed; the fracture was enhanced by a linear landslide aggravated by a severe rain storm and erosion prior to the April 1977 site visit (see figure 9, p. 62, in Berberian and Navā'i, 1977). The 1-km-long fracture is located along the Ardal fault in the meizoseismal area.

Ambraseys and Jackson (1998), referring to Ambraseys (1979), reported spurious faulting associated with this earthquake. In fact, Ambraseys (1979) stated that the ground deformation found in the epicentral area was of a purely secondary nontectonic nature.

The meizoseismal area of the earthquake is located at the southeastern extremity of the NW–SE-trending Ardal fault (Berberian and Navā'i, 1977, 1978; Berberian, 1995). However, the fault-plane solution (Jackson and McKenzie, 1984; Ni and Barazangi, 1986) shows an E–W thrusting with a centroid depth of 8 ± 3 km (Ni and Barazangi, 1986). The 1665–1666 M_s ~6.5 earthquake took place to the northwest of the 1977 event along the Ardal fault. Other known events in the area took place in 1874 (?, felt), 1880 ($M_s \gtrsim 5.5$), 26 July 1958 (*mb* 4.2), 21 September 1960 (*mb* 5.0), and 4 March 1992 (M_w 5.1) (Berberian, 1995).

13.10 THE 19 DECEMBER 1977 M_W 5.9 DARTANGAL EARTHQUAKE

This earthquake destroyed villages in the district of Zarand, killing 665 people. It was associated with a 19.5-km-long coseismic fault break at the surface along the Kuhbanān fault, with a maximum 20 cm right-lateral and 6 cm reverse movement consistent with the fault-plane solution (Figure 13.7;

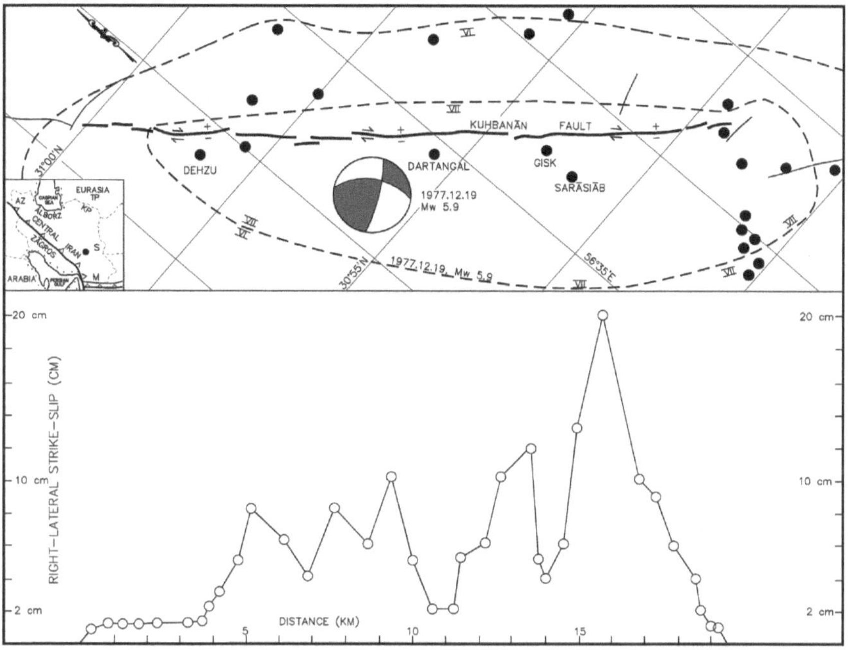

FIGURE 13.7 Meizoseismal area of the 19 December 1977 M_w 5.9 Dartangal earthquake along the Kuhbanān right-lateral strike-slip fault in the Kermān province of southeast Iran (top); and measured coseismic right-lateral displacement (bottom). Observed points at which offsets were measured have been joined by lines and do not imply that the offsets varied between those points in the jagged manner indicated by the profiles. Fault and other symbols as in Figure 11.1. See also Plate 13.10. *Modified from Berberian et al. (1979a).*

Plate 13.10). Several en-echelon tension openings with a strike of N–S to N20°E, indicative of right-lateral slip, were found along the fault line. Stream beds and wheel tracks of tractors were displaced right-laterally a maximum of 20 cm by the earthquake fault (Berberian et al., 1979a; Berberian, 1981a, 2005).

The coseismic right-lateral displacements showed variable patterns (Figure 13.7). The first 4 km of the northwestern end showed about 2 cm of right-lateral displacement. The lateral displacement increased to about 10 cm, in the area about 5–10 km from the northwestern end of the coseismic surface rupture. The maximum 20-cm measured horizontal slip was observed southeast of the Gisk village (about 16 km from the northwestern tip of the ground rupture), near the southern end of the earthquake fault. Although the horizontal movement changed inconsistently along the fault line, the displacements decreased and the fresh surface rupture gradually died out as the earthquake fault approached the extreme northwest and southeast ends of the fault segment (Berberian et al., 1979a).

PLATE 13.10 The 19 December 1977 M_w 5.9 Dartangal coseismic right-lateral strike-slip displacement along the Kuhbanān fault. See also Figure 13.7.

Ambraseys et al. (1979) presented the coseismic fault as a straight line drawn on the 1:55,000 aerial photograph of the region. Ambraseys et al. (1979) and Ambraseys and Melville (1982) reported a 9–10 km of surface rupture with predominantly oblique compression features in places with a strike-slip motion. Zohoorian Izadpanah et al. (1981) reported that no fresh fault rupture was observed at the surface; however, a consistent pattern of en-echelon fracturing for a length of 9 km was mentioned.

Several historic earthquakes are recorded along the Kuhbanān fault, especially in ca. twelfth century ($M_s \gtrsim 5.5$), ca. fourteenth century (?), May 1875 ($M_s \sim 6.0$), 22 May 1897 ($M_s \gtrsim 5.5$), 4 January 1931 (M_s 5.0), 28 November 1933 (M_w 6.1), 12 May 1937 (M_s 5.0), and 8 October 1951 (M_s 5.0). For pattern of historical seismicity along the fault, see Chapter 16 and Figures 16.19 and 16.20.

13.10.1 The Post-Neogene Regional Change in Kinematics

The 1977 earthquake mechanism showed a considerable change of the slip vector along the Kuhbanān range-front fault from predominantly reverse to primarily strike-slip since perhaps the Late Neogene–Early Quaternary. The NW–SE-trending Kuhbanān fault runs about 200 km at the southwestern foot of the steep and linear Kuhbanān fold-and-thrust Mountain belt. The Kuhbanān range-front fault has possessed a significant component of shortening and upthrusting of the Cambrian deposits onto the Kermān–Zarand plain during its history. The fault segment that moved has a gouge zone varying in width from a few meters to 30 m. The pre-earthquake movement along this segment (with a trend of N140°E and dip of 70°NE) produced new slickensides on the hanging wall with an average strike of N116°E and a plunge of 50°SE. Along this trend, the Lower Paleozoic rocks (in the northeast) are thrust over the

Quaternary alluvial deposits (in the southwest); thus, the fault was an active reverse fault in the Late Neogene–Early Quaternary (Berberian et al., 1979a).

13.10.2 Oral Earthquake Hazard Warning

Ancient place names such as Zelzeleh Sang [lit., "Earthquake Rock"] and Gaud-e-Zelzeleh Zadeh ["Playa Smitten by Earthquake"] along the Kuhbanān fault may indicate additional historical events preserved in the ancient oral earthquake hazard warnings (see Chapter 1 for details).

13.10.3 Seismology

The fault-plane solution of the 19 December 1977 earthquake along the southern part of the Kuhbanān fault (Figure 13.7) indicates almost pure right-lateral strike-slip motion on a NW–SE-trending plane dipping to the northeast (Berberian et al., 1979a).

13.11 THE 16 SEPTEMBER 1978 M_W 7.3 TABAS-E GOLSHAN EARTHQUAKE

Ten years after the 1968 Dasht-e Bayāz/Ferdows earthquakes, the seismicity seems to have migrated cross fault from the Ferdows thrust in the northeast (Figure 13.2) to the Tabas thrust fault in the southwest (Figure 13.8). The 1978 earthquake was the last of a southward-migrating progression of smaller intramontane events in 1939 (*mb* 5.5), 1974 (*mb* 4.8), and 1977 (VI); the 1977 event was near the northern end of the 1978 rupture (Figure 13.8; Berberian, 1979a).

The 1978 earthquake killed more than 20,000 people, destroyed or severely damaged about 90 villages, partially damaged another 50 villages in the region, and completely demolished the evergreen oasis town of Tabas-e Golshan, where 85% of the inhabitants (11,000 out of 13,000) were killed (Figure 13.8; Plate 13.11). The earthquake ruptured the unmapped and unknown Tabas thrust fault at the western Neogene foothills of the Shotori Mountains (Figure 13.8). Evidence of active faulting is preserved in the landscape in the form of truncated asymmetric anticlinal folding in the Neogene molase deposits, deformation of the late Quaternary alluvial fan deposits, and widespread river incision (Berberian, 1979a,d; King et al., 1981; Niazi and Kanamori, 1981; Berberian, 1982; Silver and Jordan, 1983; Shoja-Taheri and Anderson, 1988; Walker et al., 2003, 2013b; Sarkar et al., 2005).

The Shotori Mountains are composed of heavily folded and thrusted, Paleozoic–Mesozoic sediments, with the folded and faulted Neogene molasse deposits at its western foothills. The 1978 earthquake was preceded by small events on 10 June 1939 (M_s 5.5), 17 June 1974 (M_s 4.6; with a thrust mechanism), and 26 September 1977 (VI^+), showing southward migration of

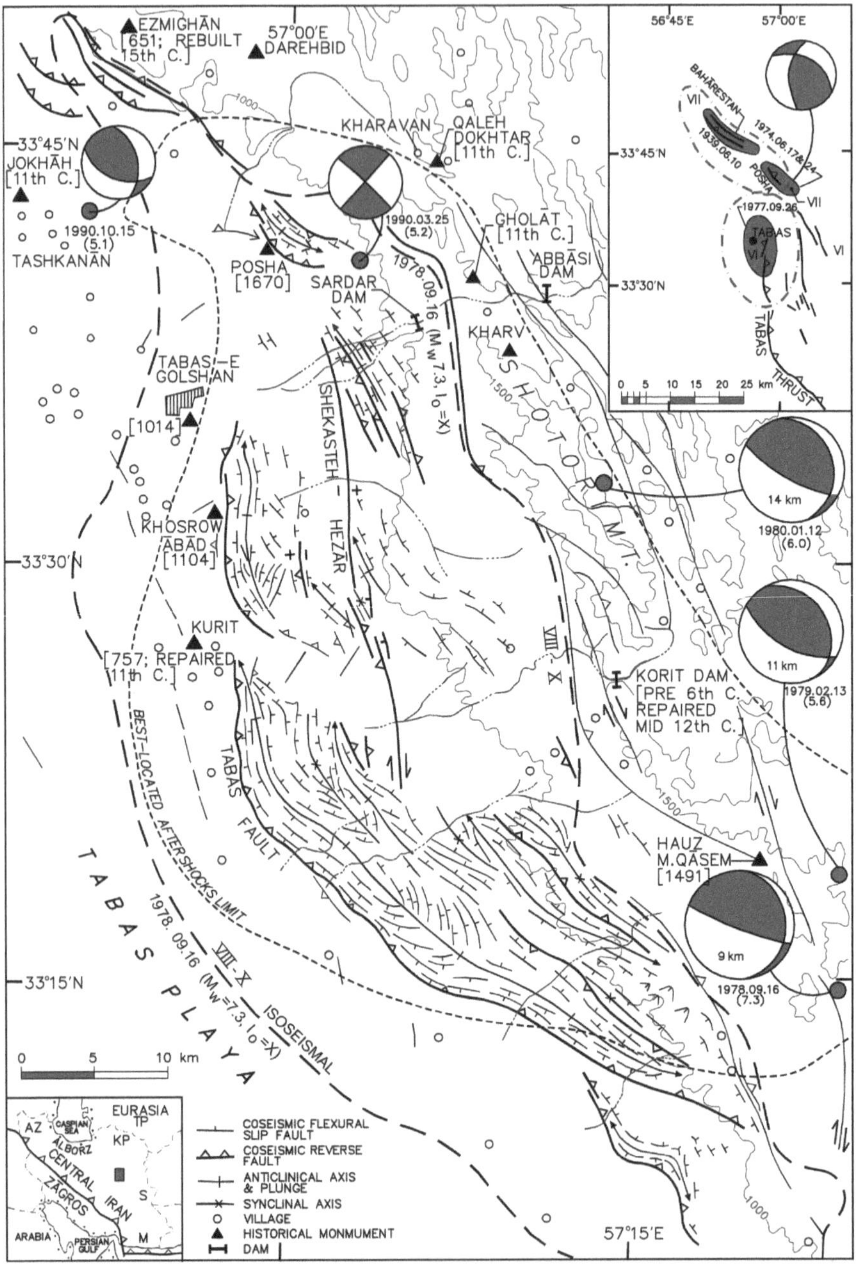

FIGURE 13.8 Meizoseismal area of the 16 September 1978 M_w 7.3 Tabas-e Golshan earthquake along the Tabas thrust in eastern Central Iran. Fault and other symbols as in Figure 11.1. The earthquake was associated with surface fault rupture and flexural-slip faulting. Focal mechanisms of the mainshock and two aftershocks constrained by body-wave modeling with centroid depths (Walker et al., 2003). Fault-plane solutions of the two small-magnitude events to the northwest (best-double-couple HRVD CMT solution). Top right: Southward migration of seismicity toward the town of Tabas-e Golshan. See also Plates 13.11–13.19. *Modified from Berberian (1979a,d, 1982).*

PLATE 13.11 Destruction of the Tabas-e Golshan oasis town by the 16 September 1978 M_w 7.3 earthquake. Only a few domed structures withstood the earthquake. See Figure 13.8 for the location. *Photographed on 28 September 1978.*

seismicity toward the town of Tabas-e Golshan (Berberian, 1979a,d, 1981a,b, 1982; Berberian et al., 1979b; King et al., 1981; see Figure 13.8).

Mohajer-Ashjai and Nowroozi (1979) and Nowroozi et al. (1980) introduced a total 40 km, N10°E-trending straight line dipping WNW (for a length of 16 km in the Tabas-e Golshan area) and ESE (for 33 km south of Tabas-e-Golshan). Mo'infar (1978) noted a 25-km surface rupture with 30 cm of vertical and 10 cm of horizontal (?) movements. Despite reproducing the fault map of Berberian (1979a) as their figure 3.57, Ambraseys and Melville (1982) stated that no proper field mapping of the fault break has so far been published. Coseismic surface fault mapping and aftershock recording by nine portable seismometers were conducted during a period of chaotic social unrest unseen in the recent history of the country; this work was halted in a premature abandonment of the research activity and the unexpected departure from the camp because of constant threats and violent, by organized protests in almost all the villages.

Aryanmanesh et al. (2007) reported a 25–35 cm left-lateral strike-slip displacement of the rows of trees in an E–W direction (?) in the area northwest of Dehshur, about 35 km northwest of Tabas-e Golshan. Aryanmanesh et al. (2007) mentioned that the trees have been growing on a farm for only 30 years. Assuming that the time of the site visit was in 2005 (2 years prior to the publication of their report), the trees might have been planted in about 1975. The authors do not mention during which of the Tabas earthquakes (16 September 1978 (M_w 7.3), 13 February 1979 (M_w 5.6), 12 January 1980

(M_w 6.0), 25 March 1990 (M_w 5.2), and 15 October 1990 (M_w 5.1)) the proposed displacement took place. No fault map or photograph of the reported displaced trees was published. Their statement cannot be confirmed.

13.11.1 Surface Deformations

The Tabas-e Golshan earthquake was associated with (i) ~75-km long, coseismic surface thrust faulting dipping east–northeast along a curved fault line (Figure 13.8; Plate 13.12) and (ii) flexural-slip faulting in the form of bedding-plane slip with a thrust mechanism on the hanging wall of the Tabas thrust (Figure 13.8; Plates 13.13 and 13.14) in a vast area (Berberian, 1979a,d, 1982; Berberian et al., 1979b; King et al., 1981).

PLATE 13.12 Coseismic thrust faulting associated with the 16 September 1978 M_w 7.3 Tabas-e Golshan earthquake, southeast of Tabas-e Golshan. Looking north. See also Figure 13.8. *Photographed in September 1978 (Berberian, 1979a).*

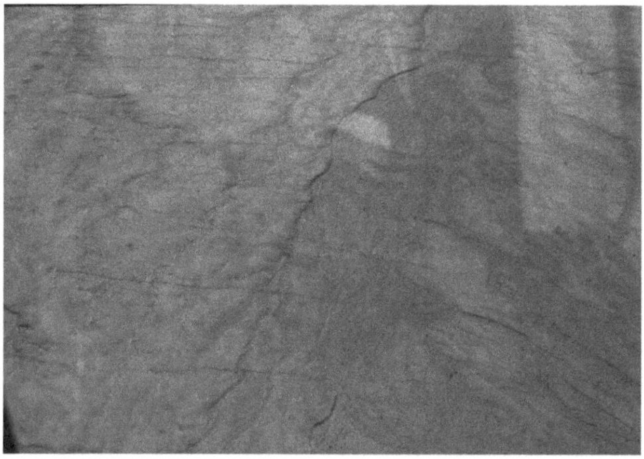

PLATE 13.13 Flexural-slip faulting showing bedding-plane slip with thrust mechanisms in the Neogene molasse deposits associated with the 16 September 1978 M_w 7.3 earthquake, southeast of Tabas-e Golshan, developed along the hanging wall of the Tabas thrust. See also Figure 13.8. The average distance between the circular access shafts of qanāt line (top left) is about 50 m. Looking east.

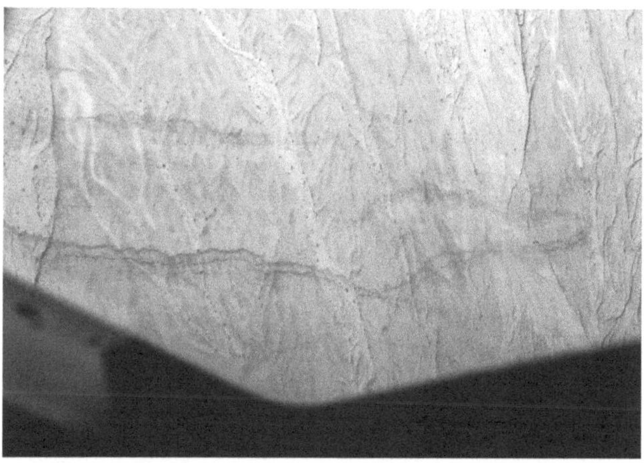

PLATE 13.14 Flexural-slip faulting showing bedding-plane slip with thrust mechanisms in the Neogene molasse deposits associated with the 16 September 1978 M_w 7.3 event southeast of Tabas-e Golshan, developed along the hanging wall of the Tabas thrust. See also Figure 13.8. The flexural-slip faults are oriented N–S at this location.

About 75 km of discontinuous coseismic thrust faulting, in several segments of arcuate form, was mapped at the surface along an existing but unrecognized foothill-front reverse fault, the Tabas active thrust. An abrupt change of nearly 60° occurs near 33.40°N latitude in the orientation of the surface ruptures from NW–SE on the southern half of the faulting to nearly N–S on the northern half, and no single surface rupture extended for more than one-third of the length of faulting (Figure 13.8). About 8 km east of the Tabas frontal thrust, a 17-km-long, nearly vertical surface faulting with the western block downthrown developed in the Shekasteh Hezār [literary thousands faults] area east of the town (Plate 13.15). This fault was originally defined as a high-angle reverse fault but could be a normal fault developed on the hanging wall of the Tabas thrust due to flexural-folding of the Neogene molasse deposits above the Tabas thrust. The observation of the surface ruptures in a complicated and segmented pattern (Figure 13.8) may suggest that the main shock was a multiple event that occurred over a system of imbricate faults at the western limbs of the Neogene molasse folds (Berberian, 1979a).

Shallow parallel thrust faulting dipping 30°NE was observed within a zone of 5 km width along the 75 km of the foothill-front. Nearly all the surface breaks followed the obvious scarps created by the historic faulting of unknown date. Overall vertical uplift (throw) and slip were measured at about 150 and 300 cm, respectively. Maximum displacement was measured to be 35 cm along one individual thrust fault. No lateral displacement was observed along the 1978 surface ruptures in the field (Plates 13.12 and 13.16; Figure 13.8). In three places, slip direction and amplitude could be determined using fresh slickensides [N60°E at Poshā; N65°E at Tabas-e Goshen; and N55°E at Kalāteh area, northwest of Pāykuh] (Berberian, 1979a). Quaternary cumulative slip of 198 cm and throw of 90 cm was measured on a thrusted Neogene clay keybed at the base of the Quaternary alluvial deposit, across the Tabas thrust, 4 km southeast of the Tabas-e Golshan town (Plate 13.16). The 1978 earthquake movement at this locality produced a 9-cm slip and a 4-cm throw. The Shotori range-front fault, located about 18–24 km to the east of the Tabas coseismic surface thrust was checked in the field and was not reactivated during this earthquake (Berberian, 1979a).

An extensive zone of coseismic bedding-plane slips in a reverse sense (flexural-slip faulting), extending for a width of >8 km and length of

PLATE 13.15 Coseismic surface rupture along the N–S-trending, Shekasteh Hezār fault on the hanging wall of the Tabas thrust. The western block (foreground) is downthrown. For the location of the fault, see Figure 13.8 (Berberian, 1979a).

PLATE 13.16 Quaternary alluvial deposits, about 4 km southeast of the Tabas-e Golshan town displaced by the late Quaternary movement of the Tabas thrust (see Figure 13.8 for the location). The early Quaternary (?) cumulative slip was 198 cm and the throw was measured about 90 cm. Note the displaced Neogene clay keybed at the base of the Quaternary alluvial deposits. The fault is N–S and dips 27°W. The early Quaternary slip vector trends N65°E and the plunge is 20°SW. The 1978 earthquake movement at this locality produced a 9-cm slip and a 4-cm throw. Looking north (Berberian, 1979a).

>90 km, was developed east of and along the main surface thrust break (Figure 13.8; Plates 13.13 and 13.14). These were developed during flexural-slip folding above the Tabas thrust fault (Berberian, 1979a; King et al., 1981). A vertical rate of ~1 mm/year of peak surface uplift above the Sardar anticline, east-northeast of the town (33°38′N–57°05′E; on the Shotori Mountain piedmont, sitting on the Tabas thrust hanging wall), and a horizontal shortening of 1.6–2.0 mm/year were obtained by Walker et al. (2013b).

13.11.2 Seismology

The highest-quality, locally recorded aftershock hypocenters occurred mainly at a depth less than 23 km with a high concentration of seismic activity between 8 and 14 km (Berberian et al., 1979b; Berberian, 1982). The best-located aftershocks with their well-constrained focal mechanisms demonstrated an active imbricate listric thrust system (covering both the hanging wall as well as the footwall blocks) with fault planes flattening into a possible basement decollement zone (Berberia et al., 1979b; Berberian, 1982). Analyses of strong motion records (Shoja-Taheri and Anderson, 1988; Hartzell and Mendoza, 1991) indicated that the rupture propagation was mostly unilateral from the southeast to the northwest at an average rupture velocity of 2.5 km/s. The majority of the slip, in at least four subevents, contributing to the strong-motion signals and the WWSSN body waves, terminated about 15 km NW of Tabas-e Golshan, giving a fault length of 90 km (Shoja-Taheri and Anderson, 1988). Maximum coseismic surface faulting of about 75 km was observed and mapped on the surface.

As with the teleseismic location (Berberian, 1979b; Berberian, 1982), the relocated epicenter of the 1978 Tabas-e Golshan mainshock (Engdahl et al., 2006; Walker et al., 2013b) is placed close to the southern end of the coseismic surface ruptures, indicating a possible unilateral northward rupture propagation (Figure 13.8). Analysis of strong-motion records (Shoja-Taheri and Anderson, 1988; Sarkar et al., 2005) indicates the presence of at least four, and probably more, discrete subevents in the mainshock, but their locations could not be constrained.

P and *SH* body-waveform inversion for the 1978 Tabas-e Golshan earthquakes (Walker et al., 2003) indicated NW–SE thrust motion dipping 16°NE with a minor right-lateral motion. Similar source parameters were obtained for the 1979 and 1980 shocks, though with slightly steeper dips (Tables 13.5 and 13.6; Figure 13.8). The obtained source time function indicates a single, simple rupture event lasting for 18 s; with a rupture velocity of ~3.5 km/s, this gives a 65-km fault length (Walker et al., 2003).

Walker et al. (2003) calculated, from scaling relationships, that the average slip at depth would be ~3.3 m. Since the maximum measured "single" thrust displacement at the surface was only 35 cm, the major part of the primary rupture is likely to have been blind. Aftershock locations from the deployment of nine local seismometers (Berberian et al., 1979b; Berberian, 1982) showed that (i) most of the aftershocks were located in the hanging-wall of the Tabas thrust; (ii) the seismicity flattened with depth toward east northeast; (iii) aftershocks concentrated in depth about 10 km; (iv) scattered seismicity located in the Tabas playa (on the footwall) indicating fault activity to the west of the surface rupture as well; and (v) very little activity was detected beneath the Shotori Mountains to the east, where the range-front fault did not show any surface activity (Berberian et al., 1979b; Berberian, 1982).

13.11.3 Oral Earthquake Hazard Warning

Despite the lack of recorded historic earthquakes in the remote desert area, place names such as Shekasteh Hezār [lit., "Thousands (numerous) Fractures," "Highly Fractured or Faulted"] and Shekasteh Kasuri ["Fractured or Faulted Kasuri"] indicate that there have been earthquakes in the area (see Chapter 1 for details).

13.11.4 Monuments and Large-Earthquake Elapse Time

No records of large-magnitude earthquakes had been entered into the annals for almost 1000 years prior to the 16 September 1978 M_w 7.3 earthquake; historical monuments were intact in the town until the earthquake (Berberian, 1979a). No specific reference exists to an earlier large-magnitude earthquake

TABLE 13.5 Source Parameters of the 16 September 1978 M_w 7.3 Tabas-e Golshan Mainshock

Source	Strike (°)	Dip (°)	Rake (°)	Slip Vector Azimuth	Focal Depth (km)	M_o (Nm) ($\times 10^{20}$)	M_w	Method
Berberian et al. (1979b); Berberian (1979b, 1982)	332	31	100	36	–	–	–	Long-period first motion
Berberian (1979b, 1982); King et al. (1981)	341	30	–	–	–	–	–	Waveform modeling
Harvard CMT	328	33	107	38	11	1.32	7.4	Centroid moment tensor
Niazi and Kanamori (1981)	330	30	110	36	–	1.5	7.4	Surface wave moment
	–	–	–	–	–	0.82	–	*P*-wave moment
Silver and Jordan (1983)	–	–	–	–	–	1.3	–	Total moment spectra
Walker et al. (2003)	355	16	155	19	09	1.052	7.28	Body-wave inversion

See also Figure 13.8.

TABLE 13.6 *P* and *SH* Body-Waveform Inversion Source Parameters of the 16 September 1978 M_w 7.3 Tabas-e Golshan Earthquakes (Walker et al., 2003)

Earthquake	Time (UTC)	Coordinate (N°–E°)[a]	Strike (°)	Dip (°)	Rake (°)	Slip Vector Azimuth	Centroid Depth (km)	M_o (Nm) ($\times 10^{18}$)	M_w
1978.09.16	15:35	33.24–57.38	355	16	155	19	09	105.2	7.28
1979.02.13	10:36	33.29–57.39	327	28	116	28	11	0.2976	5.58
1980.01.12	15:31	33.54–57.22	348	20	137	29	14	1.277	6.0
1990.03.25[b]	00:01	33.66–57.04	313	90	0		Fixed	0.65	5.2
1990.10.15[b]	19:06	33.71–56.86	335	53	118		Fixed	0.52	5.1

See also Figure 13.8.
[a]*Epicenters are from Engdahl et al. (2006).*
[b]*Harvard Centroid Moment Tensor Solution.*

in Tabas-e Golshan town (Daneshdoust, 1998); however, inhabitants believe that the location of the town has been changed twice. The cause and dates are unavailable (Ya'qub Daneshdoust, personal letter dated 9 October 1983).

The oldest monument in Tabas-e Golshan town was the 40-m high, slender Menār-e Kabir [lit., the "Great Minaret"] of the congregational mosque built in 405H/1014 CE (Plate 13.17). Photographs from 1880 (Stewart, 1911) and 1906 (Hedin, 1910) show a damaged and slightly leaning minaret, which later collapsed on 22 February 1907 (Daneshdoust, 1998). The original congregational mosque was replaced by a new mosque built by Mir Hassan Khān,

PLATE 13.17 The minaret of the congregational mosque (Menāreh-ye Kabir) at the town of Tabas-e Golshan photographed in 1906 by Sven Anders Hedin (Hedin, 1910; see Figure 13.8 for the location). The minaret, which was possibly built in 993–1014 (late Ghaznavid-early Saljuq), collapsed on 22 February 1907 due to construction of the Kariz-e Allāhābād (qanāt), which weakened the foundation of the minaret (Meshkati, 1970; Daneshdoust, 1998). The survival of the minaret for a period of >900 years indicates the lack of large-magnitude earthquakes since at least the year 1000. *Courtesy of the Sven Hedin Foundation, Stockholm, Sweden, published with permission.*

the governor of Tabas-e Golshan, in 1803; repairs were carried out in 1893, 1932, and 1976 (Daneshdoust, 1998). The Saljuq (ca. 1000–1218) shaking minarets of the Madraseh-ye Dau-Menār ("Two Minaret Seminary")—two minarets around which the Madraseh was rebuilt in 1803 (Daneshdoust, 1998)—completely collapsed during the 1978 earthquake (Plate 13.18). The Emāmzādeh Hossein (mausoleum) was another structure; it was built in 1957 (Meshkati, 1970). The historic, thick kiln brick entrance wall and vaulted gate to the city (Plate 13.19), and the Qājār period Robāt-e Gur caravanserai built with thick brick vaults and abutments (Plate 13.20) located about 47 km northwest of the coseismic surface fault, were also destroyed by strong ground shaking. The existence of the old minarets (Menār-e Kabir and Dau-Menār) in the citadel [Arg] Tabas-e Golshan indicates a relative seismic quiescence of large-magnitude earthquakes for at least 1000 years.

The outer wall of the citadel (Arg/Kohandezh), built possibly by Amir Esmā'il Gilaki, ca. 1100, is buttressed (Daneshdoust, 1998). The reason and date for adding the buttresses are not documented. The buttresses may have added to strengthen the walls after a pre-1000 earthquake(?). However, we cannot support this idea. A paleoseismic trench study may help in understanding the previous motion along the Tabas thrust.

13.12 THE 14 NOVEMBER 1979 M_W 6.6 KORIZĀN EARTHQUAKE

The shock killed 171, injured 297, destroyed qanāts of Korizān, and caused extensive damage in the remote, sparsely populated desert region of eastern Iran along the Ābiz fault (Figure 13.9). This earthquake occurred during a period of civil disturbance and increasingly violent protests, during which the Shāh of Iran was forced to abdicate his throne and accept a forced deportation from the country. The earthquake preceded the departure of the Shāh and his flight into imposed exile by a few hours. Hence, little information is available for this event as well as the next two events of 27 November 1979 and 7 December 1979 discussed below.

The 14 November 1979 earthquake, which occurred north of the 30 June 1936 Ābiz earthquake (discussed earlier in Chapter 12), produced about a 17-km- (Nowroozi and Mohajer-Ashjai, 1980, 1980/1981) to 20-km-long (Haghipour and 'Amidi, 1980) surface faulting along the northern section of the Ābiz fault, which has a N–S arcuate strike, convex to the west (Figure 13.9). Coseismic surface rupture extended from south of Estend, through Korizān, and died out near the Kāl-e Shur River (see also Figures 13.17, and 13.18). A maximum right-lateral displacement of 100 cm was observed in Korizān village, where a dried mud wall and small bush were cut by the fault and displaced in two pieces, with a maximum observed vertical movement of 60 cm (Nowroozi and Mohajer-Ashjai, 1980, 1980/1981). No displacement was recorded by Haghipour and 'Amidi (1980). About 2 km north of Korizān, the fault cut through a dry

PLATE 13.18 The eleventh-century Two-Minaret Seminary (madraseh-ye dau-Menār), also known as "the shaking minarets" (menār-e jonbān), at the town of Tabas-e Golshan, photographed in 1906 by Sven Anders Hedin (Hedin, 1910; see Figure 13.8 for the location). The two minarets with their faience pattern, flanking the portal of a possible twelfth-century seminary, were constructed during the Saljuq dynasty (ca. 1000–1218) and together with the seminary were destroyed during the 1978 M_w 7.3 earthquake (Meshkati, 1970; Matheson, 1972; Berberian, 1979a; Daneshdoust, 1998). *Published with permission from the Sven Hedin Foundation, Stockholm, Sweden; http://svenhedinfoundation.org.*

PLATE 13.19 The main entrance gate to Tabas-e Golshan, photographed by Sven Hedin in 1906. The very thick masonry structure was destroyed during the 1978 earthquake. *Published with permission from the Sven Hedin Foundation, Stockholm, Sweden; http://svenhedinfoundation.org.*

PLATE 13.20 The Robāt-e Gur caravanserai built with thick load-bearing abutments supporting the lateral thrust constructed sometime during the Qājār period (1779–1925). The structure was located about 43 km northwest of Tabas-e Golshan and 47 km to the northwest of the 1978 coseismic surface faulting, and was destroyed during the earthquake. *Photographed in 1906 by Sven Hedin. Published with permission from the Sven Hedin Foundation, Stockholm, Sweden; http://svenhedinfoundation.org.*

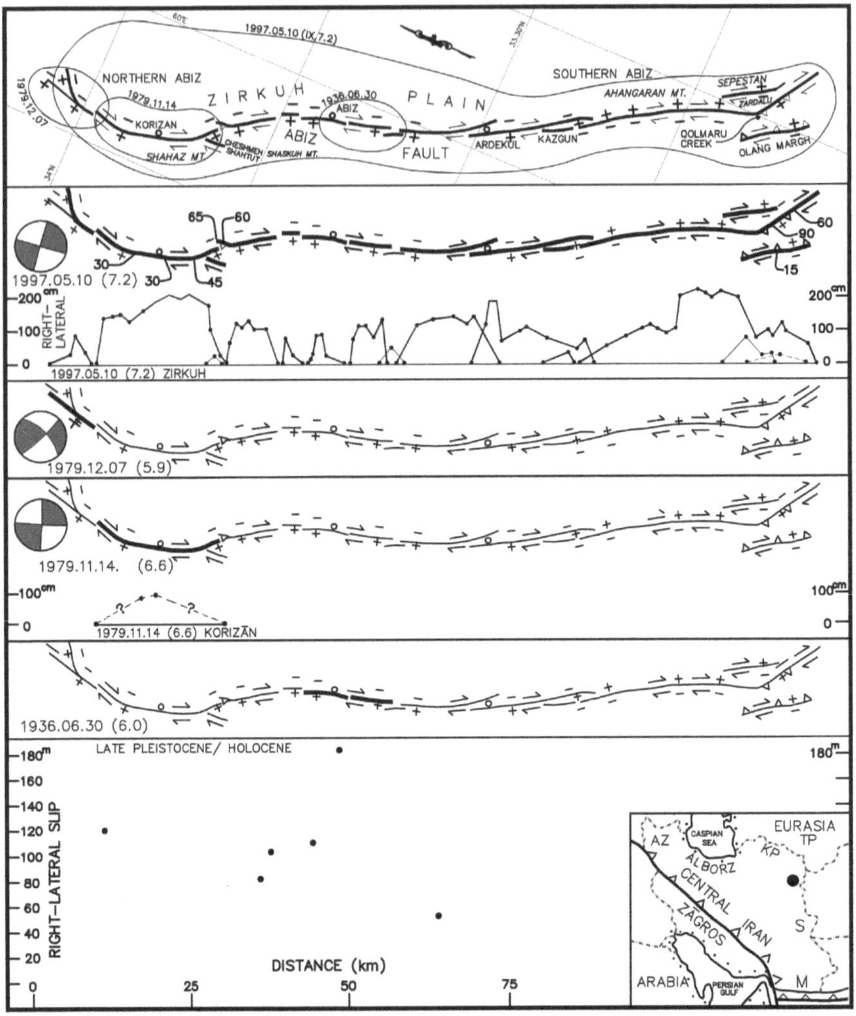

FIGURE 13.9 Coseismic slip variation along the Abiz right-lateral fault since 1936 in eastern Iran. Fault and other symbols as in Figure 11.1. The top panel shows the fault line mapped after the 10 May 1997 M_w 7.2 Zirkuh earthquake. Measured amounts of vertical displacement during the 1977 (in cm) are shown by numbers on the second panel from top. In some places, vertical displacement was not measured because it was small, or was accentuated by landslides or by differential ground settlement. The graph shows the amount of coseismic right-lateral strike-slip displacement measured along the surface rupture of the Abiz fault. Observed points at which offsets were measured have been joined by lines and do not imply that the offsets varied between those points in the jagged manner indicated by the profiles. The next three panels show the estimated surface rupture length of the 1979 (two events) and 1936 earthquakes along the Abiz fault based on the field study. The bottom panel shows estimated late Pleistocene offsets of streams along the fault. *Modified from Berberian et al. (1999) and Berberian and Yeats (1999).*

stream channel with a right-lateral displacement of 80 cm and a vertical displacement of 45 cm, with the eastern block moving upward relative to the western block (Figure 13.9). In a profile observed in the Estend ravine, the fault dips 65° toward the west near the surface; the dip changes to near vertical within a few meters and drops steeply toward the east (Nowroozi and Mohajer-Ashjai, 1980, 1980/1981; Berberian and Yeats, 1999, 2001; Berberian et al., 1999).

13.12.1 Seismology

P and *SH* body-waveform modeling indicated two right-lateral strike-slip subevents; the first striking N160°E and the second N175°E (Figure 13.9; Table 13.7). A southward propagation of the event would result in a change of strike from N–S to N160°E, which is opposite to the modeling result (Baker, 1993; Berberian et al., 1999).

For patterns of coseismic surface rupture and a historic seismicity of the area, see Chapter 16.

13.13 THE 27 NOVEMBER 1979 M_W 7.1 KOLI [EAST DASHT-E BAYĀZ] EARTHQUAKE

This earthquake killed only 20 people in sparsely populated desert villages but caused widespread damage over a large area along the eastern Dasht-e-Bayāz left-lateral, strike-slip fault from Makārem, Jabbār, Koli, and Chāh Zighān to Barkāh, Qal'eh Sorkh, Donakhi, Boniābād, and Mehrābād (Figure 13.1). Seven people were killed at Jozandar and Hātamābād, one at Chāh Mamaqāni, and four at Chāh Gachi. The miraculously low death toll can by explained by the sparse population in this remote desert area, people's awareness of the 14 November Korizān earthquake and aftershocks, and the fact that people were living in tents that had been distributed after the Korizān earthquake.

TABLE 13.7 *P* and *SH* Body-Waveform Inversion Source Parameters of the 14 November 1979 M_w 6.6 Korizān Earthquake (Baker, 1993; Berberian et al., 1999)

Earthquake	Time (UTC)	Coordinate (N°–E°)[a]	Strike (°)	Dip (°)	Rake (°)	Centroid Depth (km)	M_o (Nm) ($\times 10^{18}$)	M_w
1979.11.14	02:21	33.95–59.72	160	89	−177	10	9.8	6.57
			175					

See also Figure 13.9.
[a]*Epicenter is from Engdahl et al. (2006).*

Almost 11 years after the 31 August 1968, M_w 7.1, Dasht-e Bayāz earthquake along the western segment of the Dasht-e Bayāz fault, the 1979 M_w 7.1 Koli earthquake took place along the eastern segment of the same fault (Figure 13.1). The 1979 earthquake occurred during a period of chaotic civil disturbance. Haghipour and Amidi (1980) drew the coseismic surface fault of this earthquake as a straight line on the satellite image. Nowroozi and Mohajer-Ashjai (1980, 1980/1981) presented a fault map on a 1:250,000 scale topographic map showing one step-over about 6 km east of Chāh Zighān, with two nearly N–S cross faults in the west and one in the east (Figure 13.1).

Although the authors claimed that the 1979 earthquake produced an E–W fault break along the extension of the eastern Dasht-e Bayāz greater than 65 km, the distance from Jazandar and Makārem (34°01′N–59°14′, +1580 m; west of Koli) in the west to Boniābād in the east (34°04 N–59°52′E, +938 m) shows a surface rupture of 55 km (Figure 13.1). The 1979 coseismic, surface-fault break overlapped the one associated with the 1968 earthquake by about 20 km. Along this section of the eastern Dasht-e Bayāz fault, a maximum left-lateral displacement of 255 cm of a small channel was observed a few hundred meters northwest of the village of Koli (34°01′N–59°22′E, +1.322 m), where the vertical displacement was 190 cm (with the southern block downthrown). The maximum vertical displacement of 390 cm was observed in the limestone mountain ranges about 1.5 km west of Koli, with the northern block moving upward (Figure 13.1). The fault plane in the area north of Koli dips 65° toward the south at the surface, with slickensides plunging 35° toward the east (Nowroozi and Mohajer Ashjai, 1980, 1980/1981). Haghipour and 'Amidi (1980) only reported 60 km of surface left-lateral faulting without recording horizontal and vertical displacements.

To the west of Koli [NW of Koli and NE of Makārem (~59°20′E, NW of Koli, with 390 cm vertical and 110 cm left-lateral slip)], two subsidiary fault traces of 5 km long striking N140°E branch off the main E–W Dasht-e Bayāz fault and show vertical and horizontal displacements. Near the eastern end of the E–W surface break, in the vicinity of Qal'eh Sorkh, the vertical motion is reversed and the southern block moved upward about 45 cm with respect to the northern block. About 4 km to the west of Qal'eh Sorkh (~59°45′E, Qal'eh Sorkh), two other nearly N–S subsidiary faults with right-lateral displacements of a few cm were mapped (Figure 13.1). The shorter one has a length of nearly 6.5 km, trending N170°E. The longer fault has a length of nearly 11.5 km, trending N15°E (Nowroozi and Mohajer Ashjai, 1980, 1980/1981).

Despite having equal moment magnitude (M_w 7.1) with the 1968 Dasht-e Bayāz earthquake, the surface rupture associated with the 27 November 1979 Koli earthquake along the eastern segment of the same fault was much shorter than the length of the 31 August 1968 Dasht-e Bayāz earthquake (55 km vs. 70 km). Furthermore, the few measured amounts of 1979 fault displacements were much less than those of the 1968 earthquake (Figure 13.1).

13.13.1 Seismology

First-motion polarities (Jackson and McKenzie, 1984) show a left-lateral, strike-slip mechanism. The waveforms of the 27 November 1979 Koli earthquake are rather similar to those of the 31 August 1968 Dasht-e Bayāz mainshock (Table 13.8; Figure 13.1). The higher-frequency details in the *P* waves make it probable that the strike-slip rupture occurred in several discrete subevents. Assuming a seismogenic thickness of 15 km and a fault length of 60 km, the moment of 5×10^{19} Nm would give an average slip of 1.85 m, which is compatible to that observed in the field (Walker et al., 2004).

13.14 THE 7 DECEMBER 1979 M_W 5.9 KALĀT-E-SHUR EARTHQUAKE

This event, an aftershock of the 14 November 1979 M_w 6.6 Korizān earthquake (on the right-lateral, N–S Ābiz fault), took place 23 days afterward and 10 days after the 27 November 1979 M_w 7.1 Koli earthquake (on the left-lateral, eastern Dash-e Bayāz fault) (Figures 13.9 and 13.18). It apparently occurred further north along the Ābiz fault and produced 15 km of fresh faulting from Kalāt-e Shur in the south (Figure 13.17) to Donakhi and Chāh-e Matār in the north (Mohājer-Ashjai et al., 1980; Nowroozi and Mohājer-Ashjai, 1980). This segment of the Ābiz fault had not ruptured at the surface during the mapping of the 14 November 1970 Korizān earthquake and was later discovered during the field expedition to investigate the 27 November 1979 Koli earthquake (Mohājer-Ashjai et al., 1980; Nowroozi and Mohājer-Ashjai, 1980a; also see Berberian and Yeats, 1999, 2001; Berberian et al., 1999).

Haghipour and 'Amidi (1980) claimed that the northernmost 10 km of the Ābiz right-lateral fault was also reactivated by the 27 November 1979 Koli earthquake, but Nowroozi and Mohājer-Ashjai (1980a) and Mohājer-Ashjai et al. (1980) categorically stated that the N–S faulting was formed after the

TABLE 13.8 *P* and *SH* Body-Waveform Inversion Source Parameters of the 27 November 1979 M_W 7.1 Koli (East Dasht-e Bayāz) Earthquake (Walker et al., 2004)

Earthquake	Time (UTC)	Coordinate (N°–E°)[a]	Strike (°)	Dip (°)	Rake (°)	Centroid Depth (km)	M_o (Nm) ($\times 10^{19}$)	M_w
1979.11.27	17:10	34.05–59.75	261	82	8	08	5.05	7.1

See also Figure 13.1.
[a]*Epicenter is from Engdahl et al. (2006).*

27 November 1979 Koli earthquake, and there was no surface rupture between 11 and 27 November 1979.

Despite the reported surface ruptures along the Ābiz fault (Nowroozi and Mohājer-Ashjai, 1980a; Mohājer-Ashjai et al., 1980), based on the calibrated relocations, Walker et al. (2011) suggested that the 7 December 1979 M_w 5.9 earthquake occurred on a N–S fault to the north of the Dasht-e Bayāz fault line (about 12 km to the NE of the location claimed by Nowroozi and Mohājer-Ashjai, 1980a; Mohājer-Ashjai et al., 1981). The epicenter of the 7 December 1979 earthquake and its two aftershocks (both of *mb* 4.8, which took place on the same day) are located to the east of the Kuh-e Khaybar, and east of the northern continuation of the Ābiz fault (north of the eastern segment of the east Dasht-e Bayāz fault) near the Chāh-e Matār and Mehrābād villages (Figures 13.1, 13.9, and 13.18). Unfortunately, the area was not visited immediately after the earthquake. Walker et al. (2011) believe that the cause of surface rupture reported on the Ābiz fault (Haghipour and Amidi, 1980) is a major subevent of the 29 November 1979 M_w 7.1 Koli mainshock. In any event, it is likely that the controversy is either due to teleseismic location error of the Iranian earthquakes (very likely), or because the rupture propagated from the north was not reported.

13.14.1 Seismology

P and *SH* body-waveform modeling (Table 13.9; Figure 13.9) shows right-lateral strike-slip motion (Table 13.9; Baker, 1993).

13.15 THE 11 JUNE 1981 M_W 6.6 GOLBĀF EARTHQUAKE

This earthquake killed about 1400 people, most in the main settlement of the remote desert town of Golbāf in southeast of Iran. All the villages in the Golbāf playa were heavily damaged (Figure 13.10). No foreshocks or preceding regional seismic activity were detected by teleseismic networks, and the general level of seismicity on the Gowk fault system had been low since the last moderate shock of 2 September 1969 (*mb* 5.2). Local inhabitants reported

TABLE 13.9 *P* and *SH* Body-Waveform Inversion Source Parameters of the 7 December 1979 M_w 5.9 Kalāt-e Shur Earthquake (Baker, 1993)

Earthquake	Time (UTC)	Coordinate (N°–E°)[a]	Strike (°)	Dip (°)	Rake (°)	Centroid Depth (km)	M_o (Nm) ($\times 10^{18}$)	M_w
1979.12.07	09:23	34.07–59.84	113	84	21	10	1.0	5.9

See also Figure 13.9.
[a]*Epicenter is from Engdahl et al. (2006).*

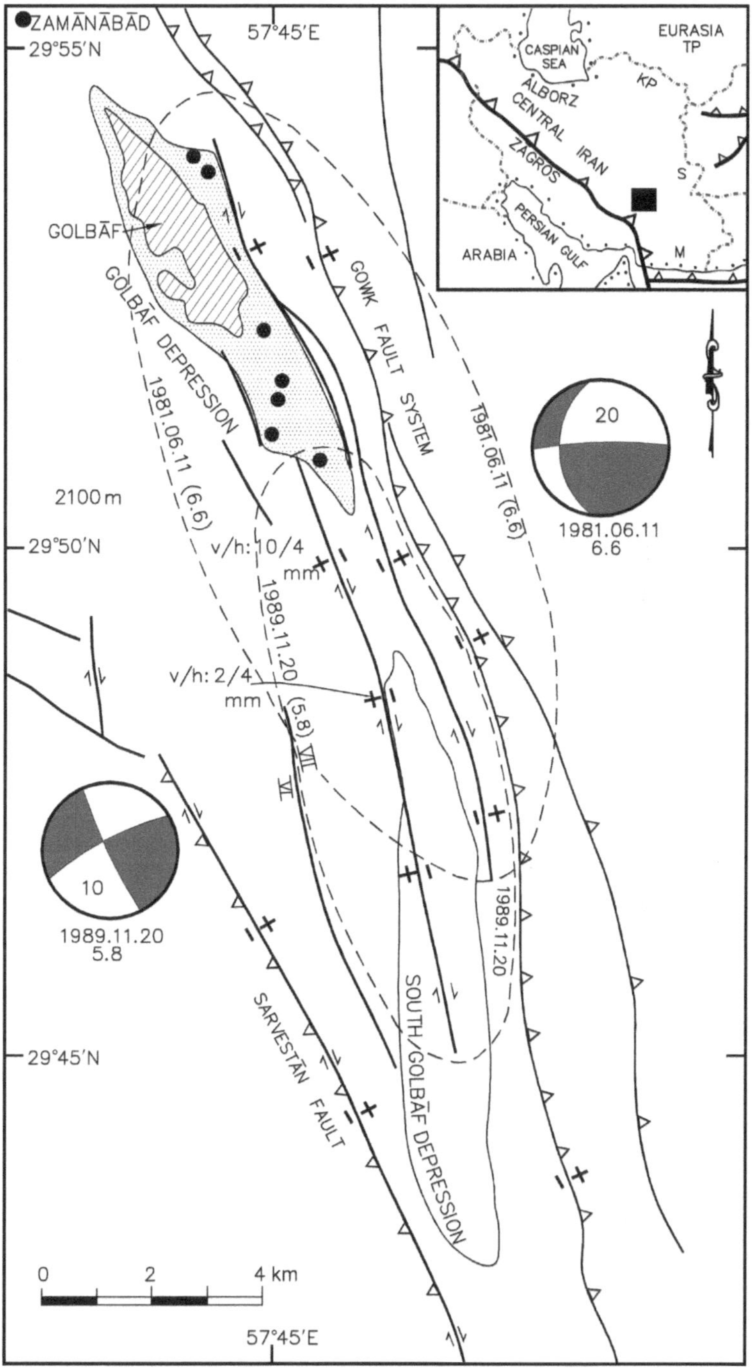

FIGURE 13.10 Meizoseismal areas of the 1981 Golbāf and 1989 South Golbāf earthquakes along the Gowk fault system in southeast Iran. Focal mechanisms constrained by body-wave modeling with centroid depths (Berberian et al., 1984, 2001). The Golbāf playa is highlighted by dotted pattern. See also Plates 13.21 and 13.22. *Modified from Berberian et al. (1984, 2001) and Berberian and Qorashi (1994).*

no indication of seismic activity in the epicentral region during the hours or days preceding the 11 June earthquake.

Modern attention was first drawn to the Gowk fault system because of the two earthquakes in 1981, when four successive earthquakes with magnitudes larger than M_w 6.0 ruptured the northern part of the Gowk fault system, exhibiting failure relationship (Figures 13.10 and 13.11). No historic (pre-1900) earthquake has been documented along the southern Gowk fault system due to its location in a remote and sparsely populated desert area. The seismic quiescence has been documented after 1900, and it will likely have its own earthquakes in the future (Berberian et al., 1984, 2001).

The 11 June 1981 earthquake was associated with right-lateral reverse displacement along faults bounding the Golbāf depression and dipping both east and west along the Gowk fault system. The eastern fault moved over a length of about 14.5 km at the surface south of Golbāf, with an average of approximately 3 cm right-lateral and 5 cm vertical motion (Figure 13.10; Plates 13.21 and 13.22). The western fault appeared to have moved over a length of about 7.5 km southeast of Golbāf with only small displacements and hair-line cracks visible at the surface (Berberian et al., 1984). The relative timing of the movement on the east- and west-dipping Gowk faults is not known from firsthand accounts, though both had moved before the 28 July 1981 M_w 7.0 Sirch earthquake; it is worth noting that there were no aftershocks of significant magnitude before 28 July 1981. Therefore, it is likely that, if both faults moved seismically rather than by creeping motion, this happened during the Golbāf mainshock of 11 June 1981 (Berberian et al., 1984, 2001).

The 11 June 1981 M_w 6.6 Golbāf earthquake produced surface ruptures for 14.5 km south of Zamānābād and was followed by the 28 July 1981 M_w 7.0 Sirch earthquake (Figures 13.10 and 13.11), which was associated with 65 km of discontinuous surface ruptures north of Zamānābād (Berberian et al., 1984). The area south of Zamānābād then ruptured again in the 20 November 1989 M_w 5.8 South Golbāf earthquake, whose 11 km of surface ruptures followed the scarps formed in the 11 June 1981 M_w 6.6 Golbāf earthquake exactly (Berberian and Qorashi, 1994). The 14 March 1998 M_w 6.6 Fandoqā earthquake again ruptured 20 km north of Zamānābād, with coseismic surface faulting following that observed after the 28 July 1981 M_w 7.0 Sirch earthquake (see Figure 13.11). Finally, a small (M_w 5.3) earthquake (18 November 1998) near Chāhār Farsakh produced minor surface cracking over about 4 km, again along the scarp of the 28 July 1981 M_w 7.7 Sirch earthquake (Berberian and Yeats, 1999; Berberian et al., 2001; Berberian, 2005). For seismic history and patterns of coseismic surface rupture along the Gowk fault, see Chapter 16.

The GPS measurements at the Gowk fault area show the direction of maximum principal horizontal stress is about N8°E with 8 mm/year slip rate. This rate is much larger than the Gowk fault slip rate of 2 mm/year derived from geomorphic data (Walker and Jackson, 2002).

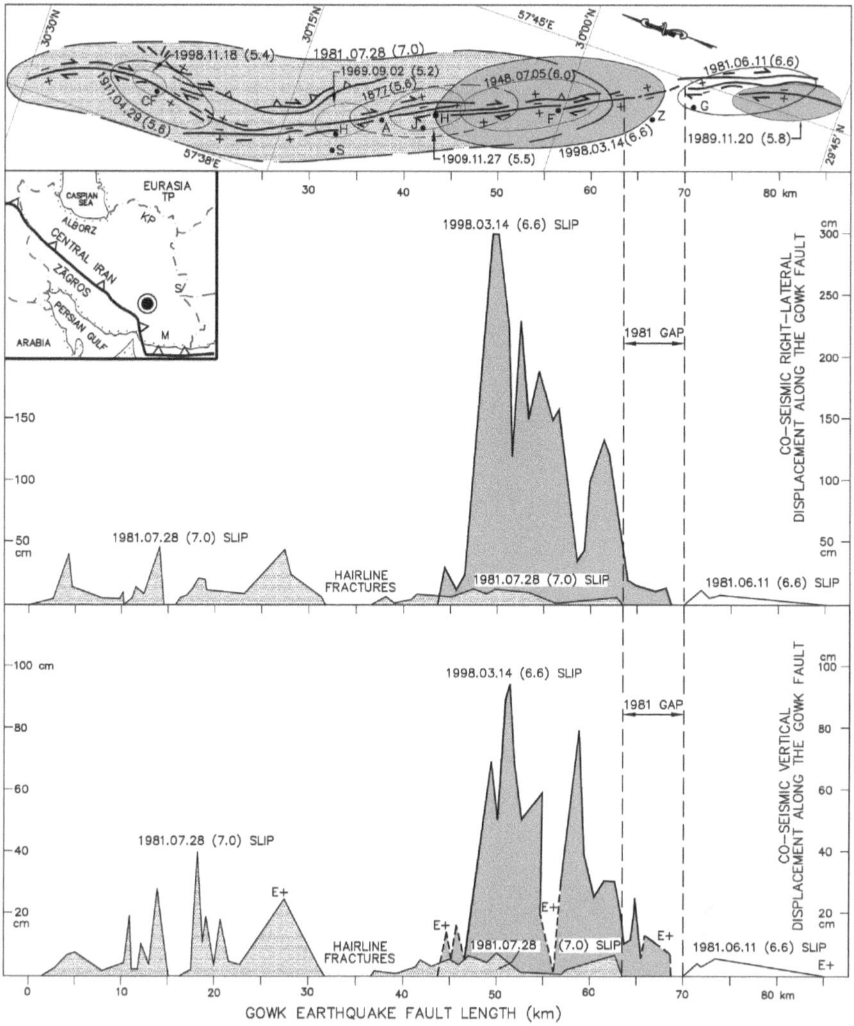

FIGURE 13.11 Top panel: Meizoseismal areas of the earthquakes along the Gowk fault system in southeast Iran. Fault and other symbols as in Figure 11.1. A, Āb-e Garm; CF, Chahār Farsakh; F, Fandoghā; G, Golbāf; H (north), Hasanābād; H (south), Hashtādān; J, Jowshān; S, Sirch; Z, Zamanābād. Middle and bottom panels: Observed amplitudes of coseismic right-lateral (middle) and vertical (bottom) offsets along the Gowk fault system following the July 28 1981 M_w 7.0 Golbāf (highlighted with dots) and 14 March 1998 M_w 6.6 Fandoqā (highlighted gray) earthquakes (the latter event was erroneously marked as 11 June 1981 in the caption to figure 8 in Berberian et al., 2001). Offsets are in centimeters plotted against distance along the fault measured on the map in the top panel. Where coseismic fault segments from the same earthquake overlapped, their cumulative value has been added. In the bottom panel, graphs marked E+, indicate that the vertical sense of slip was east-side-up. Observed points at which offsets were measured have been joined by lines to distinguish the 1981 and 1998 displacements and do not imply that the offsets varied between those points in the jagged manner indicated by the profiles. Note the large offsets from the 14 March 1998 M_w 6.6 Fandoqā earthquake compared with the much bigger July 28 1981 M_w 7.0 Golbāf earthquake. See also Plates 13.23 and 13.24. *Modified from Berberian and Yeats (1999), Berberian et al. (2001), and Berberian (2005).*

PLATE 13.21 Coseismic right-lateral surface rupture of the 11 June 1981 M_w 6.6 Golbāf earthquake along the Gowk fault system in southeast Iran (see Figure 13.10 for the location). Looking east (Berberian et al., 1984).

PLATE 13.22 Close up view of the coseismic right-lateral surface rupture of the 11 June 1981 M_w 6.6 Golbāf earthquake in southeast Iran (see Figure 13.10 for the location). The two pencils are placed parallel to the surface rupture along the Gowk fault system. Looking east (Berberian et al., 1984).

13.15.1 Seismology

SH and *P* waveform modeling indicated a right-lateral, strike-slip faulting with a small normal-faulting component, dipping west, striking NW–SE, parallel to the orientation of the Gowk fault (Figure 13.10; Table 13.10). The source was modeled as two subevents, suggesting a slip of about 75 cm for the first subevent based on the seismic moment of 4.18×10^{18} Nm (Berberian et al., 2001). As with the next Sirch earthquake, the Golbāf earthquake produced surface offset much smaller than expected from its seismic moment (Berberian et al., 2001). We will return to this issue in the section on the 14 March 1998 M_w 6.6 Fandoqā earthquake.

The calculated stress changes from both coseismic sudden movement in the upper crust and the time-dependent, viscous relaxation of the lower crust and/or upper mantle following the earthquake (Nalbant et al., 2006) shows that a coseismic, Coulomb stress load of up to 2 bars was created by the 11 June 1981 M_w 6.6 Golbāf earthquake. Apparently, this stress load had a significant influence on the time of occurrence of the 28 July 1981 M_w 7.0 Sirch earthquake (Nalbant et al., 2006), which took place 47 days later.

13.16 THE 28 JULY 1981 M_W 7.0 SIRCH EARTHQUAKE

Official estimates of the death toll of this event vary between 846 and 1300. Twenty-five villages were completely destroyed and several more severely damaged in the remote desert area of the southeastern Iran. The town of Golbāf, severely damaged by the 11 June 1981 M_w 6.6 Golbāf earthquake, was completely destroyed on July 28, though few additional casualties were sustained as most people were living in tents by then (Figure 13.11).

The Sirch earthquake was associated with 65 km of complex and discontinuous surface rupture on both east- and west-dipping faults of the Gawk Fault System north of Zamānābād (Figure 13.11). The earthquake occurred 47 days after the 11 June 1981 M_w 6.6 Golbāf earthquake (Figure 13.10)

TABLE 13.10 ***P*** **and** ***SH*** **Body-Waveform Inversion Source Parameters of the 11 June 1981** M_w **6.6 Golbāf Earthquake (Berberian et al., 1984, 2001)**

Earthquake	Time (UTC)	Coordinate (N°–E°)[a]	Strike (°)	Dip (°)	Rake (°)	Slip Vector Azimuth	Centroid Depth (km)	M_o (Nm) ($\times 10^{18}$)	M_w
1981.06.01 S1	07:24	29.85–57.68	169	52	156	184	20	4.18	6.58
S2			182	88	198	182	12	5.3	

See also Figure 13.10.
[a]*Epicenter is from Engdahl et al. (2006).*

PLATE 13.23 Right-lateral reverse displacement of ridges across the zone of coseismic surface rupture of the Gowk fault system during the 28 July 1981 M_w 7.0 Sirch earthquake (see also Figure 13.11). Looking northeast (Berberian et al., 1984).

and its southernmost extent was about 6.5 km north of the northern end of the Golbāf's coseismic surface rupture. No single fault trace extended more than one-fifth of the length of the activated system and discontinuous fractures spread over a wide zone (Berberian et al., 1984, 2001). Displacement on the surface breaks was commonly distributed over a series of en-chelon or parallel faults rather than concentrated on a single fracture. The maximum displacements were observed in the north near Chāhār Farsakh, where 43 cm of right-lateral slip and 40 cm of uplift (east side up) caused by reverse motion were recorded. Right-lateral displacement was about 25 cm in the south (Figure 13.11; Plates 13.23 and 13.24).

The overall motion was right-lateral, oblique-reverse slip with individual fault surfaces appearing to show predominantly either strike-slip, or dip-slip displacement. Slip-vector measurements show that displacements varied within individual parallel, or en-echelon segments of the fault zone; commonly being a maximum in the middle and dying out toward the ends. Individual en-echelon segments trended about 35° clockwise from the general N160°E strike of the surface ruptures (Berberian et al., 1984, 2001; Berberian and Yeats, 1999). A similar pattern was also recorded in the 28 June 1966 M_w 6.4 Parkfield, California (Brown et al., 1967), 31 August 1968 M_w 7.1 Dasht-e Bayāz, Iran (Tchalenko and Ambraseys, 1970; Tchalenko and Berberian, 1975), and 9 April 1968 M_w 6.8 Borrego Mountain (Clark, 1972) earthquakes. Nearly all significant vertical displacements occurred at locations where there was clear evidence of vertical displacement in the past,

PLATE 13.24 Right-lateral reverse displacement of a ridge across the zone of coseismic surface rupture of the Gowk fault system during the 28 July 1981 M_w 7.0 Sirch earthquake (see also Figure 13.11). The pencil indicates a right lateral displacement of about 17.5 cm. Looking northeast (Berberian et al., 1984).

generally in the form of uplifted beveled strata at the base of mountain slopes (Berberian et al., 1984, 2001; Berberian and Yeats, 1999).

The main ruptures in the 1981 Sirch earthquake probably occurred on different, deeper parts of the same fault system, producing only a minor reactivation of the shallower faults at the surface. Although the 1981 (M_w 7.0) and the 1998 (M_w 6.6) earthquakes (discussed later) apparently repeatedly ruptured parts of the same Gowk fault system at the surface with a short 17-year interval, these earthquakes had very different rupture characteristics (Figure 13.11). The regional kinematics, which involve oblique, right-lateral, and convergent motion, are evidently achieved by a complex configuration of faults with normal, reverse, and strike-slip components. Some of the complexity at the surface may be related to a ramp-and-flat fault geometry at depth, but could also be related to the large topographic contrast of ~2000 m across the Gowk fault system, which separates the high Kermān plateau from the low Dasht-e Lut desert (Berberian et al., 2001).

Following the 28 July Sirch earthquake, additional vertical motion of 1–2 cm was observed on the Golbāf section of the Gowk Fault since it had moved in the 11 June shock. If this happened during the 28 July shock and not as postseismic creep in the interval between the two mainshocks, then the total fault length in the 28 July Sirch earthquake could be considered 83 km rather than 65 km (Berberian et al., 1984, 2001).

Apparently, the almost pure right-lateral strike-slip faulting earthquake experienced 3.3-m average slip by using an empirical relation of displacement to fault length ratio (5×10^{-5}) (Berberian et al., 2001). The slip–seismic moment relation (Kanamori and Anderson, 1975) gives a 2.7 m average slip (Nalbant et al., 2006). The 6.6 km of unbroken surface fault that was left between the 11 June 1981 M_w 6.6 Golbāf and the 28 July 1981 M_w 7.0 Sirch earthquakes (the Zamānābād gap; Figure 13.11) is explained by a strong Coulomb stress decrease over most of the fault segment. Subsequent positive coseismic stress may have overcome this stress decrease, permitting the occurrence of the 14 March 1998 M_w 6.6 Fandoqā earthquake (Nalbant et al., 2006).

Without presenting an earthquake fault map, 'Adeli (1982) stated that the earthquake was associated with a fresh surface normal faulting with maximum vertical displacement of about 1 m. Gheitānchi (1999) presented the fault data from Berberian et al. (1984) without citing any reference. Finally, Zāre' and Hamzehloo (2004) stated that the 1981 Sirch earthquake started as a "*secondary faulting*" along the N–S Gowk fault or was triggered by activation of the Gowk fault in the hidden continuation of the NW–SE Kuhbanān fault, in their intersection zone. This strange statement cannot be warranted.

13.16.1 Seismology

Seismic parameters of the earthquake are summarized in Table 13.11. The earthquake produced surface offset that was much smaller than expected from its seismic moment (Berberian et al., 2001). We will return to this issue in the 1998 Fandoqā earthquake.

13.17 THE 20 NOVEMBER 1989 M_W 5.8 SOUTH GOLBĀF EARTHQUAKE

Since there was no inhabitation in the remote desert epicentral region of the earthquake, casualty figures and damage were not considerable and the meizoseismal area of the earthquakes, especially to the south, is not constrained. The earthquake killed 4 and injured 45 people. Two sets of fresh surface ruptures were developed along the Gowk fault system, south and southeast of Golbāf (Figures 13.10 and 13.11). As with the Golbāf and Sirch earthquakes of 1981 (discussed above), two faults on either sides of the South Golbāf Depression were reactivated. The surface rupture on the western side of the depression extended for 11 km (Plate 13.25) and showed 1-cm vertical versus 4-mm right-lateral motion (Figures 13.10 and 13.11). The width of the

TABLE 13.11 *P* and *SH* Body-Waveform Inversion Source Parameters of the 28 July 1981 M_w 7.0 Sirch Earthquake (Berberian et al., 1984, 2001)

Earthquake	Time (UTC)	Coordinate (N°–E°)[a]	Strike (°)	Dip (°)	Rake (°)	Slip Vector Azimuth	Centroid Depth (km)	M_o (Nm) ($\times 10^{18}$)	M_w
1981.07.28	17:22	29.97–57.76	177	69	194	176	18	36.69	7.0

See Figure 13.11.
[a]*Epicenter is from Engdahl et al. (2006).*

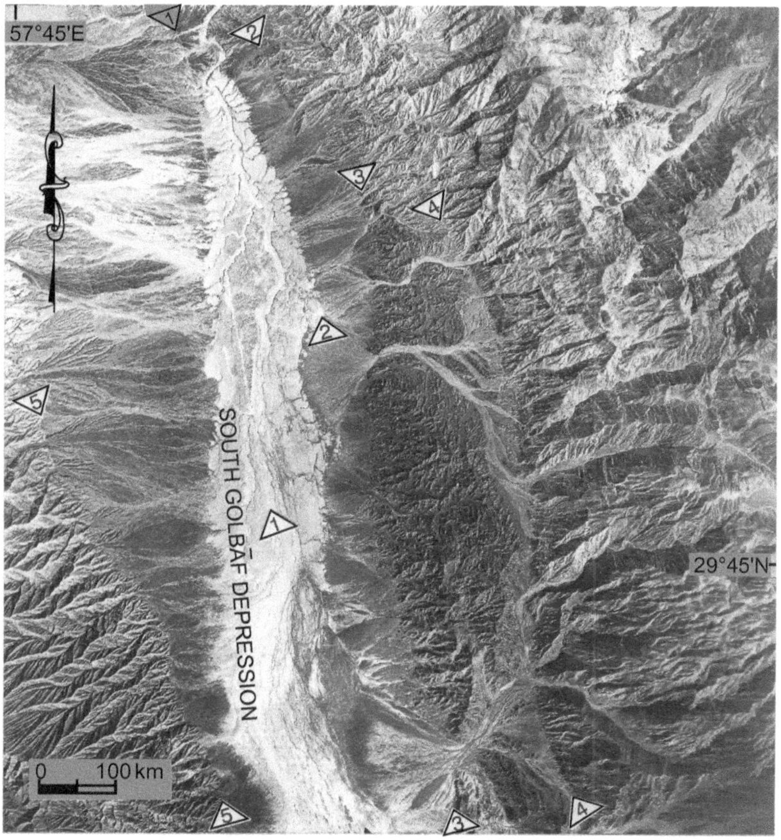

PLATE 13.25 Aerial photograph of the active fault features of the Gowk fault system at the South Golbāf playa, the meizoseismal area of the 20 November 1989 M_w 5.8 South Golbāf earthquake. See Figure 13.10 for the fault and seismicity map of the area. The central-valley west-dipping (No. 1) and east-dipping (s) segments of the Gowk fault system (3 and 4) can be seen. The white patch is the Holocene playa deposits cut by the west-dipping fault (1) and river beds. Only faults in the immediate vicinity of the playa were reactivated during the 1989 earthquake (1 and 2). No surface rupture was detected along the faults further east or west (3–5). *Photograph No. 39918, 19 September 1956 (Worldwide Aerial Surveys, Inc., Project 158; Cartographic Department of the Imperial Government of Iran, Joint Operations Graphics; Berberian and Qorashi, 1994).*

deformed zone along the western fault was up to 11 m with eight parallel hairy fault lines (Berberian and Qorashi, 1994).

The surface rupture of the eastern side was only 8 km long (Berberian and Qorashi, 1994). It should be noted that this fault segment had moved during the 11 June 1981 M_w 6.6 Golbāf earthquake (Berberian et al., 1984). The hairy surface ruptures of the eastern fault appeared to be more eroded than the western fault when the area was visited a week after the 1989 mainshock. However, it should be noted that a day after the 20 November 1989 earthquake, there was rainfall in the epicentral area (Berberian and Qorashi, 1994).

The earthquakes were also associated with coseismic folding (Berberian and Qorashi, 1994). Observation at a natural trench exposure across the surface trace of the western Gowk fault in the Holocene playa deposits at the western edge of the South Golbāf depression (Plate 13.25) showed elastic folding of the horizontally bedded Holocene clay deposits in the earthquake fault zone (Plates 13.26 and 13.27; Figure 13.10). The exposure revealed both faulting (with splayed fresh ruptures and minimum surface displacement) and folding (flexural-slip faulting on bedding planes developed in the Holocene playa deposits above the active fault tip). The clay layers dip 15–20°E toward the playa at the southern part of the western Gowk fault zone. It is clear in this case that surface folding of the horizontal Holocene clay is intimately related to earthquake faulting and could be explained by repeated motions on the steeply dipping west Golbāf fault during several historic earthquakes (Berberian and Qorashi, 1994).

PLATE 13.26 Surface deformation associated with the 20 November 1989 M_w 5.8 South Golbāf earthquake along the Gowk fault system. See Figure 13.10 for the fault and seismicity map of the area. Looking north.

PLATE 13.27 Surface deformation associated with the 20 November 1989 M_w 5.8 South Golbāf earthquake along the Gowk fault system. See Figure 13.10 for the fault and seismicity map of the area. Surface ruptures together with unknown amount of coseismic folding and flexural-slip faulting (bedding-plane slip with thrust mechanism) above the upper tips of the Gowk fault developed during the 1989 earthquake. Looking north (Berberian and Qorashi, 1994).

TABLE 13.12 *P* and *SH* Body-Waveform Inversion Source Parameters of the 20 November 1989 M_w 5.8 South Golbāf Earthquake (Berberian et al., 2001)

Earthquake	Time (UTC)	Coordinate (N°–E°)[a]	Strike (°)	Dip (°)	Rake (°)	Slip Vector Azimuth	Centroid Depth (km)	M_o (Nm) ($\times 10^{18}$)	M_w
1989.11.20	04:19	29.89–57.71	145	69	188	142	10	0.70	5.83

See also Figures 13.10 and 13.11.
[a]*Epicenter is from Engdahl et al. (2006).*

13.17.1 Seismology

An event of this size might be expected to occur on a fault about 8 km long that slipped about 0.4 m. However, the depth is not well constrained by the seismic data, and whether the small surface ruptures can be attributed to a deeper centroid or to the unconsolidated playa deposits in which most of the ruptures were seen remains uncertain (Berberian et al., 2001) (Table 13.12).

13.18 THE 20 JUNE 1990 M_W 7.3 RUDBĀR EARTHQUAKE

The earthquake killed about 40,000 people (according to official reports on July 28; later it was lowered to 13,000), injured 60,000, left more than 500,000 homeless, destroyed three cities (Rudbār, Manjil, and Lowshān) and 700 villages, and damaged another 300 villages in the area southwest of the Caspian Sea in the High-Alborz Mountains. Three discontinuous fault segments were mapped after the earthquake (Figure 13.12) along a previously unrecognized fault of 80 km length, passing north of the city of Rudbār (Berberian et al., 1992; Berberian and Walker, 2010). The 80-km-long fault was not mapped on the 1:250,000 geological maps of the area published prior to the event (Stocklin and Eftekhārnezhād, 1969; Annells et al., 1985). The western portion of the fault, west of the Sefidrud deep gorge, was also not shown on the 1:100,000-scale post-earthquake geological map of the area (Nazari and Salāmati, 1998), while only about 15 km of the eastern end of the fault is marked on the neighboring map (Ghalamghāsh and Rashid, 2002).

The Rudbār coseismic surface fault has a strike of N95°–120°E with principal left-lateral, strike-slip motion on sub-vertical fault segments (Figure 13.12; Plates 13.28 and 13.29). Maximum observed surface displacements (at the locations visited) were 100 cm left-lateral and 120 cm vertical (Berberian et al., 1992; Berberian and Walker, 2010). The heavily dense High-Alborz jungle of the region, which has no access road, was not visited. Different field interpretations and many speculative coseismic fault maps have been provided by several incompatible field reports, which has caused confusion in almost all the reports published since 1990. It is, therefore, necessary to comment on the misreported coseismic surface ruptures entered in the literature.

Mo'infar and Nāderzādeh (1990) reported 85 km of coseismic, "right-lateral" surface faulting with 20-cm right-lateral and 50-cm of vertical displacements along a straight trend of N112°E, partly covering the Manjil thrust introduced by Berberian and Qorashi (1984). Their speculative coseismic surface fault is located 12.5 km to the south of the actual coseismic surface rupture of the 1990 earthquake in the west, and 10 km to the south passing by the Sefidrud dam axis. They also mapped a questionable fault trace of 12 km length, striking N152°E in the area of Baklor to the northwest. Meanwhile, Dashti (1990) introduced several major longitudinal and transverse active faults in a crisscrossed pattern, which allegedly moved during the earthquake. Such a seismic fault pattern is not warranted (Berberian and Walker, 2010).

Following Mo'infar and Naderzadeh (1990), Zāre' (1991a,b), Zāre' and Mo'infar (1993, 1994), Maheri (1991), Niazi and Bozorgnia (1992), Haghshenas (1998), and many others, introduced "two" coseismic surface fault sets for the 1990 earthquake: (i) the "*Harzevil fault*" along the Sefidrud dam axis (the Manjil thrust of Berberian and Qorashi, 1984; see Berberian et al., 1992), with no data regarding the length and amount of displacements, and (ii) the "*earthquake fault of the Sefid Rud Dam,*" with 3-km length and 10–70-cm displacement (Zāre', 1991a) or 5-km long with 0.3-m left-lateral and 0.5-m upward movement of the northern part at the left abutment hill above of the Sefidrud Dam (Zāre', 1991b) (Berberian and Walker, 2010).

Ramazi (1991) correctly reported 15 km of the eastern, Zardgeli, segment of the faulting with maximum left-lateral displacement of 100 cm and vertical displacement of 50 cm. In their later publication, Zāre' and Mo'infar (1993,

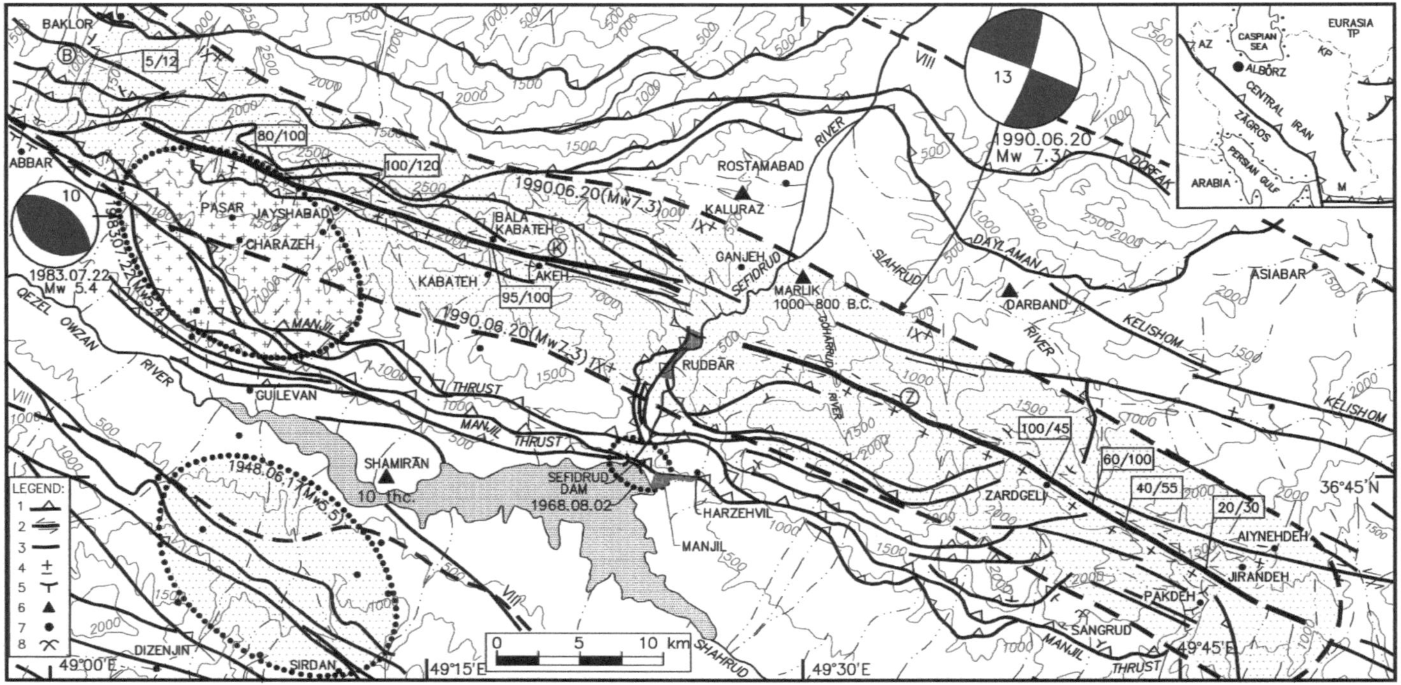

FIGURE 13.12 Seismicity, fault, and topographic map of the Rudbār area in the western Alborz Mountains. Fault and other symbols as in Figure 11.1. The 20 June 1990 M_w 7.3 Rudbār earthquake surface ruptures are drawn by a thick line. Coseismic horizontal/vertical displacements (in cm) are given in boxes inside the isoseismal lines (marked by broken lines). Three main segments of the 1990 coseismic surface rupture are labeled by encircled letters (from the NW to the SE) as B: Baklor, K: Kabateh, and Z: Zardgeli. Focal mechanism determined by waveform modeling with source parameters for the 20 June 1990 Rudbār mainshock are from Gao and Wallace (1995) and from Priestley et al. (1994) for the 22 July 1983 Charazeh event. The creation of the unique Sefidrud River gorge, which is the only river crossing the ~1000-km length of the Alborz Mountains and acted as a major barrier during the 1990 earthquake, may have been facilitated by the segment boundary in the NNE–SSW faulting in the Rudbār area. Meizoseismal areas of other earthquakes are marked by dotted lines. Topographic contour lines are in meters above mean sea level. See also Plates 13.28–13.31. *Modified from Berberian and Walker (2010), and Berberian et al., (1992).*

PLATE 13.28 North-facing coseismic fault scarp of the 20 June 1990 M_w 7.3 Rudbār earthquake along the Rudbār left-lateral strike-slip fault in the area between Kelās and Jayshābād (36.90°N–49.16°E, +1951 m) in the High-Alborz. The southern block (right) is raised against the general higher topography of the High-Alborz (left/north). Left-lateral displacement of 80 cm and vertical slip of 100 cm were recorded at this locality. See also Figure 13.12. View to the southeast. *Photographed in June 1990.*

PLATE 13.29 The same as Plate 13.26, photographed in September 2000, showing 10 years of erosion along the fault scarp. The left-lateral displacement could not be measured 10 years after the earthquake (Berberian and Walker, 2010).

1994) and Zāre' (1995) speculated "*six coseismic faults*" distributed to the north and the south, and collectively named them as the "*Earthquake Origin Zone*" (Berberian and Walker, 2010).

Ishihara et al. (1992) combined the speculated "*northern*" and "*southern*" faults together and introduced a single arcuate ground rupture of 100 km long. It extends from Baklor in the northwest to Guilévān in the south with a strike of N150°E, and then curves to an almost E–W direction from Guilévān to the Sefidrud Dam and then to Pākdeh in the east. Ishahara et al. (1992) reported maximum horizontal and vertical coseismic displacements of 20 and 50 cm, respectively. Finally, Hamzehloo et al. (1997) suggested that the 1990 Rudbār earthquake was caused by the Sefidrud Dam impoundment (Berberian and Walker, 2010).

The striking features of the surface ruptures of this earthquake (Berberian et al., 1992; Berberian and Walker, 2010; Figure 13.12) were as follows:

i. The 1990 coseismic surface rupture was composed of three main, right-stepping, en-echelon, left-lateral strike-slip fault segments of the Baklor, Kabateh (west of the Sefidrud gorge surface gap), and Zardgeli (east of the Sefidrud gorge surface gap), with nearly vertical dip. No single fault segment extended more than half of the total length of the system.

ii. The coseismic surface fault did not cross the Sefidrud deep gorge, its elevated Quaternary terraces, or the highway running along the gorge, and no surface deformation was observed during the mainshock and/or the aftershock activities in the gorge (Figure 13.12). The Sefidrud gorge, where the mighty Sefirdrud River runs, may, therefore, be located at an important discontinuity along Rudbār coseismic fault line. Slight vertical displacement observed on the highway near the tunnel, as well as in the tunnel near the Sefidrud Dam, was caused by landslides and not by tectonic fault movement.

iii. The surface rupture virtually followed the drainage divide of the "High-Alborz" very close to the peak of the western "High-Alborz," and mostly at or above elevation +2000 m amsl.

iv. The coseismic vertical displacements along the total length of the earthquake fault were consistently down to the north and northeast, opposite to the existing topography of the "High-Alborz." The coseismic faulting thus had a tendency to reverse the existing topography of the "High-Alborz" Mountains (Plates 13.28–13.30).

v. A large portion of the strike-slip deformation occurred in a narrow zone of about 6–10 m wide (Plate 13.28).

vi. The amount of vertical displacement was usually more than the amount of the left-lateral displacement in the locations visited (Plate 13.30).

vii. In addition to the main left-lateral slip, the western fault segments, the Baklor segment, and the western portion of the Kabateh segment showed local surface evidence of slight normal-slip component in the form of large open fissures (Plate 13.30). In contrast, the eastern segment of the Kabateh and most of the Zardgeli segments were associated with slight reverse-slip component in the form of push-ups (Plate 13.31).

viii. The fault plane was nearly vertical, with a slight dip toward the SSW at the surface (Plate 13.30).

ix. The subtle topographic expression of the Rudbār fault suggests that the fault does not move often enough to exert much influence on the geomorphology of the soft rocks (Berberian and Walker, 2010).

x. The mainshock was also associated with bedding plane slip with thrust mechanism showing coseismic flexural-slip folding in the High-Alborz.

PLATE 13.30 Coseismic surface rupture of the 20 June 1990 M_w 7.3 Rudbār earthquake with NNE-facing scarp along the Kabateh segment of the Rudbār left-lateral strike-slip fault, near Jayshābād (west of the Kelās ravine). See also Figure 13.12. The southern block (right) is uplifted against the northern (left) higher topography of the High-Alborz. Left-lateral displacement of 100 cm and vertical motion of 120 cm were recorded at this locality. View to the east. *Photographed in June 1990 (Berberian and Walker, 2010).*

13.18.1 The Rudbār Seismic Gap

The 1990 Rudbār earthquake struck the Rudbār–Āqdāgh seismic gap, where the surface expression of active faults was less developed than in other regions (Berberian et al., 1992). The gap was located to the immediate west of the 15 August 1485 M_s ~7.3 Upper Polrud earthquake along the Kelishom left-lateral strike-slip fault (Figure 13.13), a fault subparallel to the Rudbār fault and located to its northeast (Berberian and Walker, 2010). The 1990 Rudbār earthquake filled the Rudbār gap; however, the Āqdāgh seismic gap to the northwest remains unruptured, although the region is characterized by moderate seismicity since 1900.

Both the Rudbār and Kelishom left-lateral strike-slip faults have a similar setting in mechanism and orientation and are about 10 km apart (Figure 13.13). Considering the entirety of the Alborz Mountains as similar in response to oblique-slip convergence with uniform slip rate along strike, the absence of earthquakes along the Rudbār fault is offered as evidence that the Rudbār earthquake filled a seismic gap. This conclusion is strengthened by the lack of geomorphic evidence for activity on the Rudbār source fault, as seen on pre-1990 aerial photographs.

PLATE 13.31 A series of push-ups (mole-tracks) formed by a reverse or thrust component associated with a left-lateral strike-slip faulting along the Zardgeli segment of the Rudbār fault, near Zardgeli, during the 20 June 1990 M_w 7.3 Rudbār earthquake. See also Figure 13.12. The southern block (right) is raised and upthrust relative to the northern (left) block, against the existing topography of the High-Alborz. Left-lateral displacement of 60 cm and vertical motion of 100 cm were recorded in this area. *Photograph taken in June 1990 (Berberian and Walker, 2010).*

13.18.2 Seismicity

The distance between the 1990 mainshock epicenters relocated by Engdahl et al. (2006) and Gao and Wallace (1995) is a few kilometers, but both locations are located 10 km to the northeast of the nearly vertical coseismic surface rupture. The 10–15 km systematic distance between the post-1970s epicentral locations of large-magnitude earthquakes on vertical coseismic surface faults has been typical for large earthquakes in Iran (Ambraseys, 1978a, 2001; Berberian, 1979c; Berberian et al, 1992). We speculate that the

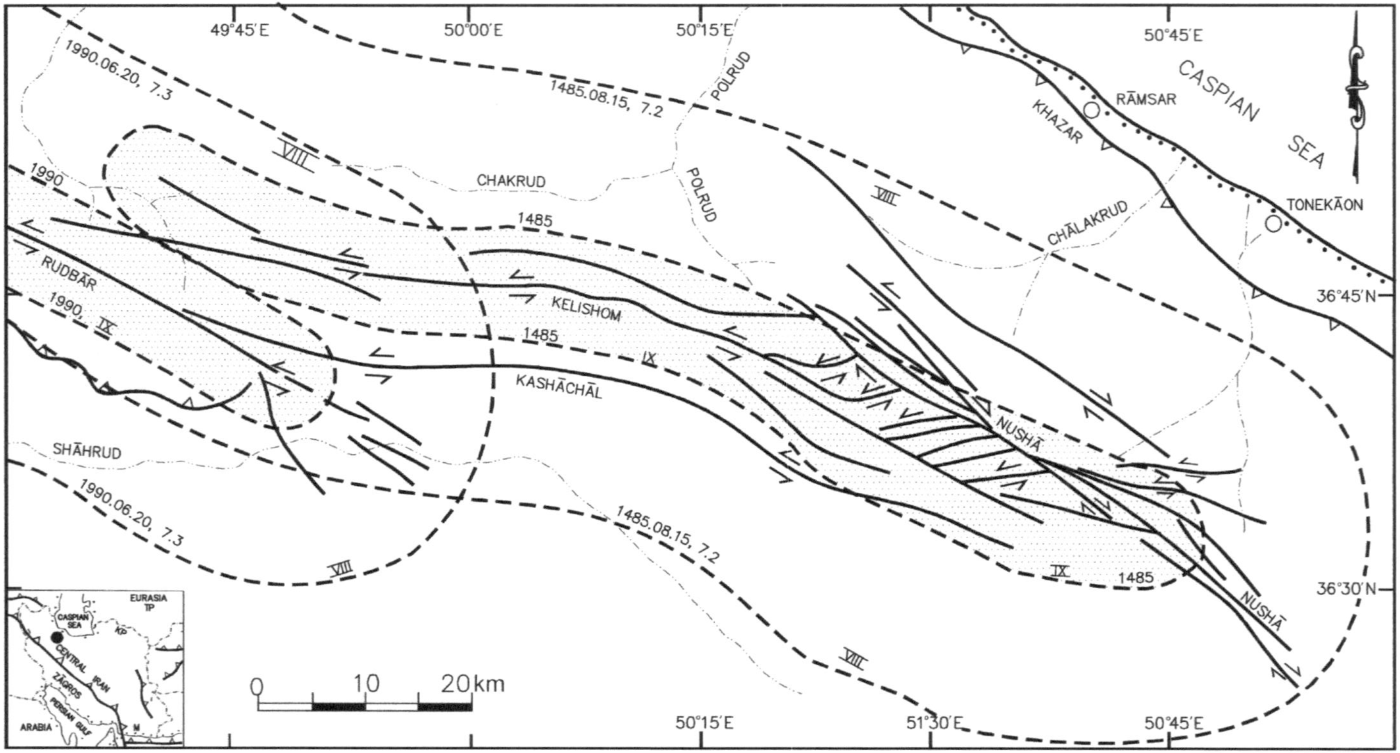

FIGURE 13.13 Estimated meizoseismal area of the 15 August 1485 M_s ~7.2 Upper Polrud earthquake along the Kelishom left-lateral strike-slip fault to the east of the 1990 M_w 7.3 Rudbār earthquake meizoseismal area, southwest of the Caspian Sea in northwest Iran. Fault and other symbols as in Figure 11.1. After 505 years, the 1990 earthquake filled the seismic gap to the west of the 1485 earthquake epicentral area. *Modified from Berberian and Walker (2010).*

systematic bias in the Iranian teleseismic earthquake locations is mainly introduced by deviations in travel times from globally averaged velocity models and asymmetric/inadequate azimuthal distribution of recording stations. The typical location error values of 10–15 km for the post 1970s large-magnitude Iranian earthquakes have precluded correct identification of the causative faults even in 1990.

Assuming a fault length of 80 km, depth of 15 km, and rigidity of 3×10^{10} N/m^2, the calculated seismic moment is able to account for an average seismologically determined slip of 240 cm, which is much more than the observed displacements in the locations visited in the field (Figure 13.12) (Berberian et al., 1992; Berberian and Walker, 2010). This discrepancy could be due to factors such as (i) the fault and the displacements were not thoroughly mapped immediately after the earthquake, especially at the heavily wooded High-Zāgros with no access roads and helicopter facility; (ii) the rupture might have propagated deeper than 15 km, though the microseismicity data of Tatar and Hatzfeld (2008) suggest that the majority of seismic deformation is limited to <16 km depth; and (iii) not all the slip reached the surface. Note that, for instance, the 28 June 1992 Landers, California, earthquake, which was of similar magnitude of M_w 7.3 and produced the same amount of surface rupture [80 km], with an 8-km source depth, accommodated 6-m right-lateral and 2.5-km vertical motion (Arrowsmith and Rhodes, 1994; Johnson et al., 1994).

In their inverted source model, Campos et al. (1994) and Virieux et al. (1994) defined nine subevents over the first 25 s with lateral heterogeneities in the source region. On the other hand, Choy and Zednik (1997) modeled a tiny precursory subevent followed after 20 s by a series of four subevents with a left-lateral, strike-slip mechanism, depth ranging from 10 to 15 km.

The earthquake was followed by several teleseismically recorded aftershocks showing both left-lateral, strike-slip, and reverse mechanisms (Berberian et al., 1992; Gao and Wallace, 1995; Berberian and Walker, 2010). Locally recorded aftershocks were studied by Tsukuda et al. (1991), Eslami (1992), and Hamzeloo et al. (1997). The relocated aftershocks defined a narrow belt oriented N125°E, with focal depths ranging from 1 to 16 km depth. In the southeast, the aftershocks are shifted to the NE from the Zardgeli segment of the surface rupture, but to the northwest, some aftershocks were located near the Kabateh segment (Figure 13.12).

About 8 years after the Rudbār earthquake, a microearthquake survey utilizing 30 portable seismographs was deployed along the fault and recorded 410 events (Tatar, 2001; Tatar and Hatzfeld, 2008). The microseismicity forms a wide zone of seismic activity trending NW–SE. As with the aftershock sequence, the blurred microearthquake activity of 1998 was clustered in (i) the Sefidrud gorge coseismic, surface fault-gap area and (ii) the southeastern termination of the 1990 surface rupture near Jirandeh (Figure 13.12). The majority of focal depths ranged from 8 to 16 km (Tatar and Hatzfeld, 2008). Very little activity was recorded along the Kabateh segment west of

TABLE 13.13 ***P*** **and** ***SH*** **Body-Waveform Inversion Source Parameters of the 20 June 1990** M_w **7.3 Rudbār Earthquake (Berberian et al., 1992; Gao and Wallace, 1995; also see Berberian and Walker, 2010)**

Earthquake	Time (UTC)	Coordinate (N°–E°)[a]	Strike (°)	Dip (°)	Rake (°)	Slip Vector Azimuth	Centroid Depth (km)	M_o (Nm) ($\times 10^{20}$)	M_w
1990.06.20	21:00	36.99–49.22	288	88	−9	112	13	1.4	7.3

See Figure 13.12.
[a]*Epicenter is from Engdahl et al. (2006).*

the Sefidrud gorge, and almost no activity was recorded along the northwestern termination of the fault (Berberian and Walker, 2010). The Baklor and the western half of the Kabateh segments were not covered by any seismic station. For the source parameters of the main shock see Table 13.13.

13.19 THE 6 NOVEMBER 1990 M_W 6.4 EAST FURG EARTHQUAKE

Preceeded by a foreshock approximately 15 s before the mainshock, the earthquake destroyed several villages in the remote Furg area of the High-Zāgros (west of Hajiābād and southeast of Dārāb). At least 22 people were killed, 100 were injured, 1535 houses and 20 schools were destroyed, 1445 houses suffered about 50% damage, and 21,000 people became homeless in 18 villages (Figure 13.14). The earthquake was associated with 15 km of E–W thrust faulting dipping 30°N with a maximum vertical displacement of 1.5 m and an average displacement of 1.3 m at the surface (Raisi, 1991; Zamani and Raisi, 1991; Walker et al., 2005a) along the southeastern section of the High-Zāgros reverse master fault (Berberian, 1995) (Figure 13.14). A few meters of ground fracture trending NE–SW in Mohammadābād village, east southeast of Furg, was also reported, however, the nature of the fracture was not specified (figure 3.3 in Nateghi Elāhi et al., 1991).

Berberian (1995) mentioned that despite the decoupling of the basement structures from the top sedimentary cover in the Zāgros Mountains and lack of coseismic surface ruptures along the Zāgros fold-and-thrust mountain belt (stated in Berberian, 1976a, 1977a; Berberian and Tchalenko, 1976b, c; Berberian and Navai, 1977, 1978; Berberian and Papastamatiou, 1978; Berberian, 1981a), medium- to large-magnitude earthquakes along the deep-seated, Zāgros master reverse faults, such as the High-Zāgros (in this case), Zāgros Mountain Front, and Zāgros Foredeep reverse faults, as well as the transverse right-lateral strike-slip faults in the Zagros (Kāzerun, Sabzpushān, Karéhbas, and Sarvestān), can propagate to the surface.

13.19.1 Seismology

The main shock of 18:46 UTC was followed by a strong aftershock on 19:30 UTC (M_s 5.7) on the same day. *P* and *SH* body-waveform inversion of the mainshock (Walker et al., 2005a) indicate that the main shock was on a thrust fault, striking approximately E–W and dipping 33°N with shallow centroid depth of 5 km (Table 13.14). The source time function suggests a rupture of between 4 and 8 s that, at a typical rupture velocity of ~3.5 km/s, would require a fault length of about 14–28 km. If the fault dips at about 30°N, it would imply that the magnitude of slip on the fault plane at the surface was ~2.6 m. With the seismic moment of 5.1×10^{18} Nm, a fault length of 15 km, down-dip width of 15 km (10 km depth at 30° dip), and rigidity modulus of 3×10^{10} N/m^2, an average slip of 0.57 m is expected, which is smaller than what was observed at the surface. This may imply that slip was relatively nonuniform and concentrated near the surface (Walker et al., 2005a).

The southeastern portion of the High-Zāgros reverse fault was associated with earthquakes on 6 November 1962 (5.4) Gahkom, 29 July 1963 (5.2) Gahkom, 27 August 1964 (*mb* 5.3), 21 June 1965 (M_w 6.1) Sarchāhān, 12 April

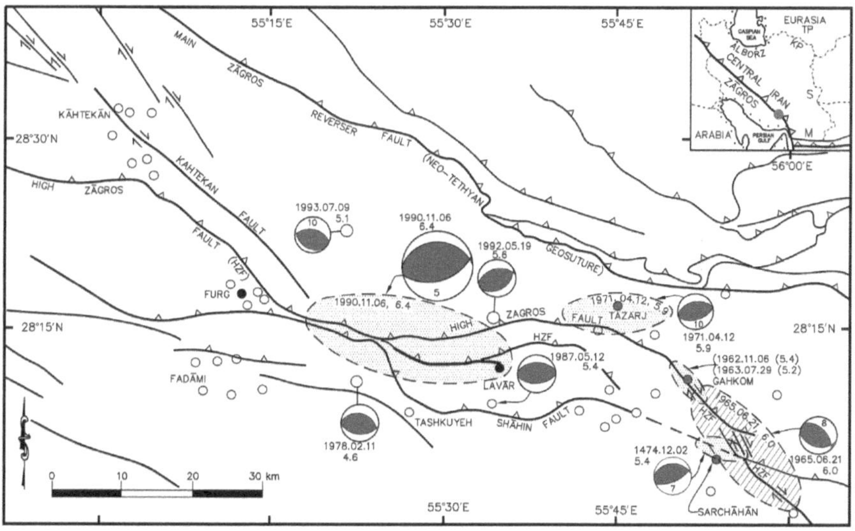

FIGURE 13.14 Meizoseismal areas of the earthquakes along the High-Zāgros fault in southeast Iran. Fault and other symbols as in Figure 11.1. The coseismic faulting associated with the 6 November 1990 M_w 6.5 East Furg earthquake (Raisi, 1991; Zamani and Raisi, 1991; Walker et al., 2005a) is marked by a thicker line in the area west of Lavār. Focal mechanism determined by waveform modeling with source parameters for the 6 November 1990 M_w 6.5 East Furg earthquakes (Walker et al., 2005a). Fault-plane solutions for the other earthquake: 12 April 1971 M_w 5.9 Tazarj (Ni and Barazangi, 1986); 2 December 1974 M_w 5.2 (Baker, 1993); 11 February 1978 (Jackson and McKenzie, 1984); 12 May 1987 M_w 5.5, and 19 May 1992 M_w 5.6 (best-double-couple HRVD CMT solution); 9 July 1993 M_w 5.1 (Talebian and Jackson, 2004). *Modified from Berberian and Tchalenko (1976c) and Berberian (1995).*

TABLE 13.14 ***P*** **and** ***SH*** **Body-Waveform Inversion Source Parameters of the 6 November 1990 East Furg** M_W **6.4 Earthquake**

Earthquake	Time (UTC)	Coordinate (N°–E°)[a]	Strike (°)	Dip (°)	Rake (°)	Centroid Depth (km)	M_0 (Nm) ($\times 10^{18}$)	M_w	Source
1990.11.06	18:45	28.24–55.46	268	33	90	05	5.17	6.4	Walker et al. (2005a)
			275	30	101	07	7.1	6.5	Maggi et al. (2000a,b)
			274	37	107	15 (fixed)		6.13	Harvard CMT

See also Figure 13.14.
[a]*Epicenter is from Engdahl et al. (2006).*

1971 (M_w 5.9) Tazarj, 2 December 1974 (5.4) Sārchāhān, 11 February 1978 (4.6), 19 May 1992 (M_w 5.6), and possibly two additional events of 12 May 1987 (M_w 5.4) and 19 May 1992 (M_w 5.6) East Furg, and 9 July 1993 (M_w 5.1) North Furg (Figure 13.14) (Berberian and Tchalenko, 1976c; Berberian, 1995).

13.19.2 Oral Earthquake Hazard Warning

Despite a lack of a pre-1900 recorded earthquake history, the place names "Kāhtekān" ["Straw Shake"; NW of Furg, see Figure 13.14] and "Tang-e Chāk Chāk" ["Sliced or Fractured Gorge"] may indicate ancient earthquakes only preserved in an oral hazard warning (see Chapter 1 for details).

13.20 THE 23 FEBRUARY 1994 M_W 6.2 SEFIDĀBEH EARTHQUAKE

The Sefidābeh earthquake sequence of 22 February–2 March 1994 took place in a sparsely populated remote desert region of the Sistān Province of East Iran (see Figure 15.1 in Chapter 15). The main shock of 23 February was blamed for six deaths (three in Sefidābeh and three in Kalāt-e Hāji Allāhdād); 10 people were injured, and an estimated 300 buildings damaged (200 were destroyed). The area is sparsely populated because of the arid climate, scarce vegetation, and its remote location. These factors together with the time of the mainshock (11:34 local time) resulted in a very low death toll for a M_w ~6.2 earthquake in the Iranian standard (see Introduction).

Based on Zāre' (1995), Ambraseys and Jackson (1998) reported a 4-km-long coseismic surface rupture with thrust and left-lateral mechanism and 30-cm vertical displacement. The earthquake was not associated with coseismic surface rupture but produced bedding-plane slips due to flexural-slip folding above the Sefidābeh blind thrust dipping SW (Berberian et al., 2000b).

13.21 THE 4 FEBRUARY 1997 M_W 6.4 NĀVEH EARTHQUAKE

The mainshock killed 88 people (official figure) and 4000 livestock, injured 1950 people, completely destroyed 4200 houses in five villages (Nāveh, Qezelqān, Surk, Sheikh, and Yekeh Shākh), and damaged about 11,000 houses in more than 170 villages in the Central Kopeh Dāgh Shear Zone of northeast Iran (Figure 13.15). Damage was estimated at more than US$1000 million (175 billion Iranian riāls; official figure).

The earthquake, which was preceded by a foreshock of M_w 5.4 about 40 min earlier, produced a ground rupture about 15 km long with a 0.5–1-m, right-lateral, strike-slip motion. North of Nāveh, en-echelon, left-stepping ruptures, up to 1 m high, ran along the mountainside (37.425°N–57.237°E) for about

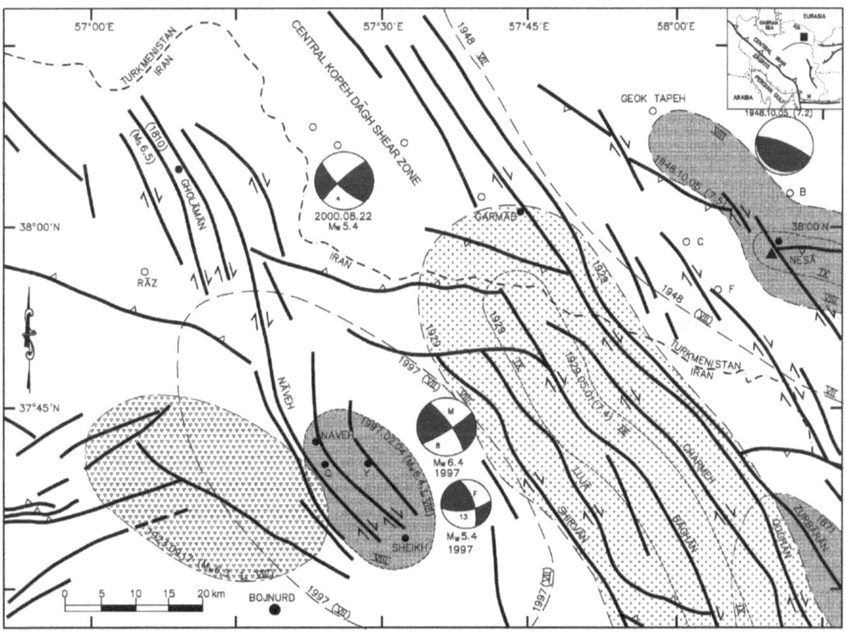

FIGURE 13.15 Meizoseismal areas of the earthquakes in the Central Kopeh Dāgh Shear Zone (CKDSZ) of northeastern Iran. Fault and other symbols as in Figure 11.1. The 4 February 1997 M_w 6.4 Nāveh earthquake was associated with surface rupture. Focal mechanism determined by waveform modeling with source parameters (Hollingsworth et al., 2006). The earthquake filled the gap between the 1923 North Bojnurd and 1929 Bāghān earthquakes. The fault-plane solution of the 5 October 1948 M_w 7.2 Ashkābād earthquake (McKenzie, 1972). Q, Qezelqān; Y, Yekeh Shāikh. *Modified from Tchalenko (1974), Afshar-Harb (1979), GSI (1986a), Berberian and Yeats (2001), Jackson et al. (2002), and Hollingsworth et al. (2006).*

2 km, and with an overall trend of about N340°E (Figure 13.15). The ruptures occurred on a steep slope and were accentuated by landslides in spots. Further south across the valley, between Qezelqān and Yekeh Shākh (37.390°N–57.275°E), villagers reported similar rupturing, directly along strike of that north of Nāveh (Hosseini and 'Asgari, 1997; Hollingsworth et al., 2006, 2007).

A 20-km surface faulting, from Nāveh to Sheikh villages, was reported by Ramazi et al. (1997), whereas Tatar et al. (1997) reported 9 km of surface faulting from Qezelqān to Qal'eh Jaq Kuchek villages.

The coseismic ground rupture took place along the Nāveh fault (Afshar-Harb, 1979; GSI, 1986a). The fault, with a general trend of NW–SE, is a right-lateral, strike-slip fault with a total length of about 25 km. The northwestern end of the fault changes its strike to a nearly N–S trend. The earthquake took place in the "Central Kopeh Dāgh Shear Zone" (see Chapter 16 for further discussion about the shear zone).

13.21.1 Earlier Earthquakes

Postdepositional soft sediment deformation structures were observed in the recent sedimentary deposits (exact age unknown) in the area north of the Bojnurd town. These structures may indicate seismically induced structures ("Seismites" of Seilacher, 1969, 1984) in the Bojnurd area. Study of the Bojnurd seismites could provide information about the earlier earthquakes of the area.

13.21.2 Seismology

Teleseismic *P* and *SH* waveform modeling indicates a nearly pure right-lateral, strike-slip mechanism (Jackson et al., 2002; Figure 13.15; Table 13.15). Apparently, the fault rupture initiated from the northwest and propagated to the southeast, causing an enhanced directivity effect resulting in an elevated level of destruction to the farther southeast (Hollingsworth et al., 2007).

TABLE 13.15 *P* and *SH* Body-Waveform Inversion Source Parameters of the 4 February 1997 M_w 6.4 Nāveh Earthquake (Jackson et al., 2002)

Earthquake	Time (UTC)	Coordinate (N°–E°)[a]	Strike (°)	Dip (°)	Rake (°)	Slip Vector Azimuth	Centroid Depth (km)	M_o (Nm) ($\times 10^{18}$)	M_w
1997.02.04	09:53	37.68–57.27	338	67	150	350	13		5.4
	10:37	37.72–57.31	326	75	173	328	8	5.75	6.4

See also Figure 13.15.
[a]*Epicenters are from Engdahl et al. (2006).*

13.22 THE 28 FEBRUARY 1997 M_W 6.0 SHIRĀN EARTHQUAKE

The earthquake shook the southeastern foothills of the Sabalān Quaternary volcano in the Ardebil province of northwestern Iran. The main shock killed 965 people (official figure; 300 were killed in Shirān village) and thousands of livestock, completely destroyed 30 villages, injured 2600 people, damaged about 13,500 houses in more than 110 villages in the epicentral area, and left 36,000 people homeless (Figure 13.16). Approximately 600 classrooms in 105 schools, 49 hospitals and health centers, and 19 telecommunication centers were completely destroyed and the water distribution systems in 30 villages were damaged beyond repair. Severe damage was caused to 347 km of secondary roads and electrical power lines. Damage was estimated at more than US$132 million (230 billion Iranian riāls; official figure). Hampering the rescue operation were heavy snow fall, landslides, rockfalls blocking the access roads, and the location of the villages in the mountainous areas, especially on the steep slopes of the Sabalān volcano (Figure 13.16).

No definite coseismic surface faulting was detected immediately after the earthquake as the area was covered by snow. Surficial secondary fractures, some associated with landslides, were developed in numerous places (Naiieri et al., 1997; Abbasi et al., 1998).

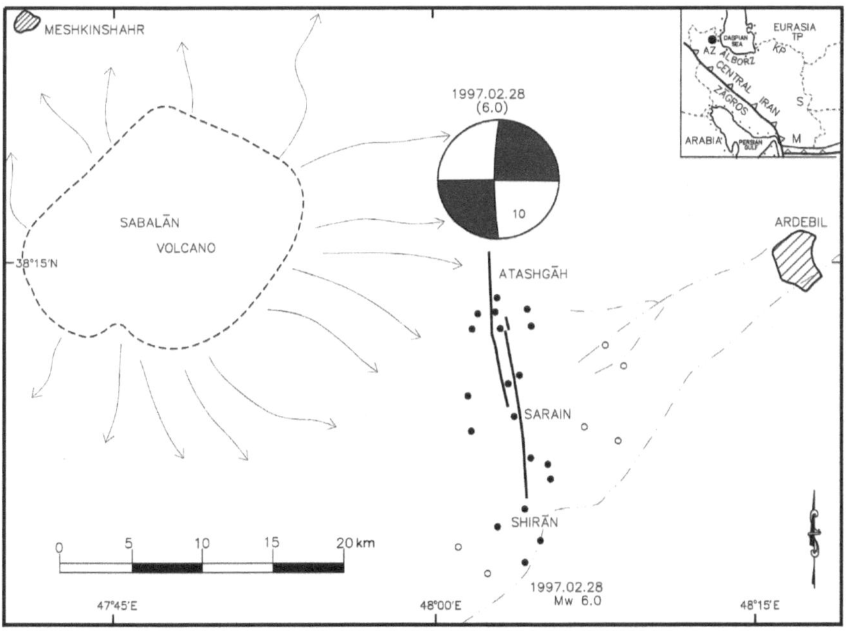

FIGURE 13.16 Meizoseismal area of the 28 February 1997 M_w 6.0 Shirān earthquake in the northwest of Iran and southeast of the Sabalān Volcano (with arrows showing the flow direction). Fault and other symbols as in Figure 11.1. Focal mechanism determined by waveform modeling with source parameters (Jackson et al., 2002).

Two left-stepping, nearly N–S fault scarps of at least 18 km length can be seen on the aerial photographs of the region (Nos. 3301 R-38 and 3337 R-39). The faults are about 700 m apart and a linear mini-graben is formed in between them. The western fault segment of about 9 km length starts from the area north of Ātashgāh (38°13′N–48°03′E, +1778 m amsl), passes the area immediate to the west of Arjestān (38°12′N–48°03′E, +1777 m amsl) and west of Bileh [Vilā] Darreh (38°10′N–48°03′E, +1801 m amsl), and dies out about 1.5 km south of the latter (Figure 13.16). The fault is in the contact between the Neogene andesitic basalt in the west and the old Quaternary tuff, volcanic ash, and lahar of the eastern foothills of the Sabalān Volcano.

The eastern fault segment of about 14 km in length starts from the area southeast of Arjestān, continues southward to Sar'ain hot spring (38°09′N–48°04′E, +1671 m amsl), then west of Varināb (38°07′N–48°04′E, +1678 m amsl), east of Shirān (38°05′N–48°03′E, +1619 m amsl), and dies out in the area south of Shirān (Figure 13.16). Several springs (including the Sar'ain hot spring) as well as numerous landslides have been developed along the fault escarpments.

Field investigation of July 1997 in selected areas (M. Qorashi and M. Talebian, personal communication, 13 July 1997) did not show fresh surface fracturing along these fault segments. However, the reported NNE–SSW ground fractures at Ātashgāh, Bileh [Vilā] Dareh, south Sar'ain, and Shirin Bulāq areas (Naiieri et al., 1997; Zar'ayan, 1997; 'Abbassi et al., 1998), accentuated by landslides in places, are aligned along the above-mentioned fault scarps (Figure 13.16).

13.22.1 Seismology

Teleseismic *P* and *SH* waveform modeling indicates a nearly pure, left-lateral strike-slip mechanism along an N–S plane (Jackson et al., 2002; Table 13.16; Figure 13.16). Copley et al. (2013) argued that absence of significant N–S left-lateral shear in GPS data from the region suggests that the earthquake had an E–W right-lateral mechanism. Damage distribution associated with this event does not support the E–W right-lateral motion (Figure 13.16).

TABLE 13.16 ***P*** **and** ***SH*** **Body-Waveform Inversion Source Parameters of the 28 February 1997** M_w **6.0 Shirān Earthquake (Jackson et al., 2002)**

Earthquake	Time (UTC)	Coordinate (N°–E°)[a]	Strike (°)	Dip (°)	Rake (°)	Centroid Depth (km)	M_o (Nm) ($\times 10^{18}$)	M_w
1997.02.28	12:57	38.10–48.07	273	89	−171	9	1.279	6.0

See also Figure 13.16.
[a] *Epicenter is from Engdahl et al. (2006).*

Nearly N–S faults have been mapped to the west and east of Ardebil (Figures 13.16 and 11.14).

13.23 THE 10 MAY 1997 M_W 7.2 ZIRKUH EARTHQUAKE

The main shock killed at least 1568 (official figure), injured 2600, completely destroyed 12,000 houses, and damaged 7000 more in 147 villages (Figures 13.17 and 13.18; Plate 13.32). The highest casualty count was in the Ābiz village, where 264 people were killed. The death toll and damage from the earthquake were relatively mild for a few reasons: this remote desert area close to the Iran–Afghanistan border has a low population density, most people were working in the field, and the main shock was preceded by a warning foreshock. Financial losses were estimated to be at least US$100 million (official estimate; 271 milliard Iranian riāls).

The earthquake produced 125 km of NNW–SSE right-lateral strike-slip surface faulting on the Ābiz right-lateral strike-slip fault northeast of the Sistān suture zone of eastern Iran (Berberian et al., 1999). This is the longest recorded surface rupture associated with an Iranian earthquake. The fault rupture was much larger than the 31 August 1968 (M_w 7.1) Dasht-e Bayāz and the 27 November 1979 (M_w 7.1) Koli earthquakes along the E–W Dasht-e Bayāz left-lateral strike-slip fault located in the same area (Figures 13.18 and 13.9).

The entire length of the Ābiz fault (Figure 13.18), including those segments that had ruptured during the 30 June 1936 M_w 6.0 Ābiz, 14 November 1979 M_w 6.6 Korizān, and 7 December 1979 M_w 5.9 Kalāt-e Shur earthquakes (Figure 13.9), ruptured during the 1997 event (Berberian et al., 1999). The 1936 earthquake produced about 12 km of surface rupture on the central segment of the Ābiz fault. The second event caused an additional 20 km of surface rupture on the fault, although about 15 km of the fault between the 1936 and 1979 surface ruptures did not rupture at the surface (Figure 13.9). The third event ruptured the remaining 15 km of the Ābiz fault up to its intersection with the left-lateral Dasht-e Bayāz fault (Figures 13.9 and 13.18). About 18 years later, the entire Ābiz fault length was ruptured. The three previous earthquakes together ruptured <50 km of the 125 km of the Ābiz fault system (Berberian et al., 1999).

The magnitude of the 1997 coseismic horizontal offset varied along the fault trace, but was typically 80–140 cm, with a maximum right-lateral strike-slip offset of around 230 cm (Plates 13.33–13.36). The maximum lateral displacements were recorded near the northern and southerns tips of the fault (Figure 13.9). The southern tip of the fault bends toward the southeast and shows a reverse mechanism (Plate 13.37). In addition, vertical displacements up to 90 cm were measured (Berberian et al., 1999) (Figures 13.9 and 13.18).

The 1997 surface ruptures on the Ābiz right-lateral fault system consist of about 10 en-echelon, stepping or overlapping segments, separated by gaps and developed in zones typically between 5 and 15 m wide (Figures 13.17, 13.18, and 13.9;

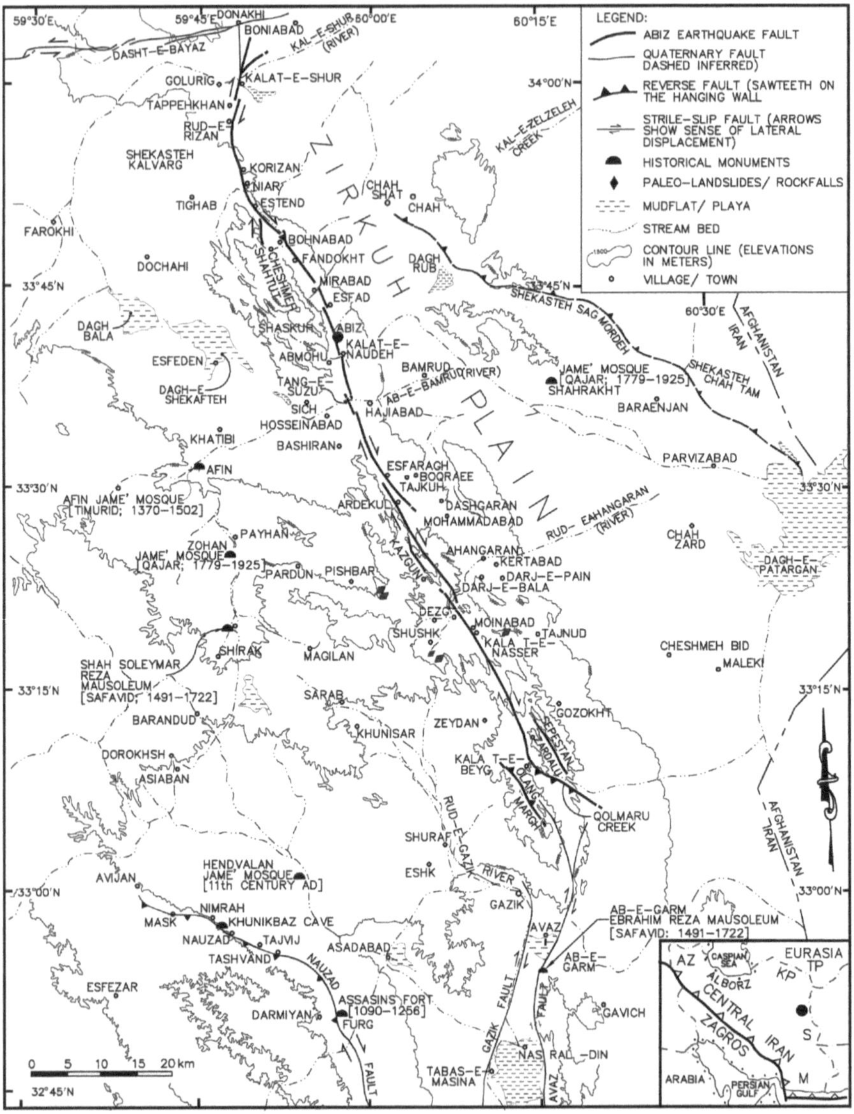

FIGURE 13.17 Coseismic surface faulting (thick lines) of the 10 May 1997 M_w 7.2 Zirkuh earthquake along the Ābiz right-lateral strike-slip fault in the Qa'enāt area of eastern Iran. Other major faults that did not move during the 1997 earthquake are shown as thin lines. Fault and other symbols as in Figure 11.1. See also Plates 13.33–13.37. *Modified from Berberian et al. (1999) and Berberian and Yeats (1999, 2001).*

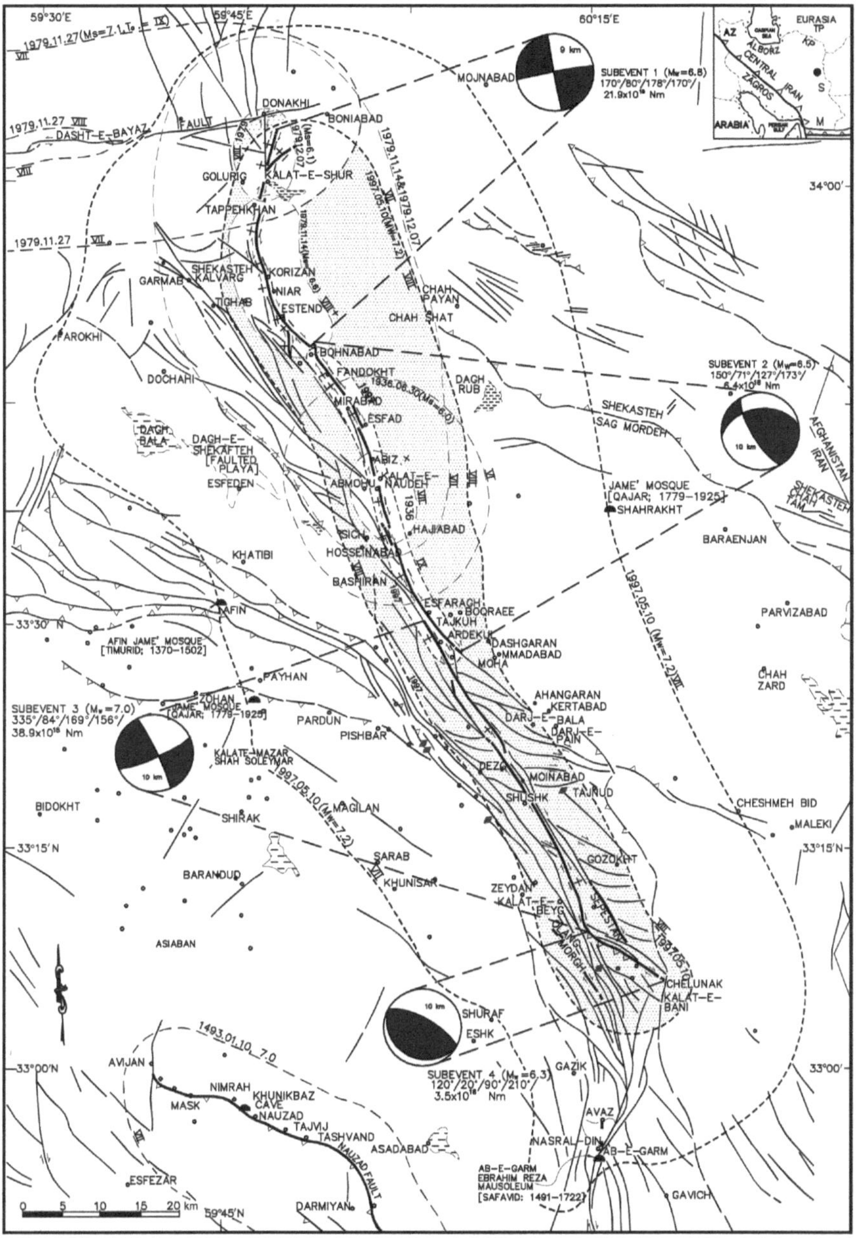

FIGURE 13.18 Meizoseismal area of the 10 May 1997 M_w 7.2 Zirkuh earthquake along the Ābiz right-lateral strike-slip fault (thick line) with focal mechanisms of the four subevents determined by waveform modeling with source parameters (Berberian et al., 1999). Other major faults that did not move during the 1997 earthquake are shown as thin lines. Meizoseismal areas of the earlier earthquakes in the area are also added. Fault and other symbols as in Figure 11.1. See also Plates 13.33–13.37. *Modified from Berberian et al. (1999) and Berberian and Yeats (1999, 2001).*

PLATE 13.32 Complete collapse of the reinforced concrete structures at Ardekul during the 10 May 1997 M_w 7.2 Zirkuh earthquake. See Figures 13.17 and 13.18 for the location, fault, and meizoseismal area of the earthquake. The collapsed buildings were constructed after the 14 November 1979 M_w 6.6 Korizān earthquake. *Photographed in May 1997. View to the north.*

PLATE 13.33 The 1997 surface rupture with 130-cm right-lateral strike-slip displacement offsetting furrows in fields at Bohnābād, along the Ābiz fault during the 10 May 1997 M_w 7.2 Zirkuh earthquake. See Figures 13.17 and 13.18 for the location, fault, and meizoseismal area of the earthquake. View to the northwest. *Photograph taken in May 1997 (Berberian et al., 1999).*

Plates 13.33–13.36). No single segment extends more than one-third of the length of the whole rupture, and discontinuous features are spread over a wide zone. The longest uninterrupted segment is in the south, with a length of ~20 km. To the south where the fault bends to the southeast, a clear reverse mechanism (Figure 13.17; Plate 13.37) was documented (Berberian et al., 1999).

PLATE 13.34 The 1997 surface rupture with 195-cm right-lateral and 30-cm vertical (east-side down) offset of the road 500 m west of old Korizān village leading to new Korizān during the 10 May 1997 M_w 7.2 Zirkuh earthquake. See Figures 13.17 and 13.18 for the location, fault, and meizoseismal area of the earthquake. View to the south. *Photograph taken in May 1997 (Berberian et al., 1999).*

PLATE 13.35 The 1997 surface rupture with 120-cm right-lateral displacement splitting a single tree at Estend during the 10 May 1997 M_w 7.2 Zirkuh earthquake. See Figures 13.17 and 13.18 for the location, fault, and meizoseismal area of the earthquake. View to the southeast. *Photograph taken in May 1997.*

PLATE 13.36 The 1997 surface rupture with 230-cm right-lateral displacement along the southeast segment of the Ābiz fault northeast of Zeydān, north of Kuh-e Zardālu during the 10 May 1997 M_w 7.2 Zirkuh earthquake. See Figures 13.17 and 13.18 for the location, fault, and meizoseismal area of the earthquake. View to the north northwest. *Photograph taken in May 1997.*

PLATE 13.37 Reverse slip of 90 cm at Qol Māru Creek along the southernmost tip of the Ābiz fault during the 10 May 1997 M_w 7.2 Zirkuh earthquake. See Figures 13.17 and 13.18 for the location, fault, and meizoseismal area of the earthquake. View to the south. *Photograph taken in May 1997 (Berberian et al., 1999).*

At least four other shorter faults, off the main trace of the Ābiz fault, were also reactivated by the Zirkuh main shock. From north to south the main faults are as follows (Figures 13.17 and 13.18):

a. The Cheshmeh Shāhtut fault at Sar Takht-e Mehdi (2.5 km west of Bohnābād), with a length of 3.5 km and 25 cm of right-lateral displacement.

b. The Kāzgun fault, 4.5 km northwest of the Āhangarān River and 1 km west of the Ābiz fault, with 4.5 km length and 40 cm of right-lateral displacement.
c. The Sepestān fault, 3.5 km east of the southern section of the Ābiz fault, with a 7 km length and 80 cm of right-lateral displacement.
d. The Olang Margh fault, 3.5 km west of the bend at the south southeast end of the Ābiz fault, with 7 km length, 30 cm of right-lateral, and 15 cm of vertical (east side up) displacement. In one place, the surface dip was estimated at 68–80 NE, and fresh slickensides were seen with a pitch of 15° to the northeast, measured on the fault plane (Berberian et al., 1999).

In addition, multiple, discontinuous rupture traces striking approximately N–S to NNE–SSW in a zone 150 m wide and 1 km long with 15 cm of right-lateral strike-slip displacement were developed in the Āvāz (lit., "song") basin (both at Āvāz and Nasr al-Din villages), about 20 km south of the end of the main Ābiz fault system (Figure 13.17). The Āvāz basin lies in a right-step (and thus "pull-apart"' location) between the Ābiz fault in the north and the Gazik and Āvāz faults in the south, which continue the right-lateral faulting of the Sistān suture zone to the south (Berberian et al., 1999).

Cumulative right-lateral displacement of Eocene green tuff key bed of 45–50 km along the Ābiz fault can be measured in a highly shear zone with several strike-slip faults in close proximity. For the pattern of coseismic surface rupture and historical earthquake clusters in the region, see Chapter 16.

13.23.1 Oral Earthquake Hazard Warning

Although no pre-1936 earthquake has been documented, several ancient place names along the fault may refer to prehistoric events along the fault. Names used in an oral seismic warning system include: *Dagh-e Shekāfteh* ("Faulted or Fractured Playa" in Persian), *Kāl-e-Zelzeleh* ("Earthquake Creek"), *Kuh-e-Shekasteh Chāhaki* ("Chāhaki Broken/Fractured/Faulted Mountain"), *Shekasteh Kālurak* ("Fractured or Faulted Creek"; Kālvarg in top left of Figure 13.17), and *Shekasteh Sag Mordeh* ("Fractured or Faulted Dead Dog") (see Chapter 1 for details, and Figure 13.17).

13.23.2 Seismology

P and *SH* body-wave modeling of the earthquake showed a complex rupture process with at least four subevents (Berberian et al., 1999; Figure 13.18; Table 13.17). Similar results were obtained from a source model derived from InSAR indicating; (i) variable dip and rake along the Ābiz fault with three rupture segments separated by regions of slow slip; and (ii) apparent slip deficit at the surface, with a modeled peak slip of 4 m beneath Ardekul village (Sudhaus and Jonsson, 2011).

TABLE 13.17 ***P*** **and** ***SH*** **Body-Waveform Inversion Source Parameters of the 10 May 1997** M_w **7.2 Zirkuh Earthquake (Berberian et al., 1999)**

Earthquake	Time (UTC)	Coordinate (N°–E°)[a]	Strike (°)	Dip (°)	Rake (°)	Slip Vector Azimuth	Centroid Depth (km)	M_o (Nm) ($\times 10^{18}$)	M_w
1997.05.10 Single-source model	07:57	33.84–59.81	156	89	200	155	13	60.5	7.2
Multiple-source model									
Subevent 1			170	88	178	170	9	21.9	6.8
Subevent 2			150	71	127	173	1	6.4	6.5
Subevent 3			335	84	169	156	10	38.9	7.0
Subevent 4			120	20	90	210	10	3.5	6.3

See also Figure 13.18.
[a]*Epicenter is from Engdahl et al. (2006).*

13.24 OVERVIEW: 1964–1997 COSEISMIC SURFACE RUPTURES

The 1968 UNESCO missions opened the systematic investigation of large-magnitude earthquakes and their coseismic faulting in Iran (mostly by N. N. Ambraseys and J.S. Tchalenko). In 1972, I established the first Tectonics and Seismotectonics Department in the country at the Geological Survey of Iran (a United Nations Special Fund Project), thus broadening the investigation and studies of the paleo- and active tectonics, geomorphic features, and ground deformations associated with both historical and modern medium- to large-magnitude earthquakes.

During 1964–1997 period, the longest ever recorded coseismic surface faulting of 125 km took place during the 10 May 1997 M_w 7.2 Zirkuh earthquake along the Ābiz right-lateral, strike-slip fault (Figures 13.9 and 13.18), and the lowest magnitude earthquake with coseismic surface rupture occurred during the 14 February 1971 M_w 5.7 Serokhi earthquake along the short Serokhi thrust fault (Figure 13.3).

Field study and body-waveform analysis of the 1997 Zirkuh earthquake along the Ābiz fault showed oblique shortening was achieved by four subevents propagating from the north to the south, with orientation of the fault segments changing in each subevents with a substantial reverse component at the southern tip of the fault. The 1997 earthquake clearly showed that the uplift of the Abiz Mountains along the fault (Shāhāz +2388, Ardekul +2532, Āhangarān +2314, and Sepestān +2380) and its folded structures were not achieved by the strike-slip faulting.

The 1997 Zirkuh earthquake was the sole example of a large-magnitude earthquake that ruptured multiple segments of a fault (the Ābiz fault in this case) in a single earthquake, despite some of the segments moving during earlier smaller earthquakes. In the case of the recorded earthquakes along other large faults, such as the Kuhbanān fault (Figures 14.8 and 16.19), seismic risk could have been underestimated if the 1997 Zirkuh earthquake had not occurred. Hence, the potential earthquake on multisegmented large faults, such as the Kuhbanān, Nāyband, Doruneh, Moshā, North Tehran, North Tabriz, Khazar, Anār, Kāshān, Kahurak, Neh, Sabzvārān, Gowk, and Zāgros Main Recent faults (Figure 9.1), is far greater if the fault segments break simultaneously during a single event in the same way as the Ābiz fault failed in 1997.

The Rudbār seismic gap was filled with the 20 June 1990 M_w 7.3 earthquake with 80 km of coseismic, left-lateral surface faulting indicating slip partitioning in the Alborz fold-and-thrust mountain belt. The 1983 thrust and 1990 strike-slip earthquakes in the western Alboz indicated that oblique convergence across the Alborz Mountains is separated into its two orthogonal components of (i) a range-normal shortening (1983) and (ii) a range-parallel strike-slip (1990) along two parallel thrust and left-lateral strike-slip faults (Figure 13.12). Except for a small area at Jaishābād (36°54′N–49°10′E), where the Rudbār fault has a clear geomorphology, the entire length of the strike-slip fault did not show visible active geomorphological features at the surface, possibly due to a long recurrence period and high rates of erosion along the High-Alborz near the Alborz-Caspian watershed. The coseismic surface rupture failed to break through a segment gap at the Sefirdrud gorge and the highway crossing the fault trend. The lesson learned from this large-magnitude event is that there should be numerous active faults that have not yet been detected due to the difficulty of tracing their active geomorphological indicators.

Despite having a larger magnitude than the 1997 Zirkuh event, the 1990 Rudbār surface faulting was 45 km shorter. The two earthquakes with M_w 7.1, which took place on the Dasht-e Bayāz fault, were accompanied by surface faulting of 70 km (31 August 1968) and 55 km (27 November 1979) in length with different amounts of maximum displacements. Cross-strike seismicity migration (1968 and 1978) as well as complex clustering of seismicity and faulting were also recorded and will be discussed in Chapter 16 (see also Tables 14.9–14.11, and 17.2).

During this period the largest reverse fault earthquake with coseismic surface faulting was documented during the 16 September 1978 M_w 7.3 Tabas-e Golshan earthquake (Figure 13.8). The 1994 Sefidābeh earthquake sequence was accompanied only by superficial flexural-slip faulting along the uplifted asymmetric ridge with steep northeast flank and gentle southwest side. Body waveform analysis showed that the sequence involved southwest-dipping, blind reverse faulting, at centroid depths of 6–10 km (Figure 15.1). Active geomorphological features were clearly developed in response to the growing hanging-wall block of the fold along the blind Sefidābeh thrust. Lessons learned in this case can be used for other similar cases in Iran and other active parts of the world.

Chapter 14

1998–2013 Coseismic Surface Faulting

The crust of the Earth cracked and formed an abyss

Vārtābed Ārākel Tāvrizetzi (1641)

The rationale for choosing this period is that, in 1998, space-based geodetic techniques with InSAR (synthetic aperture radar interferometry) for imaging earthquake ground deformation were used for the first time; the techniques were applied to the earthquakes along the Gowk fault, in addition to field studies, aerial photographs, satellite imagery, and seismological investigations (Berberian et al., 2001). The 1998 InSAR showed that the 14 March 1998 M_w 6.6 Fandoqā earthquake with a strike-slip displacement along the Gowk fault in southeast Iran triggered a seismic slip of 8 cm at a shallow depth along the Shahdād thrust located to the east (Berberian et al., 2001; Fielding et al., 2004). The InSAR technique was first used after the 28 June 1992 M_w 7.3 Landers, California, earthquake (Massonnet et al., 1993).

The Iranian National Broadband Seismic Network (INSN; THR), operated by the International Institute for Seismology and Earthquake Engineering (IIEES) in Tehran, started issuing seismological bulletins in 2004. After 2005, the Iran Telemetered Seismographic Network (ITSN; TEH), operated by the Iranian Seismological Research Center (IRSC) at the Institute of Geophysics of University of Tehran (IGUT), began to routinely report the Iranian earthquake parameters. Nonetheless, the epicentral and focal depths of earthquakes located by these two organizations show considerable shifts in the parameters of each event.

Advancement of relocation techniques (Engdahl et al., 1998, 2006) during this period also partially improved the teleseismic locations of major Iranian earthquakes by using surface reflection phases such as p^P and s^P. Teleseismic locations continue to be incorrect; for example, the mainshock epicenter location of the 26 December 2003 M_w 6.6 Bam earthquake (NEIC, HRVD, ISC) and relocation (Engdahl et al., 2006) falls about 7 km to the west of the southern tip of the N–S vertical right-lateral strike-slip coseismic surface rupture.

☆"To view the full reference list for the book, click here"

Developments in Earth Surface Processes, Vol. 17. http://dx.doi.org/10.1016/B978-0-444-63292-0.00014-4

The instrumental epicenters located by the national seismographic networks of INSN (IIEES, Tehran) and ITNS (IRSC; IGUT) for the 20 December 2010 M_w 6.5 South Rigān earthquake are 30 km apart. The distance between the epicenters located by the two national networks for the 27 January 2011 M_w 6.2 earthquake is about 20 km. In both cases the INSN (IIEES) locations are to the SSE of the INSN (IGUT)-located epicenters. Because of several factors—such as poor azimuthal distribution of the recording stations located mostly in the northwest quadrant relative to the epicentral region as well as far from it and the uncertainty about the crustal velocity structure and additional unknown errors—the epicentral locations for these two medium-magnitude earthquakes located by the two national Iranian agencies have a large magnitude of epicentral error. The location errors of the very small- to small-magnitude earthquakes located by the INSN (IIEES) and ITSN (IGUT) are much larger than the quoted location errors of the medium-magnitude earthquakes.

During this period, a report was published that introduced data about the historical and twentieth-century earthquakes and earthquake faulting as well as seismic hazards in the Neyshābur and Mashhad urban and rural areas together with fault and seismic hazard maps of the area (Berberian et al., 2000a). In 1999, the rich Iranian historical and archeological records spanning several thousand years were first used to study the pattern of historical earthquakes on the Iranian plateau. The study established recurrence intervals of 1000–5000 years on individual fault segments. Furthermore, several clusters of earthquakes provided evidence of interaction among reverse and strike-slip faults, probably because adjacent faults were loaded by individual earthquakes (Berberian and Yeats, 1999, 2001). The study also emphasized that an increased understanding of seismic hazards in Iran could be obtained by an extensive paleoseismology trench program and space-geodetic arrays in the urban and rural areas, supplementing the abundant historical and archeological record. The first analysis of archeoseismicity in Iran was also published during this period (Berberian and Yeats, 2001; Berberian et al., 2012, 2014).

Within a framework of French–Iranian cooperation (CNRS-INSU, Universite Montpellier II; NCC and IIEES, Tehran), the first large-scale Iranian Permanent Global Positioning System (IPGN) with 25 sites was implemented; it performed measurements beginning in September 1999. The space-based geodetic investigation using GPS with a network of 28 sites (25 in Iran, 2 in Oman, and 1 in Uzbekistan), combined with seismological analysis, measured the velocity fields addressing the distribution of crustal deformation on the Iranian plateau (Nilforoushan et al., 2003; Vernant et al., 2004). The resulting velocities constrained the kinematics of the regional Iranian tectonics and showed that the north–south shortening from Arabia to Eurasia is about 25 mm/year. Unlike the fast-moving fault systems in Turkey, such as the North and East Anatolian faults (McClusky et al., 2000) with numerous large-magnitude earthquakes (Ambraseys, 1970, 1989, 2009; Barka, 1996, 1999), no fast-moving fault system was detected on the Iranian plateau.

14.1 THE 14 MARCH 1998 M_W 6.6 FANDOQĀ EARTHQUAKE

The 14 March 1998 Fandoqā earthquake, which occurred in a remote and sparsely inhabited desert area, killed five people in Golbāf (official figure), injured 15, and damaged seven villages: Golbāf (VII, MMI), Zamānābād (VII), Fandoqā (VII), Hashtādān (VII), Jowshān (VI), Dehu (VI), and Deh Qanbar (VI) (Figures 13.11 and 14.1).

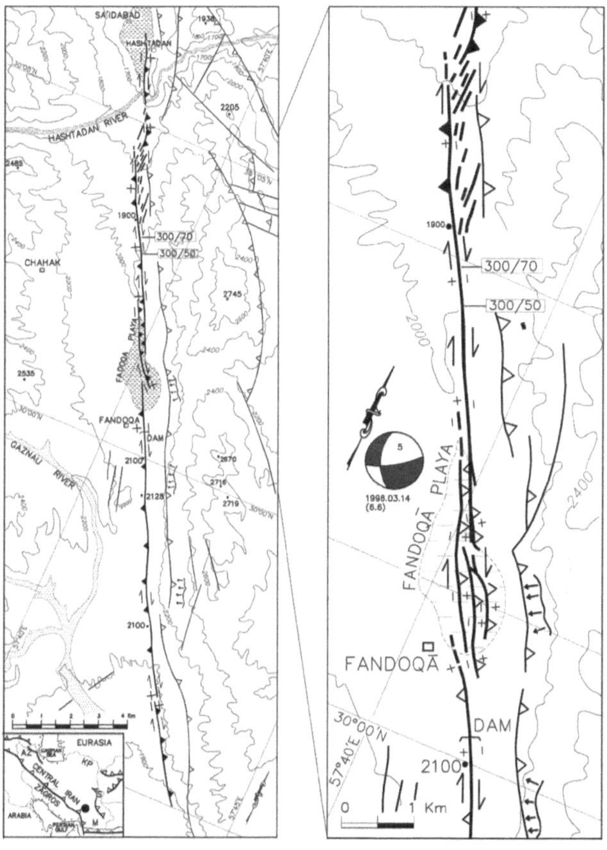

FIGURE 14.1 Detailed map of coseismic surface rupture (thick lines) along the Gowk fault system during the 14 March 1998 Fandoqā earthquake (the right map shows details of the coseismic surface rupture in the Fandoqā area). Fault and other symbols as Figure 11.1. Thin lines are other faults that showed no surface reactivation during the 1998 earthquake. Relative uplift and subsidence across the 1998 ruptures are marked by + or − signs. Groups of arrows on the east side of the valley are earthquake-triggered landslides. Numbers in the two boxes between Fandoqā and Hashtādān show the locations of maximum horizontal/vertical slip (in cm) in 1998. The Hashtādān and Fandoqā playas are marked by horizontal dashed lines. Elevations are in meters. Note the right-lateral offset and diversion of the Hashtādān River where it crosses the Gowk fault. Focal mechanism of the earthquake constrained by body-wave modeling with centroid depths (Berberian et al., 2001). See also Plates 14.1–14.10. *Modified from Berberian et al. (2001).*

The 1998 Fandoqā earthquake was the penultimate in a series of five substantial earthquakes on the Gowk fault system of southeast Iran to occur since 1981; all were associated with coseismic surface ruptures along the northern section of the Gowk fault (Figure 13.11). No significant post-1900 earthquake has been recorded along the southern segment of the Gowk fault during the instrumental time period. The 1998 Fandoqā earthquake, which occurred in the gap left between the 11 June 1981 M_w 6.6 Golbāf and the 28 July 1981 M_w 7.0 Sirch events on the Gowk fault, produced 23.5 km of surface faulting along the eastern side of the Gowk valley with up to 3-m right-lateral strike-slip and 1-m vertical offsets (Figures 13.11 and 14.1). The 23.5-km-long Fandoqā surface rupture along the Gowk fault system includes re-rupturing of 19 km of the southernmost portion of the 28 July 1981 Sirch rupture plane and the 6.6 km gap left between the 1981 Golbāf and Sirch earthquakes (Berberian et al., 2001).

Of the 1998 ruptures, about 14.25 km were east-facing and 9.25 km were west-facing. The west-facing sections were at both ends of the rupture zone and in the middle, in Fandoqā playa (Figure 14.1; Plate 14.1). The surface ruptures were expressed as either: (i) zones up to 300–400 m wide, containing

PLATE 14.1 Aerial photograph of the Fandoqā playa, the epicentral area of the 14 March 1998 M_w 6.6 Fandoqā earthquake along the Gowk fault system. The earth dam (30°00′05.13″N–57°40′51.17″E, +2071 m) about 1.7 km to the south of Fandoqā playa was fractured by the coseismic surface fault. The length of the Fandoqā playa is about 3 km. See Figure 14.1 for the location. *National Cartographic Center of Iran.*

distributed tension cracks, mole tracks, and short (<100 m) anastomosing faults, commonly displaying strike-slip motion and sometimes arranged in en-echelon patterns; or (ii) well-defined linear scarps (Plates 14.1–14.3); or (iii) small fractures near Zamānābād (Figures 14.1 and 13.11; Plates 14.1–14.10). SAR interferometry and seismic waveforms showed that the main rupture plane dipped west at about 50° and had a normal component although the surface ruptures were more complicated, being downthrown to both the east and the west on steep faults in near-surface sediments (Berberian et al., 2001).

Although the 1981 M_w 7.0 Sirch and 1998 M_w 6.6 Fandoqā earthquakes apparently ruptured parts of the same fault system repeatedly at the surface with a short 17-year interval, these earthquakes had very different rupture characteristics (Figure 13.11). The magnitude of the strike-slip and vertical

PLATE 14.2 Aerial photograph to the south of the Fandoqā Playa (south of Plate 14.1), the epicentral area of the 14 March 1998 M_w 6.6 Fandoqā earthquake along the Gowk fault system. See Figure 14.1 for the location. The rural road runs along the Gowk fault. The distance between the southern edge of the Fandoqā playa and the earth dam (white patch) is 1.7 km. *National Cartographic Center of Iran.*

PLATE 14.3 Aerial photograph of the Hashtādān area (30°07′N–57°37′E, +1652 m) located to the north of the Fandoqā playa (north of Plate 14.1), the epicentral area of the 14 March 1998 M_w 6.6 Fandoqā earthquake along the Gowk fault system. See Figure 14.1 for the location. The length of the Hashtādān playa is 5.7 km. *National Cartographic Center of Iran.*

offsets measured at the surface in the 14 March 1998 Fandoqā earthquake M_w 6.6, centroid depth 5 km were consistently greater than those measured after the much larger 28 July 1981 Sirch earthquake M_w 7.0, centroid depth 18 km. The maximum strike-slip and vertical slip observed in 1998 were 300 and 95 cm, respectively, compared with 43 and 40 cm, respectively in 1981 (Figure 13.11). Moreover, whereas a maximum of 300 cm strike-slip offset was observed in 1998, only 13 cm was observed in 1981. The main ruptures in the 1981 earthquake probably occurred on different, deeper parts of the same fault system (Figure 14.2), producing only minor reactivation of the shallower faults at the surface (Berberian et al., 2001).

SAR interferometry modeling showed that, in addition to a dominantly right-lateral strike-slip movement on the 1998 Fandoqā rupture plane, the nearby Shahdād thrust with a similar strike, but dipping at about 8° west, moved about

PLATE 14.4 The 14 March 1998 M_w 6.6 Fandoqā earthquake right-lateral strike-slip surface rupture of the earth dam (30°00′05.13″N–57°40′51.17″E, +2071 m) located about 1.7 km to the south of Fandoqā playa (for location see Plate 14.1, and Figure 14.1).View to the northeast. *Courtesy of the photograph pool of Berberian et al. (2001).*

PLATE 14.5 The 14 March 1998 M_w 6.6 Fandoqā earthquake right-lateral strike-slip surface rupture of a ridge in the area east of Fandoqā; the village is in the background. The 1998 displacement at this locality was 150 cm, whereas the 28 July 1981 M_w 7.0 Sirch earthquake resulted in 20 cm displacement at the same locality (see Figure 13.11 for the slip difference between the two events). View to the southwest. See also Figure 14.1 for the location in the lower right panel. *Courtesy of the photograph pool of Berberian et al. (2001).*

PLATE 14.6 The 14 March 1998 M_w 6.6 Fandoqā earthquake right-lateral strike-slip surface rupture of a concrete canal east of Fandoqā. The canal was earlier displaced by the 28 July 1981 M_w 7.0 Sirch earthquake and was repaired 17 years earlier, when it ruptured again by the1998 event. See also Figure 14.1. *Courtesy of the photograph pool of Berberian et al. (2001).*

PLATE 14.7 The 14 March 1998 M_w 6.6 Fandoqā earthquake right-lateral strike-slip surface rupture in the area northeast of Fandoqā playa (the white patch). View to the north. See also Figure 14.1. *Courtesy of the photograph pool of Berberian et al. (2001).*

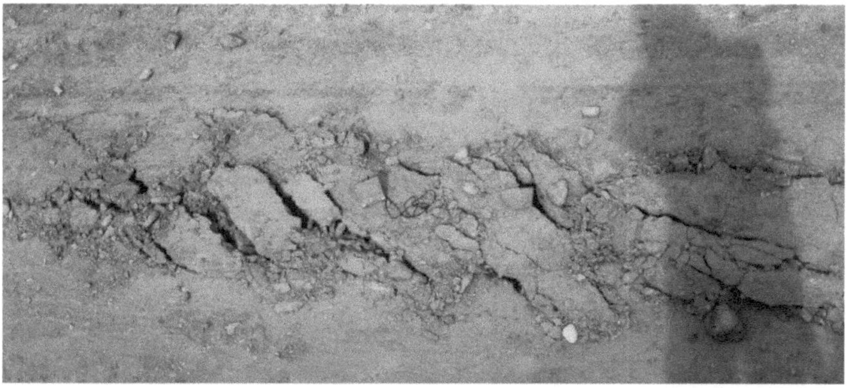

PLATE 14.8 The 14 March 1998 M_w 6.6 Fandoqā earthquake right-lateral strike-slip surface rupture in the area south of Hashtādān. See Figure 14.1 for the location. *Courtesy of the photograph pool of Berberian et al. (2001).*

PLATE 14.9 The 14 March 1998 M_w 6.6 Fandoqā earthquake right-lateral strike-slip surface rupture in the area south of Hashtādān near the dirt road. See Figure 14.1 for the location. *Courtesy of the photograph pool of Berberian et al. (2001).*

PLATE 14.10 Vertical motion along the 14 March 1998 M_w 6.6 Fandoqā earthquake right-lateral strike-slip surface rupture in the Hashtādān playa. See Figure 14.1 for the location. *Courtesy of the photograph pool of Berberian et al. (2001).*

7 cm in a time interval and in a position that makes it likely that its slip was triggered by the 1998 Fandoqā earthquake (Figure 14.3). InSAR modeling indicated that the Fandoqā earthquake transferred stress to the nearby Shahdād basal thrust and fold, triggering slip either immediately or in the following 6 months. SAR interferometry has imaged slip on at least 600 km^2 of the Shahdād basal-thrust and splay-fault network [30 × 20 km rupture area extending from 1 to 4.5 km below the surface]. Approximately 70 mm of possibly aseismic thrust motion on a 8°-dipping Shahdād basal thrust occurred 30 km to the east of the 14 March 1998 Fandoqā earthquake surface rupture along the Gowk fault (Berberian et al., 2001; Fielding et al., 2004) (Figures 14.2 and 14.3). The aseismic growth of the Shahdād anticline was not accompanied by surface rupture. However, active geomorphological features (deep incised stream beds, diverted streams around the nose of the folds) show active fold growth above the Shahdād thrust.

Coulomb-failure change study indicated that the high stress decrease from the 11 June 1981 M_w 6.6 Golbāf earthquake led to the termination of rupture of the 28 July 1981 M_w 7.0 Sirch earthquake, leaving behind the 6.6-km gap (Figure 14.3). Subsequent positive coseismic stress may have overcome this stress decrease, permitting the occurrence of the 14 March 1998 M_w 6.6 Fandoqā earthquake (Nalbant et al., 2006). Additional positive coseismic stress load due to the 1981 Sirch earthquake and postseismic loading over 17 years may have helped overcome the stress shadow (Nalbant et al., 2006). Apparently, the stress field generated by the three 11 June 1981 (6.6), 28 July 1981 (7.0), and 14 March 1998 (6.6) earthquakes on the Gowk fault system explains the triggered deformation on the Shahdād thrust (Figure 14.3) and fold zone as captured by the SAR interferometry (Nalbant et al., 2006).

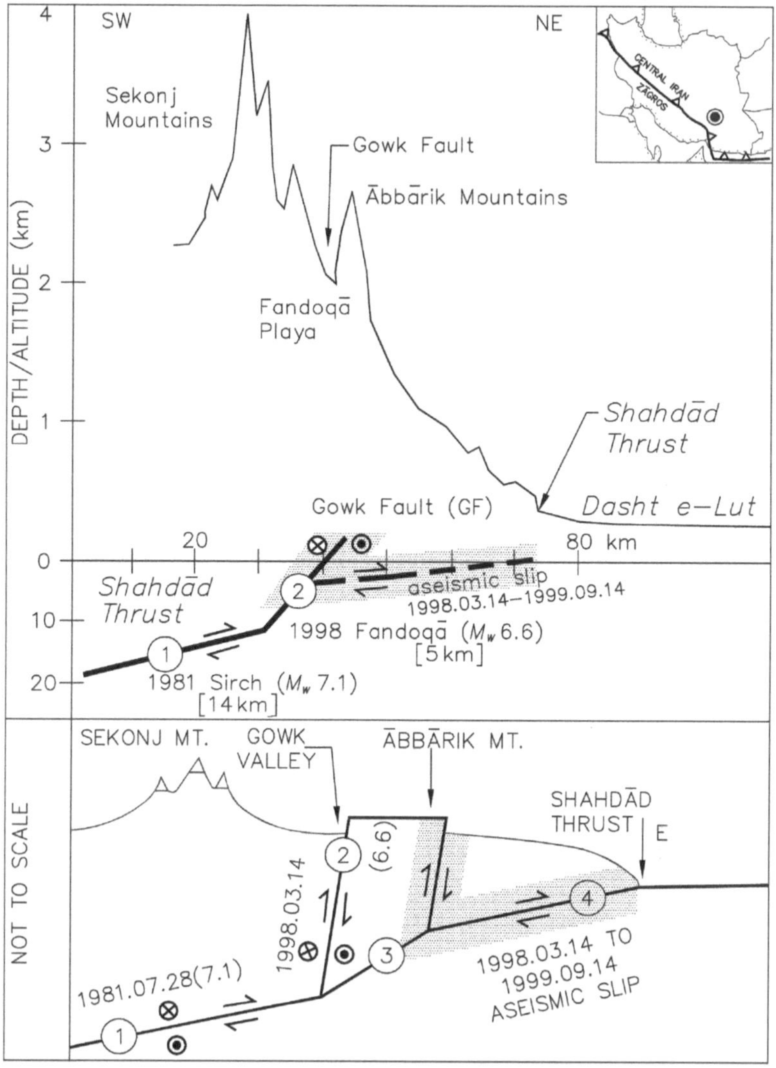

FIGURE 14.2 Topographic cross-section (top) and cartoons (middle and bottom) illustrating fault geometries in and beneath the Gowk fault system during the two earthquakes of 28 July 1981 M_w 7.0 with centroid depth of 5 km with smaller slip at the surface, and 13 March 1998 M_w 6.6 with centroid depth of 5 km with larger slip at the surface (compare with Figure 13.11 for the displacement differences). The fault system achieved slip partitioning by separating oblique slip at depth into its strike-slip and thrust components near the surface modified from (Berberian et al., 2001; Walker and Jackson, 2002).

An overall right-lateral slip-rate of about 1.5–2.4 mm/year is suggested for the Gowk active fault (Walker and Jackson, 2002). This rate is much smaller than the overall 10–20 mm/year of shear expected between Central Iran and Afghanistan (Nilforoushan et al., 2003; Vernant et al., 2004). The deficit is

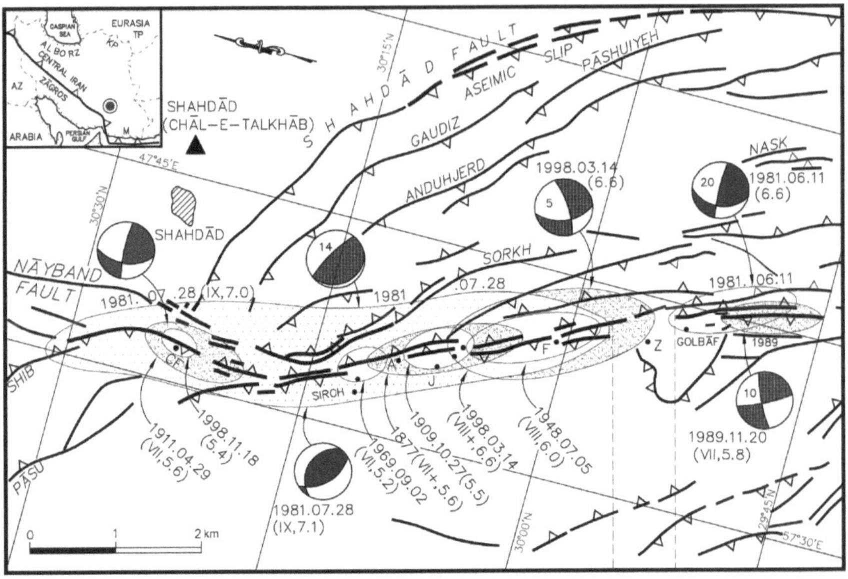

FIGURE 14.3 Meizoseismal areas (shaded, stippled ellipses) of documented earthquakes along the northern section of the Gowk fault system. Fault and other symbols as Figure 11.1. Thick lines are fault segments that were activated during the earthquakes. Focal mechanism of the earthquake constrained by body-wave modeling with centroid depths (Berberian et al., 2001). F: Fandoqā. J: Jowshān. Z: Zamānābād. *Modified from Berberian and Yeats (1999), Berberian et al. (2001), and Berberian (2005).*

likely to have been accommodated on other faults east of the Gowk fault system on the eastern side of the Lut Block along the Neh and Kahurak faults (Walker and Jackson, 2002).

14.1.1 Seismology

P and *SH* waveform modeling indicates a well-constrained right-lateral strike-slip on a fault dipping 54° WSW, with a strike of N156°E (Table 14.1) that is almost identical to the overall trend of the coseismic ruptures observed at the surface (Berberian et al., 2001).

For historical seismicity and coseismic surface rupture patterns along the Gowk fault, see Chapter 16.

14.2 THE 18 NOVEMBER 1998 M_W 5.3 CHAHĀR FARSAKH EARTHQUAKE

Eight months after the 13 March 1998 M_w 6.6 Fandoqā earthquake, an earthquake of M_w 5.3 damaged houses at Chahār Farsakh (VI$^+$) and Puzeh Bāgh (VI$^+$), 45 km north of the 14 March 1998 M_w 6.6 Fandoqā meizoseismal area, located to the northern end of the Gowk fault, which was activated during the

TABLE. 14.1 ***P*** **and** ***SH*** **Body-Waveform Inversion Source Parameters of the 14 March 1998** M_w **6.6 Fandoqā Earthquake (Berberian et al., 2001)**

Earthquake	Time (UTC)	Coordinate (N°–E°)[a]	Strike (°)	Dip (°)	Rake (°)	Slip Vector Azimuth	Centroid Depth (km)	M_o (Nm) ($\times 10^{18}$)	M_w
1998.03.14	19:40	30.13–57.58	156	54	195	147	5	9.09	6.6

See also Figures 14.1–14.3.
[a]*Epicenter is from Engdahl et al. (2006).*

28 July 1981 M_w 7.0 Sirch earthquake (Figures 14.3 and 14.4). The shock was strongly felt at Hashtādān, Fandoqā, and Golbāf.

The 18 November 1998 M_w 5.3 Chahār Farsakh earthquake was accompanied by 4 km of surface rupturing in the area between Chahār Farsakh and Puzeh Bāgh at the northern end of the Gowk fault (Figures 14.3 and 14.4; Plate 14.11). The surface faulting followed exactly the well preserved escarpment of the 28 July 1981 M_w 7.0 Sirch earthquake (Figure 14.3) with a few centimeters of horizontal and vertical displacements on the west-dipping fault segment of the northern Gowk fault system (Figure 14.4; Plate 14.12). The surface rupture broke an area where only hairline fractures had been developed in Chahār Farsakh during the 1981 M_w 7.0 Sirch event (Figure 14.3); to the north, it covered the area where the maximum 43 cm of right-lateral displacement had been measured during the 1981 event (Berberian et al., 2001).

It is difficult to prove that the minor surface ruptures in the 1998 M_w 5.3 earthquake represent a true reactivation of the Gowk fault rather than a consolidation of near-surface deposits, especially because an earthquake of M_w 5.3 is only expected to produce a slip of the order of 20 cm on a fault of dimension about 4×4 km (assuming an equidimensional fault plane with a slip-to-length ratio of 5×10^{-5}), and rupture may not have reached the surface if the hypocenter was deeper than about 5 km (Berberian et al., 2001).

14.2.1 Seismology

Table 14.2 presents the HRVD CMT solution source parameters of this event.

14.3 THE 22 JUNE 2002 M_W 6.4 CHANGUREH EARTHQUAKE

An earlier death toll for this event was reported as 500; however, this number was apparently inflated as some of the severely injured had been mistaken for dead. The official count was 261 fatalities, 1500 injured, and 25,000 left homeless. Most of the dead were women, children, and the elderly, as many of the men were working in local vineyards. An estimated 5000 buildings were damaged beyond repair in several villages.

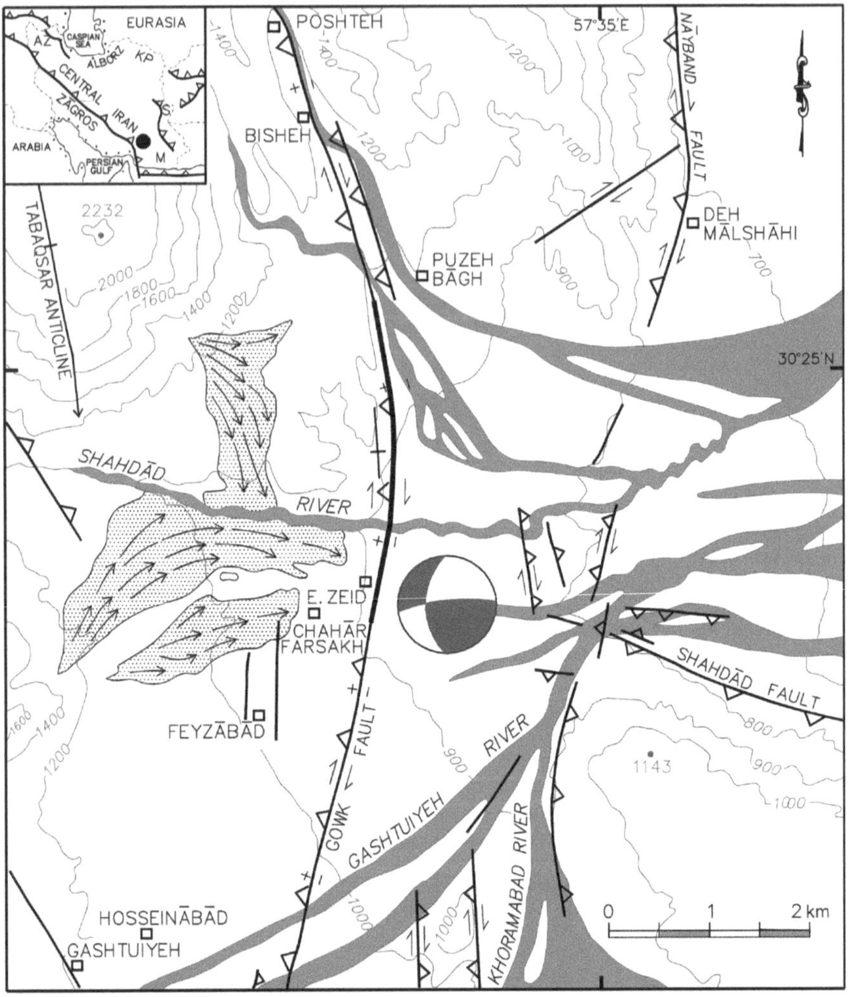

FIGURE 14.4 Detailed map of the northern end of the Gowk fault system, including the area affected by the 18 November 1998 M_w 5.3 Chahār Farsakh earthquake, and the location of surface breaks observed after the earthquake (thick fault line). Fault and other symbols as in Figure 11.1. The stippled regions with arrows to the west are old alluvial fans and their flow direction, now abandoned by the deep incision on the Shahdād River through the uplifted (west) side of the Gowk fault. Except for the Nāyband fault, the transverse fault west of Deh Mālshāhi, the Shahdād fault and the fault southwest of the Tabaqsar anticline, all other faults were reactivated at the surface during the 28 July 1981 M_w 7.0 Sirch earthquake. See also Plates 14.11 and 14.12. *Modified from Berberian et al. (2001).*

The earthquake was associated with discontinuous surface ruptures on the southwest-dipping Ābdarreh thrust fault (Figure 14.5) with an average slip of ~10 cm, and a maximum slip of 16 cm (Walker et al., 2005b). The ruptures appeared on the northeastern edge of the Ābdarreh Neogene anticline and were not mapped on the geological map of the region (Bolurchi and Hājiān, 1979).

PLATE 14.11 Gowk fault scarp at northern Chahār Farsakh area. View to the north. See Figure 14.1 for the location. *Courtesy of the photograph pool of Berberian et al. (2001).*

PLATE 14.12 Surface rupture along the Gowk fault viewed after the 18 November 1998 M_w 5.3 Chahār Farsakh earthquake, north of Chahār Farsakh village. *Courtesy of the photograph pool of Berberian et al. (2001).*

TABLE 14.2 ***P*** **and** ***SH*** **Body-Waveform Inversion Source Parameters of the 18 November 1998** M_w **5.3 Chahār Farsakh Earthquake (HRVD CMT)**

Earthquake	Time (UTC)	Coordinate (N°–E°)[a]	Strike (°)	Dip (°)	Rake (°)	Slip Vector Azimuth	Centroid Depth (km)	M_o (Nm) ($\times 10^{18}$)	M_w
1998.11.18	07:39	30.32–57.58	174	55	173	178	15 (fixed)	0.13	5.3

See also Figure 14.4.
[a]*Epicenter is from Engdahl et al. (2006).*

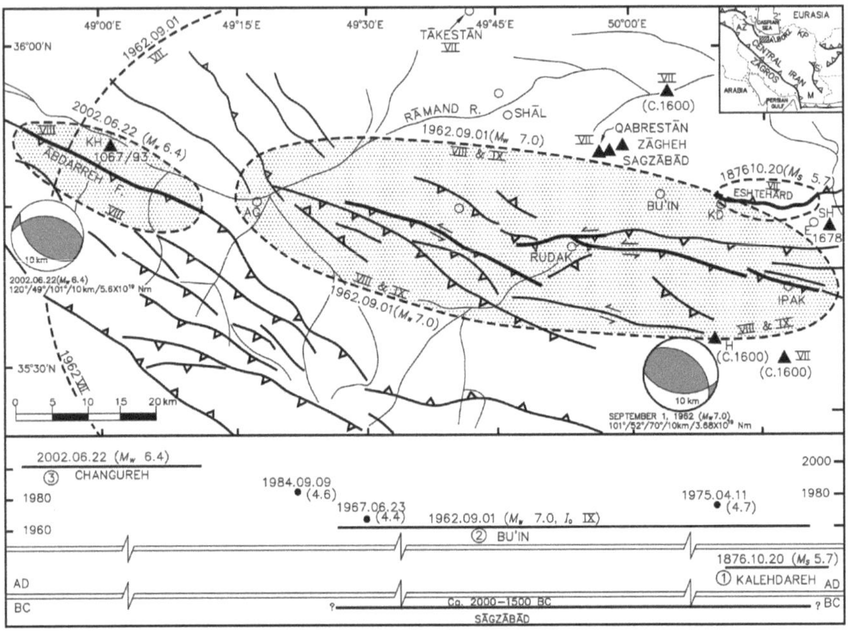

FIGURE 14.5 Top: Meizoseismal areas of three earthquakes in the Bu'in region showing westward migration illustrated by the 20 October 1876 M_s 5.7 Kaleh Dareh/Kalehdareh (KD), 1 September 1962 M_w 7.0 Bu'in, and 22 June 2002 M_w 6.4 Changureh earthquakes along the Eshtehārd, Ipak, and Ābdarreh faults. Fault and other symbols as in Figure 11.1. Focal mechanism of the earthquakes constrained by body-wave modeling with centroid depths (1962: Priestley et al., 1994; 2002: Walker et al., 2005b). Triangles: location of archeological mounds and historical monuments with date of construction. Bottom: Time–space diagram of the three recorded earthquakes in the area since 1876. See also Plates 14.13 and 14.14, Figure 12.13, and Chapter 12. *Modified from Berberian et al. (1983) and Berberian and Yeats (2001).*

Solaymani and Feghhi (2002) recorded a 14-km-long reverse fault, striking N105°–115°E, dipping 30–32° toward southwest, with a maximum vertical displacement of 65 cm. Qorashi and Ovaisi (2002) reported three parallel surface ruptures labeled as "*surface ruptures with the largest right-lateral vertical to left-slip vertical slips*" *[!]*. Their observation is very confusing.

PLATE 14.13 The two eleventh-century Kharaqān tomb-tower monuments (35°50′N–49°00′E, +1684 m), located 4 km north of the Ābdarreh thrust on the footwall (see Figure 14.5 for the location). The monuments were intact until the 22 June 2002 M_w 6.4 Changureh earthquake. *Courtesy of Wikipedia commons; zereshk; http://en.wikipedia.org/wiki/File:kharaghan.jpg.*

The 2002 M_w 6.4 Changureh earthquake meizoseismal area lays within the intensity VI (MMI) zone of the 1 September 1962 M_w 7.0 Bu'in earthquake (discussed earlier in Chapter 12; Figure 12.13). The 2002 surface rupture is located about 36 km to the west of the western end of the 1962 coseismic surface fault (Figure 14.5). The seismic pattern of the Bu'in area shows a westward propagation with left-stepping of seismicity from the 20 October 1876 $M_s \sim 5.7$ Kolā Darreh/Kalehdareh-KD in Figures 12.3 and 14.5 (Berberian et al., 1985), to the 1962 M_w 7.0 Bu'in, and the 2002 M_w 6.4 Ābdarreh events covering a 140-km-long zone along active fold-thrust belt. Unlike the 1876 [in the east] and 2002 [in the west], which took place along the foothill-ridge, the 1962 Bu'in earthquake was an intramontane area event (Figure 14.5).

The two Kharaqān tomb-tower monuments (1067 and 1093; Stern, 1966; Stronach and Cuyler Young, 1966; Meshkāti, 1970), located 4 km north of the Ābdarreh thrust on the footwall (Figure 14.5; Plate 14.13), were intact until 2002 when they were damaged (VIII) during the 2002 M_w 6.4 earthquake (Plate 14.14) indicating a minimum regional seismic quiescence of about 935 years for a medium-magnitude earthquake. The 2002 earthquake took place in a seismic gap to the west of the 1962 earthquake (Figure 14.5).

14.3.1 Seismology

Although the main rupture discontinuously propagated to the surface with small slip, the major part of the slip (75–150 cm) presumably accommodated by folding of the Ābdarreh anticline at the surface. Seventeen earthquakes in this region were determined to have focal depths between 7 and 10 km (Walker et al., 2005b).

PLATE 14.14 The eleventh-century Kharaqān tomb-tower monument, damaged (VIII) during the 22 June 2002 M_w 6.4 Changureh earthquake, indicating a minimum regional seismic quiescence of about 935 years for a medium-magnitude earthquake. See Figure 14.5 and Plate 14.13. *Courtesy of GSI Website; http://gsi.ir/General/Lang_en/Page51.*

Inversion of long-period *P* and *SH* body-wave seismograms shows rupture on a thrust fault dipping 49° to the southwest with a centroid depth of about 10 ± 3 km (Walker et al., 2005b; Table 14.3). The source–time function indicates that the earthquake rupture lasted for ~5 s. At a typical rupture velocity of ~3 km/s, the fault length would be ~15 km. Apparently, the major part of the slip probably did not propagate to the surface and was presumably accommodated by folding of the Ābdarreh anticline at the surface (Figure 14.5). Using the average slip (~1 m), fault length (~20 km), and moment (5.6×10^{18} Nm), a down-dip fault width is estimated of ~9 km. With a centroid depth of ~10 km and a fault-plane dip of ~49°, the main slip event would have occurred over a depth range of ~7–13 km, which suggests that the main slip probably failed to reach the surface (Walker et al., 2005b).

TABLE 14.3 ***P*** **and** ***SH*** **Body-Waveform Inversion Source of the 22 June 2002** M_w **6.4 Changureh Earthquake (Walker et al., 2005b)**

Earthquake	Time (UTC)	Coordinate (N°–E°)[a]	Strike (°)	Dip (°)	Rake (°)	Slip Vector Azimuth	Centroid Depth (km)	M_0 (Nm) ($\times 10^{18}$)	M_w
2002.06.22	02:58	35.59–49.02	120	49	101	13	10	5.6	6.4

See also Figure 14.5.
[a]*Epicenter is from Engdahl et al. (2006).*

14.4 THE 26 DECEMBER 2003 M_W 6.6 BAM EARTHQUAKE

During the medium-magnitude Bam urban earthquake, which completely ruined the city of Bam, between 31,000 and 43,000 lives were lost in a population of about 142,000 in Bam and surrounding areas. Officially, the initial death toll estimate was 43,200–45,000; it was later lowered to 40,000, 30,000, 26,500, and finally fixed at 31,500. On 29 March 2004 the head of the Statistical Center of Iran (SCI) announced that "*Some victims were counted more than once in the chaotic aftermath of the disaster; the earthquake killed 26,271 people and 525 people were still missing*" (Iranian Newspapers, 29 March 2004 Tehran; irna.ir; SCI, 2004; news.bbc.co.uk). However, a few months after the event, local inhabitants estimated a possibly exaggerated death toll as high as 79,000. More than 100,000 residents were trapped under collapsed buildings. Between 30,000 and 50,000 people suffered injuries in the Bam region (mainly in the city) and required hospitalization, yet all the functioning hospitals in the area were severely damaged or demolished. The surviving inhabitants were rendered homeless. The head of the Kermān province Education Department announced that more than 11,000 students (one-third of the students of Bam); 1200 teachers (one-half of the Bam teachers; this was later changed to 560 teachers, or one-fifth); and 200 health professionals perished in Bam alone. More than 6000 children were orphaned, and approximately 500 people were permanently disabled.

The Bam earthquake occurred on an unknown, blind, N–S strike-slip fault located about 4 km west of the previously mapped geologic Bam fault at Baravāt (Berberian, 1976b,e, 2005). It cut the eastern limb of an asymmetric anticline ridge with a prominent active geomorphological feature, which also showed surface rupturing (Figure 14.6). A series of discontinuous coseismic, left-stepping en-echelon, short surface-rupture segments (each of 50–100 m long) with a maximum observed right-lateral slip of 20 cm and a slip vector at N10° were mapped in a flat alluvial fan south of the city of Bam (Figure 14.6) extending directly beneath the city at its northern end (Talebian et al., 2004; Wang et al., 2004b; Berberian, 2005; Fielding et al., 2005; Jackson et al., 2006). Although the earthquake fault slipped at a shallow depth, little of this slip reached the ground surface, and the earthquake fault was unknown before the 2003 earthquake. The coseismic surface ruptures occurred in a region devoid of active morphotectonic features. Additional small-scale surface ruptures developed to the north of Bam on four subparallel faults, as well as some fractures along the Bam-Baravāt segment of the Bam fault system (Fu et al., 2004; Nakamura et al., 2004; Suzuki et al., 2004; Talebian et al., 2004; Wang et al., 2004b; Fielding et al., 2005; Funning et al., 2005; Peyret et al., 2007).

The Envisat Advanced Synthetic Aperture Radar (ASAR; Funning et al., 2005) showed that the observed deformation pattern cannot be explained by slip on a single planar fault. The authors found out that slip occurred on

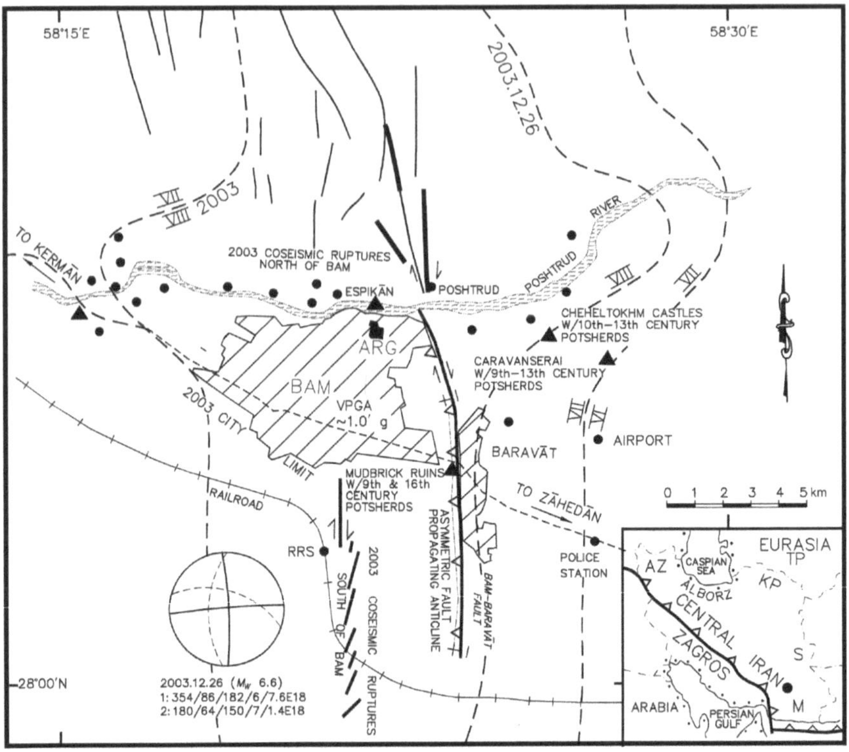

FIGURE 14.6 Meizoseismal area of the 26 December 2003 M_w 6.6 Bam earthquake with coseismic surface faulting. Fault and other symbols as in Figure 11.1. The city of Bam and town of Baravāt are crossed hatched and the location of the historic Arg (citadel) of Bam is marked by a filled square to the north of the city. The Bam fault at Baravāt (Berberian 1976b,e); the Bam fault to the south of the city (Talebian et al. 2004). The reactivated faults are shown by thicker lines. Maximum displacement was recorded on the Bam fault just south of the city. Isoseismal lines (in MMI) are shown by dashed lines. Focal mechanism of the earthquake constrained by body-wave modeling with centroid depth (Jackson et al., 2006). The dashed lines on the focal sphere show the nodal planes for the second thrust subevent with right-lateral motion. Filled triangles are ruined archeological sites (Fehrevari and Caldwell, 1967). VPGA: Location of Bam governor's building accelerogram with peak horizontal and vertical ground acceleration of ~1.0 g. *Modified from Berberian (2005).*

two subparallel faults: (i) 85% of moment release occurred on the previously unknown strike-slip Bam fault south of the city (with peak slip of over 2 m occurring at depth of ~5 km); and (ii) the remainder of the moment release occurred as a combination of strike-slip and thrusting on a southward extension of the Bam fault at Baravāt mapped by Berberian (1976b,e; Figure 14.5).

The Envisat radar interferometry and seismic body-wave inversion showed right-lateral strike-slip motion on a nearly N–S striking vertical fault to the south of the city. The coseismic surface fault was formed of four discontinuous ruptures each with a few cm right-lateral slip. Interferometry slip modeling and aftershock distribution indicated that the length of the rupture was about 20 km with a fault width of 15 km and a maximum slip of 2.5 m (Nakamura et al., 2004; Talebian et al., 2004; Wang et al., 2004b).

The total rupture length estimated by Wang et al. (2004b) is about 24 km in three segments (13 km in the south, 6 km in the north, and a middle connecting segment of 5 km). The southern and northern segments are in agreement with those reported by Talebian et al. (2004) and Fielding et al. (2005). The middle segment introduced by Wang et al. (2004b), which connects between the southern and the northern fault segments below the city of Bam, seems arbitrary and provides minimal insight with respect to comprehending the situation underneath the city. Wang et al. (2004b) also maintained that the southern and the middle fault segments dip 75–80°E, while the northern segment dips 55°W. More than 80% of the seismic moment was released from its southern segment of about 13 km, where slip reached a maximum of up to 20 cm resulting in a stress drop of at least 6 MPa (see also Jackson et al., 2006).

The nearly N–S strike of the Bam, Gowk, and Nāyband faults (Figures 14.3 and 14.7) are nearly aligned to the direction of the regional stress in eastern Iran (Nilforoushan et al., 2003; Vernant et al., 2004), which may cause very little strain accumulation and, therefore, may lead to a much longer earthquake recurrence period (Nalbant et al., 2006). Unlike the northern segment of the Gowk fault, no historical or post-1900 instrumental period earthquakes have been recorded along the Nāyband and South Gowk faults (Figures 14.7 and 14.3).

Some factors led to the strong ground shaking with a peak vertical ground acceleration of nearly 1.0 g/980 cm/s^2 and horizontal acceleration of 0.8 g/779 cm/s^2 (BHRC, 2003, 2004; Shoja-Taheri et al., 2005). Furthermore, the poor quality of the physical infrastructure and a lack of adequate risk management by the Iranian government led to the complete destruction of the city of Bam with only a medium-magnitude earthquake. The three geophysical factors leading to complete destruction of Bam were: (i) the Rayleigh-like speed of the Bam rupture propagation over the Bam earthquake fault immediately south of the city; (ii) the high-slip velocity, which exceeded 2 m/s over a large part of the earthquake fault; and (iii) the strong directivity, resulting from the location of the hypocenter opposite the city relative to the rupture area, which focused the elastic energy released directly toward the city of Bam. The Rayleigh-like speed of the Bam rupture implies that the Bam earthquake fault was a weak and mature seismogenic fault that broke easily (Bouchon et al., 2006).

Despite the published detailed analysis of the Bam earthquake and the N–S coseismic surface fault ruptures, unsubstantiated and conflicting claims regarding the source of this earthquake abound (Zāre', 2003; Andalibi, 2004; Fatemi Aghda, 2004; Ghafory-Ashtiany, 2004; Hessāmi et al., 2004; Zāre' and Hamzehloo, 2004). Zāre' (2003), and Zāre' and Hamzehloo (2004) introduced the Bam fault as a single fault, 170 km long extending in a NW–SE direction. They also claimed that the intersection zone between the N–S (Nāyband, Gowk, and Bam) and the NW–SE (Kuhbanān) faults is the main source for disastrous earthquakes in the region. Their statement cannot be warranted.

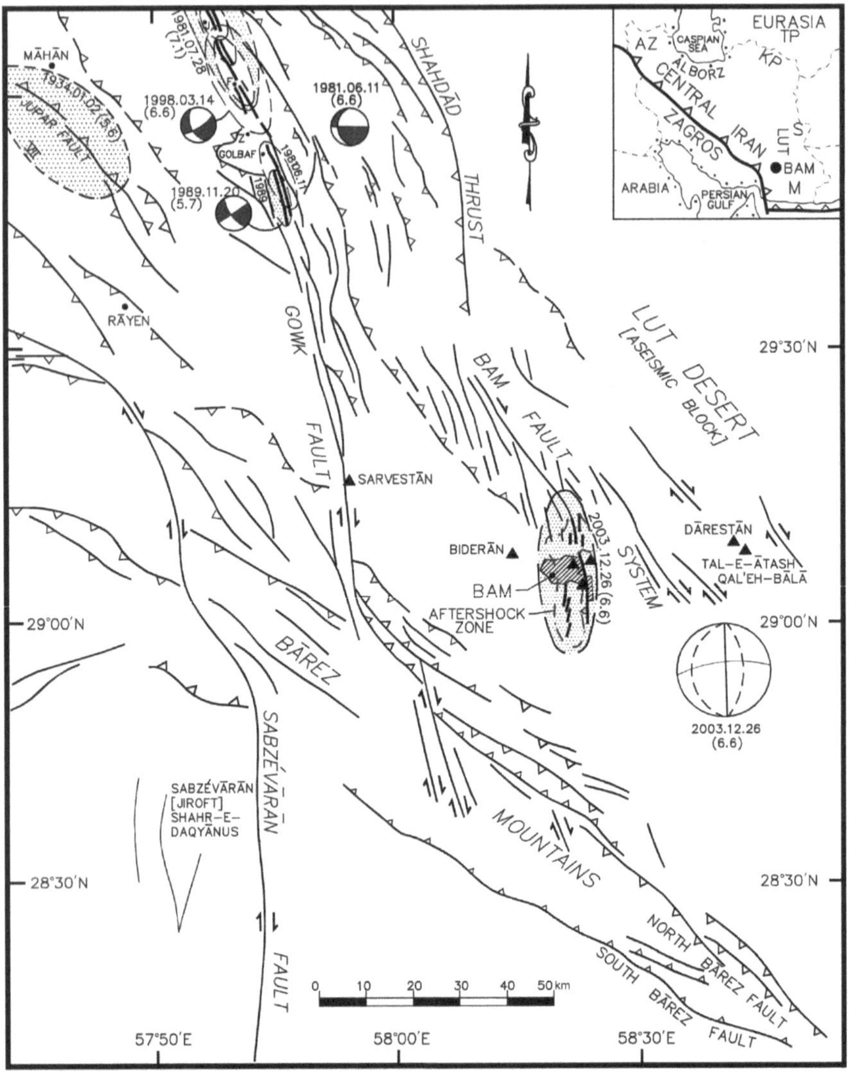

FIGURE 14.7 Fault and recorded seismicity map of the area south of the Gowk fault system, west of the rigid Lut Block in southeast Iran. Fault and other symbols as in Figure 11.1. Filled triangles show location of the ruined historic sites. Focal mechanism of the earthquakes constrained by body-wave modeling with centroid depths (Berberian et al., 2001; Jackson et al., 2006). Aftershock zones recorded at Bam (Tatar et al., 2004). The strike-slip deformation along the southern end of the Gowk fault and the Bam fault systems, and the northern end of the Sabzévārān fault is absorbed by active shortening along the Jupār and Bārez Mountains. No historical record along the southern segment of the Gowk or the Sabzévārān faults has survived. *Modified from Berberian (2005).*

Hessami et al. (2004) stated that the fault segment responsible for the Bam earthquake is difficult to locate, as there is no direct surface faulting associated with the earthquake. They also claimed that the qanāt lines are displaced right-laterally by the Bam fault about 11 ± 1 m, which cannot be substantiated. By speculating the maximum age of the qanāts as 3000 years, they implied a minimum horizontal slip rate of 3–4 mm/year. Ghafory-Ashtiany (2004) went a step further, claiming that the whole system of fresh ruptures associated with the main event is not a direct manifestation of the earthquake faults but they are secondary structures.

Fatemi Aghda (2004) mentioned that the earthquake was caused by movement along a new "secondary" fault. Finally, Andalibi (2004) reported that the earthquake occurred on the confluence point of three faults: (i) the N–S Arg fault (?) with a left-lateral sense (?); (ii) the N–S Bam fault with a left-lateral sense (?); and (iii) the W–E Tahrood fault (?), a distal part of the Nāyband segment by the complex thrust and right-lateral mechanisms. The speculative statement cannot be warranted with the well-studied field and seismological data.

14.4.1 Seismology

Teleseismic data show the location of the mainshock in the area to the southwest of Bam, far from the coseismic N–S strike-slip surface rupture and the locally recorded aftershock zone (Figures 14.6 and 14.7), which is due to location errors of the Iranian teleseismic data (Berberian 1979c), despite the use of the national seismographic networks of INSN (IIEES) and ITSN (IGUT). The locally recorded aftershocks in January 2004 indicated strike-slip faulting that extended approximately 30 km N–S (much longer than the mapped coseismic surface fault) underneath the city of Bam, with focal depths located 6–20 km below the surface (Tatar et al., 2005). There was a conspicuous absence of locally recorded aftershocks from surface to 6 km below the ground.

The major fault slip that caused the 2003 earthquake occurred at depths below 1–2 km, beneath the surface ruptures south of Bam, and extended northward beneath the center of the city (Talebian et al., 2004; Fielding et al. 2005). Waveform inversion analysis of the mainshock shows a two-source mechanism with an interval of approximately 8–10 s: (i) a right-lateral strike-slip faulting with larger moment release and centroid depth of 6 km and (ii) a thrust mechanism, dipping 64°W, with some right-lateral motion with centroid depth of 7 km and smaller moment release (Figure 14.6). Waveform modeling of the mainshock suggests that the rupture propagated northward along the seismogenic fault (Talebian et al. 2004; Jackson et al., 2006).

During the Bam earthquake mainshock, more than 80% of the moment release occurred on a near-vertical right-lateral strike-slip fault extending from the city of Bam southward for about 15 km, with slip of up to 2 m but mostly restricted to the depth range 2–7 km (Jackson et al., 2006). Little of

TABLE 14.4 *P* and *SH* Body-Waveform Inversion Source Parameters of the 26 December 2003 M_w 6.6 Bam Earthquake (Jackson et al., 2006)

Earthquake	Time (UTC)	Coordinate (N°–E°)[a]	Strike (°)	Dip (°)	Rake (°)	Centroid Depth (km)	M_o (Nm) ($\times 10^{18}$)	M_w
2003.12.26 (single Source)	01:56	28.95–58.26	354	86	182	6	7.6	6.6
Two sources			354	86	182	6	7.6	
			180	64	150	7	1.4	

See also Figure 14.6.
[a]*Epicenter is from Engdahl et al. (2006).*

the slip in the mainshock reached the surface and, more importantly, aftershocks revealed that a ~12 km vertical extent of a deeper part of the fault system remained unruptured beneath the coseismic rupture plane, at depth of 8–20 km. Some oblique-reverse slip (up to 2 m, and less than 20% of the released seismic moment) occurred at a restricted depth of 5–7 km on a blind west-dipping fault that projects to the surface at the Bam-Baravāt escarpment mapped by Berberian (1976b) earlier (Jackson et al., 2006) (Table 14.4).

14.5 THE 22 FEBRUARY 2005 M_W 6.4 DĀHUIYEH EARTHQUAKE

At least 612 people were killed and 1411 injured in the mountainous area of the Kermān province of southeastern Iran. An estimated 8000 homes in 40 villages were damaged or destroyed. The preliminary damage was estimated at US$ 79 million (700 billion riāls).

The earthquake was associated with an intramountain surface rupture with an ~E–W strike and a thrust mechanism dipping 60°N (Figure 14.8). Vertical displacement of up to 105 cm was recorded [raise to the north], with surface ruptures crossing the road between Darbidkhun (Darbidkhān) and Kotkan over a total distance of about 13 km along a mapped thrust fault north of Darbidkhun, east of Dāhuiyeh (GSI, 1992b; Talebian et al., 2006a; Figure 14.8). The coseismic fault has an oblique angle to the NW–SE Kuhbanān range-front right-lateral strike-slip fault and seems to be one of the thrust splays that come off the Kuhbanān strike-slip fault to the east. It is associated with the termination of the Kuhbanān strike-slip fault and may indicate block rotation about vertical axes (Talebian et al., 2006a).

The apparent episodic bursts of activity in the region started in 1981 along the Gowk fault (discussed earlier in Chapter 13 and above), which migrated southward to Bam in 2003 (Figures 13.11, 14.3, and 14.7) and reached the

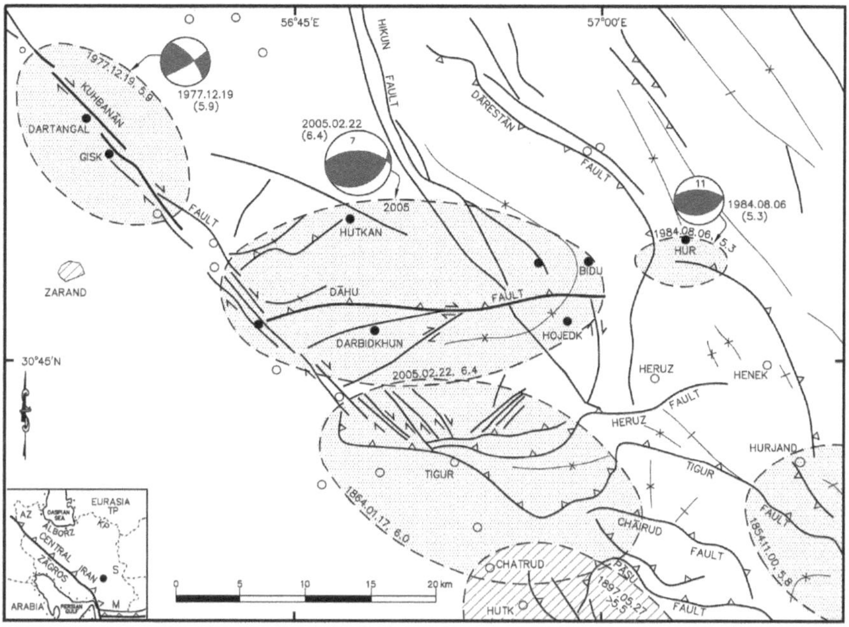

FIGURE 14.8 Meizoseismal areas of the recorded earthquakes along the southern tip of the range-front Kuhbanān right-lateral strike-slip fault and the oblique thrust splays. Fault and other symbols as in Figure 11.1. Focal mechanism of the earthquakes constrained by body-wave modeling with centroid depths (1984: Baker, 1993; 2005: Talebian et al., 2006a). Focal mechanism of the 1977 earthquake (Berberian et al., 1979a). *Modified from Berberian (2005).*

north Kermān area in 2005 (Figure 14.8). The 2005 earthquake took place to the west of the meizoseismal area of the 6 August 1984 M_w 5.3 Hur earthquake (Figure 14.8) with E–W thrust fault plane solution (Berberian, 2005). The combined Coulomb stress changes with a maximum value of 0.9 bars at the hypocentral region of the 2005 Dāhuiyeh earthquake appear to have advanced it toward failure (Nalbant et al., 2006).

Despite the mountain-bordering reverse faults on the Iranian plateau and the fact that the geological fault was mapped prior to the earthquake (GSI, 1992b), no pre-2005 earthquake active fault features were noticed along the 2005 intramountain coseismic surface rupture line. The lack of geomorphological active fault features of these types of faults may create serious seismic hazard evaluation (Talebian et al., 2006a).

14.5.1 Seismology

Inversion of long-period *P* and *SH* body-wave seismograms (Table 14.5; Figure 14.8) together with radar interferometry show rupture on an E–W thrust fault dipping 60° to the north (Talebian et al., 2006a).

TABLE 14.5 *P* and *SH* Body-Waveform Inversion Parameters of the 22 February 2005 M_w 6.4 Dāhuiyeh Earthquake (Talebian et al., 2006a)

Earthquake	Time (UTC)	Coordinate (N°–E°)	Strike (°)	Dip (°)	Rake (°)	Slip Vector Azimuth	Centroid Depth (km)	M_o (Nm) (×10^{18})	M_w
2005.02.22	02:25	30.774–56.736	270	60	104		7	4.83	6.4

See also Figure 14.8.

14.6 THE 31 MARCH 2006 M_W 6.1 CHĀLĀNCHULĀN EARTHQUAKE

At least 70 people were killed, more than 1300 people injured, 40 villages completely destroyed, and 300 villages damaged in the Silākhor plain, south of the city of Borujerd. The earthquake took in the seismic gap left after the 23 January 1909 M_w 7.4 Silākhur earthquake along the northwestern part of the Dorud fault segment of the Zāgros Main Recent fault (see Chapter 12 and Figure 12.1). Talebian et al. (2006b) reported short en-echelon fractures indicating right slip along the fault (Figure 2 in their short Persian report; gsi.ir).

14.6.1 Seismology

Inversion of the long-period *P* and *SH* body-wave seismograms (Table 14.6; Figure 12.1) together with radar interferometry showed right-lateral, strike-slip rupture along the Dorud fault. The InSAR uniform slip dislocation model excluded any significant rupture close to surface and indicated an average right-lateral slip of 1.07 m of 13.5 km long at a depth of 4 km along the Dorud fault. The InSAR variable slip model gave an average slip of 0.4 m displacement of 20-km-long faulting (Peyret et al., 2008).

TABLE 14.6 *P* and *SH* Body-Waveform Inversion Parameters of the 31 March 2006 M_w 6.1 Chālānchulān Earthquake (Peyret et al., 2008)

Earthquake	Time (UTC)	Coordinate (N°–E°)	Strike (°)	Dip (°)	Rake (°)	Centroid Depth (km)	M_o (Nm) (×10^{18})	M_w
2006.03.31	01:17	33.58–48.79	318	63	174	6	1.522	6.1
InSar uniform slip			320	60	180		1.70	
InSAR variable slip			320	60	180		1.58	

See Figure 12.1.

14.7 THE 20 DECEMBER 2010 M_W 6.5 AND 27 JANUARY 2011 M_W 6.2 SOUTH RIGĀN EARTHQUAKES

At least 7 people were killed, 25 injured, and 3 villages destroyed in the southeast of Kermān. The reason for so few casualties and little damage was that the earthquake took place in a remote, very low, and sparsely populated rural desert region with a mostly nomadic and seminomadic population living in tents about 45 km south of Rigān (Figure 14.9).

The 2010 and 2011 earthquakes were accompanied by coseismic minor surface fractures associated with a NE–SW right-lateral strike-slip (2010) and NW–SE left-lateral (2011) faulting (Figure 14.9). The main slip of both

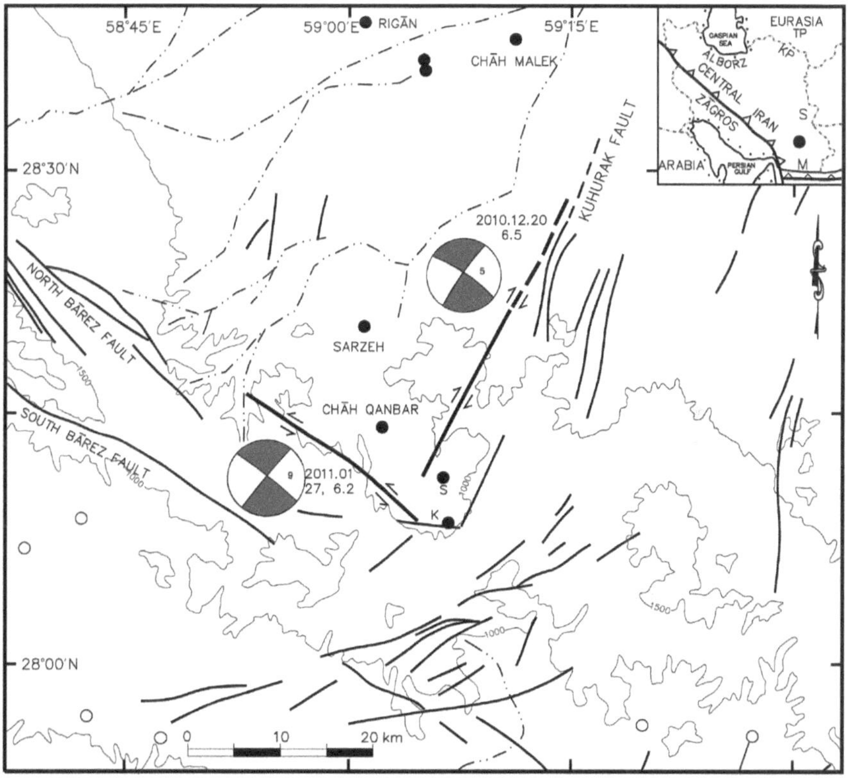

FIGURE 14.9 Surface ruptures of the south Rigān earthquakes (thicker lines) with their focal mechanism of the earthquakes constrained by body-wave modeling with centroid depths (Walker et al., 2013c). Fault and other symbols as in Figure 11.1. The southern tip of the Kahurak right-lateral fault was reactivated during the 2010 earthquake, whereas the southeastern tip of the North Bārez left-lateral fault reactivated 38 days later in 2011. The area is located to the south of the Lut Block and north of the Jāz Muriān Depression in southeast Iran, north of the Makrān accressionary prism.

earthquakes failed to reach the surface. Like the causative fault of the 25 December 2003 M_w 6.6 Bam earthquake, the 2010 and 2011 earthquake faults did not have any clearly identifiable effect on the geomorphology (Walker et al., 2013c).

Seismological investigation, SAR interferometry, and field investigations revealed that the 2010 earthquake involved an average of ~1.3 m right-lateral slip on a vertical fault trending ~210°. The 2011 earthquake resulted from an ~0.6 m slip on a conjugate left-lateral fault striking ~310° and confined within the zone of increased Coulomb stress from the 2010 event (Walker et al., 2013c). The 2010 fault is the southern tip of the Kahurak right-lateral fault (Figure 14.10); whereas the 2011 earthquake with a left-lateral mechanism took place on a conjugate fault located at the southeastern tip of the northern Bārez fault (Figure 14.10).

The head of the Fahraj District County (Figure 14.11), Mohammad Khorshid, reported that a fracture of about 2 km long and a width of up to 40 cm was developed at the edge of the Hosseinābād-e ‘Alamkhān village near the village houses (irna.ir, 22 December 2010). The nature of this fracture has not been investigated.

Foroutan et al. (2010) described en-echelon, left-stepping, right-lateral, strike-slip surface ruptures striking N70°–80°E, with maximum horizontal slip of 5 cm in the area north of Konārak village for a length of >2 km. They added that the surface fractures are the southwestern portion of the Kahurak fault as the southern termination of the West Neh fault in the damaged area. The authors drew the Kahurak fault as a continuous curved line to the southwest with a bending toward WSW, which cannot be warranted.

14.7.1 Seismology

Body-waveform modeling for both events show almost pure strike-slip faulting, with either left-lateral faulting on a northwest-southeast fault or right-lateral slip on a northeast–southwest fault (Figure 14.9; Table 14.7). The centroid depth of the 2010 event is at 5 km. The 2011 event is deeper, with a centroid depth of 9 km (Walker et al., 2013c).

Based on the dislocation grid search method, Ashtari Jafari (2012) suggested that the 20 December 2010 event was nearly strike-slip with a reverse component (strike 115°, dip 50°, and rake −5°), whereas the 27 January 2011 earthquake had some amount of reverse component (strike 125°, dip 75°, and rake −5°). Based on a synthetic Green's function deconvolution and IGUT/IRSC aftershock distribution, Ashtari Jafari (2012) assumed a downdip-propagated rupture migrated toward the southeast for both events along the continuation of the Bam fault in a NW–SE direction, which cannot be true.

Maleki et al. (2012), by using data from the national permanent seismogenic stations (IGUT/IRSC and IIEES) with the unknown velocity model,

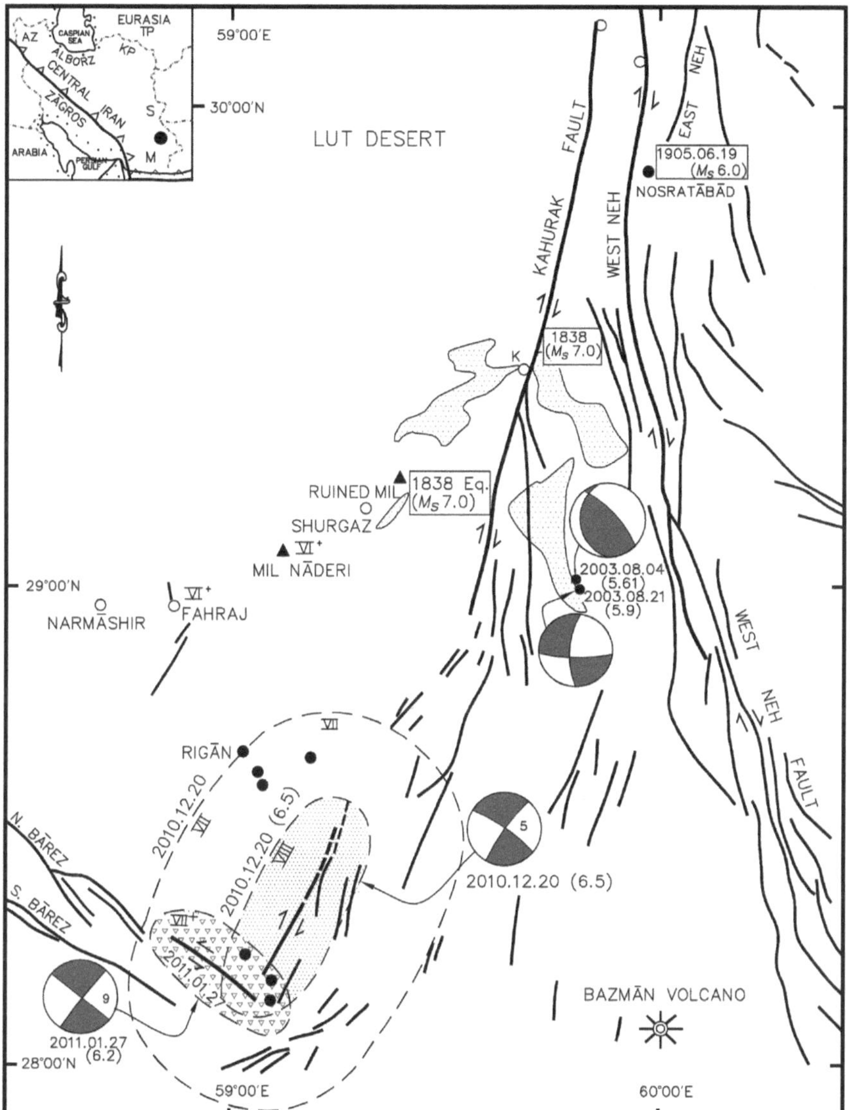

FIGURE 14.10 Recorded seismicity along the southern segment of the Kahurak fault since 1838, with unconstrained meizoseismal areas of the 2010 and 2011 South Rigān earthquakes. Fault and other symbols as in Figure 11.1. References as in Figure 14.9. The 2003 focal mechanism solutions (best-double-couple HRVD CMT solution).

introduced a 40-km-thick seismic zone, which seems unrealistic. By deployment of a six-station portable seismographic network, Rezapour and Mohsenpur (2013) found the aftershocks' depth range to be 1–12 km with ml ranging from 1.8 to 4.1 (the latter took place on 28 December 2010).

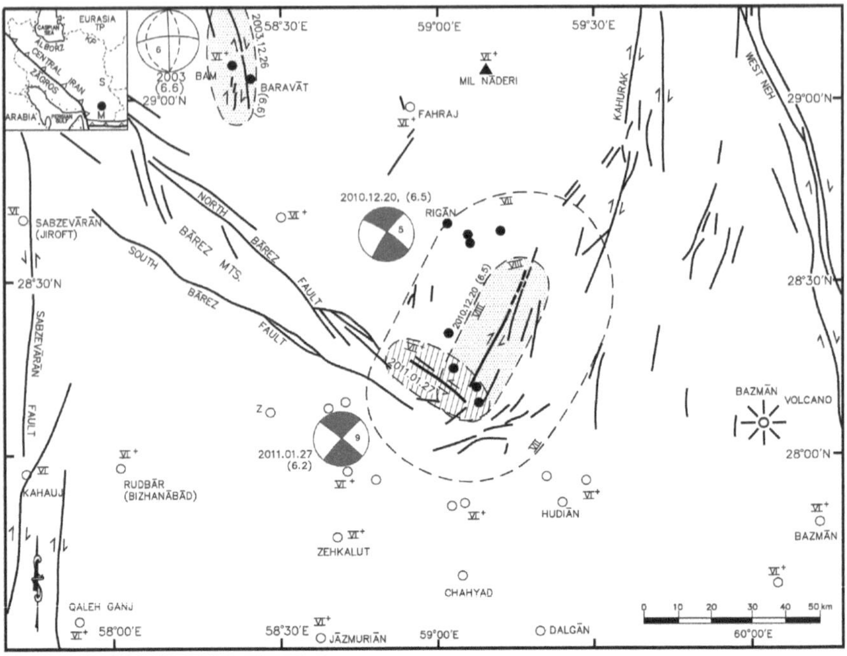

FIGURE 14.11 Recorded seismicity of the Bam-Rigān area to the south of the Lut Block and north of the Jāz Muriān Depression with unconstrained meizoseismal areas of the 2010 and 2011 South Rigān earthquakes. Fault and other symbols as Figure 11.1. References as in Figures 14.6 and 14.9.

TABLE 14.7 ***P*** **and** ***SH*** **Body-Waveform Inversion Parameters of the 20 December 2010** M_w **6.5 and 27 January 2011** M_w **6.2 South Rigān Earthquakes**

Earthquake	Time (UTC)	Coordinate (N°–E°)	Strike (°)	Dip (°)	Rake (°)	Centroid Depth (km)	M_o (Nm) ($\times 10^{18}$)	M_w	Method (Ref.)
2010.12.20	18:41	28.325–59.188	213	85	173	05	5.39	6.5	Body-wave (Walker et al., 2013c)
2011.01.27	08:39	28.169–59.044	221	87	176	09	1.448	6.2	Body-wave (Walker et al., 2013c)
2011.01.28	04:21	28.199–59.021	133	74	−14	12 fixed		5.2	CMT (HARV)

See also Figure 14.11.
HRVD: http://www.globalcmt.rog.

14.8 THE 11 AUGUST 2012 M_W 6.4 AND 6.3 SOUTH AHAR TWIN EARTHQUAKES

The double earthquakes, which took place within an 11-min interval (Figure 14.12), killed 327 people, injured more than 3000, left about 30,000 homeless, totally destroyed more than 20 villages (including Zangābād, Gurdeh, Dino, Hājebāj, Sarand, and Shāhsavār), and damaged 425 other villages from 50% to 90%. Hospitals in the stricken areas as well as the 540 km of rural (130 km) and secondary (410 km) roads and bridges were damaged. Forty percent of the 713 schools were damaged. About 25% of other public buildings were also severely damaged. Buildings in the Ahar and Varzeqān towns were also damaged. Several industrial facilities suffered economic losses.

The earthquake caused a total break in communications lines, electricity, and water supplies. Electricity was also lost in the city of Tabriz, located about 70 km to the southwest of the epicentral area. On 10 October 2012 an unrated faulty electric wire set fire to 24 tents of the survivors still living outside under the cloth despite the very cold nights and earlier housing promises; one 70-year-old woman was killed in the fire. Heavy snow on 13 and 14 November made life very difficult for the survivors who were still living in

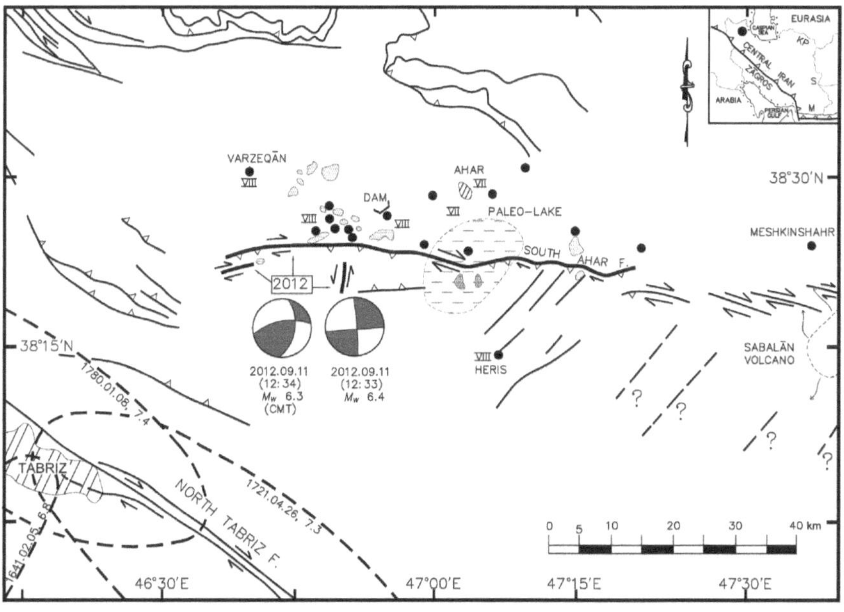

FIGURE 14.12 Epicentral area of the 11 September 2012 South Ahar twin earthquakes (12:33 UTC M_w 6.4, and 12:34 UTC M_w 6.3) in the area northeast of the North Tabriz right-lateral strike-slip fault in northwest Iran. Fault and other symbols as Figure 11.1. The South Ahar fault (Faridi et al., 2006; Baron et al., 2013). Focal mechanisms (Copley et al., 2013 for the first and HRVD CMT for the second events). Locations of a paleo-lake (Baron et al., 2013) and paleo-landslides are added.

tents for nearly 3 months after the earthquake. Maximum uncorrected accelerations of 0.425 g (first event) and 0.534 g (second event) were recorded in the BHRC Varzeqān station, located to the northwest of the South Ahar fault (38°30′N–49°39′E). Recorded acceleration at Ahar (38°28′N–47°04′E) and Heris (38°14′N–47°06′E) were 0.038 and 0.018 g, respectively (bhrc.ac.ir). Numerous paleo-slope failures are present to the east and southeast of Varzeqān (Figure 14.12).

The earthquakes took place on previously unrecognized active faults with a lack of distinctive tectonic geomorphological indicators. Faridi (2012) introduced more than a 65-km-long surface faulting along the South Ahar fault system dipping to the south with a right-lateral slip component. The fault is located to the south of the Ahar town (Faridi et al., 2006). Baron et al. (2013) showed E–W right-lateral coseismic surface rupture along the South Ahar reverse fault dipping south. They also showed a short N–S coseismic surface rupture with left-lateral strike-slip displacement to the south of the South Ahar fault (south of Qara Darreh, to the west of the Ahar-Tabriz main road) in the Cretaceous deposits. No slip measurements were provided by the authors.

Copley et al. (2013) reported a 13-km-long, E–W, dominantly right-lateral, strike-slip surface faulting with a maximum horizontal slip of 100 cm, with left-stepping en-echelon, surface fissures on scales of meters to tens of meters in length. No proper field mapping of the surface faulting or horizontal and vertical slip measurements were prepared. The reported surface ruptures along the South Ahar fault took place near the crests of two broad topographic ridges and crossed an intervening valley. The surface ruptures associated with the second event were not observed in the visited areas during the short site visit, or the event was associated with an unknown blind fault.

14.8.1 Paleoseismicity

Baron et al. (2013) found several phases of deformation, some of which might have been initiated by ancient earthquakes and earthquake-induced slope failure:

- **i.** Opening of fractures around 42,072 ±497 cal. BP;
- **ii.** Calcite-filled joints (31,682 ±510 cal. BP) were subsequently deformed by thrust faulting;
- **iii.** Ahar basin dammed about 32,100 ±3300 cal. BP, possibly by an *earthquake-triggered slope failure* or during the Last Glacial Maximum;
- **iv.** Calcite-filled joints (23,935 ±503–24,346 ±582 cal. BP) were subsequently deformed by thrust faulting;
- **v.** Fluvial gravels (8200 ±2200 cal. BP) were deformed by a thrust phase; and
- **vi.** Paleosol (9612 ±89 cal. BP) was deformed *by minor strike-slip faults.*

TABLE 14.8 *P* and *SH* Body-Waveform Inversion Parameters of the 11 August 2012 South Ahar Earthquake (Copley et al., 2013 for the First Event, and HRVD CMT for the Second and Third Events)

Earthquake	Time (UTC)	Coordinate (N°–E°)	Strike (°)	Dip (°)	Rake (°)	Slip Vector Azimuth	Centroid Depth (km)	M_o (Nm) (×10[18])	M_W
2012.08.11	12:23		265	90	175	85	07	3.595	6.4
2012.08.11	12:34		255	63	134	100	f		6.3
2012.08.11	22:34		254	89	149		f		5.2

See Figure 14.12. f, fixed

Some indications of younger earthquakes in the Ahar city area were also detected (Baron et al., 2011). Despite contacting the authors, I was not able to review the report.

14.8.2 Seismology

P and *SH* waveform modeling of the first earthquake showed mostly a pure strike-slip mechanism (Copley et al., 2013), whereas the CMT solution of the second event indicated an oblique combination of reverse and strike-slip motion (HARD CMT, USGS W-phase solution) (Table 14.8; Figure 14.12).

14.9 THE 16 APRIL 2013 M_W 7.7 GOSHT-SARĀVĀN SLAB EARTHQUAKE

About 40 people were killed, 300 injured, and 35,000 left homeless in a remote and sparsely populated area of southwest Pakistan. About three people were reportedly injured in southeast Iran. The earthquake was felt IV (MMI) at Bandar ʻAbbās port (560 km to the southwest) and III at Shirāz (860 km to the west). An aftershock of M_w 5.6 took place on 17 April 2013.

Ground shaking of this intermediate slab event (Barnhart et al., 2014) apparently triggered slip on shallow faults in the Iranian Makrān. Sharifi and Qasabian (2013) reported a 5-km-long surface rupture striking N150°E with maximum 8-cm left-lateral displacement. The reported fracture is located about 500 m to the SW of the Sarāvān fault. It apparently runs parallel to the fault from west of Gosht village [in the NW] to Dampani village [in the SE].

14.10 THE 11 MAY 2013 M_W 6.1 GOHARĀN EARTHQUAKE

The earthquake killed 2 and injured about 20 people in a remote, mountainous area of the western Makrān. It was felt with the intensity of VI$^+$ at Irar, VI at

Goharān and Sardasht, V at Jakdān and Bolbolābād, IV at Jāsk and Mināb, and III at Bandar ‘Abbās Port. An alarming foreshock warned the people to stay out of their houses.

An 8-km-long surface rupture trending N60°–70°E, dipping steeply south with 20-cm left-lateral displacement cutting the Oligo-Miocene deposits was reported to extend from Irar village in the west to Kumestān in the east (gsi.ir, 2013). In another report, Bolurchi et al. (2013) introduced a nearly E–W (N55°–65°E) surface rupture dipping 80° south with a left-lateral displacement at least from Dormush to Irar and beyond cutting all the regional structural trends. Zaré and Ansari (2013) reported extensional fractures of 6–20 m in a NNW–SSE and E–W direction. The exact location of the observed fractures was not reported.

14.11 OVERVIEW: COSEISMIC SURFACE FAULTING OF THE HISTORICAL AND MODERN EARTHQUAKES

The purpose of Chapters 11–13 and this chapter and Tables 14.9–14.11 is (i) to search for and identify seismogenic faults of the historical earthquakes on the Iranian plateau by utilizing the current and near-current accounts and available geomorphological, remote sensing, geological and geophysical data backed with field investigation; and (ii) to create an improved data set to help determine the earthquake hazard, risk minimization, active tectonics studies, and future paleoseismologic trenching for interested groups of earthquake geologists, seismologists, engineers, planners, and authorities. The summary of the latest data on historical and modern coseismic surface faulting of the Iranian plateau is presented in Tables 14.9–14.11. The historical data collection and analysis of earthquakes and active faulting, which was initiated in the early 1970s (Berberian, 1976a, 1977a, 1981a, 1994, 1995, 1997a, 2005) and has benefitted from numerous individual investigations cited in the aforementioned chapters, can also be used to define zones of future hazards and in risk minimization efforts.

Tables 14.9 and 14.10 show cases of coseismic surface faulting covering the period from ca. 280 BCE through 2014 [though for the most recent cases, the results of the ongoing research have not yet been published]. Almost all the faults that showed pre-1900 seismic activity with large-magnitude earthquakes appear to be relatively quiescent during the short period of 1900–2014. Almost all the documented coseismic surface faults could have been predicted from the active geomorphological features observed along the faults. During the period of 1900–1962, about 16 cases of documented coseismic surface faulting were collected, whereas from 1962–2014, about 24 surface faulting cases were recorded (Table 14.9). Table 14.11 presents several cases inferred mainly from elongated shape of the meizoseismal areas along known active faults. The cases in this table require further detailed research and paleoseismic trench studies.

TABLE 14.9 Earthquakes in Iran Associated with Documented Coseismic Surface Fault Ruptures (ca. 943–2013) Deduced Mainly from Contemporary or Near-Contemporary Accounts with Meizoseismal Areas of Large Earthquakes

Eq. Date	Local Time	Epicenter (00°N–00°E)	Location	Fault Name	$\sim M_s$	M_w	$\sim I$ (MMI)	CD_w	FT MEC.	FT Length (km)	FT Width[a] (km)	FT Strike (°)	FT Dip (°)	FT Rake (°)	H (cm)	V (cm)	Mean Slip[b] (cm)	Moment ($\times 10^{18}$ Nm)
943.08.06–09.09			Samalqān	Shoqān	7.4		IX$^+$		R	?					?	++		
1042.11.04	19:30		Tabriz	N. Tabriz	>7.0		IX		RL	?					?	?		
1493.01.10	Midday		Nauzād	Nauzād	7.0		IX		R	>12					?	?		
1550.01.20–1551.01.08			Tabriz	N. Tabriz	>6.0		>VII$^+$		RL	?					?	?		
1721.04.26	07:48		Shebli	North Tabriz	7.3		IX		RL	>35					?	?		
1780.01.08	19:00		Tabriz	North Tabriz	7.4		IX		RL	>42					?	400		
1838.03.27–1839.03.17			Nosratābād	Kahurak	7.0		IX		RL	>70					?	?		
1877.00.00			Āb-e Garm, Sirch	Gowk	5.6		VII		RL	Few					?	?		

Eq. Date	Time (UTC)	Epicenter (00°N–00°E)	Location	Fault Name	M_s	M_w	I (MMI)	CD_w	FT MEC.	FT Length (km)	FT Width[a] (km)	FT Strike (°)	FT Dip. (°)	FT Rake (°)	H (cm)	V (cm)	Mean Slip[b] (cm)	Moment ($\times 10^{18}$ Nm)
1909.01.23	02:48		Silākhor	Dorud (ZMRF)	7.4	7.4	IX$^+$		RL	>70					?	>100		
1923.09.22	20:47	29.20–56.90	Lālehzār	Chamanrang	6.7	6.7	VIII$^+$		R	?					?	?		
1929.05.01	15:37	37.89–57.58	Bāghān	Bāghān	7.3	7.1	IX$^+$		RL	>70					?	>200		

Continued

TABLE 14.9 Earthquakes in Iran Associated with Documented Coseismic Surface Fault Ruptures (ca. 943–2013) Deduced Mainly from Contemporary or Near-Contemporary Accounts with Meizoseismal Areas of Large Earthquakes—Cont'd

Eq. Date	Time (UTC)	Epicenter (00°N–00°E)	Location	Fault Name	M_s	M_w	I (MMI)	CD_w	FT MEC.	FT Length (km)	FT Width (km)	FT Strike (°)	FT Dip. (°)	FT Rake (°)	H (cm)	V (cm)	Mean Slip (cm)	Moment ($\times 10^{18}$ Nm)
1930.05.06	07:03		Salmās Foreshock	Salmās	5.4		VII		RL	>1					?	?		
1930.05.06	22:34	38.17–44.66	Salmās	Salmās, Derik	7.2	7.1	IX		RL RL?	>43 >5					400 ?	>500 100		
1933.11.28	11:09		NW. Behābād	Kuhbanān	6.2	6.2	VIII		RL	>12					?	100		
1936.06.30	19:26		Ābiz	Ābiz	6.0	6.0	VII^+		RL	>8								
1940.05.04	21:01		Estāyesh	Zelzelehkhiz	6.2	6.4	6.4	VIII	R	?					?	?		
1941.02.16	16:39		Kumirān	Chāhak	6.2	6.1	VIII		RL	>12					?	50		
1947.09.23	12:18	33.55–58.69	Charmeh	Changrakhsh	6.9	6.8	VIII^+		RL	>20					100	80		
1947.09.26	03:04		Dusrābād	Ferdows	6.0	6.0	VII^+		R	>6					?	?		
1957.12.13	01:45	34.34–47.63	Fārsinaj	Sahneh (ZMRF)	6.7	6.8	VIII		RL	?					?	?		
1958.08.14	11.27		Givaki	Garrin (ZMRF)	5.7		VII^+		RL	>15					?	200		
1958.08.14	15:26		Kalādeh	Sahneh	5.5		VII^+		RL	>8					?	?		
1958.08.16	19:13	34.25–47.84	Firuzābād	Nahāvand (ZMRF)	6.6	6,6	VIII^+		RL	>20					?	10		

1958.09.21	16:18		Kargsar	Dinevar (ZMRF)	5.2		VII		RL	?					?	10		
1962.09.01	19:20	35.55–49.83	Bu'in	Ipak	7.2	7.0	IX		R, LL	>70					>50	>76		36.8
1968.08.31	10:47	34.04–58.95	Dasht-e-Bayāz	West Dasht-e Bayāz	7.2	7.1	IX	17 10	LL	70		254 320	84 70	5 90	450	250		54.78 4.57
1971.02.14	16:27	36.55–55.68	Serokhi	Serokhi	5.3	5.5	5.7	11	R	>2					–	Few		0.40
1976.11.25	12:22	39.08–44.03	Chālderān	Chālderān-Khoi	7.2	7.0	IX		RL	>50					350	Few		
1977.12.19	23:34	30.90–56.41	Dar Tangal	Kuhbanān	5.7	5.9	VII	07	RL	19.5		328	58		20	6		
1978.09.16	15:35	33.24–57.38	Tabas-e Golshan	Tabas	7.4	7.3	IX^+	09	T	75		355	16	155	–	150		105
1979.11.14	02:21	33.95–59.72	Korizān	Ābiz	6.7	6.6	VIII^+	10	RL	20		160	89	−177	100	60		9.8
1979.11.27	17:10	34.05–59.75	Koli	East Dasht-e Bayāz	7.2	7.1	IX	08	LL	55		261	82	8	390	190		50.5
1979.12.07	09:23	34.07–59.84	Kalāt-e-Shur	Ābiz	6.1	5.9	VIII	10	RL	15		113	84	21	?	?		1.0
1981.06.11	07:24	29.85–57.68	Golbāf	Gowk	6.6	6.6	VIII^+	20 12	RL	14.5	15	169 182	52 88	156 198	3	5	140	9.48
1981.07.28	17:22	29.97–57.76	Sirch	Gowk	7.0	7.0	IX	18	RL	65	16	177	69	194	43	40	270	36.69
1989.11.20	04:19	28.89–57.71	S. Golbāf	Gowk	5.7	5.8	VII	10	RL	11 08		145	69	188	4	1		0.70
1990.06.20	21:00	36.99–49.22	Rudbār	Rudbār	7.4	7.3	IX^+	13	LL	80		288	88	−9	100	120		140
1990.11.06	18:45	28.24–55.46	East Furg	High Zāgros	6.7	6.5	VIII^+	05	T	15		268	33	90	–	150		5.17
1997.02.04	10:37	37.73–57.31	Nāveh	Nāveh	6.7	6.6	VIII^+	08	RL	15–20		326	75	173	100			5.75

Continued

TABLE 14.9 Earthquakes in Iran Associated with Documented Coseismic Surface Fault Ruptures (ca. 943–2013) Deduced Mainly from Contemporary or Near-Contemporary Accounts with Meizoseismal Areas of Large Earthquakes—Cont'd

Eq. Date	Time (UTC)	Epicenter (00°N–00°E)	Location	Fault Name	M_s	M_w	*I* (MMI)	CD_w	FT MEC.	FT Length (km)	FT Width (km)	FT Strike (°)	FT Dip. (°)	FT Rake (°)	H (cm)	V (cm)	Mean Slip (cm)	Moment ($\times 10^{18}$ Nm)
1997.05.10	07:57	33.84–59.81	Zirkuh-e-Qā'enāt	Ābiz	7.2	7.2	IX	10	RL	125		156	89	200	230	90		60.5
1998.03.14	19:40	30.13–57.58	Fandoqā	Gowk	6.6	6.6	VIII$^+$	05	RL	23	12.4	156	54	195	300	95	170	9.09
1998.11.18	07:39	30.32–57.58	Chahār Farsakh	Gowk	5.1	5.3	VI$^+$		RL	04		174	55	178	<1	<1		0.13
2002.06.22	02:58	35.59–49.02	Changureh	Ābdarreh	6.4	6.4	VIII	10	R	14		120	49	101		16		5.6
2003.12.26	01:56	28.95–58.27	Bam	Bam and Baravāt	6.6	6.6	VIII$^+$	5.5 6.7	RL T	12 15	9 1	355 180	86 64	−178 150	20	?	2.1 2.0	7.6 1.4
2005.02.22	02:25	30.76–56.80	Dāhuiyeh	Dāhu	6.5	6.4	VIII	07	T	15	11	270	60	104		105	08	4.8
2006.03.31	01:17	33.58–48.79	Chalānchulān	Dorud (ZMRF)	5.9	6.1	VIII	06	RL	?		310	63	174	<1	<1		1.52
2010.12.20	18:41	28.37–59.15	South Rigān	Kahurak	6.7	6.5	VIII$^+$	05	RL	15		213	85	173	1.3	–		6.5
2011.01.27	08:38	28.15–59.05	South Rigān	North Bārez	6.2	6.2	VIII	09	LL	7.5		221	57	176	0.6	–		6.2

Complete references for each event are cited in the text and the related figure captions and event tables.
CD_w, centroid depth, waveform modeling (conceptual point in space representing a weighted center of rupture surface during earthquakes); FT Length (in km): coseismic surface fault length (in km); FT Mec, fault mechanism: LL, left-lateral strike-slip; R, reverse; RL, right-lateral strike-slip; T, thrust; and ZMRF, Zāgros Main Recent fault. M_s, surface wave magnitude. M_w, moment magnitude. *I*, intensity (Modified Mercalli scale). *H*, horizontal displacement. *V*, vertical displacement.
[a]*Calculated based on the slip-seismic moment relation of Wells and Coppersmith (1994).*
[b]*Based on empirical relation of Kanamori and Anderson (1975).*

TABLE 14.10 Early Earthquakes in Iran Associated with Possible Coseismic Surface Fault Ruptures Deduced Mainly from Late-Contemporary Accounts and Limited Macroseismic Data

Eq. Date	Local Time	Epicenter (00°N–00°E)	Location	Fault	$\sim M_s$	M_w	$\sim I$ (MMI)	FT MEC.
ca. 1200 BCE	–	?	Komesh (Dāmghān)	Dāmghān	–	–	–	LL
ca. 280 BCE	–	35.48–51.82	Rhagae (Ray)	Pārchin	>7.0	>VIII		R, RL
856.12.22	23:00	36.23–54.14	Komesh (Dāmghān)	Dāmghān	7.2	IX⁺	IX⁺	LL

Complete references for each event are cited in the text and the corresponding figure captions.
LL, left-lateral strike-slip; R, reverse; RL, right-lateral strike-slip.

Except for the greater part of the Zāgros Mountains (where the uppermost sedimentary cover is mechanically decoupled along numerous tectonically incompetent beds) and the Makrān subduction zone (which is offshore and not accessible), active intracontinental faults on the surface of the Iranian plateau are superbly exposed in semi-arid and arid conditions and can easily be seen in the field and with remote-sensing tools. Nonetheless, a comparison of Tables 14.9–14.11 with the historical and modern medium- to large-magnitude earthquakes (Tables 17.10–17.12) clearly demonstrates that little is known about the responsible faults of the major historical earthquakes on this part of the planet.

Despite causing stronger shaking and large casualties, the reverse-fault earthquakes have rarely produced surface ruptures on the Iranian plateau. Only a few historical earthquakes, such as the 1962 Bu'in, 1971 Serokhi, 1978 Tabas-e Golshan, 1990 East Furg, 2002 Changureh, and 2005 Dāhuiyeh, were expressed by surface ruptures with slips much smaller than the calculated subsurface slips. Hence, the observed coseismic surface ruptures make up only part of the subsurface rupture length and displacement. In most cases, the rupture failed at shallow depths and was accompanied by secondary ruptures associated with folding. Hence, the ruptures showed the poorest correlation between earthquake magnitude, length of surface rupture, and displacement (if any). During reverse earthquake faulting, the maximum principal compressive stress is due to horizontal tectonic loading, whereas the load stress is the minimum principal stress, which decreases more rapidly than horizontal stress as the surface is approached. The resultant increase in the differential stress and maximum compressive shear stress leads to the coseismic rupture failure at the surface. Therefore, in most cases, reverse faults are unlikely to be identified in paleoseismic trench excavations. However, they show valuable active geomorphologic indicators related to folding.

TABLE 14.11 Earthquakes in Iran Associated with Coseismic Surface Fault Ruptures (ca. 874–2014) Inferred Mainly from Elongated Shape of the Meizoseismal Areas Along Known Active Faults

Eq. Date	Local Time	Epicenter (00°N–00°E)	Location	Fault	~M_s	M_w	~I (MMI)	FT MEC.
874.10.16–11.4	–	33.16–55.28	Jorjān (Gonbad Kāvus)	Khazar	>6.0		VII$^+$	R
912.08.18–913.08.07	–	34.61–47.50?	Dinévar	ZMRF	?	?	?	RL
943.08.06–09.03	–	37.50–56.75	Samalqān	Shoqān	7.4		IX$^+$	R
958.02.23	–	?	Ruyān	Moshā	7.1		X	R, LL
1008.04.27	Night	34.61–47.50	Dinévar	ZMRF	7.0		VIII	RL
1052.06.02	–	36.22–57.68	Baihaq (Sabsevār)	Sabsevār	7.0		IX	R
1119.12.10	Night	36.38–50.00	Qazvin	N. Qazvin	6.5		VIII	R
1127.01.17–1128.01.06	–	36.33–53.48	Farim (Parim)	N. Alborz	6.8		VIII$^+$	R
1177.05.01–05.30	–	36.10–50.43	Ray-Qazvin	N. Tehran	7.0		IX	R
1208.07.16–1209.07.06	Day	36.22–59.00	SE. Neyshābur	N. Neyshābur	7.3		IX$^+$	R
1237.08.24–1238.08.13	–	34.28–58.62	Gonābād	Gonābād or Bidokht	>6.0		>VII$^+$	
1251.03.26–1252.03.13	–	36.20–58.88	Shādyākh	Neyshabur system	>6.0		>VII$^+$	R
1270.10.07	Morning	36.35–58.45	W. Neyshābur	Neyshābur system	7.1		IX	R
1300.09.16–1301.09.05	–	36.10–53.10	Farim (Parim)	Larzaneh	6.7		VIII$^+$	R
1316.01.05	–	34.08–48.40	Borujerd	Nahāvand (ZMRF)	>6.5		>VIII	RL
1319.00.00	–	39.00–44.40	St. Taddeus	Siah Cheshmeh-Khoi	>6.5		>VII$^+$	RL
1336.10.19[a]	Dawn[a]	34.85–59.74	Jizd[a]	South Roshtkhār[a]	>7.0[a]		IX[a]	RL, R

1336.10.21[a]	Dawn[a]	34.65–59.55	Zuzan[a]	Jangal[a]	>7.0[a]	IX[a]	R
1366.09.07–1367.08.27	–	?	Kuchesfahān	Khazar?	?	VII	R
1389.01.30–02.27	Dawn	36.32–58.93	Neyshābur System	N. Neyshābur or Rayvand	7.3	IX[a]	R
1405.11.23	–	36.44–58.52	Neyshābur System	Binālud	7.4	IX[a]	R
1435.07.27–1436.07.15	–	37.16–55.28	Jorjān (Gonbad Kāvus)	Khazar	>5.5	>VII	R
1485.08.15	Sunset	36.75–50.00	Upper Polrud	Kelishom	7.2	IX	LL
1497.08.30–1498.08.18	–	37.18–55.30	Jorjān (Gonbad Kāvus)	Khazar	6.5	VIII	R
1549.02.15	Night	32.40–59.12	Birjand	Birjand	6.7	VIII+	R
1608.04.20	AM	36.38–50.50	Alamutrud	Alamutrud	7.4	IX[a]	
1641.02.05	20:00	37.95–46.05	Dehkhārqān	Talkhehrud	6.8	IX	?
1665.06.15–07.13	–	35.75–52.08	Damāvand	Moshā	6.5	VIII	R, LL
1665.07.14–1666.07.03	–	32.15–50.55	NW. Ardal	Ardal	6.5	VIII	R, RL
1673.07.30	–	36.45–59.25	Shāndiz	Shāndiz	6.6	VIII+	RL
1678.02.03	05:30	37.20–50.00	Lāhijān	Khazar	6.5	VIII	R
1675 or 1678 (?) Winter	Night	34.30–58.70	Gonābād	Gonābād	>6.5	>VIII	
1695.05.11	Dawn	36.98–57.28	Esfarāyen	Esfarāyen, Rayvand, or Jovain	7.0	IX	R
1755.06.07	AM	33.78–51.50	S. Kāshān	Kāshān	5.9	VII+	RL
1778.12.15	Dawn	34.04–51.30	Kāshān	Kāshān	6.2	VIII	RL
1804	–	36.32–57.22	Mehr (NW Sabsevār)	Sabsevār	>5.6	>VII	R
1809	AM	36.34–52.60	Āmol	Khazar	6.5	VIII	R

Continued

TABLE 14.11 Earthquakes in Iran Associated with Coseismic Surface Fault Ruptures (ca. 874–2014) Inferred Mainly from Elongated Shape of the Meizoseismal Areas Along Known Active Faults—Cont'd

Eq. Date	Local Time	Epicenter (00°N–00°E)	Location	Fault	$\sim M_s$	M_w	$\sim I$ (MMI)	FT MEC.
1824.06.02	–	29.70–51.50	Bishāpur	Kāzerun	>6.0		VIII	RL
1824.06.25	Dawn	29.75–52.40	Guyum (NW Shirāz)	Sabzpushān	6.4		VIII	RL
1830.03.27	Morning	35.81–51.76	Lavāsānāt	Moshā	7.1		IX	R, LL
1840.07.02	19:00	39.52–43.95	Ārārāt	Gāilātu (Pāmbukh)-Khoi	7.4		IX	RL
1843.04.18	08:00	38.50–44.80	Khoi	Khoi	5.9		VIII	RL
1844.05.12	17:00	33.55–51.35	Kāmu (SW Kāshān)	Joshaqān	6.4		VIII	RL
1851.06.00	08:30	36.83–58.52	Sarvelāyat	Kalidar	6.9		IX	RL
1863.12.30	21:48	38.12–48.52	Hir	Sangāvar	6.1		VIII	LL
1865.06.00	–	29.62–53.05	Dāryān	High Zāgros	6.0		VII$^+$	R
1871.12.23	Night	37.40–58.25	N. Quchān	Zurbārān?	7.2		IX	RL
1872.01.06	–	37.38–58.58	Darbandān (NNE Quchān)	Darbandān?	6.3		VIII	RL
1879.03.22	03:42	37.80–47.85	SE. Bozqush	S. Bozqush or Germirud?	6.7		VIII$^+$	
1880.00.00	–	32.00–50.68	Ardal	Ardal				R, RL
1890.02.07	06:00	34.18–51.22	Nassirābād	Kāshān	>5.5		>VII	RL
1890.07.11	05:55	36.62–54.60	Tāsh	Shāhkuh	7.2		IX	R
1891.12.14	Night	29.90–51.08	N. Kāzerun	Kāzerun	>5.5		>VII	RL
1893.11.17	19:36	36.96–58.35	S. Quchān	Chakāneh?	7.0		IX	R, RL

1894.02.26	01:30	29.50–53.30	Kharāmeh	High Zāgros	5.9	VII	R
1895.01.17	11:30	37.08–58.38	Quchān	Chakāneh?	6.8	VIII^{+}	R, RL
1895.00.00	17:30	34.18–51.23	Sāruq	Kāshān	>5.5	>VII	RL
1896.01.04	18:28	37.98–48.38	Sangāvar	Sangāvar	6.7	VIII^{+}	LL

Eq. Date	Time (UTC)	Epicenter (00°N–00°E)	Location	Fault	M_s	M_w	*I* (MMI)	FT MEC.
1900.02.24	00:30	38.42–44.81	Khoi	Khoi	5.4		VII	RL
1913.04.16	06:00	39.10–48.60	Lankaran	Tālesh	5.1		VII	R
1925.12.18	05:53	28.88–51.30	Ahram	Kāzerun	5.4		VII	RL
1934.03.13	23:33	30.95–51.40	S. Maymand	High Zāgros	5.3			R
1935.03.05	10:26		S. Doāb	Orim?	5.8		VII^{+}	LL
1935.04.11	23:14	36.48–53.59	Kusut	N. Alborz or Khazar	6.2	6.6	VIII	R
1944.04.05	18:05	36.78–54.46	Gorgān	Khazar	5.2		VI^{+}	R
1945.09.27	12:30	36.42–50.28	Herian	Alamutrud	–	–	VII	R
1957.05.06	15:06	36.13–52.08	Namār	Namarostāq	5.5		VII	R
1957.07.02	00:42	36.06–52.66	Band-e pay	Larzaneh	6.8	7.1	VIII^{+}	R
1960.09.21	23:05	31.90–50.80	Ardal	Ardal	5.0		VII	R, RL
1963.03.31	02:27	36.91–57.78	Dahaneh Ojāq	Esfarāyen	5.0		VI^{+}	R
1967.01.15	00:03	29.22–51.48	Kamāraj	Kāzerun	4.7		VI	RL
1968.04.29	17:01	39.22–44.19	Gol	Siah Cheshmeh	5.5		VII	RL
1968.08.02	03:59	36.43–50.32	Manjil	Manjil	4.7		VI	R

Continued

TABLE 14.11 Earthquakes in Iran Associated with Coseismic Surface Fault Ruptures (ca. 874–2014) Inferred Mainly from Elongated Shape of the Meizoseismal Areas Along Known Active Faults—Cont'd

Eq. Date	Time (UTC)	Epicenter (00°N–00°E)	Location	Fault	M_s	M_w	I (MMI)	FT MEC.
1969.01.03	03:16	37.06–57.82	Dahaneh Ojāq	Esfarāyen	5.5	5.5	VII	R
1970.03.14	01:51	38.59–44.77	Badalān	Khoi	5.2		VI^+	RL
1970.07.11	22:41	37.72–49.10	Kapurchāl	Tālesh	5.2		VI^+	R
1971.08.09	02:54	36.20–52.57	Bābol Kenār	Khazar	5.3		VII	R
1971.10.23	11:49	29.60–51.48	Kamaraj	Kāzerun	4.5		VI^+	RL
1977.04.06	13:36	31.95–50.64	Nāghān	Ardal	6.1	5.9	VIII	R, RL
1977.05.26	01:35	38.89–44.32	Mokhur	Chālderān-Khoi	5.4	5.7	VII	RL
1978.11.04	15:22	37.67–48.91	Siāhbil	Tālesh	6.0	6.1	VII^+	R
1980.05.04	18.35	38.04–49.01	Shirābād	Tālesh	6.1	6.3		R
1983.03.25	11:57	36.03–42.29	Baijān	Baijān	4.9	5.5	VII	LL
1983.03.26	04:07	35.98–52.24	Baijān	Baijān	4.9	5.4	VII	LL
1983.07.22	02:04	36.94–49.22	Charazeh	Manji	5.0	5.5	VII	R
1986.07.12	07:54	29.91–51.56	Golgun	Kāzerun	5.6	5.5	VII	RL
1986.12.20	23:47	29.90–51.57	Golgun	Kāzerun	5.0	5.3	VI^+	RL
1987.01.11	12:31	29.93–51.79	Golgun	Kāzerun	4.1		V	RL
1987.04.10	06:43	37.16–57.68	Esfarāyen	Esfarāyen	4.9	5.1	VII	R
1988.08.11	16:00	29.94–51.58	Doshman Ziāri	Kāzerun	6.1	5.5	VIII	RL

1988.08.11	16:04	29.88–51.65	Doshman Ziāri	Kāzerun	5.9	5.8	VII	RL
1989.10.01	02:59	30.59–51.40	Sisakht	High Zāgros	4.7		VI	R
1990.01.20	01:27	35.90–52.97	Gaduk	Firuzkuh	5.8	6.0	VII$^+$	LL
1992.09.08	00:38	29:13–52.15	Dārénjān	Karébas	4.8		VI	RL
1992.09.22	14:05	36.29–52.72	Bābol Kenār	Khazar	4.3	5.1	VI	R
1993.01.08	22:51	29.04–52.13	Dārénjān	Sabzpushān	5.3	5.4	VII	RL
1993.08.04	19:46	36.90–57.90	Bām	Esfarāyen	4.4		VI	R
1994.03.01	03:49	29.14–52.64	Mook	Karébas	6.0	5.9	VII$^+$	RL
1994.06.20	09:09	29.05–52.67	Ebrāhimābād	Sabzpushān	5.9	5.8	VII$^+$	RL
1998.08.04	11:41	37.22–57.33	Esfarāyen	Esfarāyen	4.9		VI$^+$	R
1999.04.24	09:22	37.15–57.03	Esfarāyen	Esfarāyen	4.3			R
2002.12.24	17:03	34.59–47.45	Dinévar	ZMRF	4.6	5.2		RL
2003.01.11	17:45	29.62–51.53	Kāzerun	Kāzerun	5.2	5.2	VI$^+$	RL
2003.07.03	14:59	35.41–60.76	Yakhak	N. Torbat Jām	5.3	4.9	5.2	R
2004.05.28	12:38	36.26–51.56	Firuzābād-e Kojur	Khazar	6.4	6.2	VIII	R
2010.07.31	06:52	29.58–56.75	Negār, Lālehzār	Chamanrang	4.9	5.4	VII	R

The faults were not investigated after the earthquakes; in need of authentication by archaeoseismic trench study. Complete references for each event are cited in the text and the corresponding figure captions and the text.
LL, left-lateral strike-slip; R, reverse; RL, right-lateral strike-slip; ZMFF, Zāgros Mountain Front fault; ZMRF, Zāgros Main Recent fault.
[a]*See discussion in Chapter 11.*

Since at least the 1970s, we have been aware that, unlike the rest of the Iranian plateau, the top sedimentary cover in the Zāgros fold-and-thrust belt is mechanically decoupled from the underlying pre-Neoproterozoic basement (where the majority of earthquakes happen) along several incompetent beds, including the thick Upper Vendian-Lower Cambrian Hormoz Salt décollement (Berberian, 1976a, 1977a, 1981a, 1995). As with the Andes of Peru (Suarez et al., 1983) and Papua New Guinea (Hamilton, 1979; Abers and McCaffrey, 1988; Hill, 1991), the basement earthquakes in the Zāgros occur along high-angle reverse faults (Jackson and Fitch, 1981; Maggi et al., 2000a); they do not reach the surface except for those that occur along the major, deep-seated reverse faults, such as the High-Zāgros, Zāgros Mountain Front, Zāgros Foredeep faults as well as the transverse reverse faults, such as the Kāzerun, Karébas, and Sarvestān (Berberian, 1976a, 1977a, 1981a, 1995).

Thrust-fault earthquakes in other parts of the Iranian plateau either produced coseismic ruptures at the surface with a reduced amount of displacements or failed to reach the surface and instead produced coseismic folding and flexural-slip faulting above the blind thrusts at the surface. Examples are the 1953 Torud (Abdalian, 1953), 1978 Tabas-e Golshan (Berberian, 1979a), and 1994 Sefidābeh earthquakes (Berberian et al., 2000b). In the latter case, active geomorphological indicators as well as seismology helped recognize the blind thrust.

In the case of the 1998 Fandoqā earthquake along the Gowk strike-slip fault system, the SAR interferometry analysis showed aseismic, blind-thrust movement beneath the Shahdād foreberg ridges with active geomorphological features to the east of the Gowk fault with no surface break. The complex interaction between an active strike-slip fault and shortening along the Shahdād folds above the Shahdād thrust within a system of oblique convergence was revealed by detailed multidisciplinary study for the first time on the Iranian plateau. Likewise, the 1994 Northridge, California, earthquake consisted of three subevents that ruptured successively shallower parts of a blind reverse fault.

Strike-slip faults produced much longer coseismic surface ruptures with larger displacements than the reverse faults. The 1997 M_w 7.2 earthquake created the longest recorded coseismic surface rupture of 125 km along the Ābiz right-lateral strike-slip fault.

Detailed study of earthquakes along the Gowk right-lateral, strike-slip fault system revealed that the main ruptures in the 1981 M_w 7.0 earthquake probably occurred on different, deeper parts of the same fault system, producing only minor reactivation of the shallower faults at the surface comparing with much larger displacement that the 1998 M_w 6.6 earthquake created (Berberian et al., 2001). The 1981 M_w 7.0 earthquake on the deeper part of the reverse fault loaded the shallower part of that fault in the 1998 M_w 6.6 earthquake, which ruptured the surface 17 years after the first event. The latter triggered aseismic slip on the nearby Shahdād blind thrust. Similarly, the 1971

San Fernando, California, earthquake ruptured the Sierra Madre reverse fault, which triggered a shallower earthquake on the San Fernando fault that resulted in surface rupture.

The two earthquakes of 1968 and 1979 with identical magnitude of M_w 7.1 produced 70- and 55-km coseismic surface faulting along the Dasht-e Bayāz left-lateral strike-slip fault, respectively.

It is clear that the cases described here as well as the accompanying figures and tables are works in progress; they are subject to modifications and corrections by additional research, especially paleoseismic trench investigations.

Chapter 15

Coseismic, Blind, Reverse-Fault-Related, Flexural-Slip Folding and Faulting at the Surface

Upheavals of the crust of the Earth
might occur during a violent earthquake.

Avicennā (Pur Sinā) (1037)

Rapid topographic modifications, base-level and river course changes, and deep incisions on the Iranian plateau have been recorded since at least the eleventh century. Despite a lack of coseismic surface fault ruptures, some related features were assumed to be coseismic surface folding above buried (blind) reverse faults. This can easily be detected by the effect on the surface topography and drainage patterns, indicating elastic response to dip-buried, reverse faults and the formation of flexural-slip faulting. In most of these cases—especially in the Zāgros active fold-and-thrust mountain belt, where generally coseismic surface ruptures in the basement are decoupled from the surface sediments and do not reach the surface—paleoseismic trench studies to obtain past seismic history are not helpful at the present time. Therefore, the best tool seems to be an investigation into the constrained meizoseismal areas of past earthquakes and active tectonics indicators along with waveform modeling and In-SAR studies.

15.1 UPLIFT AND INCISION

Bruin Khārazmi (1025) wrote about an interesting observation of active uplift and incision by rivers in the eleventh century: "*If the land somewhere had risen up and filled up the adjoining region, then the waters in that region would have been diminished, the springs would have been sunk, the valleys would have been made deeper, and the district would have been rendered uninhabitable. Its inhabitants would have moved to another district; and people would have ascribed that destruction to old age of the ground in the former district, and the building up of the desolate land in the latter would have been ascribed to its youth.*" Biruni

☆"To view the full reference list for the book, click here"

Developments in Earth Surface Processes, Vol. 17. http://dx.doi.org/10.1016/B978-0-444-63292-0.00015-6

Khārazmi (1025) continued: "*He* [Abu al-'Abbās al-Irānshahri (Neyshāburi), 9th century] *also added that, when the level of that locality* [possibly at Qal'eh Baizā/White castle] *rose, many kārizes* [qanāts][1] *and rivers, that had been flowing, were sunk*" (ed. Aram, 1973; ed. Jamil 'Ali, 1967).

The location of the white castle is given as "*one farsakh (6 km) from Shirjān, a town in the district of Kermān.*" The old Shirjān (present Sirjān) was located about 5 miles east of the modern Sa'idābād at a location known as *Qal'eh Sang* or *Qal'eh Baizā* (29°22′N–55°45′E; see Sykes, 1902; Le Strange, 1905). The location is 5 km southeast of modern Sirjān [29°26′N–55°40′E]. Qal'eh Sang or Qal'eh Baizā [lit., 'Stone-' or 'White-Fort'] is on an arid plain, and there is no river in the area. Another famous Baizā (lit., "white" in Arabic) occurred in the ancient Parthian city of Neysāyak,[2] located in the district of Kāmfiruz, Sepidān County, western Fārs province. The important ancient Elamite city of Anshān (Anzān; the present Maliān archeological mound: ca. 3500–1000 BCE; 30°00′41″N–52°24′29″E, north of Baizā) was located on the same plain [Dasht-e Baizā, Sepidān] (Hansman, 1985, 2011; Miller and Abdi, 2003). The reference was likely made to this locality.

The geomorphological, crustal movement and hydrological observation at Baizā (in old Sirjān or in the Zāgros) is very interesting. This seems to be one of the oldest references to a change in water table and in the base level of river beds caused by crustal uplift. Although Biruni Khārazmi (1025) did not mention any earthquake when quoting from the ninth-century Abu al-'Abbās al-Irānshahri (Neyshāburi), it is likely that the site was on a growing fold above a blind fault.

15.2 THE 23 SEPTEMBER 1947 M_W 6.8 CHARMEH EARTQHUAKE

See Section 12.12 and Figure 12.8.

15.3 THE 12 FEBRUARY 1953 M_W 6.5 TORUD EARTHQUAKE

See Section 12.14 and Figure 12.9.

15.4 THE 10 APRIL 1972 M_W 6.7 KĀRZIN EARTHQUAKE

The 10 April, 1972, Kārzin earthquake was associated with minor and discontinuous fractures on the alluvial surface in zones up to 160 m wide and many

---

1. Kāriz, kahriz, qanāt: A traditional, ancient Iranian technology of a series of well-like vertical shafts, connected by a gently sloping tunnel, that supplied a reliable source of water for settlements in semiarid and arid areas.

2. Neysāyak: lit., "bright, shining"; later changed to Nasātak and Nesā' in the district of Kāmfiruz, Sepidān County, western Fārs province (29°58′N–52°24′E; west of Naqsh-e Rostam and the Kur River; 36 km NW of Shirāz; Moqaddasi, 985; Yāqut, 1225).

hundreds of meters long, striking either (i) NW–SE (~N120°E; in the center of meizoseismal area) or (ii) NNW–SSE (~N145°E; mostly located to the east of the meizoseismal area). Individual cracks were invariably oriented in an en-echelon pattern, indicating right-lateral displacements of a few centimeters (Ambraseys et al., 1972; Ambraseys and Melville, 1982). However, Haghipour et al. (1972) only reported 32 to 71 km of deformation consisting of parallel or subparallel surface fractures with an azimuth of 120°.

The short surface ruptures striking ~N145°E with right-lateral slip are located along the Sabzpushān right-lateral shear zone (with dextral-dragged anticline axes) and may indicate reactivation of the shear zone during a subevent or aftershock (Figure 13.4; Plate 13.7). In contrast, the ~N120°E discontinuous fractures are located on the flanks of anticlines with minor vertical displacement, indicating coseismic folding above the Surmeh reverse fault (flexural-slip faulting above the Surmeh thrust). Microseismicity study of the area (Tatar et al., 2004a) showed that the earthquakes are located between 8 and 15 km deep, with the maximum number of events at a depth of 11 km. Almost all the focal mechanism solutions located to the eastern part of the meizoseismal area (i.e., the Sabzpushān right-lateral shear zone) show a right-lateral strike-slip mechanism. On the other hand, the rest of the mechanisms show thrusting mainly dipping to the north (Berberian, 1995).

Berberian (1981a, 1995) suggested that the 1440 M_s~6.7–6.9, 14/15 November 1903, 10 April 1972 M_w 6.7, 14 September 1968 M_w 5.8, and 2 February 1985 M_w 5.6 earthquakes took place along the Surmeh master basement fault, along which the Ordovician rocks are thrust over Pliocene deposits (Figure 13.4).

15.4.1 Seismology

The source–time function of this earthquake shows two distinct ruptures by two subevents: the second one occurring about 13.7 s after the onset of the first subevent with a different strike (Table 15.1). Both subevents involved high-angle reverse faulting (Baker et al., 1993).

TABLE 15.1 *P* and *SH* Body-Waveform Inversion Source Parameters of the 10 April 1972 Kārzin Earthquake (Baker et al., 1993) (See Figure 13.4)

Earthquake	Time (UTC)	Coordinate (N°–E°)[a]	Strike (°)	Dip (°)	Rake (°)	Centroid Depth (km)	M_o (Nm) ($\times 10^{18}$)	M_w
1972.04.10 SE-1	02:06	28.41–52.78	288	49	99	9	8.44	6.7
SE-2			322	40	98	10	3.704	

[a]*Epicenter from Engdahl et al. (2006).*

15.5 THE 2 JULY, 1972 M_W 5.3 MISHĀN EARTHQUAKE

See Section 13.5 and Figure 13.5.

15.6 THE 16 SEPTEMBER 1978 M_W 7.3 TABAS-E GOLSHAN EARTHQUAKE

Extensive bedding-plane slip with a thrust mechanism (flexural-slip faulting) developed in the hanging-wall block of the Tabas thrust in the Neogene molasse deposits, indicating coseismic folding above the Tabas thrust (Berberian, 1979a; See also Section 13.11, Figure 13.8, and Plates 13.13 and 13.14).

15.7 THE 20 NOVEMBER 1989 M_W 5.8 SOUTH GOLBĀF EARTHQUAKE

In addition to the two fresh surface faults, folding was also developed along the Gowk fault system, south and southeast of Golbāf (Berberian and Qorashi, 1994). The Holocene playa deposits at the western edge of the South Golbāf depression show coseismic elastic folding of the horizontal clay-flat deposits in the earthquake fault zone (Figure 13.10; Plates 13.25 and 13.27).

The horizontal clay layers dip 15–20° east (toward the playa) at the southern part of the western fault zone (Figure 13.25). In a cross-section (Figure 13.27), both earthquake faulting (with splayed fresh ruptures) and earthquake folding (slip along bedding planes) are clearly visible. It is clear in this case that surface folding of the horizontal Holocene clay is intimately related to earthquake faulting and could be explained by repeated motions on the steeply dipping west Gowk fault during the past earthquakes. The south Golbāf depression was therefore formed by earthquake faulting as well as coseismic folding, and its present topography was controlled by these two elastic-brittle deformations (Berberian and Qorashi, 1994).

15.8 THE 20 JUNE 1990 M_W 7.3 RUDBĀR EARTHQUAKE

Bedding-plane slip with reverse mechanism (flexural-slip faulting) was also observed at the surface in the Jurassic Shemshak molasse, Jurassic Dalichāi Formation (in the Jirandeh area), and the Neogene molasse deposits (north of the Qezel Owzan River; figures 4 and 10 in Berberian et al., 1992), showing coseismic folding in the Alborz Mountains.

15.9 THE 23–28 FEBRUARY 1994 M_W 5.5–6.2 SEFIDĀBEH EARTHQUAKES

The 1994 Sefidābeh earthquake sequence, with a NW–SE reverse mechanism and centroid depths of 6–10 km, occurred in a desert area of eastern

Iran. The sequence was associated with coseismic ruptures at the surface involving bedding-plane slip on a growing hanging-wall anticline displaying a geomorphological evidence of uplift and lateral propagation (figures 5a–h in Berberian et al., 2000b). The ruptures were all within the vertically dipping and isoclinally folded Upper Cretaceous–Paleocene flysch and volcanoclastic rocks of the Sefidābeh Ridge (Figures 15.1 and 15.2; Plates 15.1 and 15.2).

Body-waveform analysis of the earthquake sequence indicated uplift of the Sefidābeh Ridge above the Sefidābeh blind reverse fault (Berberian et al., 2000b). The Ridge shows asymmetric folding with a steep northeast flank close to the reverse fault tip near the surface and gentle southwest dip away from the near-surface fault tip. Analyses of a SAR interferogram determined the precise location and amount of coseismic surface displacements and indicate that the earthquake sequence probably occurred on en-echelon fault segments associated with three stepping ridges. The center of the cluster lies about 10.4 km away on an azimuth of 170° from the center of the area of significant fault slip. The regions of maximum slip on the fault segments appear to be aligned, suggesting that slip on one segment may then have triggered slip on its neighbor (Parsons et al., 2006).

As mentioned, the dominant characteristic of the surface ruptures was bedding-plane slip, producing extensional features on the ridge. From the seismological and geomorphological evidence, it appears that these ruptures formed in response to the growth of an anticline above the Sefidābeh blind thrust (Berberian et al., 2000b). Despite the abovementioned fact, Razaqi-Azar et al. (1994) wrote that the earthquake was associated with more than 10 km of surface faulting with a normal mechanism.

Progressive uplift of the Sefidābeh Ridge during the prehistoric and historical earthquakes has resulted in the formation of the dammed paleo-Sefidābeh gorge; it formed a small ephermal lake, which was later abandoned (Figure 15.3; see also figures 3, 4, 5, and 13 in Berberian et al., 2000b). The paleo-Sefidābeh River was later diverted about 7 km to the northwest at the end of the present growing Sefidābeh Ridge; there it flows about 70 m below the uplifted abandoned lake deposits (Berberian et al., 2000b).

The 1994 Sefidābeh earthquakes were associated with a NW-trending blind thrust system that splays off the northern termination of the Zāhédān N–S right-lateral strike-slip fault (Figures 15.1 and 15.4). The system accommodates oblique NE–SW shortening across the N–S-deforming zone, probably by anticlockwise rotation about a vertical axis. This style of fault kinematics may be transitional, and move to a more evolved state that involves partitioning of the strike-slip and convergent motion onto separate subparallel faults (Freund, 1970; Berberian et al., 2000b). U/Th dates of ~100 ka were obtained from the Sefidābeh uplifted lake deposits above the growing Ridge, indicating that about 120 earthquakes, the size of those in

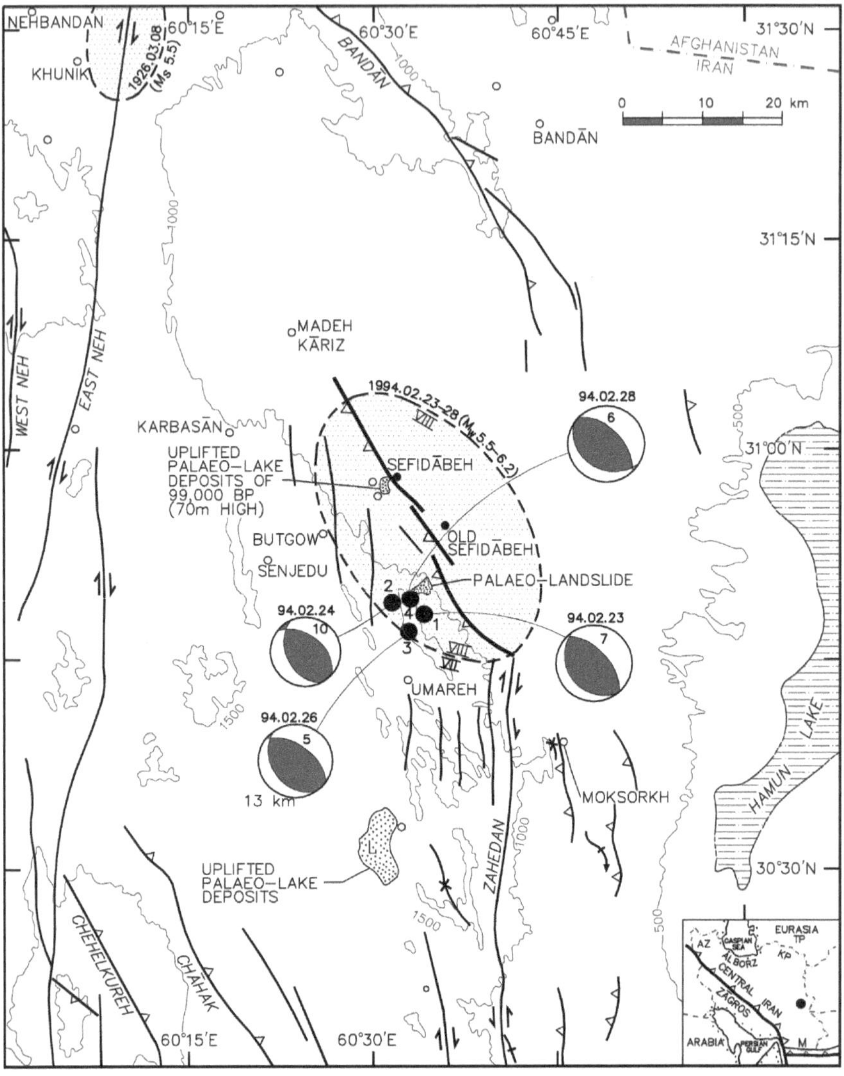

FIGURE 15.1 Fault map and meizoseismal area of the 23–28 February 1994 M_w 5.5–6.2 Sefidābeh reverse-blind fault earthquake sequence at the northern end of the Zāhedān right-lateral strike-slip fault in southeast Iran. Fault and other symbols as in Figure 11.1. Focal mechanisms constrained by body-wave inversion with centroid depths (Berberian et al., 2000b). Locations of the uplifted paleo-lake deposits of Sefidābeh (west of Sefidābeh) and the Palangkuh paleo-rockfall (southwest of ruins of the Old Sefidābeh village) are stippled.

1994, occurred in the past 100 ka with an average interval of 830 years and an average convergence rate of 1.5 mm/years on the Sefidābeh blind reverse fault. Apparently, the Sefidābeh buried reverse fault formed by coalescence of many small fault segments and has grown in length at about 2 cm/year in the past 100 ka (Parsons et al., 2006).

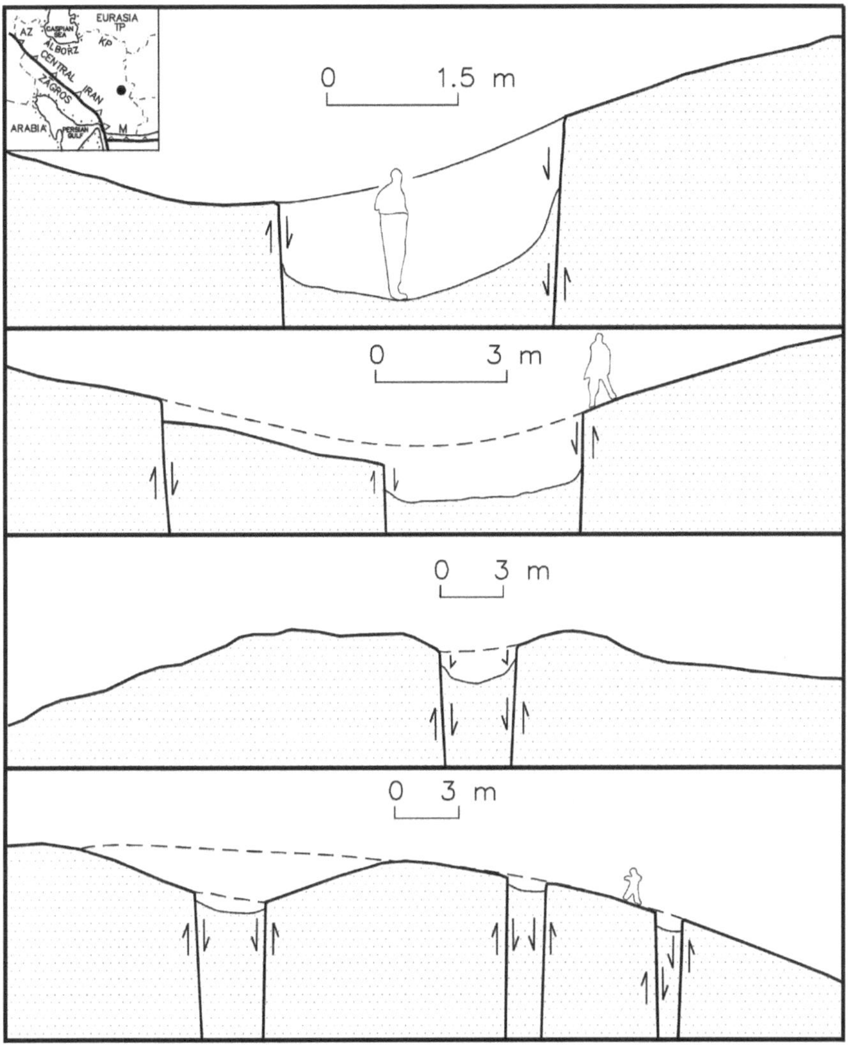

FIGURE 15.2 Coseismic surface flexural-slip faulting involved slip on subvertical bedding planes in Paleocene flysch deposits forming mini-grabens on the Sefidābeh ridge crest about 1 km southwest of Sefidābeh village and the buried reverse fault. For location see Figure 15.1.

15.9.1 Seismology

P and *SH* body-wave modeling of the four events showed similar mechanisms involving reverse faulting with a NW strike and a nodal plane dipping SW at 29–45° (Berberian et al., 2000b; Figure 15.1; Table 15.2).

15.10 THE 4 FEBRUARY 1997 M_W 6.4 NĀVEH EARTHQUAKE

East–west bedding-plane slip due to coseismic folding developed in the Yekeh Shākh syncline, east of the Nāveh fault, at (i) the contact between the Cretaceous

PLATE 15.1 Coseismic surface flexural-slip faulting on the Sefidābeh ridge crest formed by slip on subvertical bedding planes in the Paleocene flycsh beds. View looking southeast (Berberian et al., 2000b). For location see Figure 15.1.

PLATE 15.2 Coseismic surface flexural-slip faulting on the Sefidābeh ridge crest formed by slip on subvertical bedding planes in the Paleocene flycsh beds. View looking southeast 1 km southwest of Sefidābeh village. For location see Figure 15.1.

marl of the Sarcheshmeh Formation and (ii) along the bedding planes of the Lower Cretaceous Sangāneh Formation. The flexural-slip faults developed for a distance of 6 km and were clearly visible in the area between Yekeh Shākh and Qezelqān in a general ENE–WSW direction (Tatar et al., 1997). See additional information about this event in Chapter 13 and Figure 13.15.

15.11 THE 14 MARCH 1998 M_W 6.6 FANDOQĀ EARTHQUAKE

After the 1998 Fandoqā earthquake, which produced 23 km of coseismic faulting (see Chapter 14 for details), SAR interferometry showed that the

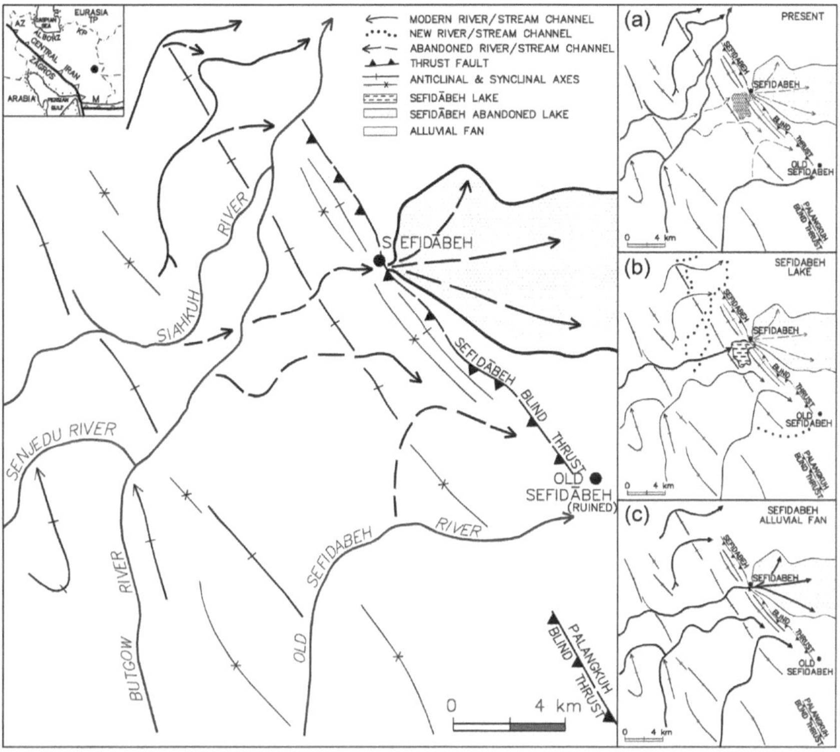

FIGURE 15.3 Left: Modern- (continuous lines with arrow showing the flow direction) and paleo-stream channels (broken lines with arrows) on the hanging-wall block of the Sefidābeh blind reverse fault. Right: Tentative evolution of the stream system near Sefidābeh in response to the growing hanging-wall anticline above the Sefidābeh active blind reverse fault and the northward propagation of the reverse fault tip. (a) The present-day configuration, showing the abandoned Sefidābeh paleo-lake bed near the village. (b) The situation just prior to the abandonment of the lake, with dotted future (i.e., present) stream channels. (c) An older Quaternary configuration (exact age unknown), when the hanging-wall ridge was lower and the alluvial fan at Sefidābeh was still active. This figure illustrates the nature of the changes that occurred to the drainage system rather than the timing. *Modified from Berberian et al. (2000b).*

nearby Shahdād thrust, located 30 km to the east of the Gowk fault (Figures 14.2 and 14.3), slipped along a plane dipping at ~6°W for about 8 cm in a time interval and in a position that makes it likely that its aseismic slip was triggered by the Fandoqā earthquake (Berberian et al., 2001).

15.12 27 NOVEMBER 2005 M_W 5.8 CENTRAL QESHM ISLAND EARTHQUAKE

This earthquake was associated with two sets of surface fractures striking N40°E parallel to the Ramkān anticline axis (see figures 2, 3 and 4 in

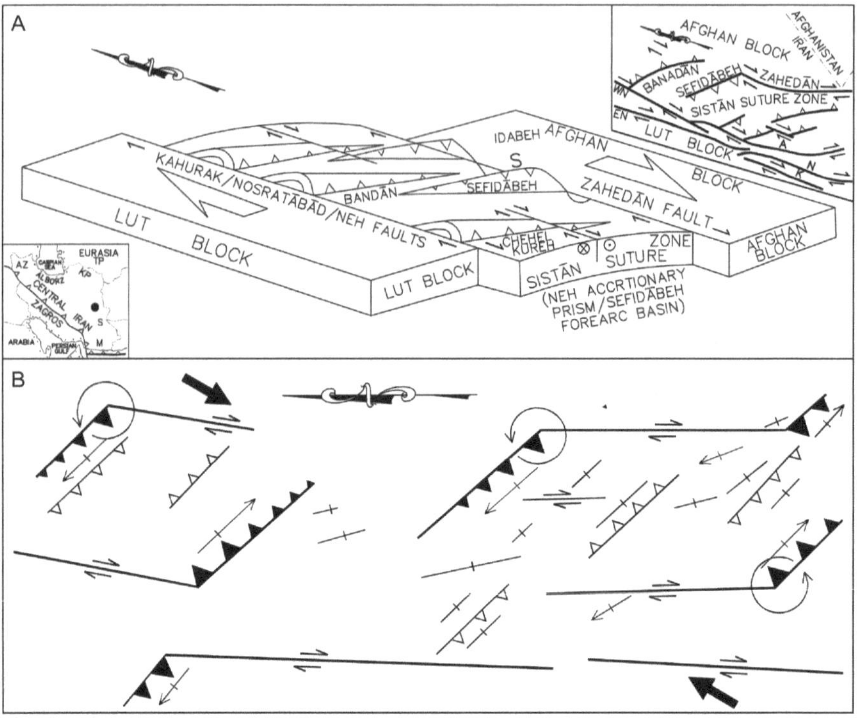

FIGURE 15.4 Schematic block diagram and cartoon plan view of a deforming zone that accommodates oblique overall NE–SW shortening (large solid arrows) partly by reverse faulting (open teeth) and partly by counterclockwise rotating, strike-slip faults about a vertical axis that end in reverse faults whose displacements die away from their intersections with the strike-slip faults (solid teeth). This allows the counterclockwise rotating, strike-slip faults (deforming zone between two rigid blocks) to be spatially associated with other reverse faults that do not need to rotate.

Nissen et al., 2007). The surface cracks are not a direct continuation of the seismic faulting at depth, because In-SAR and body-wave modeling showed a reverse mechanism along a nearly E–W strike dipping ~50°N. The surface ruptures represent flexural-slip faulting during coseismic folding of the surface sediments in the Zāgros (Nissen et al., 2007) (Table 15.3).

15.12.1 Overview

Most of the medium- to large-magnitude, reverse-fault, crustal earthquakes on the Iranian Plateau have not ruptured the surface. However, their surface expression is a broad anticlinal warp, some of which is associated with flexural-slip faulting, above a blind reverse fault. As with the Coalinga (Stein and Ekström, 1992) and Whittier Narrows (Dolan et al., 2003) earthquakes, the coseismic warping can be compared with long-term warping of

TABLE 15.2 ***P*** **and** ***SH*** **Body-Waveform Inversion Source Parameters of the 23–28 February 1994 Sefidābeh Earthquake sequence (Berberian et al., 2000b) (See Figure 15.1)**

Earthquake	Time (UTC)	Coordinate (N°–E°)[a]	Strike (°)	Dip (°)	Rake (°)	Slip Vector Azimuth	Centroid Depth (km)	M_o (Nm) (×10^{18})	M_w
1994.02.23	08.02	30.78–60.53	143	29	96	46	7	1.51	6.05
1994.02.23[b]	11:54	30.81–60.54	108	31	62	49	15 fixed (HRVD)	0.22	5.49
1994.02.24	00:11	30.79–60.51	155	45	110	38	10	2.54	6.20
1994.02.26	02:31	30.79–60.53	146	36	107	35	5	1.06	5.95
1994.02.28	11:13	30.90–60.62	122	33	78	78	6	0.22	5.49

[a]*Epicenters are from Engdahl et al. (2006).*
[b]*CMT Harvard (HRVD; http://www.globalcmt.org).*

TABLE 15.3 ***P*** **and** ***SH*** **Body-Waveform Inversion Source Parameters of the 27 November 2005 Central Qeshm Island Earthquake (Nissen et al., 2007, 2010)**

Earthquake	Time (UTC)	Coordinate (N°–E°)	Strike (°)	Dip (°)	Rake (°)	Centroid Depth (km)	M_o (Nm) (×10^{18})	M_w
2005.11.27	10:22	26.75–55.82	259	50	95	9	0.64	5.8
2005.11.27 AFS	16:30	26.80–55.77	212	89	358	10	0.158	5.4

AFS, aftershock.

terraces of antecedent streams crossing the folding or of Holocene sediments in the forelimbs of the folds.

Earthquakes in the Zāgros, East Iran, Central Iran, and Alborz showed evidence of coseismic flexural-slip folding and faulting of the surficial formations by slip along bedding planes or other structural features above seismic blind reverse faults. In all these regions, the coseismic reverse fault rupture at depth fails to reach the surface. The slip dissipates toward the surface and produces flexural-slip folding and faulting by sliding on bedding planes with a thrust mechanism and results in growth of the folds. In most cases, active geomorphological indicators, in the form of recent changes in drainage incision and deflection, as well as asymmetric young folds and topographic drop, show evidence of slip along active buried reverse faults (Tables 15.4 and 15.5).

TABLE 15.4 Documented Earthquakes in Iran Associated with Blind Reverse-Fault-Related Folding and Bedding-Plane Slips (Flexure-Slip Faulting), or General Folding at the Surface (1947–2005)

Eq. Date	Time (UTC)	Epicenter 00°N–00°E	Location	M_s	M_w	*I* (MMI)	Blind Fault Name
1947.09.23	12:28	33.77–58.66	Charmeh	6.9	6.8	VIII^+	Ferdows
1953.02.12	08:15	35.40–55.00	Torud	6.5	6.5	VIII^+	Morghāb
1972.04.10	02:06	28.41–52.78	Kārzin	6.9	6.7	VIII	Surmeh
1972.07.02	12:56	30.06–50.85	Mishān	5.3	6.3	VII	?
1978.09.16	15:35	33.24–57.38	Tabas-e Golshan	7.4	7.3	IX^+	Tabas
1989.11.20	04:19	29.89–57.72	South Golbaf	5.7	5.8	VII	Gowk
1990.06.20	21:00	36.99–49.22	Rudbār	7.4	7.3	IX^+	–
1994.02.23	08:02	30.79–60.53	Sefidābeh	6.0	6.1	VII^+	Sefidābeh
1994.02.23	11:54	30.81–60.54	Sefidābeh	4.9	5.5		Sefidābeh
1994.02.23	22:45	30.90–60.55	Sefidābeh	5.0			Sefidābeh
1994.02.24	00:01	30.79–60.51	Sefidābeh	6.1	6.2	VIII	Sefidābeh
1994.02.26	02:31	30.79–60.53	Sefidābeh	5.8	6.0	VII^+	Sefidābeh
1994.02.28	11:13	30.90–60.62	Sefidābeh	5.5	5.5	VII	Sefidābeh
1997.02.04	10:37	37.73–57.31	Nāveh	6.7	6.6	VIII^+	-
1998.03.14	19:40	30.13–57.59	Fandoqā	6.6	6.6	VIII^+	Triggered Shahdād
2005.11.27	10:22	26.84–55.93	Central Qeshm Island	5.8	5.8	VII^+	Qeshm

For details and references see the text of Chapters 12–15.

TABLE 15.5 Blind Thrust Earthquakes in the Zagros Fold-and-Thrust Mountain Belt Associated with Flexural-Slip Folding and Faulting (839–2008) Deduced Mainly from Meizoseismal Areas of Large Earthquakes

Eq. Date	Local Time	Epicenter 00°N–00°E	Location	Fault Name	$\sim M_s$	$\sim I$ (MMI)	FT. MEC.	References
839.11.12–840.10.20	–	31.32–48.67	Ahvāz	ZFF	6.5	VIII	R	Berberian (1994, 1995)
872.06.22	–	33.11–47.15	Saimareh	Kabirkuh	6.8	VIII^{+}	R	Berberian (1994, 1995)
1008.05.10–06.08	–	27.68–52.37	Sirāf	ZMFF	6.5	VIII	R	Berberian (1994, 1995)
1052.05.03–1053.04.22	–	30.65–50.32	Arrājān [Argān]	ZFF	6.8	VIII^{+}	R	Berberian (1994, 1995)
1085.04.29–05.28	–	30.65–50.32	Arrājān [Argān]	ZFF	5.8	VII^{+}	R	Berberian (1994, 1995)
1150.04.01	Morning	34.53–45.78	Sar-e Pol-e Zahāb [Holvān]	ZMFF	5.9	VII^{+}	R	Berberian (1994, 1995)
1336.05.08	Midday	26.82–55.92	Central Qeshm Island	Qeshm	>6.0	VII^{+}	R	Berberian (1994, 1995)
1440.06.02–1441.05.21	–	28.42–53.08	Kārzin	Surmeh	7.1	IX	R	Berberian (1994, 1995)
1665.07.14–1666.07.03	–	32.15–50.55	NW Ardal	?	6.5	VIII		Berberian (1994, 1995)
1883.10.16	Night	27.40–52.20	Kangān	ZMFF	5.8	VII^{+}		Berberian (1994, 1995)

Continued

TABLE 15.5 Blind Thrust Earthquakes in the Zagros Fold-and-Thrust Mountain Belt Associated with Flexural-Slip Folding and Faulting (839–2008) Deduced Mainly from Meizoseismal Areas of Large Earthquakes—Cont'd

Eq. Date	Time (UTC)	Epicenter 00°N-00°E	Location	Fault Name	M_s	M_w	I (MMI)	CD_w (km)	FT. MEC.	CD_w References
1903.11.14/15	–	–	Kārzin	Surmeh			VII^+	–	R	
1913.03.24	10:34	26.80–53.70	Chiru	ZMFF	5.8		VII^+	–	R	
1929.07.15	07:44	32.15–49.60	Andika	ZMFF	6.0		VII^+	–	R	
1949.04.24	04.22	27.22–56.38	Nakhl-e Nākhodā	ZFF	6.3		VIII	–	R	
1950.01.19	17:27	27.38–52.78	Dehnau 'Assaluyeh	ZMFF	5.5		VII	–	R	
1950.01.22	04:07	27.38–52.78	Dehnau 'Assaluyeh	ZMFF	5.3			–	R	
1956.10.31	14:03	27.26–54.42	Gowdeh	ZMFF	6.2	6.6	VIII	–	R	
1960.04.24	12:14	27.75–54.40	Lār	Lār	6.0	5.8	VII^+	–	R	
1968.06.23	09:16	29.75–51.26	Khesht	?	5.3	5.5		09	R	Baker et al. (1993)
1968.09.14	13:48	28.33–53.17	Tang-e Ru'in	Surmeh	6.0	5.8	VII^+	–	R	
1970.02.23	11:22	27.79–54.48	Dehkuiyeh	Beriz	5.5	5.6		–	R	
1971.12.09	01:42	27.30–56.37	Sarkhun	ZFF	5.7	5.8	VII	–	R	
1972.04.10	02:06	28.41–52.79	Kārzin	Surmeh	6.9	6.7	VIII	10	R	Baker et al. (1993)
1973.06.09	20:36	27.78–52.18	Kangān	ZMFF	4.7		VI^+	–	R	
1974.12.02	09:05	28.09–55.86	Sarchāhān	?	5.1	5.2		07	R	Baker (1993)

1975.03.07	07:04	27.47–56.22	Sarkhun	ZFF	6.0	6.1		–	R	
1976.04.22	17:03	28.68–51.12	Farrashband	ZFF	5.7	5.7		07	R	Baker et al. (1993)
1977.02.14	05:30	27.22–53.06	Gāvbandi	ZMFF	4.6			-	R	
1977.03.21	21:18	27.58–56.36	Khurgu	ZMFF	6.8	6.7	VIII+	12	R	Jackson and Fitch (1981)
1977.03.21	22:42	27.60–56.46	Khurgu	ZMFF	6.0	6.1			R	
1977.03.22	11.57	27.59–56.42	Khurgu	ZMFF	6.9	6.0		12	R	Maggi et al. (2000a)
1977.03.23	23:51	27.59–56.55	Khurgu	ZMFF		5.5		09	R	Jackson and Fitch (1981)
1977.03.24	04:42	27.62–56.58	Khurgu	ZMFF		5.3			R	
1977.04.01	13:36	27.55–56.27	Khurgu	ZMFF	6.0	6.0	VII+	12	R	Jackson and Fitch (1981)
1977.04.06	13:36	31.95–50.64	Nāghān	Ardal?	6.1	5.9	VIII	06	R	Baker (1993)
1977.07.14	15:23	26:83–53.56	Nakhl-e Jamal	ZMFF	4.3			–	R	
1978.12.14	07:05	32:13–49.63	Andikā	ZMFF	6.2	6.2	VIII	–	R	
1980.10.19	17:24	32.70–48.65	Shahbāzān	ZMFF	5.7	5.6		17	R	Maggi et al. (2000a)
1983.07.12	11:34	27.61–56.40	Khurgu	ZMFF	5.8	5.9		17	R	Maggi et al. (2000a)
1985.02.02	20:52	28:35–52.97	Gavaki, SE Qir	Surmeh	5.3	5.6		11	R	Maggi et al. (2000a)
1987.04.29	01:45	27.42–56.11	Bandar 'Abbās	ZMFF	5.4	5.6		10	R	Maggi et al. (2000a)
1988.03.30	02:12	30.84–50.18	Tashān	ZFF	5.7	5.9	VII	–	R	

Continued

TABLE 15.5 Blind Thrust Earthquakes in the Zagros Fold-and-Thrust Mountain Belt Associated with Flexural-Slip Folding and Faulting (839–2008) Deduced Mainly from Meizoseismal Areas of Large Earthquakes—Cont'd

Eq. Date	Time (UTC)	Epicenter 00°N-00°E	Location	Fault Name	M_s	M_w	I (MMI)	CD_w (km)	FT. MEC.	CD_w References
1991.05.22	16:29	27.38–55.77	Gishu		5.1	5.4		13	R	Talebian and Jackson (2004)
1991.11.04	01:50	30.69–50.25	Tashān	MFF	5.4	5.8	VII	05	R	Talebian and Jackson (2004)
1993.06.22	16.32	30.18–50.83	Gachsārān	ZMFF	4.9	5.4		05	R	Maggi et al. (2000a)
1994.07.31	05:15	32.68–48.42	Badreh	ZMFF	5.3	5.5		14	R	Maggi et al. (2000a)
1994.04.22	00:21	30.97–49.93				5.1		14	R	Talebian and Jackson (2004)
1996.05.24	06:35	27.85–53.57	Khonj			5.0		06	R	Talebian and Jackson (2004)
1997.10.03	11:28	27.79–54.73	Lār	Lār	4.8	5.3		04	R	Talebian and Jackson (2004)
1998.06.15	01:14	31.71–50.77			4.9	5.0		–	R	Talebian and Jackson (2004)
1998.08.01	23:38	27.74–56.51	Lengeh	ZFF	4.8	5.1		09	R	Talebian and Jackson (2004)
1998.11.13	13:01	27.76–53.62	Khonj	Lār	5.1	5.4		07	R	Talebian and Jackson (2004)

2000.03.05	09:40	27.95–56.47	Ahmadi		5.2	5.4		12	R	Talebian and Jackson (2004)
2000.05.03	09:01	29.66–50.80	Chehel Zari		4.6	5.1		05	R	Talebian and Jackson (2004)
2005.11.27	10:22	26.84–55.93	Central Qeshm	Qeshm	5.8	5.8	VII^+	09	R	Nissen et al. (2007, 2010)
2005.11.27	16:30	26.86–55.87	Central Qeshm	Qeshm	5.3	5.4	VI^+	10	R	Nissen et al. (2007, 2010)
2006.03.25	07:29	27.55–55.66	Fin		5.6	5.7		08	R	Roustaei et al. (2010)
2006.03.25	09:55	27.58–55.76	Fin		5.0	5.5		04	R	Roustaei et al. (2010)
2006.06.03	07:15	26.85–55.97	Central Qeshm	Qeshm		5.0		09	R	Nissen et al. (2007, 2010)
2006.06.28	21:02	26.93–55.92	Central Qeshm	Qeshm	5.5	5.6		11	R	Nissen et al. (2007, 2010)
2008.09.10	11:00	26.84–55.92	Central Qeshm	Qeshm	5.9	5.9		08	R	Nissen et al. (2007, 2010)
2008.09.17	17:43	26.96–56.10	Central Qeshm	Qeshm		5.1		06	R	Nissen et al. (2007, 2010)
2008.12.07	13:36	26.96–55.94	Central Qeshm	Qeshm	5.1	5.4		04	R	Nissen et al. (2007, 2010)
2008.12.08	14:41	26.98–55.95	Central Qeshm	Qeshm		5.1		06	R	Nissen et al. (2007, 2010)

CD_W, centroid depth deduced from *P* and *SH* body-waveform analysis; *FT*. Mec: Fault Mechanism; *I*, maximum intensity in Modified Mercalli Intensity Scale; ZFF, Zāgros Foredeep fault; ZMFF, Zāgros Mountain Front fault; ZMRF, Zāgros Main Recent fault.

Chapter 16

Patterns of Historical Earthquake Ruptures on the Iranian Plateau

When sorrows come,
They come not single spies,
But in battalions.

Claudius, in Shakespeare's *Hamlet*, IV, Scene 5, 1602

Quantifying the medium- to large-magnitude historical earthquakes and their causative faults in different active parts of the world has increased our ability to forecast the probability of occurrence, location, approximate magnitude, depth, mechanism, and rupture length of future earthquakes. It has consequently improved our general understanding of continental tectonics and intraplate earthquake-fault hazard for urban and rural areas as well as reduced the destruction and casualty of earthquakes that occur in industrialized countries. Despite the complexity of the Earth's crust and its faults and deformation characteristics away from the faults, we may be able to forecast the next event along a seismic fault or in an active region by studying the historical pattern of faulting and earthquakes.

Unlike the numerous paleoseismic studies and dense seismogenic networks of industrialized nations, earthquake risk minimization is in the early stages on the Iranian Plateau, where limited paleoseismic data with unconstrained dates are available. However, owing to the long history of ancient civilizations in Iran and Armenia, the extensive seismic record (Berberian, 1976a, 1977b, 1981a, 1994, 1995, 1997a, 2005; Ambraseys and Melville, 1982; Berberian and Yeats, 1999) and archeoseismic indicators (Berberian, 1994; Berberian and Yeats, 2001; Berberian et al., 2012, 2014) stretch back several millennia. This ancient data bank, containing many large earthquake cycles, could help our understanding of earthquake patterns and rupture behavior. This has important implications for seismic hazard assessment.

☆"To view the full reference list for the book, click here"

Developments in Earth Surface Processes, Vol. 17. http://dx.doi.org/10.1016/B978-0-444-63292-0.00016-8

The comprehensive seismotectonic studies of Iranian earthquakes and active faulting thus far have provided us with a narrow and unique window into different earthquake-fault behaviors in space and time and offered some preliminary understanding of the stress accumulation–release cycles in the past, which can be used as models for future earthquakes throughout the world. The observed long-term fault behaviors and earthquake patterns in the complex tectonic setting of the Iranian Plateau show faults in seismically active periods separated by periods of relative quiescence (Berberian, 1981a, 1994, 1995, 1997a, 2005; Berberian and Yeats, 1999, 2001). Such faulting and seismic behaviors and patterns have been observed in theoretical and dynamic models of individual faults because of postseismic reloading and model switching (Ben-Zion et al., 1999; Kenner and Simons, 2005).

Seismogenesis in the Iranian Plateau and other active parts of the planet, such as California, New Zealand, Venezuela, and Argentina, shows similar earthquake sequences and interplay between strike-slip and reverse faults, with intraplate shear strain partitioning induced by local changes in the regional stress field. Thousands of years of historical and archeological records from the Iranian Plateau could provide mitigating strategies for seismic hazards in the metropolitan Los Angeles and Ventura basins, Istanbul, Yerevān, Ashkābād, and other earthquake-prone areas.

The short-term variations of the slip rates during temporally clustered earthquakes with large rupture displacements along the faults could have resulted in variable/chaotic (noncharacteristic) seismic behaviors of the fault slip rates and recurrence intervals throughout the Holocene on the Iranian Plateau. In order to generate a long series of earthquake intervals and fault rupture displacements and study the variability in recurrence intervals of large-magnitude earthquakes, we need more detailed paleoseismic trench investigations across the faults, especially in the urban areas, and particularly in the mega-city of Tehrān. Despite the paramount importance of such studies in seismic risk minimization of urban and rural areas, they have regrettably not been conducted in countries such as Iran since 1971 (see the data and discussion in Berberian, 2005).

The discussion covered in this chapter is critical for seismic hazard assessment of countries such as Iran. Most of the estimated fault slip rates in Iran mentioned here are not based on direct dating of the offset morphological features; hence, some rates are speculative and might be greater or smaller than the assessed values. It should be noted that the loose and scattered GPS network in Iran has not been designed to ascertain interseismic accumulation of strain across the numerous faults on the vast plateau. For seismic parameters of the earthquakes discussed below, see Tables 17.10–17.12.

16.1 POST-NEOGENE CHANGE OF SENSE OF MOTION OF ACTIVE FAULTS; TECTONIC REACTIVATION OF INHERITED STRUCTURES

The Kuhbanān range-front fault in southeastern Iran (Figures 9.1 and 16.19) was introduced as a 300-km long, NW–SE-trending, range-front active, high-angle reverse fault by Huckriede et al. (1962) and Berberian (1976c,e, 1977i). The fault is composed of several deep-seated, en-echelon segments along which the Late Precambrian and Lower Paleozoic rocks (in the NE) are thrust over the Quaternary alluvial deposits (in the SW) with a significant component of geological shortening in the Kuhbanān fold-and-thrust mountain belt. The pre-1977 earthquake movement along the fault produced new slickensides on the hanging-wall, with an average strike of N116°E and a plunge of 50°SE (see Chapter 13 and Figure 13.7). The fault thus had a predominant active reverse or oblique-reverse motion during the post-Neogene period.

Despite the post-Neogene reverse indications along the Kuhbanān fault, field study of coseismic surface rupture and focal-mechanism solution of the 19 December 1977, M_w 5.9 Dartangal earthquake surprisingly indicated almost pure, right-lateral strike-slip motion without any significant vertical motion (Berberian et al., 1979a; Figure 13.7). This was the first documented evidence of a change of sense of motion from a predominantly reverse/oblique reverse to right-lateral slip along the faults on the Iranian Plateau, and it indicates tectonic reactivation of inherited deformation zones induced by changes in the regional stress field (see also Figure 9.4).

Meyer and LeDortz (2007) suggested that the ongoing strike-slip tectonics of Central Iran might have originated between 8 and 22 Ma ago (in the Early to Late Miocene). The last marine reefal limestone deposits in Central Iran (the Qom Formation) and the Zāgros (the Āsmāri Formation) is dated Burdigalian in the late Early Miocene (~15 Ma; James and Wynd, 1965; Bozorgnia, 1966). Hence, from the time of the early Late Miocene, the marine condition had ended in Iran because of contractional movements of the Arabian plate. Because the diachronous contraction deformation of the outer Zāgros took place in the Late Miocene, 8 Ma (Homke et al., 2004), the inferred fault inception and change in sense of motion might have taken place beginning in ca. 8 Ma and younger (Figures 9.1 and 9.4).

16.2 LATE NEOGENE–QUATERNARY MIGRATION OF ACTIVE FAULTING AWAY FROM THE MOUNTAIN RANGE-FRONT TO THE PIEDMONT AREA

Immediately after the 16 September 1978, M_w 7.3 Tabas-e Golshan (Berberian, 1979a), 28 July 1981, M_w 7.0 Sirch (Berberian et al., 1984), and 10 May 1997, M_w 7.2 Zirkuh strike-slip (Berberian et al., 1999) earthquakes, the mountain range-front faults were checked and did not show any coseismic surface deformation (Figures 13.8, 13.10, 13.11, and 13.18). In the case of the 1978 Tabas-e Golshan earthquake (Figure 13.8), the Shotori range-front reverse fault with geomorphic evidence of reverse and right-lateral strike-slip (located at the northern end of the Nāyband right-lateral strike-slip fault system) did not undergo coseismic surface faulting; hence, the active faulting appeared to show Quaternary migration away from the range-front, possibly in response to stresses produced by the elevated Shotori range topography (+2838 m amsl) versus the playa (+650 m) (Berberian, 1979a; Walker et al., 2003).

In all these cases, the coseismic deformations took place along the active faults with frontal escarpments located at the edge of the recent asymmetric folds, with playa-facing, steeper truncated limbs in the piedmont area. In the case of the Tabas-e Golshan thrusting event (Figure 13.8), the migration of activity varies from a few kilometers to 25 km away from the Shotori range-front fault. In the case of the Gowk (Figures 13.10 and 13.11) and Zirkuh with a strike-slip mechanism, the distance is less than a kilometer to 1.5 km (Figure 13.18).

Usually active geomorphological features such as active folding, bedding-plane slips with a thrust mechanism, faulting, and uplift of the Neogene molasse deposits and Quaternary alluvial terraces are visible in the piedmont area. This is a clear indication of migration of the present-day activity away from the mountain range-front to the piedmont area, possibly in response to stresses produced by the elevated topography of the active fold-thrust mountain belts (Walker et al., 2003). However, in the case of the 1 September 1962, M_w 7.0 Bu'in earthquake (thrust with slight left-lateral motion), only parts of the range-front, intermountain, and piedmont faults were reactivated (Figure 12.13).

16.3 TOPOGRAPHIC RIDGE-TOP STRIKE-SLIP FAULTING

Two earthquakes, the 20 June 1990, M_w 7.3 Rudbār earthquake in the western High Alborz Mountains (Berberian et al., 1992; Berberian and Walker, 2010; Figure 13.12) and the 11 August 2012, M_w 6.4 South Ahar earthquake (Copley et al., 2013) along the ridge south of Ahar (Figure 14.12), showed very weak active topographic indicators that were not detected beforehand. Coseismic surface ruptures of both events were developed close to the crest lines of the mountain belts in both events (>+2000 and ~+2000 m, respectively); they also involved oblique strike-slip shortening and slip partitioning.

The 1990 coseismic surface ruptures showed unusually large vertical displacements in the opposite sense of the existing topography of the High Alborz near the High Alborz–Caspian watershed. Unfortunately, no slip measurements were carried out for the 2012 twin earthquakes, and the case has not been properly studied in the field.

As mentioned, both earthquake faults have subtle expression in the pre-earthquake active geomorphology. Apparently, both faults do not move often enough to exert much influence on the topography. It is highly likely that there are other similar cases that have not yet been detected, which will have a direct effect on the assessment of earthquake hazards.

16.4 SPATIAL SHEAR-STRAIN PARTITIONING ON THE IRANIAN PLATEAU FAULTS

The partitioning—or special separation of oblique convergence into parallel strike-slip and reverse faulting—induced by local changes in the regional stress field along inherited structures (Figures 9.1 and 9.4) has been documented in the analyses of earthquakes and their faulting in transpressional belts in the Alborz, NW Iran, Kopeh Dāgh, Central Iran, and Zāgros (Jackson and McKenzie, 1984; Berberian et al., 1992, 2000b, 2001; Jackson, 1992, 2002; Jackson et al., 2002; Talebian and Jackson, 2002; Copley and Jackson, 2006; Masson et al., 2007; Tatar and Hatzfeld, 2008; Berberian and Walker, 2010).

The Alborz Mountains of northern Iran (Figures 9.1–9.4) behave as a transitional orogenic belt, accommodating the differential motion between Central Iran in the south and the South Caspian Basin in the north, which seems to involve oblique left-lateral shortening (see also Chapter 9). Strain is partitioned onto the subparallel strike-slip and reverse faults with orthogonal slip vectors. Examples of this include the 22 July 1983, M_w 5.5 Charazeh thrust event and the 20 June 1990, M_w 7.3 Rudbār left-lateral strike-slip earthquake on the subparallel Manjil thrust and the Rudbār strike-slip faults in the western Alborz (Figure 13.12; Berberian et al., 1992; Jackson, 1992, 2002; Jackson et al., 2002; Vernant et al., 2004; Berberian and Walker, 2010).

In the northwest of the Zāgros (Figures 9.1–9.4), where the overall N–S convergence is oblique to the trend of the Zāgros, the strain is partitioned into its orthogonal strike-slip (the Zāgros Main Recent fault; Figures 12.1, 12.11, 12.12, and 16.12) and shortening components on separate parallel basement reverse faults (Jackson and McKenzie, 1984; Berberian, 1995; Talebian and Jackson, 2004).

Along the Gowk fault system on the western side of the Lut Block (Figure 9.1), where it is oblique to the convergence, the overall motion is spatially partitioned into strike-slip faulting along the Gowk fault and shortening on the Shahdād thrust-and-fold system to its east (Figure 14.3). Both the Gowk strike-slip and Shahdād thrust faults moved during the 1998 Fandoqā earthquake; the latter occurred either immediately after or within 6 months (Berberian et al., 2001; Fielding et al., 2004). The Gowk fault system itself has evolved into an underlying ramp-and-flat, thrust geometry (1981 M_w 7.1, $h \sim 14$ km, Sirch earthquake) with the strike-slip fault at a higher-level, steeper splay off a structural ramp (1998, M_w 6.6, $h \sim 5$ km; Figure 14.2) (Berberian et al., 2001; Walker and Jackson, 2002; Jackson et al., 2006). This will be discussed later.

Study of the 26 December 2003 M_w 6.6 Bam earthquake also showed evidence of spatial partitioning of slip on two subparallel faults (Figures 14.6 and 14.7). About 85% of moment release occurred on a previously unknown, N–S strike-slip fault running into Bam city, with peak slip of over 2 m occurring at a depth of ~5 km; the remaining 15% of the moment release occurred as a combination of strike-slip and thrusting motion on the 1976-mapped Bam fault at Baravāt (Berberian, 1976b), located about 5 km to the east of the former fault (Fielding et al., 2005; Funning et al., 2005; Jackson et al., 2006).

Slip partitioning of strike-slip and thrust components of motion in obliquely converging belts is common in oceanic island arcs, especially in the Java–Sumatra subduction zone (Fitch, 1972; McCaffrey, 1991, 1992) and on the continents. It is an important mechanism by which oblique convergence is achieved on the continents (Jackson et al., 1992, 2002). Slip partitioning has also been documented in California (Mount and Suppe, 1987; Ekstrom et al., 1992), the Caucasus and the North Anatolian faults (Jackson, 1992; Jackson and Ambraseys, 1997; McClusky et al., 2000), the Himalayas (McCaffrey and Nabelek, 1998), Pakistan (Nakata et al., 1990), Mongolia (Bayasgalan et al., 1999; Bendick et al., 2000), and Irian Jaya, Indonesia (Abers and McCaffrey, 1988).

16.5 FAULT-BOUNDED BLOCK ROTATION

As with eastern Tibet (England and Molnar, 1990) and the Western Transverse Ranges, California (Jackson and Molnar, 1990; Molnar and Gipson, 1994), the nearly E–W left-lateral (Doruneh, Dasht-e Bayāz, etc.), N–S, right-lateral (Nāyband, Gowk, Neh, Kahurak, etc.), and NNW–SSE, right-lateral (west Āzarbāijān, Kāshān, Dehshir, Anār, Kuhbanān, Rafsanjān, etc.) strike-slip faults in Central and East Iran (Figure 9.1), all accommodate the NNE convergence of Arabia–Eurasia by block rotation about vertical axes (Freund, 1970; Jackson and McKenzie, 1984, 1988; Jackson et al., 1995; Berberian et al., 2000b; Jackson, 2002; Walker and Jackson, 2004; Allen et al., 2011).

The block rotations are also documented by paleomagnetic studies from Oligocene–Miocene sedimentary units in Central Iran (Mattei et al., 2012). The inherited structures within the broad deforming zone of the Iranian Plateau are, therefore, being reactivated in the present deformation kinematics (Figure 9.4) by strike-slip, block rotation (Figure 15.4). Apparently, the nearly E–W to WSW–ENE, left-lateral strike-slip faults and the N–S to NNW–SSE, right-lateral faults accommodate the present nearly N–S shortening caused by the Arabian plate motion and indentation (Figure 9.4) by counterclockwise and clockwise rotations about vertical axes, respectively. Hence, there is evidence that the fault pattern on the Iranian Plateau may change with time in response to block rotation (Jackson, 1988, 2001; Jackson and McKenzie, 1999; Mattei et al., 2012).

16.6 THRUST FAULT EARTHQUAKES AT THE TERMINATION OF STRIKE-SLIP FAULTS

Some intracontinental strike-slip faults terminate at one or both ends with splay thrust faults that die away with distance from their intersection with the strike-slip faults (Baljinnyam et al., 1993; Berberian, 1995; Bayasgalan et al., 1999; Berberian et al., 2000b; Talebian and Jackson, 2002). This seems to be a common fault configuration on the Iranian Plateau, as well as other places, that accommodates oblique shortening across the N–S deforming zones, probably by counterclockwise rotation about a vertical axis (Figures 9.1, 13.4, 13.15, 13.17, 14.7, 14.8, 15.1, 15.4, 16.1, and 16.2). This might be a transition to a more evolved state that involves partitioning of the strike-slip fault and convergent motion onto separate subparallel faults (Freund, 1970; Jackson, 1992a,b, 2002; Bayasgalan et al., 1999). Some of the examples on the Iranian Plateau are as follows:

i. The central Zāgros transverse strike-slip faults: Kāzerun, Karéhbas, Sabzpushān, and Sarvestān (Figures 9.1, 13.4, 16.1, and 16.2; Berberian, 1995; Talebian and Jackson, 2002).
ii. The Ābiz fault (Figure 13.18; Berberian et al., 1999; Berberian, 2005).
iii. The Zāhedān and Sefidābeh faults (Figures 15.1 and 15.4; Berberian et al., 2000b).
iv. The Kuhbanān fault (Figure 14.8; Berberian, 1995, 2005; Talebian et al., 2006a).
v. The East and West Neh faults (Figure 9.1, Berberian and Yeats, 1999).
vi. The Nāyband, Gowk, and Ābiz faults (Figures 9.1, 13.18, 14.3, and 14.7; Berberian and Yeats, 1999; Berberian et al., 1999, 2001; Berberian, 2005).
vii. The North Tabriz fault (NTF; Figures 11.7 and 11.12; Berberian and Yeats, 1999).
viii. The Rudbār fault (Figure 13.12; Berberian et al., 1992; Berberian and Walker, 2010).
ix. The Central Kopeh Dāgh Shear Zone (CKDSZ; Figure 16.3).

16.7 DECOUPLED EARTHQUAKES IN THE ZĀGROS FOLD-AND-THRUST MOUNTAIN BELT

Medium-magnitude earthquakes (maximum recorded M_w 6.7 1972 earthquake) in the Zāgros active fold-and-thrust mountain belt (Figures 9.1–9.4) of southern and southwestern Iran, northern Iraq, northern Syria, southern Asia-Minor, and the Salt Range of Pakistan, generally do not show surface faulting because of the special salt-tectonics character of the region (Berberian, 1976a, 1977b, 1981a, 1995; Berberian and Tchalenko, 1976b,c; Berberian and Papastamatiou, 1978). The apparent lack of agreement between

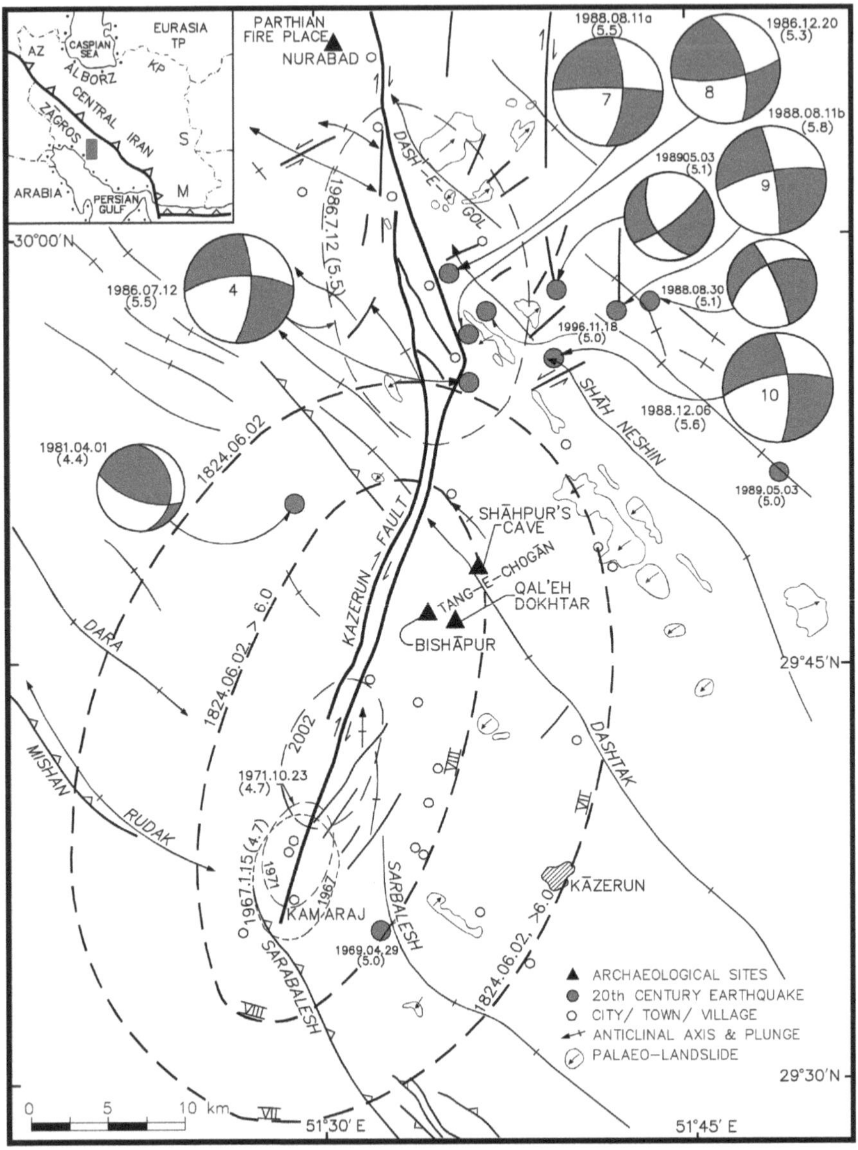

FIGURE 16.1 Meizoseismal area of the recorded historical earthquakes along the central section of the Kãzerun right-lateral strike-slip transverse fault in the Zãgros since 1824. Fault and other symbols as Figure 11.1. Focal mechanism of the earthquake constrained by body-wave modeling with centroid depths (Baker et al., 1993). Fault-plane solutions of smaller events (best-double-couple HRVD CMT solution). *Modified from Berberian (1995).*

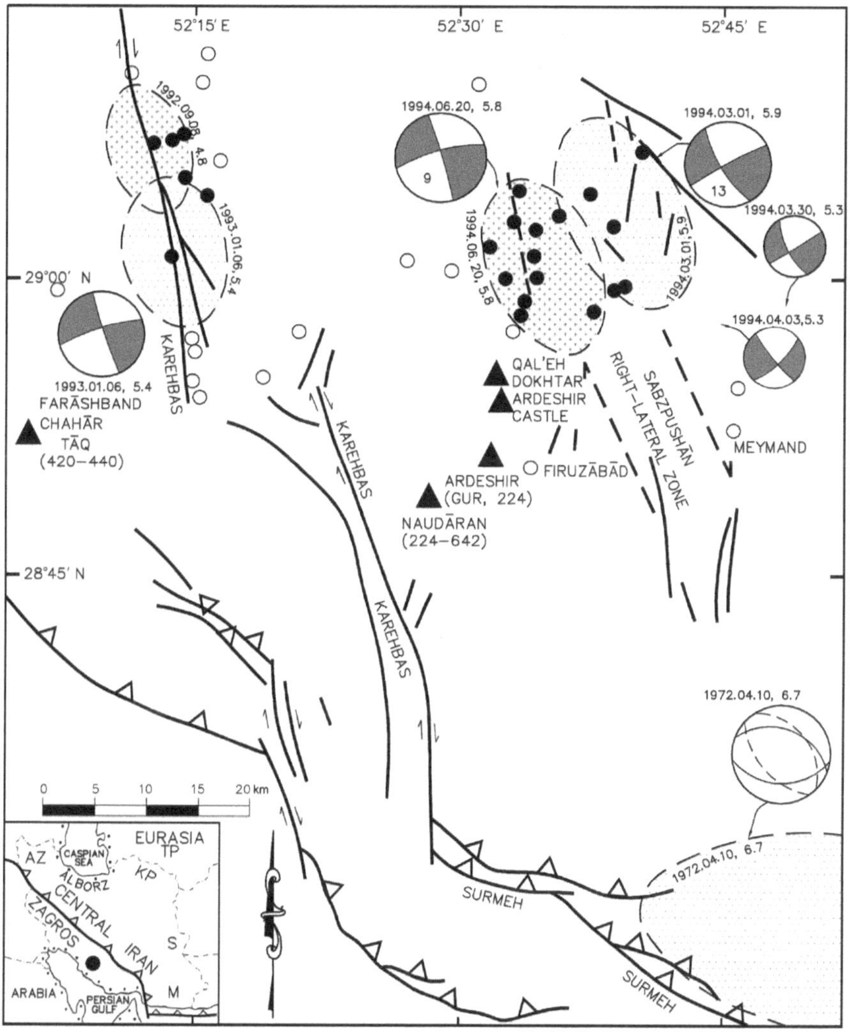

FIGURE 16.2 Recorded earthquakes and their meizoseismal areas along the Karehbas right-lateral transverse fault and Sabzpushān right-lateral strike-slip zone in the central Zāgros. Fault and other symbols as Figure 11.1. Destroyed sites are shown by filled circles, and historical monuments with their dates by filled triangles. No earthquake has yet been recorded along the southern segment of the Karehbas fault. Focal mechanism of the earthquakes constrained by body-wave modeling with centroid depths (20 April 1994: Maggi et al., 2000a; 1 March 1994: Talebian and Jackson, 2004; 10 April 1972: Baker et al., 1993). Fault plane solutions of smaller events (best-double-couple HRVD CMT solution).

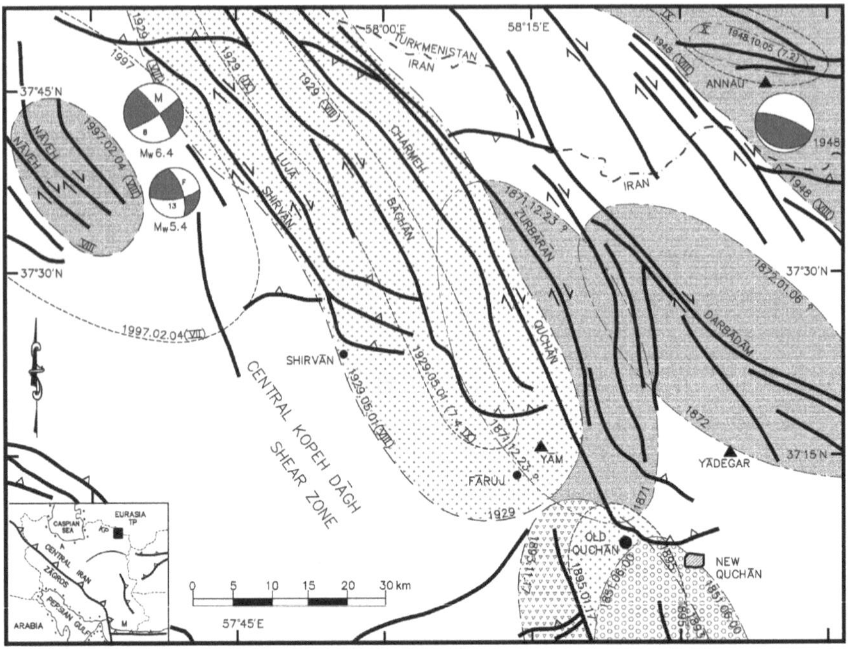

FIGURE 16.3 Meizoseismal areas and fault map of the Central Kopeh Dāgh Shear Zone (CKDSZ) of northeast Iran. Fault and other symbols as Figure 11.1. Focal mechanism of the earthquake constrained by body-wave modeling with centroid depths (Hollingsworth et al., 2006). Fault plane solution of the 1948 Ashkābād earthquake (McKenzie, 1972).

earthquakes and surface faults in the Zāgros arises from the relation between the late Neoproterozoic metamorphosed basement at depth (where most earthquakes occur) and the 8–14 km of Phanerozoic top sedimentary cover, where the geological structures are observed at the surface. A highly plastic salt layer (the Upper Vendian–Lower Cambrian Hormoz Salt; Stocklin, 1968b) acts as a "zone of slippage and decollement," disconnecting the top sedimentary cover structures from those in the basement.

Furthermore, within the Zāgros sedimentary cover there are plastic layers other than the Hormoz Salt: (i) the Permian evaporites (Nat member) of the Dālān Formation, (ii) the Lower Triassic Dashtak evaporites, (iii) the Lower Jurassic Adaiyah and Alan anhydrite, (iv) the Upper Jurassic Gotniā and Hith anhydrite, (v) the Lower Eocene Kashkān-Sāchun gypsum, (vi) the Lower Miocene Kalhur gypsum, and finally, (vii) the Miocene salt–gypsum layers of the Gachsārān Formation (the Lower Fārs in Iraq and Kuwait) (see table 1, p. 215 in Berberian and King, 1981a) (James and Wynd, 1965; Setudehnia, 1972; Szabo and Kheradpir, 1978; Berberian, 1981a, 1995). Each of these incompetent evaporite horizons (see also figure 6, pp. 46 and 47 in Berberian, 1981a), together with other semiductile beds (such as the Jurassic Neyriz and the Eocene Pābdeh Formations), mechanically decouples the overlying structures from the underlying ones, behaving as zones of decollement.

These tectonically incompetent units, characterized by extreme mobility, are responsible for remarkable structural changes and complications: synclines at the surface correspond to anticlines at 1 km depth; surface faults change from high-angle reverse to near-horizontal (listric) thrust, or die out altogether (Berberian, 1976a, 1977a, 1981a, 1995; Berberian and King, 1981a).

The Zāgros decoupling idea was later confirmed by P-wave modeling of the Zāgros earthquakes (Jackson, 1980; Jackson and Fitch, 1981; Jackson et al., 1981; Maggi et al., 2000a,b; Talebian and Jackson, 2004; and many more) and show that some events take place in the sedimentary cover as well. The only exception to the decoupling idea is the master basement faults cutting the Hormoz and other salt layers and reaching the surface (Berberian, 1981a, 1995). These are (i) the High Zāgros, Zāgros Mountain Front, and Zāgros Foredeep longitudinal reverse faults; (ii) the Zāgros Main Recent longitudinal right-lateral strike-slip fault at the northeast border; and (iii) the central Zāgros transverse right-lateral strike-slip faults such as the Kāzerun, Karébas, Sabzpushān, and Sarvestān faults (Berberian, 1995; Figure 9.1). These Zāgros master fault sets show clear escarpments and large horizontal and vertical displacements at the surface, with a concentration of linear zones of meizoseismal areas of historical and modern large earthquakes along their trends (Berberian, 1981a, 1995; see figure 1 in Berberian, 1995).

Due to the presence of the Upper Vendian–Lower Cambrian Hormoz Salt bed (Rāvar or Desu Series in the Kuhbanān Mountains; Huckriede et al., 1962; Stocklin, 1961, 1968b, 1972) in the Kuhbanān fold-and-thrust belt in southeast Iran (Figures 9.1 and 14.8), similar decoupling may be present in the area north of Kermān (see also figure 10, p. 228, and table 1, p. 215 in Berberian and King, 1981a).

16.8 GEOMORPHOLOGICALLY FEATURELESS ACTIVE FAULTS

The coseismic surface ruptures of the 20 June 1990 M_w 7.3 Rudbār (Figure 13.12; Berberian et al., 1992; Berberian and Walker, 2010), 15 December 2003 M_w 6.6 Bam (Figure 14.6; Talebian et al., 2004; Jackson et al., 2006), 22 February 2005 M_w 6.4 Dāhuiyeh (Figure 14.8; Talebian et al., 2006a), 20 December 2010 M_w 6.5, 27 January 2011 M_w 6.2 South Rigān (Figure 14.9; Walker et al., 2013c), and 11 September 2012 M_w 6.4 and 6.3 (twin shocks) (Figure 14.12; Copley et al., 2013) earthquakes did not have clear pre-earthquake features on the local geomorphology. The main slip in some of the aforementioned cases failed to reach the surface. In the case of Bam, at least the Baravāt segment of the Bam fault system shows a significant active morphology (Berberian, 1976b), whereas in the case of the two 2011 South Rigān earthquakes, such active morphologic indicators were absent and the area was covered by Recent volcanic products. Some of the aforementioned earthquakes occurred in seismically quiet regions (or without records of seismicity), where it is difficult to predict future events. This may partly be due to the earthquakes' long return period.

Although this issue makes seismic hazard studies difficult, a glance at the geological map and aerial photos of the South Rigān/Jahānābād area (GSI, 1992c) clearly reveals some short fault splays at the southern tip of the Kahurak active strike-slip fault as well as the southeastern tip of the North and South Bārēz faults (Figure 14.11), which makes it possible to expect future seismic activity in the area at the tips of active faults.

The cases of the 1990 Rudbār and 2012 South Ahar earthquake faults, with pre-earthquake, subtle geomorphologic expressions, have been addressed in the previous chapters (Figures 13.12 and 14.12).

16.9 FAULTS WITH LARGE AREAS OF INTERVENING UNRUPTURED GAPS

Several fault segments with a temporary lack of recorded historical seismicity, or low level of activity, have been observed along some active faults on the Iranian Plateau (Figures 11.2–11.6, 11.8–11.11, 11.13–11.15, 12.6–12.12, 13.14, 13.15, 14.7–14.11, 16.2, 16.4, 16.11, 16.12, 16.15, 16.18, and 16.19). The seismic gaps indicate unruptured segments of the fault zones that will move in future earthquakes. Some of the cases are addressed below.

16.9.1 Rudbār Seismic Gap

The 20 June 1990 M_w 7.3 Rudbār earthquake struck a region that was a seismic gap, where the surface expression of active faults was less developed than in other regions (Berberian et al., 1992; Berberian and Walker, 2010). The gap was located to the immediate west of the 18 August 1485 $M_s \sim 7.3$ Upper Polrud earthquake along the Kelishom left-lateral strike-slip fault (Figures 13.12 and 13.13), a fault subparallel to the Rudbār fault and located to its northeast (Berberian and Walker, 2010). The 1990 Rudbār earthquake filled the Rudbār gap, but the seismic gap to the northwest remains unruptured, although the region is characterized by moderate seismicity. Both the Rudbār and Kelishom faults have a similar setting in mechanism and orientation and are about 10 km apart (Figure 13.13). Considering the entirety of the Alborz Mountains as similar in response to oblique-slip convergence with uniform slip rate along strike, the absence of earthquakes along the Rudbār fault is advanced as evidence that the Rudbār earthquake filled a seismic gap. This conclusion is strengthened by the lack of geomorphic evidence for activity on the Rudbār source fault, as seen on pre-1990 aerial photographs.

16.9.2 Changureh Seismic Gap

The Changureh seismic gap is another recent case located to the west of the 1 September 1962 M_w 7.0 Bu'in earthquake that occurred during the 22 June 2002 M_w 6.4 Changureh earthquake (Figure 14.5).

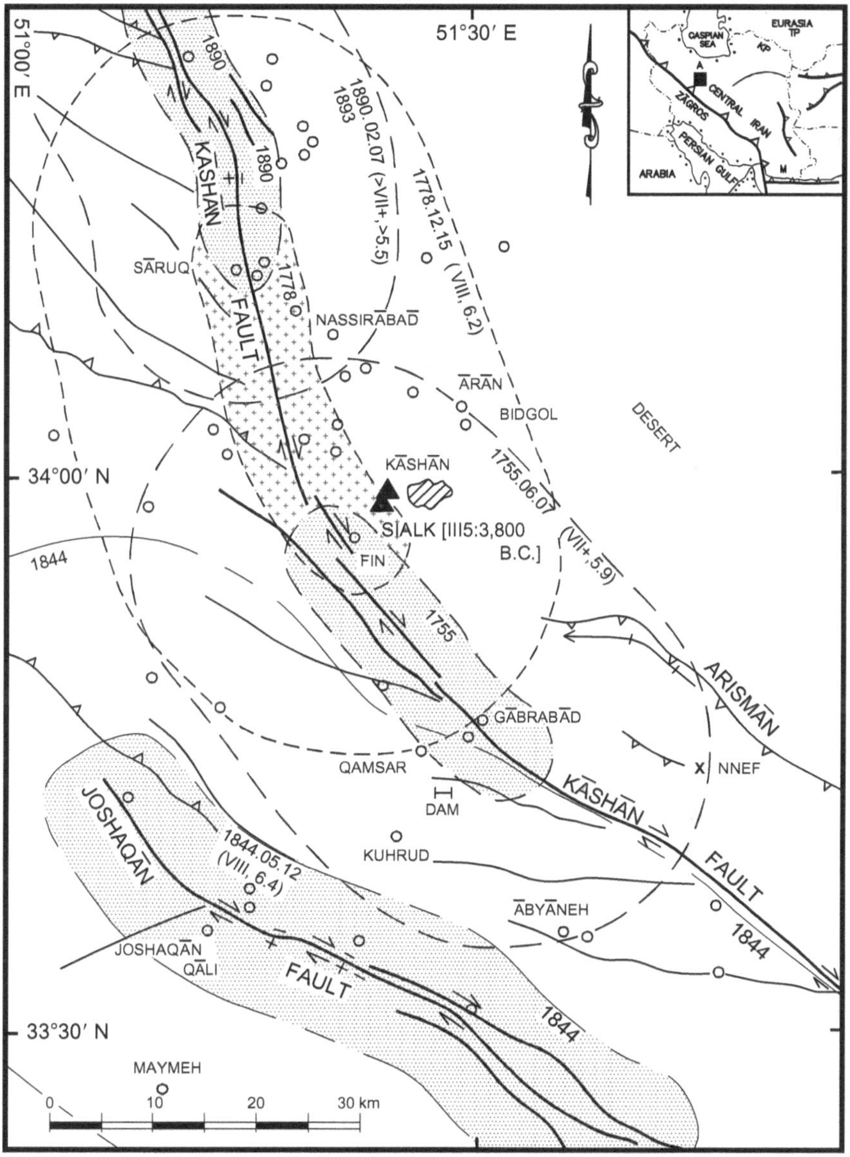

FIGURE 16.4 Meizoseismal areas of recorded earthquakes along the Kāshān and Joshaqān faults in Central Iran since 1775. Fault and other symbols as Figure 11.1. Locations of the Sialk archeological mounds to the west of the city of Kāshān are shown by filled triangles. Meizoseismal area of the 1844 earthquake is not constrained. *Modified from Berberian et al. (2012).*

16.9.3 Tehrān Seismic Gap

The North Tehrān fault system (the North Tehrān thrust, Niāvarān and Darakeh left-lateral strike-slip faults, Mahmudiyeh, Dāvudiyeh, and ʻAbbāsābād thrusts: Berberian et al., 1985; Berberian and Yeats, 2014) has not generated historical or modern earthquakes during recorded seismic history (Figure 11.2). Faults to the southeast (Pārchin, ca. 280 BCE?), northeast and north (Moshā, 958,1665, and 1830), and northwest (western North Tehrān thrust, 1177?) have apparently slipped during large-magnitude earthquakes. Hence, the mega-city of Tehrān is located in a seismic gap section of the aforementioned faults and has a greater chance for strain in the future. Since the controversial nature of the Kahrizak and North and South Ray escarpments has not yet been resolved, I do not include them in this discussion (see Berberian and Yeats, 2014).

In recent years, there has been a rapid increase in population and area of Tehrān. Because the city is located in a seismic gap region with its current demography, a catastrophe is expected to happen at any minute (Berberian and Yeats, 2014).

16.9.4 Additional Gap Cases

Some of the seismic gaps on the Plateau are summarized below.

i. The Kuhbanān fault (Figures 12.6, 14.8, 16.19, and 16.20).
ii. The Gāilātu–Khoi fault in northwest Iran, along the Western Āzarbāijān Shear Zone, where the 2 July 1840 $M_s \sim 7.4$ Ārārāt and 29 April 1968 M_s 5.5 Gol earthquakes took place along the northwestern segment, and the 18 April 1843 M_s 5.9 Khoi, 1900 and 1971 earthquakes took place along the southeastern segment with an intervening unruptured 70-km long, central segment (Figure 13.6; Berberian, 1997a).
iii. The Chālderān–Khoi fault, along the Western Āzarbāijān Shear Zone, between the 24 November 1976 M_w 7.0 Chālderān and the 18 April 1843 M_s 5.9 Khoi earthquakes (Figure 13.6; Berberian, 1997a).
iv. The 200-km long Kāshān fault, with ruptured northern segment during the 1755, 1778, 1890, 1893 earthquakes; in contrast, the southern Natanz segment has not shown seismic activity during the historical period (Figure 16.4; Berberian et al., 2012).
v. The 600-km long Khazar reverse fault (Figures 9.1, 11.6, and 12.10; Berberian, 1981a, 1983a), with recorded earthquakes scattered along its length: 874, 1367, 1400, 1436, 1498, 1678, and 1809, 5 April 1944, 22 July 1980, 3 December 1980, and 24 May 2004.
vi. The 400 Tālesh reverse fault bordering the western margin of the South Caspian basin (Figures 9.1, 11.14; Berberian, 1981a, 1983a), with recorded events on 11 July 1970 (5.2), 4 November 1978 (6.1), and 4 May 1980 (6.2) (Berberian, 1997a).
vii. The Sabzévār reverse fault, with recorded earthquakes of 2 June 1052 (7.0) and 1804 (Berberian, 1976c, 1981a; Figure 9.1).

viii. The Esfarāyen reverse fault, with earthquakes on 11 May 1695 (7.0), 31 March 1963 (5.0), 3 January 1969 (5.5), 10 April 1987 (4.9), 4 August 1998 (4.9), and 24 April 1999 (4.3) (Berberian, 1981a; Figure 9.1).

ix. The Shoqān reverse fault, which possibly moved in August 943 (~7.4) (Figure 11.5).

x. The unruptured northern (Yāsuj) and southern (Borāzjān) segments of the Kāzerun fault (Figures 9.1 and 16.1; Berberian et al., 2014).

xi. The Karébas fault (Figure 16.2; Berberian, 1995).

xii. The Zāgros Main Recent, High Zāgros, Zāgros Mountain Front, and Zāgros Foredeep faults (Figures 9.1, 11.4, 12.1, 12.11, 12.12, and 13.14; Berberian, 1995).

xiii. The western segment of the Moshā fault (Figures 11.2 and 11.6; Berberian and Yeats, 1999, 2001, 2014).

16.10 TEMPORAL EARTHQUAKE CLUSTERING AND COMPLETE RUPTURING OF MULTISEGMENTED INDIVIDUAL FAULTS

We know that active faults do not show unique forms of behavior and, therefore, recurrence intervals of large-magnitude earthquakes along faults are complicated and highly variable in different structural provinces. The earthquakes may form temporal clusters along a single fault, subparallel faults, and in a zone covered by several interacting faults striking in variable directions (Kagan and Knopoff, 1976; Wallace, 1987; Kagan and Jackson, 1991; King et al., 1994; McCalpin and Nishenko, 1996; Xu and Deng, 1996; Stein et al., 1997; Berberian and Yeats, 1999, 2001; Rockwell et al., 2000; Berberian, 2005).

Cases of temporal clustering of large-magnitude earthquakes along a single fault have been recorded on the Iranian Plateau. Some faults show alternating periods of medium-magnitude seismic activity followed by a large-magnitude earthquake rupturing the complete fault length or a major long segment. Examples are the Ābiz (1936, 1979, 1979, 1997; Figures 13.9 and 13.18) and Gowk (1877, 1909, 1911, 1948, 1961, 1969, 1981; Figure 13.11) faults. Other faults show rupture progressing along a single fault, such as the North Tabriz (1721, 1780, 1786; Figures 11.7 and 11.12), Sangāvar (1863, 1896; Figure 11.14), Moshā (958, 1665, 1830; Figure 11.2), Larzaneh (1301, 1957; Figure 12.10), Neyshābur fault system (1209, 1270, 1389, 1405; Figure 16.15), Chāhak (1941, 1962; Figure 12.8), northern Kāshān (1755, 1778; Figure 16.4), Zāgros Main Recent (1909, 1957, 1958, 2006; Figure 16.12), central Kāzerun (1824, 1987; Figure 16.1), Southern Ardal (1666, 1880, 1960, 1977), Surmeh (1440, 1903, 1968, 1972, 1973, 1985; Figure 13.4), Central Karéhbas (1992, 1993; Figure 16.2), Sabzpushān (1506, 1591, 1752, 1765, 1784, 1862, 1812, 1824, 1853, 1862, 1892, 1994; Figures 13.4 and 16.2), Eastern Makrān Subduction Zone (1765, 1851, 1945, 1947), and many more.

16.10.1 The Ābiz Fault Earthquake Sequence Stress Build-Up and Triggering

An important feature of the 10 May 1997 M_w 7.2 Zirkuh earthquake, which ruptured the complete 125 km of the Ābiz right-lateral strike-slip fault (Figure 9.1), is that virtually all the northern segment of the rupture, from Ardekul to Donakhi, had already ruptured relatively recently in medium-magnitude earthquakes on 30 June 1936 (M_w 6.0; >12 km surface rupture) and in two earthquakes on 14 November 1979 (M_w 6.6; >20 km rupture) and 7 December 1979 (M_w 5.9; >15 km rupture). The epicenter of the 1997 event was apparently along the unruptured part of the Ābiz fault between the 1936 and 1979 surface ruptures (Figures 13.9 and 13.18). It is rare to definitively observe this behavior in modern times; it clearly has implications for the nature of stress build-up, release, and triggering during the earthquake cycle (Berberian et al., 1999).

We know of two other places where similar behavior has been documented in modern times, also on strike-slip faults. One is on the Imperial Valley Fault in California, where a section that ruptured on 19 May 1940 (M_w 7.2) ruptured again on 15 March 1979 (M_w 6.7) (Ulrich, 1941; Sharp et al., 1982). In this case, unlike at Zirkuh, the larger earthquake preceded the smaller one. The other example is at the western end of the North Anatolian Fault, where an overlap occurred in the rupture zones of two earthquakes of M_s 7.0 and 7.1 on 26 May 1957 at Abant and 22 July 1967 at Mudurnu (Ambraseys and Zatopek, 1969; Barka, 1992; Stein et al., 1997).

16.10.2 The Tabriz Clustered Earthquake Sequence of 1641–1786

The city of Tabriz, located to the immediate south of the North Tabriz Fault (NTF) in northwest Iran (Figures 9.1 and 16.5), is a highly seismic region in which most historical monuments have been eliminated by several historical, large-magnitude earthquakes (Figure 16.6). The NTF is mainly a segmented, right-lateral strike-slip fault with a considerable vertical component (Berberian and Arshadi, 1976; Berberian, 1981a, 1997a), and the fault can be divided into a few segments in a right-stepping, en-echelon geometry (Berberian, 1997a; Berberian and Yeats, 1999). The fault is bounded at the northwestern and southeastern tips by north-dipping reverse faults (the Tasuj—in the NW—and the South Bozqush—in the SE—reverse faults). Locally recorded microseismic activity along the fault shows predominant right-lateral strike-slip focal mechanisms (Moradi et al., 2011).

Since 858, the region of Tabriz has been shaken by at least 10 moderate- to large-magnitude earthquakes; occurring in 858, 1042, 1273, 1304, 1440, 1641, 1717, 1721, 1780, and 1786, they devastated the urban and rural areas along the fault (Figures 11.12 and 16.7). Of these 10 recorded earthquakes, three (1721, 1780, and 1786) can be assigned to the motion being ruptured along different

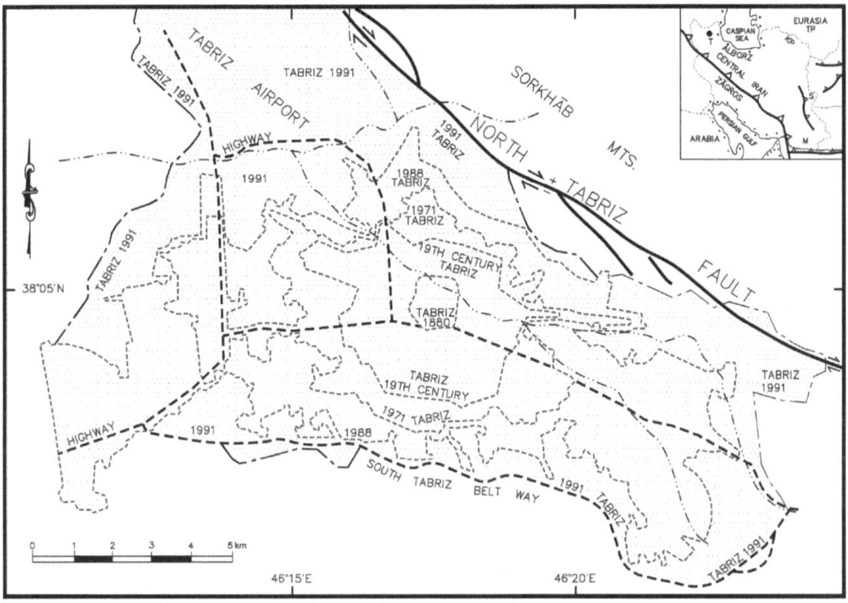

FIGURE 16.5 Growth of the city of Tabriz from 1880 through 1991. Since 1990, the northern city limit surpassed the North Tabriz strike-slip fault and buildings have been built on the fault line and beyond on the slopes of the Sorkhāb Mountain in the north and northeast. *City limits modified from Arseh (1988).*

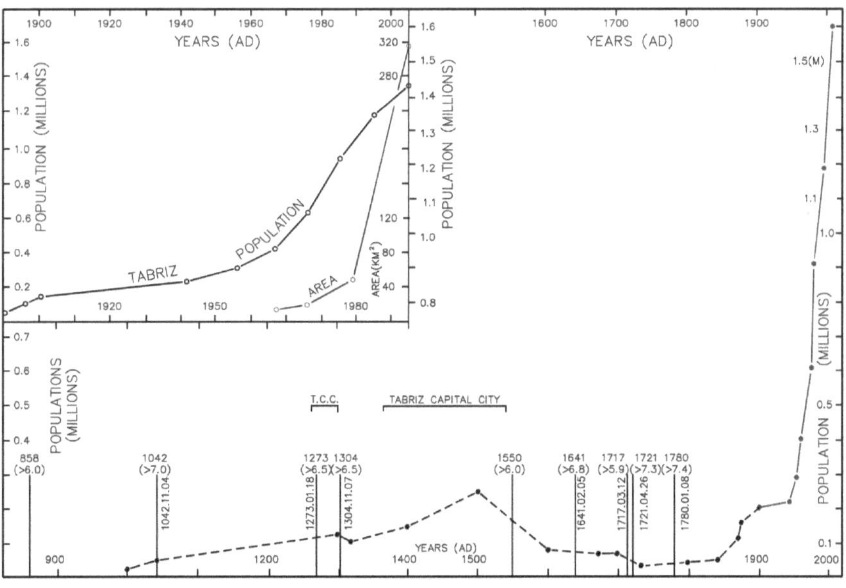

FIGURE 16.6 Recorded seismicity of Tabriz (vertical lines) with population and area growth of the city. Demography based on data from Chardin (1673), Lockhart (1960), Issawi (1971), Bharier (1972), Arseh (1988), Azimi (1995), and amar.org.ir.

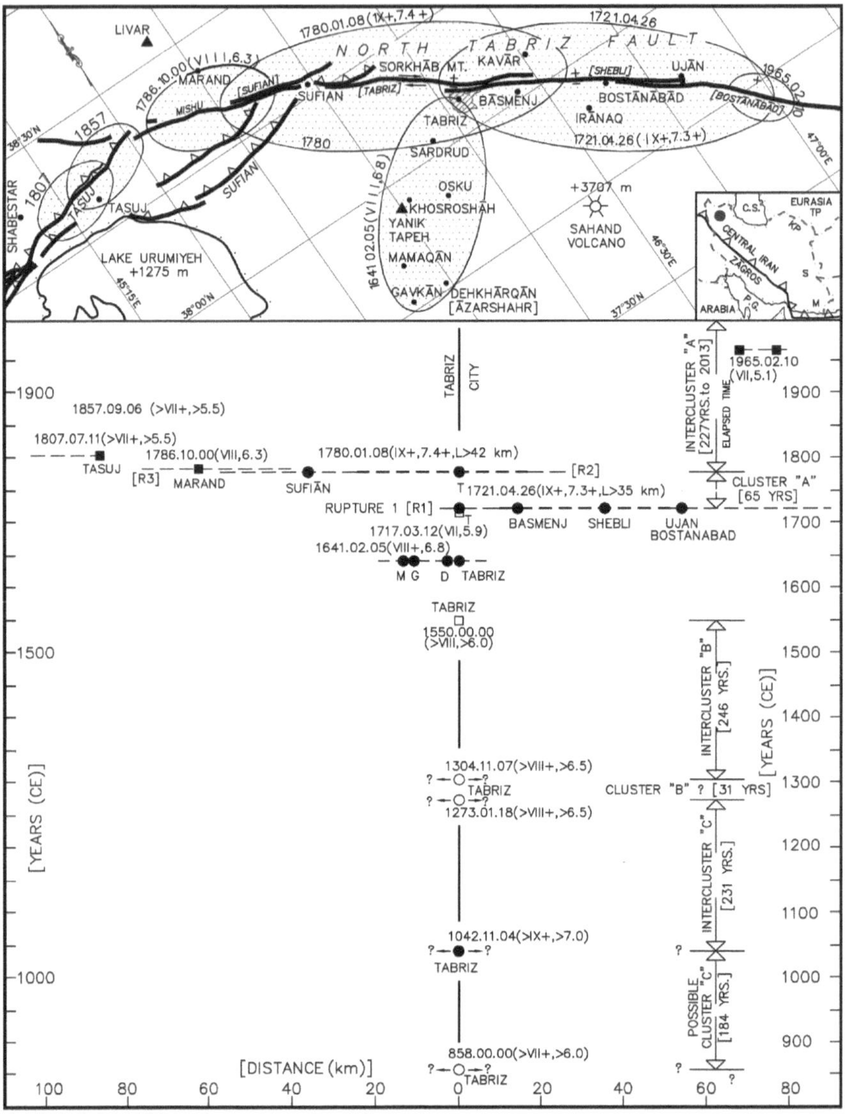

FIGURE 16.7 Meizoseismal area (stippled) of medium- to large-magnitude earthquakes, active fault map (top), and the space-time diagram of 1156 years of seismicity of the Tabriz region (bottom) striking the city of Tabriz. Fault and other symbols as in Figure 11.1. The Yanik Tapeh archeological mound is marked as a filled triangle in the meizoseismal area of the 1641 earthquake. Zone of extensive damage is shown by horizontal dashed line, queried where uncertain in the bottom diagram. Specific sites damaged or destroyed shown in solid circles ($M_s \geq 7.0$), blank circles (M_s 6.0–6.9). Where date of the earthquake is shown, it is given by year.month.day. Distances are along strike with respect to the city of Tabriz. *Modified from Berberian and Yeats (1999).*

segments of the northwestern part of the NTF during seismic clusters (Berberian and Yeats, 1999). Data on the earlier events are not well constrained, and certainly more earthquakes have occurred than survived in the historical annals (Daulatabadi, 1964; Mashkur, 1973; Berberian and Arshadi, 1976; Berberian, 1981a, 1994, 1997a; Melville, 1981; Ambraseys and Melville, 1982; Zokā', 1989; Berberian and Yeats, 1999). The recorded data do not show the occurrence of a very large earthquake that broke the entire fault length of about 200 km through the geometrical barriers; however, this does not indicate that it may not occur in the future (Figures 16.7 and 11.12).

An anomalous cluster of large earthquakes within a 145-year interval (1641–1786, excluding the 1550 event) devastated the city of Tabriz separated by 337 years of pre-1641 and 227 years of post-1786 quiescence periods (cluster "A" in Figure 16.7). Prior to the 1641–1786 clustered earthquake sequence, Tabriz was struck by earthquakes in 858 ($M_s \sim 6.0$), 1042 ($M_s \sim 7.3$; with ground fracture witnessed by Qatrān Tabrizi, 1042), 1273 ($M_s \sim 6.5$; mentioned by Bar-Hebraeus, 1286), and 1304 ($M_s \sim 6.7$; Figure 16.7; also see Chapter 11 and Figure 11.12). However, the meizoseismal zones of these older earthquakes are not known well enough to speculate on which fault segments were responsible for them. The 1273 and 1304 earthquakes are a temporal cluster (Cluster "B" in Figure 16.7) like the 1721–1786 events, but the direction of propagation is not known (Berberian and Yeats, 1999). If the 858 and 1042 events were part of an older cluster (cluster "C" in Figure 16.7), and if we assume that all the three clusters were associated with the NTF, it yields an approximate 450-year recurrence period of seismic clusters along the fault (Figure 16.7; Berberian and Yeats, 1999).

The northwestern section of the NTF system and adjacent reverse faults ruptured from southeast to northwest in three earthquakes during a 65-year interval: (i) the 26 April 1721 $M_s \sim 7.3$ Shebli earthquake on the southeastern NTF, with a minimum recorded surface rupture >35 km long (along the Shebli segment); (ii) the 8 January 1780 $M_s \sim 7.4$ Tabriz earthquake on the northwestern NTF, with recorded surface rupture >42 km long and 2–4-m vertical separation (along the Marjān segment); and (iii) the October 1786 $M_s \sim 6.3$ Marand-Mishu earthquake along the Sufiān segment of the NTF (Figures 11.12 and 16.7; Berberian and Yeats, 1999).

The abovementioned events show a northwestward propagation of seismic activity along the NTF (Figure 16.7). The fourth and fifth earthquakes struck the Tasuj reverse fault farther west on 11 July 1807 ($M_s \sim >5.5$) and 6 September 1857 (>VII) between Tasuj and Marand-Mishu on the Tasuj fault (Figure 16.7). A sixth earthquake of M_s 6.7 took place, possibly along the South Bozqush reverse fault farther southeast on 22 March 1879 (Figure 11.14). The earthquake segment boundaries correspond to geometric discontinuities such as stepovers (overlaps, offsets) and abrupt changes in strike (Berberian and Yeats, 1999).

The short-term variations of the NTF slip rates during the 65-year temporally clustered earthquakes (1721–1786), with large rupture displacements and strain release (more than the long-term average rate) along the fault, as well as other recorded seismic cycles (Figures 11.12 and 16.7). These could result in variable/chaotic (noncharacteristic) seismic behavior of the fault slip rates and recurrence intervals throughout the Holocene, as observed along the San Andreas fault (Weldon et al., 2004).

Prior to this sequence and outside the aforementioned seismic clustered events along the NTF, the Tabriz–Dehkhārqān earthquake of 25 February 1641 $M_s \sim 6.8$ took place off the NTF to the southwest of the city, south of the boundary between the 1721 Shebli and 1780 Tabriz earthquakes (Berberian and Yeats, 1999; Figures 11.12 and 16.7). Peculiarly, the well-constrained meizoseismal area of this event (by the contemporary elites who studied the earthquake; Mohammad Nassir, 1641; Ārākel Vārtābed Tāvrizetsi, 1670; see Chapter 11) is not oriented along the NW–SE-trending NTF, but nearly at right angles to the NTF, as though the source fault was oriented in a NNE–SSW direction along the northwestern foothills of the Sahand Quaternary volcano and parallel to the Talkhehrud River course (Figures 11.12 and 16.7). This alignment does not correspond to any mapped surface fault.

The 1641 off-NTF earthquake might have occurred on (i) a NNE–SSW-tending cross blind thrust, similar to the 1968 Dasht-e Bayāz and Ferdows sequence (Berberian, 1981a; Berberian and Yeats, 1999; Figure 13.2); (ii) a conjugate strike-slip fault to the NTF (like the 1986 North Palm Spring and the 1992 Joshua Tree, Mojave Desert, California conjugate strike-slip earthquakes [Yeats, 2012], or the South Rigān earthquakes of 20 December 2010 M_w 6.5 and 27 January 2011 M_w 6.2 [Walker et al., 2013c; Figure 14.11]); or (iii) a normal fault responsible for the creation of the Tabriz plain as a pull-apart-basin covered by the Sahand volcanic lava and ash flow. In any case, the 1641 earthquake might have increased the Coulomb failure stress along the adjacent NTF and triggered the clustered sequence of 1721–1786 along the NTF (Figures 11.12 and 16.7).

It is interesting to note that the earliest known isoseismal map for the 4 October 1856 small-magnitude earthquake felt in Tabriz (not causing damage but triggering a Cacciatore-type seismometer in Tabriz) drawn by Khanikoff (1858), has an elongated shape trending ENE–WSW [N70°E; the seismometer showed the shock had a direction of E23°16′N; followed by a second shock in the W31°12′S; Khanikoff (1858)]. Moradi et al. (2011) showed a short ENE–WSW-trending, normal fault to the southwest of Tabriz; however, neither their recorded seismicity nor the focal-mechanism solutions correlate with the introduced short fault and the nature of a pull-apart-basin southwest of Tabriz. Contact with the first author regarding the regency of movement along this short fault and its active fault features was not constructive.

As discussed, the NTF system ruptured from southeast to northwest in three clustered earthquakes from 1721 to 1786; a previous cluster may have struck this region in 855–958 (Figures 16.7 and 11.12). The destruction of almost all chronicles during the invasions of Alexander III of Macedonia (336–330 BCE) and the Moslem Arabs (636–652 AD) makes it impossible to broaden the historical earthquake window. Unfortunately, the paleoseismological trench studies with unconstrained dates could not give a better picture.

Limited paleoseismological trenches in separate areas across the NTF indicated at least three strong earthquakes since 33.5 ka (Figure 11.7). The first trench was dug in the area to the northwest of the city in the meizoseismal area of the 1780 earthquake (Hessami et al., 2003); the second trench was located to the southeast of the city in the epicentral area of the 1721 earthquake (Solaymani Azad, 2009; Solaymani Azad et al., 2009).

Based on intrinsic uncertainties from the lack of properly datable materials, poorly constrained ages of the events, and the lack of correlation between the two nearby trenches [100 m apart; 20 km NW of Tabriz], Hessami et al. (2003) defined four surface faulting paleo-earthquakes during the past 36.0 ka with unconstrained dates along the northwestern segment of the NTF (Figure 11.7):

- 1780 CE [1190–1780 CE] (the 1780 date is based on speculation that the youngest event observed in the trench was the historically documented 1780 earthquake).
- 910 ± 250 AD [660–1320 CE].
- 320 ± 320 AD [0–640 CE].
- 700 ± 920 BCE [1620 BCE–220 CE].

With no radiometric dating of the displaced alluvial deposits, Hessami et al. (2003) reported 4 ± 0.5 m horizontal slip/event, 3.1–6.4 mm/year minimum horizontal slip rate along the fault, and 0.5–0.8 mm/year vertical slip rate along the NTF, which cannot be warranted because of their speculative approach to the dates of the displaced alluvial deposits. With great uncertainty, average recurrence intervals of 350–1430 years, with a mean recurrence interval of 821 ± 176 years, were also estimated (Hessami et al., 2003). This value is a much larger value than 250 years (taking into account the whole set of recorded earthquakes; Berberian and Yeats, 1999) and 700 years (taking into account the strongest events; see Berberian and Yeats, 1999; Figure 16.7). Hessami et al. (2003) concluded that the northwestern segment of the NTF does not appear to present major seismic potential for the near future. This conclusion may rest on the last clustered earthquakes having taken place to the northwest of the NTF (Figures 11.12 and 16.7).

The second paleoseismological study in the southeast of the city suggested that three earthquakes occurred during the past 3300 years (Solaymani Azad, 2009; Solaymani Azad et al., 2009), implying a longer recurrence interval for earthquakes along the NTF than those calculated from historical seismic data (Berberian and Yeats, 1999).

GPS data and slip rate: Active current deformation in the Āzarbāijān province of northwestern Iran between the Central Iranian block (in the south) and the Caucasus (in the north) is mainly accommodated by right-lateral strike-slip motion along the NTF (Figure 16.7) and along the Western Āzarbāijān Shear Zone (WASZ; Figure 13.6). GPS data show that the N–S shearing rate between the Central Iranian Block and the Eurasian Plate is about 14 ± 2 mm/year (Nilforoushan et al., 2003; Vernant et al., 2004; Masson et al., 2006; Djamour et al., 2011). The GPS velocity measurements along two profiles normal to the NTF suggest a right-lateral slip rate of 7.7 ± 0.9 and 7.2 ± 1.2 mm/year, with an estimate of 0–3 mm/year of the fault-normal component of possible extension normal to the fault (Djamour et al., 2011; Rizza et al., 2013). The permanent-scatters (PS-InSAR) or persistent-scatters (PSI) radar interferometry resulted in slip rates of 7.3 ± 1.3 and 6 ± 3 mm/year (Rizza et al., 2013).

Morphotectonic analyses estimated a mean cumulative right-lateral offset of 320 ± 40 m since 43–49 ka (luminescence dating of the displaced alluvial fan) and resulted in a Late Quaternary slip rate of 6.5 and 7.3 mm/year, which is consistent with the geodetic slip rate (Rizza et al., 2013). This long-period estimate definitely spans numerous clustered events, such as those mentioned above. Considering the minimum right-lateral offset of 4 m and assuming a characteristic slip for the past earthquakes along the NTF with 6.5–7.3 mm/year slip rate, Rizza et al. (2013) estimated a recurrence interval of 480–715 years.

Despite the introduction of the NTF since at least 1976 (Berberian and Arshadi, 1976; Berberian, 1981a), the high seismic risk for the city of Tabriz associated with faulting along the North Tehran Fault system (NTF has not been taken seriously by the authorities (Table 16.1; Figures 11.12 and 16.5–16.7). Unfortunately, as with the case of the city of Tehran, the expansion of the city has included buildings that cover the NTF line that are not reinforced against strong ground motion.

16.10.3 The Northern Gowk Clustered Earthquake Sequence of 1877–1989

Unlike the Nāyband fault (to the north) and the Shahdād and Bam faults (to the south; Figure 9.1), for which there has been no historical seismic record for the past millennia, the northern Gowk fault has been associated with earthquakes since at least 1877 (Figure 16.8). During the 28 July 1981 M_w 7.1 Sirch earthquake, the whole northern segment of the Gowk fault reactivated with coseismic surface rupture of at least 60 km (Berberian et al., 1984, 2001; Berberian and Yeats, 1999, 2001; Berberian, 2005). No pre-1877 earthquake record has been preserved in the annals and the historic behavior of the fault cannot be known unless proper paleoseismic study is carried out. During

TABLE 16.1 Major Recorded Earthquakes at the City of Tabriz

Eq. Date	Eq. Name or Major Event	No. Killed	$\sim i^a$	$\sim M_s{}^a$	$\sim i^b$	$\sim M_s{}^b$	I^c	$M_s{}^c$	I^d	$M_s{}^d$	Fault
18 April 2013	West Marand	–								(*mb* 4.3)	?
01 December 2007										(*mb* 4.8)	?
10 February 1965	Chamkhorān	5	–	–	VII	5.1	–	–	VII	5.1	NTF
GAP											
00 October 1786 (1201 end of Autumn)	Marand	K	2 (VIII$^+$)	6.3	VIII	6.3	–	–	VIII	6.3	NTF
20 February 1780 (14 February 1194)	Tabriz AFS										NTF
12 February 1780 (06 February 1194)	Tabriz AFS								>VI		NTF
08 January 1780 (01 January 1194)	Tabriz	0.0.50?	1 (X)	7.7	IX$^+$	7.4	XI	7.4	XI	7.4	NTF
07 January 1780 (30 December 1193)	Tabriz FRS								>VI		NTF
26 April 1721 (28 June 1133)	Shebli (SE Tabriz)	0.0.40?	1 (X)	7.7	IX$^+$	7.3	XIe	7.4^e	IX	7.3	NTF
00.00.1720			–	–	–	–			?	?	?
12 March 1717 (09 April 1129)		>700	2$^+$ (VIII$^+$)	5.9	VII	5.9	–	5.9	VII$^+$	5.9	?

Continued

TABLE 16.1 Major Recorded Earthquakes at the City of Tabriz—Cont'd

Eq. Date	Eq. Name or Major Event	No. Killed	$\sim i^a$	$\sim M_s^a$	$\sim i^b$	$\sim M_s^b$	I^c	M_s^c	I^d	M_s^d	Fault
08 February 1641 (25 October 1050)	Dehkhārqān AFS								>VII	>5.5	?
05 February 1641 (22 October 1050)	Dehkhārqān (SW Tabriz)	3.6.12	2 (VIII$^+$)	6.8	VIII$^+$	6.8	IX	6.8	IX	6.8	?
20 January 1550–08 January 1551 (00.00.957)	Tabriz	K	3 (VIII)	–	>VII	>5.3	IX	7.1	>VII$^+$	>6.0	?
00 December 1385 (00 November 787)	Tabriz and R'ab-e Rashidi were sacked and pillaged by Tuqtamish Khān of Qipchāq [Turkic Tātār tribe confederation] (Hāfez Abru, 1414; Mashkur, 1973)										
1635	Ottoman Turks sacked the city of Tabriz										
07 November 1304 (07 April 707)	Tabriz	K	2 (VIII$^+$)	6.7			VIII$^+$	6.4	>VIII	>6.7	?
ca. 1300 (ca. 699)	Construction of R'ab-e Rashidi (Rashidi Quarter/Estate/Foundation) University City by Minister Rashid al-Din Fazl Allāh (Wilber, 1955)										
18 January 1273 (26 June 671)	Tabriz	250 in Tabriz	–	–	VIII$^+$	6.5			>VIII	>6.5	?
1221	Invasion of the Mongol Hordes										
04.11 October 2 (17 March 434)	Tabriz	0.0.40?	1 (X)	7.6	IX$^+$	7.3			>IX	>7.0	?

19 April 858–07 April 859 (00.00.244)	Tabriz	0.0.40?	2^+ (VIII$^+$)	6.0	VII$^+$	6.0		>VII$^+$	>6.0	?
642	Invasion of the Moslem Arabs. Almost all the pre-636 written documents were destroyed. Destruction of infrastructure of the country									
GAP										

See also Figures 11.12 and 16.7.
Eq. Date: earthquake date: the original lunar dates are added in parenthesis for consulting the contemporary annals.
I: Highest intensity estimated within the meizoseismal area based on written accounts (in MMI).
i: intensity used by Ambraseys and Melville (1982); the equal MMI values are added in parenthesis.
K: People killed.
Mech: Fault mechanism (R: reverse; RL: right-lateral strike-slip).
M_s: The equivalent surface-wave magnitude derived from macroseismic information embedded in written accounts calibrated against instrumental M_s values. Therefore, they present poorly constrained estimates.
NTF: North Tabriz fault.
[a] *Ambraseys and Melville (1982).*
[b] *Berberian and Yeats (1999).*
[c] *Utsu (2002).*
[d] *This study.*
[e] *Utsu (2002) added two new entries as 26 September 1721 with I: VIII–IX, M 6.0, and 18 November 1727 with I: X, M 7.2, which are duplicates for the 26 April 1721 event.*

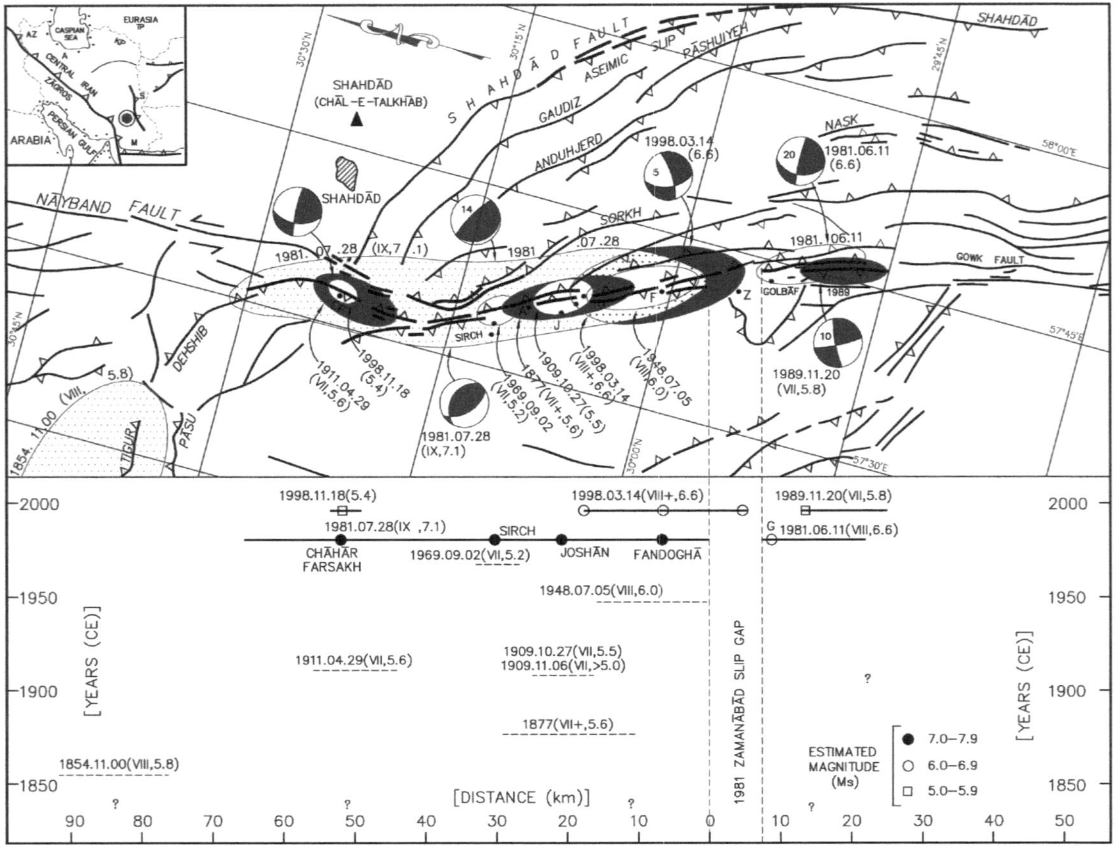

FIGURE 16.8 Meizoseismal area (shaded, stippled) of medium- to large-magnitude earthquakes (top) and space-time diagram of 160-year of seismicity (bottom) along the Gowk fault, southeast Iran. Symbols as in Figures 11.1 and 16.7. Faults with documented surface ruptures are shown by thicker line. The thickened part of the Shahdād thrust slipped aseismically after the 1998 earthquake along the Gowk fault. A, Āb-e Garm; CF, Chahār Farsakh; F, Fandoqā; J, Jowshān; Z, Zamānābād. Focal mechanism of the earthquake constrained by body-wave modeling with centroid depths (Berberian et al., 2001). *Modified from Berberian et al. (1984, 2001), Berberian and Yeats (1999), and Berberian (2005).*

this time period, the southern segment of the fault did not show any activity (Figure 14.7).

During the 104 years of recorded history from 1877 to 1981, seven scattered sections along the northern segment of the Gowk fault that total about 70 km ruptured (Figure 16.9). The 28 July 1981 M_w 7.1 event re-ruptured previously ruptured sections. During 1877–1998, a total of 92 km of the northern segment of the fault ruptured (about 44% of the total fault length), yet no earthquake was detected along the 115-km-long, southern segment of the Gowk fault during this period. Limited data make it difficult and unwise to draw broader conclusions about the southern segment of the fault (Berberian et al., 2001).

The 11 June 1981 M_w 6.6 Golbāf earthquake produced 15 km of right-lateral coseismic surface ruptures south of the Zamānābād gap, with very small surface displacements of up to 2 cm (Figures 16.8 and 16.9). This event was followed by the 28 July 1981 M_w 7.1 Sirch earthquake, with 65 km of discontinuous, right-lateral surface ruptures north of the Zamānābād gap with a maximum displacement of less than 50 cm (Berberian et al., 1984). The area south of the Zamānābād gap ruptured again during the 20 November 1989 M_w 5.8 South Golbāf earthquake, with 11 km of surface ruptures following the identical path of the scarps formed on 11 June 1981 M_w 6.6 Golbāf earthquake (Berberian and Qorashi, 1994). The 14 March 1998 M_w 6.6 Fandoqā earthquake again ruptured 23 km north of the Zamānābād gap, with an average 1.3-m and a maximum 3-m right-lateral coseismic faulting, following that which was observed after the 28 July 1981 M_w 7.1 Sirch earthquake (Berberian et al., 2001). The last event in this sequence was the 11 November 1988 M_w 5.4 Chahār Farsakh earthquake in the north, which produced minor surface cracking over approximately 4 km (Figures 16.8 and 16.9), again along the 28 July 1981 Sirch earthquake (Berberian et al., 2001). During the 17 years of the 1981 to 1998 earthquake sequence, about 92 km of the northern segment of the total 160 km Gowk fault system ruptured at the surface (Berberian and Yeats, 1999; Berberian, 2005).

The 28 July 1981 M_w 7.1 Sirch and the 14 March 1998 M_w 6.6 Fandoqā earthquakes along the northern segment of the Gowk fault showed contrasting coseismic displacements despite their related magnitudes (Berberian et al., 2001). The 1981 M_w 7.1 Sirch earthquake nucleated at an 18-km centroid depth and produced a coseismic surface rupture on a strike-slip fault averaging 0.4 m, whereas the smaller 1998 M_w 6.6 Fandoqā earthquake, with a nucleation depth of 5 km, produced offsets of up to 3 m (Figure 16.10). Analysis of long-period, seismic body waves and InSAR data showed that the 1998 M_w 6.6 Fandoqā earthquake ruptured a fault dipping west at ~50° to a depth of 7–10 km, with a small normal component and a slip-vector azimuth of ~147° (Berberian et al., 2001). Berberian et al. (2001) suggested that the overall rupture of the Gowk fault system is possibly that of a ramp-and-flat

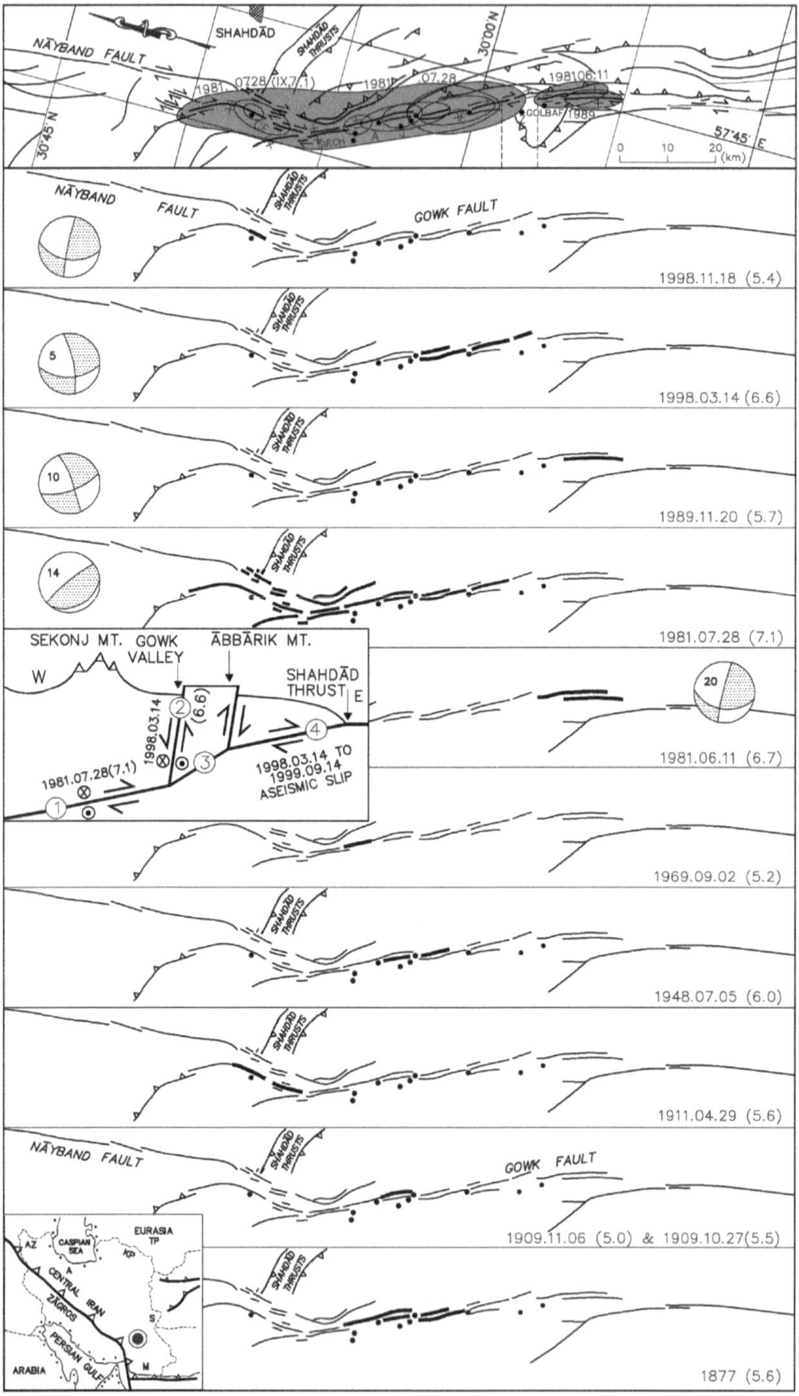

FIGURE 16.9—CONT'D

thrust, but with strike-slip motion superimposed. They suggested that the 1981 earthquake ruptured a deeper, flatter part of the system, which later failed on a steeper, shallower fault in 1998 (Figure 16.10).

Apparently, the deeper 1981 M_w 7.1 Sirch earthquake loaded the shallower part of the fault and triggered the shallower 1998 M_w 6.6 Fandoqā earthquake 17 years later (Figure 16.10). Five years after the last earthquake on the northern section of the Gowk fault, on 26 December 2003, an earthquake of M_w 6.6 destroyed the city of Bam south of the Gowk fault (Figure 14.7; Berberian, 2005).

In addition, synthetic aperture radar interferograms (InSAR) showed evidence of deformation on the nearby Shahdād thrust underlying the Shahdād fold-and-thrust system 35 km east of the 1988 ruptures in the Gowk valley, moving aseismically about 8 cm on a fault dipping 6° west between depths of ~1 and 5 km, in a time interval of either immediately after or within 6 months of the 1981 earthquake (Berberian et al., 2001; Figures 14.3 and 16.10). Apparently, the slip on the Shahdād thrust was triggered by the 1998 strike-slip earthquake (Figure 16.10). The slip-vector azimuth on the thrust was ~62°, nearly perpendicular to that of the Gowk strike-slip fault. The oblique shortening across the Gowk fault system, required by its general trend of N155°E, therefore, appears to be achieved by a spatial partitioning of the strike-slip and reverse component onto adjacent, subparallel faults (Berberian et al., 2001; Walker and Jackson, 2002; Fielding et al., 2004; Berberian, 2005).

The gently dipping decollement surface with blind faults is a common feature in the regions of continental shortening and has been inferred beneath many fold-and-thrust belts (Lettis et al., 1997). Shallow thrust faults are also believed to be present beneath urban areas such as Los Angeles, California (Davis and Namson, 1994) and possibly Tehrān (Berberian and Yeats, 1999, 2001, 2014).

The Gowk fault case seems to be similar to the 9 February 1971 M_w 6.7 San Fernando, California, earthquake at 13 km depth with coseismic surface rupture up to 2.5 m, and the 17 January 1994 M_w 6.7 Northridge earthquake at 17 km depth with only secondary coseismic flexural-slip faulting observed at the surface (Sharp, 1975; Mori et al., 1995).

FIGURE 16.9—CONT'D Sequence of the recorded rupture propagation during different earthquakes along the Gowk fault system. Focal mechanism of the earthquake constrained by body-wave modeling with centroid depths (Berberian et al., 2001). Inset center left: Structure of the Gowk fault system as proposed by Berberian et al. (2001) and Walker and Jackson (2002). Gently dipping parts of the system (1 and 4 in the inset left) are separated by a steeper "ramp" (3) that causes uplift of the Ābbārik mountains and localized oblique normal faulting (2) in the Gowk valley. Sections (3) and (4) show purely thrust motion. A Coulomb failure function will help understand the rupture propagation along the Gowk fault system. *Modified from Berberian et al. (2001) and Berberian (2005).*

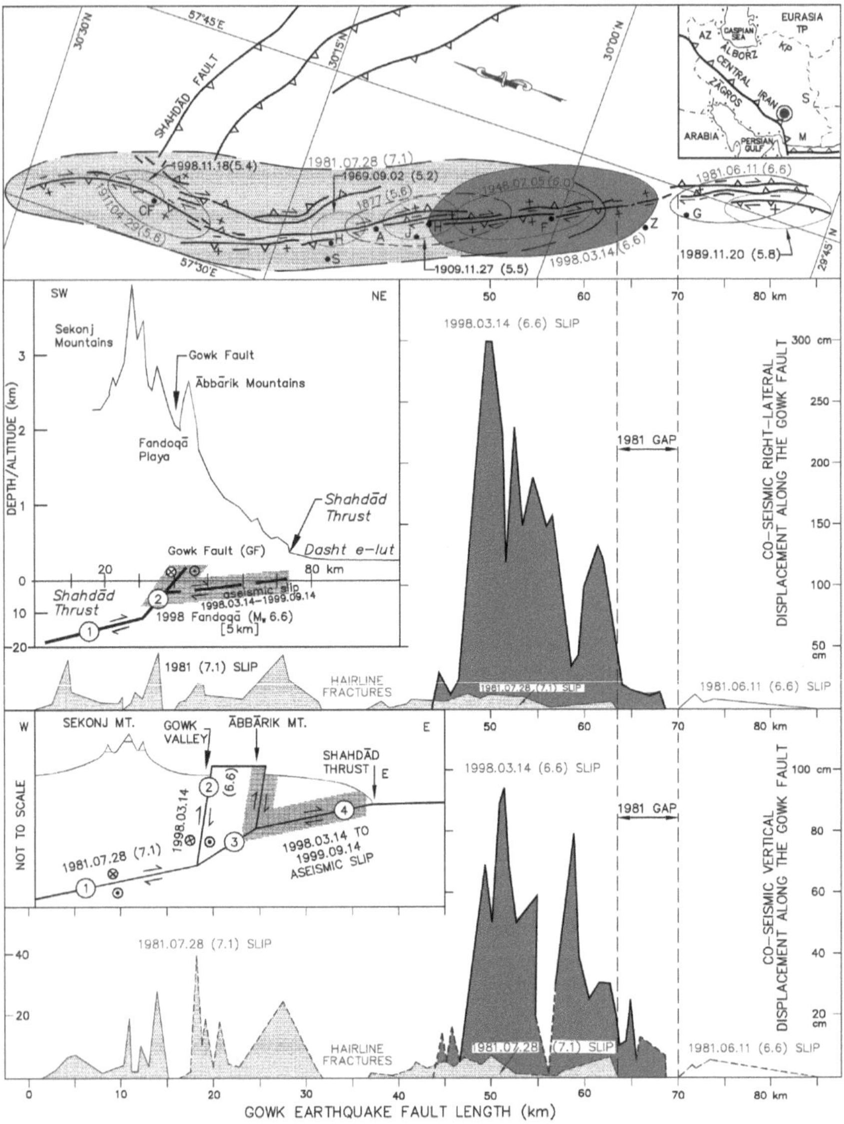

FIGURE 16.10 Seismicity-fault map (top) observed amplitudes of coseismic right-lateral (middle panel), and vertical (bottom) offsets on the Gowk fault system following the 28 July 1981 M_w 7.1 Sirch (stippled) and the 14 March 1998 M_w 6.6 Fandoqā (gray shaded) earthquakes. Offsets are in centimeters, plotted against distance along the fault measured on the map in the top panel. Where coseismic fault segments from the same earthquake overlapped, their cumulative value has been plotted. Observed points at which offsets were measured have been joined by lines to distinguish the 1981 and 1998 displacements and do not imply that the offsets varied between those points in the jagged manner indicated by the profiles. Note the large offset from the 1998 M_w 6.6 Fandoqā earthquake compared with the much shorter (M_w 7.1) 1981 Sirch earthquake. A, Āb-e Garm; CF, Chahār Farsakh; F, Fandoqā; G, Golbāf; H (in the north), Hasanābād; H (in the south), Hashtādān; J, Jowshān; S, Sirch. Inset left, Structure of the Gowk fault system as proposed by Berberian et al. (2001) and Walker and Jackson (2002). Gently dipping parts of the system (1 and 4 in the inset left) are separated by a steeper "ramp" (3) that causes uplift of the Ābbārik mountains and localized oblique normal faulting (2) in the Gowk valley. Sections (3) and (4) show purely thrust motion. *Modified from Berberian et al. (2001).*

A difference occurs in the Gowk fault right-lateral slip rates inferred from geologic measurements of 1.5–2.4 mm/year (Walker and Jackson, 2002, 2004) or 0.9–1.9 mm/year (Nāyband-Gowk; Walker et al., 2009) and the GPS data (10–20 mm/year) of shear expected between Central Iran and Afghanistan. The deficit is likely to have been accommodated on other faults east of the Gowk fault system on the eastern side of the Lut Block such as the Neh and Kahurak faults (Walker and Jackson, 2002; Figure 9.1). It is surprising to have so many earthquakes on the northern Gowk fault system with a low slip rate. With an average slip rate of ~2 mm/year on the Gowk fault, the recurrence time for events that slip ~5 m on the fault 100 km long is likely to be on the order of 2000 years.

16.10.4 The Moshā Clustered Earthquake Sequence of 958, 1665, and 1830

The Moshā fault located about 12 km to the northeast of the present northern mega-city of Tehrān (Figures 9.1 and 11.2) shows active fault features along its length. Macroseismic data show that at least the three earthquakes of 23 February 958 M_s~7.1 Ruyān, 15 June–13 July 1665 M_s~6.5 Damāvand and 27 March 1830 M_s~7.1 Lavāsānat (the latter damaging Tehrān with casualties) took place along the fault (Figure 16.11; see also Chapter 11). (Berberian et al., 1985; Berberian and Yeats, 1999, 2001, 2014; Berberian, 2005). The most recent earthquake creating concentrated damage with loss of life on the fault was the 2 October 1930 M_s 5.2 Āh earthquake and its aftershock of 7 October 1930 M 5.0 at 35.76°N–52.00°E (Berberian et al., 1985).

Four medium- to large-magnitude earthquakes struck the Tehrān/Ray region from ca. 280 BCE to 1177 CE (Figure 16.11), but only one that large has struck the area since 1177. Despite the earthquake's location close to the mega-city of Tehrān and its destructive effects on the city during the 1830 M_s~7.1 earthquake, no proper paleoseismic trench study has been carried out along different segments of the Moshā and North Tehrān faults.

16.10.5 The Zāgros Main Recent Fault

The multisegmented Zāgros Main Recent fault (Figures 9.1 and 16.12) is a major NW–SE-trending right-lateral strike-slip seismic fault zone of more than 800 km long located between the Zāgros fold-and-thrust belt in the southwest and Central Iranian range-and-basin in the northeast (Tchalenko and Braud, 1974; Berberian, 1995; Berberian and Yeats, 2001). The fault more or less follows the trend of the Zāgros Mountains and the Main Zāgros reverse fault (the Neo-Tethys geosuture; Berberian and King, 1981a; Berberian, 1995). The fault has been the source of frequent significant earthquakes of up to M_w 7.4 during the twentieth century (in 1909) and should have been during the earlier historical record (Figure 16.12) as well.

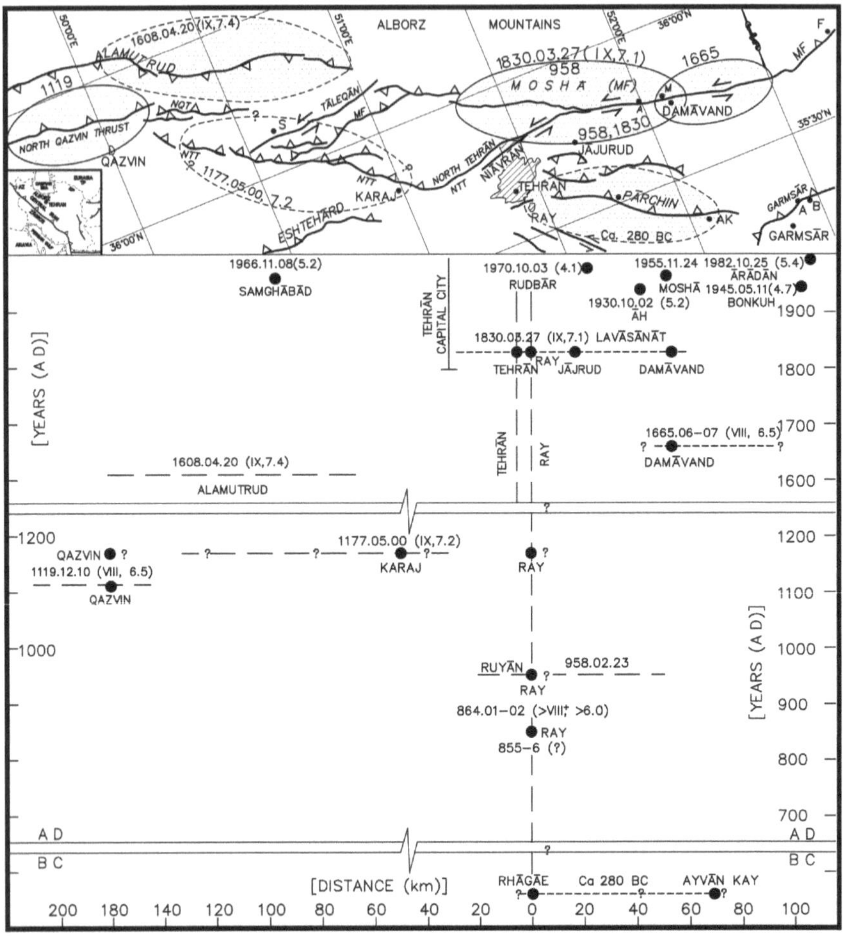

FIGURE 16.11 (Top) Meizoseismal areas of medium-to large-magnitude earthquakes striking the ancient city of Ray of the southern Tehrān metropolitan area (hatched). Symbols as in Figures 11.1 and 16.7. (Bottom) Time–space diagram of earthquakes in the Tehrān region. Queried where meizoseismal areas are not well defined and constrained. Distances are along strike with respect to the capital city of Tehrān. *Modified from Berberian and Yeats (1999).*

We are aware of a cluster of events destroying Dinévar in 913, 1008, and 1107 (Figures 11.4 and 16.12). During the twentieth century, the 23 January 1909 M_w 7.4 Silākhor earthquake along the Dorud segment of the Zāgros Main recent fault was followed by the 31 March 2006 M_w 6.1 Chālānchulān earthquake; the latter filled the gap left on the northwestern segment of the Dorud segment by the 1909 event (Figures 12.1 and 16.12). The northwestern Nahāvand and the Sahneh segments of the fault ruptured in 1957 and 1958, leaving a twentieth-century seismic gap to the southeastern part of the Nahāvand segment, which itself was ruptured during the 1316 earthquake

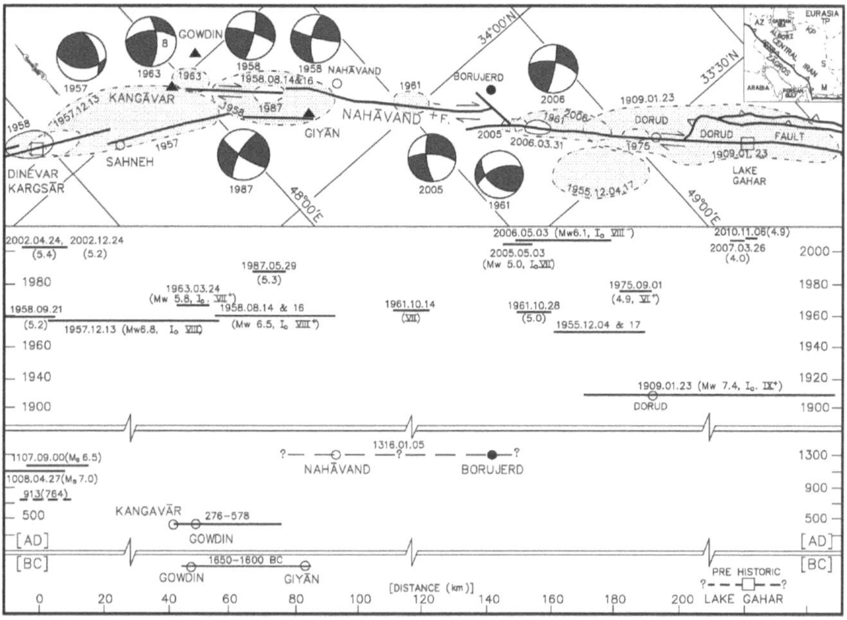

FIGURE 16.12 (Top) Meizoseismal areas of medium-to large-magnitude earthquakes along the Zāgros Main Recent fault. (Bottom) Time–space diagram of earthquakes along the Zāgros Main Recent fault. Distances are along strike with respect to Dinévar/Kargsār. Symbols as in Figures 11.1 and 16.7. The 1957 and 1958 focal mechanisms are from Shirokova (1962); the 1963 solution is from Ni and Barazangi (1986) with waveform modeling indicating a centroid depth of 8 km; the 1987 is the CMT solutions from Harvard; the 2006 earthquake with a centroid depth of 6 km is from Peyret et al. (2008). The early fault plane solutions were based on the published short-period polarities, and it is not possible to assess the reliability of these early solutions.

(Berberian et al., 2014; Figures 12.11, 12.12, and 16.12). Other segments of the Zāgros Main Recent fault show seismic gaps that will be activated in the future.

16.10.6 The High-Zāgros Fault

The 1962, 1963, 1965, 1971, 1974, 1990, and 1992 earthquakes were clustered along the southeastern segment of the High Zāgros fault at the Gahkom-Furg area (Figures 9.1 and 13.14). To the northwest, the fault shows large areas of intervening unruptured gaps (see figure 4–10 in Berberian, 1995).

16.10.7 The Zagros Mountain Front Fault

The 978 and 1008 earthquakes clustered along the Sirāf segment; the 1929 and 1978 earthquakes clustered along the Andikā segment; the 1883, 1950, and 1954 along the ‘Assaluyeh segment; and the 1949, 1971, and 1975 clustered along the Bandar 'Abbās segment of the Zāgros Mountain Front fault (Figure 9.1), with numerous areas of intervening unruptured gaps (see figures 5–11 in Berberian, 1995).

16.11 TEMPORALLY CLUSTERED EARTHQUAKE SEQUENCES ALONG DENSELY POPULATED, INTERACTING SUBPARALLEL FAULT ZONES

Cases of temporal clustering along interacting subparallel fault zones are documented in Neyshābur (1209–1405), Quchān (1833–1895), and the Central Kopeh Dāgh Shear Zone, CKDSZ (Chapter 11; Figure 9.1). Understanding the long-term interaction between such faults is critical in seismic risk assessment. The Bam and Bam-Baravāt subparallel faults ruptured during the 2003 earthquake (Figure 14.6; Funning et al., 2005; Fielding et al., 2005; Berberian, 2005; Jackson et al., 2006).

The nearly simultaneous rupture of side-by-side parallel faults had been previously documented in the 23 December 1972 M 6.2 Managua, Nicaragua, earthquake where four small (up to 10 km long), equally spaced, parallel faults, within a 4-km wide zone, developed, traversing the capital city of Managua. The 1972 surface ruptures stepped to the right of the 31 March 1931 Managua rupture (Sultan, 1931; Brown et al., 1973). During the 24 November 1987 M_w 6.2 Elmore Ranch, California, earthquake, five parallel faults developed in a zone that was 8-km wide and 1-km long (Kahle et al., 1988; Sharp et al., 1989; Hudnut et al., 1989). Near its southwestern termination, left-lateral faults of the Elmore Ranch fault zone intermingled with right-lateral splays of the Superstition Hill faults, which produced a larger 24 November 1987 M_w 6.6 earthquake, 12 h later (Kahle et al., 1988; Sharp et al., 1989).

16.11.1 The Neyshābur–Binālud Catastrophic Thrust Clustered Earthquakes Followed by Long Periods of Relative Quiescence

Active tectonics in the northeastern boundary belt of the Arabian–Iranian collision zone of the Iranian Plateau (along the Binālud Mountains and the Kopeh Dāgh fold-thrust Mountain belt of northeast Iran; Figures 9.1 and 9.2) is principally accommodated by active faulting and related seismicity (Tchalenko, 1975; Berberian and Yeats, 1999; Berberian et al., 2000a; Shabanian et al., 2009a,b, 2012a; Hollingsworth et al., 2010). The ~130-km-long and ~+2500-m-high Binālud Mountains, separating Central Iran from the Kopeh Dāgh (to the south of the Paleo-Tethys geosuture), are apparently taking up the motion between Central Iran and Eurasia at a rate of about 4.0 ± 1.3 mm/year (Shabanian et al., 2012a). The Binālud Mountains are being deformed by the southwestern range-front reverse faulting of the Neyshābur fault system (Berberian and Yeats, 1999; Berberian et al., 2000a) as well as strike-slip faulting to the northeast and northwest (Shabanian et al.,

2009a,b, 2012a). The unfortunate ancient city of Neyshābur (Nishāpur), famous for its turquoise mine and the Zoroastrian Borzin-Mehr fire temple (Bosworth, 2010) and one of the four great cities in the Khorāsān province and the capital of the ancient province, is located on the southwestern foothills of the Binālud Mountain, adjacent to the Neyshābur fault system (Figure 16.13).

Cumulative geological offset along the Neyshābur Fault System yielded slip rates of 2.4 ± 0.5 mm/year for horizontal (right-lateral), and 2.8 ± 0.6 mm/year

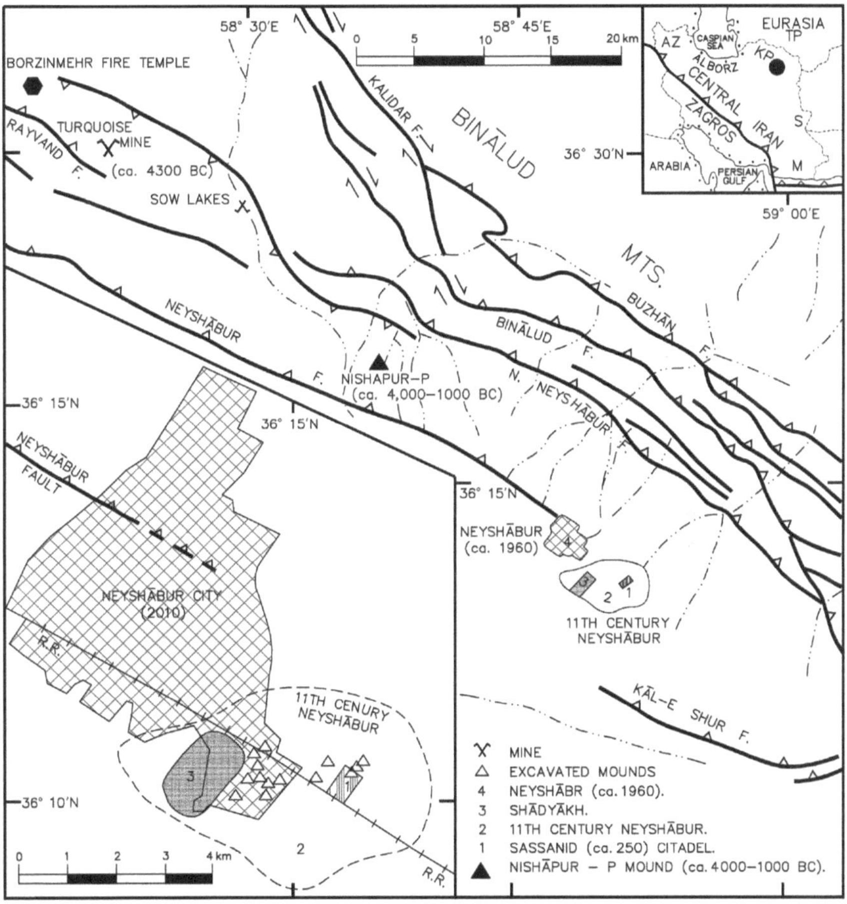

FIGURE 16.13 The Neyshābur fault system located on the southwestern Binālud Mountain front. Faults and symbols as in Figure 11.1. The location of the prehistoric Nishāpur (ca. 4000–1000 BCE) is marked by a filled triangle. Inset bottom left: The 2010 Neyshābur city limits (cross-hatched) crossing the Neyshābur active fault with construction of unreinforced buildings upon and beyond the fault line. The locations of the historic city of Neyshābur are also included with archeological mounds marked by open triangles (see legend). See Figures 16.14 and 16.15 for the seismicity of the area.

for the uplift (reverse) component of active faulting (averaged over 105 ± 0.6 ka), corresponding to an oblique rate of 3.6 ± 1.0 mm/year (Shabanian et al., 2012a,b). An overall rate of 3.7 ± 0.6 mm/year for the range-parallel displacement and an uplift rate of about 2.8 mm/year due to the range-normal shortening (1.6–2.2 mm/year) was estimated during the Late Quaternary (Shabanian et al., 2012a,b).

The ancient Sassanid city of Neyshābur (ca. 250 CE) is located along the Silk Road to the southwest of the Binālud Mountains (Figure 16.13). The Binālud Mountains area is delimited to the northeast by the Shāndiz right-lateral strike-slip fault; to the southwest by the Neyshābur range-front reverse fault system (the Neyshābur, North Neyshābur, Binālud, Rayvand, and Buzhān thrusts; Figure 16.14); and to the northwest by right-lateral and left-lateral strike-slip faults (GSI, 1986b, 1992a,b, 1993, 2000; Berberian and Yeats, 1999; Berberian et al., 2000a; Shabanian et al., 2009a,b, 2012a,b). The overall Binālud structure seems to be similar to the western Transverse Ranges, California, with the Red Mountain, San Cayetano, Santa Susana, San Fernando, and Sierra Madre reverse faults, and the San Gabriel right-lateral fault to the southwest of the San Andreas fault (Yeats et al., 1981; Tsutsumi and Yeats, 1999).

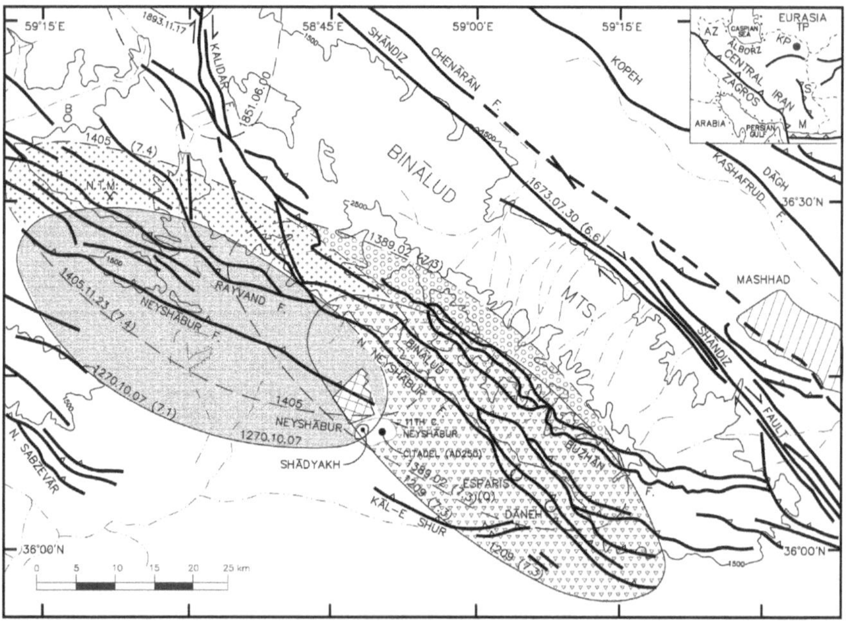

FIGURE 16.14 Approximate estimated meizoseismal areas of earthquakes (not well constrained; see the text) devastating the ancient city of Neyshābur along the Neyshābur fault system at the foot of the Binālud Mountains. Faults and other symbols as in Figure 11.1. See also Figures 16.13 and 16.15. After the 1209–1405 earthquake cluster, seismicity migrated to the northeast where the 30 July 1673 M_s ~6.6 earthquake took place along the Shāndiz fault. *Modified from Berberian et al. (2000a).*

Within a period of 196 years (1209–1405; excluding the 1145 event), four large-magnitude, clustered earthquakes struck the Neyshābur–Binālud range-front, closely spaced, reverse fault system (the Neyshābur, North Neyshābur, Rayvand, Binālud, and Buzhān faults), destroying the ancient city of Neyshābur on the eastern continuation of the Alborz Mountains in northeast Iran (Figure 16.14). These reverse faults are spread along a narrow belt of 7 to 10-km wide that runs along the southwestern foot of the Binālud Mountains. The intensity of active reverse faulting along the Neyshābur–Binālud fault system in this tectonically active region is expressed by (i) rapid erosion of the uplifting Binālud Mountains and (ii) subsequent high rates of sedimentation in the Neyshābur plain, which buried and preserved the archaeological material of the Sassanid city of Nivshāpur (Meysam Labbaf Khaniki, personal communication, 21 December 2012; Fouache et al., 2011; Rante and Collinet, 2013).

Remnants of the Sassanid Nivshāpur citadel (Arg/Kohandezh; ca. 224–642) was excavated in the area about 6.5 km to the southeast of the modern city of Neyshābur near the village of Tarbābād (Kervran and Labbaf Khaniki, 2008a,b; Labbāf Khāniki and Labaf Khaniki, 2012; Rante and Collinet, 2013), in the area marked by Bulliet (1976) as *Kohandezh* (Citadel) and *Shārēstān* (City Proper) (Figure 16.13). The rampart of the Sassanid citadel was constructed of 40 × 40 × 12-cm bricks with 5–6-m thickness and 8-m height (Labbāf Khāniki and Labaf Khaniki, 2012). Figure 7 on page 343 of the rampart in Labbāf Khāniki and Labaf Khaniki (2012) shows a well-developed “Y”-shaped, through-going fracture cutting the Sassanid bricks of the northern rampart above the arrow slot. This is clearly an archeoseismic indicator of a post-third-century earthquake; however, it cannot be dated at this stage because no archeoseismic investigation was included during the excavation project.

Except for the questionable earthquake of 25 July 454, no recorded pre-1209 strong earthquake had struck the Neyshābur area since the late seventh century, and no strong earthquake has struck since 1405; the magnitude of the 856–857 earthquake is not known (Figure 16.15). Monuments such as the congregational mosque of Neyshābur (1494), the mausoleum of Mahruq (mid-sixteenth century AD), and the mausoleum in Esparis (later, Qadamgāh; 1680) are extant (Meshkāti, 1970). Obviously, earthquakes struck the region in 454, the seventh century, 856–857, 1145 (with unknown magnitudes), 1209, 1270, 1389, 1405, 1928 (5.2), and 2012 (5.3) (Table 16.2; Figure 16.15). The Neyshābur region is highly seismic, and historical data on the earlier events are not well constrained; certainly more earthquakes have occurred in the region than have been recorded (Melville, 1980; Ambraseys and Melville, 1982; Berberian, 1994; Berberian and Yeats, 1999; Berberian et al., 2000a).

The meizoseismal areas of the 1209, 1270, 1389, and 1405 earthquakes are so entangled with each other that Ambraseys and Melville (1982) presented a

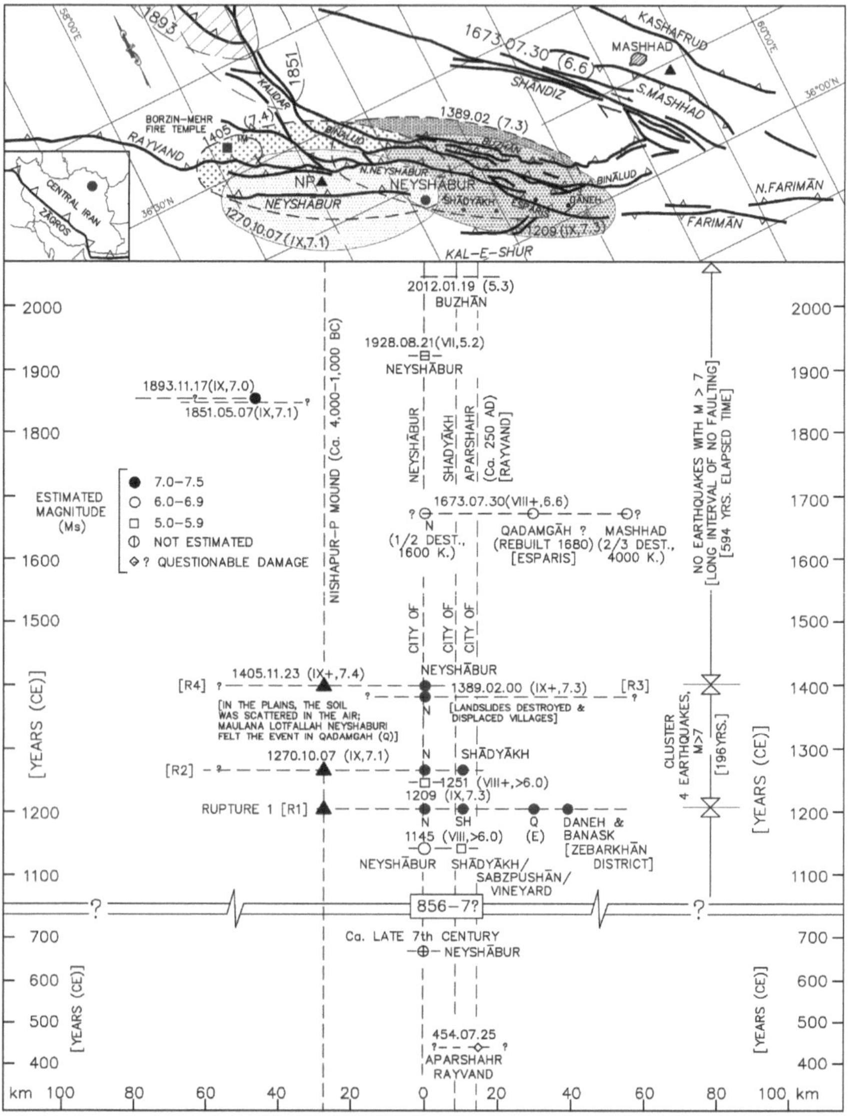

FIGURE 16.15 (Top) Meizoseismal areas of medium-to large-magnitude earthquakes along the Neyshābur reverse fault system (south); Shāndiz right-lateral strike-slip fault (NE, near the city of Mashhad); and the Quchān right-lateral and reverse fault system (NW) striking the cities of Neyshābur, Mashhad, and Quchān. (Bottom) Time–space diagram of earthquakes in the Neyshābur, Mashhad, and Quchān areas. Zone of extensive damage shown by horizontal dashed line, queried where uncertain. Distances are along strike with respect to the city of Neyshābur, the ancient and old capital city of the Khorāsān province. Symbols as in Figures 11.1 and 16.7. *Modified from Berberian and Yeats (1999) and Berberian et al. (2000a).*

TABLE 16.2 Major Earthquakes and Calamitous Events at the City of Neyshābur

Eq. Date & Major Event	Eq. Name	Archao-Seismicity	No. Killed	$\sim i^a$	$\sim M_s^a$	$\sim i^b$	$\sim M_s^b$	$\sim I^c$	$\sim M_s^c$	$\sim M_s^d$	$\sim I^e$	$\sim M_s^e$	M_w	Refs.
19 January 2012	Buzhān		1	–	–	–	–	–	–	–	VII	–	5.3	
21 August 1928	Neyshābur		30–40	VIII	5.4	–	–	VII	5.2		VI$^+$	–		
May 1857	Neyshābur Turquoise Mine			VII	5.1						F	?		Khanikoff (1861)
21 March 1855	Neyshābur Turquoise Mine			VII	5.1						F	?		Khanikoff (1861)
30 July 1673 (15 April 1084)	Shāndiz		M:4000 N:1600	IX	6.2	2 (VIII$^+$)	6.6	VIII$^+$	6.6	7.1	VIII$^+$	6.6		Melville (1980), Ambraseys and Melville (1982), Berberian and Yeats (1999)
1598	Uzbeks ravaged and seized the Khorāsān province. Neyshābur suffered.													Bosworth (2010)
23 November 1405 (30 May 808)	Neyshābur		0.0.30	IX	6.7	1 (X)	7.6	IX$^+$	7.4		IX$^+$	7.4		Hāfez Abru (1414), Melville (1980), Ambraseys and Melville (1982), Berberian and Yeats (1999)
1404	Ruy Gonzales de Clavijo visited Neyshābur.													Clavijo (1405)
30.01.1389 –02.27 (00.02.791)	Neyshābur		Many	VIII/IX	5.8	1 (X)	7.6	IX$^+$	7.3		IX$^+$	7.3		al-Maqrizi (1426), Melville (1980), Ambraseys and Melville (1982), Berberian and Yeats (1999)

Continued

TABLE 16.2 Major Earthquakes and Calamitous Events at the City of Neyshābur—Cont'd

Eq. Date & Major Event	Eq. Name	Archao-Seismicity	No. Killed	$\sim i^a$	$\sim M_s{}^a$	$\sim i^b$	$\sim M_s{}^b$	$\sim I^c$	$\sim M_s{}^c$	$\sim M_s{}^d$	$\sim I^e$	$\sim M_s{}^e$	M_w	Refs.
1332	Ibn Battuta visited Neyshābur.													Ibn Batuta (1355)
07 October 1270 (19.02.669)	Shādyākh W. Neyshābur		0.0.10	VIII/IX	5.8	1^+ (X^+)	7.1	IX	7.1		IX	7.1		Rashid al-Din Fazlollāh Hamédāni (1304), Hāfez Abru (1431), Wilkinson (1975), Melville (1980), Ambraseys and Melville (1982), Berberian and Yeats (1999)
[19 May 1270 (26 September 668)]	Margha'ul Mongol sacked Neyshābur 5 months before the earthquake (Rashid al-Din Fazlollāh Hamédāni, 1304).													
26 March 1251 –13 March 1252 (649)	Shādyākh		Many	IX	6.2	3 (VIII)	5.3				>VII^+	>6.0		Melville (1980), Ambraseys and Melville (1982), Berberian and Yeats (1999)
[10 April 1221 (15.02.618)]	Mongol Hordes calamitously sacked Neyshābur and massacred the population of 1.7 M. Total destruction of the city, the city ploughed. Only 400 survived (Yāqut, 1225; Jovaini, 1260; Rashid al-Din Fazlollāh Hamédāni, 1304; Esfezāri, 1493; Eqbāl Āshtiāni, 1986).													
1216	Yāqut stayed in Shādyākh and observed the damage done by the earthquake and by the Quzz Turks													Yāqut (1225)
16 July 1208 –06 July 1209 (605)	SE. Neyshābur (Dāneh-Shādyākh)		2000–0.0.10	IX	6.2	1 (X)	7.6	IX	7.3		IX^+	7.3		Jovaini (1260), Melville (1980), Ambraseys and Melville (1982), Berberian and Yeats (1999)

Eq. Date & Major Event	Eq. Name	Archao-Seismicity	No. Killed	$\sim I^a$	$\sim M_s^a$	$\sim I^b$	$\sim M_s^b$	$\sim I^c$	$\sim M_s^c$	$\sim M_s^d$	$\sim I^e$	$\sim M_s^e$	M_w	Refs.
[1153 (548)]	Quzz/Oghuz Turkish nomads sacked Neyshābur. Total destruction of the city, including Mani'i mosque, colleges, and libraries (Hakem Neyshāburi, 998; Mojmal al-Tavārikh val Qesās, 1126; Ibn al-Athir, 1231; Al-Rāwandi [Rāvandi], 1238; Zakariyā Qazvini, 1263; Bondāri Esfahāni, 13th century; Rashid al-Din Fazlollāh Hamédāni, 1304; Hāfez Abru, 1414)													
24 June 1145 –13 June 1146 (540)	Neyshābur	Vineyard, Sabz Pushān, and Madreseh mounds		–	–	3 (VIII)	5.3	VII	5.3		>VII$^+$	>6.0		Wilkinson (1975), Melville (1980), Ambraseys and Melville (1982), Berberian and Yeats (1999)
856–7	Neyshābur		Many								>VII$^+$	>6.0		Ibn al-Jauzi (1181), Zakariyā Qazvini (1263)
Late seventh century (Late first century)	Neyshābur (Shahr-e Kohneh)		Many								>VII$^+$	>6.0		Melville (1980), Ambraseys and Melville (1982), Berberian and Yeats (1999)
[650 (30)]	Invasion of the Moslem Arab raiders. The Zoroastrian fire temples were demolished (al-Balādhuri, 890; Tabari, 915).													
25 July 454	Aparshahr													Ghazar Parpetsi (500, tr. Thomson, 1991; Yeghishé (last decade of the sixth century; tr. Boyajian, 1952)
?														
[309–379]	Shāpur-II Sāssānid rebuilt the city of Nivshāpur.													Joneidi (1987)
270–309	Earthquake?													
[242–270]	Shāpur-I Sāssānid founded the city of Nivshāpur/Aparshahr.													

Continued

TABLE 16.2 Major Earthquakes and Calamitous Events at the City of Neyshābur—Cont'd

Eq. Date & Major Event	Eq. Name	Archao-Seismicity	No. Killed	$\sim i^a$	$\sim M_s{}^a$	$\sim i^b$	$\sim M_s{}^b$	$\sim I^c$	$\sim M_s{}^c$	$\sim M_s{}^d$	$\sim I^e$	$\sim M_s{}^e$	M_w	Refs.
														Shahrestān-hāyeh Irānshahr (Joneidi, 1987; Tafazzoli, 1989; Daryāee, 2002), Bundahishn, Bosworth (2010)
GAP														
[330 BC]	Invasion of Alexander III of Macedonia													
[550–330 BC]	The city of Raévanta (modern Rayvand)													The Avestā; Joneidi (1987), Bosworth (2010)

See Figures 16.14 and 16.15.
~: Approximate value for the pre-1900 events.
F: Felt.
I: Highest intensity estimated within the meizoseismal area based on written accounts (in MMI).
i: intensity used by Ambraseys and Melville (1982); the approximate equal MMI values are given in parenthesis.
M: Mashhad.
M_s: The equivalent surface-wave magnitude derived from macroseismic information embedded in written accounts calibrated against instrumental M_s values. Therefore, they present poorly constrained estimates.
M_w: Moment magnitude.

135-km-long meizoseismal area trending NW–SE for the 1209 event. The limited available historical data of the 1209–1405 large-magnitude, quadruple earthquake cluster suggest that the city of Neyshābur (and/or its relocated spot, Shādyākh) was completely destroyed and suffered with a large number of casualties (Figure 16.14). Therefore, the meizoseismal areas of these four earthquakes should cover the city. This, together with the 7–10 km narrow zone of the four faults (Figures 16.14 and 16.13) making up the Neyshāur fault system, have made difficult to identify the respective causative fault of each event (Berberian and Yeats, 2001).

We are certain that the city and the villages of Dāneh and Banask (in the Zebarkhān district, about 40 km to the southeast of Neyshābur) were destroyed during the 1209 earthquake (Jovaini, 1260). This gives a semiconstrained >40 km-long meizoseismal area along the southeastern segment of the Binālud fault system, with the possible reactivation of the North Neyshābur fault (Figure 16.14). The 1405 earthquake that also destroyed the city was felt without reported damage by Maulānā Lotfallāh Neyshāburi (Hāfez Abru, 1414), who was residing in the village of Esparis/Qadamgāh (28 km southeast of the city; Figure 16.14). Hence, that meizoseismal area of the 1405 earthquake should cover the city and the area to its west-northwest. This event might have happened along the North Neyshābur-Rayvand fault system (Figure 16.14).

Al-Maqrizi (1426) wrote that the destructive 1389 earthquake was not confined to the city of Neyshābur. Outside the city, one village was transferred from its location down onto another village. We know that several villages at the foot of the Binālud mountains to the southeast are located between the North Neyshābur and Binālud faults (from Kharv to Fakhrābād for a distance of 40 km), where the Neyshābur plain becomes narrow; the villages to the west are scattered on a larger plain. It is possible that the 1389 event took place to the southeast of Neyshābur along either the North Neyshābur of Binālud faults (Figure 16.14).

Both historical (Rashid al-Din Fazlollāh Hamédāni, 1304; Hāfez Abru, 1431) and archaeological data (Wilkinson, 1975) documented that the Neyshābur-Shādyākh area and villages were completely destroyed during the 7 October 1270 earthquake. No other data are available to constrain the meizoseismal area of this event. We have tentatively located the event to the west because two earthquakes of 1029 and 1389 took place to the east (Figure 16.14) within a short period. A detailed paleoseismologic trench study would clear up the ambiguities and provide a window into the source faults and the nature of clustering sequences along the densely populated, subparallel Neyshābur fault zone.

Uncertainties aside, it is clear that since the late seventh century AD, most moment release has occurred in the 196 years along the narrow fault zone of the Neyshābur system. The 1209–1405 Neyshābur earthquake sequence may resemble reverse fault pairs of earthquakes such as the 17 June 1929 M_s 7.8

Murchison–24 May 1968 M_s 7.4 Inangahua in New Zealand (Anderson et al., 1994; Yeats, 2000); the 15 January 1944 M_s 7.4 San Juan–23 November 1977 M_s 7.4 Caucete in the Sierras Pampeanas, Argentina (Kadinsky Cade et al., 1985; Siame et al., 2002); and the 9 February 1971 M_w 6.7 Sylmar/San Fernando–17 January 1994 M_w 6.7 Northridge in the San Fernando Valley, California (Tsutsumi and Yeats, 1999).

Clearly, with the short return period of 180 years (in the east) and 135 years (in the west) for the 1209–1405 quadruple cluster, the four earthquakes did not occur on the same fault segments. Simply stated, with the slip rates of about 2.1–2.4 mm/year (Shabanian et al., 2009a,b, 2012a,b), not enough time occurred for stress accumulation on a single fault or fault segment (Table 16.3). To resolve this issue, we may think of the following possibilities:

i. The 1029–1405 large-magnitude, quadruple-earthquake cluster occurred along different segments of the parallel range-front northeast-dipping thrusts (Neyshābur, North Neyshābur, Rayvand, and Binālud faults; Figures 16.14 and 16.16);
ii. The Neyshābur reverse fault, which dies out near the city, may continue for another 40 km to the southeast beneath the alluvial deposits of the plain (Figure 16.14);
iii. The Kāl-e Shur reverse fault to the southeast of Neyshābur, which dies out approaching the south Neyshābur area, may continue northwestward toward south of the city as a blind thrust (Figure 16.14);
iv. Or, similar to the Central San Fernando Valley (the 1971 M_w 6.7 Sylmar and the 1994 M_w 6.7 Northridge, California, earthquakes; Tsutsumi and Yeats, 1999), south-dipping foreberg thrust(s) underneath the Neyshābur–Binālud north-dipping, range-front thrust system was activated (?) (Figures 16.14 and 16.16).

Unlike the Tehrān plane, at the foot of the Touchāl Mountain of the central Alborz fold-and-thrust belt (Figure 11.2; Berberian et al., 1985; Berberian and Yeats, 1999, 2001, 2014; Landgraf et al., 2013), or the western Transverse Range, California (Yeats and Huftile, 1995), no south-dipping, foreberg ridge or foreberg thrust fault (and/or oblique faults in the case of SE-dipping Oak Ridge fault lying beneath the active north-dipping Santa Susana fault, Southern California) has been detected in the Neyshābur plain topography or as reaching the surface in the neighboring areas—unless it is in the early stages of development and covered by the rapid rate of alluvial sedimentation in the Neyshābur plain.

The available teleseismic data from national and international sources were studied in order to test some of the alternatives mentioned above. The instrumental epicenters of the Neyshābur area (Ml > 2.0) located by the Iran National Seismograph Network (INSN; denoted as THR) for the period 2003–2013 and the Iran Telemetered Seismograph Network (ITNS; denoted as TEH) for the period 2005–2013 were plotted on the Neyshābur fault map. No obvious trend was noted to improve our understanding of the active, buried structures of the area.

TABLE 16.3 Summary of the Fault Parameters in the Neyshābur Area[a]

Fault Name[b]	Mech.	Min. Fault Length (in km)[b]	M_w Max.[c]	*I* Max.	Max. Displ. (in m)[c]	Slip Rate (mm/year)[d]	Ave. Slip Rate (mm/year)[d]	Slip Rate (mm/year)[e]	Return Period (years)
Binālud	R	105	7.4	IX⁺	2.0	*v*: 1.2–1.6 *h*: 2.1–2.7 *o*: 2.4–3.1	2.1	–	952
Buzhān	R	100	7.2	IX	1.7		–	–	?
Chakāneh	RL/R	50	7.0	IX	1.5	1.1; 1.2	–	–	1363
Kāl-e Shur	R	25	6.7	VIII⁺	1.2		–	–	?
Kalidar	RL/R	60	7.1	IX	1.6	0.8–1.2; 2.0	2.0	–	800
Neyshābur	R	60	7.1	IX	1.6	–	–	0.5–0.7	3200–2285
North Neyshābur	R	95	7.4	IX⁺	2.0	*v*: 3–3.2 *h*: 2.2–2.8 *o*: 3.3–4.3	2.4 ± 0.5 (RL during 2.8 ± 0.5 kyr) 2.8 ± 0.6 SH	0.3–1.0	833
Rayvand	R	150	7.6	X	2.3	–	–	–	?
Shāndiz	RL	120	7.5	X	2.1	*h*: 0.9–1.8	1.3		1615

See also Figures 16.14 and 16.15.
I: Highest intensity estimated within the meizoseismal area based on written accounts (in MMI).
Mech: Fault mechanism (R: reverse; RL: right-lateral strike-slip).
M_w: Moment magnitude.
R: Reverse.
RL: Right-lateral strike-slip.
SH: Shortening.
[a]*Rigid block modeling of recent GPS data (Mousavi et al., 2013) showed 3.5 mm/year right-lateral slip and 3.4 mm/year shortening along the Binālud–Neyshābur fault system.*
[b]*Modified after Berberian et al. (2000a), Berberian and Yeats (1999), GSI (1986a,b, 1992a–d, 1993, 2000), Shabanian et al. (2009a,b, 2012), aerial photographs and satellite imagery.*
[c]*Wells and Coppersmith (1994).*
[d]*Shabanian et al. (2009a,b, 2012a,b).*
[e]*Hollingsworth et al. (2010).*

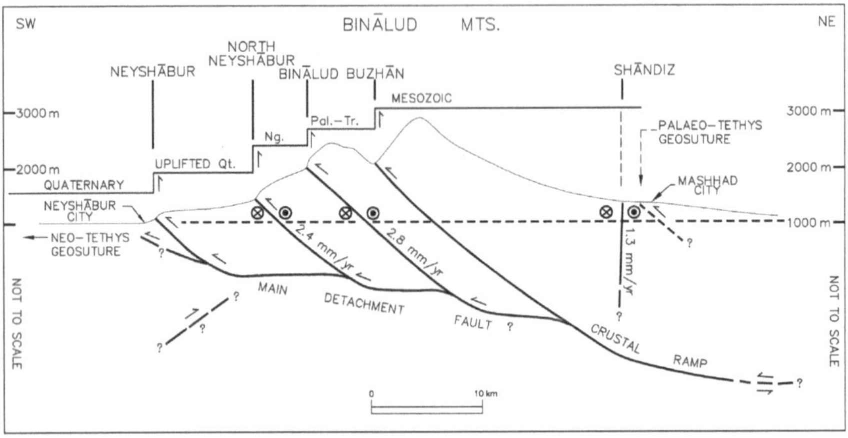

FIGURE 16.16 NE–SW schematic cross-section of the Binālud Mountains (with above-ground scaled topography and possible structure below the ground—not to scale) showing the Neyshābur–Binālud thrust system dipping NE with possible growing blind thrust underneath the city of Neyshābur in the southwest and speculated foreberg thrust possibly dipping to the southwest (?). Except for the hypothetical foreberg thrust, the Neyshābur–Binālud thrust faults and rows of the uplifted Neogene folds (Ng.) at the foothills of the Binālud Mountains may be interpreted as joining a decollement zone at depth; a structure similar to the Tien Shan at the Jungger Basin (Avouac et al., 1993; Wang et al., 2004a), the Western Foothills of Taiwan, east of the Taiwan Strait and the 1999 M_w 7.6 Chi-Chi, Taiwan, earthquake (Rubin et al., 2001; Hsu et al., 2009), or the Garhwal Himalaya (Yeats and Thakur, 2008). Detailed microearthquake survey, waveform modeling of the future earthquakes, and their InSAR study will give a clear picture of the structures at depth in the future. Bl, Binālud; Bu, Buzhān; NE, Neyshābur; NNE, North Neyshābur; and SH, Shāndiz faults. Ng, Neogene; Pal-TR, Paleozoic–Triassic; Qt, Quaternary. Slip rates are after Shabanian et al. (2009a,b, 2012a,b).

The epicentral locations for these two national Iranian agencies have a large magnitude of epicentral error for several reasons, including poor azimuthal distribution of the recording stations, uncertainty about the crustal velocity structure, and unknown errors. The epicenters of the very small- to small-magnitude earthquakes located by the INSN (IIEES) and ITSN (IGUT) are poorly located and have larger location magnitude errors than the agencies claim. Comparison of the TEH, THR, and NEIC epicenters of the 19 January 2012 M_w 5.3 earthquake with the damaged villages near Neyshābur show that the epicenters are shifted toward the west-southwest. We then used the ISC-located epicenters from 1981 to 2010 with *mb* >3.0 for the Neyshābur region. Due to the low number of recorded events and epicentral location errors (Ambraseys, 1978a; Berberian, 1979c), no seismic trend was noticed in the Neyshābur plain.

Quoting a 1317 marginal note of a copy of al-'Utbi's treatise *Tārikh al-Yamini* (1036), Khalifeh Neyshāburi (1291–1349), in his treatise *Tārikh-e Neyshābur* [*The History of Neyshābur*; an abridged Persian translation of Hākem Neyshāburi's 388/998 lost *History of Neyshābur*], wrote that since its first foundation (possibly by King Shāpur-I Sassanid: r. 242–270, or

Shapur-II: 309–379), the city of Neyshābur has been destroyed by 18 earthquakes, possibly from 350 to 1317. Unfortunately, al-'Utbi's text (1036) itself (Reynolds, 1858; Shafi'ee-Kadkani, 1996) does not mention an earthquake in Neyshābur. Within this 967-year period (350–1317), we have been able to find limited records for only six earthquakes (Table 16.2; Figure 16.15). If the statement is correct, much of the seismicity record (for 12 large earthquakes) has vanished. Furthermore, al-Maqrizi (1426) referred to a later tradition among the people of Neyshābur that the city had been destroyed seven times by earthquakes (Melville, 1980). Both statements indicate a high seismicity rate for the city of Neyshābur (Table 16.2; Figure 16.15).

Despite the intense activity during the clustered earthquakes of 1209–1405 in Neyshābur, the long-term slip rate on the Neyshābur fault system is 0.3–2.4 mm/year, and earthquake recurrence interval is about 800–3200 years (Table 16.3). Rupture occurred in a zone about 120-km-long and 7–10 km wide at the Binālud mountain-foot (Figures 16.14 and 16.15). Each earthquake might have increased the Coulomb failure stress along the adjacent faults. Due to the lack of historic seismic data, we are uninformed about the character of the previous earthquakes.

According to records, earthquake cluster activity seems to alternate in northeast Iran. Almost 268 years after the 1209–1405 Neyshābur large-magnitude, quadruple reverse faulting clustered earthquakes ($M_s > 7.0$), the seismicity migrated to the northeast, triggering the Shāndiz right-lateral strike-slip fault on 30 July 1673 ($M_s \gtrsim 6.6$) (Figure 16.15). This event was followed by other strike-slip faulting clustered events in the northwest along the Kalidar, Chakāneh, and the Central Kopeh Dāgh Shear Zone, CKDSZ, faults with the 1851–1895 M_s ~6.0–7.2 earthquake cluster, followed by the 1929 M_w 7.1 earthquake (see Section 16.11.2 below). The seismicity then migrated to the northeastern margin of the CKDSZ with the 1948 M_w 7.2 Ashkābād thrust earthquake (Figure 12.4; see Berberian and Yeats, 2001).

The alternation in earthquake cluster activity observed in the Neyshābur and the Quchān area (described below) might have happened earlier, but no additional recorded historic seismic data are available (especially prior to 1831 for the Quchān area) and no paleoseismic trenching has been performed in the area. Nonetheless, we are confident that Neyshābur has been in a lull for the past 609 years, since the 1405 earthquake (except for the 1673 Shāndiz earthquake, which did not occur along the Neyshābur fault system; and the 2 May 1928 and 19 January 2012, 5.3 earthquakes near Neyshābur; Figure 16.15). Alternation in earthquake activity has been well documented in the Eastern California Shear Zone (ECSZ) and Los Angeles Basin (Dollan et al., 2007; Yeats, 2007) and at Hog Lake along the San Jacinto fault and Wrightwood along the San Andreas Fault zone (Rockwell et al., 2004).

The severe earthquake-fault hazard and risk associated with the faulting at Neyshābur should be taken seriously. Unfortunately, despite the publication of the seismicity (Melville, 1980; Ambraseys and Melville, 1982; Berberian, 1994; Berberian et al., 2000a) and active fault maps of the area (Berberian and

Yeats, 1999; Berberian et al., 2000a; Shabanian et al., 2009a,b, 2012a,b), the city has been expanded and buildings cover the Neyshābur fault (Figure 16.13, lower-left inset). As with the mega-city of Tehrān and the city of Tabriz, unreinforced buildings have been constructed on the Neyshābur fault and the city has become closer to the North Neyshābur fault (Figure 16.13); these changes were approved by the authorities without upholding building seismic codes to withstand strong ground motion or enforcing a setback zone along the active fault lines.

Neyshābur has had a long archeological record since at least 4300 BCE (Table 16.4). Therefore, detailed archeoseismic investigation and paleoseismic trench study can advance our knowledge of seismic history and faulting patterns along the Neyshābur fault system. More than 240,000 people live in the city of Neyshābur and more than 500,000 in the Neyshābur County (2011 census of Iran, www.amar.org.ir).

16.11.2 The Quchān Temporal Earthquake Clusters (NW Binālud Faults and the Central Kopeh Dāgh Shear Zone)

The Kopeh Dāgh Mountain belt of the northeast Iranian Plateau is bound to the north by the Ashkābād fault (Figure 12.4) and to the south by the Paleo-Tethys suture line (Figures 9.1 and 9.2). Short-term GPS data (Nilforoushan et al., 2003; Vernant et al., 2004; Masson et al., 2007; Tavakoli, 2007) indicate a convergence rate of 6.0 ± 2 mm/year (range-parallel right-lateral shear of 2–4 mm/year) for the Kopeh Dāgh, which is close to the N–S geological convergence slip rate of 8.0 ± 2 mm/year (Shabanian et al., 2009a). Shabanian et al. (2009a) showed that the northward motion of Central Iran relative to Eurasia is principally accommodated by strike-slip faulting (6.0 ± 2 mm/year geological rate) localized along the Quchān fault system. The shear faults in this zone (the Central Kopeh Dāgh Shear Zone; CKDSZ; Figure 13.15) are capable of generating large-magnitude ($M_w > 7.0$) earthquakes (Table 16.5), the latest of which was the 1 May 1929 M_w 7.1 Bāghān earthquake (Figure 12.4), which took place to the NW of the unfortunate city of Quchān (Figure 16.17).

During a short period of 62 years (1833–1895), six medium- to large-magnitude earthquakes occurred in the Quchān area of the CKDSZ of northeast Iran (Table 16.5; Figure 16.17). The clustered earthquakes took place in close proximity in space and time with the intervening period of relative quiescence lasting between 20 days and 21 years (1832–1833, 1851, 1871, 1872, 1893, and 1895) along a dense network of closely related active faults (Figure 16.17; Table 16.5). The CKDSZ is an approximately 70 km wide and >120-km-long deformation zone of predominantly NW–SE subparallel right-lateral strike-slip faulting that cuts across the central Kopeh Dāgh fold-and-thrust mountain belt (Tchalenko, 1975; Afshar-Harb, 1979; GSI, 1986a,b; Shabanian et al., 2009a,b; Hollingsworth et al., 2010). The CKDSZ bears a resemblance to the ECSZ, California (Sauber et al., 1986, 1994) and the Marborough strike-slip fault system of northeastern South Island, New Zealand (Van Dissen and Yeats, 1991).

TABLE 16.4 Archaeological Chronology of Neyshābur

Site	~Date	~Ceramicware Period	Major Social Event	Natural Event
Shādyākh/ Kohandezh	11th century–1165	Period III_B[a]		
Shādyākh/ Kohandezh	2nd half of 8th–early 11th centuries	Period III_A[a]		
	Late 7th century			Earthquake
Neyshābur	Late 4th–late 8th centuries	Period II[a]		
	651 CE		Invasion by the Moslem Arabs	
Aparshahr/ Rayvand Sassanid city	224–642 CE			
	336–330 BCE		Invasion by Alexander III of Macedonia	
	450–150 BCE	Period I[a]		
Nishāpur-P Mound[b]	1000 BCE	–	Site abandonment and shift of settlement	Natural disaster?
	1500–1000 BCE	Yaz I/Iron Age I		
	2500–1750 BCE	Namazga V/VI		
	2700–2500 BCE	Namazga IV		
	3100–2700 BCE	Namazga III		
	3500–3100 BCE	Namazga II		
	4000–3500 BCE	Namazga I	Exploitation of the Neyshābur Turquoise Mine	
	~4300 BCE	?	Turquoise used at Sialk II_3 (Ghirshman, 1938)	

See NP in Figure 16.15, and Nishapur-P in Figure 16.13. 26 km Northwest of the Modern City of Neyshābur.
[a]*Rante and Collinet (2013) (Figure 16.13).*
[b]*Hiebert and Dyson (2002). The Nishāpur-P Mound (Figures 16.13 and 16.15) should be investigated for archeoseismic indicators.*

TABLE 16.5 Summary of the Seismic Activity in the Quchān Area

Eq. Date	Eq. Name or Major Event	No. Killed	$\sim i^a$	$\sim M_s^a$	$\sim i^b$	$\sim M_s^b$	$\sim M_s^c$	I^d	M_s^d	Fault
1895	Relocation of the town		The Old Quchān (Khabushān) was relocated ~12 km to the SE in the Nazarābād/Haihai area (Shākeri, 1986).							
17 January 1895 [20 July 1312]	Quchān	>1000	IX	6.0	2 (VIII$^+$)	6.8	6.8	VIII$^+$	6.8	
12 January 1894a		Many	–	–	–	–	–	>VII	?	
19 November 1893a [10 May 1311]	Kalukhi	Many	VI	4.5	–	–	–	>VII	?	
18 November 1893a [09 September 1311]	Gabrābād	Many	–	–	–	–	–	>VII	?	
17 November 1893 [08 May 1311]	S. Quchān	>0.0.1	IX–X	6.6	2 (VIII$^+$)	7.1	7.1	IX	7.0	Chakāneh
06 January 1872 [24 October 1288]	Darbādām, NNE Quchān	Large No.	VIII–IX	6.3	–	–	6.3	VIII	6.3	Darbādām
23 December 1871 [09 October 1288]	N. Quchān	Large No.	VII–VIII	5.6	2 (VIII$^+$)	7.2	7.2	IX	7.2	Quchān
00 May 1857	F. Mine (f)	?	VII	5.1	–	–	–	F	?	?
21 March 1855 [02 July 1271]	F. Mine (f)	?	VII	5.1	–	–	–	F	?	?
22 February 1852[e]	–		–	–	–	–	5.8[e]			–
00 June 1851 [00 July 1267]	Sarvelāyat (SE. Quchān)	>2000	VIII–IX	5.8	2 (VIII$^+$)	6.9	6.9	IX	6.9	Kalidar
31 May 1832–20 May 1833 [1248]	Quchān-F. Mine (c)	?	–	–	3 (VIII)	6.2	–	VII	6.2	?
?										

Seventeenth century	Old Quchān (Khabushān)	Sani' al-Dauleh (1883–1886)
?		
Pre-sixteenth century	Bizhanyurt	Sani' al-Dauleh (1883–1886)
GAP		

See also Figure 16.17.
~: Approximate value for the pre-1900 events.
a: Strong aftershock.
c: Collapsed.
F. Mine: Firuzeh (Turquoise) mine of Neyshābur.
f: Felt.
I: Highest intensity estimated within the meizoseismal area based on written accounts (in MMI).
i: instensity scale used by Ambraseys and Melville (1982); the equivalent MMI values are given in parenthesis.
M_s: The equivalent surface-wave magnitude derived from macroseismic information embedded in written accounts calibrated against instrumental M_s values. Therefore, they present poorly constrained estimates.
[a] *Kondorskaya and Shebalin (1977, 1982). The magnitude–intensity values do not match.*
[b] *Ambraseys and Melville (1982).*
[c] *Utsu (2002).*
[d] *This study.*
[e] *Spurious entry by Utsu (2002); not valid.*

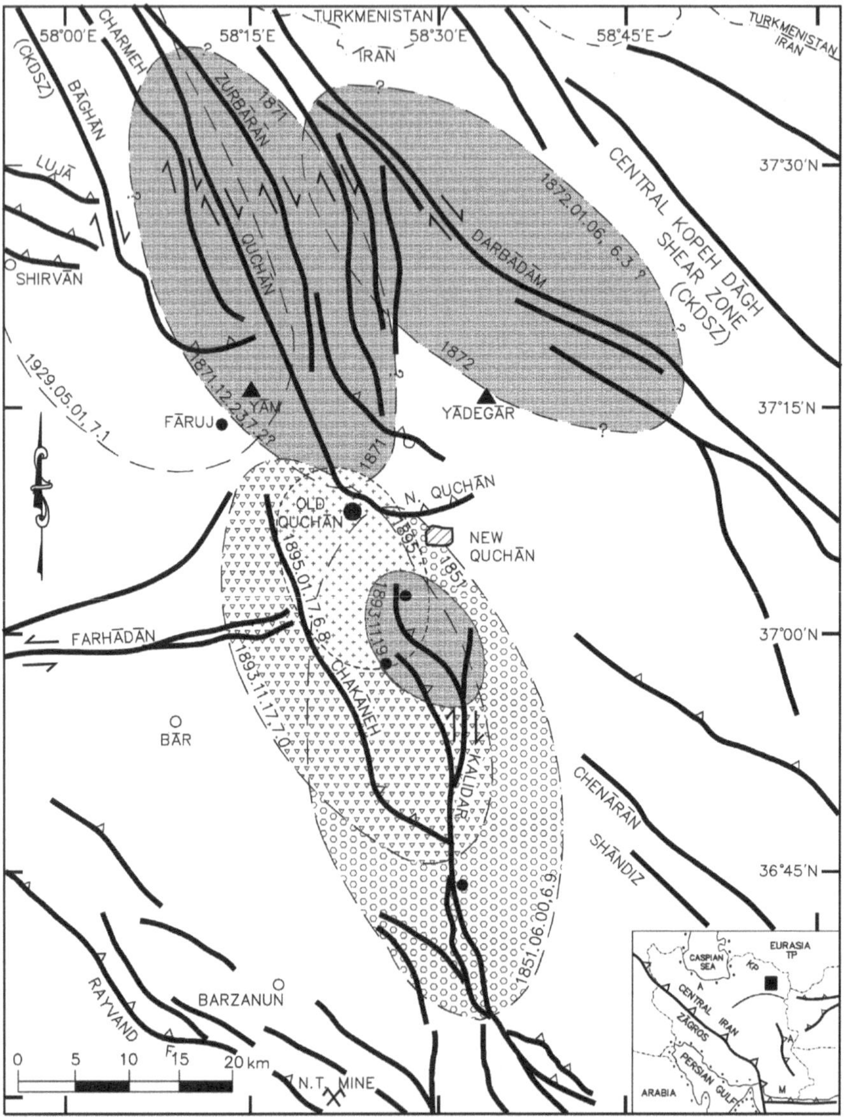

FIGURE 16.17 Meizoseismal areas of medium- to large-magnitude earthquakes in the Quchān area (NW of Neyshābur) and the Central Kopeh Dāgh Shear Zone (CKDSZ) in northeast Iran. Queried where uncertain. Faults from Berberian et al. (2000a), GSI (1986a,b), Shabanian et al. (2009a,b). Faults and symbols as in Figure 11.1.

Despite the intense activity of the 1833–1895 Quchān clustered earthquakes, the long-term slip rates on the nearby faults are estimated at 1.0–2.8 mm/year, and earthquake recurrence intervals are ~900–2000 years (Figure 16.17; Tables 16.6A and 16.6B). Rupture occurred across 40-km-wide areas of the

TABLE 16.6A Summary of the Fault Parameters in the Quchān Area

Fault Name	Mech.	Min. Fault Length (in km)	M_w Max.[a]	*I* Max.	Max. Displ. (in m)[a]	Slip Rate (mm/year)[b]	Slip Rate (mm/year)[c]	Slip Rate (mm/year)[d]	Slip Rate (mm/year)[e]	Slip Rate (mm/year)[f]	Slip Rate (mm/year)[g]	Return Period (years)
Bāghān	RL	80	7.3	IX	3.0			1.0	2.8			1071
Chakāneh	RL/R	50	7.0	IX	1.5					1.2		1250
Charmeh	RL	55	7.1	IX	1.9					1.1		1727
Darbādām	RL	100	7.4	IX⁺	3.9			1.5	1.5			2600
Farhādān	LL	60	7.1	IX	1.9					1.8		1055
Kalidar	RL/R	60	7.1	IX	1.9					0.8-1.2	2.0	950
Lujā	RL	55	7.1	IX	1.9							
Nāveh	RL	60	7.1	IX	1.9				0.5			3800
North Quchān	R	15	6.4	VIII	1.0							
Quchān	RL	100	7.4	IX⁺	3.9	9.0±2	4.0±1	1.5	4.3			906
Shirvān	RL	55	7.1	IX	1.9				1.0			1900
Zurbārān	RL	100	7.4	IX⁺	3.9							

See also Figure 16.17.
I: Highest intensity estimated within the meizoseismal area based on written accounts [in MMI].
Mech: Fault mechanism (LL: left-lateral strike-slip; R: reverse; RL: right-lateral strike-slip).
M_w: Moment magnitude.
Faults based on: GSI (1986a,b, 1987), Berberian et al. (1387/1999), Shabanian et al., 2009a,b, 2012a; aerial photographs and satellite imagery.
[a]*Wells and Coppersmith (1994).*
[b]*Tavakoli (2007).*
[c]*Masson et al. (2007).*
[d]*Hollingsworth et al. (2008).*
[e]*Shabanian et al. (2009a).*
[f]*Shabanian et al. (2009b).*
[g]*Shabanian et al. (2012a).*

TABLE 16.6B Summary of Long-Term and Current Fault Slip Rates Along the Central Kopeh Dagh Shear Zone

Fault Name	Geodetic Slip (mm/year) (Mousavi et al., 2013)	Block Modeling (mm/year) (Mousavi et al., 2013)	Geologic Slip (mm/year) (Hollingsworth et al., 2006)	Geologic Slip (mm/year) (Shabanian et al., 2009a)	Geologic Time (kyr) (Shabanian et al., 2009a)
Bāghān	4.4±0.8 RL 1.2±0.7 SH	5.4±1 RL 2.1±1 SH	1 RL (without age constraint)	2.8±1 RL	280+16
Quchan	–	–	1.5 RL (without age constraint)	4.3±0.6 RL	83±4

See also Figure 16.17.
RL, right-lateral; SH, shortening.

southern CKDSZ and the northwest Binālud Mountains (Figures 16.15 and 16.17). Each earthquake might have increased the Coulomb failure stress along the adjacent faults of this densely faulted area with unstable tectonic situation of NNW–SSE strike-slip and NW–SE thrust faulting. Lack of a pre-1832 seismic record as well as an archeoseismic trench study prevents us from seeing if the previous earthquakes were also clustered in the Quchān region as well as examining the interaction between the strike-slip and reverse faults.

The effects of the medium- to large-magnitude earthquake cluster (that occurred during a short period of 62 years (1833–1895)—their concentrated meizoseismal areas and responsible faults—are difficult to disentangle due to a high density of faulting and lack of proper recorded historical data. After the 1895 earthquake, the old city of Quchān (Khabushān) was relocated about 12 km to the southeast in the Nazarābād area (Shakeri, 1986; Figure 16.17). We know that the city was relocated at least three times in recorded history because of earthquakes (Sani' al-Dauleh, 1883–1886) (Table 16.5). Recorded historical seismicity indicates that the city was heavily damaged or destroyed by the 1832–1833, 1851, 1871, 1893, and 1895 earthquakes (Table 16.5; Figure 16.17).

The damage areas of the two large-magnitude earthquakes of 23 December 1871 and 6 January 1872 in the north and northeast Quchān area, in the CKDSZ, are so entangled that it is not clear which shock was stronger (Figure 16.17). Kondorskaya and Shebalin (1977, 1982), Tchalenko (1975), and Ambraseys and Melville (1982) treated them as foreshock and mainshock, or mainshock and aftershock sequences. The latter presented a single meizoseismal area for the 1871–1972 earthquakes (covering the damage zones of the both 1871 and 1872 events) with a NNE–SSW -trending ellipse, covering several NW–SE -trending active right-lateral strike-slip faults of the Kopeh Dāgh (see figure 3.23, p. 65, in Ambraseys and Melville, 1982).

Alison (1872) received a report that 2000 inhabitants were killed by the first shock (23 December 1871) and 4000 by the second (6 January 1872). Napier (1876) stated that no lives were lost in the city of Quchān because the population had been warned by foreshocks in December 1871 and left town. The *Iran Newspaper* (1 February 1872) reported that only two people were killed in the city of Quchān. Apparently, Quchān was half destroyed (Napier, 1876).

The difference in the reported casualties in the two events could be the result of distinct magnitudes or the population density. The fertile lands and valleys north of Quchān are much more populated (Ja'farābād, Khabushān, Esfejir, and many more) than the mountainous region to the northeast of Quchān with fewer small villages (Shamkhāl, Darbādām, and Emān Qoli). However, the reported sequence of earthquakes and casualty figures does not match the population density. The destroyed villages lie along two different NW–SE trending strike-slip faults of the CKDSZ and may indicate that two different faults (the Quchān and the Darbādām faults) were reactivated during the two events (Figure 16.17).

The Quchān seismic pattern resembles the 28 June 1992 M_w 7.3 Landers and the 16 October 1999 M_w 7.1 Hector Mine, California, earthquakes, along two nearby parallel strike-slip faults in the central Mojave Desert (Harris and Simpson, 2002; Pollitz and Sacks, 2002). Both earthquakes occurred within the active ECSZ, where earthquake return periods are estimated to be about 4000 years (Sauber et al., 1994; Rubin and Sieh, 1997). The fault system and earthquake pattern north of Quchān in the CKDSZ (Figure 16.17) seem to be also analogous to the northern South Island, New Zealand, subparallel right-lateral strike-slip faults in an oblique convergence at the Pacific–Australian plate boundary, off the Alpine fault, with the 6 October 1848 M 7.5 earthquake on the Awatere fault, 1 September 1888 M 7.0–7.3 earthquake on the Hope fault, and the 9 March 1929 M 7.0 earthquakes on the Hope fault (Yeats and Berryman, 1987; Cowan, 1990; Van Dissen and Yeats, 1991; Yeats, 2012).

At the end of the 1833–1895 Quchān-clustered earthquakes, the authorities realized that there was something wrong with the location of the devastated town. Hence, the site of the town was relocated about 12 km to the southeast (Figure 16.17). Nonetheless, the new site is still close to the active faults and the modern buildings of the town of Quchān, with population of 104,000, (2011 census of Iran, www.amar.org.ir) are not reinforced against strong ground shaking.

16.11.3 Possible Alternating Seismic Clusters Between the Neyshābur Reverse Fault System (1209–1405) and the CKDSZ (1833–1895 and 1929–1997)

Almost 268 years after the 1029–1405 quadruple large-magnitude thrust seismic cluster of the Neyshābur area (Figures 16.14 and 16.15), the seismicity migrated to the northeast, triggering strike-slip faulting in 1673 ($M_s \sim 6.6$) along the Shāndiz fault to the northeast (Figure 16.15). Later on, the Quchān/CKDSZ, northwest of Neyshābur, experienced increased earthquake activity in 1851–1895 (Quchān) and 1929–1997 (CKDSZ; Figure 16.17). The cluster of strike-slip faulting earthquakes in the Quchān area devastated the old city of Quchān (Khabushān) several times (Figure 16.17). During the 1851–1895 Quchān cluster and the 1929–1997 CKDSZ cluster (Figures 12.4, 13.15, and 16.17), no seismic activity was recorded in the Neyshābur area (Figure 16.15). It is not clear if the mechanism of alternating seismic clusters had taken place earlier. Thus, variations in strain accumulation and release over different time scales are unclear.

A similar mechanism of alternating seismic clusters between the San Andreas/Los Angeles region/Garlock faults and ECSZ was suggested by Dollan et al. (2007). Paleoseismologic study is needed to better understand the mechanics of temporal and spatial clustering, alternating seismic clusters between different regions, and seismic risk measurements.

16.11.4 Coseismic Noncontemporaneous Ruptures of the Zāgros Main Recent Parallel Fault Segments

Analyses of the 14 August 1958 (two events) and 16 August 1958 earthquake sequence in Chapter 12 showed that the damaging 14 August 1958 (11:27 UTC) M_s 5.7 earthquake was associated with coseismic surface faulting along the Garrin fault segment (Figure 12.12). The second M_s 5.5 earthquake that same day (15:25 UTC) ruptured the Sahneh fault segment of the Zāgros Main Recent fault to the northwest of the first event. Finally, the 16 August 1958 M_w 6.5 earthquake was associated with coseismic surface fault rupture along the Nahāvand fault segment (Figure 12.12). The distance between the two subparallel segments of the Zāgros Main Recent fault is about 8 km. The first earthquake triggered the second event along the same fault trend, and the third event was the cross-strike jump for 8 km from the southwest to the northeast (Figure 12.12).

16.12 SPATIALLY CLUSTERED SEQUENCES OF EARTHQUAKES ON TECTONICALLY RELATED FAULTS WITH VARIABLES STRIKES AND MECHANISMS

Several clusters of earthquakes in the Iranian Plateau provide evidence of interaction among reverse and strike-slip faults, probably due to adjacent faults being loaded by individual earthquakes. In contrast, other historical events did not occur as parts of a sequence (Berberian, 1994, 1997a, 2005; Berberian and Yeats, 1999, 2001).

A few rare examples exist of densely populated interacting faults with different seismic characters and mechanisms showing spatial and temporal clustering of major earthquakes and reloading of neighboring faults with thrust and strike-slip mechanisms striking in different directions. Earthquake activity on one fault or fault segment interacts with other adjacent faults, some of which are many rupture-lengths away from the original fault. Understanding such clustering, with episodic bursts of seismic activity, and long-term interaction between faults in regions with complex active faulting patterns and shorter historic seismic data, has some critical consequences in earthquake hazard evaluation and risk-mitigation efforts.

16.12.1 The Qohestān Sequence of 1936–1997: Earthquake Clustering, Seismic Interaction, and Triggering Between the Adjacent Conjugate Faults of Dasht-e Bayāz, Ferdows, Ābiz, Chāhak, Chang Raqsh, and Āvash

One of the rare and well-pronounced earthquake clustering history on adjacent conjugate faults took place during 61 years (between 1936 and 1997) in the Dasht-e Bayāz, Ābiz, and Ferdows area in the eastern Iranian Plateau (Figures 9.1 and 16.18; Table 16.7). The earthquake cluster consisted of 11 earthquakes of M_w 6.0–7.2 (16 events of M_w 5.5–7.2) in an area of 160×110 km, approximately the same size as the metropolitan area of the Los Angeles basin, California (Berberian and Yeats, 1999, 2001; Berberian et al., 1999). During this short period, stored energy along several faults was released in an area of $\sim$17,600 km^2 crisscrossed with several major faults. Due to the remoteness of the area and the lack of paleoseismic trench study, we are unaware how long the interim quiet period was in this region.

We know that earthquakes release accumulated elastic strain along active faults and load faults in nearby (and even more distant) regions that will be the location of future events (Stein and Lisowski, 1983; Hudnut et al., 1989; Reasenberg and Simpson, 1992; Stein et al., 1992; Du and Aydin, 1993; King et al., 1994; Bennett et al., 1995; Harris and Simpson, 1996; Jacques et al., 1996). The geometry of interactions between groups of active faults is critical for seismic risk evaluation.

The clustered sequence of 16 medium- to large-magnitude earthquakes (Table 16.7; Figure 16.18) that began in 1936 is likely to represent the triggering of motion on successive faults through stress enhancement caused by motion on nearby faults (King et al., 1994). This is clear where successive strike-slip ruptures occur along a strike of the same system (as in 1968 and 1979 on the E–W Dasht-e Bayāz fault, or in 1936, 1979 (twice), and 1997 on the N–S Ābiz fault), but is less obvious in the case of interaction between the conjugate Dasht-e Bayāz and Ābiz faults, as their ends lie in each other's compressional quadrants where Coulomb failure stress should be decreased rather than enhanced (Walker et al., 2011). However, the Ābiz fault has a high length/width ratio of about 8 (125/15 km) and the off-fault lobes of Coulomb stress change might be suppressed, allowing other factors such as the curvature of the fault in plain view and the gradient of slip along it to be important (King et al., 1994; Stein et al., 1997; Berberian and Yeats, 1999, 2001; Berberian et al., 1999; Walker et al., 2011).

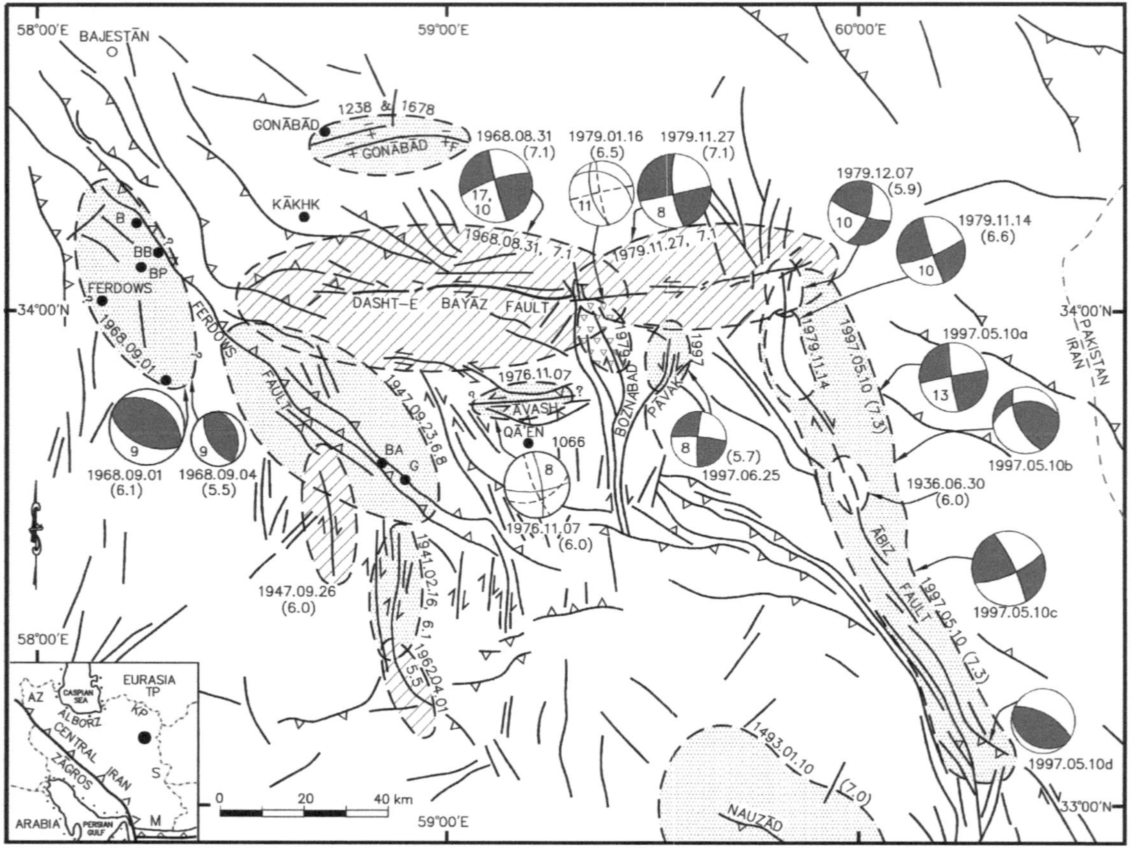

FIGURE 16.18 Meizoseismal areas of medium- to large-magnitude earthquakes in the Qohestān region of the Dasht-e Bayāz/Ābiz in eastern Iran. Queried where uncertain. Faults and symbols as in Figure 11.1. Focal mechanism of the earthquakes constrained by body-wave modeling with centroid depths: 31 August 1968 (Walker et al., 2004); 1 and 4 September 1968 (Walker et al., 2003); 7 November 1976, 16 January 1979, 14 November 1979 (Baker, 1993; Jackson, 2001); 27 November 1979 (Walker et al., 2004); 7 December 1979 (Baker, 1993); 10 May 1995, 25 June 1997 (Berberian et al., 1999). *Modified from Berberian and Yeats (1999–2001) and Berberian et al. (1999).*

TABLE 16.7 The Clustered Earthquake Sequence in the Qohestān Area

Date	Time (UTC)	Coordinate 00.00°N–00.00°E	Earthquake Name	M_s	M_w	CD_w (km)	Mechanism	Fault Name
30 June 1936	19:26		Ābiz	6.0	6.0	–	N–S right-lateral	Ābiz (middle)
16 February 1941	16:39		Kumirān	6.2	6.1	–	N–S right-lateral	Chāhak
23 September 1947	12:28		Charmeh	6.9	6.8	–	NW–SE thrust	Ferdows
26 September 1947	03:04		Dustābād	6.0	6.0	–	N–S right-lateral	Chang Raqsh
01 April 1962	00:45		Chāhak	5.5	5.5	–	N–S right-lateral	Chāhak
31 August 1968	10:47	34.04–58.95	Dasht-e Bayāz	7.2	7.1	09	E–W left-lateral	Dasht-e Bayāz (west)
01 September 1968	07:27	34.04–58.20	Ferdows	6.4	6.3	10	NW–SE Thrust	E. or W. (?) Ferdows
04 September 1968	23:24	34.03–59.31	Ferdows		5.5	09	NW–SE Thrust	E. or W. (?) Ferdows
11 September 1968	19:17	33.97–59.53	Dasht-e Bayāz	6.0	5.6	06	E–W left-lateral?	?
07 November 1976	04:00	33.82–59.18	Vondik	6.3	6.0	08	E–W left-lateral	Āvash
16 January 1979	09:50	33.90–59.46	Boznābād	6.7	6.5	11/13	Reverse/N–S right-lateral	Boznābād
14 November 1979	02:21	33.95–59.72	Korizān	6.7	6.6	10/6	N–S right-lateral	Ābiz (north)
27 November 1979	17:10	34.05–59.75	Koli	7.2	7.1	08	E–W left-lateral	Dasht-e Bayāz (east)
07 December 1979	09:23	34.07–59.84	Kalāt-e Shur	6.1	5.9	10	N–S right-lateral	Ābiz (north)
10 May 1997	07:57	33.84–59.81	Zirkuh-e Qā'enāt	7.2	7.2	13	N–S right-lateral	Ābiz (complete fault length)
16 June 1997	03:00	33.24–60.20		5.0	5.0	–	Reverse	
20 June 1997	12:57	32.29–59.99	Chākhu	5.0	5.4	–	(N–S right-lateral)	?(Purang)
25 June 1997	19:38	33.91–59.43	Boznābād	5.8	5.7	08	(N–S right-lateral)	Pāvak

See also Figure 16.18. CD_w, centroid depth from waveform modeling (see the text for references).

The twentieth-century earthquake sequence began with the 30 June 1936 M_w 6.0 Ābiz earthquake, producing >12 km of coseismic surface rupture on the central segment of the N–S right-lateral strike-slip Ābiz fault (Figures 13.18 and 16.18; Table 16.7; see also Chapters 12 and 13). Five years later, the 16 February 1941 M_w 6.1 Kumirān earthquake caused >12 km surface rupture along the Chāhak N–S right-lateral strike-slip fault, about 100 km to the west-southwest of the first event (Figures 12.8 and 16.18; Table 16.7; Chapter 12). This event was followed by the 23 September 1947 M_w 6.8 Charmeh earthquake along the NW–SE Ferdows thrust, about 40 km to the north-northwest of the 1941 earthquake (Figures 12.8 and 16.18). Three days later, the 26 September 1947 M_w 6.0 Dustābād earthquake took place along the N–S Chang Raqsh strike-slip fault, with an unknown length of surface rupture about 20 km to the south of the previous earthquake (Figure 12.8). The last event in the southwestern area was the 1 April 1962 M_w 5.5 Chāhak earthquake along the southern part of the Chāhak N–S right-lateral strike-slip fault, with an unknown length of surface rupture (Figures 12.8 and 16.18; Table 16.7; Berberian and Yeats, 1999, 2001).

Six years later, the seismicity shifted about 80 km to the north when the 31 August 1968 M_w 7.1 Dasht-e Bayāz earthquake produced 70 km of surface faulting along the western segment of the E–W Dasht-e Bayāz left-lateral, strike-slip fault (Figures 13.1 and 16.18; Table 16.7; Chapter 13). This earthquake was the sixth earthquake in the 1936–1997 earthquake-cluster sequence. The 1968 earthquake on the west segment of the Dasht-e Bayāz fault was followed 20 h later by the 1 September 1968 M_w 6.3 Ferdows earthquake on the NW–SE Ferdows thrust system, possibly due to loading of that fault by the west Dasht-e Bayāz earthquake (Figures 13.2 and 16.18; Table 16.7; Berberian and Yeats, 1999, 2001).

Eight years later, the seismicity migrated to the central area with the 7 November 1976 M_w 6.0 Vondik earthquake along the E–W left-lateral Āvash fault (Figures 11.8 and 16.18; Table 16.7; Chapter 13). This event was followed by the 16 January 1979 M_w 6.5 Boznābād earthquake along the N–S, right-lateral strike-slip Boznābād fault about 20 km to the northeast of the 1976 earthquake (Figure 11.8). About 10 months later, the 14 November 1979 M_w 6.6 Korizān earthquake, with >20 km coseismic surface rupture, took place along the N–S right-lateral strike-slip Ābiz fault to the north of the 1936 event (Figure 13.18). Thirteen days later, the 27 November 1979 M_w 7.1 Koli earthquake produced 55 km of coseismic surface rupture along the eastern segment of the E–W Dasht-e Bayāz left-lateral strike-slip fault. With this earthquake, the entire Dasht-e Bayāz fault system had ruptured (Figures 13.1 and 16.18; Berberian and Yeats, 1999, 2001). Ten days later, the seismicity jumped to the Ābiz fault in the east when the 7 December 1979 M_w 5.9 Kalāt-e Shur earthquake ruptured >15 km of the northern part of the fault (Figures 13.18 and 16.18; Table 16.7; Chapter 13).

The clustered sequence ended a decade later with two earthquakes: (i) 10 May 1997 M_w 7.2 Zirkuh earthquake, producing 125 km of N–S right-lateral strike-slip faulting along the Ābiz fault, with its epicenter being along the unruptured part of the fault between the 1936 and 1979 surface ruptures, and (ii) the 25 June 1997 M_w 5.7 Boznābād earthquake along the Pāvak N–S, right-lateral strike-slip fault (Figures 11.8, 13.18 and 16.18; Table 16.7; Berberian and Yeats, 1999, 2001; Berberian et al., 1999). No medium- to large-magnitude earthquake has occurred since 1997, and we do not know if the cluster sequence has ended. Nonetheless, there are still some seismic gap areas in this crisscrossed faulted area likely to be activated in the future (Figure 16.18).

Regarding the individual faults reactivated during the 1936–1997 cluster, the Dasht-e Bayāz E–W left-lateral strike-slip fault consists of a 70-km-long west segment that ruptured in 1968 and a 50-km-long east segment that ruptured 11 years later in 1979 with two M_w 7.1 earthquakes occurring in a west-to-east progression of strike-slip earthquakes (Figures 13.1 and 16.18). The two segments of the Dasht-e Bayāz fault are separated by the north–south-trending, Mahyār right-lateral strike-slip fault. The intersection is marked by structural complexity, including a local change of strike, a zone of splays of the Dasht-e Bayāz fault, and a right-step of about 1 km on the Mahyār fault. The Mahyār fault extends farther south, with a left-stepover to the Pāvak fault (Figures 11.8, 13.1, and 16.18). So within 11 years, the whole E–W left-lateral strike-slip fault was ruptured (Figure 16.18).

Whereas the Ābiz N–S right-lateral strike-slip fault was ruptured in 1936 (>12 km along the central part), 1979 (>15 km along the northern part close to the Dasht-e Bayāz fault), and 1997, when the complete 125 km of the fault length was ruptured (Figures 13.18 and 16.18). The NW–SE Ferdows thrust was ruptured in 1947 and 1969, with unruptured segments remaining along the fault (Figure 13.2). The total fault length of the Chāhak N–S right-lateral strike-slip fault was ruptured in 1941 and 1962 (Figure 12.8). The Chang Raqsh N–S right-lateral and Āvash E–W left-lateral faults ruptured in 1947 and 1976, respectively (Figure 12.8, 11.8, and 16.18), whereas only the northern parts of the Boznābād and Pāvak faults were ruptured in 1979 and 1997, respectively (Figures 11.8 and 16.18). The southern parts of these two faults, as well as the rest of the crisscrossed faults of the region, remained as seismic gaps. In the last nine centuries, only the twentieth-century earthquakes showed clustering in this region (Berberian and Yeats, 1999) (Figure 16.18).

The gradually accumulated strain of the crisscrossed faults in the area was released in a series of medium- to large-magnitude clustered earthquakes along eight faults. The twentieth-century initial earthquake of 1936 ruptured a small segment of the Ābiz fault, but apparently it transferred more stress onto segments of Chāhak, Ferdows, Chang Raqsh, Dasht-e Bayāz, Boznābād, and Pāvak faults. During these loading and triggering processes, the entire length of five faults was ruptured (Figure 16.18).

Similar intercontinental clustered earthquakes on adjacent conjugate faults are documented in other active regions:

i. The Superstition Hills, Southern California, along the Superstition Hill right-lateral fault (24 November 1987, M_w 6.5) and Elmore Ranch left-lateral fault (24 November 1987, M_w 6.6) (Hudnut et al., 1989; Sharp et al., 1989).

ii. The Al Hoceima, Morocco, along the Boussekkor left-lateral fault (26 May 1994 M_w 6.0) and a right-lateral fault (24 February 2004, M_w 6.4) (Bezzeghoud and Buffon, 1999; Biggs et al., 2006; Cakir et al., 2006).

iii. The 1 May 2003 M_w 6.4 Bingol earthquake and its aftershocks along a NNW-striking, right-lateral fault, which is perpendicular to the left-lateral East Anatolian fault, eastern Turkey (Milkereit et al., 2004).

iv. The 6 May 1930 M_w 7.1 Salmās, NW Iran, earthquake sequence along the Salmās (right-Lateral), Derik (left-lateral), Tasuj (thrust), and Mir 'Omar (?, Rāviān left-lateral, or Goharān right-lateral) faults (see Figure 12.5 and Chapter 12).

v. The 10 December 1977 M_w 5.9 Dartangal (right-lateral strike-slip Kuhbanān fault) range-front and the 6 August 1984 M_w 5.3 (Figure 13.7), and 22 February 2005 M_w 6.4 Dāhuiyeh (thrust with slight left-lateral motion; Figure 14.8) intramountain earthquakes (Berberian et al., 1979a; Talebian et al., 2006a).

vi. The 20 December 2010 M_w 6.5 (right-lateral) and 27 January 2011 M_w 6.2 (left-lateral) South Rigān earthquakes, in southeast Iran (Walker et al., 2013c; Chapter 14.9).

vii. The Tabriz 1721, 1780, and 1786 earthquakes along the right-lateral strike-slip fault and the 1641 earthquake perpendicular to it (Figures 11.12 and 16.7; Berberian and Yeats, 1999).

However, none of the abovementioned cases were as remarkable or spectacular as the Qohestān (Dasht-e Bayāz area) 1963–1997 clustered earthquake sequence of 16 M_w 5.5–7.3 earthquakes (Table 16.7; Figure 16.18). Obviously, the documented 1936–1997 general system behavior with the most recent cluster of intense seismic activity along right-lateral and left-lateral strike-slip as well as reverse faults could be very dangerous if located in an urban area, especially in the vicinity of mega-cities like Tehrān.

Because we lack historic seismic data and paleoseismological trench studies, we do not know if the seismic activity in the Qohestān area has been governed by temporary and spatial clustering and loading of adjacent faults in the past. We also do not know if this complex cluster pattern will be a general characteristic of rupture process in this and other structurally similar regions. Furthermore, the ruptured areas (Figure 16.18) show several seismic gaps along other faults of the area. These may be the sites of future earthquakes. Obviously, in such regions, we have to extend the earthquake record back to the pre-1936 period.

The left-lateral offset of about 10 m across the Dasht-e Bayāz fault has been measured along old qanāts with unknown ages of displacement (Ambraseys and Tchalenko, 1969). One of the oldest known qanāts in Iran was constructed in the second millennium BCE in the Semnān area at the northern edge of the central Kavir depression (Mehryar and Kabiri, 1986). Assuming the qanāts offset by the Dasht-e Bayāz fault are at least 4000 years old (?), the minimum horizontal slip rate on the Dasht-e Bayāz fault would be ~2.5 mm/year. The 250-cm left-lateral slip released during the 1968 earthquake (Tchalenko and Berberian, 1975) could have accumulated in about 1000 years at a rate of 2.5 mm/year. Cumulative Holocene offsets of 8–28 m are documented at the Khidbas Creek as well as other drainages crossing the fault (Berberian and Yeats, 1999).

As mentioned previously, the length of the pre-twentieth-century interseismic interval in this area is uncertain. We are aware of the 1233 and 1678 earthquakes that destroyed the town of Gonābād, northwest of Dasht-e Bayāz (Figure 11.11), and the 1066 earthquake that destroyed the city of Qā'en, south of Dasht-e Bayāz (Figure 11.8). It is probable that the 1238 Gonābād earthquake was the penultimate earthquake before the twentieth-century cluster occurring on the Dasht-e Bayāz fault, an interpretation consistent with the approximately 1000-year average recurrence interval discussed earlier. However, details about the 1847 event damaging the Qā'en congregational mosque are not known.

The occurrence of a very large earthquake that ruptured the entire 120-km length of the Dasht-e Bayāz fault cannot be demonstrated during the last 776 years because the historical monuments at Kākhk and Ferdows have survived for the last 423 and 776 years, respectively (Figure 16.18). It seems, therefore, that unlike the twentieth-century cluster, the earlier events were possibly isolated in time. The available macroseismic data do not allow the estimation of intensity, magnitude, constrained meizoseismal areas, and thus the seismic sources for these earlier earthquakes (Berberian and Yeats, 1999).

We do not know when the previous cycle of clustered events occurred in east Iran and are not aware of the physical crustal processes that might be responsible for the observed 1941–1997 clustering.

16.12.2 The Interacting Kuhbanān Range-Front Strike-Slip and the Intermountain Cross-Reverse Faults

The 200-km-long Kuhbanān range-front right-lateral strike-slip fault (Huckriede et al., 1962; Berberian, 1976c,e, 2005; Berberian et al., 1979a) is a major active fault striking N140°E in the vicinity of the provincial capital city of Kermān (Figure 9.1). The fault consists of a series of right-lateral strike-slip segments that step to the right in an en-echelon pattern (Figure 16.19). Toward the southern tip of the fault, splay intermountain, cross-reverse faults (Dāhu, Heruz, Tigur, Chāirud, Pāsu, Bāzargān, and

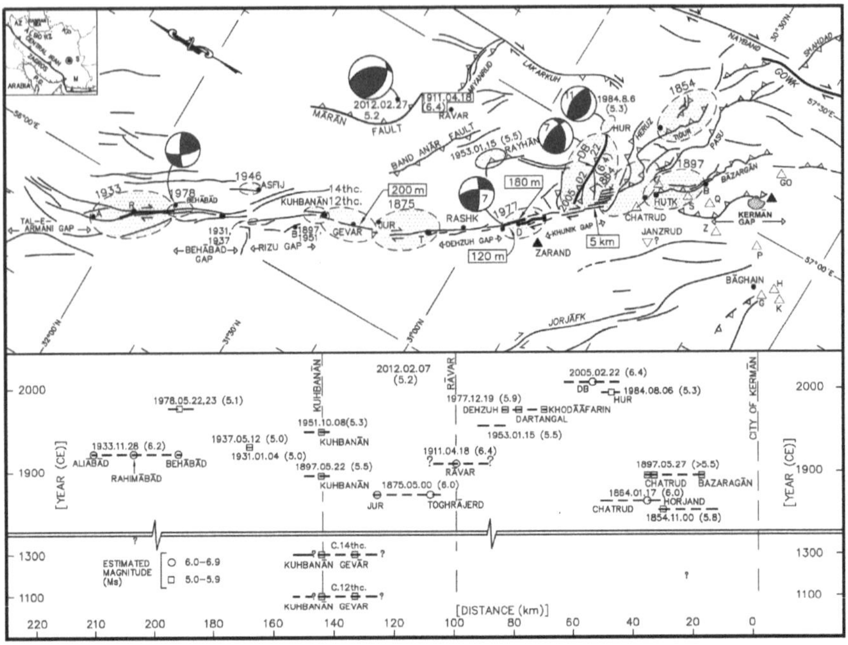

FIGURE 16.19 Meizoseismal areas of medium-magnitude earthquakes (top) and space–time diagram (bottom) of nearly 160 years of recorded seismicity along the Kuhbanān right-lateral strike-slip and cross-thrust system in the Kuhbanān fold-and-thrust Mountain belt north-northwest of the city of Kermān in southeast Iran. Faults and other symbols as in Figures 11.1 and 16.7. Thicker lines show faults with documented surface ruptures. Dashed horizontal lines in the lower panel indicate possible rupture area. Distances are along strike with respect to the provincial capital city of Kermān. Focal mechanisms: 1977 (Berberian et al., 1979a; Baker, 1993); 1985 (Baker, 1993); 2005 (Talebian et al., 2006a; 1978, 2011 [HRVD]). Triangles show location of standing (filled) and ruined (blank) archeological sites. The reversed triangle with "?," the approximate location of the lost tenth-century town of Janzrud. DB, Darbidkhun; G, Tapeh Gajin; GO, Goshkin; H, Qal'eh Hosseinābād; P, Panhe Parān; Q, Qal'eh; Z, Zangiābād. *Modified from Berberian (2005).*

Kermān; Figure 14.8) and folds are well developed at almost a right angle (Berberian, 2005). An active slip rate of ~1.0–2.0 mm/year is estimated for this fault (Walker and Allen, 2012).

During the 123 years of recorded seismic history (from 1854 to 1977), five scattered sections on the 190-km length of the Kuhbanān fault that total about 100 km in length have ruptured with medium-magnitude earthquakes; this is about 52% of the total known length of the fault (Figure 16.19). Unlike the Gowk and Ābiz fault cases (Figures 13.18 and 16.8), no large-magnitude earthquake has yet been recorded during the last 160 years of recorded seismic history along the Kuhbanān fault (Figure 16.19). Therefore, the largest earthquakes in the current cluster may not have yet occurred. Although references are made to the twelfth- and fourteenth-century earthquakes at Kuhbanān (Figure 16.19), no macroseismic data have survived to estimate the intensity and the magnitude of these events (Berberian, 2005). Limited macroseismic data make it difficult and unwise to draw broader conclusions regarding the recurrence period of large-magnitude earthquakes along the Kuhbanān fault.

The recorded preinstrumental period seismic history of the Kuhbanān strike-slip and cross-thrust fault system began with the November 1854 $M_s{\sim}5.8$ Hurjand earthquake possibly along the Tigur cross-thrust in the south (Figure 16.20). Ten years later, the seismicity migrated 30 km to west-northwest along the same cross-thrust, approaching the Kuhbanān fault where the 17 January 1864 $M_s{\sim}6.0$ Chatrud earthquake took place. Eleven years after the cross-thrust event in the south, the May 1875 $M_s{\sim}6.0$ earthquake took place 70 km to the northwest along the Kuhbanān range-front, strike-slip fault. Twenty-two years after this event, the seismicity migrated about 30 km to the northwest along the Kuhbanān fault with the 22 May 1897 $M_s{\sim}5.5$ Kuhbanān earthquake. Five days later, the seismicity propagated 110 km to the southeast, where the 27 May 1895 $M_s{\gtrsim}5.5$ event occurred on the Bazargān cross-thrust (Figures 16.19 and 16.20; Berberian, 2005).

After 14 years of relative seismic quiescence, the 18 April 1911 M_s 6.4 earthquake destroyed the city of Rāvar in the northeast on a so far, undetected causative fault (Figures 12.2 and 16.19). Twenty-two years later, the 28 November 1933 M_s 6.2 Behābād earthquake took place along the northern part of the Kuhbanān fault (Figures 12.6 and 16.19). The seismicity then shifted almost 95 km to the southeast with the 15 January 1953 M_s 5.5 Rayhān earthquake. Twenty-four years later, the 19 December 1977 M_w 5.9 Dartangal earthquake created 20 km of coseismic surface rupture along the right-lateral strike-slip of the Kuhbanān fault (Figures 13.7 and 16.20). This was followed a year later by the 22 May 1978 M_w 5.1 Behābād earthquake, about 110 km to the northwest of the previous event. The last earthquake was the 22 February 2005 M_w 6.4 Dāhuiyeh, cross-thrust earthquake (Figure 14.8), about 120 km to the southeast of the latter (Berberian, 2005; Talebian et al., 2006a). Nonetheless, there are still some intervening unruptured gaps between the ruptured zones (Figure 16.19).

Both strike-slip earthquakes (twelfth and fourteenth century, 1875, 1897, 1933, 1951, 1977, 1978) along the Kuhbanān range-front and reverse fault and cross reverse earthquakes along the intermountain thrusts (1854, 1864, 1897, 1984, and 2005) have been documented as interaction and triggering between two diverse fault sets. The pattern indicates the strain partitioning along the Kuhbanān Mountains of southeast Iran.

The limited period of 160-year seismic data along the Kuhbanān fault system is insufficient to place bounds on the recurrence behavior of large-magnitude earthquakes. Furthermore, the available seismic data may underestimate the occurrence of maximum credible earthquake on the individual fault segments, which will release virtually all of the accumulated strain energy (Berberian, 2005).

A review of the very brief seismic history along the Kuhbanān fault also shows that M 5.0–6.4 earthquakes left several seismic gaps in between. These gaps have not been the site of earthquakes since at least 1875 (Figure 16.19).

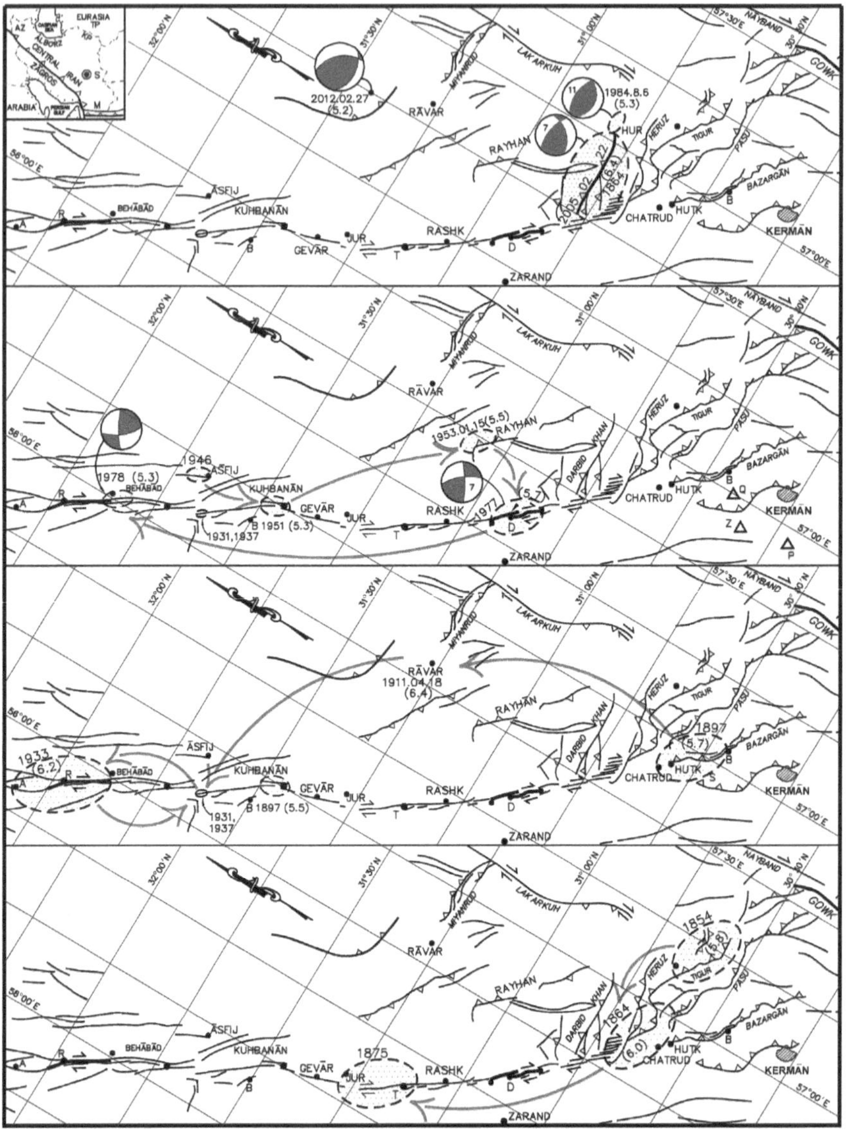

FIGURE 16.20 Sequence of the recorded earthquake pattern and rupture propagation during medium-magnitude earthquakes along the Kuhbanān right-lateral strike-slip and cross-reverse fault system. Based on data from Figure 16.19. *Modified from Berberian (2005).*

The small- to moderate-magnitude Kuhbanān earthquakes may presumably represent local readjustment of strain during interseismic cycles and have different recurrence mechanisms than those of large events. The pattern is similar to the seismic pattern observed along the northern Gowk fault prior to the

28 July 1981 M_w 7.1 Sirch earthquake (Figure 16.8). Unlike the northern Gowk fault, no maximum credible earthquake has been documented in the last 160 years of recorded seismic history of the Kuhbanān strike-slip fault. The probability of simultaneous rupturing of contiguous segments with MCE increases with the passage of time since the previous MCE of unknown date (if any). This is the result of continuing tectonic deformation, which further stresses the locked fault segments. Hence, the likelihood of an MCE along the Kuhbanān strike-slip fault, next to the provincial capital city of Kermān, with an estimated population of 535,000 (2011 census of Iran, www.amar.org.ir), is presumably higher than that along the northern segment of the Gowk fault after the 1891 earthquake (Berberian, 2005).

As with the Superstitions Hills earthquake of 24 November 1987 M_w 6.6 (Kahle et al., 1988; Hudnut et al., 1989; Sharp et al., 1989), cross-fault movement in the Kuhbanān fold-thrust mountain belt may control both the timing and extent of a major devastating earthquake along the Kuhbanān strike-slip fault toward the provincial capital city of Kermān.

16.13 CROSS-STRIKE EARTHQUAKE INTERACTION AND TRIGGERED MIGRATION

Some earthquakes showed a pattern of cross-strike propagation, in which one followed another in a time span of decades. Examples are (Figure 9.1):

- **i.** The 1 September 1968 M_s 6.3 Ferdows and 16 September 1978 M_w 7.3 Tabas-e Golshan thrust earthquakes (Figures 13.2 and 13.8).
- **ii.** The 1493 M_s 7.0 Nauzād and 1549 M_s 6.7 Birjand thrust earthquakes (Figure 11.10).
- **iii.** The 14 March 1998 M_w 6.6 Fandoqā strike-slip and Shahdād thrust case with strain partitioning (Figures 14.2 and 14.3, 16.8–16.10).
- **iv.** The 4 August 2003 M_w 5.6 and 24 August 2003 M_w 5.9 earthquakes along the Kahurak fault east of the Lut Block, with the 26 December 2003 M_w 6.6 Bam earthquake on the western side of the Lut Block (Figures 14.6, 14.7, and 14.10).
- **v.** The 2 July 1840 M_s 7.4 Ārārāt and 24 November 1976 M_w 7.0 Chālderān right-lateral strike-slip earthquakes along the Western Āzarbāijān Shear Zone (Figure 13.6).

Other examples are (i) the 17 June 1929 M 7.6 Murchison and 24 May 1968 M 7.1 Inangahua earthquakes in northwest Nelson, New Zealand (Henderson, 1937; Lenson and Suggate, 1968; Berryman, 1980); (ii) the 9 February 1971 M_w 6.7 Sylmar/San Fernando and 17 January 1994 M_w 6.7 Northridge earthquakes in the San Fernando Valley, California (Sharp, 1976; Tsutsumi and Yeats, 1999); (iii) the earthquakes along the San Andreas and San Jacinto faults, California (Sanders, 1993); and (iv) the 15 January 1944 M 7.4 San Juan and 23 November 1977 M 7.4 Caucete earthquakes in the Pampean

Ranges of northwest Argentina (Volponi, 1979; Bastias, 1985; Smalley et al., 1993; Costa et al., 2000). These examples occurred in thick-skinned range-and-basin provinces, where, apparently, earthquakes on one fault stimulate earthquakes on the other fault (Sanders, 1993; Yeats et al., 1997; Berberian and Yeats, 1999).

16.13.1 The 1968 Ferdows and 1978 Tabas-e Golshan Thrusts Case

Ten years after the 1 September 1968 M_w 6.3 Ferdows thrust earthquake (which itself was triggered a day after the 1968 M_w 7.1 Dasht-e Bayāz earthquake), the 16 September 1978 M_w 7.3 Tabas-e Golshan thrust earthquake ruptured the entire foothill-front of the Shotori Mountains, about 140 km southwest of and across strike from the Ferdows thrust (Figures 9.1, 13.2, and 13.8). The Tabas-e Golshan earthquake might have been triggered by the Ferdows earthquake (Berberian and Yeats, 1999).

16.13.2 The Kāzerun and Karéhbas Strike-Slip Faults Case

The 1986, 1988, 1989 earthquake sequence along the Kāzerun right-lateral strike-slip fault in the Zāgros was followed by the 1992, 1993, and 1994 earthquake cross-strike along the Karébas strike-slip fault (Figures 9.1, 16.1, and 16.2).

16.13.3 The 10 January 1493 $M_s\sim$7.0 and 15 February 1549 $M_s\sim$6.7 Birjand Thrusts Case

About 56 years after the 10 January 1493 $M_s\sim$7.0 Nauzād earthquake along the Nauzād reverse fault, the 15 February 1549 $M_s\sim$6.7 Birjand earthquake took place along the Birjand thrust located about 40–60 km to the west of the former (Figures 9.1 and 11.10).

16.13.4 The Neyshābur 1209, 1270, 1389, and 1405 Thrusts Case

The case of the clustered earthquakes in the Neyshābur region (discussed above), which accounts for all moment release in 196 years since the seventh century AD, may be considered as cross-strike earthquake pairs (Figure 16.14).

16.13.5 The 30 December 1863 $M_s\sim$6.1 Hir and 22 March 1879 $M_s\sim$6.7 Bozqush Case

Sixteen years after the 30 December 1863 $M_s\sim$6.1 Hir earthquake along the Sangāvar fault, the 22 March 1879 $M_s\sim$6.7 Bozqush earthquake took place about 70 km to the southwest of the former event either on the Germirud or the South Bozqush fault (Figure 11.14).

16.13.6 The 1810, 1871, 1872, 1923, 1929, 1948, 1997, and 2000 CKDSZ Case

See Section 16.11.2 and Figures 16.17 and 13.15.

16.13.7 The Kāshān and Joshaqān Faults Case

See Figures 9.1 and 16.4.

16.14 ALONG-STRIKE MIGRATION OF SEISMICITY

Along-strike strain migration of large crustal earthquakes has been documented along plate boundary faults. Examples include the westward migration of seismicity along the North Anatolian strike-slip fault zone since 1939 (Barka, 1996; Stein et al., 1997), and northward progression on the Japanese Honshu Island since 1923 (Mogi, 1985). The following are examples from the Iranian Plateau (Figure 9.1):

- **i.** Northwestward progression of seismicity along the NTF (1641–1786; Section 16.10.2; Figure 16.7).
- **ii.** Westward migration of seismicity along the Moshā fault (1665–1830; Section 16.10.4; Figure 16.11).
- **iii.** Migration of seismicity along the Zāgros Main Recent fault (Section 16.10.5; Figure 16.12).
- **iv.** Northward seismic migration along the Sangāvar (1863–1896; Figure 11.14).
- **v.** Northwestward migration along the Kāshān (1755, 1778, 1890, 1893; Figure 16.4);
- **vi.** Migration along the Karébas fault (1992, 1993, Figure 16.2).
- **vii.** Migration of seismicity along the Tālesh faults (1970, 1978, 1980).
- **viii.** Along the Larzaneh fault (1301, 1957; Figure 12.10).
- **ix.** The Bu'in Earthquake Sequence Propagation: There seems to have been a westward propagation of the recorded seismic activity clustered in the Bu'in area of south Qazvin since 1876. The 20 October 1876 M_s 5.7 Kaleh Dareh earthquake (Figure 12.13) took place in the east along the Eshtéhārd thrust (Berberian et al., 1983; Berberian, 1994). This event might have brought the Ipak fault closer to rupture in 1962, as 86 years later, the 1 September 1962 M_w 7.0 Bu'in earthquake ruptured the Ipak oblique-slip reverse fault to the west of the 1876 event (Kaleh Dareh village was destroyed during the both earthquakes; Figure 12.13). The third earthquake in the Bu'in sequence took place 40 years later with the 22 June 2002 M_w 6.4 Changureh earthquake (Walker et al., 2005b) to the west on the Ābdarreh reverse fault (Figure 14.5). The whole sequence released the stored elastic energy in three consecutive steps along a nearly E–W area of 150 km long within a period of 126 years. Apparently, like

the domino effect, each energy release transferred stress to the west from the Eshtèhārd thrust (1876) to the Ipak reverse fault (1962) and then to the Ābdarreh reverse fault (2002). The historical monuments of the area show a minimum seismic quiescent period of 935 years in the area of the 2002 event in the west (Figures 12.13 and 14.5).

Due to the low number of earthquakes along each of the aforementioned faults on the Iranian Plateau and a lack of detailed studies, it is not clear if the few directed progressions are part of the physical mechanism of the faults that will be continued in the future or not.

16.15 POSSIBLE LESSER-KNOWN CLUSTERS

Less-known possible clustered earthquakes need further investigation:

i. The 734, 806, and 816 Sistān (possibly Zarang) earthquakes.
ii. The 863, 893, and 906 Dvin/Hāyotz-dzor earthquakes.
iii. The 1844, 1863, 1879, 1883, 1896 Ardébil/Miyāneh earthquakes.

16.16 FAULTS WITH GAPS IN THE HISTORICAL SEISMIC RECORDS

Active faults such as the Doruneh, Nāyband, Sabzvārān, Dehshir, Anār, Jorjāfk, West Neh, East Neh, Zāhedān, and many more have no record of historical and instrumental period large-magnitude earthquakes (Figure 9.1). Of course, a lack of historical seismic record does not indicate a lack of seismic activity along these faults. Gaps in the historic records of earthquakes could either be due to their location in remote, arid, sparsely populated areas or a long recurrence period. If earthquake recurrence intervals on these faults are several thousand years long, then we conclude that they have been accumulating strain because the plate tectonic slip deficit has not been relieved by earthquakes (Berberian and Yeats, 1999). As with the 26 December 2003 M_w 6.6 Bam earthquake, which occurred along the Bam fault with no historical and modern earthquake record, these active faults will pose a major threat in the future.

16.16.1 The Doruneh Fault

The Doruneh left-lateral strike-slip fault of 600 km length in central east Iran (Berberian, 1976b,e) has not produced a large-magnitude earthquake during the recorded historical or instrumental period (Figure 9.1). Medium-magnitude earthquakes on 25 September 1903 M_w 6.0 West Turshiz (possibly to the south of the fault), 25 May 1923 M_w 5.9 Kāj Derakht, 9 December 1979 M_w 5.5, and 2 February 2000 M_w 5.3 occurred during the instrumental period.

Despite the lack of recorded historical earthquakes along the Doruneh fault, the survival of place names such as Shekasteh [Broken/Faulted]

Malakhi (35°08′N–59°09′E), *Shekasteh Chāhuk* (35°11′N–59°14′E), *Shekasteh Galéhband* (35°06′N–59°34′E), *Shekasteh Gushmir* (to the south of the latter), and *Kuh-e Kamar Rikhteh* (NW of Kāshmar) indicate an occurrence of historical earthquakes along the fault.

A Holocene left-lateral slip rate of ~2.5 mm/year, with average recurrence time of ~2000 years between large-magnitude earthquakes along the Doruneh fault were estimated (Fattahi et al., 2006, 2007). A GPS block model yields ~2 mm/year of left-lateral slip along the eastern Doruneh fault, rising to ~3.5 mm/year in the west (Mousavi et al., 2013). Pezzo et al (2012) estimated a characteristic recurrence interval between 630 and 1400 years along the fault.

In discussing the legendary character of the Kāshmar Cypress-tree, Mostaufi Qazvini (1340) wrote that: "*In the village of Kāshmar no earthquake is ever felt, although, in the various other places, of the entire neighbourhood round and about, earthquakes are common.*" This fourteenth-century statement shows that the surrounding regions were hit by historic earthquakes. The recorded earthquakes around Kāshmar are presented in Table 16.8. The earliest known earthquake is the Dughābād earthquake of May 1619, which took place about 85 km to the east of ancient town of Kāshmar, and to the south of the Doruneh fault. The Turshiz [modern Kāshmar] earthquake of 25 September 1903, which took place to the south of the Doruneh fault and to the northeast of the Ancient Kāshmar, may not have originated on the Doruneh fault. The meizoseismal area of the 1903 earthquake is located on the northern limbs of actively deforming Neogene-Pleistocene folds, possibly located on the hanging-wall of north-dipping, blind thrust fault. The 1903, 1994, and 2000 earthquakes (Table 16.8) took place along the northern limb of the aforementioned active folds. The rest of the seven post-1903 earthquakes took place to the north, northwest, and northeast of ancient Kāshmar, possibly on other active faults.

Except for the 25 May 1923 M_s 5.9 Kāj Derakht [SSW of Torbat Haydariyeh] earthquake, no medium- to large-magnitude earthquake has taken place along the Doruneh fault in the Kāshmar region since at least the thirteenth century. The 18-m-high brick tower and minaret at Firuzābād [Ancient Kāshmar mound, 17.5 km to the south of the Doruneh fault] and at Kashmer [located exactly on the Doruneh fault line] are intact since their construction in the thirteenth century AD. Therefore, the minimum return period of 630 years (Pezzo et al., 2012) may not be realistic.

16.16.2 The Nāyband Fault

With a N–S trend and right-lateral strike-slip mechanism, this fault is located on the western margin of the Lut Block (Berberian, 1976b,e; Figures 9.1 and 16.8). Despite the occurrence of numerous earthquakes along the northern Gowk fault located to the south of the Nāyband fault (Figure 16.8), the

TABLE 16.8 Recorded Earthquakes and History Time Line of the Ancient (Now Firuzābād) and the Modern Kāshmar (Turshiz, Soltānābād, Soltāniyeh) Region, to the South of the Doruneh Fault

Date	Monument	Major Events	Earthquakes	References
02 December 2000	SE Bardaskan, 17 km NE of Ancient Kāshmar, 34 km W of Modern Kāshmar. 1 killed, >10 villages damaged		Jaubuzjān [M_w 5.2]	
14 December 1994	8 km S of Kāshmar, damaged (immediate S of the 1903 meizoseismal area)		Saghābād [M_w 5.2]	
01 December 1972	34 km NW of Ancient Kāshmar. 12 houses collapsed		Khanjari [M_s 5.2]	
26 May 1971	26 km N of modern Kāshmar		Rivash [M_s 5.4]	
04 May 1940	50 km NNE of modern Kāshmar		Estāyesh [M_w 6.4]	See Figure 12.7
05 October 1933	45 km N of Ancient Kāshmar		Sebeh [M_s 6.1]	See Figure 12.7
25 May 1923	Kāj Derakht, SSW of Torbat Haydariyeh destroyed		Kāj Derakht [M_s 5.9; I VII$^+$]	Ambraseys and Moinfar (1977b), Tchalenko (1973), Tchalenko et al. (1973)
1909 Nov.		Sykes visited Turshiz (now Kāshmar); 7000–8000 inhabitants.		Sykes (1911)
25 September 1903	Ayvān of mosque collapsed; bricks detached from the parapets of the mosque; minaret slightly damaged	Turshiz damaged, 100 killed.	Turshiz [M_s 5.9; I VIII]	Almazov (1905), Tchalenko (1973), Tchalenko et al. (1973), Ambraseys and Moinfar (1975), Ambraseys and Melville (1982)

TABLE 16.8 Recorded Earthquakes and History Time Line of the Ancient (Now Firuzābād) and the Modern Kāshmar (Turshiz, Soltānābād, Soltāniyeh) Region, to the South of the Doruneh Fault—Cont'd

Date	Monument	Major Events	Earthquakes	References
1894		Turshiz has 6000 inhabitants and 1200 dwellings covering 0.5 km^2, w/80–120 villages.		Baumgarten (1896)
1880 Nov.		Soltānābād (Turshiz) flourishes w/5000 inhabitants.		Stewart (1881)
1857		The town is in ruins w/ruined remains of villages.		Clark (1861)
Early 1800		Uzbeks plundering the villages.		Malcom (1829)
~1822		Turshiz w/3000–4000 inhabitants is in ruins.		Fraser (1825)
1783 Dec.		Foster visited Turshiz.		Foster (1798)
1736–1739	Turshiz (Modern Kāshmar)	Present Turshiz founded.		Sykes (1911)
1619 May	Dughābād [35°05′N–38°51′E; 40 km SE of modern Kāshmar and 85 km E of ancient Kāshmar]	Dughābād region destroyed.	Dughābād [*I* VIII, M_s 6.5]	
1381	Ancient Kāshmar	Besieged by Tamerlane hordes. Ancient Turshiz (now Firuzābād village) ruined.		Sykes (1911)
Thirteenth century	'Aliābād/Keshmar Tomb Tower [32 km WNW of Kāshmar; 35°17′N–58°06′E] (18 m high tower)	The tomb Tower was not damaged during the 1903 earthquake.		Sykes (1911), Wilber (1955), Meshkāti (1970)

Continued

TABLE 16.8 Recorded Earthquakes and History Time Line of the Ancient (Now Firuzābād) and the Modern Kāshmar (Turshiz, Soltānābād, Soltāniyeh) Region, to the South of the Doruneh Fault—Cont'd

Date	Monument	Major Events	Earthquakes	References
Thirteenth century	Firuzābād (Ancient Kāshmar) Mināret/Watch Tower [49 km SW of modern Kāshmar] (18 m high; top is missing); and Fort	The Minaret was not damaged during the 1903 earthquake.		Sykes (1911), Meshkāti (1970)
1256	Ancient Kāshmar	Onslaught of Hulagu Monqol.		Sykes (1911)
1218	Ancient Kāshmar	Invasion of the Mongol Hordes.		Sykes (1911)
Twelfth century	Ancient Kāshmar	The district was seized by the Isma'ilis.		Sykes (1911)
861 winter	Ancient Kāshmar	Cypress-tree was cut by the order of Caliph al-Mutawakkil 'Abbāsid.	Sudden lack of earthquake immunity in Kāshmar	Mostaufi Qazvini (1340). See the text (Section 3.1.1)
Pre-861	Ancient Kāshmar		Earthquakes around Kāshmar, but not in Kāshmar	Mostaufi Qazvini (1340). See the text and Section 3.1.1
636–642 CE	Destruction of the fire temples and infrastructure	Invasion of Moslem Arabs.		
?	Fort of Barkala in Kondor [28 km WSW of Kāshmar]; ruined			Sykes (1911)
?	Shahr-e Sefid [White City], Doruneh; ruined			Sykes (1911)
?	Qal'eh Ātashgāh [Fire temple Fort]			Sykes (1911)
330 BCE		Invasion of Alexander III of Macedonia.		330 BCE

Nāyband active fault does not have a recorded large-magnitude historical earthquake along its 400 km length. The fault may have the potential to generate a magnitude >8.0 earthquake if the entire length of the fault ruptures with a single earthquake (Berberian and Yeats, 1999); however, no intracontinental earthquake of such magnitude has yet been documented in Iran. A geological offset of ~12–15 km and a slip rate of ~1.4 ± 0.5 mm/year (~0.9–1.9) averaged over ~2.25 Ma is estimated along the Nāyband fault (Walker et al., 2009).

16.16.3 The Sabzvārān Fault

This 150-km-long active fault (Berberian, 1976b,e) is located to the east of the southeastern Zāgros in southeast Central Iran (Figures 9.1 and 14.11). A measured Holocene slip rate of 4.0–7.4 is estimated along the fault (Regard et al., 2005) in an area with the GPS slip rate of 0.6–5.6 mm/year (Bayer et al., 2006).

16.16.4 The Dehshir fault

The Dehshir right-lateral strike-slip active fault (Berberian, 1976b,e, 1981a) is the westernmost prominent NNW–SSE -striking right-lateral faults of Central Iran with a total length of about 450 km (Figure 9.1). A geological offset of 65 ± 15 km (with inferred fault inception of ~20 Ma) is measured along the fault resulting in a very long-term slip rate of ~2.5–4.0 mm/year (Meyer et al., 2006). A 2.0 mm/year right-lateral short-term Holocene strike-slip rate (Meyer et al., 2006; Meyer and LeDortz, 2007) and 0.9–1.5 mm/year (Le Dortz et al., 2012) are estimated for the Dehshir fault.

A preliminary paleoseismic trench study showed three earthquakes; the oldest two events apparently took place over the last 10–30 ka, and the last event at 2.8 ± 1.4 ka (Nazari et al., 2009). Unfortunately, the dates of the events are not properly constrained. Fattahi et al. (2010) estimated that the most recent earthquake causing slippage along the Dehshir fault took place around 2.0 ± 0.2 ka.

The exact location of the 8 August 848–27 July 849 destructive Herat (Harāt) earthquake reported by Ibn al-Jauzi (1181) and al-Suyuti (1499) is not known. It either took place at Harat along the Dehshir fault in modern Iran, or at the town of Herat (Harāt) in modern Afghanistan. Mickey Kādjār (personal communication, 4 February 2013) indicated a totally destroyed Seljuk/Ilkhanid [1000–1334] edifice to the west of the Dehshir village and the fault, where only the basal structure and pottery shards remained. The cause of complete destruction of this structure is not known.

16.16.5 The Anār Fault

The ~200-km-long, N–S -striking, active, right-lateral strike-slip Anār fault (Berberian, 1976b,e) is located in Central Iran subparallel to the Dehshir fault to its west and Kuhbanān fault to its east (Figure 9.1). A measured geological offset of 25 ± 5 km with an inferred fault inception of ~20 Ma (Meyer and LeDortz, 2007; Le Dortz et al., 2009) is reported. The fault also has an inferred, very long-term slip rate of ~1.0–1.5 mm/year and a short-term Holocene slip rate of 0.5–0.75 mm/year (Meyer and LeDortz, 2007), with an estimated minimum slip rate of 0.8 ± 0.1 mm/year (Le Dortz et al., 2009). Later, the slip-rate estimate was reduced to ~0.55 (Fattahi et al., 2011). A preliminary paleoseismic trench study showed three earthquakes within 9.8 ± 2.0, 6.8 ± 1.0, and 4.4 ± 0.8 ka (Foroutan et al., 2012). The dates are not properly constrained and details about these events are not known.

16.16.6 The Jorjāfk fault

The ~180-km-long, NW–SE -trending fault, which was first identified as a major active fault (Berberian, 1976e; Berberian et al., 1984), is located to the southwest of the Kuhbanān fault and runs subparallel to it (Figure 9.1). Despite its young active fault features and its location between two major cities of Kermān and Rafsanjān, no historical or instrumental period earthquake is recoded along the fault.

16.16.7 The Rafsanjān Fault

The 200-km-long fault (Figures 9.1 and 12.3), with a right-lateral strike-slip motion, has a NW–SE strike and is located to the south of the Anār and southwest of the Jorjāfk faults (Berberian, 1976b,e, 1981a; Berberian and Qorashi, 1984, 1989a; Berberian et al., 1984). A slip rate of ~0.4 mm/year is estimated for this fault (Fattahi et al., 2011).

Archeoseismic study of numerous archeological mounds (including Tal-e Eblis and Ghobayrā) in the area together with paleoseismic trench study can help us understand the past history of the Rafsanjān fault. Tal-e Eblis (Iblis) mound (Figure 12.3) is the oldest recorded copper smelting complex on the Iranian Plateau, dating approximately from 5500 to 3500 BCE (Caldwell and Malek Shahmirzadi, 1966; Caldwell, 1967, 1968; Ehrich, 1992). The site is located about 10 km to the north of the southeastern segment of the Rafsanjān fault system. During the 22 September 1923 M_w 6.7 Lālēhzār earthquake, the Tal-e Eblis site underwent a seismic event of intensity VII (MMI) (Figure 12.3).

It is reported that probably shortly after the "Iblis V Period" (ca. 2750–2500 BCE), the old channel of the Lālēhzār (Āb Bakhshā) River ceased to be a major source of water and became abandoned (Caldwell, 1967, 1968; Chase et al., 1967; Hume, 1967). The reason for the abandonment is unknown and has not been investigated. Tal-e Eblis is located to the east and downstream of the Āb Bakhshā Riverbed (Figure 12.3). About 10 km south-southwest of Tal-e Eblis, the Lālēhzār River at the upgradient crosses the southeastern segment of the Rafsanjān, Ākhurak, Morghāb, and Chamanrang active faults with very young deformation clearly visible on the 1955 aerial photographs (Worldwide Aerial Surveys, Inc., Project 158, scale 1:55,000). Moreover, local tradition favors that some generations ago landslides triggered by a past earthquake buried the wealth of villages and formed the present *Tal-e Zelzeleh* [lit., "the Earthquake Mound"], beneath which lie buried treasures (Mickey Kādjār interviewed the inhabitants of a small village located about 10 km to the north of Bardsir road crossing the Rafsanjān fault; personal communication, 7 December 1998).

The Ghubayrā mound (third millennium BCE to CE) is another promising mound for archeoseismic investigation located to the east of Tal-e Eblis (Figure 12.3). The excavations at Chahārdaru's (Site D) circular tomb tower (Phase 1) of the mound (eleventh/twelfth century CE) revealed rows of a brick structure (part of an old arch, arches, or dome) fallen in one direction (Pl. 27b in Bivar, 2000). Bivar (2000) wrote that the fallen structure with larger bricks was not part of the standing structure (eleventh/twelfth century CE), but must have belonged to an earlier structure of which only the foundations have been preserved. The cause of the fallen old brick structure is not known.

16.16.8 The West Neh Fault

The active West Neh, right-lateral strike-slip fault (Berberian, 1976b,e) is located on the eastern margin of the Lut Block and within the highly deformed Sistān accretionary prism (Figures 9.1, 11.13, and 14.10). A geological offset of ~10 km (possibly since ~5 Ma) is measured along the fault (Walker and Jackson, 2004), with an inferred Holocene rate of ~1.0–5.0 mm/year (Meyer and LeDortz, 2007).

16.16.9 The East Neh Fault

As with the West Neh fault, the active East Neh right-lateral strike-slip fault (Berberian, 1976b,e) is located on the eastern margin of the Lut Block and within the highly deformed Sistān accretionary prism (Figures 9.1 and 14.10). A geological offset of ~50 km is measured along the fault (Walker and Jackson, 2004), with a slip rate of ≳1.2 mm/year, averaged over ~1.7 Ma estimated along the fault (Walker et al., 2009). Meyer and LeDortz (2007) estimated a Holocene slip rate of ~1.75-2.5 mm/year for the East Neh fault. The 8 March 1928 M_w 5.7 Nehbandān earthquake, which destroyed Nehbandān (located near the West Neh and 12 km to the west of the East Neh faults), was possibly generated along the East Neh fault (Figures 11.13 and 14.10).

16.16.10 The Asagie Fault

A Holocene (~12 ka) slip rate of ~1.0–2.5 mm/year is estimated for this fault (the southern continuation of the East Neh fault; Figure 9.1) located to the east of the Lut Block (Meyer and LeDortz, 2007).

16.16.11 The Sarvestān Fault

This right-lateral strike-slip fault (Berberian, 1976b,e, 1995) is located to the east of the Kāzerun and the Sabzpushān strike-slip cross faults in the central Zāgros (Figure 9.1; also see figure 12, p. 211 in Berberian, 1995). The remains of the Sassanid Palace (420–440 CE) and other ruins scattered in the surrounding fields (see figures 3 and 4 in Bier, 1986) are located about 12 km to the south of the town of Sarvestān near the fault. The cause of the destruction of the surrounding structures is not known.

16.16.12 The Zāhedān Fault

This N–S -trending, right-lateral strike-slip fault (Berberian, 1976b,e; Berberian et al., 2000b) is located near the Afghanistan border and within the highly deformed Sistān accretionary prism in southeast Iran (Figures 9.1 and 15.1). No historic or instrumental period earthquake is recorded along the fault, however, the NW–SE splay, thrust fault to its northern end became active during the 1994 Sefidābeh earthquake sequence (Figures 15.1 and 15.4; Berberian et al., 2000b). A geological offset of ~13–20 km (possibly since ~5 Ma) was measured along the fault (Walker and Jackson, 2004), and an inferred Holocene slip rate of 1.75–2.50 mm/year is estimated along the fault (Meyer and LeDortz, 2007).

16.16.13 The Mināb-Zendān Fault System

This right-lateral fault borders the southeastern Zāgros fold-and-thrust mountain belt and the western Makrān accretionary wedge (Figure 9.1; Berberian, 1976b,e). No historical earthquake is documented along this fault system. The epicenter of the 7 February 1983 M_w 6.0 event, with a synthetic waveform showing east-dipping thrust mechanism with right-lateral slip (Ni and Barazangi, 1986), is located between the coast line and the southern section of the Zendān fault; no macroseismic data are available for this event. The fault possibly accommodates 5.6 ± 2.3 or 7.4 ± 2.7 mm/year of slip (Regard et al., 2004, 2005).

The Old Hormoz Port [Shahr-e Kohneh-ye Hormoz/old city of Hormoz/Hormozia/or Moghestān; on the mouth of the Mināb River, the main port of Sistān and Kermān] on the mainland (Morgan, 1991) was located about 10 km to the southwest of the modern town of Mināb, 11 km west of the Mināb fault, and 15 km west of the Zendān fault. The reason for the decline and abandonment of the Old Hormoz Port on the mainland is not clear. The port was described by Estakhri (951), Moqqadasi (985), and Idrisi (1154) and was visited by Marco Polo in 1272 and 1293. Ibn Batuta (1355) described both the Port (Old city of Hormoz on the mainland) and the city of Hormoz on Hormoz Island (45 km to the west of the Old Hormoz port).

16.17 OVERVIEW

The Iranian Plateau is a diffuse plate-boundary zone of active strike-slip and reverse faults, analogous to southern California. Unlike California, we can draw on several thousand years of historical and archeoseismic data to work out the earthquake histories and faulting patterns in Iran. Analysis of the historical earthquake sequences show that earthquakes were triggered by other earthquakes, with interplay among left-lateral, right-lateral, and reverse faults.

The long-term seismic pattern and active faulting behavior of the Iranian Plateau can be categorized as follows:

- **i.** Temporal active seismic cycles (clustered earthquake sequence) followed by a long period of seismic quiescence in a densely populated, subparallel fault zone (Neyshābur, 1209–1405; Quchān, 1833–1895; Tehrān, 855–958; Dasht-e Bayāz, 1968–1997).
- **ii.** Segmented rupturing and along-strike migration of seismicity of a single fault zone (North Tabriz, 1641–1786; Ardal, 1666–1977; Moshā, 958–1665–1830; Kāshān, 1755–1778; Kāzerun, 1824–1986; Zāgros Main Recent fault, 1909–2006; Dasht-e Bayāz, 1968–1997; Ipak, 1876–2002; Surmeh, 1968–1972; Kuhbanān, twelfth century–1978; Khoi-Chālderān, 1843–1976; Makrān Subduction zone, 1483–1954; Karéhbas, 1992–1993; Sabzpushān, 1752–1994).
- **iii.** Complete rupturing of a multisegmented single fault within a short time period of decades to hundreds of years (Sangāvar, 1863–1896; Gowk, 1877–1981; Zirkuh, 1936–1997).
- **iv.** Seismic migration and interaction along multiple parallel faults (15 August 1485 Upper Polrud, 20 April 1608 Alamutrud, and 20 June 1990 Rudbār earthquakes).
- **v.** Interplay between strike-slip and reverse faults within the seismogenic layer; transverse reverse faulting off the main longitudinal strike-slip fault (Tabriz, 1641; Ferdows, 1968; Manjil-Rudbār 1983, 1990 and its aftershock sequences; Kuhbanān, 2005).
- **vi.** Contrasting surface active faults and deeper seismogenic sources (Gowk 28 July 1981 and 14 March 1998).
- **vii.** Slip partitioning onto separate reverse and strike-slip faulting (Rudbār, 1983 and 1990 and its aftershocks; Nayshābur; etc.).
- **viii.** Active conjugate strike-slip faulting (South Rigān, 2010–2011).
- **ix.** Major strike-slip fault with no recorded great earthquakes (Nāyband; Doruneh, etc.).
- **x.** Aseismic slip on the Shahdād thrust (14 March 1998–14 September 1999; post-slip of the 14 March 1998 M_w 6.6 Fandoqā earthquake on the Gowk fault).
- **xi.** Blind-reverse basement faulting (Kārzin, 1972; Mishān, 1972; Khurgu, 1977; Sefidābeh, 23–28 February 1994; and many more).

xii. The Central Kopeh Dāgh and the Western Āzarbāijān Shear Zones: The NW–SE -striking subparallel right-lateral strike-slip faulting of north-western (1810, 1871, 1872, 1923, 1929, 1997, 2000; Berberian, 1997a) and northeast Iran (363, 1319, 1696, 1840, 1843, 1900, 1976, 1977). These zones show strain partitioning and clockwise rotation of blocks (Jackson, 1992a,b; Copley and Jackson, 2006; Hollingsworth et al., 2006).

Chapter 17

Earthquake History of Iran

They sow the Wind,
And they shall reap Whirlwind.

Hosea, 8:7

Earthquakes have created severe humanitarian crises and tragedies throughout the history of Iran, and critical, emergency-response issues have never been properly addressed (See Introduction and Chapters 1–8). These disasters have always generated short periods of attention for discussions of earthquake risk and relocation plans for affected areas for rich countries with plenty of resources (money, engineers, scientists, and planners), where people and investments are concentrated in settlements located along active fault zones (Chapters 9–16). A few months after each disaster-causing medium- to large-magnitude earthquake, earthquake-risk-minimization plans and awareness by news media, authorities, and citizens are forgotten. Even complete destruction of urban and rural areas has not stimulated any serious action in reinforcing buildings, retrofitting existing public structures, or avoiding construction of buildings on active faults in cities like Tehrān, Tabriz, and Neyshābur. Since ancient times, natural disaster hazards have been disregarded; earthquakes have officially been accepted as acts of God, and people believe that to minimize their risks is beyond the power of the citizens and the state (Chapters 2 and 3). Hence, improvements in building materials, workmanship, construction methods, and enforcement of seismic codes have never been seriously implemented (Introductions and Chapters 11–16).

The 2003 M_w 6.6 Bam earthquake (Figure 14.6), with a death toll ranging from 26,500 to 43,200 in the first years of the twenty-first century, is an unfortunate and unacceptable reminder of tragedies that happened centuries ago at Tabriz, Ray/Tehrān, Komesh/Dāmghān, Neyshābur, Quchān, Kāshān, Shirāz, and other major urban and rural areas of the Iranian plateau (Chapters 11–16). The large number of casualties from Iranian earthquakes (Figures 1–3 in Introduction), as well as other earthquake-prone developing countries throughout the world, has been mainly due to: (i) the immediate, total

☆"To view the full reference list for the book, click here"

destruction of adobe and other improperly built buildings and a lack of reinforcement of aseismic building codes or any proper control; (ii) time of the earthquake (day or night); (iii) a delay in emergency action and lack of plans for a proper and immediate search-and-rescue operation following an earthquake; (iv) demography; (v) underestimating geological hazards; (vi) endemic poverty and vulnerability; and (vii) corruption of irresponsible authorities.

Despite these facts, there is unfortunately no comprehensive regional way to adequately analyze and assess the potential for human casualties, injuries, and social disruptions (homeless, recovery, orphans) from future earthquakes. In most earthquakes, the high death tolls and total destruction of buildings reflect the extremely poor construction of the buildings rather than high intensity. Furthermore, because of the vulnerability of the buildings, there is no private earthquake insurance available and the governments are not interested in earthquake-related business.

17.1 POPULATION DENSITY AND CHANGES THROUGH TIME

Population density and its changes over time have considerable bearing on the extent of the recorded material available about earthquakes from which we can make assessments. For certain areas, such as the central zone of the Dasht-e Kavir (Central Desert salt playa in northern Central Iran) and Dasht-e Lut (the Lut Desert, in the east) there is a lack of evidence of early seismic information. However, this lack should not be interpreted to mean that these areas have been free from large-magnitude earthquakes. Obviously, there have been fewer casualties in sparsely inhabited regions than in populated areas. For example, the 23 September 1947 M_s 6.8 Charmeh earthquake in a remote semiarid area killed about 500 people (Figure 12.8), while the 20 December 2010 M_w 6.7 South Rigān earthquake, which occurred in the remote and arid areas of southeast Iranian with an extremely low population density, killed only six (Figure 14.11).

However, this has not been always a hard and fast rule. The 31 August 1968 M_w 7.1 Dasht-e Bayāz earthquake (Figure 13.1) occurred in a rather isolated and thinly populated rural area on the abandoned old trade route to Baluchestān and to Bandar ‘Abbās port. The earthquake affected an area of 10,000 km^2 with a population density of just under nine people per km^2 (about one-third the average density of Iran in 1968). Nonetheless, the earthquake killed about 10,000 people. Over 12,000 housing units were destroyed or damaged beyond repair, and 180 villages were affected, leaving over 70,000 people homeless (Ambraseys and Tchalenko, 1969). Had the earthquake happened in an urban area, the casualties would have been doubled, if not quadrupled.

17.1.1 Historical (Pre-1900) Demography

Knowledge about the pre-1900 population of Iran is very limited and based mostly on documentation by historians and travelers, which is itself based on the "guesses" of local inhabitants. Ya'qubi (892) reported that 200,000 people were killed during the 856 Komesh (modern Dāmghān) earthquake (Figure 11.3). Later, Tabari (915) wrote that 45,096 people were killed by the earthquake, mostly in the city of Dāmghān. It is important to remember that the population of the city of Dāmghān during its zenith in the first half of the eleventh century was about 25,000. During the Timurid (1370–1502) and Safavid (1491–1722) eras, the population dropped to about 2000–3000. The population climbed to 5000 in 1929; 8900 in 1956; 13,175 in 1966; 25,500 in 1984; 34,057 in 1986; and 58,770 in 2011 ('Adle, 1971, 1993, 2001; amar.org.ir). We include these population estimates as evidence that the number of casualties—40,000 in the city and 200,000 in the area—in 856 has been highly exaggerated and cannot be true (Berberian and Yeats, 2014).

17.1.2 Post-1900 Demography

The first modern national census of Iran was conducted in November 1956 (MI/DPS, 1961, 1962; amar.org.ir). Prior to this date, two official population records existed of: (i) an urban head count held between June 1939 and August 1941 (PO/SCI, 1967), and (ii) the Civil Registration Office, which started operations in 1928 (Bharier, 1968, 1972). Naturally, population and its density have grown in the twentieth century in Iran. Until possibly the early 1960s, the rate of investment in large urban and industrial developments in Iran was minimal, at best. From 1900 to 1926, the average rate of population growth was less than 1%. After World War II, the rate accelerated. Between 1941 and 1956, the annual rate of population growth averaged 2.2%. Between 1956 and 1966, the average annual rate of growth increased to 3.1%. The growth rate decreased to 2.7 in 1976, but after the regime change, it increased to 3.8% in 1986 (Bharier, 1968, 1972; Aghajanian, 1991; sci.org.ir; amar.org.ir).

During the first half of the twentieth century, the population of Iran doubled from ~10 million (in 1900) to ~20 million (in 1955). During the regime change in 1979, the population was about 35 million (sci.org.ir; amar.org.ir). By 2006, the population had doubled to 70.5 million (Table 17.1). Within 40 years, the urban population had increased from 6 million (in 1956) to 37 million (in 1996) (with a building boom in the cities, especially Tehrān); the rural population increased from 13 million (in 1956) to 23 million (in 1996) (Tables 17.2 and 17.3). Hence: (i) a sixfold increase in the population of Iran took place only in one century, and (ii) a sixfold increase in rural population took place within the last 40 years of the twentieth century (Hourcade, 1994; SCI, 2004; sci.org.ir; amar.org.ir).

TABLE 17.1 Post-1900 Population of Iran

Year (Christian/Persian)	Estimated Population (in Millions) (Bharier, 1968)	Official Census (in Millions) (SCI)	Average Annual Growth Rate During the Preceding Period (%) (Aghajanian, 1991)	Population Density (km^2) (SCI)	Literacy (% of Population) (SCI)	Major Political Event
1900/1279	9.8	–	–	6.0		
1901/1280	9.9					D'Arcy's Colonial Oil Concession (1901–1961) given by Mozafar al-Din Shāh Qājār, an inept and corrupt Iranian despot
1905/1284	10.2	–				Iranian Constitutional Revolution (1905–1911)
1906/1285	10.3	–				First Iranian Parliament established
1908/1287	10.4					First Iranian oil well hit oil
1909/1288	10.5	–			1–5% (Amirahmadi, 2012)	Ahmad Shāh Qājār (r. 1909–1925) Russian occupation of Tabriz The Anglo-Iranian Oil Co. (APOC) was established
1910/1289	10.6	–				
1911/1290	10.7	–				War with Russia. Russians temporarily occupied Tehrān
1916/1295	11.0	–				

1920/1299	11.4	–			Soviet occupation of Guilān
1925/1304	11.8	–			Rezā Shāh Pahlavi (r. 1925–1941)
1926/1305	11.9	–			
1927/1306	12.0	–	0.8		
1930/1309	12.6	–			
1935/1314	13.5	–	1.5		
1936/1315	13.7	–			
1940/1319	14.5	–		8.8	
1941/1320	14.8	–	1.5		Anglo-Soviet Occupation of Iran (1941–1946) Abdication and Deportation of Rezā Shāh Pahlavi by the UK Mohammad Rezā Shāh Pahlavi (r. 1941–1979)
1946/1325	16.0	–			
1950/1329	17.6	–			
1951/1330	18.0				Iranian oil nationalization by Premier Dr. Mosaddeq, Esq.
1953/1331	18.9				CIA/MI6 Coup d'état and overthrow of the democratic government of Dr. Mosaddeq, Esq.

Continued

TABLE 17.1 Post-1900 Population of Iran—Cont'd

Year (Christian/ Persian)	Estimated Population (in Millions) (Bharier, 1968)	Official Census (in Millions) (SCI)	Average Annual Growth Rate During the Preceding Period (%) (Aghajanian, 1991)	Population Density (km^2) (SCI)	Literacy (% of Population) (SCI)	Major Political Event
1954.08.05	–	–	–	–	–	Signing of the 25-year oil extraction agreement (1954–1979) with British Petroleum
1956/1335	20.3	18.9	2.2	11.5	14.9	
1960/1339	22.9	–				
1966/1345	27.0	25.0	3.1	15.4	29.4	
1976/1355	–	33.7	2.7	20.5	47.5	
1978/1357	–	–	–	–	–	Negotiations between the Shah's government and British Petroleum for renewal of the 25-year-old extraction agreement (signed in 1954) collapsed The British government-owned BBC Persian language program gave Ayatollah Khomeini a full propaganda platform inside Iran

1979/1358	–	–					Guadeloupe Summit Meeting (1979.01.05) Abdication and deportation of Mohammad Rezā Shāh Pahlavi (1979.01.19) Establishment of Islamic regime, resulting in capital flight, a serious brain drain, and 8-year war with Iraq
1980–1988							Islamic regime of Iran–Iraq War with staggering damages of US$ 650 billion (1980.09.24–1988.08.25)
1986/1365	–	49.4	3.8		30.3	61.7	
1996/1375	–	60.0					
2006/1385	–	70.5					
2009							Mass protest of people against fraudulent presidential election of the Islamic regime of Iran
2011/1390		75.1					

Bharier (1968), Aghajanian (1991), Statistical Centre of Iran (SCI), Plan and Budget Organization of Government of Iran [later, Ministry of the Interior], Markaz-e Āmār-e Irān, **sci.org.ir**, amar.org.ir.

TABLE 17.2 Population Data of the Iranian Cities Based on Official Census

Town/City	1956/1335	1966/1345	1976/1355	1986/1365	1991/1370	1996/1375	2006/1385	2011/1390
Ābādān	226,083	272,962	294,068	6 [War]	84,774	206,073	219,772	212,744
Ahvāz	120,098	206,375	334,399	579,826	724,653	804,980	985,614	1,112,021
Āmol	22,251	40,076	68,963	118,242	139,923	159,092	199,698	219,915
Andimeshk	7324	16,195	32,085	56,288	73,759	106,923	120,177	126,811
Anzali [Pahlavi] Port	3134	41,785	55,481	87,063	94,697	98,544	110,643	116,664
Arak	58,998	71,925	116,832	265,349	331,345	380,755	446,760	484,212
Ardébil	65,742	83,596	147,865	281,973	310,022	340,386	418,262	482,632
Bābol	36,194	49,973	68,059	115,320	137,348	158,346	201,335	219,467
Bam	15,735	21,761	30,422					107,131
Bandar 'Abbas Port	17,710	34,627	87,981	201,642	249,504	273,578	379,301	435,751
Behbahān	29,886	39,874	49,378	78,694	85,846	88,213	101,178	107,412
Birjand	13,934	25,845	46,943	81,798	101,177	127,608	166,138	178,020
Bojnurd	19,253	31,248	47,719	93,392	112,426	134,834	176,726	199,791
Borujerd	49,186	71,486	101,342	183,879	201,016	217,804	229,541	240,654
Bukān	5307	9357	20,579	67,938	83,401	120,020	150,703	170,600
Bushehr Port	18,412	23,547	58,965	120,787	132,824	143,641	169,966	195,222
Damāvand				82,177				37,315

Dezful	52,121	84,499	121,251	151,420	181,309	202,639	235,819	248,380
Dorud	7088	14,060	27,621	62,517	77,299	88,152	101,219	99,499
Esfahān	254,708	424,045	661,510	986,753	1,122,703	1,266,072	1,602,110	1,756,126
Eslāmshar (SW Tehrān)			50,292	215,129	230,183	265,450	357,389	389,102
Ferdows					18,468	21,784	24,703	25,968
Golestān [Soltānābād]	669	827	1074	9927	45,892	112,554	231,905	
Gonābād					24,708	30,149	36,996	36,367
Gonbad Kāvus	18,347	40,667	60,721	87,100	102,768	111,253	129,167	144,546
Gorgān	25,380	51,181	88,033	139,430	162,468	188,710	274,438	329,536
Hamédān	99,909	124,167	165,785	272,499	349,653	401,281	497,640	525,794
Ilām	8346	15,493	32,476	89,035	116,428	126,346	160,355	172,213
Irānshahr	3618	5386	11,386	40,027	56,581	76,959	100,642	97,012
Izeh	1983	5115	10,257	46,042	64,072	81,288	104,364	111,093
Jahrom	29,169	38,236	48,530	77,174	88,693	94,185	105,285	114,108
Karaj	14,526	44,243	137,926	611,510	806,363	940,968	1,386,030	1,614,626
Kāshān	45,955	58,468	84,863	138,599	155,188	201,372	253,509	275,325
Kāshmar					58,522	69,177	83,667	90,200

Continued

TABLE 17.2 Population Data of the Iranian Cities Based on Official Census—Cont'd

Town/City	1956/1335	1966/1345	1976/1355	1986/1365	1991/1370	1996/1375	2006/1385	2011/1390
Kāzérun					74,048	81,713	84,594	89,685
Kermān	62,157	85,404	140,761	257,284	311,643	384,991	515,114	534,441
Kermānshāh	125,439	187,930	290,600	560,514	624,084	692,986	794,863	851,405
Khorramābād	38,676	59,578	104,912	208,952	249,258	272,815	333,945	348,216
Khorramshahr	43,850	88,536	140,490	0 [War]	34,750	105,636	125,859	129,418
Khoi	34,491	47,648	70,357	115,343	137,885	148,944	181,465	200,958
Lowshān				9056				15,193
Mahābād	20,332	28,610	44,067	75,238	81,987	107,799	135,780	147,268
Māhshar Port	15,694	16,594	29,940	71,808	74,248	88,394	111,448	153,778
Malārd	1369	2596	5970	9160	32,364	88,118	228,713	290,817
Malāyer	21,105	28,434	47,117	103,640	130,458	144,373	156,289	159,848
Manjil				11,107				17,396
Marāgheh	36,551	54,106	65,172	100,679	117,388	132,318	149,929	162,275
Marand	13,822	23,818	36,108	71,394	85,253	96,369	114,841	124,323
Marvdasht	8987	25,498	50,446	79,132	92,013	103,579	124,350	138,649
Mashhad	241,989	409,616	667,770	1,463,508	1,759,155	1,887,405	2,427,316	2,749,374
Masjed Soleymān	44,651	64,488	77,098	104,787	107,539	116,883	108,682	103,369

Miyāndoāb	14,796	18,767	27,739	59,551	71,273	90,141	114,153	123,081
Nahāvand					59,307	65,164	73,141	75,445
Najafābād	30,422	43,384	75,276	129,058	160,004	178,498	206,647	221,814
Narmāshir [Akbarābād]	67	91	202	13,750	49,945	85,124	135,846	6167
Neyshābur	25,820	33,482	59,562	109,258	135,681	158,847	208,860	239,185
Pākdasht	1704	2668	5459	18,308	33,273	49,220	126,937	206,490
Qā'en								40,226
Qazvin	66,420	88,106	139,258	248,591	278,826	291,117	355,338	381,598
Qom	96,499	134,292	247,219	543,139	681,253	777,677	964,706	1,074,036
Quchān	21,250	29,133	40,301	66,531	74,919	85,750	101,313	103,760
Rafsanjān	9212	21,425	36,025	66,498	79,926	98,257	139,219	151,420
Rasht	109,491	143,557	188,957	290,897	340,637	417,748	557,366	639,951
Ray		174,066		431,846		213,161		
Rudbār				14,834				10,926
Sabzévār	30,545	42,415	69,562	129,103	148,065	170,738	214,582	231,557
Salmās					57,441	65,416	81,342	88,196
Sanandaj	40,641	54,578	95,872	204,537	244,039	277,808	316,862	373,987

Continued

TABLE 17.2 Population Data of the Iranian Cities Based on Official Census—Cont'd

Town/City	1956/1335	1966/1345	1976/1355	1986/1365	1991/1370	1996/1375	2006/1385	2011/1390
Saqqez	12,729	17,834	30,661	81,351	98,933	115,349	133,331	139,738
Sāri	26,278	44,547	70,753	141,020	167,602	195,882	261,293	296,417
Sāveh	14,537	17,565	25,751	64,081	87,621	111,245	180,548	
Sedeh [Homāyunshahr/ Khomeinishahr]		46,836	65,495	104,647	118,348	165,888	223,071	244,696
Semnān	29,036	31,058	38,786	64,891	75,131	91,045	126,780	153,680
Shāhi [Qā'émshahr]	23,055	38,898	63,377	109,288	123,684	143,286	174,768	196,050
Shāhinshahr	4195	6089	8321	49,312	62,592	84,827	127,412	143,308
Shahr-e Kord	15,476	23,575	40,359	75,080	89,253	100,477	131,612	159,775
Shahrezā	29,311	34,220	46,956	73,367	84,666	89,779	109,601	123,767
Shahriyār	3748	6626	11,697	22,439	32,459	40,058	189,421	249,473
Shāhrud	17,058	30,767	49,783	78,950	92,195	104,765	132,379	140,474
Shirāz	170,659	269,865	425,813	848,289	965,117	1,053,025	1,227,331	1,460,665
Sirjān	12,160	19,568	39,464	90,072	107,887	135,024	170,916	185,623
Tabas-e Golshan			~13,000				30,681	39,620
Tabriz	289,996	403,413	597,976	971,482	1,088,985	1,191,043	1,398,060	1,494,998
Tajrish/Shémirānāt	26,525			40,617		30,397		

Tehrān (including Ray and Shemirānāt/Tajrish)	1,797,429	3,094,563	4,689,497	6,530,670	6,758,845	7,037,907*	7,797,520	8,154,051–11,305,832
Tehrān (excluding Tajrish and Ray)	1,512,082	2,719,730	4,530,223	6,042,584	6,475,527	6,758,845	7,088,287	8,154,051
Torbat Heydariyeh	19,830	30,106	43,259	72,068	81,781	94,647	121,300	131,150
Urumiyeh	67,605	110,749	164,419	300,746	357,399	535,200	583,255	667,499
Varāmin	5205	11,183	25,792	58,311	77,624	107,233	208,996	218,991
Yāsuj	34	931	4524	29,991	48,957	69,133	100,544	108,505
Yazd	63,502	93,241	135,925	230,483	272,298	362,776	432,194	486,152
Zābol	12,221	18,806	29,404	75,105	91,041	100,887	136,956	137,722
Zāhédān	17,495	39,732	93,740	281,923	361,623	419,518	567,449	560,725
Zanjān	47,159	58,714	100,351	215,261	254,100	286,295	349,713	386,851

Statistical Centre of Iran, Plan and Budget Organization of Government of Iran [later, Ministry of the Interior], sci.org.ir, amar.org.ir, Markaz-e Āmār-e Irānr, Khosravi (1990).

Dates: Christian/Persian

*Despite the official population figure of 7,037,907 for Tehrān (the Iranian Statistical Center, Plan and Budget Organization of Government of Iran), on December 11, 1999, Mr. 'Asghar Zebardasht, the deputy governor general of Tehrān Province for planning, announced that: "...Before the revolution [1977], population of Tehrān city was 5,000,000, but with unchecked immigration, currently [1999], the figure is about 11 million ..." (IRNA, 1999.12.11).

TABLE 17.3 Population of the Country

	1956/1335		1966/1345		1976/1355		1986/1365		1996/1375	2006/1385	2010/1389
Population	*People*	%	*People*	%	*People*	%	*People*	%	*People*	*People*	*People*
Urban	5,953,563	31	9,794,246	39.04	15,854,680	47.03	26,844,561	54.29	36,817,789	48,259,964	53,637,652
Rural	13,001,141	69	15,284,677	60.96	17,854,064	52.97	22,600,449	45.71	23,237,699	22,235,818	21,095,578
Total	18,954,704	100	25,078,923	100	33,708,744	100	49,445,010	100	60,055,448	70,495,782	74,733,230

Statistical Centre of Iran, Plan and Budget Organization of Government of Iran, sci.org.ir, amar.org.ir.
Country area: 630,577 miles^2/1,633,188 km^2.
Dates: Christian/Persian
The 1996 census: population in Iran = 67,283,000.
Population growth rate: 4.2% (1995).
Population density: 41 km^2 (1995).
Urban population: 57.5% (1993).

In 2012, following the order of Iran's supreme religious leader āyatollāh 'Ali Khāmeneh'i, the country's ministry of health inappropriately announced that it was completely suspending all programs and procedures related to family planning and control in Iran. Āyatollāh Khāmeneh'i called for the country's population to grow to 150–200 million people! According to Iran's latest census, the population of the country today stands at about 75 million. Earlier, on 27 July 2010, Khameneh'i had supported Mahmoud Ahmadinezhād's policies to increase the country's population. In a meeting with some authorities of the Islamic Regime in early September 2011, he publicly said: "*I believe that with the resources that we have, we can have a population of 150 million people.*" He further added that any family planning aimed at stopping population growth should be exercised after the country's population reached that figure. Ahmadinezhād had announced his desire to increase the population of Iran soon after coming to office in 2005 (payvand.com/news, 27 August 2012). This decision in an earthquake-prone country with no earthquake-risk minimization plan will have catastrophic consequences. During the first days of the Iran–Iraq war, Khomeini ordered the people to increase the population of the country in order to produce more soldiers.

Although earthquakes took a toll of more than 24,750 lives during the first 60 years of the twentieth century, damage in economic terms remained relatively low (Figure 1 in Introduction). However, with the massive investments of the past five decades in new urban and industrial centers, future earthquakes in Iran are likely to result in serious economic loss and casualty. There has been a 3.5-fold increase in the Iranian population since 1956, and a two-fold increase since 1976 (SCI, 2004). Population growth has been concentrated in the mega-city of Tehrān and in the large provincial capital cities throughout the country, most of which are located in active fault zones (Chapters 10–16).

17.1.3 Demographic Density and Distribution

Due to the geographic and climatological features of the country, the distribution of the Iranian population has always been very uneven. The heaviest concentration is in the western and northern part of the country, where 64% of Iranians live on 27% of the territory (72.4 km^{-2}). This high density reflects stability among the large rural population, rather than growth of the cities, with the exception of the region around Tehrān (308 km^{-2}). Population density in the Hamédān province is 77 km^{-2}; in Māzandarān 73 km^{-2}; and in Gilān 140 km^{-2}. The rest of the country—73% of the territory—is largely desert (14.9 km^{-2}), with a population concentrated around cities and on several irrigated plains. The southern regions (Fārs, Baluchestān, and Kermān) have experienced higher population growth than those of the north and west (the central plateau, the Caspian provinces, Āzarbāijān, and even Kordestān). The 1986 census provided evidence of the effects of regional wars, with heavy increases in Khorāsān and Baluchestān, where Afghan refugees have settled, and, on the other hand, abandonment of the areas along the Iraqi frontier, especially those of Ābādān and Khorramshahr (Hourcade, 1994).

17.2 TIME OF THE EARTHQUAKE AND DEATH RATIO

There should of course be a clear difference in the death ratio of earthquakes that occur during the day and at night. The greatest number of fatalities occurs during earthquakes that take place at night, when the majority of people are at home. However, because of a lack of accurate earthquake casualty data, it is difficult to arrive at a reasonable empirical relation (Tables 17.4 and 17.5). A high ratio of death toll to injured people also occurs in Iranian earthquakes due to sudden destruction of indigenous dwellings and unreinforced masonry buildings.

17.3 EXTREME CLIMATE CONDITIONS

Intense cold weather (−20 to −30 °C) during the 5 February 1641 $M_s \sim 6.8$ Dehkhārqān earthquake (Figure 11.12; Vārtābed Ārākel Tāvrizetsi, ed. Bournoutian, 2006) and 17 January 1895 $M_s \sim 6.8$ Quchān earthquake (Figure 16.17; Tchalenko, 1975) led to additional victims. After the 13 December 1957 $M_w \sim 6.8$ Fārsinaj earthquake (Figure 12.11), about 20 people and many animals perished from exposure to below-freezing temperatures, particularly during the snowstorm of the 21 December 1957 earthquake (Ambraseys et al., 1973).

The 28 February 1997 $M_w \sim 6.2$ Shirān earthquake (Figure 13.16) help-and-rescue mission as well as the relief operations were hampered by: (i) extreme climate conditions in the form of freezing temperatures and major snowfalls on 7 March 1997, when more than 1 m of snow blocked roads to nearly all of the affected villages; (ii) temperature dropped to −18 °C during the day and −28 °C at night; and (iii) difficult mountainous terrain. An Iranian army plane crashed several days after the earthquake as a result of snowstorms and fog. The heavy loss of life was to some extent attributable to the fact that many remote villages could not be easily accessed and to the difficulties in surviving the aftermath of the earthquake (Iranian Newspapers; Iranian Red Crescent Society; IFRC; reliefweb.net, 7 January 1998).

17.4 DIFFICULT MOUNTAINOUS TERRAIN

Even during recent earthquakes such as the 22 July 1983 M_w 5.5 Charazeh, 20 June 1990 M_w 7.3 Rudbār (Figure 13.12), and 28 February 1997 M_w 6.2 Shirān (Figure 13.16), relief operations were hampered by the lack of proper rural road access in difficult mountainous terrains.

TABLE 17.4 Earthquake Fatality Ratio in Iran

Earthquake	M_w	I (MMI)	Local Time (UTC)	People Killed (Injured)	Regional Demographic Density	Climate Zone	Bldg. Type	~Fatality Ratio (%) ~*Range*	~Fatality Ratio (%) ~*Mean*	References
1968.09.31 Dasht-e Bayāz	7.1	IX	14:17 (10:17)	10,000 (4000)	Low	Arid	Adobe	40–100	86	Ambraseys et al. (1969), Tchalenko and Ambraseys (1973)
1972.04.10 Kārzin	6.7	VIII	**05:36** (02:06)	5010 (1340)	Moderate	Semiarid	Adobe, Masonry	20–67	48	Ambraseys et al. (1972)
		VII						1–20	18	
1978.09.16 Tabas-e Golshan	7.3	IX	**19:35** (15:35)	20,000 (6400)	Low	Arid	Adobe, Masonry	67–90	84	Berberian (1979a,d)
		VIII						20–43	17	
		VII						1–20	6	
1981.06.11 Golbāf	6.6	VIII⁺	10:54 (07:24)	1400 (3000)	Low	Arid	Adobe		10	Berberian et al. (1984)
1981.07.28 Sirch	7.0	IX	**20:52** (17:22)	1300 (1000)	Low	Arid	Adobe	33–58	48	Berberian et al. (1984)
		VIII						2–10	8	
1990.06.20 Rudbār	7.3	IX⁺	**00:32** (21:00)	40,000 (105,000)	High	Subtropical	Adobe, Masonry	66–90	80	Berberian et al. (1992); Berberian and Walker (2010)
		VIII						10–40	18	
1997.05.10 Zirkuh	7.2	IX	12:27 (07:57)	1568 (5059)	Low	Arid	Adobe	20–46	38	Berberian et al. (1999)
		VIII						2–20	12	
2003.12.26 Bam	6.6	VIII⁺	**05:26** (01:56)	26,500–43,200 (30,000)	Low	Semiarid	Adobe, Masonry	50–90	78	Berberian (2005)

Bold: Time of earthquake during night, evening, and early morning, when majority of people were at home.
~: Approximate estimated value.

TABLE 17.5 Earthquake Fatalities as a Percentage of Population

Earthquake Date	Earthquake Time (Local)	M_s	M_w	*I* (MMI)	CD_w (km)	Location	*I* (MMI)	Population at the Time of Earthquake	No. of People Killed	Fatality Ratio (%)*
1909.01.23	14:17	7.4	7.4	IX^+	–	Budineh	IX^+			**100**
						Miyān Rudān	IX^+			**100**
						Nasir al-Din Bālā	IX^+			**100**
						Sanbāl	IX^+			**100**
						Sangar	IX^+	400	397	**99**
						Chamanār [Chamnār]	IX^+			>**90**
						Choghā Bahrām	IX^+			>**90**
						Qal'eh Qāsem	IX^+			>**90**
						Qal'eh Rostam	IX^+			>**90**
						Bahrain [Dorud] Town	IX^+			>**50**
						45 Villages	IX^+			>**50**
1941.02.16		6.2	6.1	VIII	–	Mohammadābād	VIII	920	680	**74**
1957.12.13	05:15	6.7	6.8	VIII	–	Fārsinaj	VIII	1400	702	**50**
1962.09.01	22:50	7.2	7.0	IX	10	Rudak	IX	3500	2310	**66**
						Amirābād Kohneh		285	162	**57**

						Ipak	IX	444	289	**56**
						Indrejin	IX	400	200	**50**
						Hesār	IX	400	200	**50**
						Amirābād Nau	IX	550	242	44
						Rostamābād	IX	850	357	42
						Ildarjin		327	137	42
						Kodik		300	120	40
						Āghcheh Mazār	IX	502	176	35
						Chenār Soflā	IX	200	70	35
						Morādbeglu	IX	203	71	35
						Cheskin	IX	1450	464	32
						Gonbadak		378	117	31
						Kambadak		370	115	31
						Qurqurak	IX	70	20	29
						'Abbasābād	IX	745	164	22
						Keshmarz		486	97	20
						Bu'in Zahrā	VIII	2515	478	19
1968.08.31	14:17	7.2	7.0	IX	10	Kareshk	IX	670	570	**85**
						Dasht-e Bayāz	IX	1670	1230	**73.6**
						Miām	IX	302	120	39.7
						Kākhk Town	IX	4380	1379	31

Continued

TABLE 17.5 Earthquake Fatalities as a Percentage of Population—Cont'd

Earthquake Date	Earthquake Time (Local)	M_s	M_w	I (MMI)	CD_w (km)	Location	I (MMI)	Population at the Time of Earthquake	No. of People Killed	Fatality Ratio (%)
1972.04.10	05:36	6.9	6.7	VIII	10	Qir Town in the Kārzin borough	VIII	4068 5068	3069 3399	**75** **67**
						Gāvaki	VIII	872	400	46
						Shāhābād	VIII	195	80	41
						Sharaf Khalil	VIII	182	64	35
						Biān/Bayān (Kārzin)	VIII	941	300	32
						Manganuiyeh		154	40	26
						Ābdeh		242	60	25
						Berijān (Berikhun)	VIII	160	40	25
						Sekeh Ravān		627	150	24
						Tang-e Ru'in	VIII	238	56	23
						Haft Āsiāb	VIII	365	80	22
						Marand		182	40	22
						Rizjān (Zirjun)		10	2	20
						Fathābād		17	3	18
						'Azizābād		183	31	17
						Khoshābjān (Khosh Kalleh)		29	5	17

						Mozaffari	VII	456	80	17
						Tan-e Bādi	VII	26	4	16
						Bang-e Nau		467	70	15
						Otruiyeh	VIII	81	12	15
						Ābābād	VII	59	8	14
						Kakruiyeh		268	31	12
						Kordshul		90	11	12
						Dasht-e Shur		144	16	11
						Khairābād (Kandun)		332	36	11
						'Aliābād Bahman (Pahn)		160	16	10
						Haidarābād		20	2	10
						Manāl		254	25	10
						Sarcheshmeh	VIII	699	70	10
1977.12.19	03:34	5.7	5.9	VII	07	Sarbāgh (Sarāsiāb)	VII	200	90	45
						Gisk	VII	1200	160 175	13 14.5
						Dartangal	VII	1995	250 382	12 19
1978.09.16	19:35	7.4	7.3	IX^+	21	Tabas-e Golshan Town	IX	~13,000	~11,000	**85**
						Fahālanj	IX	3521	2500	**71**

Continued

TABLE 17.5 Earthquake Fatalities as a Percentage of Population—Cont'd

Earthquake Date	Earthquake Time (Local)	M_s	M_w	I (MMI)	CD_w (km)	Location	I (MMI)	Population at the Time of Earthquake	No. of People Killed	Fatality Ratio (%)
						Bāghestān-e Korit	IX	500	300	**60**
						Korit	IX	3335	2000	**60**
						Mir 'Omaru	IX	151	80	**53**
						Deheshk [Dehestak]	IX	3846	2000	**52**
						Khosrowābād	IX	1000	500	**50**
						Gharān		6060	2000	33
						Dasht-e Gharān	VIII	1000	300	30
1990.06.20	00:39	7.4	7.3	IX^+	13	Fatalak (landslide)				**100**
						Fisham				**100**
						Manjil Town	IX	11,107	7000?	**63?**
						Bivarzin		688	316	46
						Pākdeh	IX	1462	460	31
						Rudbār Town	IX	14,834	4000?	27?
						Ābbar	VIII	2249	500	22
						Lowshān Town +Iraqi Camp	VIII	11,107	2000?	22?

**:Organized in descending order of fatality ratio for each event.*
?:uncertain values.
Bold: fatality ratio greater than 50%.

17.5 NOMADS IN THE TENTS

During the 23 January 1909 M_w 7.4 Silākhor earthquake (Figure 12.1), most nomads who were living in their tent camps suffered fewer casualties from the earthquake; however, some were killed by rockfalls and rock avalanches. This factor reduced the number of casualties in certain places. The combined casualty figure for the two earthquakes of 20 December 2010 M_w 6.5 and 27 January 2011 M_w 6.2 South Rigān at a remote desert environment where people were living in tents was six dead (Figure 14.11).

17.6 ALARMING FORESHOCKS

Numerous alarming foreshocks helped people staying outdoors during the devastating mainshocks, reducing the number of deaths. About 2514 people died during the 6 May 1930 M_w 7.1 Salmās earthquake in the fertile Salmās plain (Figure 12.5) because the majority of people spent the night in the fields after experiencing an M_s 5.4 damaging foreshock that took place 10 hours before the devastating mainshock. The army troops at Salmās (Dilmān) also abandoned their barracks and camped in tents set outside the buildings. This saved the lives of the soldiers, who became of vital help in rescuing and evacuating the trapped and injured inhabitants in the ruined cities of Salmās and Kohnehshahr as well as in the numerous devastated villages.

17.7 STATISTICAL DATA ON HUMAN CASUALTIES AND STRUCTURAL DAMAGE

Accurate statistics of earthquake-related social loss and building damage are either not available or are inaccessible for the general public and earthquake researchers. The data are needed for vulnerability assessment of earthquake impacts and consequences. The following is an attempt to collect some of the existing historic earthquake data scattered throughout the literature.

17.7.1 15 August 1485 M_s~7.2 Upper Polrud Earthquake

See Table 17.6 and Figure 13.13.

17.7.2 20 April 1608 M_s~7.6 Alamutrud Earthquake

See Table 17.7 and Figure 11.6.

17.7.3 22 March 1879 M_s~6.7 SE Bozqush Earthquake

See Table 17.8 and Figure 11.14.

TABLE 17.6 **Summary of the 15 August 1485 [$I \sim$IX, $M_s \sim 7.2$] Upper Polrud Earthquake Macroseismic Data Along the Kelishom Fault (Berberian and Walker, 2010)**

Location	Casualty	Earthquake Destruction/ Damage	Estimated Intensity
(I) Northern Meizoseismal Area Data			
Tonékābon [District between Upper Polrud and Upper Sehezār Valleys]		High buildings (palaces, mosques, and shrines) and public baths collapsed, the rest fissured	VIII
Shakvar [modern Eshkévar, Middle Polrud River area]	70	Many villages affected; old buildings collapsed	
Gorjiān [upper Chālakrud River area] and Golijān [lower Chālakrud River area]	106		VIII
Castle of Sardābsar in Gorjiān [lower Chālakrud River area]	2	Collapsed	VII
Jandeh Rudbār of Gorjiān [Jennat Rudbār of Chālakrud?]	–	A pig in panic leapt from top of mountain into river below and perished	
Daylamān [36.88°–49.90°, +1449 m]	?	Many old buildings collapsed	VIII
Daylamestān [Daylamān area; Chalakrud River area, west of Polrud River]	?	Numerous large rockfalls killed livestock	VIII
Rānku [lower Polrud area, south of Rudsar]	–	Strongly felt, minor damage. A part of an old palace fell down [possibly built by Seyyed Rāzikiyā in 820/1417 at Tamijān town, WSW of Rudsar (35.40°N–37.20°E)]	VII
Hashtpar-e Kohneh of Rānku	–	Partly collapsed	VII
Biyehpish [37.00°–49.60°]	–	Strongly felt, minor damage	VI
Lāhijān [37.20°–50.00°]	–	Strongly felt, minor damage	VI
Kissum [district between Lāhijān and Sefidrud]	–	Strongly felt, minor damage	VI
Gukeh [district between Lāhijān and Sefidrud, south of Kissum]	–	Strongly felt, minor damage	VI
Pāshijā [E. Sefidrud Delta]	–	Strongly felt, minor damage	VI

TABLE 17.6 Summary of the 15 August 1485 [$I \sim$IX, $M_s \sim 7.2$] Upper Polrud Earthquake Macroseismic Data Along the Kelishom Fault (Berberian and Walker, 2010)—Cont'd

Location	Casualty	Earthquake Destruction/ Damage	Estimated Intensity
Lashteh-neshā [W. Sefidrud Delta]		Strongly felt, minor damage	VI
(II) Southern Meizoseismal Area Data			
Pālisān Castle [in Rudbārāt of Alamut; exact location unknown]	78		VIII
Rudbārāt [Alamutrud Valley, north of east Shāhrud-Alamutrud Rivers]	Many perished, number unknown		VIII

Note that the width of intensity VIII between the Northern and the Southern meizoseismal area is over 40 km (Berberian and Walker, 2010). See Figures 11.6 and 13.13. Coordinates are in 00°N–00°E.

TABLE 17.7 Summary of the 20 April 1608 Alamutrud Earthquake [$I \sim X^+$, $M_s \sim 7.4$] Limited Macroseismic Data Along the Alamutrud Fault (Berberian and Walker, 2010). See Figure 11.6

Location	Coordinates (N–E)*	Casualty	Earthquake Destruction/ Damage	Estimated Intensity
Darband Castle	?	+	Ruined	X
Five Villages NW of Chāhār Kalāteh				
(i) Ovān	36°29′31″–50°27′01″, +1845 m	+	Ruined, landslides	X
(ii) Varbon	36°29′14″–50°27′15″, +1882 m	+	Ruined, landslides	X
(iii) Zarābād	36°29′23″–50°25′52″, +1789 m	+	Ruined, landslides	X
(iv) Zavārdasht	36°29′45″–50°25′56″, +1814 m	+	Ruined, landslides	X

Continued

TABLE 17.7 Summary of the 20 April 1608 Alamutrud Earthquake [$I \sim X^{+}$, $M_s \sim 7.4$] Limited Macroseismic Data Along the Alamutrud Fault (Berberian and Walker, 2010). See Figure 11.6—Cont'd

Location	Coordinates (N–E)*	Casualty	Earthquake Destruction/ Damage	Estimated Intensity
(v) 300 m N of Zavārdasht	?	+	Ruined, landslides	X
Chāhār Kalāteh				
(i) Shams Kālāyeh	36°27′–50°28′, +1700 m	+	Ruined, landslides	X
(ii) Mo'alem Kālāyeh	36°26′–40°28′, 1541 m	+	Ruined, landslides, springs w/ red water	X
(iii) Estalbar	36°27′–50°28′, +1567 m	+	Ruined, landslides	X
(iv) Anādeh [possibly Nāveh, south of Mo'alem Kālāyeh]	36°26′–50°28′, +1629 m	–	Ruined, landslides	X
Caspian Sea Shore [50 km NNE]	Southern shore of the Caspian Sea	–	Seiche	VII
Damāvand [150 km SE]	35°43′–52°04′, +1964 m	–	Repair of the Jame' Mosque [inscription dated 1615]	Possibly associated w/ the 1608 eq. damage
Āmol [195 km NE]	36°28′–52°20′, +87 m	–	Chimneys collapsed, buildings fractured	VI
Sāri [230 km NE]	36°33′–53°03′, +38 m	–	Buildings fractured	V
Ashraf [modern Behshahr; 275 km NE]	36°41′–53°32′, +30 m	–	Buildings fractured	V
Miyān Kāleh Peninsula [300 km NE]	SE Caspian Sea	–	Strongly felt	IV+
Shamakhi [500 km NW]	40°38′–48°38′, +705 m	–	Felt	IV

*, due to proximity of some small sites, an exact coordinate is given.
?, location unknown. See Figure 11.6.

TABLE 17.8 Statistics of the 22 March 1879 SE Bozqush Earthquake (Iran, 1296, nos. 382–3, 1974; Nabavi, 1972; Ambraseys, 1974a; Musketoff and Orloff, 1893), Figure 11.14

Village/Town	No. of People Killed	No. Animals Perished	No. Houses Destroyed	*I* (MMI)
Ardebil	3		Some	
Armudāgh	2 survived		All 150	VIII+
Bāshkand				VIII+
Dāshanli	50		Ruined	VIII+
Daushānjaq				VIII+
Dizaj, Yehgijeh, and Miyāneh (town)	1000			VIII+
Dursun Haj 'Ali	Many		Ruined	VIII+
Hadili	Many		Ruined	VIII+
Karandāgh				VIII+
Khanavand				VIII+
Manak	600		Ruined	
Meshkijik	All	1000 sheep, 100 horses, 50 camels	All 100	VIII+
Mishānjaq				VIII+
Miyāneh (town) ⇒ Dizaj			Several	VII
Niār (Nir)	Many		Half ruined	VII
Qaraqayeh				VII
Qarashirān	100		Ruined	VIII+
Qezelyātāq				VII
Sagānchi	50			VIII+
Saqezchi				VIII+
Sārighamish	All		Ruined	VIII+
Tark	All (500); many killed by landslides		Ruined	VIII+
Yengijeh ⇒ Dizaj				VIII+
Total: 21 Villages totally destroyed. 54 Villages heavily damaged.	>922	2660 sheep, 1125 oxen, 124 horses, 55 camels		

17.7.4 The 11 July 1890 $M_s{\sim}7.2$ Tash Earthquake

See Table 17.9 and Figure 11.15.

17.8 COMPREHENSIVE LIST OF THE RE-EVALUATED IRANIAN EARTHQUAKES

Numerous earthquake catalogs cover the Iranian earthquakes, including works by Perrey (1845–1875), Mallet (1850–1858), Abich (1857, 1858, 1882),

TABLE 17.9 Statistics of the 11 July 1890 Tāsh Earthquake (Musketoff, 1891; Musketoff and Orloff, 1893; Ambraseys, 1974a,b; Kondorskaya and Shebalin, 1977, 1982), Figure 11.15

Town/Village	Population	No. of People Killed	*I* (MMI)	Damage/Destruction
Astarābād [Estārehābād/Gorgān]		+	VII	Damaged
Avarsin		+	VIII	
Bastām		+	VII	
Chahār Bāgh		?		
Gorgān ⇒ Astarābād		?		
Kalāt-e Khān		?	VI	
Marta' Ganju		+	IX	Completely destroyed
Mojen		7	VIII	Completely destroyed
Niknāmdeh		+	VII	
Qāle'hnau Kharaqān		+	VII	Damaged, a few houses collapsed
Shāhkuh Bālā		14	IX	Completely destroyed
Shāhkuh Pā'in		+	IX	
Shāhrud		+	VII	Many houses collapsed
Siyāh Khāneh		+	VIII	
Siyāh Marzkuh		+	IX	Completely destroyed
Tāsh	200	140	IX	Completely destroyed
Tazareh		+	VIII	

+: people killed; ?: data not available.

Schmidt (1879), Fuchs (1886), Musketoff (1891, 1899), Musketoff and Orloff (1893), Ballore (1900, 1906, 1924), Stahl (1911, 1962), Milne (1911), Wilson (1930), Sieberg (1932), Savaresnki et al. (1962), Rothé (1969), Karnik (1971), Nabavi (1972, 1977), Alsinawi and Ghalib (1975), Kondorskaya and Shebalin (1977, 1982), Ambraseys (1978b, 1979, 1989, 1990, 2001), Ambraseys et al. (1979), Poirier and Taher (1980), Berberian (1976a, 1977a, 1981a, 1994, 1997, 1996, 2005), Ambraseys and Melville (1977, 1982), Alsinawi (1988), Zokā' (1989), Karapetian (1991), Mansuri (1991), Guidoboni and Traina (1995), Ambraseys and Finkel (1987, 1995), Utsu (2002), and Babayan (2006).

Except for the overestimation of the size of some historical earthquakes reevaluated by Ambraseys and Melville (1982) and used by the others, which are beyond the capability limits of the causative faults, their catalog is still very valuable. However, previous catalogs suffer from numerous shortcomings: being incomplete and/or inhomogeneous; using secondary sources with erroneous, dubious, and duplicate events; and including incorrect seismic parameters and/or mislocations.

An inclusive list of the Iranian medium- to large-magnitude earthquakes is prepared in three separate tables covering: (i) the prehistoric period (BCE), (ii) the historical period (10–1900 CE), and (iii) the instrumental period (1900–2014). As much as possible, the list includes the corrected seismic parameters, damage and destruction zones, casualty figures, and direct economic losses of the earthquakes (Tables 17.10–17.12). The tables are based on rigorous reassessment of primary source data in the literature. The most recent available data scattered in numerous publications have been critically analyzed. Overestimation of earthquake size and damage zone beyond the capability of the causative faults (when detected) caused by amalgamation of historic data compiled through the centuries has been avoided in this work. Duplicate and dubious evens have also been filtered throughout the list. Nonetheless, the assessed intensity and magnitude of historical events should be used carefully for critical risk evaluation. For specific local usage, the contemporary primary sources should be reevaluated for reliability and completeness of data, and the process should be backed by detailed trench investigation of the stratigraphy and dating of paleoseismicity.

Data for many historical earthquakes are not sufficient enough to permit the assessment of intensity and hence the size of the damage zones and magnitude of the earthquakes. Some of the earthquakes were only felt in one or a few sites, and there are catalogs that assigned magnitude for such events. Thus, small-magnitude earthquakes are deleted from this list. Clearly, the data are highly biased to the urban areas, and a lack of information is characteristic about the remote rural and desert areas.

The list is of course not complete; more work (which is in progress) is needed to complete the data set. Nonetheless, it is good enough to be used by earth scientists, earthquake engineers, planners, and decision makers for analyses of active tectonics and earthquake-fault hazards assessments.

TABLE 17.10 Selected Prehistoric (Pre-1 BCE) Earthquakes in Iran [25°00′–41°00′N and 43°00′–64°00′E]

Date (Approximate)	Approximate Location (00°N–00°E)	Area	~*I* (MMI)	~*M*	Active Fault	Cities Destroyed/ Damaged (D)	Towns Destroyed/ Damaged (D)	Deformation	Source
1.8 Ma?	36.11–50.78	Tāléqān			Tāléqān, Moshā?	–	–	Paleo-landslides	Berberian et al. (1983), Berberian (1994)
1.5–0.1 Ma?	32.37–52.35	Gavkhuni			?	–	–	Soft-sediment deformation, earthquake-induced liquefaction, and sedimentary dikes in playa deposits at seven layers	Price (1978)
??	28.88–50.86	Bushehr			?	–	–	Earthquake-induced liquefaction and sedimentary dikes	Berberian (1997c, 1998)
??	33.30–49.26	Lake Gahar			Dorud (ZMRF)	–	–	Earthquake-triggered landslide and rock avalanche	Sawyer (1891, 1894), Harrison and Falcon (1937), Tchalenko and Braud (1974), Nabavi (1985), Berberian (1994)
??	33.29–49.64	Tang-e Zireh, Shimbār Valley			ZMFF	–	–	Paleo-landslides of 280 million m^3, forming a dam on the Kārun River	Layard (1846, 1887), Busk (1926), Berberian (1994)
??	30.81–60.55	Palangkuh			Sefidābeh, Palangkuh	–	–	Paleo-landslides/rock avalanche	Berberian et al. (2000b)
24,000–14,000 BP?	35.82–52.11	Lāsem/ Damāvand			Moshā?	–	–	Earthquake/volcanic-induced soft sediment deformational structures in ancient Lāsem Lake deposits	Berberian et al. (1985), Berberian (1994)

9600–4600 BCE	36.25–54.00	Āstāneh		Āstāneh	–	–	Paleoseismic trench	Hollingsworth et al. (2010)
9000 BCE	33.00–47.60	Saimareh/ Kabirkuh		Kabirkuh	Saimareh district	Darreh shahr	Saimareh landslip w/ a debris volume of 20,000 million m^3	Harrison (1936, 1946), Harrison and Falcon (1934, 1937, 1938), Stein (1940), Oberlander (1965), Watson and Wright (1969), van Zeist and Bottema (1977), Berberian (1994)
4600–600 BCE	36.25–54.00	Āstāneh		Āstāneh	–	–	Paleoseismic trench*	Hollingsworth et al. (2010)
3470–1540 BCE	36.01–51.02 (36°07′24.75″N–51°12′48.28″E, +2995 m)	Tāléqān		Tālēqān	–	–	Paleoseismic trench*	Nazari et al. (2009a)
3000 BCE?	35.50–51.47	Kahrizak?	7.0–7.4	Kahrizak?	–	–	Paleoseismic trench*	De Martini et al. (1998), De Martini (2007)
1620 BCE–AD 220	35.75–51.00 (35°45′05.17″N–51°05′05.51″E, +1311 m)	Tehrān	6.1–7.2	North Tehrān	?	?	Paleoseismic trench*	Nazari et al. (2007, 2008, 2011), Ritz et al. (2012)
1200 BCE	?	Komesh Mt.		Dāmghān	Dāmghān	?	Landslide, rock avalanche	Bondeheshn; see Chapter 11, Figure 11.1
1263 BCE	56.13–43.01 35.45–43.05	Nineveh Assur			Ninveh, Assur			See Chapter 8, Figure 8.1
1170 BCE	36.13–43.01	Nineveh			Ninveh			See Chapter 8, Figure 8.1
680–669 BCE	36.05–43.40	Nimrud					Felt at Calh/Nimrud	See Chapter 8, Figure 8.1
647 BCE	36.50–43.02	Dur-Sharrukin					Felt at Dur-Sharrukin	See Chapter 8, Figure 8.1

Continued

TABLE 17.10 Selected Prehistoric (Pre-1 BCE) Earthquakes in Iran [25°00′–41°00′N and 43°00′–64°00′E]—Cont'd

Date (Approximate)	Approximate Location (00°N–00°E)	Area	~*I* (MMI)	~*M*	Active Fault	Cities Destroyed/ Damaged (D)	Towns Destroyed/ Damaged (D)	Deformation	Source
600–500 BCE	32.18–48.15	Shush (Suza)							Sieberg (1932)
600 BCE–1300 CE	36.25–54.00	Āstāneh			Āstāneh			Paleoseismic trench*	Hollingsworth et al. (2010)
550 BCE?	39.78–44.36?	Ārārāt (Māsis)							Tarasov (1902), Stepanyan (1942), Kondorskaya and Shebalin (1977, 1982), Karapetian (1991), Guidoboni and Traina (1995), Babayan (2006)
336–330 BCE Invasion of Alexander III of Macedonia Almost all manuscripts, including the 21 Nasks of the Sacred Avesta, were either destroyed or transferred to Greece. Loss of macroseismic data.									
280 BCE	35.48–51.82	Rhagae [Ray]	>VIII	>7.0	Pārchin	Rhagae [Ray]	Numerous counties and cities	Ground rupture and diversion of a river course	See Chapter 11
200 BCE–AD 200	31.00–53.90	E. Abarqu playa		7.0	Dehshir			Paleoseismic trench*	Nazari et al. (2009b), Fattahi et al. (2010)

Clearly, the list is not complete and more work (which is in progress) is needed to complete the data set.
I: Highest intensity (MMI) estimated within the meizoseismal area based on written accounts.
M: Approximate magnitude.
ZMFF: Zāgros Mountain Front Fault;
ZMRF: Zāgros Main Recent Fault.
*: Paleoseismic trench dates in Iran have large error bars that are not properly constrained; some dates are speculative and in general cannot be correlated properly.

TABLE 17.11 Historical (10–1900) Earthquakes in Iran and the Surroundings [25°00′–41°00′N and 43°00′–64°00′E] (See Also the Notes and Comments at the End of the Table and Chapters 10 and 11)

Date[a] (Christian) (Year.Month. Date)	Arabic Lunar Date[b]	Time (Local)/Day[c]	Macroseismic Epicenter[d] 00°N–00°E/Q[e]	Location[f]	~I (MMI)	~M_s	People Killed/ Injured	Cities Destroyed/ Damaged (d)	Towns Destroyed/ Damaged (d)	Villages Destroyed/ Damaged (d)	No. of Houses Destroyed	Direct Economic Loss (Million Tumān; Original Values)
~10 AD	–	–	37.95–58.18/2	Mithrādātkart								
139	–	–	39.80–44.40/1	Ārārāt				Nisibus [Mtsbin or Mtsurn]				
276–578	–	–	34.45–47.93/2	Kangāvar								
293–303	–	–	29.30–51.51/2	Bishāpur								
363	–	–	39.52–44.10/2	Ārshākāvān [Dogubāyazid]	>IX?			Ārshākāvān				
454.07.25	–	–	36.20–58.85/1	Aparshahr [Neyshābur]	?	?				Tegharkuni?		
461	–	–	39.25–43.20/1	Āpāhunik [Mānāvāzākert; Manāzkert; Malāzgert]	>IX							
531–590	–	–	29.30–51.51/2	Bishāpur								
629.06?	6–7	–	–/0	Karkuk	F (IV)							
632 or 639?	11	–	–/0	N. Vān region								

Continued

TABLE 17.11 Historical (10–1900) Earthquakes in Iran and the Surroundings [25°00′–41°00′N and 43°00′–64°00′E] (See Also the Notes and Comments at the End of the Table and Chapters 10 and 11)—Cont'd

Date (Christian) (Year.Month. Date)	Arabic Lunar Date	Time (Local)/Day	Macroseismic Epicenter 00°N–00°E/Q	Location	~I (MMI)	~M_s	People Killed/ Injured	Cities Destroyed/ Damaged (d)	Towns Destroyed/ Damaged (d)	Villages Destroyed/ Damaged (d)	No. of Houses Destroyed	Direct Economic Loss (Million Tumān; Original Values)
636–652: Invasion of the Muslim Arabs Almost all the pre-636 written documents were destroyed. Destruction of the infrastructure. Gap in historic data												
658	37–38	–	–/0	Basra	F (IV)	?	–	–	–	–		
Late seventh century?	Late first century?	–	36.20–58.85/1	Neyshābur (Shahr-e Kohneh)	>VII⁺	>6.0	K					
713–762	–		39.30–51.51/2	Bishāpur								
734	116	–	–/2	Sistān Province	>VIII	>6.5	K	Zarang (Shahr-e Sistān)?				
735	117	–	39.70–45.60/2	Vāyots Dzor	>VIII	>6.5	10,000	Moz	+	+		
737	120	–	37.70–45.60/2	Vāyots Dzor AFS								
743, Late Spring	125	–	–/0	Caspian Gates [i. Tang-e Sardareh; ii. Greater Caucasus; iii. Derbent]?	?	?	?	Iran?, Georgia?, or Dāghestān?	?	?	?	?
763.03–764.03	146.00.00	–	–/0	Khorāsān Province	>VIII	>7.0	?	?				
805.12.02	190.01.06	–	–/2	Sistān Province	>VIII	>6.5		Zarang?				

815	199	–	–/0	Sistān Province	>VIII	>6.5	?	?			
840.07.?	225?	–	–/0	Khorāsān Province?	>VIII	>6.5	K		+	+	+
839.11.12–840.10.20	225.00.00	–	31.32–48.67/2	Ahvāz	VIII	6.5	K	Ahvāz			+
846.08.28–847.08.16?	232?	–	–/0	Mosul (al-Mousel)		?					
848.08.05–849.07.25	234.00.00	–	–/0	Harāt (modern Iran or Afghanistan?)	>VII	>5.7					
855.05.22–856.05.11	241.00.00	–	36.60–51.50/2	Ray	IX	7.1	K^{+}	Ray		+	+
856.12.22	242.08.18	23:00/ Tuesday	36.23–54.14/2	Komesh (Damghān)	IX^{+}	7.2	45,096	Komesh, Dāmghān	Bastām??	+	+
856.05.10–857.04.29?	242.00.00?	–	–/0	Fārs Province	?	?					
858.04.19–859.04.07	244.00.00	–	38.12–46.31/2	Tabriz	>VII^{+}	>6.0	40,000?	Tabriz		+	+
859.04.08–860.03.27	245.00.00	–	–/0	Khorāsān Province	>VII	>5.7	K	?			
859.04.08–860.03.27	245.00.00	–	–/0	Baghdād, Ctesiphon	F (IV)?	?					
863.02.13	248.12.19	–/Saturday	40.06–44.58/2	Dvin	>VII^{+}	>6.0	K; 120,000?	Dvin		+	+
864.01.15–02.12	249.12.00	–	36.60–51.50/2	Ray	>VII^{+}	>6.0	K	Ray		+	+
872.06.21	258.08.10	–/Saturday	33.11–47.15/1	Saimareh FRS	>VII	>5.7				+	
872.06.22	258.08.11	–/Sunday	33.11–47.15/2	Saimareh	VIII^{+}	6.8	20,000	Saimareh, Dareh Shahr	Sirvān	+	+

Continued

TABLE 17.11 Historical (10–1900) Earthquakes in Iran and the Surroundings [25°00′–41°00′N and 43°00′–64°00′E] (See Also the Notes and Comments at the End of the Table and Chapters 10 and 11)—Cont'd

Date (Christian) (Year.Month. Date)	Arabic Lunar Date	Time (Local)/Day	Macroseismic Epicenter 00°N–00°E/Q	Location	~I (MMI)	~M_s	People Killed/ Injured	Cities Destroyed/ Damaged (d)	Towns Destroyed/ Damaged (d)	Villages Destroyed/ Damaged (d)	No. of Houses Destroyed	Direct Economic Loss (Million Tumān; Original Values)
874.10.16–11.14	261.01.00	–	33.16–55.28/2	Gorgān/Jorjān [Gonbad Kāvus]	>VII$^+$	>6.0	>2000	Gorgān/Jorjān		+	+	
880.09.29–10.28	268.03.00	–	–/0	Baghdād	F	?						
893.12.28	280.10.15	Night	40.10–44.50/2	Dvin	>VIII	>6.1	30,000	Dvin		+	+	
902.06.11–07.10	289.07.00	–	–/0	Baghdād	F	?						
906.04.00	293.06-07	–	39.70–45.00/3	Vāyots-Dzor	>VIII	>6.1	K					
912.04.21–05.20	299.09.00	–	–/0	Kufa	>VII	?	K					
912.08.18–913.08.07	300.00.00	–	–/0	Dinévar	?	?						
943.08.06–09.03	331.12.00	–	37.50–56.75/3	Samalqān	IX$^+$	7.4	>5000			>30	+	
956.04.15–957.04.05	345.00.00	–	–/2	Hamédān	>VII$^+$	>6.0	K	Hamédān	Asadābād	+	+	
958.02.23	346.12.01	–	–/1	Ruyān	X	7.1	K$^+$	Ray		>150	+	
958.04.00	347.01.08–02.07	–	34.53–45.78/3	Sar-e Pol-e Zahāb [Holvān]	VIII	6.4	K	Holvān	Qasr-e Shirin	+	+	

972	361–2	–	40.30–44.19/2	Aragocotn?	>VII	>5.4					
973.10.02–974.09.20	363	–	–/0	Wāsit [Wāset]	F (V)	?					
977.10.18–11.16	367.00.00	–	–/0	Baghdād	F (IV)	?		–	–	–	–
978.06.17	367.11.07	–/Sunday	27.68–52.37/2	Sirāf	>VIII	>6.5	>2000	Sirāf port		+	+
978.08.09–979.07.28	368.00.00	–	–/0	Iraq?	?	?					
986.11.07–12.06	376.07.00	–	–/0	Mosul (al-Mousel)	?	?	K				
1008.04.27	398.08.16	Night/Sunday	34.61–47.50/2	Dinévar	IX	7.0	>16,000	Dinévar	?	–	–
1008.05.10–06.08	398.09.00	–	27.68–52.37/2	Sirāf	VIII	6.5	K	Sirāf port		+	+
1013.03.18–1014.03.17	–	–	41.00–44.70/2	Shirākāvān	>VII	>5.5		Hālbāt	Kelargom		
1031.01.01–12.18	422.00.00	–	36.33–43.01/2	Mosul (al-Mousel)	>VII	>5.5	4		Mosul (d)		
1040?	432?	–	–/0	Gold mine, Zelzeleh Kuh, E. Ghaznain, NE Sistān				Gold mine destroyed and disappeared			
1042.11.04	434.03.17	19:30/Thursday	37.95–46.62/3	Tabriz	>IX	>7.0	40,000?	Tabriz			
1052.06.02	444.02.01	–	36.22–57.68/2	Baihaq [Sabzévār]	IX	7.0	K	Baihaq			

Continued

TABLE 17.11 Historical (10–1900) Earthquakes in Iran and the Surroundings [25°00′–41°00′N and 43°00′–64°00′E] (See Also the Notes and Comments at the End of the Table and Chapters 10 and 11)—Cont'd

Date (Christian) (Year.Month. Date)	Arabic Lunar Date	Time (Local)/Day	Macroseismic Epicenter 00°N–00°E/Q	Location	~I (MMI)	~M_s	People Killed/ Injured	Cities Destroyed/ Damaged (d)	Towns Destroyed/ Damaged (d)	Villages Destroyed/ Damaged (d)	No. of Houses Destroyed	Direct Economic Loss (Million Tumān; Original Values)
1052.05.03–1053.04.22	444.00.00	–	30.65–50.32/2	Arrajān [Argān]	VIII^+	6.8	K	Arrajān	Izeh	+	+	
1058.12.08	450.10.18	Dusk/ Tuesday	–/1	Karkuk?	>VII^+	>6.0	K	–	–	–	–	
1063.11.25–12.24	455.12.00	–	–/0	Wāsit [Wāset]	F	?						
1064.06.19–07.18?	456.07.00?	–	–/0	Āni (?)	–	–						
1066.04.30–05.28	458.06.00	–	33.72–59.20/2	Qā'en	>VIII	>6.5	K^+	Qā'en	+	+		
1069?	461?	–	–/0	Khurshāh (?)	–	–						
1072.01.20	464.04.25	–/Friday	–/0	Baghdād	F	?						
1085.04.29–05.28	478.01.00	–	30.65–50.32/2	Arrajān [Argān]	VII^+	5.8	K	Arrajān		+	+	
1087.11.01–11.29	480.08.00	–	–/2	Hamédān	>VII^+	>6.0	K	Hamédān		+	+	
1094.01.21–02.19	487.01.00	Early AM	–/0	Baghdād	F	?						
Eleventh century	Fifth century	–	–/0	Sistān Province [E. Ghaznain]								

Twelfth century	Sixth century	–	–/0	Bishāpur							
Twelfth century	Sixth century	–	31.30–56.28/1	Gevār [Jevār]	>VII	>5.5		Kuhbanān			
1101.10.26–1102.10.14 (?)	495.00.00 (?)	–	–/0	Gerdkuh (Dāmghān)	?	?					
1102.02.28	495.05.08	Night/Friday	34.35–62.18/2	Harāt		5.3		Harāt			
1107.08.22–1108.08.11	501.00.00	–	34.60–47.50/2	Kargsār	VIII	6.5	K	Dinévar			+
1118.04.03	511.12.09	–	–/0	W. Zāgros	F	5.9?	?				
1119.12.10	513.09.05	Night/ Wednesday	36.38–50.00/2	Qazvin	VIII	6.5	K	Qazvin			+
1121.03.22–1122.03.11	515.00.00	–	–/0	Mosul (al-Mousel)	F	?					
1127.01.17–1128.01.06	521.00.00	–	36.33–53.48/3	Farim (Parim)	VIII+	6.8	K			+	+
1130.02.27	524.03.16	Early Evening	–/0	Baghdād/W. Zāgros?	?	6.8					
1131.11.28		–	40.50–43.57/2	Āni	>VII	>5.7					
1135.03.19–04.20	529.06.00	–	–/0	Baghdād	F	?					
1135.07.25	529.10.11	–/Thursday	–/0	W. Kordestān	>VIII	>6.5				+	+
1135.08.13	529.11.01	–/Thursday	–/0	W. Kordestān AFS	>VIII	>6.5	K			+	+
1139.09.30	534.02.04	Night/ Saturday	40.66–43.31/3	Ganjeh	XI	>7.0	100,000–300,000#	Ganjeh	Hālbāt, Tātev	+	+
1144.05.29	538.11.24	Night/ Tuesday	–/0	Baghdād	F(IV)	?					

Continued

TABLE 17.11 Historical (10–1900) Earthquakes in Iran and the Surroundings [25°00′–41°00′N and 43°00′–64°00′E] (See Also the Notes and Comments at the End of the Table and Chapters 10 and 11)—Cont'd

Date (Christian) (Year.Month. Date)	Arabic Lunar Date	Time (Local)/Day	Macroseismic Epicenter 00°N–00°E/Q	Location	~I (MMI)	~M_s	People Killed/ Injured	Cities Destroyed/ Damaged (d)	Towns Destroyed/ Damaged (d)	Villages Destroyed/ Damaged (d)	No. of Houses Destroyed	Direct Economic Loss (Million Tumān; Original Values)
1145.06.24–1146.06.13	540.00.00	–	36.33–58.75/2	Neyshābur	>VII$^+$	>6.0		Neyshābur	Vineyard, Sabzpushān, and Madreseh mounds			
1150.04.01	544.12.01	Morning/ Saturday	34.53–45.78/2	Sar-e pol-e Zahāb [Holvān]	VII$^+$	5.9	K	Holvān		+	+	
1153/548: Invasion of the Ghuzz (Oghuz) Turk Nomads—gap in historic data												
1156.01.26–02.24	550.12.00	–	–/0	Baghdād	F	?						
1177.05.01–05.30	572.11.00	–	36.10–50.43/2	Ray-Qazvin	IX	7.0	K$^+$	Ray-Qazvin?	+	+	+	
1179.04.29	574.11.12	–	36.25–44.30/2	Arbil (Irbil)	VIII	6.6		Arbil		+	+	
Late 1191–Early 1192	Late 587–Early 588	–	–/0	Hamédān	F	?						
1194.02.24–03.25	590.03.00	–	32.00–44.30/1	Najaf	VII	>5.5		Najaf (d)				
1208.07.16–1209.07.06	605.00.00	Day	36.22–59.00/3	SE Neyshābur (Dāneh-Shādyākh)	IX$^+$	7.3	2,000–10,000	Shādyākh		+	+	
1220–1221: Invasion of the savage mongol hordes: People slaughtered and cities devastated—gap in historic data												
1225.03.04	622.02.21	Night	–/0	Mosul (al-Mousel)	F	?						

1226.11.18	623.11.25	Early AM/ Wednesday	35.40–45.90/2	Shahr-e Zur [Yāsin Tapeh]	VIII	6.5		Shahr-e-Zur	+	+
1237		–	–/1	Mosul (al-Mousel)	>VII	>5.5		Mosul (d)		
1237.08.24–1238.08.13	635.00.00	–	34.28–58.62/2	Gonābād	>VII^{+}	>6.0				
1251.03.26–1252.03.13	649.00.00	–	36.20–58.88/2	Shādyākh	>VII^{+}	>6.0		Shādyākh	+	+
1252.03.14–1253.03.02	650.00.00	–	–/0	Baghdād	F (IV)	?				
1255–1258: Invasion of Hulāku Mongol: People slaughtered and cities devastated—gap in historic data										
1261.11.24–1262.11.14	660	–	–/0	S. Iraq	F	?				
1262.11.15–1263.11.03	661	–	–/0	Mosul (al-Mousel)	?	?		Mosul (d)		
1270.10.07	669.02.19	Morning/ Tuesday	36.35–58.45/2	W. Neyshābur	IX	7.1	>10,000	Shādyākh	+	+
1273.01.18	671.06.26	21:00/ Wednesday	38.13–46.30/2	Tabriz	>VIII	>6.5	250 in Tabriz	Tabriz (d)	+	+
1300.09.16–1301.09.05	700.00.00	–	36.10–53.10/2	Farim [Parim]	VIII^{+}	6.7		Farim	+	+
1304.11.07	704.04.07	Night/Friday	–/0	Tabriz?	>VIII?	>6.7?				
1305.04.16	704.09.20	Day/ Saturday	–/0	Tabriz AFS	F	?				
1310.05.31–1311.05.19	710.00.00	–	–/1	Shahr-e Zur [Yāsin Tapeh]	>VII?	>5.5?	K	Shar-e Zur	+	+

Continued

TABLE 17.11 Historical (10–1900) Earthquakes in Iran and the Surroundings [25°00′–41°00′N and 43°00′–64°00′E] (See Also the Notes and Comments at the End of the Table and Chapters 10 and 11)—Cont'd

Date (Christian) (Year.Month. Date)	Arabic Lunar Date	Time (Local)/Day	Macroseismic Epicenter 00°N–00°E/Q	Location	~*I* (MMI)	~M_s	People Killed/ Injured	Cities Destroyed/ Damaged (d)	Towns Destroyed/ Damaged (d)	Villages Destroyed/ Damaged (d)	No. of Houses Destroyed	Direct Economic Loss (Million Tumān; Original Values)
1316.01.05	715.10.08	–	34.08–48.40/2	Borujerd	>VIII	>6.5	K	Borujerd		>20	+	
1319.00.00	718.12.01–719.11.30	–	39.00–44.40/2	St. Thāddeus	>VII$^+$	>6.5	75 in the monastery			+	+	
1320.00.00	719.11.19–720.11.28	–	40.50–43.56/2	Āni	VIII	>6.0	K	Āni				
1336.05.08	736.09.26	Mid-day/ Wednesday	26.82–55.92/2	Kusheh, Central Qeshm Island	>VII$^+$	>6.0				Kusheh, +	+	
1336.10.19[a]	737.03.12	Dawn/ Saturday	34.85–59.74/2[a]	Jizd[α]	>IX	>7.0	10,000?	Jizd		+	+	
1336.10.21[a]	737.03.14	Dawn/ Monday	34.65–59.55/2[a]	Zuzan[α]	>IX	>7.0	10,000?	Zuzan				
1344.05.15–1345.05.03	745.00.00	–	–/1	Esfahān	>VII	>5.7	20 in Esfahān					
1345.05.04–1346.04.23	746.00.00	–	–/0	Tabriz	F	?						
1364.02.10	765.05.06	–	34.90–61.70/1	Harāt	>VII	5.8	Harāt	?				
1366.09.07–1367.08.27	768.00.00	–	–/0	Kuchesfahān	VII	?						
1383–1385: Invasion of Timur [Tamburlaine]: People slaughtered and cities devastated—gap in historic data												
1389.01.30–02.27	791.02.00	Dawn	36.32–58.93/2	Neyshābur	IX$^+$	7.3	K$^+$	Neyshābur		+	+	

Fourteenth century	Eighth century	–	31.30–56.28/1	Gevār [Jevār]	?	?		Kuhbanān			
Early 1400	802	–	–/0	Kutam [Rudsar]	VII	?		Kutam [Rudsar]?			
1400?	802–3	–	27.66–54.33/1	Lār	>VII	>5.3		Lār			500
1405.11.23	808.05.30	–	36.44–58.52/2	Neyshābur	IX^{+}	7.4	>30,000	Neyshābur		+	+
1405	–	–	–/0	Heshāt							+
1406.11.29	809.06.17	–	39.38–46.23/1	Tātev [Siunik]	>VII	>5.5	K		Tātev, Jolfā and Shamb (d)?		
1427.10.22–1428.10.10	831.00.00	–	–/0	Tāléqān	>VII	6.5	K	Iran or Afghanistan?			
1429.09.30–1430.09.15	833.00.00	–	–/0	Hamédān	$>VII^{+}$	>6.0					
1429.09.30–1430.09.15	833.00.00	–	–/0	Wāsit [Wāset]	?	5.3					
1435.07.27–1436.07.15	839.00.00	–	37.16–55.28/2	Jorjān [old Gorgān, modern Gonbad Kāvus]	>VII	>5.5	50				
1435.07.27–1436.07.15	839.00.00	–	–/0	Āzarbāijān Province	>VII	?	K			+	+
1440.06.02–1441.05.21	844.00.00	–	28.42–53.08/2	Kārzin	VIII	6.7?	10,000?	Kārzin Borough			
1456.11.29–1457.11.18	861.00.00	–	–/0	Amāra	?	?					
1458.11.08–1459.10.27	863.00.00	–	–/0	Central Zāgros	?	>6.5					

Continued

TABLE 17.11 Historical (10–1900) Earthquakes in Iran and the Surroundings [25°00′–41°00′N and 43°00′–64°00′E] (See Also the Notes and Comments at the End of the Table and Chapters 10 and 11)—Cont'd

Date (Christian) (Year.Month. Date)	Arabic Lunar Date	Time (Local)/Day	Macroseismic Epicenter 00°N–00°E/Q	Location	~*I* (MMI)	~M_s	People Killed/ Injured	Cities Destroyed/ Damaged (d)	Towns Destroyed/ Damaged (d)	Villages Destroyed/ Damaged (d)	No. of Houses Destroyed	Direct Economic Loss (Million Tumān; Original Values)
1458.11.08–1459.10.27	863.00.00	–	–/0	Āzarbāijān Province	?	?						
1470.06.30–1471.06.19	875.00.00	–	–/0	Jorjān [old Gorgān, modern Gonbad Kāvus]	>VII	>5.5						
1482.11.03	877.09.21	–	–/0	W. Makrān FRS	?	?						
1483.02.18	888.01.10	–	–/0	W. Makrān	?	7.7		Jārun [Hormoz Island]				
1485.08.15	890.08.03	Sunset/ Sunday	36.75–50.00/3	Upper Polrud	IX	7.2	K^+	Tonékābon, Daylamān		+	+	
1486.07.03	891.07.01	–/Monday	–/1	Upper Polrud AFS								
1493.01.10	898.03.21	Mid-day/ Friday	32.88–59.88/3	Nauzād	IX	7.0	K^+			+	+	
1494.10.02–1495.09.21	900.00.00	–	–/0	Jebāl Province [Media], Hamédān	>VII$^+$	>6.0	?					

1497.04.04–05.02	902.08.00	–	–/0	Hormoz-Gamberon [modern Bandar ʻAbbās]	VIII	6.5	K	Jārun, New Hormoz, or Gamberon?		
1497.08.30–1498.08.18	903.00.00	–	37.18–55.30/2	Jorjān [old Gorgān, modern Gonbad Kāvus]	VIII	6.5	1000	Jorjān [Gorgān]		
1502.07.07–1503.06.25	908.00.00	–	–/0	Hakkāri	>VII	6.9		Mosul (d)		
1505.07.06	911.02.03	–	34.60–68.93/3	Paghmān	IX	7.3		Paghmān, NW Kābol		
1506.05.24–1507.05.12	912.00.00	–	–/0	Shirāz	>VII			Shirāz (d)		
1549.02.15	956.01.17	Night/ Wednesday	32.40–59.12/4	Birjand	VIII^{+}	6.7	2000–3000	Birjand	+	+
1550.01.20–1551.01.08	957.00.00	–	38.10–46.30/2	Tabriz	$>VII^{+}$	>6.0	K			
1566.07.19–1567.07.07	974.00.00	–	38.80–47.40/1	Arasbārān [Qaracheh Dāgh]	VIII?		3 at Qahqaheh castle/prison			
1590.10.20–1591.10.08	999.00.00	–	29.80–52.40/2	NW (?) Shirāz	$>VII^{+}$	5.9		Shirāz	+	+
1593.06.08	1001 Summer	–	37.88–47.52/2	Sarāb	VIII	6.1		Sarāb	+	+
1593.09.00	1001 Late Summer	–	27.65–54.34/2	Lār	VIII	6.5	3000	Lār	+	1200
1598 (?)	1006.00.00?	–	–/0	Mashhad	?	?				
1608.04.20	1017.01.04	AM	36.38–50.50/3	Alamutrud	IX^{+}	7.4	K^{+}			
1619.05.00 (?)	1028.06.00	AM	35.12–58.88/2	Dughābād	VIII	6.5	800		+	+

Continued

TABLE 17.11 Historical (10–1900) Earthquakes in Iran and the Surroundings [25°00′–41°00′N and 43°00′–64°00′E] (See Also the Notes and Comments at the End of the Table and Chapters 10 and 11)—Cont'd

Date (Christian) (Year.Month. Date)	Arabic Lunar Date	Time (Local)/Day	Macroseismic Epicenter 00°N–00°E/Q	Location	~I (MMI)	~M_s	People Killed/ Injured	Cities Destroyed/ Damaged (d)	Towns Destroyed/ Damaged (d)	Villages Destroyed/ Damaged (d)	No. of Houses Destroyed	Direct Economic Loss (Million Tumān; Original Values)
1620	1029–30	–	–/0	Mosul	F	?						
1621.05.21	1030.07.10	–	–/0	Miyāneh	F	?						
1622.10.04	1031.12.08	AM	27.20–56.42/2	Hormoz Island, Bandar 'Abbās	>VII	>5.5		Hormoz, Bandar 'Abbas				
1622	1031–2	–	–/0	Syunik	F	?						
1623	1032–3	–	29.92–52.92/2	Pārsé	>VII	>5.5		Persepolis, Estakhr, Marvdasht		+	+	
1641.02.05	1050.11.04	20:00/Friday	37.95–46.05/4	Dehkhārqān	IX	6.8	12,613	Tabriz	Osku, Dehkhārqān	+	12,620	
1641.02.08	1050.11.07	Monday	37.95–46.05/2	Dehkhārqān AFS	>VII	>5.5						
1643.03.12–1644.02.28	1053.00.00	–	30.35–57.20/1	Kermān	>VII	>5.5		Kermān (d)				
1646.04.07	1056.03.01	Midnight/ Holy Friday	38.37–43.50/4	Hayots Dzor, SE Vān	IX	6.7	2000	Vān (Vāsporākān)	Khoshāp, Echmiādzin			
1659	–	–	–/0	Tātev	?	?						
1664.05.29	1074.11.04	Early Evening/ Friday	–/0	Ārārāt, felt in Tabriz	?	?						

1665.06.15–07.13	1075.12.00	–	35.75–52.08/2	Damāvand	VIII	6.5	K			+	+
1665.07.14–1666.07.03	1076.00.00	–	32.15–50.55/2	NW Ardal	VIII	6.5					
1666.09.22	1077.04.02	–/Saturday	36.70–43.90/2	N. Mosul	IX	6.9					
1668.01.14	1078.08.09	10:00	40.50–48.50/3	Shemākhā	XI	7.0	10,000–80,000?	Shemākhā		+	+
1673.07.30	1084.04.15	–	36.45–59.25/2	Shāndiz	VIII+	6.6	5600 in Mashhad. 1600 in Neyshābur.	66% of Mashhad. 50% of Neyshābur.		+	+
1675 or 1678 (?) Winter	1086 or 1089 (?)	Night	34.30–58.70/2	Gonābād	>VIII	>6.5	K	Gonābād, Bidokht		+	+
1677.02.24–1678.02.12	1088	–	27.84–54.18/1	NW Lār	VIII	6.4	?	Lār			
1678.02.03	1088.12.10	05:30/Wednesday	37.20–50.00/2	Lāhijān	VIII	6.5		Lāhijān			
1678.04 (?)	1089	–	–/0	Mashhad (?)							
1679.06.14	1090.05.05	10:30	40.14–44.76/4	Gārni	X	6.7	>7600	Gārni, Yerevān		+	+
1679.06.22	1090.05.13	10:00	–/1	Gārni AFS							
1679.09.18	1090.08.10	09:00	–/1	Gārni AFS							
1686.04.22	1097.05.28	14:00	–/0	Gārni AFS							
1686.11.07–1687.10.27	1098.00.00	–	–/0	Mazandarān Province	?	?				+	+
1687.04.00	1098.06.00	–	–/0	Mashhad	?	?					
1692.04.13	1103.07.26	–	–/1	Vān (Vāsporākān)					Ādelcivās Castle (d)	+	+

Continued

TABLE 17.11 Historical (10–1900) Earthquakes in Iran and the Surroundings [25°00′–41°00′N and 43°00′–64°00′E] (See Also the Notes and Comments at the End of the Table and Chapters 10 and 11)—Cont'd

Date (Christian) (Year.Month. Date)	Arabic Lunar Date	Time (Local)/Day	Macroseismic Epicenter 00°N–00°E/Q	Location	~I (MMI)	~M_s	People Killed/ Injured	Cities Destroyed/ Damaged (d)	Towns Destroyed/ Damaged (d)	Villages Destroyed/ Damaged (d)	No. of Houses Destroyed	Direct Economic Loss (Million Tumān; Original Values)
1692.10.27	1104.02.16	Evening/ Thursday	–/1	Vān AFS								
1695.05.11	1106.09.27	Dawn	36.98–57.28/3	Esfarāyen	IX	7.0	K^+			+	+	
1696.04.14	1107.09.21	–	39.05–44.00/3	Chālderān	IX	>7.0	K^+		St. Thāddeus	+	+	
1696.05.00	1107.09.00	–	39.00–44.40/2	Chālderān AFS					St. Thāddeus	+	+	
1701.03.26	1112.10.27	Evening/ Friday	38.38–43.35/2	Vān	>VII	>5.5			Pertek [Berdāk]			
1701.11.30	1113.07.10	–	–/0	Vān	F	?						
1703.05.13–1704.05.05	1115.00.00	–	–/0	Turiyān (Qeshm Island)	>VII?	>5.5?				Turiyān		
1705.01.13	1116.09.17	–	–/0	Vān	F	?						
1705	1116–7	–	–/0	Basra	F	?						
1707.07.05	1119.04.04		–/0	Vāsporāgān (Vān)	F	?						
1709.03.13–1710.03.11	1121.00.00	–	–/0	Rasht	F	?						
1713.01.18–1714.01.16	1125.00.00	–	–/0	Rasht	F	?						
1714.01.07–12.26	1126.00.00	–	–/0	Arbil	F	?						

1715.03.08	1127.01.13	Dawn	38.40–43.90/3	E. Vān	VIII	6.6	K^+			+	+
1716.01.23	1128.02.09	–	–/0	Echmiādzin	F	?					
1717.03.12	1129.04.09	Midnight	38.12–46.32/2	Tabriz	VII^+	5.9	>700				4000
1720		–	–/0	Tabriz	F	?					
1721.04.26	1133.06.28	07:48/ Sunday	37.88–46.72/3	Shebli	IX	7.3	>40,000?	Tabriz [3/4th]		+	45,000
1721.07.00	1133.09–10	–/Monday	–/1	S. Qazvin?	>VII?	>5.5?				+	+
1727.06.05	1139.10.15	–	–/0	Echmiādzin	F	?					
1752	1165.03.01– 1166.02.29	–	–/0	Shirāz	F	?					
1755.06.07	1168.08.24	AM	33.78–51.50/3	S. Kāshān	VII^+	5.9	>1200	Kāshān, Fin		+	3000
1759.11.13	1173.03.22	–	–/0	Nākhijevān					St. Stephānos		
1764.07.01– 1765.06.19	1178	Evening	–/0	Mosul	F	?					
1765.04.23	1178.10.29	–	29.60–52.50/2	Shirāz	>VII	>5.5		Shirāz (d)			
1765?	1178–9	–	–/0	E. Makrān (R'as Kuchāri)	?	?					
1766.06.09– 1767.05.19	1180.00.00	–	–/0	Lār	?	?					
1769.05.01	1182.12.24	AM	–/0	Baghdād	F	?					
1778.12.15	1192.11.25	Dawn/ Tuesday	34.04–51.30/3	Kāshān	VIII	6.2	>8000	Kāshān, Fin		+	+
1780.01.07	1193.12.30	–/Friday	–/1	Tabriz FRS	>VI						
1780.01.08	1194.01.01	19:00/ Saturday	38.20–46.18/3	Tabriz	IX	7.4	50,000?	Tabriz	Sufiān, Marand	+	+
1780.02.12	1194.02.06	13:00/ Saturday	–/1	Tabriz AFS	>VI	?				+	+

Continued

TABLE 17.11 Historical (10–1900) Earthquakes in Iran and the Surroundings [25°00′–41°00′N and 43°00′–64°00′E] (See Also the Notes and Comments at the End of the Table and Chapters 10 and 11)—Cont'd

Date (Christian) (Year.Month. Date)	Arabic Lunar Date	Time (Local)/Day	Macroseismic Epicenter 00°N–00°E/Q	Location	~I (MMI)	~M_s	People Killed/ Injured	Cities Destroyed/ Damaged (d)	Towns Destroyed/ Damaged (d)	Villages Destroyed/ Damaged (d)	No. of Houses Destroyed	Direct Economic Loss (Million Tumān; Original Values)
1780.02.19	1194.02.13	21:30/ Sunday	–/1	Tabriz AFS	>VI	?				+	+	
1780	1194		–/0	Khorāsān Province	?	6.5?	3000					
1784.03.01	1198.04.09	Night/ Monday		SE Shirāz	>VII	?						
1784.07.06	1198.08.17	Mid-day/ Friday	–/0	Mosul	F	?						
1785.04.15	1200.06.15	–	–/1	Ray	?	?	Few				+	
1785.08.07	1199.10.01	–	37.10–58.58/1	Quchān	?	?		Khabushān (d)				
1786.07.28	1200.10.02	02:00/Friday	–/0	Basra								
1786.10.00	1201 end of Autumn	Night, before sleep	33.36–45.72/3	Marand	VIII	6.3		Marand, Tabriz	Sufiān	+	+	
1802	1216–1217	–	–/0	Damāvand	F	?			Damāvand	+	+	
1803?	1217–1218?	–	36.43–48.83/2	Soltāniyeh	>VII	>5.6		Soltāniyeh (d)		+	+	
1804	1218–1219	–	36.32–57.22/2	Mehr (WNW Sabzévār)	>VII	>5.6			Mehr	+	+	
1805.04.01–1806.03.20	1220.00.00	–	36.32–52.55/2	Bārforush (Bābol)	>VII			Bārforush (d)		+		

1806.05.00	1221.02–03	–	38.12–46.31/1	Tabriz	>VI	?				40
1806	1220–1221	–	–/0	Bushehr	F	?				
1807.07.11	1222.05.05	Early Morning/ Saturday	38.32–45.20/2	Tasuj	>VII^{+}	>5.5		Tasuj		
1808.02.23	1222.12.25	–	–/0	Jahrom	F	?				
1808.06.00	1223.04–05	–	–/1	Tasuj AFS	?	?				
1808.06.26	1224.06.02	PM/Friday	35.30–54.50/2	Reshm	VIII^{+}	6.6	Few		+	+
1808.10.09	1223.08.18	Night/ Sunday	–/0	Gilān Province	F	?				
1808.12.16	1223.10.29	PM	–/1	NE Qazvin	VII^{+}	5.9		Qazvin (d)	+	+
1809	1224	AM	36.34–52.60/3	Āmol	VIII	6.5		Āmol, Bārforush (d), Ashraf (d)	+	+
1810	1225	–	38.10–57.15/2	Gholāmān	VIII?	6.5?			Gholāmān	+
1811.06.20	1226.05.28	–	–/0	Damāvand	F	?				
1812.05.14	1227.06.02	Night	–/0	Tabriz	F	?				
1812.06.23	1227.06.13	14:00/ Tuesday	–/1	Tabriz	F	?				
1812.00.00	1227	–	–/0	Jolfā	F	?				
1812.00.00	1227	–	–/0	Shirāz	>VII	?				+
1815.06.00	1230.06–07	–	–/0	Damāvand	F	?				
1819.01.00	1234.01.00	–	–/0	Tabriz	F	?				+
1820.06.00	1235.08–09	–	–/0	Tabriz	F	?				
1823.12.00	1239.03–04	–	–/0	Tabriz	F	?				
1823.09.07–1824.08.25	1239.00.00	–	35.98–52.30/2	Bāijān	?	?	Many		+	+

Continued

TABLE 17.11 Historical (10–1900) Earthquakes in Iran and the Surroundings [25°00′–41°00′N and 43°00′–64°00′E] (See Also the Notes and Comments at the End of the Table and Chapters 10 and 11)—Cont'd

Date (Christian) (Year.Month. Date)	Arabic Lunar Date	Time (Local)/Day	Macroseismic Epicenter 00°N–00°E/Q	Location	~I (MMI)	~M_s	People Killed/ Injured	Cities Destroyed/ Damaged (d)	Towns Destroyed/ Damaged (d)	Villages Destroyed/ Damaged (d)	No. of Houses Destroyed	Direct Economic Loss (Million Tumān; Original Values)
1824.06.02	1239.10.04	–/ Wednesday	29.70–51.50/3	Bishāpur	VIII	>6.0	>150 in Kāzerun	Kāzerun		+	+	
1824.06.25	1239.10.27	Dawn	29.75–52.40/3	Guyum (NW Shirāz)	VIII	6.4	>1000	Shirāz (d)		+	+	0.5–1.5
1824.08.28	1240.01.02	–/Saturday	–/1	Guyum AFS	>VI	?						
1824.12.30	1240.05.09	–/Thursday	–/1	Guyum AFS	>VI	?						
1825.10.00	1241.02–03	–	–/1	Shirāz (Guyum AFS?)	VII	?		Shirāz (d)				
1826.06.14	1241.11.08	–	–/1	Kāzerun	F							
1827?	1242–3?	–	33.20–46.10/2	Badra (Badreh)	>VII	?		Badra, Zurbātiya (d)				
1827.10.20	1243.03.28	–/Saturday	40.53–44.84/3	Tsāghkādzor	>VII	5.6						
1828.08.06	1244.01.24	Night	–/0	Stepānākert	F	?						
1829.03.00	1244	Night	–/0	W. Quchān	F	?						
1829.03.00	1244.08–09	Night	–/0	Basaidu (W. Qeshm)	F	?						
1829.04.24	1244	–	–/0	Khoi	F	?						
1829.08.14	1244.02.02	AM	–/0	Stepānākert	F	?						

1830.03.27	1245.10.02	Morning/ Saturday	35.81–51.76/3	Lavāsānāt	IX	7.1	K; 500 in Damāvand; 30 in Tehrān	Tehrān (d)		70	+	0.5 in Tehrān
1830.04.06	1245.10.12	Midnight/ Tuesday	35.76–51.96/1	Lavāsānāt AFS	>VII^{+}	>6.0				+	+	
1831	1246–7	–	–/0	Tabriz	F	?						
1832.12.08	1248.07.15	–	–/0	Lankarān	F	?						
1832.05.31–1833.05.20	1248.00.00	–	37.00–58.20/2	Quchān-Neyshābur Turquoise mine	VIII	6.2		Quchān, Shirvān (d), Turquoise mine (d)		+	+	
1834	1249–50	Morning	39.70–43.70/3	Pāmbuk	VIII	6.0			Dogubāyazit	+	+	
1837.06.00	1253.02–03	–	–/1	Salmās	>VII	>5.5						
1838.03.27–1839.03.17	1254.00.00	–	29.50–59.71/3	Nosratābād	IX	7.0				+	+	
1839.01.29	1254.11.14	–	–/0	Mosul	F	?						
1840.01.22	1255.11.27	Late afternoon	–/0	Lankarān	F	?						
1840.07.02	1256.05.02	19:00/ Thursday	39.52–43.95/4	Ārārāt	IX^{+}	7.4	K^{+}	Dogubāyazit		+		
1840.07.06	1256.05.06	–/Monday	–/1	Ārārāt AFS	VII	?						
1840.07.08	1256.05.08	–/ Wednesday	–/1	Ārārāt AFS		?						
1840.07.26	1256.05.26	Midnight/ Sunday	–/1	Ārārāt AFS	VII	?						
1840.07.27	1256.05.27	16:00/ Monday	–/1	Ārārāt AFS	VII	?	K					
1840.08.06	1256.06.07	–/Thursday	–/1	Ārārāt AFS	VII	?						

Continued

TABLE 17.11 Historical (10–1900) Earthquakes in Iran and the Surroundings [25°00′–41°00′N and 43°00′–64°00′E] (See Also the Notes and Comments at the End of the Table and Chapters 10 and 11)—Cont'd

Date (Christian) (Year.Month. Date)	Arabic Lunar Date	Time (Local)/Day	Macroseismic Epicenter 00°N–00°E/Q	Location	~I (MMI)	~M_s	People Killed/ Injured	Cities Destroyed/ Damaged (d)	Towns Destroyed/ Damaged (d)	Villages Destroyed/ Damaged (d)	No. of Houses Destroyed	Direct Economic Loss (Million Tumān; Original Values)
1840.08.14	1256.06.15	07:00/Friday	–/1	Ārārāt AFS	VI	?						
1840.11.29	1256.10.04	–/Sunday	–/1	Ārārāt AFS		?						
1840.12.07	1256.10.12	06:40/ Monday	–/1	Ārārāt AFS	VII	?						
1841.05.08	1257.03.16	13:00/ Saturday	–/1	Ārārāt AFS		?		Dogubāyazit				
1841.05.17	1257.03.25	19:00/ Monday	–/1	Ārārāt AFS	VII$^+$	5.7						
1841.09.22	1257.08.05	16:00/ Wednesday	–/1	Ārārāt AFS	VII	?						
1842.01.02	1257.11.19	22:00/ Sunday	40.50–50.00/1	Mashtaqi	>VII	?				+		
1843.04.18	1259.03.18	08:00	38.60–44.80/3	Khoi	VIII	5.9	1000	Khoi (d), Māku (d)		+	+	
1843.04.18	1259.03.18	08:30	–/1	Khoi AFS	>VI	?						
1843.04.25	1259.03.25	01:02	–/1	Khoi AFS	>VI	?						
1843.04.25	1259.03.25	15:01	–/1	Khoi AFS	>V	?						
1843.04.26	1259.03.26	00:29	–/1	Khoi AFS	>VI	?						
1843.04.26	1259.03.26	05:00	–/1	Khoi AFS	>VI	?						

1843.04.28	1259.03.28	09:37	–/1	Khoi AFS	>VI	?				
1843.04.29	1259.03.29	13:03	–/1	Khoi AFS	>VI	?				
1843.04.30	1259.03.30	21:00	–/1	Khoi AFS	>V	?				
1843.05.11	1259.04.11	21:02	–/1	Khoi AFS	>V	?				
1843.12.05	1259.11.13	17:56	–/1	Khoi AFS	>VI	?				
1844.05.12	1260.04.23	17:00/ Sunday	33.55–51.35/3	Kāmu (SW Kāshān)	VIII	6.4	1500	Esfahan (d)	+	+
1844.05.12	1260.04.23	18:00	–/1	Kāmu FRS	>VI	?				
1844.05.12	1260.04.23	19:04	–/1	Germirud FRS	>VI	?				
1844.05.13	1260.04.24	Evening/ Monday	37.40–48.85/3	Germirud	IX	6.9	K+		+	+
1845.07.09	1261.07.03	00:01/ Wednesday	–/1	WSW Marand	VII	?				
1846.01.11	1262.01.13	07:00	–/0	Tabriz (F)	VI	?				
1847.09.07	1263.09.26	01:25	–/0	Tehrān	F	?				
1847.00.00	1263	–	–/1	Qā'en	VII?	5.5?		Qā'en (d)		
1848.02.03	1264.02.27	20:08	–/0	Tabriz (F)	>V	?				
1848.02.20	1264.03.15	17:00	–/0	Tabriz (F)	VI	?				
1848.02.20	1264.03.15	19:06	–/1	Tabriz (F)	VI	?				
1848.02.20	1264.03.15	21:14	–/1	Tabriz (F)	VI	?				
1848.03.07	1264.04.01	05:10	–/1	Tabriz (F)	V	?				
1848.03.07	1264.04.01	05:50	–/1	Tabriz (F)	V	?				
1848.07.19	1264.08.17	21:00	–/1	Tabriz (F)	V	?				
1848.08.13	1264.09.13	04:48	–/1	Kuhrud (Qohrud)	F	?				

Continued

TABLE 17.11 Historical (10–1900) Earthquakes in Iran and the Surroundings [25°00′–41°00′N and 43°00′–64°00′E] (See Also the Notes and Comments at the End of the Table and Chapters 10 and 11)—Cont'd

Date (Christian) (Year.Month. Date)	Arabic Lunar Date	Time (Local)/Day	Macroseismic Epicenter 00°N–00°E/Q	Location	~I (MMI)	~M_s	People Killed/ Injured	Cities Destroyed/ Damaged (d)	Towns Destroyed/ Damaged (d)	Villages Destroyed/ Damaged (d)	No. of Houses Destroyed	Direct Economic Loss (Million Tumān; Original Values)
1848.10.00	1265	–	–/1	Ardebil	F	?						
1849.08.07	1265.09.18	10:30	–/1	Tabriz (F)	VI	?						
1849	1265–1266	–	–/0	S. Kahnuj	F	?						
1851.02.16	1267.04.14	03:07	–/1	Tabriz (F)	VI	?						
1851.03.14	1267.05.11	14:20	–/1	Vān	>VII	>5.5						
1851.03.23	1267.05.20	09:01	–/1	Vān AFS	VI	?						
1851.04.09	1267.06.07	16:00	40.00–47.30/2	Pāidākārān	VIII	6.2						
1851.04.13	1267.06.11	15:00	–/1	Pāidākārān AFS	VI	?						
1851.04.19	1267.06.17	17:00	25.10–62.00/1	Gwadur (Gavāder)	?	?						
1851.06.00	1267.07.00	08:30/ Saturday	36.83–58.52/3	Sarvelāyat (SE Khabushān/ Quchān)	IX	6.9	>2000	Quchān		+	+	
1851.10.30	1268.01.04	07:52	–/0	Tabriz (F)	>VI	?						
1851.10.31	1268.01.05	00:07	–/0	Tabriz (F)	>V	?						
1852.01.00	1268	–	–/1	Sarvelāyat AFS	>VII			Quchān (d)				
1852.04.25	1268.07.05	Night	–/0	Shāhrud	F	?						
1852.08.03	1268.10.16	11:40	–/0	Tabriz (F)	>VI	?						

1852.09.19	1268.12.04	–	–/0	Dogubāyazit	F	?				
1853.05.04	1269.07.25	Dawn	–/1	Shirāz FRS	F	?				
1853.05.04	1269.07.25	06:00	–/1	Shirāz FRS	>VII		K	Shirāz (d)		
1853.05.05	1269.07.26	11:45	–/1	Shirāz FRS	>VII		K	Shirāz (d)		
1853.05.05	1269.07.26	12:00	29.60–52.50/3	Shirāz	VIII	6.2	9000–11,000	Shirāz	+	+
1853.06.02	1269.08.24	–	–/1	Izadkhāst FRS	F	?				
1853.06.05	1269.08.27	–	31.26–57.00/2	Izadkhāst	VII	5.5				
1853.06.11	1269.09.04	–/Monday	32.57–50.20/2	Periā (SE Faridan)	VII	5.5			+	+
1853.08.28	1269.11.23	07:07	–/1	Tabriz (F)	VI	?				
1854.09.22	1270.12.29	23:48	–/1	Tabriz (F)	>VI	?				
1854.09.22	1270.12.29	23:49	–/1	Tabriz (F)	IV–V	?				
1854.09.22	1270.12.29	23:51	–/1	Tabriz (F)	IV–V	?				
1854.09.23	1270.12.30	00:46	–/1	Tabriz (F)	IV	?				
1854.09.23	1270.12.30	01:16	–/1	Tabriz (F)	IV	?				
1854.09.27	1271.01.04	20:15	–/1	Tabriz (F)	IV	?				
1854.09.28	1271.01.04	00:40	–/1	Tabriz (F)	IV	?				
1854.10.01	1271.01.08	15:00	–/0	SW Caspian Sea	?	5.9?				
1854.10.13	1271.01.20	23:48	–/1	Tabriz (F)	IV	?				
1854.10.26	1271.02.03	00:07	–/1	Tabriz (F)	IV	?				
1854.11.01	1271.02.09	14:15	–/1	Rasht	F	?				
1854.11.04	1271.02.12	04:00	–/1	Tabriz (F)	F	?				
1854.11.00	1271.02–03	–	30.54–57.35/2	Horjand	VII^+	5.8			+	+

Continued

TABLE 17.11 Historical (10–1900) Earthquakes in Iran and the Surroundings [25°00′–41°00′N and 43°00′–64°00′E] (See Also the Notes and Comments at the End of the Table and Chapters 10 and 11)—Cont'd

Date (Christian) (Year.Month. Date)	Arabic Lunar Date	Time (Local)/Day	Macroseismic Epicenter 00°N–00°E/Q	Location	$\sim I$ (MMI)	$\sim M_s$	People Killed/ Injured	Cities Destroyed/ Damaged (d)	Towns Destroyed/ Damaged (d)	Villages Destroyed/ Damaged (d)	No. of Houses Destroyed	Direct Economic Loss (Million Tumān; Original Values)
1855.01.30	1271.05.11	17:30	–/1	Tabriz (F)	F	?						
1855.03.21	1271.07.02	–/ Wednesday	–/0	Neyshābur Turquoise mine	F	?						
1856.04.06	1272.07.30	–/Sunday	–/0	Tabriz	F	?						
1856.10.04	1273.02.04	19:49/ Saturday	–/0	Tabriz-Vān	F	?						
1856.10.05	1273.02.06	00:41	–/0	Tabriz	F	?						
1857.04.18	1273.08.23	18:30	–/0	Rasht	F	?						
1857.05.00	1273	–	–/0	Neyshābur Turquoise mine	F	?						
1857.06.26	1273.11.04	Lunch time/ Friday	–/0	Tabriz (Tasuj FRS?)	F	?						
1857.08.23	1274.01.02	17:00/ Sunday	–/0	Tabriz (Tasuj FRS?)	F	?						
1857.08.30	1274.01.09	Afternoon	–/0	Tabriz (Tasuj FRS?)	F	?						
1857.09.06	1274.01.15	13:50/ Wednesday	38.35–45.35/2	Tasuj	>VII	?	20 in Dizaj		Tasuj	+	+	
1857.09.07	1274.01.01	06:48 and 07:18/ Thursday	–/1	Tasuj AFS	F	?						

1857.09.14	1274.01.23	16:13	–/1	Tasuj AFS	F	?				
1857.09.15	1274.01.24	16:12 and night	–/1	Tasuj AFS	F	?				
1857.09.16	1274.01.25	04:30	–/1	Tasuj AFS	F	?				
1857.11.19	1274.04.01	21:00	–/1	Tasuj AFS	F	?				
1857.11.27	1274.04.09	–	–/1	Tasuj AFS	F	?				
1858.04.23	1274.09.09	04:30	–/0	Āzarbāijān	VI	?				
1858.06.13	1274.11.01	–	–/0	Bushehr, Genāveh	F	?				
1859.04.11	1275.09.08	–	–/0	Khoi	F	?				
1859.06.04	1276.02.06	Night	–/0	Tabriz (Erzrum AFS?)	F	?				
1860.05.13	1276.10.22	18:30	–/1	Kurā	VI	?				
1861.02.28	1277.08.17	17:00	39.20–47.90/2	Kurā	VII	5.6				
1861.05.24	1277.11.14	16:00	39.40–47.50/2	Kurā	VII	6.0				
1862.10.01	1279.04.06	–	–/0	Rasht, Anzali	F	?				
1862.11.07	1279.05.14	21:00	–/0	Karkuk	F	?				
1862.12.19	1279.06.26	05:00	39.30–47.80/2	Kurā	VII	6.1				
1862.12.21	1279.06.28	10:00	29.50–52.50/3	Shirāz	VIII	6.2	K	Shirāz	+	+
1863.01.01	1279.07.10	–	–/1	Shirāz AFS	F	?				
1863.01.04	1279.07.13	–	–/1	Shirāz AFS	F	?				
1863.01.21	1279.07.30	–	–/1	Shirāz AFS	F	?				
1863.12.30	1280.07.19	21:48	38.12–48.52/3	Hir	VIII	6.1	>1000		+	+
1864.01.02	1280.07.22	–	–/1	Hir AFS	VII	>5.5	K			
1864.01.03	1280.07.23	–	–/1	Hir AFS	F	?				

Continued

TABLE 17.11 Historical (10–1900) Earthquakes in Iran and the Surroundings [25°00′–41°00′N and 43°00′–64°00′E] (See Also the Notes and Comments at the End of the Table and Chapters 10 and 11)—Cont'd

Date (Christian) (Year.Month. Date)	Arabic Lunar Date	Time (Local)/Day	Macroseismic Epicenter 00°N–00°E/Q	Location	~I (MMI)	~M_s	People Killed/ Injured	Cities Destroyed/ Damaged (d)	Towns Destroyed/ Damaged (d)	Villages Destroyed/ Damaged (d)	No. of Houses Destroyed	Direct Economic Loss (Million Tumān; Original Values)
1864.01.04	1280.07.24	–	–/1	Hir AFS	F	?						
1864.01.17	1280.08.07	Night	30.70–57.00/3	Chatrud	VII$^+$	6.0	K	Kermān (d)		+	+	
1864.08.25	1281.03.20	12:45	–/0	Gwadur	F							
1864.12.07	1281.07.08	20:00	33.30–45.90/2	Tursaq	VIII	6.4						
1865.06.00	1282	–	29.62–53.05/2	Dāryān (E. Shirāz)	VII$^+$	6.0						
1865.00.00	1281–2	–	27.20–53.10/2	Maqām	VII	5.6						
1866.05.08	1282.12.22	–	–/0	Chālderān	>VII	>5.5						
1867.07.23	1284.03.21	11:52	40.33–46.64/3	Zurnābād	VII	5.8						
1868.01.20	1284.09.25	–	–/0	Alhāk (E. Shāhrud)	F	?						
1868.03.18	1284.11.24	15:05	40.00–47.00/2	Kārābāgh	>VII	6.5	6			+	+	
1868.04.19	1284.12.26		–/0	Bushehr	F	?						
1868.07.29	1285.04.08	04:00/ Wednesday	–/0	Shiraz	F	?						
1868.08.01	1285.04.11	22:30/ Saturday	–/1	Kavir FRS?	F							
1868.08.30	1285.05.11	23:15/ Sunday	–/1	Kavir?	F	6.4?						

1868.09.11	1285.05.23	19:50/Friday	–/1	Kavir AFS?	F					
1869.03.01	1285.11.17	–	–/0	Bushehr	F	?				
1869.12.26	1286.09.22	19:00	40.70–44.24/3	Ārgāts		4.4				
1870.01.26	1286.10.23	–	–/0	Tabriz	F	?				
1871.03.17	1287.12.24	–	38.00–43.00/2	Hakkāri		6.8				
1871.08.04	1288.06.29	14:00	30.70–56.90/2	Chatrud	VII⁺	6.0			+	+
1871.09.06	1288.06.20	–	–/0	Bushehr	F	?				
1871.12.11	1288.09.28	13:45	39.92–43.54/3	Āgri	>VII	>5.5				
1871.12.23	1288.10.09	Night	37.40–58.25/3	N. Quchān	IX	7.2	30,000 w/ AFSs	Quchān	+	+
1872.01.06	1288.10.24	–/Sunday	37.38–58.58/2	Darbandān (NNE Quchān)	VIII	6.3	4000			
1872.01.28	1288.11.17	–	40.60–48.70/3	Shemākhā		5.7	118			
1872.06.20	1289.04.13	–	–/1	Sonqor/Farsinaj	VIII?	6.1?	1500		+	+
1873.06.20	1290.04.23	21:30	–/0	Baghdād	F	?				
1874.02.00	1290.12–1291.01	–	–/0	Choghākhor Plain	F	?				
1874.06.12	1291.04.26	–	–/0	Baghdād	F	?				
1875.03.21	1292.02.13	15:00	–/0	Dogonbadān	F	5.7?				
1875.05.00	1292	–	31.15–56.32/2	Jur	VII⁺	6.0			+	+
1876.09.28	1293.09.08	03:00	33.00–49.60/2	Periā (NW Faridan)	F	5.8				
1876.10.20	1293.10.01	15:00	35.78–50.20/3	Kaleh Dareh	VII	5.7	35		+	+
1877.00.00	1294	–	30.15–57.60/3	Āb-e Garm, Sirch	VII	5.6			+	+

Continued

TABLE 17.11 Historical (10–1900) Earthquakes in Iran and the Surroundings [25°00′–41°00′N and 43°00′–64°00′E] (See Also the Notes and Comments at the End of the Table and Chapters 10 and 11)—Cont'd

Date (Christian) (Year.Month. Date)	Arabic Lunar Date	Time (Local)/Day	Macroseismic Epicenter 00°N–00°E/Q	Location	~I (MMI)	~M_s	People Killed/ Injured	Cities Destroyed/ Damaged (d)	Towns Destroyed/ Damaged (d)	Villages Destroyed/ Damaged (d)	No. of Houses Destroyed	Direct Economic Loss (Million Tumān; Original Values)
1878 (?)	1295 (?)	–	28.03–57.20/2	Golāshgerd Castle	>VII	>5.5						
1878.04.07	1295.04.14	Daytime/ Wednesday	30.42–53.42/2	Simakān	>VI	>5.0	3				+	
1878.05.11	1295.05.09	Morning/ Saturday	–/0	Shirāz	F	?						
1878.12.21– 1879.01.20	1295.12.26– 1296.01.26	–	–/0	Behbahān	F	?						
1879.03.22	1296.03.28	–	–/1	SE Bozqush FRS	F	?						
1879.03.22	1296.03.28	03:42/ Saturday	37.80–47.85/3	SE Bozqush	VIII^{+}	6.7	>2000	Miyāneh and Ardebil (d)		21, 54(d)	+	
1879.03.22	1296.03.28	04:00/ Saturday	–/1	SE Bozqush AFS	F	?						
1879.03.22	1296.03.28	09:00/ Saturday	–/1	SE Bozqush AFS	F	?						
1879.03.22	1296.03.28	17:00/ Saturday	–/1	SE Bozqush AFS	F	?						
1879.03.22	1296.03.28	21:00/ Saturday	–/1	SE Bozqush AFS	F	?						
1879.03.25	1296.04.01	–	–/1	SE Bozqush AFS	F	?						

1880.07.04	1297.07.26	–	36.45–47.55/3	SE Takht-e Solaymān	VII	5.6	60	+	+
1880.07.05	1297.07.27	–	–/1	SE Takht-e Solaymān AFS	F	?			
1880.08.00	1297	12:00	27.00–54.20/3	Jonāh	>VII	>5.5	>100	+	+
1880.00.00	1297	–	32.00–50.68/3	Ardal	>VII	>5.5			
1881.08.17–09.24	1298.09.21–10.29	–	–/0	Kāzerun	F	?			
1881.08.28	1298.10.02	–	–/1	Khoi	>VI	>5.3			
1881.09.13	1298.10.18	–	–/1	Khoi	F	?			
1881.09.25–10.21	1298.10.30–11.27	–	–/0	Shirāz	F	?			
1881.09.30–10.12	1298.11.06–18	–	–/0	Khoi, 2 shocks in 2 nights	F?				
1882.03.30	1299.05.10	–	–/0	Estārehābād (Gorgān)	F	?			
1883.04.28	1300.06.20	17:00	36.88–56.30/2	Garmeh (W. Esfarāyen)	VII⁺	5.8			
1883.05.03	1300.06.25	12:00	38.84–47.10/2	Duzduzān	VIII	6.2		+	+
1883.10.16	1300.12.14	Night	27.40–52.20/3	Kangān	VII⁺	5.8	50	+	+
1883.12.26	1301.02.25	01:00/ Wednesday	–/0	Mashhad	F	?			
1884.05.19	1301.07.23	Night/ Monday	26.81–55.91/4	Central Qeshm Island	VIII	>6.0	238/500	31	
1884.10.27	1302.01.07	AM	–/0	Karkuk	F	?			

Continued

TABLE 17.11 Historical (10–1900) Earthquakes in Iran and the Surroundings [25°00′–41°00′N and 43°00′–64°00′E] (See Also the Notes and Comments at the End of the Table and Chapters 10 and 11)—Cont'd

Date (Christian) (Year.Month. Date)	Arabic Lunar Date	Time (Local)/Day	Macroseismic Epicenter 00°N–00°E/Q	Location	~I (MMI)	~M_s	People Killed/ Injured	Cities Destroyed/ Damaged (d)	Towns Destroyed/ Damaged (d)	Villages Destroyed/ Damaged (d)	No. of Houses Destroyed	Direct Economic Loss (Million Tumān; Original Values)
1885.00.00	1302–3	–	–/0	Mālamir (Izeh)	F	?						
1885.10.10–12.08	1303.01–02	–	–/0	Kāzerun	V	?						
1887.11.14–24	1305.02.27–03.08	–	–/0	Bushehr	F	?						
1888.08.13–1888.09.17	1305.12.05–1306.01.11	–	–/0	Shiraz	F	?						
1890.02.07	1307.06.16	06:00	34.18–51.22/3	Nassirābād (NW Kāshān)	>VII	>5.5						
1890.03.25	1307.08.03	–	28.78–53.65/2	S. Fasā	VIII	6.4	30					
1890.07.11	1307.11.23	00:55/Friday	36.62–54.60/4	Tāsh	IX	7.2	K+	Bastām, Shāhrud, and Gorgān (d)		+	+	
1890.07.11	1307.11.23	01:33/Friday	–/1	Tāsh AFS	>V	?						
1890.07.11	1307.11.23	01:47/Friday	–/1	Tāsh AFS	>IV	?						
1890.07.12	1307.11.24	–/Saturday	–/1	Tāsh AFS	F (many)	?						
1890.07.13	1307.11.25	–/Sunday	–/1	Tāsh AFS	F (11)	?						
1890.07.14	1307.11.26	07:22/ Monday	–/1	Tāsh AFS	>IV	?						

1890.07.14	1307.11.26	10:25/ Monday	–/1	Tāsh AFS	>V	?
1890.07.14	1307.11.26	15:00/ Monday	–/1	Tāsh AFS	>IV	?
1890.07.15	1307.11.27	04:45/ Tuesday	–/1	Tāsh AFS	>V	?
1890.07.15	1307.11.27	20:00/ Tuesday	–/1	Tāsh AFS	>IV	?
1890.07.16	1307.11.28	–/ Wednesday	–/1	Tāsh AFS	F>IV (many)	?
1890.07.17	1307.11.29	14:45/ Thursday	–/1	Tāsh AFS	>V	?
1890.07.17	1307.11.29	15:05/ Thursday	–/1	Tāsh AFS	>V	?
1890.07.17	1307.11.29	16:10/ Thursday	–/1	Tāsh AFS	>IV	?
1890.07.17	1307.11.29	17:10/ Thursday	–/1	Tāsh AFS	>IV	?
1890.07.18	1307.11.30	00:45/Friday	–/1	Tāsh AFS	>VII	?
1890.07.18	1307.11.30	03:00/Friday	–/1	Tāsh AFS	>IV	?
1890.07.18	1307.11.30	05:00/Friday	–/1	Tāsh AFS	>IV	?
1890.07.18	1307.11.30	05:45/Friday	–/1	Tāsh AFS	>IV	?
1890.07.18	1307.11.30	06:45/Friday	–/1	Tāsh AFS	F	?
1890.07.18	1307.11.30	09:00/Friday	–/1	Tāsh AFS	F	?
1890.07.18	1307.11.30	11:05/Friday	–/1	Tāsh AFS	F	?
1890.07.18	1307.11.30	23:00/Friday	–/1	Tāsh AFS	F	?
1890.07.19	1307.12.01	01:30/ Saturday	–/1	Tāsh AFS	F	?

Continued

TABLE 17.11 Historical (10–1900) Earthquakes in Iran and the Surroundings [25°00′–41°00′N and 43°00′–64°00′E] (See Also the Notes and Comments at the End of the Table and Chapters 10 and 11)—Cont'd

Date (Christian) (Year.Month. Date)	Arabic Lunar Date	Time (Local)/Day	Macroseismic Epicenter 00°N–00°E/Q	Location	~I (MMI)	~M_s	People Killed/ Injured	Cities Destroyed/ Damaged (d)	Towns Destroyed/ Damaged (d)	Villages Destroyed/ Damaged (d)	No. of Houses Destroyed	Direct Economic Loss (Million Tumān; Original Values)
1890.07.19	1307.12.01	03:00/ Saturday	–/1	Tāsh AFS	F	?						
1890.07.19	1307.12.01	08:18/ Saturday	–/1	Tāsh AFS	F	?						
1890.07.19	1307.12.01	08:23/ Saturday	–/1	Tāsh AFS	F	?						
1890.07.25	1307.12.07	23:56/Friday	–/1	Tāsh AFS	>VI	?						
1890.07.27	1307.12.09	13:28/ Sunday	–/1	Tāsh AFS	>VII	?						
1890.08.02	1307.12.15	11:53/ Saturday	–/1	Tāsh AFS	F	?						
1890.08.06	1307.12.19	03:00/ Wednesday	–/1	Tāsh AFS	F	?						
1890.08.11	1307.12.24	11:40/ Monday	–/1	Tāsh AFS	F	?						
1890.08.11	1307.12.24	11:42/ Monday	–/1	Tāsh AFS	F	?						
1890.08.14	1307.12.27	13:30/ Thursday	–/1	Tāsh AFS	F	?						
1890.08.16	1307.12.29	08:52/ Saturday	–/1	Tāsh AFS	F	?						

1890.08.16	1307.12.29	13:30/ Saturday	–/1	Tāsh AFS	F	?
1890.08.17	1308.01.01	08:24/ Sunday	–/1	Tāsh AFS	>VI	?
1890.08.17	1308.01.01	18:20/ Sunday	–/1	Tāsh AFS	>VI	?
1890.08.19	1308.01.03	13:30/ Tuesday	–/1	Tāsh AFS	F	?
1890.08.19	1308.01.03	16:00/ Tuesday	–/1	Tāsh AFS	F	?
1890.08.20	1308.01.04	00:30/ Wednesday	–/1	Tāsh AFS	F	?
1890.08.20	1308.01.04	03:00/ Wednesday	–/1	Tāsh AFS	F	?
1890.08.23	1308.01.07	02:15/ Saturday	–/1	Tāsh AFS	F	?
1890.08.23	1308.01.07	02:30/ Saturday	–/1	Tāsh AFS	F	?
1890.08.23	1308.01.07	05:00/ Saturday	–/1	Tāsh AFS	F	?
1890.08.24	1308.01.08	13:30/ Sunday	–/1	Tāsh AFS	F	?
1890.08.26	1308.01.10	20:25/ Tuesday	–/1	Tāsh AFS	F	?
1890.08.27	1308.01.11	02:00/ Wednesday	–/1	Tāsh AFS	F	?
1890.08.27	1308.01.11	06:42/ Wednesday	–/1	Tāsh AFS	F	?
1890.08.31	1308.01.15	17:18/ Sunday	–/1	Tāsh AFS	F	?

Continued

TABLE 17.11 Historical (10–1900) Earthquakes in Iran and the Surroundings [25°00′–41°00′N and 43°00′–64°00′E] (See Also the Notes and Comments at the End of the Table and Chapters 10 and 11)—Cont'd

Date (Christian) (Year.Month. Date)	Arabic Lunar Date	Time (Local)/Day	Macroseismic Epicenter 00°N–00°E/Q	Location	~I (MMI)	~M_s	People Killed/ Injured	Cities Destroyed/ Damaged (d)	Towns Destroyed/ Damaged (d)	Villages Destroyed/ Damaged (d)	No. of Houses Destroyed	Direct Economic Loss (Million Tumān; Original Values)
1891.02.06	1308.06.26	–	–/1	Adelcivās (N. Lake Vān)								
1891.12.14	1309.05.12	Night	29.90–51.58/2	N. Kāzerun	>VII	>5.5						
1892.08.15	1310.01.21	AM/Monday	–/1	Kavār?	VIII?	>5.5				+	+	
1893.10.20	1311.04.09	Midnight/ Friday	–/1	S. Quchān FRS	>VI	?						
1893.11.17	1311.05.08	19:36/Friday	36.96–58.35/4	S. Quchān	IX	7.0	12,000–18,000 5000–7000 in Quchān			+	+	
1893.11.18	1311.05.09	Night/ Saturday	–/1	S. Quchān AFS (Gabrābād)	>VII	?	K			+	+	
1893.11.19	1311.05.10	Dawn/ Sunday	36.99–58.48/2	S. Quchān AFS (Kalukhi)	>VII	?	K			+	+	
1894.01.12	1311.07.04	Evening/ Friday	–/1	S. Quchān AFS	>VII	?	K			+	+	
1894.02.26	1311.08.19	01:30/ Monday	29.50–53.30/3	Kharāmeh	VII$^+$	5.9	K			+	+	
1894.05.08	1311.11.03	11:30/ Tuesday	–/0	Tabriz	F	?						
1894.06.20	1311.12.16	–/ Wednesday	–/0	Salmās	F	?						

1895.01.17	1312.07.20	11:30/ Thursday	37.08–58.38/4	Quchān	VIII⁺	6.8	>1000 700 in Quchān	+	+
1895.01.22	1312.07.25	–/Tuesday	–/1	Quchān AFS	>VII	>5.5		+	+
1895.07.08	1313.01.15	02:30 [07/09, Tuesday] (21:30 UTC: 07:08)/ Monday	40.00–52.80	Krasnovodsk	IX	7.5			
1895.10.26	1313.05.07	22:35	–/0	Karkuk	F	?			
1895.12.18	1313.07.01	–/ Wednesday	–/1	Sangāvar FRS	>VI	?		+	+
1895.12.24	1313.07.07	–/Tuesday	–/1	Tehrān	>VI	?			+
1895.00.00	1312.08.01–1313.07.30	17:30	34.18–51.23/2	Sāruq	>VII	>5.5			
1896.01.02	1313.07.16	Night/ Thursday	–/1	Sangāvar FRS	>VII	>5.5	300 in Sangābād	+	+
1896.01.04	1313.07.18	18:28/ Saturday	37.98–48.38/4	Sangāvar	VIII⁺	6.7	>1100	+	+
1896.01.05	1313.07.19	10:51	–/1	Sangāvar AFS	>VII	>5.5			
1896.01.08	1313.07.22	10:50	–/0	Mashhad and Kalāt	F	?			
1896.01.14	1313.07.28	–/Tuesday	–/1	Sangāvar AFS	>VII	>5.5	K	+	+
1896.02.03	1313.08.18	18:25	38.10–43.40	SE Vān	~VII	~5.5			
1896.04.03	1313.10.19	21:48	–/1	Vān AFS	F	?			
1896.08.23	1314.03.14	–/Sunday	–/1	Sangāvar AFS	>VI	?			
1896.09.14	1314.04.06	–/Monday	–/1	Sangāvar AFS	F	?			

Continued

TABLE 17.11 Historical (10–1900) Earthquakes in Iran and the Surroundings [25°00′–41°00′N and 43°00′–64°00′E] (See Also the Notes and Comments at the End of the Table and Chapters 10 and 11)—Cont'd

Date (Christian) (Year.Month. Date)	Arabic Lunar Date	Time (Local)/Day	Macroseismic Epicenter 00°N–00°E/Q	Location	~I (MMI)	~M_s	People Killed/ Injured	Cities Destroyed/ Damaged (d)	Towns Destroyed/ Damaged (d)	Villages Destroyed/ Damaged (d)	No. of Houses Destroyed	Direct Economic Loss (Million Tumān; Original Values)
1896.10.01	1314.04.23	06:30	–/0	Zāngézur FRS	F	?						
1896.10.01	1314.04.23	11:00	39.50–46.40/3	Zāngézur	>VII	>5.5						
1897.01.10	1314.08.06	21:00 (UTC)/ Sunday	26.95–56.26/3	Qeshm Town	VIII	6.4	1600			+	+	
1897.02.07	1314.09.05	Night/ Sunday	–/0	Quchān	F	?						
1897.05.22	1314.12.20	–/Saturday	31.32–56.28/2	Kuhbanān	VII	5.5	K	Kermān (d)				
1897.05.27	1314.12.25	23:30/ Thursday	30.70–57.00/3	Chatrud	>VII	>5.5	K			+	+	
1897.07.20	1315.02.09	18:20	–/0	Shirāz	F	?						
1897.07.27	1315.02.26	18:24	–/0	Shirāz	F	?						
1897.12.20	1315.07.25	Night/ Monday	–/1	Sangāvar AFS	F	?						
1898.01.15	1315.08.21	03:42/ Saturday	36.60–54.70/2	Shāhkuh	>VII	>5.5		Estārehābād [Gorgān] (d)				
1898.03.04	1315.10.10	21:30/Friday	–/0	Shirāz	F	?						
1898.03.04	1315.10.10	22:00/Friday	–/0	Sarāb	F	?						
1898.12.01	1316.07.17	12:18	–/0	Shirāz	F	?						

1898.12.05	1316.07.21	–	–/0	Shirāz	F	?
1899.01.13	1316.09.01	–	–/0	Hakkāri	F	?
1899.05.09	1316.12.28	–	–/0	Hakkāri	F	?
1899.05.09	1316.12.28	AM	–/0	Shirāz	F	?

Revised and modified after Heuckroth and Karim (1970), Nabavi (1972), Ambraseys (1974a), Berberian (1976a, 1977a, 1979b, 1981a, 1983a, 1994, 1995, 1997a, 2005), Ambraseys and Melville (1982), Kondorskaya and Shebalin (1977, 1982), Berberian et al. (1983, 1984, 1985, 1992, 1996, 2000a), Bommer and Ambraseys (1989), Guidoboni (1994), Guidoboni and Traina (1995), Ambraseys and Finkel (1987, 1995), Ambraseys (1997), Berberian and Yeats (1999, 2001), Ambraseys and Bilham (2003a), Guidoboni and Comastri (2005), Babayan (2006), Ambraseys (1997, 2009), Berberian and Walker (2010); travel books; contemporary and near-contemporary Persian, Arabic, and Armenian treatises, colophons, and newspapers; and personal databank collected since 1971. The list is not complete and more work (which is in progress) is needed to complete the data set.

The macroseismic data are commonly incomplete and inhomogeneous and, for some regions, lacking entirely. This is due to a lack of historic data in the remote, relatively lowly populated desert and mountainous regions far from the main routes of historic communications. More work is needed to complete the macroseismic data of historical earthquakes.

In order to avoid duplication and find the possible origin of the felt data, earthquakes outside the modern borders of Iran are mentioned; they have been felt in the country.

AFS: (Damaging) aftershock. Number of aftershocks in brackets.

F: Earthquake felt. They are included since they may possibly represent fringe effects of medium- to large-magnitude distant earthquakes for which no additional data have yet been identified in the historical documents.

FRS: (Damaging) foreshock.

I: Highest intensity (MMI) estimated within the meizoseismal area based on written accounts.

K: People killed.

K^+: Unknown large fatalities.

M_s: The equivalent surface-wave magnitudes of the historical (pre-1900) earthquakes mentioned throughout the text are estimates taken from Ambraseys et al. (1979), Berberian (1981a), Ambraseys and Melville (1982), Ambraseys and Finkel (1987, 1995), Ambraseys and Adams (1989), Berberian (1994), and Ambraseys and Bilham (2003a), unless otherwise stated. These pre-instrumental magnitudes were derived from macroseismic information embedded in written accounts calibrated against instrumental M_s values based on the regional twentieth-century earthquakes. Therefore, they may represent poorly constrained magnitude estimates for some historical events.

~: Approximate estimation.

`: 840-Hollingsworth et al. (2010) erroneously located this event at Torbat-e Jām.

#: The fatality figures are exaggerated and do not fit the population levels of the area. Therefore, the numbers should be taken as an indication that the death toll was very high.

$^\alpha$: See discussion in Chapter 11.

?: Events or values queried where more speculative.

Comments on some earthquake dates and the felt data:

242–245: The Antioch earthquake was felt in Iran (Sieberg, 1932).

276–578: The earliest date of AD 224–549 (Berberian and Yeats, 2001) was bracketed on data by Kāmbakhsh-Fard (1994). The present date is bracketed on new data by Āzarnoush (2009).

363: An earthquake on 363.05.18–19 took place in Palestine-Jordan in this year (Guidoboni, 1994; Sbeinati et al., 2005).

531–534: Sieberg (1932) reported that the Syrian [Allepo-Homs] earthquake was felt (IV) in Mesopotamia (Sbeinati et al., 2005).

881.05.16: The Syrian earthquake of 267 AH [880.08.12–881.07.31] was also felt in Mesopotamia (Ibn al-Athir, 1231; Guidoboni, 1994; Sbeinati et al, 2005). It is not clear yet if the 880.09.29–10.28 earthquake for Baghdād is related to this event.

893.12.18: 893.12.24 (sic) in Ambraseys and Melville (1982, p. 158).

1013.03.18–1014.03.17 (Armenian year of 462): The exact date is not known. Ambraseys (2009) gives 1016.03.17–1017.03.16 (the Armenian year of 465). See also Guidoboni and Traina (1995) and Guidoboni and Comastri (2005).

1042.11.04: Ambraseys and Melville (1982, p. 39) give 17 Rabi' II (sic). It is not clear yet if the strong earthquake felt in Tabriz (III) in 1042.08.21 (reported by Sieberg, 1932; Ben-Menahem, 1979) was one of the aftershocks of the 434H [1042.08.21–1043.08.09] Palymra earthquake (see Sbeinati et al., 2005).

1135.05.25 and 1135.08.13: Sieberg (1932) reported that an earthquake in 1135 was felt in Syria (see also Sbeinati et al., 2005).

Continued

1137.10.19–11.16: The Syrian earthquake of 532.02.00H was felt in al-Mousel, Mesopotamia, and Iraq (Ibn al-Athir, 1231; Abu'l-Fida, 1329; Sieberg, 1932; Ben-Menahem, 1979; Sbeinati et al., 2005).
1170.06.29: The 565H al-Shām (Syria) earthquake was felt in al-Mousel, al-Jazira, Iraq, and Mesopotamia (Ibn al-Athir, 1231; Ben-Menahem, 1979).
1202.05.20: The Kittany valley, M. Lebanon-Baalbek earthquake was felt in Mesopotamia, Lesser Armenia, and NW Iran (Ibn al-Athir, 1231; Sibt Ibn al-Jauzi, 1256; Ambraseys et al., 1994; Sbeinati et al., 2005).
1287.03.22: According to Sieberg (1932) an earthquake [the Lattakia, Syria, earthquake; Sbeinati et al., 2005] was felt in Armenia (IV). Ben-Menahem (1979) added that it was destructive in northern Syria and Armenia (Willis, 1928; Wills, 1933; Amiran, 1950, 1951; Alsinawi and Ghalib, 1975; Plassard and Kogoj, 1981).
1549.02.15: 1549.02.16 (sic) by Utsu (2002, p. 692) in his "List of the Deadly Earthquakes in the World: 1500–2000."
1574: 1574 [Fin, Kashan, 1200 killed] (sic) in Utsu (2002, p. 692) should be 1755.06.07.
1577: The northern Syrian earthquake was felt (IV) in Armenia (Sieberg, 1932; Sbeinati et al., 2005).
1619.06: Duplicated by Utsu (2002, p. 693) in his "List of the Deadly Earthquakes in the World: 1500–2000" as: (i)1618.12.19 [800 killed at Dughabad] (sic), and (ii) 1619.05. [800 killed at Dughabad] (sic).
1640: Perrey (1850), Sieberg (1932), Plassard and Kogoj (1981), and Sbeinati et al. (2005) reported that the Damascus earthquake was felt in Tabriz.
1646.04.07: Tchalenko (1977), Ambraseys and Melville (1982), Brommer and Ambraseys (1989) reported the event on 1648.03.31. Contemporary source Ārākel Vārtābed of Tabriz (Bournoutian, 2006) mentioned April 2 in the Armenian Calendar 1097 [1648], on the night of Holy Friday and the Eve of Holy Sunday. Bosset (1874–6) stated that Easter occurred on April 2nd and Holy Friday was March 31th. Ambraseys and Finkel (1995) and Ambraseys (2009) gave 1646.04.07.
1665: Utsu (2002, p. 694) gives 1665.06.15 (06.15–07.13) [many killed at Damavand] (sic).
1666.09.22: The al-Mousel earthquake was felt (V) in Tabriz and Van (Theatrum Europaeum, 1617–1721; Hammer-Purgstal, 1832; Fiey, 1965; Ambraseys, 1989, 2009).
1673.07.30: Utsu (2002, p. 694) gives M 7.1.
1679.06.14: Ambraseys and Melville (1982, p. 51) give June 04 (sic).
1693–4: Al-'Umari (1793) and Ambraseys and Finkel (1995) reported an earthquake being felt in NW Iraq.
1721.04.26: Triplicated recently by Utsu (2002, p. 695) in his "List of the Deadly Earthquakes in the World: 1500–2000" as: (i) 1721.04.26 [*M* 7.4, 40,000 killed in Tabriz], (ii) 1721.09.26 [*M* 6.0, many killed in Tabriz] (sic), and (iii) 1721.11.18 [*M* 7.2, 77,000 killed in Tabriz] (sic).
1820: Utsu (2002, p. 698) presents an earthquake in 1820 (10.19?) Shiraz, Babol (sic) many killed, without mentioning his source.
1823: Utsu (2002, p. 698) gives an earthquake on 1823.09.07 [many killed in Mazandaran], without mentioning the source.
1827.10.20: Babayan (2006) gives 1827.10.08. See Brommer and Ambraseys (1989), Ambraseys (2009).
1830.03.27: Utsu (2002, p. 699) gives duplicate of (i) 1830.03.27 [530 killed in Damavand, Shamiranat, Tehrān], and (ii) 1830.09.09 [500 killed in Tehrān, Damavand, Semnan].
1847.00.00: Ambraseys and Melville (1977) give "some years before 1882."
1853.06.11: Ambraseys (1971) gives July 11. See Ambraseys and Melville (1982).
1864.01.02: Utsu (2002, p. 699) gives 1864.01.03 [500 killed in Shiraz].
1879.03.22: Utsu (2002, p. 700) triplicated this event and added an earthquake somewhere else: (i) 1879.03.13 [many killed in Uremia Lake, Marand], (ii) 1879.03.22 [2000 killed in Bozqush, Gertrud, Ardabil], (iii) 1879.03.22 [992 killed in Mianeh], and (iv) 1879.04.02 [700 killed in Bodzhnurd].
1890.07.11: Utsu (2002, p. 701) duplicates this event: (i) 1890.06.28 [many killed in Tasch, Meshed, Tehrān (6.27–28)], (ii) 1890.07.11 [100s killed in Tash, Gorgan, Shahrud].
1893.11.27: Despite numerous attempts and correspondences with the scientists in Moscow, Russia, and Yerevan, Armenia, I have not yet been able to consult the detailed list of the aftershocks of this event, which appeared in the Russian report by Baumgarten (1896).
1894.02.26: Utsu (2002, p. 702) gives 1894.02.27 [many killed in Karameh, Shiraz].

[a] *Date: Dates are in the Gregorian System (New Style).*
[b] *Arabic Lunar Date: The Arabic Lunar Date is used to facilitate checking with the original historical Arabic and Persian sources dealing with each earthquake.*
[c] *Time (local)/Day: Given in a 24-h clock. The Iranian local time was 4 h:30 m ahead of the GMT (Greenwich Mean Time)/UTC (Coordinated Universal Time).*
[d] *Macroseismic Epicenter: Center of affected (meizoseismal) area.*
[e] *Q: Estimated geometric center of the historically reported sites in the meizoseismal area that experienced the maximum effects of the earthquake. The quality of the macroseismic epicenter is classified as: 0 = Unidentified location. No details of earthquake effects are available. 1 = Shock felt or occurred with slight damage in a few sites. It may indicate a small local event or an effect of a medium- to large-magnitude distant earthquake not yet identified. 2 = Significantly documented earthquake destroying one or two sites. 3 = Significantly documented historical earthquake at a number of sites with rough epicentral assessment. 4 = Well-documented historical earthquake with effects at several sites with good epicentral assessment.*
[f] *Location: Name of the earthquake, not necessarily used as macroseismic epicenter.*

TABLE 17.12 Instrumental Period Medium- to Large-Magnitude Earthquakes in Iran (1900–2014)

Date (Christian)	Time (UTC)	Epicenter 00°N–00°E	Location	*mb*	M_s	M_w	*I* (MMI)	CD_w (km)	PGA (H/V)	People Killed/ Injured	Official Death Toll/ Injured	Cities Destroyed/ Damaged (d)	Towns Destroyed/ Damaged (d)	Villages Destroyed/ Damaged	No. of Houses Destroyed/ Damaged	Direct Economic Loss (U.S$ m, Original values)
1900–1917: The British Association for the Advancement of Science (BAAS) Seismic Data Period																
1900.02.24	00:30	38.42–44.81*	Khoi		5.4		VII	–	–	++		Khoi (d)		5	++	
1901.05.20	12:29	36.39–50.48*	Alamut		5.4		VII	–				Qazvin (d)		+	+	
1901.08.05	18:35	–			6.1		–	–								
1902.02.13	09:40	40.65–48.60*	Shemākhā		6.5		VIII			100						
1902.07.09	03:38	26.95–56.28*	Qeshm town		6.4		VIII	–	–	10		–	Qeshm			
1903.02.09	05:18	36.58–47.67	Angurān area		5.6		VII	–								
1903.03.22	14:35	33.16–59.71*	Dorakhsh?		6.2	6.1	–	–	–			–	Dorakhsh			
1903.06.24	16:56	–	Anzali?		6.0		VII+	–	–			–				
1903.09.25	01:20	35.23–58.25*	W. Turshiz (Kāshmar)	5.3	5.9	6.0	VIII	–	–	>350		Turshiz		25		
1903.11.14	–	–	Kārzin				VII+	–								
1904.11.09	03:28	36.94–59.77*	Kaakhkā-Klāt-e Nāderi?		6.4	6.3	VIII	–								
1904.11.09	03:30	–			5.3	5.5		–								
1905.01.09	06:17	37.00–48.68*	Darrām		6.0		VII+	–	–			–		Darrām?		
1905.04.25	14:01	27.42–56.28*	Issin-Genu		5.8		VII+	–	–			–		>2		
1905.06.19	01:27	29.87–59.92*	Nosratābād	6.8	6.0	6.0	VII+	–	–			–		>1		

Continued

TABLE 17.12 Instrumental Period Medium- to Large-Magnitude Earthquakes in Iran (1900–2014)—Cont'd

Date (Christian)	Time (UTC)	Epicenter 00°N–00°E	Location	*mb*	M_s	M_w	*I* (MMI)	CD_w (km)	PGA (H/V)	People Killed/ Injured	Official Death Toll/ Injured	Cities Destroyed/ Damaged (d)	Towns Destroyed/ Damaged (d)	Villages Destroyed/ Damaged	No. of Houses Destroyed/ Damaged	Direct Economic Loss (U.S$ m, Original values)
1907.04.17	08:36	37.74–57.85*	NW Bāghān		5.8			–								
1907.07.04	09:12	27.22–56.28*	Bandar 'Abbās area	6.0	5.7			–								
1908.09.28	06:28	–	Bash Kaleh?		6.0			–								
1909.01.23	**02:48**	**33.38–49.28***	**Silākhor**	**7.2**	**7.4**	**7.4**	**IX$^+$**	–	–	**8000**		**Dorud [Bahrain]**		**130**	**++**	
1909.04.11	04:02	33.10–49.52*	Arjanak, Silākhor AFS		5.6		VII$^+$							+	+	
1909.10.27	18:45	30.10–57.63*	Jowshān		5.5		VII	–	–			–		>3		
1910.12.04	14:02	39.35–47.26*	Moghān		5.4											
1911.01.20	04:05	29.30–51.20*			5.2											
1911.04.18	18:14	31.26–56.88*	Rāvar	6.7	6.4		VIII	–	–	700		Rāvar		>9		
1911.04.29	05:33	30.35–57.65*	Feyzābād	6.4	5.6		VII	–	–	++		–		>2		
1911.06.07	23:58	40.00–51.00		6.4	6.1			–								
1911.09.13	03:22	29.65–54.45*	Bastak-Lār		5.5		VII	–	–	10		Bastak		>2	>150	
1912.02.24	14:36	37.00–59.50^K			5.3	5.6		–								
1913.03.24	10:34	26.80–53.70*	Chiru		5.8		VII$^+$	–								
1913.04.16	06:00	39.10–48.60*		5.2			VI$^+$									
1913.04.25	04:13	30.59–56.95*	Chatrud		5.2		VI							+	+	
1914.02.06	11:42	28.65–64.75*	Rās Kuh, Makrān	5.9			–	–	–	–						
1915.08.29	13:00	37.40–58.00*			5.7											

1916.11.14	13:54	40.80–44.40*			5.3		VI$^+$						
1917.07.15	17:58	33.50–45.95*	Tursaq	6.3	5.6			–					
1917.07.15	21:22	35.50–46.50	Tursaq AFS		5.5								
1917.07.24	16:12	35.50–46.50	Tursaq AFS		5.5								
1917.08.29	13:00	37.38–58.10*	Bāghān		5.7	5.8		–					
1917.10.24	11:–	–	Gorgan area		5.3			–					
1917.11.28	14:42	37.10–57.90*	Kaliān, NE Esfarāyen		5.9	5.9	VII$^+$	–	–		–	+	++
1917.11.28	17:43	37.38–58.10*	Bāghān		5.2								
1918–1963 International Seismology Summary (ISS) Seismic Data Period													
1918.03.24	23:14	35.35–60.70*	Torbat Sheikh Jām	6.3	6.0	6.0	VII$^+$	–	–		–	+	++
1919.05.12	22:30	36.20–44.00	Arbil, Iraq		5.7			–		2		+	+
1919.10.24	20:32	26.10–62.00	Mand, Makrān		5.6			–					
1922.03.21	16:56	31.11–51.11*	Vastégān	5.5			VII						
1923.05.25	22:21	35.22–59.11*	Kāj Derakht		5.8	5.9	VII$^+$	–	–	2200	–	>7	
1923.09.14	08:10	27.97–59.33			5.6	5.8		–					
1923.09.17	07:09	37.66–57.20*	N. Bojnurd	6.1	6.4	6.3	VIII	–	–	157	–	32	
1923.09.22	20:47	29.62–56.60*	Lālehzār	6.9	6.7	6.7	VIII$^+$	–	–	290	–	>12	
1923.09.23	03:18	29.62–56.60*	Lālehzār AFS		5.4		VII						
1923.11.29	03:36	35.40–59.20*	SE Qā'en		5.3	5.5	VII	–	–		–		++

Continued

TABLE 17.12 Instrumental Period Medium- to Large-Magnitude Earthquakes in Iran (1900–2014)—Cont'd

Date (Christian)	Time (UTC)	Epicenter 00°N–00°E	Location	*mb*	M_s	M_w	*I* (MMI)	CD_w (km)	PGA (H/V)	People Killed/ Injured	Official Death Toll/ Injured	Cities Destroyed/ Damaged (d)	Towns Destroyed/ Damaged (d)	Villages Destroyed/ Damaged	No. of Houses Destroyed/ Damaged	Direct Economic Loss (U.S$ m, Original values)
1924.01.18	14:56	29.62–56.60*	Lālehzār AFS		5.4		VII									
1924.02.19	07:01	39.27–47.65*	Moghān	6.8	5.9		VII+	–								
1924.06.30	03:41	27.50–53.80			5.8			–								
1925.09.24	04:38	26.60–55.40*	W. Qeshm Island	6.1	5.5			–								
1925.12.10	04:59	37.50–58.00	Shirvān-Bāghān		5.4											
1925.12.18	05:53	28.88–51.30*	Ahram	5.2	5.4		VII			2						
1926.04.23	01:31	27.50–55.00			5.3											
1927.05.09	10:31	28.05–56.72		6.4	5.8			–								
1927.07.07	20:06	26.96–62.08		6.4	5.7	6.3		–								
1927.07.22	03:55	34.90–52.90	Dasht-e-Kavir	6.9	6.3			–	–	?		–	–	+	+	
1927.07.22	08:37	34.90–52.90	Dasht-e-Kavir AFS		5.3											
1927.07.23	20:17	34.90–52.90	Dasht-e-Kavir AFS		5.8											
1927.07.23	22:40	34.90–52.90	Dasht-e-Kavir AFS		5.8											
1927.11.12	14:46	32.15–47.40*	Dehlorān area	6.0	5.6		VII	–								
1928.03.08	18:14	31.14–60.15*	Nehbandān		5.5	5.7	VII	–	–	4		–		>2		
1928.04.14	13:16	35.71–54.74*	Satveh	5.2												
1928.08.21	19:01	36.25–58.75*	Neyshābur	5.2	5.4		VI+	–	–	>30		–				

1929.05.01	**15:37**	**37.60–57.98* 37.89–57.58**	**Bāghān**	**7.1**	**7.3**	**7.1**	**IX⁺**	–	–	**>3800/1121**	**Shirvān (d)**		**300**	
1929.05.03	16:20	38.04–57.83	Bāghān AFS	5.2	5.1									
1929.05.13	13:27	38.10–57.50*	Bāghān AFS	6.5	5.8								+	+
1929.07.13	07:36	37.30–58.28*	Fāruj, Bāghān AFS		6.0		VII⁺	–		21			10	+
1929.07.15	07:44	32.15–49.60*	Andikā	6.3	6.0		VII⁺	–	–	++	–		>4	
1929.07.25	00:17	37.60–57.90*	Bāghān AFS		5.7								+	+
1929.09.03	12:07	25.77–62.65		5.9	5.6	6.3		–						
1930.05.06	07:03	38.15–44.75*	Salmās FSK		5.4		VII	–	–	25	–		4	
1930.05.06	**22:34**	**38.15–44.67***	**Salmās**	**7.0**	**7.2**	**7.1**	**IX**	–	–	**2514**	**Dilmān, Kohneh Shahr**		**>60**	
1930.05.08	15:05	28.32–44.68*	Salmās AFS; Shekaryāzi	5.4			VII			>4		Shekaryāzi	+	+
1930.05.08	15:35	38.48–44.48*	Salmās AFS; South Qotur		6.2		VIII	–				>7	+	+
1930.05.11	22:35	27.60–55.25	–	5.8	5.8			–						
1930.05.29	17:14	37.50–45.50	Salmās AFS	5.3	5.3								+	+
1930.08.08	08:54	37.00–59.00^K			5.3									
1930.08.23	10:53	27.88–55.02	–	6.2	6.1			–	–		–			
1930.10.02	15:32	35.78–51.96*	Āh		5.2		VII⁺	–	–	Few	–		>6	
1931.04.27	16:50	39.25–46.00* 39.27–45.93	Zāngézur	6.3	6.3	6.3	VIII	–		2890		+	+	+
1931.04.27	18:05	39.25–46.00*	Zāngézur AFS		5.4									

Continued

TABLE 17.12 Instrumental Period Medium- to Large-Magnitude Earthquakes in Iran (1900–2014)—Cont'd

Date (Christian)	Time (UTC)	Epicenter 00°N–00°E	Location	*mb*	M_s	M_w	*I* (MMI)	CD_w (km)	PGA (H/V)	People Killed/ Injured	Official Death Toll/ Injured	Cities Destroyed/ Damaged (d)	Towns Destroyed/ Damaged (d)	Villages Destroyed/ Damaged	No. of Houses Destroyed/ Damaged	Direct Economic Loss (U.S$ m, Original values)
1931.05.12	10:25	39.25–46.00*	Zāngézur AFS		5.3											
1932.02.04	21:18	26.44–62.50	–	5.4?												
1932.09.08	07:25	30.83–36.40	–		5.6	5.7		–								
1933.10.05	13:29	35.55–57.88*	Sebeh	6.2	6.1		VIII	–	–			–		+	++	
1933.11.28	11:09	32.00–55.88*	NW Behābād	6.4	6.2	6.2	VIII	–	–	+		–		>10	++	
1933.12.14	18:51	32.55–55.88*	NW Behābād AFS		5.6											
1934.01.02	20:55	29.90–57.38*	Sekonj		5.6		VII	–	–	+		Māhān (d)		>5	+	
1934.02.04	13:27	30.25–51.53	Kāzerun-Shirāz area		6.3			–	–			–		?		
1934.02.22	08:07	38.75–45.95*	Ordubād area	6.1	5.7			–								
1934.03.13	23:33	30.95–51.40*	S. Maymand	5.3										+	+	
1934.06.13	**22:10**	**27.38–62.56**	**Makrān**	**6.9**	**6.6**	**7.0**		–								
1934.10.29	16:15	39.90–47.80*	Kurā		5.7			–								
1935.03.05	10:26	35.90–53.08*	S. Dauāb		5.8		VII^+	–	–	>60		–		8	+	
1935.04.11	23:14	36.35–53.32* 36.48–53.59	Kusut	6.8	6.2	6.6	VIII	–	–	400		–		106	+	
1935.04.12	01:06	36.35–53.32*	Kusut AFS	5.5										+	+	

1935.04.12	12:44	36.35–53.32*	Kusut AFS	5.7						?		+	+
1935.04.12	22:23	36.35–53.32*	Kusut AFS	5.5								+	+
1935.04.12	22:31	36.35–53.32*	Kusut AFS	5.3									
1936.04.21	02:14	26.29–55.28	–	6.2	5.5			–					
1936.06.10	03:29	26.50–64.00	–		5.7			–					
1936.06.30	19:26	33.70–59.97*	Ābiz	6.2	6.0	6.0	VII+	–	–	>12	–	>5	+
1938.01.26	03:40	33.45–47.80*	Tursaq	5.7	5.4			–					
1938.02.14	02:54	40.50–53.50	–		6.1		VIII	–					
1938.04.23	09:26	27.22–53.28	–		5.5			–					
1938.09.19	03:23	38.40–57.30		5.9	5.5								
1938.12.19	18:56	36.65–58.52*	W. Binālud		5.6	5.7		–					
1939.04.06	04:08	35.15–54.56	Satveh	5.6						?			
1939.06.10	08:36	33.82–56.90*	Bahārestān		5.5		VII	–	–		–	>2	+
1939.09.19	03:23	37.80–57.70*	Garmāb area	6.0	5.6			–					
1939.11.04	10:15	32.40–48.50* 32.55–49.08	Khoramābād area	6.0	5.7			–					
1939.11.08	17:21	36.30–57.95*	NE Sabzevār	6.3	5.5	5.7		–					
1940.05.04	21:01	35.68–58.52*	Estāyesh	6.2	6.4	6.4	VIII	–	–	?	–	+	++
1940.07.19	04:53	37.50–57.50K			5.3								
1940.07.19	05:53	37.50–57.50K			5.8								

Continued

TABLE 17.12 Instrumental Period Medium- to Large-Magnitude Earthquakes in Iran (1900–2014)—Cont'd

Date (Christian)	Time (UTC)	Epicenter 00°N–00°E	Location	*mb*	M_s	M_w	*I* (MMI)	CD_w (km)	PGA (H/V)	People Killed/ Injured	Official Death Toll/ Injured	Cities Destroyed/ Damaged (d)	Towns Destroyed/ Damaged (d)	Villages Destroyed/ Damaged	No. of Houses Destroyed/ Damaged	Direct Economic Loss (U.S$ m, Original values)
1941.02.16	16:39	33.40–58.88*	Kumirān	6.4	6.2	6.1	VIII	–	–	>680		–		9	+	
1941.06.10	20:38	33.51–46.95	–		5.5			–								
1941 August: Allied Forces (Great Britain and Soviet Union) occupied Iran: Macroseismic data lost during WWII																
1943.02.06	02:35	24.68–63.20	Makrān	6.2	5.9			–								
1944.04.05	18:05	36.78–54.46*	Gorgān		5.2		VI+	–	–	>20/>2		Gorgān (d)		+	+	
1944.04.05	18:29	36.78–54.46*	Gorgān	5.4												
1944.07.23	12:00	29.85–56.80	Negār		5.5		VII	–	–	Few			Negār	+	++ at Negār	
1945.01.15	17:21	27.00–54.85	–		5.6			–								
1945.03.12	01:38				5.4											
1945.09.27	12:30	36.42–50.28*	Heriān				VII							>2	+	
1945.11.27	**21:56**	**25.20–63.42**	**Pāsni, Central Makrān**	**7.6**	**8.0**	**7.9**	**X**	–	–	**>300, 45 in Pāsni**		**Pāsni, Qwadur, Ormārā, + Tsunami**		**300**	+	
1946.02.10	13:13	32.52–59.37*	Giv		5.2	5.5	VI+	–		3				>2		
1946.03.12	02:21	29.72–51.72	Kāzerun	5.7	5.7		VII	–	–	Few		Qal'eh Kāli/ Kāzerun (d)		+	+	

1946.06.20	00:37	–	–		5.8			–							
1946.07.27	16:25	35.60–45.92*	Penjevin		5.5			–		6	Penjevin (d)		+	+	
1946.08.17	09:48	36.67–45.99*	Penjevin		5.5		VII				Penjevin (d)				
1946.11.04	**21:47**	**39.30–55.20* 39.77–54.56**	**Kāzāndzhik**	**7.4**	**7.0**	**7.1**	**IX**	–			**Kāzāndzhik**				
1947.08.05	14:24	25.01–63.38	Central Makrān	7.6	7.0	6.9	IX	–	–		Pāsni				
1947.09.23	12:28	33.77–58.66*	Charmeh	6.4	6.9	6.8	VIII+	–	–	500	–		>6	+	
1947.09.26	03:04	33.58–58.70*	Dustābād		6.0	6.0	VII+	–		?			+	+	
1947.10.03	06:13	25.99–57.41	Makrān	5.8	5.8			–							
1947.10.06	15:18	33.77–58.66*	Charmeh AFS		5.4	5.6		–							
1948.01.30	08:43	25.14–63.78	Makrān		6.5			–							
1948.06.17	14:08	36.68–49.14*	Āltinkosh	5.5			VII+	–	–	2	–		>6	+	0.07
1948.06.18	18:44	37.50–58.00*	N. Shirvān-Baghān		5.5	5.7						Shirvān (d)	+	+	
1948.07.05	13:53	29.90–57.71*	Gowk	5.9	6.0		VII+	–	–		–		+		
1948.07.30	03:30	31.21–49.12	Ahvāz area		5.5										
1948.10.05	**20:12**	**37.90–58.50* 37.95–58.32**	**Ashkābād**	**7.3**	**7.2**	**7.2**	**IX**	–	–	**176,000 in USSR; 352 in Iran**	**Ashkābād**		**>30 in Iran**	+	
1948.10.06	01:24	37.70–58.60	Ashkābād AFS		5.3	5.5		–							
1949.04.24	04:22	27.22–56.38*	Nakhl-e Nākhodā	6.1	6.3		VIII	–	–	–	–		12	80	
1949.05.09	11:16		Ashkābād AFS		5.8										

Continued

TABLE 17.12 Instrumental Period Medium- to Large-Magnitude Earthquakes in Iran (1900–2014)—Cont'd

Date (Christian)	Time (UTC)	Epicenter 00°N–00°E	Location	mb	M_s	M_w	I (MMI)	CD_w (km)	PGA (H/V)	People Killed/ Injured	Official Death Toll/ Injured	Cities Destroyed/ Damaged (d)	Towns Destroyed/ Damaged (d)	Villages Destroyed/ Damaged	No. of Houses Destroyed/ Damaged	Direct Economic Loss (U.S$ m, Original values)
1949.07.04	03:04	27.22–56.38*	Nakhl-e Nākhodā AFS	6.0												
1949.07.06	02:30	27.22–56.38*	Nakhl-e Nākhodā AFS	5.5												
1949.11.22	15:21	28.00–57.00	Chāh Dormeh	5.4												
1950.01.19	17:27	27.38–52.78*	Dehnau, 'Assaluyeh	5.8	5.5		VII^+	–	–	30		–		>20	+	
1950.01.22	04:07	27.38–52.78*	Dehnau, Assaluyeh AFS	5.3												
1950.04.06	02:43	37.90–58.40*	Ashkābād AFS	5.4	5.3	5.5		–								
1950.05.02	16:43	38.00–58.50	–		5.3	5.5		–								
1950.05.09	11:16	37.90–58.40*	Ashkābād AFS	6.1	6.4	6.3		–								
1950.09.24	22:56	34.50–60.70	–		5.5	5.7		–								
1951.08.16	23:52	28.00–57.00	Mināb area	5.7												
1951.12.30	18:21	28.00–57.00	Mināb area	5.9	5.5			–								
1952.09.15	04:31	37.80–58.30*		5.4	5.1											
1953.01.15	20:06	31.10–56.75*	Rayhān	5.6	5.5		VII	–	–	>13		–		+		

1953.02.12	08:15	35.40–55.00*	Torud	6.9	6.5	6.5	VIII+	–	–	>930	Torud	>8	1800	[1 million F. Francs] US$0.285
1954.11.01	21:10	37.10–57.30*		5.2	5.5									
1955.12.04	14:02	33.50–48.80*	Razan	6.0						3		7	50	
1955.12.17	08:06	33.50–48.80*	Razan	5.5								+	+	
1956.02.15	15:49	27.80–53.10		5.4	5.3		VII							
1956.04.12	22:34	37.30–50.20	Langérud area	5.5										
1956.10.31	14:03	27.28–54.60* 27.27–54.42	Gowdeh	5.9	6.2	6.6	VIII	–	–	410	–	>14		2.8
1956.10.31	14:22	27.28–54.60*	Gowdeh AFS											
1957.02.03	19:27	32.28–55.62	Chāh Dormeh		5.3									
1957.03.16	00:43	43.34–52.93		5.5										
1957.04.23	–	36.15–52.06	Kachu Mesqāl				VII	–	–	16	–		210	
1957.05.06	15:06	36.13–52.08*	Namār	5.5			VII			Few		4	+	
1957.07.02	**00:42**	**36.06–52.48*** **36.06–52.66**	**Band-e-Pay**	**7.0**	**6.8**	**7.1**	**VIII+**	–	–	**>1500**	–	**120**		**25**
1957 July: Establishment of the Institute of Geophysics, University of Tehrān														
1957.09.05	11:36	28.51–53.61	Jahrom area	5.5						1/many		+	+	
1957.12.13	01:45	34.53–47.76* 34.34–47.63	Fārsinaj	6.5	6.7	6.8	VIII	–		1200/900	–	211	5000	
1958.08.14	11:27	34.27–48.10*	Givaki		5.7							+	+	

Continued

TABLE 17.12 Instrumental Period Medium- to Large-Magnitude Earthquakes in Iran (1900–2014)—Cont'd

Date (Christian)	Time (UTC)	Epicenter 00°N–00°E	Location	*mb*	M_s	M_w	*I* (MMI)	CD_w (km)	PGA (H/V)	People Killed/ Injured	Official Death Toll/ Injured	Cities Destroyed/ Damaged (d)	Towns Destroyed/ Damaged (d)	Villages Destroyed/ Damaged	No. of Houses Destroyed/ Damaged	Direct Economic Loss (U.S$ m, Original values)
1958.08.14	15:26	34.33–48.02*	Kalādeh		5.5									+	+	
1958.08.16	19:13	34.35–48.12* 34.25–47.84	Firuzābād	6.2	6.6	6.5	VIII$^+$	–	–	>132/948		–		170	2500	
1958.09.21	16:18	34.63–47.46*	Kargsār (Firuzābād AFS)		5.2		VI$^+$	–	–	16/57		–		7	+	
1958.10.06	09:29	37.40–54.50*		5.5	5.0											
1959.05.01	08:24	36.45–51.23	Lāhijān	5.7								Lāhijān (d)				
1959.12.08	12:50	31.10–56.25*	Rayhān		5.4					Few			Rayhān	+	+	
1960.03.07	00:51	34.41–55.39?	Kargsār (Firuzābād AFS)		5.1											
1960.04.02	22:36	34.29–48.19	Kargsār (Firuzābād AFS)		5.0											
1960.04.02	23:33	34.18–48.14														
1960.04.24	12:14	27.75–54.39	Lār	6.0	5.8		VII$^+$	–	–	420		Lār		+	+	
1960.05.03	06:58	27.63–54.44	Lār AFS	5.2								Kurdeh (d)		+	+	
1960.05.18	08:41	27.28–53.01														
1960.05.20	04:14	27.32–53.02														

1960.06.02	07:22	33.18–60.30	–		5.1										
1960.06.02	12:42	33.04–48.79	–	5.5											
1960.07.31	22:26	27.89–54.39	Lār AFS	5.2											
1961.04.06	18:12	27.77–56.61	Dowlatābād area	5.7	5.3			–							
1961.06.11	05:10	27.88–54.55	Dehkuyeh	6.4	6.5	6.6	VIII+	–	–	60	–		10	500	
1961.06.11	05:30	27.88–54.81	Dehkuyeh AFS												
1961.06.11	06:19	27.80–54.90	Dehkuyeh AFS	5.4											
1961.06.11	12:31	27.74–54.50	Dehkuyeh AFS		6.1										
1961.06.11	13:57	27.87–54.73	Dehkuyeh AFS	5.2											
1961.06.23	16:36	27.71–54.98	Dehkuyeh AFS		5.8										
1961.10.14	07:00	34.00–48.51*	Mogh				VII			2			+	+	
1961.10.28	10:46	33.54–48.49	Heydarābād		5.0		VI+					Few			
1962.04.01	00:45	33.26–58.87* 33.23–58.86	Chāhak	5.5	5.5		VII	–	–	20	–		>5	+	
1962.06.23	05:05	29.70–49.20		5.6	5.4										
1962.09.01	**19:20**	**35.55–49.83**	**Bu'in**	**6.9**	**7.2**	**7.0**	**IX**	**10**	–	**12,200/3000**	–		**91**	+	**30**
1962.09.04	13:30	35.56–49.72	Bu'in AFS	5.6	5.6										
1962.09.04	22:59	40.00–44.00			5.2		VI+								
1962.09.29	06:53	28.20–57.40	–	5.5											
1962.10.01	12:13	27.89–54.73	Evaz Lār area	5.8	5.5	6.3									

Continued

TABLE 17.12 Instrumental Period Medium- to Large-Magnitude Earthquakes in Iran (1900–2014)—Cont'd

Date (Christian)	Time (UTC)	Epicenter 00°N–00°E	Location	*mb*	M_s	M_w	*I* (MMI)	CD_w (km)	PGA (H/V)	People Killed/ Injured	Official Death Toll/ Injured	Cities Destroyed/ Damaged (d)	Towns Destroyed/ Damaged (d)	Villages Destroyed/ Damaged	No. of Houses Destroyed/ Damaged	Direct Economic Loss (U.S$ m, Original values)
1962.10.05	20:02	35.20–58.71	Ahmadābād, Rikhteh		5.0		VI$^+$		–	6		–		3	+	
1962.10.13	10:23	35.61–50.05	Bu'in AFS		5.7		VII							+	+	
1962.11.06	00:09	28.09–55.50	Gahkom	5.6	5.4								Gahkom (d)	+	+	
1962: Establishment of the Geological Survey of Iran [United Nations Special Fund Project]																
1963.03.24	12:44	34.45–47.98* 34.43–47.89	Kārkhāneh	5.5	5.8	5.8	VII$^+$	08	–	–		–		22	5000	
1963.03.31	02:27	36.91–57.78*	Dahaneh Ojāq	5.0	5.0		VI$^+$		–	4		–		+	+	
1963.05.29	00:47	28.05–52.42														
1963.05.29	08:35	27.11–59.41		5.4												
1963.07.29	06:10	28.11–55.67	Gahkom AFS	5.2				05		5/8				+	+	
1964: The first fully operational year of the World-Wide Standard Seismographic Network (WWSSN)																
1964.01.19	09:13	26.72–53.94	Lengeh area	5.6	5.3											
1964.02.16	00:17	29.96–51.11	–	5.3	5.9											
1964.08.12	19:26	30.85–49.66		5.2	5.4		VII									
1964.08.19	09:33	28.11–52.59	–	5.5	5.5											
1964.08.19	15:20	28.09–52.60	–	5.5												

1964.08.20	05:08	28.05–52.58	–	5.4											
1964.08.20	05:39	28.10–52.55	–	5.6											
1964.08.27	12:56	28.07–55.83	Gahkom	5.3									+	+	
1964.11.08	10:33	29.58–50.94	–	5.3	5.7										
1964.11.10	15:47	32.78–49.16	–	5.3											
1964.12.22	04:36	28.16–56.90	Faduyeh	5.7	6.1	5.5		18	–	?	–				
1965.06.21	00:21	28.09–55.84	Sarchāhān	5.8	5.4	6.0		08							
1966.09.18	20:43	27.81–54.27	Dehkuyeh	5.9	5.4	5.9		12					+	+	
1966.09.24	10:00	27.38–54.46	Bastak	5.3											
1967.01.09	01:55	27.64–54.48	Lār area	5.3	6.1										
1967.01.11	11:20	34.06–45.65	Naft-e Shāh area	5.6		5.6		10							
1967.01.29	07:56	26.49–55.26	Lengeh area	5.1	5.5		VII		–					+	
1968.04.29	17:01	39.22–44.19	Gol	5.3	5.5		VII		–	61/100	–		10		1
1968.05.30	01:10	27.83–54.01	Evaz	5.2						0/15		Evaz (d)	+	+	
1968.06.09	00:56	39.19–46.14*	Zāngézur	5.0	5.0		VII						+	+	
1968.06.23	09:16	29.74–51.25	Khesht	5.3		5.5		09					+	+	
1968.08.02	13:30	27.55–60.89	Makrān	5.7		5.7									
1968.08.31	**10:47**	**34.04–58.95**	**Dasht-e Bayāz**	**6.0**	**7.2**	**7.1**	**IX**	**09**	–	**10,000/4050**	**Kākhk**		**157**	**9356**	**>50**
1968.08.31	11:34	33.97–59.18	Dasht-e Bayāz AFS	5.4											
1968.09.01	07:27	34.04–58.20	Ferdows	5.9	6.4	6.3	VIII	10	–	1000/300	Ferdows		+	2100	

Continued

TABLE 17.12 Instrumental Period Medium- to Large-Magnitude Earthquakes in Iran (1900–2014)—Cont'd

Date (Christian)	Time (UTC)	Epicenter 00°N–00°E	Location	*mb*	M_s	M_w	*I* (MMI)	CD_w (km)	PGA (H/V)	People Killed/ Injured	Official Death Toll/ Injured	Cities Destroyed/ Damaged (d)	Towns Destroyed/ Damaged (d)	Villages Destroyed/ Damaged	No. of Houses Destroyed/ Damaged	Direct Economic Loss (U.S$ m, Original values)
1968.09.04	23:24	34.03–58.31	Ferdows AFS	5.4		5.5		09								
1968.09.11	19:17	33.97–59.53	Dasht-e Bayāz AFS		6.0	5.6	VII$^+$	06								
1968.09.14	13:48	28.33–53.17	Tang-e Ru'in	5.8	6.0	5.8	VII$^+$	07	–	–		–		>4	+	
1968.09.14	19:20	28.38–53.19	Mobarakābād	5.1	5.9					1/3				+	+	
1968.11.15	06:25	38.10–58.30*	Ashkābād area	5.7	6.0	5.4	VII$^+$	18						+	+	
1969.01.03	03:16	37.06–57.82	Dahaneh Ojāq	5.4	5.5	5.5	VII	07	–	5		–		6	300	
1969.04.29	04:37	29.56–51.52	–	5.5	5.0											
1969.09.01	23:16	30.83–49.66	Āghājāri area	4.9	5.4											
1969.09.02	13:30	30.22–58.74	Sirch	5.2	5.3									+	+	
1969.11.07	18:34	27.82–59.98	Bazmān area	6.1	6.5	6.1								+	+	
1969: Implementation of the first Iranian Code for Seismic Resistant Design of Buildings (ISIRI Code # 519)																
1970.02.23	11:22	27.79–54.48	Dehkuyeh	5.4	5.6	5.6		09						+	+	
1970.02.28	19:58	27.81–56.31	Kohneh Kharāb	5.4	5.2	5.5								+	+	
1970.03.14	01:51	38.59–44.77	Badalān	5.2	5.2		VI$^+$		–	4		–		>3		
1970.04.04	10:58	37.00–59.40	Archigān	4.8	5.3									+	+	

1970.07.30	00:52	37.70–55.90* 37.84–55.88	Karnāveh	5.7	6.5	6.4	VIII⁺	11	–	200	–		40	2000	8
1970.10.25	11:22	36.75–45.18	Pasveh	5.3	5.2	5.5				0/21			>6	158	
1970.11.09	17:41	26.50–56.78	Sirjān area	5.4	4.7	5.5		100							
1971.02.14	16:27	36.55–55.68	Serokhi	5.3	5.5	5.7	VII	11	–	1	–		>2	+	
1971.04.06	06:49	29.78–51.88	Dasht-e Arzhan	5.2	5.1	5.2		06							
1971.04.12	19:03	28.26–55.61	Tazarj	6.0	5.8	5.9	VII⁺	10	–	1	–		>3	>100	
1971.05.26	02:41	35.53–58.20	Rivash	5.4	5.6	5.6	VII	13	–	?	–				
1971.08.09	02:45	36.20–52.75	Bābol Kenār	5.2	5.3		VII		–	1	–				
1971.09.08	12:53	29.14–60.00	Zāhedān area	5.3	5.6										
1971.11.08	03:06	27.01–54.45	Bastak	5.6	6.0	5.9	VII⁺	09	–	–	–	Bastak (d)			
1971.12.09	01:42	27.30–56.37	Sarkhun	5.3	5.7	5.8	VII		–	–	–		+		
1971: Establishment of the Tectonics and Seismotectonics Research Department, Geological Survey of Iran, by Manuel Berberian															
1972.04.10	02:06	28.41–52.78	Kārzin	6.0	6.9	6.7	VIII	10	0.40	5010/1332	–		50		8
1972.06.12	13:33	33.03–46.26	Mehrān-e Gharb	5.3	5.0	5.0		11					+	+	
1972.07.02	12:56	30.06–50.85	Mishān	5.4	5.3	5.3	VII	09					+	700	
1972.08.06	01:12	24.99–61.14	Gavāter, Makrān	5.4	5.3										
1972.08.08	19:09	25.01–61.14	Gavāter, Makrān	5.4	5.0	5.0									
1972.11.17	09:09	27.37–59.11	Gavāter, Makrān	5.2		5.4		65							
1972.12.01	11:39	35.43–57.92	Bardaskan area	5.2	5.2	5.4		08							
1973.08.25	14:58	28.11–56.75	–	5.3											
1973.11.11	07:14	30.53–52.95	Qeshlāq	5.4	5.5	5.5	VII		–	1	–		3	++	

Continued

TABLE 17.12 Instrumental Period Medium- to Large-Magnitude Earthquakes in Iran (1900–2014)—Cont'd

Date (Christian)	Time (UTC)	Epicenter 00°N–00°E	Location	*mb*	M_s	M_w	*I* (MMI)	CD_w (km)	PGA (H/V)	People Killed/ Injured	Official Death Toll/ Injured	Cities Destroyed/ Damaged (d)	Towns Destroyed/ Damaged (d)	Villages Destroyed/ Damaged	No. of Houses Destroyed/ Damaged	Direct Economic Loss (U.S$ m, Original values)
1973: Establishment of the Iranian Strong Motion Network by A.A. Mo'infar																
1974.03.07	11:36	37.70–55.70		5.4	5.3											
1974.08.02	15:06	42.31–45.93	–	5.4												
1974.12.02	09:05	28.03–55.87	Sarchāhān	5.4	5.1	5.2		07						+	+	
1975.03.07	07:04	27.47–56.22	Sarkhun	5.8	6.0	6.1	VIII	11	0.39	7		–		>2		
1975.12.24	11:48	27.00–55.47	Pol-e Angur	5.7	5.5	5.5	VII	08						+	60	
1976.03:16	07:28	27.30–54.98	–	5.4		5.2		09								
1976.04.22	17:03	28.68–52.12	Farāshband area	6.0	5.7	5.7		07								
1976.11.07	04:00	33.82–59.18	Vondik	5.8	6.3	6.0	VIII	08	0.17	16		–		>10		
1976.11.07	11:07	33.19–47.92	Pol-e Dokhtar area	5.5		4.8										
1976.11.24	**12:22**	**39.08–44.03**	**Chālderān**	**6.2**	**7.2**	**7.0**	**IX**			**6 in Iran**				**45**	**1000 in Iran**	
1976.11.24	12:36	39.13–44.27	Chālderān AFS	5.5												
1977.01.05	05:44	27.44–56.22	Genu	5.5	5.2	5.1				0/10				+	54	
1977.03.21	21:18	27.58–56.36	Khurgu	6.2	6.8	6.7	VIII+	12	0.41	152		–		>20	1500	
1977.03.21	22:42	27.60–56.46	Khurgu AFS	5.7	6.0	6.1										

1977.03.22	11:57	27.59–56.42	Khurgu AFS	5.7	5.9	6.0		12					2	+	
1977.03.23	23:51	27.59–56.55	Khurgu AFS	5.7	5.4	5.5		09							
1977.03.24	04:42	27.62–56.58	Khurgu AFS	5.2		5.3									
1977.04.01	13:36	27.55–56.27	Khurgu AFS	5.9	6.0	6.0	VII^+	12							
1977.04.06	13:36	31.95–50.64	Nāghān	5.6	6.1	5.9	VIII	06	0.52	366		–	55	2100	
1977.04.26	16:25	32.64–48.90	Shahbāzān area	5.4	4.8	5.5		20							
1977.05.25	11:01	34.84–52.01	Siyāh Kuh	5.3	4.3									+	
1977.05.26	01:35	38.89–44.32	Mokhur	5.2	5.4	5.7	VII			3		–	19	55	
1977.06.05	04:45	32.61–48.08	Andimeshk area	5.8	5.7	6.1		12							
1977.06.05	08:25	32.58–48.11	Andimeshk area AFS	5.1		5.4		09							
1977.09.17	05:25	31.12–56.60	Dartangal FRS	4.8	5.6										
1977.10.19	06:35	27.80–54.89	E. Dehkuyeh area	5.5	5.2	5.6									
1977.12.10	05:46	27.68–56.58	Khurgu AFS	5.1	5.0	5.6		18							
1977.12.19	23:34	30.90–56.41	Dartangal	5.8	5.7	5.9	VII	07		665/260		–			
1978.02.10	20:50	25.26–62.33	Makrān Coast	5.1	5.5		–								
1978.05.22	06.18	31.25–55.73	Behābād	5.0	4.9	5.1	VI								
1978.09.16	**15:35**	**33.24–57.38**	**Tabas-e Golshan**	**6.7**	**7.4**	**7.3**	**IX^+**	**09**	**0.91**	**20,000**	**6363**	**Tabas-e Golshan**	**90**	**15,000**	**11**
1978.11.04	15:22	37.67–48.91	Siāhbil	6.0	6.0	6.1	VII^+	21	0.28	26		–	20	541	1.2
1978.12.06	17:18	33.24–57.13	Tabas-e Golshan	5.3	5.2										

Continued

TABLE 17.12 Instrumental Period Medium- to Large-Magnitude Earthquakes in Iran (1900–2014)—Cont'd

Date (Christian)	Time (UTC)	Epicenter 00°N–00°E	Location	*mb*	M_s	M_w	*I* (MMI)	CD_w (km)	PGA (H/V)	People Killed/ Injured	Official Death Toll/ Injured	Cities Destroyed/ Damaged (d)	Towns Destroyed/ Damaged (d)	Villages Destroyed/ Damaged	No. of Houses Destroyed/ Damaged	Direct Economic Loss (U.S$ m, Original values)
1978.12.14	07:05	32.12–49.63	Andikā	5.9	6.2	6.2	VIII		–	60/52		–		12		
1979.01.10	01:26	26.55–60.90	Pārud, Makrān	5.6	6.0	5.8	VII⁺		–			–		+	++	
1979.01.10	15:05	26.49–60.99	Pārud, Makrān	5.5	6.2	5.9	VIII		–					28	990	
1979.01.16	09:50	33.90–59.46	Boznābād	6.0	6.7	6.5	VIII⁺	11		129/200		–		>3	1700	
1979.01.16: Abdication and deportation of the Shāh from Irān *1979.02.11: Change of regime in Irān: Loss of macroseismic data*																
1979.02.13	10:36	33.29–57.39	Tabas-e Golshan	5.4	5.5	5.6		11								
1979.11.14	02:21	33.95–59.72	Korizān	6.0	6.7	6.6	VIII⁺	10		171		–		7	216	
1979.11.27	07:12	34.01–59.83	Koli FRS	5.1	5.3											
1979.11.27	**17:10**	**34.05–59.75**	**Koli**	**6.1**	**7.2**	**7.1**	**IX**	**08**		**20/23**		–		**>10**		
1979.12.07	09:23	34.07–59.84	Kalāt-e-Shur	5.9	6.1	5.9	VIII	10						>3		
1979.12.09	09:12	35.10–56.82	W. Doruneh	5.2	5.5	5.6		09								
1980.01.01	02:45	27.34–60.34	Makrān	5.3	5.1	5.5										
1980.01.12	15:31	33.54–57.22	Tabas-e Golshan	5.3	6.0	6.0		14								
1980.05.04	18:35	38.04–49.01	Shirābād	5.3	6.1	6.3		15								

1980.07.22	05:17	37.32–50.26	Lāhijan area	5.3	5.1	5.6				1/Few				+	+	
1980.10.19	17:24	32.70–48.56	Shahbāzān area	5.2	5.7	5.6		17								
1980.11.28	21:15	27.61–56.53	Khurgu	5.5	5.1	5.4									+	
1980.12.18	12:34	35.88–44.63	–	5.4	5.8	6.2										
1980.12.19	01:16	34.47–50.64	Giv	5.5	5.8	6.0	VII^+	14	–	16		–		11		
1980.12.22	12:51	34.42–50.63	Giv	5.4	5.2	5.5	VI^+	15	–	3/100						
1981.04.01	10:16	29.81–51.47	Kāzerun area	5.5	4.4	5.2										
1981.06.11	07:24	29.85–57.68	Golbāf	6.0	6.6	6.6	VIII^+	20		1400/3000		–				5
1981.07.23	00:05	37.08–45.19	Oshnaviyeh	5.6	5.6	5.8	VII			10		–		>7		
1981.07.28	**17:22**	**29.97–57.76**	**Sirch**	**5.9**	**7.0**	**7.0**	**IX**	**18**	**0.35 at Golbāf**	**1300/1000**	**796**	–	**Golbāf**	**25**		**1000**
1981.08.04	18:35	38.15–49.37	Āstāarā area	5.4	5.2	5.6		20								
1982.10.25	12:54	35.20–52.35	Garmsār	4.6	5.4											
1982.12.19	15:40	30.56–57.51	Sirch AFS	5.1	6.2											
1983.02.07	15:06	26.86–57.56	W. Makrān	5.5	5.7	6.0										
1983.02.18	07:40	27.89–53.82	Lār area	5.2	4.3	5.2		06								
1983.03.05	14:22	32.46–49.34	Lāli area	5.3	5.3	5.7										
1983.03.14	12:12	29.32–54.53	Kum Dāgh area	5.2	5.6											
1983.03.25	11:57	36.03–52.29	Bāijān	5.1	4.9	5.5	VII			100/61				+	+	
1983.03.26	04:07	35.98–52.24	Bāijān	5.4	4.9	5.4	VII		–			–		7	+	
1983.04.18	10:58	27.76–62.05	Makrān	6.4	6.3	6.7		65								

Continued

TABLE 17.12 Instrumental Period Medium- to Large-Magnitude Earthquakes in Iran (1900–2014)—Cont'd

Date (Christian)	Time (UTC)	Epicenter 00°N–00°E	Location	*mb*	M_s	M_w	*I* (MMI)	CD_w (km)	PGA (H/V)	People Killed/ Injured	Official Death Toll/ Injured	Cities Destroyed/ Damaged (d)	Towns Destroyed/ Damaged (d)	Villages Destroyed/ Damaged	No. of Houses Destroyed/ Damaged	Direct Economic Loss (U.S$ m, Original values)
1983.05.28	11:35	32.58–48.58	Dez dam area	5.6	5.0	5.6										
1983.07.12	11:34	27.60–56.40	Khurgu	5.7	5.8	5.9		17								
1983.07.22	02:40	36.94–49.22	Charazeh	5.6	5.0	5.5	VII	10		4/41		–		7	75	5
1984.02.22	05:44	39.45–54.04	–	5.1	5.8	5.8		27								
1984.08.06	11:14	30.80–57.16	Hur	5.6	5.3	5.3	VII	11	–	–		–		>4		
1985.02.02	20:52	28.35–52.97	Gāvaki, SE Qir	5.1	5.3	5.6	VII	11	–	2/80		–		++	1500	
1985.08.07	15:43	27.85–53.04	Khonj area	5.4	5.4	5.4		17		0/2				+	+	
1985.08.16	10:46	37.08–59.30	Tirgān area	5.4	5.2	5.5		09								
1985.10.29	13:13	36.68–54.77	Nomal	6.0	6.0	6.1	VII		–	1		–		8		
1985.10.29	14:23	36.90–54.90	Nomal	5.1	5.7	6.2	VII	13						+	+	
1986.03.06	00:05	40.33–51.59	Caspian	6.1	6.3	6.4		31								
1986.03.25	01:32	26.69–54.81		4.9	5.4											
1986.05.02	03:18	28.00–53.30	Khonj area	5.5	5.1	5.6										
1986.05.03	10:37	27.97–53.33	Khonj area	5.4	4.7	5.2										
1986.07.12	07:54	29.91–51.56	Golgun	5.7	5.6	5.5	VII	07	–	2/4		–		8	300	
1986.12.20	23:47	29.90–51.57	Golgun AFS	5.4	5.0	5.3	VI^+	08							80	

1987.01.11	12:31	29.93–51.79	Golgun AFS	4.8	4.1									300
1987.04.10	06:43	37.16–57.68	Esfarāyen area	5.0		5.1								
1987.04.29	01:45	27.42–56.10	Bandar ʻAbbās area	5.8	5.4	5.6		10						
1987.05.12	07:15	28.14–55.57	Bandar ʻAbbās area	5.2	4.9	5.5								
1987.05.29	06:27	34.06–48.28	Kahriz	5.0	4.6	5.3	VI		–	2/50	–			
1987.09.07	11:32	39.40–54.79	–	5.5	5.5	5.5		30						
1987.11.24	11:23	32.65–59.10	Kariz-e-Nau	5.4	4.7	5.3	VII						2/30	728
1987.12.18	16:24	28.15–56.66	–	5.7	5.5	5.8		10						
1988 FEBRUARY: Implementation of the second Iranian Code for Seismic Resistant Design of Buildings (ISIRI Code # 2800)														
1988.03.30	02:12	30.84–50.18	Tashān	5.4	5.7	5.9	VII			0/24		140/422d	+	700
1988.04.20	03:50	39.07–44.11	Chālderān	5.1	5.1	5.5								
1988.08.11	16:00	29.94–51.58	Doshman Ziāri	5.4	6.1	5.5	VIII	07	–	6	–		45	2000
1988.08.11	16:04	29.88–51.65	Doshman Ziāri	5.6	5.9	5.8	VIII	09						
1988.08.22	21:23	35.31–52.34	Garmsār	5.0	5.0	5.3	VII						+	+
1988.08.23	05:30	35.37–52.24	Garmsār AFS	5.0	4.8	5.0								
1988.08.30	17:30	29.95–51.72	Doshman Ziāri	4.9	4.7	5.2		16		0/Few				
1988.12.06	13:20	29.89–51.62	Doshman Ziāri	5.5	5.7	5.6	VII	10					21	
1988.12.07	07:41	40.92–44.11	Spitak, Armenia	6.0	6.7	6.8	VIII^{+}			25,000/ 19,000		4/20	342	++
1988.12.07	07:45	40.94–44.23	Spitak, Armenia AFS	5.8	6.2									
1989.04.02	06:42	28.17–57.28	Chāh Dormeh area	5.2	4.8	5.4								

Continued

TABLE 17.12 Instrumental Period Medium- to Large-Magnitude Earthquakes in Iran (1900–2014)—Cont'd

Date (Christian)	Time (UTC)	Epicenter 00°N–00°E	Location	*mb*	M_s	M_w	*I* (MMI)	CD_w (km)	PGA (H/V)	People Killed/ Injured	Official Death Toll/ Injured	Cities Destroyed/ Damaged (d)	Towns Destroyed/ Damaged (d)	Villages Destroyed/ Damaged	No. of Houses Destroyed/ Damaged	Direct Economic Loss (U.S$ m, Original values)
1989.04.02	21:24	32.66–47.77	Kabir Kuh area	5.5	5.0	5.4										
1989.05.03	09:12	29.90–51.80	Doshman Ziāri	5.0	4.9	5.2	VII							7		
1989.05.27	20:08	30.14–50.89	Dogonbadān	5.6	5.8	6.0	VII$^+$					–		9		
1989.09.16	02:05	40.35–51.58	Caspian Sea	6.3	6.5	6.4		31								
1989.09.17	00:53	40.21–51.78	Caspian Sea AFS	6.0	6.1	6.2		35								
1989.10.01	02:59	30.95–51.40	Sisakht	5.2	4.7		VI								1000	
1989.11.20	04:19	28.89–57.71	S. Golbāf	5.5	5.7	5.8	VII	10	–	4						
1989.12.07	12:59	25.91–58.96	Makrān	5.7	5.8	6.0						–		+	+	
1989: Establishment of the International Institute of Earthquake Engineering and Seismology (IIEES), Tehrān																
1990.01.20	01:27	35.90–52.97	Gaduk	5.5	5.8	6.0	VII$^+$							+	–	
1990.06.20	**21:00**	**36.99–49.22**	**Rudbār**	**6.2**	**7.4**	**7.3**	**IX$^+$**	**13**	**0.65, ~11 km from the fault**	**~40,000/ 105,000**	**13,000/ 60,000**	**Rudbār, Manjil, Lowshān, Harzévil/ Rasht (d)**		**700**	**100,000**	**7200**
1990.06.21	02:08	36.78–49.75	Rudbār AFS	5.4	4.9			12								
1990.06.21	09:02	36.63–49.78	Rudbār AFS	5.8	5.3	5.6	VII	10		21				+	+	
1990.06.21	12:17	36.86–49.44	Rudbār AFS	5.1	5.3			06								

1990.06.24	09:45	36.84–49.41	Rudbār AFS	5.1	4.8	5.5	VI	10				+	+	
1990.09.26	11:32	29.14–60.90	Zāhedān area	5.4	5.4	5.6		08						
1990.11.06	18:45	28.24–55.46	East Furg	6.0	6.7	6.5	VIII+	07		23/90	–	18	1445/1200	2.5
1990.11.06	19:30	28.23–55.37	East Furg AFS	5.3	5.7									
1990.12.16	22:18	29.02–51.31	Kalameh	5.3	5.4	5.7	VII		–	2/25	–	5	+	0.01
1991.05.22	16:29	27.38–55.77	–	5.7	5.1	5.4		13						
1991.07.24	09:45	36.60–44.07	Arbil area	5.4	5.1	5.5		11		20				
1991.11.04	01:50	30.69–50.25	Tashān	5.3	5.4	5.8	VII	05		0/66		15	2000	
1991.11.28	17:19	36.82–49.59	Rudbār	5.6	5.0	5.7	VI+	08		1		+	+	
1991.12.07	14:22	25.15–62.97	Makrān	5.2	5.1	5.6								
1991.12.19	18:55	28.04–57.30	–	5.3	4.8	5.4								
1991: Establishment of the Natural Disaster Headquarters, Ministry of Interior														
1992.01.16	10:20	32.12–59.93	Shurak	4.8	4.2		VI					8		
1992.03.04	11:57	31.60–50.76	Dareh Yās	5.0	4.8	5.1	VI		–	6/50			160	
1992.05.19	12:24	28.27–55.60	East Furg AFS	5.6	5.1	5.6								
1992.09.08	00:38	29.13–52.15	Dārenjān	4.2	4.8		VI		–	1/11	–	5	200	
1992.09.11	12:06	29.92–51.13	Mishān area	5.1	5.0	5.4								
1992.09.11	18:24	30.00–60.73	Zāhedān area	5.2	4.8	5.3								
1992.12.17	10:39	25.91–61.45	Central Makrān	5.7	5.3	5.7		43						
1993.01.06	22:51	29.04–52.13	Dārenjān	5.4	5.3	5.4	VII					16	+	
1993.01.27	10:27	32.08–60.06	Shirshotor	5.0	4.8		VI+					11	200	
1993.06.22	16:32	30.18–50.83	Gachsārān	5.4	4.9	5.4	VI	05					70	

Continued

TABLE 17.12 Instrumental Period Medium- to Large-Magnitude Earthquakes in Iran (1900–2014)—Cont'd

Date (Christian)	Time (UTC)	Epicenter 00°N–00°E	Location	*mb*	M_s	M_w	*I* (MMI)	CD_w (km)	PGA (H/V)	People Killed/ Injured	Official Death Toll/ Injured	Cities Destroyed/ Damaged (d)	Towns Destroyed/ Damaged (d)	Villages Destroyed/ Damaged	No. of Houses Destroyed/ Damaged	Direct Economic Loss (U.S$ m, Original values)
1993.07.09	10:29	28.39–55.36	N. Furg	5.2	4.6	5.1		16								
1994.02.23	08:02	30.79–60.53	Sefidābeh	6.0	6.0	6.1	VII^+	07		6		–		6	300	1.4
1994.02.23	11:54	30.81–60.54	Sefidābeh	5.3	4.9	5.5										
1994.02.23	22:45	30.50–60.55	Sefidābeh	5.3	5.0											
1994.02.24	00:01	30.79–60.51	Sefidābeh	6.0	6.1	6.2	VIII	10								
1994.02.26	02:31	30.79–60.53	Sefidābeh	5.7	5.8	6.0	VII^+	05								
1994.02.28	11:13	30.90–60.62	Sefidābeh	5.5	5.5	5.5	VII	06								
1994.03.01	03:49	29.14–52.64	Mook	5.8	6.0	5.9	VII^+	13		2/31				15		
1994.03.29	07:56	29.19–51.35	Borāzjān area	5.3	4.6	5.1		07								
1994.03.30	19:55	28.97–52.80	Firuzābād area	5.3	5.2	5.4				0/30						
1994.06.20	09:09	29.05–52.67	Ebrahimābād	5.8	5.9	5.8	VII^+	09		3/100				12		
1994.07.01	10:12	40.18–53.35	–	5.9	5.2	5.6		41								
1994.07.01	19:50	40.20–53.36	–	5.6	4.5	5.2		41								
1994.07.31	05:15	32.68–48.42	Badreh	5.2	5.3	5.1		14								
1994.12.14	20:43	35.09–58.60	Saghābād	5.2	4.6	5.3	VII							+	200	
1995.04.22	00:21	30.97–49.93	–	5.1	5.0	5.3		14								

1995.10.29	06:27	39.54–51.83	–	5.3	4.9	5.3		61						
1996.02.25	16:14	35.68–56.98	–	4.8	5.0	5.4								
1996.02.25	17:42	35.60–57.01	–	4.9	4.9	5.3								
1996.02.26	08:08	28.27–57.03	–	5.2	5.1	5.5								
1996.05.24	06:35	27.81–53.56	Khonj area	4.8	4.7	5.0		06		0/20		+	+	
1997.02.04	09:53	37.68–57.27	Nāveh FRS	5.2	5.4	5.4		13						
1997.02.04	10:37	37.73–57.31	Nāveh	5.8	6.7	6.4	VIII+	08	0.3	90/2000	–	183	14,356	100
1997.02.28	12:57	38.11–48.07	Shirān	5.5	6.1	6.0	VIII	09	0.2	954/2600	–	112	13,500	132
1997.03.02	18.29	37.99–47.89	Shirān AFS	5.0	4.8	5.3								
1997.04.19	05:53	27.94–56.86	–	5.3	5.3	5.5		19						
1997.05.07	16:16	40.30–51.63	–	5.2	4.7	5.2								
1997.05.10	**07:57**	**33.84–59.81**	**Zirkuh-e Qā’enāt**	**6.2**	**7.2**	**7.2**	**IX**	**10**	**0.7**	**1568/5059**	–	**259**	**17,630**	**150**
1997.06.20	12:57	32.29–59.99	Chāhkhu	4.9	5.4	5.4	VII	02				>4	>70	
1997.06.25	19:38	33.91–59.43	Boznābād	5.4	5.8	5.7	VII+	08				>4	5000	
1997.10.03	11:28	27.76–54.70	Lār area	5.1	4.8	5.3		04		0/6				
1997.10.20	06:09	28.44–57.28	–	5.5	5.0	5.2		28						
1998.03.14	19:40	30.13–57.59	Fandoqā	5.8	6.6	6.6	VIII+	05		5/50	–	7	2000	
1998.04.10	15:00	32.44–60.06	Gazān	5.2	5.7	5.7	VII	07		12/30	–		600	
1998.06.10	08:30	28.15–58.47	Bardsir area	5.0		5.2		85						
1998.06.15	01:14	31.60–50.83	–	4.9	4.9	5.0		05						

Continued

TABLE 17.12 Instrumental Period Medium- to Large-Magnitude Earthquakes in Iran (1900–2014)—Cont'd

Date (Christian)	Time (UTC)	Epicenter 00°N–00°E	Location	*mb*	M_s	M_w	*I* (MMI)	CD_w (km)	PGA (H/V)	People Killed/ Injured	Official Death Toll/ Injured	Cities Destroyed/ Damaged (d)	Towns Destroyed/ Damaged (d)	Villages Destroyed/ Damaged	No. of Houses Destroyed/ Damaged	Direct Economic Loss (U.S$ m, Original values)
1998.07.09	14:19	38.73–48.53	Āstārā area	5.8	5.5	5.7		27								
1998.08.01	23:38	27.69–56.53	Lengeh area	5.0	4.8	5.1		09								
1998.08.04	11:41	37.22–57.33	Esfarāyen area	4.9	5.1	5.3	VII							+	+	
1998.08.05	14:26	33.18–46.26	Ilam area	4.9	4.9	5.6										
1998.10.04	00:42	33.29–47.22	Dareh Shahr	5.2	4.8	5.2		09								
1998.10.05	02:20	33.28–47.27	Dareh Shahr	5.2	4.9	5.3	VI	07						>6	100	
1998.11.13	13:01	27.79–53.64	Khonj	5.3	5.1	5.4	VI^+	07		5/105					850	
1998.11.18	07:39	30.32–57.58	Chahār Farsakh	4.8	5.1	5.3	VI^+							2		
1999.03.04	05:38	28.27–57.20	Āshub	6.0	6.5	6.2	VIII^+	28		1				+	517	
1999.03.04	05:47	28.31–57.13	Āshub	5.5												
1999.05.06	23:00	29.52–51.91	Kuhmareh Sorkhi	5.7	6.3	6.1	VIII	07		26/100				9	800	
1999.05.06	23:13	29.43–51.93	Kuhmareh Sorkhi AFS	5.2	5.7		VII							+	+	
1999.08.10	19:33	36.15–54.65	Momenābād	4.4	4.3		V^+			1				+	+	
1999.09.24	19:17	28.65–51.34	Tangestān area	5.2	4.7	5.3				0/14				+	+	
1999.10.31	15:09	29.37–51.85	Kāzerun area	5.0	4.9	5.2										

1999.11.08	21:37	35.70–61.22	Yekeh Tut	5.4	5.2	5.5	VI+	09		>5		
1999.11.26	04:27	36.95–54.89	ʻAliābād	5.0	4.8	5.3	VI			>6	50	
2000.01.26	23:00	39.92–52.87	Gorgān area	5.3	4.7	5.3		40				
2000.02.02	22:58	35.23–58.21	Jaubuzjān	4.8	5.3	5.2	VII		1/15	>10	400	
2000.03.01	20:06	28.34–52.84	Hengām	5.0	4.6	5.0						
2000.03.05	09:40	27.92–56.45	Ahmadi	5.4	5.2	5.4		12				
2000.03.21	14:07	40.03–48.22	Parsābād, Moghān	5.0		5.2						
2000.05.03	09:01	29.59–50.83	Chehel Zari	4.9	4.6	5.1		05				
2000.08.22	16:55	38.14–57.37	Rāz, N. Bojnurd	5.1	5.8	5.6	VII	04		0/9		
2000.10.23	06:54	31.54–59.81	Nehbandān	5.3	4.9	5.3						
2000.11.25	18:09	40.23–49.95	Āzarbāijān	5.7	6.4	6.8			6/298	+	+	
2000.11.25	18:10	40.23–49.94	Āzarbāijān	6.2	6.2	6.5						
2000.12.06	**17:11**	**39.53–54.80**	**Turkmenistan**	**6.7**	**7.4**	**7.0**		**31**				
2001.03.13	01:04		–			5.1						
2001.03.23	05.24	32.98–46.63	Pahle, Dehlorān area	5.1	4.9	5.4						
2001.06.10	01:52	39.80–53.86	–	5.4	5.2	5.4		31				
2002.02.17	13:03	28.08–51.79	Shonbeh	5.5	5.0	5.4	VI+	15?	1/30	Bāghān	300	
2002.04.24	19:48	34.60–47.40	Armanijān	5.3	5.2	5.4	VI+		2/56	10		
2002.06.22	02:58	35.59–49.02	Changureh	6.2	6.4	6.4	VIII	10	261/1300	15	>1000	91
2002.09.02	01:00	35.68–48.81	Changureh AFS	5.1	4.6	5.2						

Continued

TABLE 17.12 Instrumental Period Medium- to Large-Magnitude Earthquakes in Iran (1900–2014)—Cont'd

Date (Christian)	Time (UTC)	Epicenter 00°N–00°E	Location	*mb*	M_s	M_w	*I* (MMI)	CD_w (km)	PGA (H/V)	People Killed/ Injured	Official Death Toll/ Injured	Cities Destroyed/ Damaged (d)	Towns Destroyed/ Damaged (d)	Villages Destroyed/ Damaged	No. of Houses Destroyed/ Damaged	Direct Economic Loss (U.S$ m, Original values)
2002.09.25	22:28	32.07–49.32	Masjed Soleymān	5.5	5.1	5.6										
2002.10.16	09:20	31.37–56.44	Tarz, Rāvar area	4.7	4.5	5.3										
2002.12.24	17:03	34.55–47.48	Dinévar	5.1	4.6	5.2				2/56				10/57	+	
2003.01.11	17:45	29.62–51.53	Kāzerun area	5.2	5.0	5.2	VI^+								2000	
2003.01.14	14:13	27.98–62.32	Sarāvān	5.4	4.8	5.4										
2003.02.14	10:29	28.00–56.79	Ahmadi	5.3	5.3	5.6										
2003.06.24	06:52	27.29–60.95	Irānshahr, Makrān	5.4	4.7	5.5										
2003.07.03	14:59	35.41–60.76	Yakhak	5.3	4.9	5.2	VI^+							3	150	
2003.07.10	17:06	28.31–54.16	Gajjiābād	5.8	5.5	5.8	VII	11?		1/25					3500	21.6
2003.07.10	17:40	28.29–54.08	Gajjiābād	5.7	5.5	5.8	VII									
2003.08.04	03:28	29.03–59.74	SE Lut	5.3	5.3	5.6										
2003.08.21	04:02	29.02–59.74	SE Lut	5.5	5.8	5.9										
2003.12.26	01:56	28.95–58.27	Bam	5.9	6.6	6.6	VIII^+	06	0.8/ 0.98	32,000–43,200/ 30,000	31,828/ 17,500	Bam	Barāvāt	+	+	1500
2004.01.14	16:58	27.68–52.36	'Assaluyeh	5.4	4.9	5.2										
2004.05.28	12:38	36.26–51.56	Firuzābād-e Kojur	6.2	6.4	6.2	VIII	22		35/278					3880	15.4
2004.10.07	21:46	37.14–54.46	Āq Qalā, Gorgān	5.6	5.4	5.6				0/60						

2005.01.10	18:47	37.21–54.51	Āq Qalā, Gorgān	5.2	5.0	5.4								
2005.02.22	02:25	30.76–56.80	Dāhuiyeh	5.9	6.5	6.4	VIII	07		612/1411	Zarand (d)	16	8000	79
2005.03.13	03:31	27.08–61.90	Sibrurān, Makrān	5.8	5.4	6.0								
2005.05.03	07:21	33.49–48.63	Choghālvandi, S. Borujerd	4.8	4.2	4.9	VI^{+}			1/26				
2005.11.27	10:22	26.75–55.82	Suzā, Central Qeshm Island	6.0	5.8	5.8	VII^{+}	09	0.12/ 0.33	10/100		12	2600	
2005.11.27	16:30	26.80–55.77	Suzā, Central Qeshm Island AFS	5.3	5.3	5.4	VI^{+}	10						
2006.02.28	07:31	28.10–56.83	Ahmadi	5.8	6.1	6.0								
2006.03.25	07:28	27.55–55.66	Fin	5.6	5.6	5.7		08						
2006.03.25	09:55	27.58–55.76	Fin AFS	5.2	5.0	5.5		04						
2006.03.31	01:17	33.58–48.79	Chālānchulān	5.7	5.9	6.1	VIII^{-}	06		68/1246			30,000	
2006.06.28	21:02	26.86–55.82	Central Qeshm Island	5.7	5.5	5.6		11						
2006.07.18	23:27	26.28–61.22	Baftān, Makrān	5.1	4.3	5.3								
2006.10.12	17:08	39.87–54.75		5.4	4.8	5.3								
2007.06.18	14:29	34.49–50.86	Kahak	5.3	5.1	5.5	VII					+	+	
2007.11.22	03:44	26.80–54.71		5.2	4.2									
2008.08.27	21:52	32.32–47.38	Musiān	5.3	5.4	5.8								

Continued

TABLE 17.12 Instrumental Period Medium- to Large-Magnitude Earthquakes in Iran (1900–2014)—Cont'd

Date (Christian)	Time (UTC)	Epicenter 00°N–00°E	Location	*mb*	M_s	M_w	*I* (MMI)	CD_w (km)	PGA (H/V)	People Killed/ Injured	Official Death Toll/ Injured	Cities Destroyed/ Damaged (d)	Towns Destroyed/ Damaged (d)	Villages Destroyed/ Damaged	No. of Houses Destroyed/ Damaged	Direct Economic Loss (U.S$ m, Original values)
2008.09.10	11:00	26.77–55.83	Tombān, Central Qeshm Island	6.0	5.9	5.9		8		7/45						
2008.10.25	20:17	26.58–55.02	Tombān AFS, Central Qeshm Island	5.1	4.8	5.4										
2008.12.07	13:36	26.91–55.90	Tombān AFS, Central Qeshm Island	5.5	5.1	5.4		4								
2009.07.22	03:53	26.79–55.76	Tombān AFS, Central Qeshm Island	5.4	4.8	5.2		4								
2010.07.20	19:38	27.03–53.90	Lamerd, Ashkanān	5.7	5.5	5.8				4/32				Lamerd		
2010.07.30	13:50	35.26–59.38	Torbat Haydariyeh	5.4	5.0	5.4				–/274						
2010.07.31	06:52	29.58–56.75	Negār, Lālehzār	5.4	4.9	5.4										
2010.08.11	17:26	37.81–51.08	Āshkhāneh	4.9	4.0											
2010.08.27	19:23	35.48–54.49	Kuh Zar, Torud	5.6	5.5	5.8			4						700	
2010.09.27	11:22	29.65–51.69	NE Kāzerun	5.9	5.5	5.9				1/3				Konar Takhteh		
2010.11.26	12:33	28.06–52.54	S. Hengām	5.6	5.1	5.6										

2010.12.20	18:41	28.37–59.15	S. Rigān	5.9	6.7	6.5		05	7/25	1	5		
2011.01.05	05:55	30.16–51.81	Sepidān	5.3	5.0	5.4							
2011.01.27	08:38	28.15–59.05	S. Rigān (Sarzeh)	5.7	6.2	6.2		09					
2011.03.05	11:24	30.02–51.15	Bābā Monir	5.2	4.8	5.3							
2011.06.15	01:05	27.78–57.76	Kahnuj	5.4	5.0	5.5							
2011.10.19	02:52	28.05–54.28		5.5	5.2	5.2							
2011.11.08	00:24	30.10–51.67		5.0	4.5	5.3							
2012.01.19	12:35	36.28–58.89	N. Neyshābur	5.6	5.3								
2012.02.27	18:48	31.43–56.78	N. Rāvar	5.0		5.2			–/6	Ravar (d)	+	+	
2012.05.03	10:09	32.75–47.72	Murmuri, NE Dehlorān	5.6		5.4							
2012.07.01	22:01	34.49–59.88	N. Zuzan	5.3		5.1							
2012.07.24	06:50	31.70–50.96		5.2		4.7							
2012.08.11	12:23	38.38–47.13	South Ahar	6.2	6.7	6.4	VIII	07	327/3000		300		1 billion
2012.08.11	12:34	38.62–47.14	South Ahar	6.3	6.3	6.3							
2012.08.11	22:24	38.45–46.71	South Ahar	5.1		5.2							
2012.08.15	17:49	38.41–46.67	South Ahar	5.3									
2012.11.07	06:26	38.41–46.62	South Ahar	5.4	5.4	5.6							
2012.12.05	17:08	33.50–59.57	Zohān	5.6	5.6	5.8							

Continued

TABLE 17.12 Instrumental Period Medium- to Large-Magnitude Earthquakes in Iran (1900–2014)—Cont'd

Date (Christian)	Time (UTC)	Epicenter 00°N–00°E	Location	*mb*	M_s	M_w	*I* (MMI)	CD_w (km)	PGA (H/V)	People Killed/ Injured	Official Death Toll/ Injured	Cities Destroyed/ Damaged (d)	Towns Destroyed/ Damaged (d)	Villages Destroyed/ Damaged	No. of Houses Destroyed/ Damaged	Direct Economic Loss (U.S$ m, Original values)
2013.01.12	03:25	31.82–50.95		5.4		5.0										
2013.01.21	19:48	30.35–57.45	NW Sirch	5.3	5.2	5.5	VII									
2013.04.09	11:52	28.50–51.59	Shonbeh	6.2	6.2	6.3	VIII			37/850				92	+	
2013.04.09	12:05	28.46–51.56	Shonbeh	5.4												
2013.04.10	01:58	28.35–51.49	Shonbeh	5.6	5.3	5.4				40?/300				1000		
2013.04.10	07:59	28.39–51.73	Shonbeh	5.2												
2013.04.16	**10:44**	**28.10–62.17**	**E. Baluchestān**	**7.0**		**7.7**				**>40/300**					**>1000**	
2013.04.17	03:15	28.00–62.12	E. Baluchestān	5.7		5.6										
2013.05.11	02:08	26.58–57.88	Goharān	5.9	6.2	6.1				1/16				+	+	
2013.05.11	03:09	26.54–57.92	Goharān	5.3	5.4	5.5										
2013.05.12	00:06	26.40–57.52	Goharān	5.5	5.4	5.5										
2013.05.12	10:54	26.71–57.76	Goharān	5.4	5.3	5.5										
2013.05.18	10:03	26.60–57.78	Goharān	5.5	5.4	5.5										
2013.05.18	10:57	26.50–57.71	Goharān	5.5	5.4	5.4										
2013.11.22	18:31	34.34–45.65		5.7	5.6											

2013.11.23	23:26	34.26–45.69		5.2				
2013.11.24	18:05	34.21–45.58		5.4				
2013.11.28	13:51	29.49–51.41	Borāzjān	5.6	7/>200	Borāzjān, Ābpākhsh, Vahdatiyeh, Dālaki	19	350 in Borāzjān

Earthquakes with $M_w \geq 7.0$ are in bold characters.

Revised and modified after Nabavi (1972, 1977), Tchalenko et al. (1974a), Berberian (1976a, 1977a, 1979a, 1981a, 1983a,c, 1995, 1997a, 2005), Berberian and Tchalenko (1976a–c), Ambraseys and Melville (1982), Kondorskaya and Shebalin (1977, 1982), Berberian et al. (1983, 1984, 1985, 1992, 1996, 1999, 2000a,b, 2001), Ambraseys and Adams (1989), Ambraseys and Finkel (1995), Ambraseys (1988, 1997, 2009), Berberian and Yeats (1999, 2001), Ambraseys and Bilham (2003a,b), Babayan (2006), Berberian and Walker (2010), contemporary and near-contemporary Persian and Armenian articles and newspapers, and personal databank collected since 1971. Telesiemic data for the early Iranian earthquakes suffer from large-magnitude errors in epicentral locations (see the text). The list is not complete and more work (which is in progress) is needed to complete the data set See also Chapters 12 through 16.

Epicenter:

*: Macroseismic epicenter assessed in this study from published and unpublished data.

Instrumental Epicenters:

Engdahl et al. (2006) covering the period 1923–2004; Nabavi (1972), ISC, and NEIC filling the gaps. K: Kondorskaya and Shebalin (1977, 1982).

Magnitude:

Instrumental period magnitudes are mainly taken from ISS, ISC, Ambraseys (2001), Engdahl et al. (2006), and USGS earthquake Hazard Program.

+: Few.

++: Many.

CD_w: Centroid depth, waveform analyses (Ni and Barazangi, 1986; Berberian et al., 1992, 1999, 2000b, 2001; Baker, 1993; Baker et al., 1993; Priestley et al., 1994; Gao and Wallace, 1995; Maggi et al., 2000a,b; Jackson et al., 2002; Walker et al., 2003, 2004, 2005a,b; Talebian and Jackson, 2004; Peyret et al., 2008; Talebian et al., 2006a; Nissen et al., 2007, 2010; Tatar et al., 2007; Roustaei et al., 2010; Walker et al., 2013a,b). See the text for the individual events and references in Chapters 12 through 16.

AFS: Aftershock.

FSK: Foreshock

I: Maximum intensity (MMI).

mb: Body wave magnitude.

M_s: Surface wave magnitude.

M_w: Moment magnitude.

?: Events or values queried where more speculative.

Together with the tables prepared in this study for the coseismic surface ruptures (Tables 14.9 and 14.11), they allow the derivation of empirical relations between the magnitude, intensity, surface fault length, and displacement for intracontinental earthquakes, providing that accurate focal depths are used.

Comparison of the tables (17.10–17.12 and 14.9 and 14.11) shows that some regions that were active during the historical period were also active during the instrumental period. This could probably be due to a shorter recurrence period, continuous activity, and the long-term seismicity of a region. Examples are the 1440 and 1972 Kārzin earthquakes in the Fārs province of the Zāgros (Figure 13.4); and the 1336, 1703, 1884, 1879, 2005, 2006, 2008, and 2009 earthquakes at the Hormoz Island in the Persian Gulf. Faults such as Ipak, Gowk (northern segment), Kuhbanān, Zāgros Main Recent, Khazar, and Central Kopeh Dāgh Shear Zone showed seismic activities both before and after 1900 (Chapters 11–16).

Regions also exist that were active during the historical period but have so far been quiescent during the short instrumental period of the last 114 years. These regions have definitely longer recurrence periods and are capable of generating large-magnitude earthquakes in the future. Examples are the Ray/Tehrān, Karaj, Qazvin, Soltāniyeh, Miyāneh, Bozqush, Tabriz, Khoi, Kāshān, Dāmghān, Samalqān, Sabzévār, Quchān, Neyshābur, Mashhad, Gonābād, Birjand, Nauzād, Dughābād, Zuzan, Saimareh/Darreh Shahr, Arrajān, Kāzerun, Shirāz, Ahvāz, Sirāf, Tāléqan, and Alamut regions (Chapters 9–16). Many faults such as the North Tehrān, Moshā, Pārchin, Dāmghān, Larzaneh, North Alborz, North Tabriz, Nauzād, and Birjand faults were active during the historical period (Chapter 11).

Some areas that showed seismic activity since 1900 do not have recorded historical data either due to long recurrence periods or a lack of documentation. Examples include the 1930 Salmās earthquake along the Salmās right-lateral fault (Figure 12.5); the 1968 and 1979 earthquakes along the Dasht-e Bayāz left-lateral strike-slip fault (Figure 13.1); the 1978 Tabas-e Golshan earthquake along the Tabas thrust (Figure 13.8); the 1990 Rudbār earthquake along the Rudbār left-lateral strike-slip fault (Figure 13.12), and many more.

Faults also occur that lack both historical and instrumental period seismic activity. Examples include Nāyband (Figures 9.1 and 12.2), Southern Kāshān (Figure 16.4), Southern Gowk (Figure 14.7), Dehshir, Anār (Figure 9.1), and several other faults. Lack of historically documented seismic data along these faults is due to their locations in remote desert areas and their long recurrence periods.

We also see that some large-magnitude earthquakes tend to occur in episodic clusters (Chapter 16). Consequently, we may expect such episodic bursts of seismic activity that have occurred earlier will recur in the future.

Finally, the effects of earthquakes have been similar since historic times: high death tolls, widespread destruction and damage of structures and infrastructure, with heavy economic losses. The earthquake history of Iran since 1962 is an ugly and excruciating reminder that states among developing countries are not enforcing the aseismic building codes and thus not reducing the seismic risk. Governments could have averted much of the destruction and casualties that have occurred since at least 1969, when the first Iranian code for seismic resistant design of buildings (ISIRI code No. 519; revised as the second code No. 2800 in 1988) was implemented (see Figure 1 in Introduction). Regrettably, unfathomable destruction and high death tolls are a system that the Iranians and other nations have been living in, rather needlessly.

17.8.1 Epilogue

Planet Earth was born and evolved through catastrophic geological and environmental events within the framework of the partially cataclysmic nature of the cosmos as it exists in different places and times at least since 4.54 billion years ago. Traces of this evolution can be seen in active tectonics and global disasters caused by earthquakes, tsunamis, climate change, superstorms, massive floods, pervasive droughts, meteor impacts, and gamma-ray bursts of dying stars. In the case of medium- to large-magnitude earthquakes, the defense systems of several countries have been shattered numerous times by the complete destruction of infrastructure and the decline of societies living along active fault zones since prehistoric times. Nonetheless, by understanding the geological and engineering fundamentals of planet Earth, industrialized countries have achieved proper scientific and engineering defense mechanisms and have been able, thereby, to minimize the earthquake risk.

The precious lives of 427.5 million people living in the vicinity of hazardous active fault zones on the Greater Iranian plateau[1] should be safeguarded against earthquake destruction, and their unfathomable death tolls should be immediately minimized. The people should not be content with the dangerous and destructive status quo and the doctrine of fatalism. Fundamental scientific, engineering, social, and political corrective actions must be taken immediately if the vicious circle of the earthquake is to be stopped. The goal cannot be achieved unless we learn from recorded history and rectify our thoughts, deeds, and words. We should definitely avoid constructing vulnerable structures and infrastructures in hazardous active fault zones, and replace the status quo with better protection policies for critical facilities, stringent enforcement of building codes, and more appropriate land-use practices. By adopting correct policy before earthquakes strike, property damage and death tolls will be reduced, businesses will be able to recover faster, the local economy will be

1. Iran, 77 million; Afghanistan, 31 million; Armenia, 3.1 million; Azerbaijan, 9.4 m; Georgia, 4.2 million; Iraq, 34.5 million; Pakistan, 183.4 million; Turkey, 79.7 million; and Turkmenistan, 5.2 million.

maintained, and governments' disaster aid will be minimized. A disaster can strike any moment and crumple the nation; we should act now.

On 11 October 1890 strong wind blew in Astarābād *[Estārehābād, the modern city of Gorgān]*; some buildings collapsed, including a stable where two horses were killed.

Iran Newspaper, 30 October 1890, No. 731

References

Abbasi, M.R., Feghhi, K., Tavakoli, B., 1998. Earthquake of Golestan (Ardebil) February 1997. In: Proc. 1st Iran-Japan Workshop on Recent Earthquakes in Iran & Japan, May 16–18, Tehran, pp. 85–91.

Abbot Zākariā, 1329. Inscription on a stone panel regarding the 1319 earthquake repairs at the Armenian Monastery of Surp Thādei Vānk (St. Thāddeus) in Her (modern Māku) signed by Abbot (Vanahāyr) Zākariā/Zechariah (see Plate 8.2).

Abdalian, S., 1951. Larzesh-hāyeh Qeshr-e Zamin va Zelzeleh-hāyeh Keshvar-e Irān [Tremors of the Earth's crust and the earthquakes of Iran]. Yamani Publ., 1330, Tehran, Iran, 40 pp. (in Persian).

Abdalian, S., 1953. Le tremblement de terre de Torude en Iran. Nature 81 (3222), 314–319, Paris.

Abdalian, S., 1961. Les grande unités tectoniques de l'Iran. Proc. Tehran Conf. Seism. Earthq., vol. 10. Inst. Geophys., Tehran University, pp. 49–52.

Abdalian, S., 1962. La tectonique de l'Iran, vol. 8. Inst. Geophys., Tehran University, 76 pp, with Tectonic Map of Iran.

Abdalian, S., 1963a. Seismo-tectonique de l'Iran, vol. 16. Inst. Geophys., Tehran University, 105 pp, with Seismotectonic Map of Iran.

Abdalian, S., 1963b. Le séisme du premier September 1962, l'histoire geologique et tectonique. In: Report on the Great Buyin-Zahra Earthquake of Sep. 1st, 1962. Inst. Geophys., Tehran University, vol. 15, Chapter III, pp. 43–72, with isoseismal and earthquake faulting map of the area on a seismotectonic and geological background.

Abel Zaya. See Zaya.

Abers, G., McCaffrey, R., 1988. Active deformation in the New Guinea fold-and-thrust belt: seismological evidence for strike-slip faulting and basement-involved thrusting. J. Geophys. Res. 93, 13332–13354.

Abich, H., 1857. Sur les derniers tremblements de terre dans la Perse septentrionale et dans le Caucase, ainsi que sur des eaux et des gaz s'y trouvant en rapport avec ces phenomenes, par M. Abich (lu le 16 mars 1855). Bull. Cl. Phys. Math. Acad. Imp. Sci. 14 (4, 5), 49–80, St. Petersburg.

Abich, H., 1858. Tremblement de terre observe' a' Tabriz en Septembre 1856, Notice physiques et geographiques de M. Khanykof sur l'Azerbeidjan, communiquees par H. Abich (lu le 17 Janvier, 1857). Bull. Cl. Phys. Math. Acad. Imp. Sci. 16 (22), 337–352, St. Petersburg.

Abich, H., 1882. Geologische Forschungen in den Kaukasischen Landern. In: *A. Holder*. Wien, Vienna, pp. 1878–87, 2 volumes.

Abrahamian, E., 2013. The Coup: 1953, the CIA, and the Roots of Modern US-Iranian Relations. The New Press, 340 pp.

Abrishami, M.J., 1990. Zelzeleh-Shenāsi dar motun-e Kohan-e Irān, [Earthquakes in Old Iranian Chronicles]. Sakhtemān, 19, Tir 1369, p. 39-43, in Persian.

Abu Dulaf, Mas'ar ebn Mohalhel Kharzaji, 950. Safarnāmeh-ye Abu Dolaf dar Irān (Abu Dolaf's Travel Diary in Iran, 341/950). Translated into Persian by Sayyed Abolfazl Tabātabā'i, Zavvār Publ., Tehran, second ed., 1354/1975, 184 pp. Also Abu Dolaf, Ed. & Trans. V. Minorsky, Cairo, 1955. Also, Abu Dulaf Mis'ar ibn Muhalhil's Travels in Iran (c. AD 950), Ed. & Trans. V. Minorsky, Cairo, 1955.

Abu'l-Fida [Abolfeda] Isma'il Ibn'Ali, 1329. Al-Mukhtasar fi Akhbar al-Bashar (Tarikh-e-Abolfeda: A History of Mankind, events from the earliest times to the time of the writer's death, 721/1321), Ed. Istanbul, 1286/1869. Mimeograph. Tehran, 4 vols. Also Hossainiah Publ. House, Egypt, 1907, 2 vols., 4 parts.

Adamia, S., et al., 1974. Tectonic Map of the Caucasus. Academy of Sciences of the Georgian SSR Institute of Geology, Tbilisi.

Adeli, H., 1982. The Sirch (Kerman, Iran) earthquake of 28 July 1981: a field investigation. Bull. Seismol. Soc. Am. 72, 841–861.

'Adle, Ch., 1971. Contribution a la Geographie Historique du Damghan. Le Monde Iranien et l'Islam I. 69–104.

'Adle, Ch., 1988. Masjed va madraseh-ye Zauzan [The mosque and the religious school of Zauzan], Athar, 15 & 16, Sazeman Mirath Farhangi Keshvar, Winter 1367, 231-248 (in Persian).

'Adle, C., 1993 & 2011. Dāmghān. Encyclopaedia Iranica VI (6), 632–638. and updated 2011, www.Iranicaonline.org.

Afsar Kermāni, Mirzā Mehdi Qoli Khān, Afsar al-Sho'arā. Collection of Poems. Ed. R. Afsari, 1977.

Afsari, R., 1977. Zelzeleha-ye Ma'ruf-e-Kerman (The famous earthquakes of Kerman), 1356. Vahid, Tehran. with poems by Afsar Kermāni, Mirzā Mehdi Qoli Khān, Afsar al-Sho'arā; Collection of Poems.

Afshar-Harb, A., 1979. The stratigraphy, tectonics and petroleum geology of Kopeh Dagh region, northern Iran. Ph.D. thesis, Petroleum Geology Section, Royal School of Mines, Imperial College of Science and Technology, London, 316 pp.

Aghajanian, A., 1991. Population change in Iran, 1966–86: a stalled demographic transition. Population Dev. Rev. 17, 703–715.

'Ajā'ib al-Makhluqāt. See Tusi.

Ajā'ib al-Makhluqāt wa Gharā'ib al-Maujudāt. See Zakariyā Qazvini.

Akkermanns, P.M.M.G., van der Plicht, J., Nieuwenhuyse, O.P., Russel, A., Kaneda, A., Buitenhuis, H., 2010. Weathering climate change in the near East: dating and Neolithic adaptations 8200 years ago. Antiquity 84 (325), Archive Project Gallery, 2p, antiquity.ac.uk.

Āl-Ahmad, J., 1962. Gharbzadegi. First ed. Tehran, 1536/1977 (in Persian). Second ed. Ravāq Publ., Tehran, 1341/1962.

Alam, H., 1993. Cypress. Encyclopaedia Iranica VI, 505–508, Updated 2 November 2011, iranicaonline.org.

Alipoor, R., Zaré, M., Ghassemi, M.R., 2012. Inception of activity and slip rate on the main recent fault of zagros mountains, Iran. Geomorphology. 175–176, 86-97. http://dx.doi.org/10.1016/j.geomorph.2012.06.025.

Alirezaei, S., Hassanzadeh, J., 2012. Geochemistry and zircon geochronology of the Permian A-type Hasanrobat granite, Sanandaj–Sirjan belt: a new record of the Gondwana break-up in Iran. Lithos 151, 122–134.

Ālishān, L., 1895. Ancient Faith of the Armenians, Venice.

Alison, C., 1872. Mr. C. Alison to the Right Hon. Earl Granville, Tehran, February 15, 1872. The Foreign Office Archives, Great Britain.

Allen, M., Vincent, S., Alsop, G., Ismail-zadeh, A., Flecker, R., 2003. Late Cenozoic deformation in the South Caspian region: effects of a rigid basement block within a collision zone. Tectonophysics 366, 223–239.

Allen, M., Jackson, J., Walker, R., 2004. Late Cenozoic reorganization of the Arabia-Eurasia collision and the comparison of short-term and long-term deformation rates. Tectonics. 23http://dx,doi.org/10.1029/2003TC001530, TC2008.

Allen, M.B., Kheirkhah, M., Emami, M.H., Jones, S.J., 2011. Right-lateral shear across Iran and kinematic change in the Arabia-Eurasia collision zone. Geophys. J. Int. 184, 555–574.

Almazov, V., 1905. Spisok zemletriasenii, 1903. Izv. Post. Tsentral. Seism. Kom, vol. 2, St. Petersburg, pp. 14, 114, 115, 157, & 232.

Alpago-Novello, A., 1988. Sorhul. Documents of Armenian Architecture, No. 20. OEMME Edizioni, Milano, Italy, 109 pp.

Alsinawi, S.A., 1988. The historical seismicity of the Arab countries. In: Al-Furaih, A.A.F. (Ed.), Proc. 3rd Arab Symp. Earthq. Seism., 8–10 March 1986, Geophysical-Seismological Observatory. King Saud University, Riyadh, pp. 11–33.

Alsinawi, S., Ghalib, H., 1975. Historical seismicity of Iraq. Bull. Seismol. Soc. Am. 65, 541–547.

Amar.org.ir. Statistical Center of Iran. Census of Iran. amar.org.ir.

Ambraseys, N.N., 1961. On the seismicity of south-west Asia. Data from XV century Arabic manuscript. Revue pour l'Etude des Calamites 37, 18–30, Geneva.

Ambraseys, N.N., 1962. On the regional conditions and damage during the Buyin-Zahra (Iran) earthquake: UNESCO Mission Report NS/1585/62. 2 vols, December 1962.

Ambraseys, N.N., 1963. The Buyin-Zahra earthquake of September 1962: a field report. Bull. Seismol. Soc. Am. 53 (4), 705–740.

Ambraseys, N.N., 1965. An earthquake engineering study of the Buyin-Zahra, Iran, earthquake of September 1962. In: Proc. 3rd World Conf. Earthq. Eng., Auckland, vol. 3, pp. 2–21.

Ambraseys, N.N., 1968. Early earthquakes in North-central Iran. Bull. Seismol. Soc. Am. 58 (2), 485–496.

Ambraseys, N.N., 1970. Some characteristics features of the North Anatolian fault zone. Tectonophysics 9, 143–165.

Ambraseys, N.N., 1971. Value of historical records of earthquakes. Nature 232, 375–379.

Ambraseys, N.N., 1974a. Historical seismicity of North-Central Iran. In: Materials for the study of seismotectonics of Iran; North Central Iran. 29. Geol. Surv. Iran, pp. 47–96.

Ambraseys, N.N., 1974b. The seismicity of Iran. The Silakhor (Lurestan) earthquake of 23rd January, 1909 (part II). Annali di Geofisica XXVI (3–4), 399–427.

Ambraseys, N.N., 1975. Studies in historical seismicity and tectonics. Geodynamics Today, Royal Society, London, pp. 7–16.

Ambraseys, N.N., 1978a. The relocation of epicentres in lran. Geophys. J. R. Astron. Soc. 53, 117–121.

Ambraseys, N.N., 1978b. Studies in historical seismicity and tectonics. In: Bruce, W.C. (Ed.), The Environmental History of the Near and Middle East Since the Late Ice Age. Academic Press, London, pp. 185–210, (Chapter 122).

Ambraseys, N.N., 1979. A test case of historical seismicity: Isfahan and Chahar Mahal, Iran. Geogr. J. 145 (1), 56–71.

Ambraseys, N.N., 1988. Engineering seismology, Mallet-Milne Lecture. Special Issue. Earthq. Eng. Struct. Dyn. 17, 1–105.

Ambraseys, N.N., 1989. Temporary seismic quiescence: SE Turkey. Geophys. J. Int. 96, 311–331.

Ambraseys, N.N., 1990. Uniform magnitude re-evaluation of European earthquakes associated with strong-motion records. Earthq. Eng. Struct. Dyn. 19, 1–20.

Ambraseys, N.N., 1997. The Krasnovodsk (Turkmenistan) earthquake of 8 July 1895. J. Earthq. Eng. 1, 293–317.

Ambraseys, N.N., 2001. Reassessment of earthquakes, 1900–1999, in the Eastern Mediterranean and the Middle East. Geophys. J. Int. 145, 471–485.

Ambraseys, N.N., 2009. Earthquakes in the Mediterranean and Middle East. Cambridge University Press, 947 pp.

Ambraseys, N.N., Adams, R.D., 1989. Long-term seismicity of North Armenia. Eos 70 (10), March 7, 145, 152–154.

Ambraseys, N.N., Bilham, R., 2003a. Earthquake in Afghanistan (with supplement). Seismol. Res. Lett. 74 (2), 107–123. electronic supplement, 1–27, http://www.seismosoc.org/publications/SLR/SLR_74-2_ambraseys-esupp2.html.

Ambraseys, N.N., Bilham, R., 2003b. Earthquakes and associated deformation in northern Baluchistan 1892-2001. Bull. Seismol. Soc. Am. 93 (4), 1573–1605.

Ambraseys, N.N., Bilham, R., 2011. Corruption kills. Nature 469, 153–155.

Ambraseys, N.N., Finkel, C.F., 1987. Seismicity of Turkey and neighbouring regions. Annales Geophysicae 5B (6), 701–725.

Ambraseys, N.N., Finkel, C.F., 1995. The Seismicity of Turkey. M.S. Eren Press (Beyoğlu, Istanbul), ISBN: 9757622389, 223 pp.

Ambraseys, N.N., Jackson, J.A., 1998. Faulting associated with historical and recent earthquakes in the eastern Mediterranean region. Geophys. J. Int. 133, 390–406.

Ambraseys, N.N., Melville, C.P., 1977. The seismicity of Kuhistan. Iran. Geogr. J. 143, 179–199.

Ambraseys, N.N., Melville, C.P., 1982. A History of Persian Earthquakes. Cambridge University Press, London, 219 pp.

Ambraseys, N.N., Mo'infar, A., 1973. The seismicity of Iran. The Silakhor, Lurestan earthquake of 23rd January, 1909. Annali di Geofisica XXVI (4), 659–678.

Ambraseys, N.N., Moinfar, A., 1974. The Firuzabad earthquake of 16 August 1958. Annali di Geofisica 27, 1–21.

Ambraseys, N., Moinfar, A., 1975. The Turshiz earthquake of 25th September 1903. Annali di Geofis. 28, 253–269.

Ambraseys, N.N., Moinfar, A., 1976. Iran Earthquakes 1969, vol. 63. Publ. Plan & Budget Organisation, Tehran, p. 22.

Ambraseys, N.N., Moinfar, A., 1977a. The Torud earthquake of 12th February 1953. Annali di Geofisica 30, 185–200.

Ambraseys, N.N., Moinfar, A., 1977b. The Kaj Darakht earthquake of 25th May 1923. Annali di Geofisica 30, 3–18.

Ambraseys, N.N., Tchalenko, J.S., 1969. Dasht-e-Bayaz earthquake of 31 August 1968. UNESCO Publ. No. 1214/BMS, Paris.

Ambraseys, N.N., Zatopek, A., 1969. The Mudurnu Valley, West Anatolia, Turkey, earthquake of 22 July 1967. Seismol. Soc. Am. Bull. 59, 521–589.

Ambraseys, N.N., Anderson, G., Bubnov, S., Crampin, S., Shahidi, M., Tassios, T.P., Tchalenko, J.S., 1969. Dasht-e-Bayaz Earthquake of 31 August 1968. UNESCO Serial No. 1214/BMS.RD/SCE, Paris.

Ambraseys, N.N., Moinfar, A.A., Tchalenko, J.S., 1972. Ghir Earthquake of 10 April 1972. UNESCO Publ. No. 2789/RMO, Paris, 130p.

Ambraseys, N., Moinfar, A., Peronaci, F., 1973. The Farsinaj, Kermanshah, earthquake of 13th December 1957. Annali di Geofis. 26, 679–692.

Ambraseys, N.N., Arsovski, A., Moinfar, A.A., 1979. The Gisk Earthquake of 19 December 1977 and the Seismicity of the Kuhbanan Fault-Zone. UNESCO FRM/SE/GEO/79/192, Paris.

Ambraseys, N.N., Melville, C.P., Adams, R.D., 1994. Seismicity of Egypt, Arabia and the Red Sea. Cambridge University Press, 223 pp.

Ambraseys, N.N., Jackson, J.A., Melville, C.P., 2002. Historical seismicity and tectonics: the case of the Eastern Mediterranean and the Middle East. In: Lee, W.H.K., Hanamori, H., Jennings, P.C., Kisslinger, C. (Eds.), International Handbook of Earthquake and Engineering Seismology. Academic Press, pp. 747–763, International Association of Seismology and physics of the Earth's Interior (IASPEI) Committee on Education.

Amiran, D.H.K., 1950. A revised earthquake-catalogue of Palestine 1. Isr. Explor. J. 1, 223–246.

Amiran, D.H.K., 1951. A revised earthquake-catalogue of Palestine 2. Isr. Explor. J. 2, 48–62.

Amir-Mo'ezzi, M.A., 2005. Shahrbanu. Encyclopaedia Iranica. Iranicaonline.org.

Ananikian, M.H., 1925. Armenian mythology. In: The Mythology of All Races. V.III Armenian. Cooper Square Publ., Inc., New York, pp. 1–100. pp. 379–440, http://rbedrosian.com/ananik.html.

Andalibi, M.J., 2004. Systematic analysis of Bam earthquake on Dec 26th 2003. Geological Survey of Iran, Shiraz Center, P.O. Box 71365-1183, www.adrc.or.jp/counterpart_report/analyses_BamEQ.pdf, 4 pp.

Anderson, A.R., 1928. Alexander at the Caspian Gates. Trans. Proc. Am. Philol. Assoc. 59, 130–163.

Anderson, H., Beanland, S., Blick, G., Darby, D., Downes, G., Haines, J., Jackson, J., Webb, T., 1994. The 1968 May 23 Inangahua, New Zealand, earthquake: an integrated geological, geodetic, and seismological source model. N. Z. J. Geol. Geophys. 37, 57–86.

Anklesaria, B.T., 1935. Kar-nama-i Ardakhsir-i Papakan, Bombay. Also see, avesta.org.

Anklesaria, B.T., 1956. Zand-Akasih; Iranian or Greater Bundahishn. English transliteration and translation, Bombay. Digital version by J.P. Peterson available at http://www.avesta.org/mp/grb.htm.

Anonymous, 982. Hudud al-'Alam men al-Mashreq el al-Maghreb [The Limits of the World from the East to the West], 372/982; a Persian geography written by an anonymous writer, 372/982. Ed. Manuchehr Sotudeh. Tāhuri Publ., Tehran, 1362/1983 (in Persian). Also Hudud al-'Alam, Trans. & comments V. Minorsky, second ed., Bosworth, London, 1970.

Ansāri Dameshqi, Shams al-Din Muhammad ibn Abu Taleb, 1300. Kitāb Nokhbat al-Dahr fi 'Ajā'ib al Barr wa al-Bahr (ca. 700/1300). Translated into Persian by S. Hamid Tayebian, Farhangestān Adab va Honar Irān, Tehran, 1357/1978 (in Persian).

Annells, R.N., Arthurton, R.S., Bazley, R.A.B., Davies, R.G., Hamedi, M.A.R., Rahimzadeh, F., 1985. Geological map of the Qazvin-Rasht 1:250,000 quadrangle map. Geol. Surv. Iran, E3,E4.

al-Antāki, Yahyā ebn Sa'id [Yahyā of Antioch], 1066. Dhail Tārikh Sa'id ibn Bitriq (458/1066). Trans. Kratchovsky & Vasiliev, Patrologia Orientalis, v. 18, Paris 1924, 1932. English Trans. John Harper Forsyth, 1977, The Byzantine-Arab Chronicle (930–1034) of Yahya b. Said al-Antāki, Ph.D. Thesis, University of Michigan, 2 vols.

Ārākel of Tabriz. See Ārākel Vārtābed Tāvrizetzi.

Ārākel Vārtābed Tāvrizetzi (Ārākel of Tabriz; 1594–1670). Ārākel Vārtābed Dāwrizec'I, Pātmut'iwn Hāyoc' (History of Armenia, 1574–1665). *Vāghārshāpāt*, 1896. Also, Livre d'histoires (Armenian History, 1574–1665). Trans. Brosset, Collection d'Historiens Armeniens, I, St. Petersburg, 1874. Ed. George A. Bournoutian, The History of Vartabed Arak'el of Tabriz, Mazda Publishers, Inc., Costa Mesa, CA, 2 vols. (2005, 298 pp & 2006, 618 pp), George A. Bournoutian.

'Arib ibn Sa'ad al-Qurtubi, 932. Silat Tārikh al-Tabari [Tabari Continuanus, 320/923]. Ed. J.M. de Goeje, Leiden, 1897.

Aristotle, 350 B.C. Meteorologica. Translated from the Greek by H.D.P. Lee, Cambridge University Press, Cambridge, MA, 1952, Classic.mit.edu/Aristotle/meteorology.html.

Arrowsmith, J.R., Rhodes, D.D., 1994. Original forms and initial modifications of the Galway Lake Road scarp formed along the Emerson Fault during the 28 June 1992 Landers, California, earthquake. Bull. Seismol. Soc. Am. 84 (3), 511–527.

Arseh, 1988. Development plan of the city of Tabriz. Arseh Consulting Engineers and Office of Housing and Urban Development of the Eastern Āzarbāijān Province, Tabriz. Project A 3-21, scale 1:20,000.

Aryanmanesh, M., Zare, M., Kusky, M., 2007. Neotectonics control on the migration of the rivers by using remote sensing imagery—a case study from Tabas area-eastern Iran. JSEE, IIEES 9 (3), 99–109, Fall 2007.

Ashtari Jafari, M., 2012. Teleseismic source parameters of the Rigan county earthquakes and evidence for a new earthquake fault. Pure Appl. Geophys.. 169, 1655–1661. http://dx.doi.org/10.1007/s00024-011-0433-9.

Austermann, J., Iaffaldano, G., 2013. The role of the Zagros orogeny in slowing down Arabia-Eurasia convergence since 5 Ma. Tectonics. 32 (2), 351–363. http://dx.doi.org/10.1002/tect.20027.

Authemayou, C., Chardon, D., Bellier, O., Malekzadeh, Z., Shabanian, E., Abbassi, M., 2006. Late Cenozoic partitioning of oblique plate convergence in the Zagros fold-and-thrust belt (Iran). Tectonics. 25http://dx.doi.org/10.1029/2005TC001860, TC3002.

Authemayou, C., Bellier, O., Chardon, D., Benedetti, L., Malekzade, Z., Claude, C., Angeletti, B., Shabanian, E., Abbassi, M.R., 2009. Quaternary slip rates of the Kazerun and Main Recent Faults: active strike-slip partitioning in the Zagros fold and thrust belt. Geophys. J. Int.. 178 (1), 524–540. http://dx.doi.org/10.1111/j.1365246X.2009.04191.x.

Avanessian, A.S., 1994. Migration of active faults in space and time as a factor of seismic hazard. In: Bolt, B.A., Amirbekian, R. (Eds.), Proc. Int. Conf. Continental Collision Zone Earthquakes and Seismic Hazard Reduction, Yerevan-Sevan, Armenia, 1–6 October 1993, IASPE/IDNDR, 1 September 1994, 79–82. Proc. Int. Conf. Continental Collision Zone Earthquakes and Seismic Hazard Reduction, Yerevan-Sevan, Armenia, 1–6 October 1993, Ed. B.A. Bolt, R. Amirbekian, IASPE/IDNDR, 1 September 1994, 25–30.

Avestā. See Poure Davoud, 1964, 1968, and 1977a,b. Also see Zoroastrian Archives. Avesta.org.

Avicenna. See Ibn Sinā (also known as Pursinā).

Avouac, J.-P., Tapponnier, P., Bai, M., You, H., Wang, G., 1993. Active thrusting and folding along the northern Tien Shan and late Cenozoic rotation of the Tarim relative to Dzungaria and Kazakhstan. J. Geophys. Res. 98, 6755–6804.

Axen, G.J., Lam, P.S., Grove, M., Stockli, D.F., Hassanzadeh, J., 2001. Exhumation of the west-central Alborz Mountains, Iran, Caspian subsidence, and collision-related tectonics. Geology 29, 559–562.

Axworthy, M., 2013. Revolutionary Iran. Revolutionary Iran: A History of Islamic Republic. Oxford University Press, New York, 528 pp.

Azar, Lotf ‘Ali Bayg Bigdeli Shāmlu, 1722–1780. Tazkereh Ātāshkadeh Āzar. Ed. Hassan Sādāt Nāseri, Tehran, 1338/1959. Also lith. ed., Bombay.

Āzarnoush, M., 2009. New evidence on the chronology of the Anahita temple. Iranica Antiqua XLIV, 393–404.

Āzar Shāmlu, 18th century. See Qaffāri Kāshāni, 1796; Zarrābi, 1870; Nakhjavāni, 1964.

Azimi, F., 1995. Iran: The Crisis of Democracy, 1941–1953. St. Martin’s Press, New York.

Babayan, T.H., 2006. Atlas of Strong Earthquakes of the Republic of Armenia, Ārtsākh and Adjacent Territories from Ancient Times Through 2003. Tigrān Mets Publishing House, Gyumri, Armenia, 139 pp.

Bābā-ye Hamédāni. In Zokā’ (1989).

Babur, Emperor of Hindustan, 1483–1530. The Babur-nama (Memoirs of Babur). Trans. A. Beveridge, 1922. Luzac, London; also archive.org.

Bachmanov, D.M., Trifinov, V.G., Hessami, Kh.T., Kozhurin, A.I., Ivanova, T.P., Rogozhin, E.A., Hamedi, M.C., Jamali, F.H., 2004. Active faults in the Zagros and central Iran. Tectonophysics 380, 221–241.

Badalyan, H.S., 1970. The history of calendar. Istoriya Kalendarya, Erevan (in Armenian).

Bagheri, S., Stampfli, G.M., 2008. The Anarak, Jandaq and Posht-e-Badam metamorphic complexes in central Iran: new geological data, relationships and tectonic implications. Tectonophysics 451, 123–155.

Bahār, M., 1966. Vāzhehnāmeh Bondeheshn (Dictionary of Bondiheshn). Bonyād Farhang Irān Publ., 166 pp., 1345 (in Persian).

Bahār, M., 1983. A Research in Iranian Mythology. Tus Publ., Tehran, vol. 1, 326 pp., 1362 (in Persian).

Bahār, M., 1990. See Bundahishn.

Bahmanyār ibn-e Marzbān, d. 1066. Jām-e Jahān Namāi. Ed. 'Abdollāh Nurāni, Mohammad Taqi Dānesh-Pazhuh. Mo'assesseh Motāle'āt Eslāmi, Tehran, 1362/1983 (in Persian).

Bahrami, E., Joneidi, F., 1990. Dictionary of the Avesta. Balkh Publ., Neyshabur Foundation, 1369, Tehran, 4 vols. (in Persian).

Baihaqi, Abol Hassan 'Ali ebn-e Zaid (Ebn-e Fandoq-e Baihaqi/Ibn Funduq), 1170. Tārikh-e Baihaq [The History of Baihaq]. Ahmad Bahmanyār (Ed.). Bongāh-e Dānesh Publication, Tehran 1317/1938 (in Persian). Reprinted in 1982.

Bailey, H.W., 1930–1932. Review of J. Markwart, a catalogue of the Provincial Eränshahr. Bull. Sch. Orient. Afr. Stud. VI, 781–782.

Bailey, H.W., 1972. Hyaona-, Indo-Celtica. In: Pilch, H., THurow, J. (Eds.), Gedächtnisschrift für Alf Somerfelt, pp. 18–28, München.

Baird-Smith, R., 1842. On the Zelzele Namah. In: Proceedings, Asiatic Soc. Bengal, XI, 1201, Calcutta.

Baker, C., 1993. Active Seismicity and Tectonics of Iran. Ph.D. Thesis, University of Cambridge, UK.

Baker, C., Jackson, J., Priestley, K., 1993. Earthquakes on the Kazerun Line in the Zagros Mountains of Iran: strike-slip faulting within a fold and-thrust belt. Geophys. J. Int. 115, 41–61.

al-Balādhuri (Balāzuri), Ahmad ibn Yahyā, 890. Kitāb Futuh al-Buldān [Book of the Conquests of Lands]. Trans. P.K. Hitti (vol. 1) and E.C. Murgotten (vol. 2). The Origins of the Islamic State, New York, 1916–1924. Persian Trans. Āzarmāsh Āzarnush, Bonyād Farhang Irān Publ., Tehrān 1346/1967; Soroush Publ., 1364/1985, Tehran (in Persian). Also, Liber Expugnationis Regionum, Ed. De Goeje, Brill, Leiden, 1866.

Bal'ami, Abu 'Ali Mohammad ibn Mohammad ibn, ca. 996. Tārikh-e-Bal'ami [Bal'ami; translation of Tārikh-e Tabari]. Edāreh Koll Negāresh Vezārat Farhang, 1341, Ed. Mohammad Parvin Gonābādi, Tehran University Publ., 1232 pp. Ed. Bahār, 1962.

Balassanian, S.Y., Arakelian, A., et al., 1995. Retrospective analysis of the Spitak earthquake. Annali di Geofisica XXXVIII (3–4), 345–372.

Balini, M., Nicora, A., Berra, F., Garzanti, E., Levera, M., Mattei, M., Muttoni, G., Zanchi, A., Bollati, I., Larghi, C., Zanchetta, S., Salamati, R., Mossavvari, F., 2009. The Triassic stratigraphic succession of Nakhlak (Central Iran), a record from an active margin. Geol. Soc.. 312, 287–321. http://dx.doi.org/10.1144/SP312.14, London, Special Publications.

Baljinnyam, I., Bayasgalan, A., Borisov, B.A., Cisternas, A., Dem'yanovich, M.G., Ganbaatar, L., Kochetkov, V.M., Kurushin, R.A., Molnar, P., Philip, H., Vashichilov, Y.Y., 1993. Ruptures of major earthquakes and active deformation in Mongolia and its surroundings. In: Geological Society of America Memoir, vol. 181, ISBN: 0-8137-1181-9, 62 pp.

Ballato, P., Nowaczyk, N.R., Landgraf, A., Strecker, M.R., Friedrich, A., Tabatabaei, S.H., 2008. Tectonic control on sedimentary facies pattern and sediment accumulation rates in the Miocene foreland basin of the southern Alborz mountains, northern Iran. Tectonics. 27http://dx.doi.org/10.1029/2008TC002278 (3.413), TC6001.

Ballato, P., Uba, C.E., Landgraf, A., Strecker, M.R., Sudo, M., Stockli, D.F., Friedrich, A., Tabatabaei, S.H., 2011. Arabia-Eurasia continental collision: insights from late Tertiary foreland-basin evolution in the Alborz mountains, northern Iran. Geol. Soc. Am. Bull. 123, 106–131.

Ballato, P., Stockli, D.F., Ghassemi, M.R., Landgraf, A., Strecker, M.R., Hassanzadeh, J., Friedrich, A., Tabatabaei, S.H., 2013. Accommodation of transpressional strain in the

Arabia-Eurasia collision zone: new constraints from (U-Th)/He thermochronology in the Alborz mountains, north Iran. Tectonics. 32, 1–18. http://dx.doi.org/10.1029/2012TC003159.
Ballore, Montessus de, 1900. Sismicite de la peninsula Balkanique et de l'Anatolie. Bull. Com. Geol. Russie 19, 31–53, St. Petersburg.
Ballore, Montessus de, 1906. Le geogrphie seismologique. A. Colin, Paris.
Ballore, Montessus de, 1924. La Geologie Sismologique. A. Colin, Paris.
Barber, E.W., Barber, P.T., 2004. When They Severed Earth from Sky; How the Human Mind Shapes Myth. Princeton University Press, 290 pp.
Bar Hebraeus, Girgis (Gregory) Abu'l Faraj Ahrun (ibn al-Ebri bin Hārun al-Malati), 1286. The Chronography of Gregory Abu'l Faraj, 1225–1286, the Son of Aaron, the Hebrew Physician, Commonly Known as Bar Hebraeus; Being the First Part of His Political History of the World, Translated from Syriac by E.A. Willis Budge, 2 vols. London, 1932: Oxford University Press. Reprinted Apa-Philo Press, Amsterdam, 1976. Reprinted Gorgias Press, LLC, 2003. Also Tarjomeh-ye Tārikh-e Mokhtasar al-Doval, 685/1286. Translated into Persian by Mohammad 'Ali Tājvar, Heshmatolāh Riyāzi, Etela'āt Publ., Tehran 1364/1985. Also Tārikh Mukhtasar al-Duwal, Ed. Sabhāni, Beirut, 1958. Also rbedrosian.com/BH/bh24.htm.
Barka, A.A., 1992. The North Anatolian fault zone. Annales tectonicae Sp. Issue VI (Suppl.), 164–195.
Barka, A.A., 1996. Slip distribution along the North Anatolian fault associated with the large earthquakes of the period 1939 to 1967. Bull. Seismol. Soc. Am. 86, 1238–1254.
Barka, A.A., 1999. The 17 August 1999 Izmit earthquake. Science 285, 1858–1859.
Barnhart, W.D., Hayes, G.P., Samsonov, S.V., Fielding, E.J., Seidman, L.E., 2014. Breaking the oceanic lithosphere of a subducting slab: The 2013 Khash, Iran earthquake. Geophys. Res. Lett. 41, 32–36. http://dx.doi.org/10.1002/2013GL058096.
Baron, I., Bubík, M., Otava, J., Babůrek, J., Ambrozek, V., Kernstocková, M., Milovský, R., 2011. Final Report on capacity building of Geological Survey of Iran's staff in geological mapping at 1:25 000 scale of selected map sheets in the Eastern Azarbayjan Province of Iran: explanatory notes to geohazard maps, area of Ahar and Ghalandar, MS, Final Report, Czech Geological Survey, Praha, 34 pp. Also Geol. Surv. Iran, Inter. Report.
Baron, I., Kernstocková, M., Faridi, M., Bubík, M., Milovský, R., Melichar, R., Sabouri, Babůrek, J., 2013. Paleostress analysis of a gigantic gravitational mass movement in active tectonic setting: The Qoshadagh slope failure, Ahar, NW Iran. Tectonophysics 605, 70–87.
Bartsiokas, A., 2003. Shanidar Cave, Iraq: Archaeology. Encyclopedia of Caves and Karst Science, Routledge.
Bastani Parizi, M.E., 1965. Khātun-e Haft Qal'eh, third ed, Tehran, 1344/1965, 1984, Tehran (in Persian).
Bastias, H., 1985. Fallamiento Cuaternario en la region sismotectonica de precordillera: San Juan, Argentina. Facultad de Ciencia Exactas, Fisicas y Naturales, Universidad Baciobal de San Juan, Tesis Doctoral inedita, 160 pp.
Batto, B.F., 1992. Slaying the Dragon: Mythmaking in the Biblical Tradition. Westminster John Knox Press.
Baumgarten, V., 1896. Poyezdka po vostochnoi Persii L. Gv. Volynskavo polka pouchika Baumgartena v 1894g (Geografichesko-torgovoye isledovabie). Sbornik Mater Po Azii 63, 1–367.
Bausani, A., 1959. Persia Religiosa. Milan, Il Saggiatore, 143.
Bayasgalan, A., Jackson, J., Ritz, J.-F., Cartier, S., 1999. Field examples of strike-slip fault terminations in Mongolia and their tectonic significance. Tectonics 18, 394–411.
Bayer, R., Chéry, J., Tatar, M., Vernant, P., Abbassi, M., Masson, F., Nilforoushan, F., Doerflinger, E., Regard, V., Bellier, O., 2006. Active deformation in Zagros-Makran transition zone inferred from GPS measurements. Geophys. J. Int. 165, 373–381.

Behzadi, 1989. See Bundahishn.

Bell, C.M., 1840. Geological notes on part of Mazandaran. Trans. Geol. Soc. Lond. 5, 577–581, Series 2.

Bendick, R., Bilham, R., Freymuller, J., Larson, K., Yin, G., 2000. Geodetic evidence for a slow rate in the Altyn Tagh system. Nature 404, 69–72.

Benjamin of Tudela. Contemporaries of Marco Polo. Ed. M. Komroff, 1927, London.

Ben-Menahem, A., 1979. Earthquake catalogue for the Middle East (92BC to 1980 AD). Boll. Geofis Teor. Appl. 21, 245–313.

Ben-Menahem, A., 1995. Review: a history of mainstream seismology: origins, legacy, and perspectives. Bull. Seismol. Soc. Am. 85 (4), 1202–1225.

Bennett, R.A., Reilinger, R.E., Rodi, W., Li, Y., Toksöz, M.N., Hudnut, K., 1995. Coseismic fault slip associated with the 1992 M_w 6.1 Joshua Tree, California, earthquake: implications for the Joshua Tree-Landers earthquake sequence. J. Geophys. Res.. 0148-0227100http://dx,doi.org/10.1029/94JB02944.

Ben-Zion, Y., Dahmen, K., Lyakhovsky, V., Ertas, D., Agnon, A., 1999. Self-driven mode switching of earthquake activity on a fault system. Earth planet. Sci. Lett. 172, 11–21.

Berberian, F., Berberian, M., 1981. Tectono-plutonic episodes in Iran. In: Gupta, H.K., Delany, F.M. (Eds.), Zagros-Hindu Kush-Himalaya Geodynamic Evolution, Am. Geophys. Union, Geodynamics Series, vol. 3, pp. 5–32.

Berberian, F., Muir, I.D., Pankhurst, R.J., Berberian, M., 1982. Late Cretaceous and early Miocene Andean-type plutonic activity in northern Makran and Central Iran. J. Geol. Soc. Lond. 139 (5), 605–614.

Berberian, M., 1976a. Contribution to the Seismotectonics of Iran (Part II). In: Berberian, M. (Ed.), vol. 39. Geol. Surv. Iran, 518 pp. (in English).

Berberian, M., 1976b. Quaternary faults in Iran. In: Berberian, M. (Ed.), Contribution to the Seismotectonics of Iran, Part II. vol. 39. Geol. Surv. Iran, pp. 187–258, (in English).

Berberian, M., 1976c. Documented earthquake faults in Iran. In: Berberian, M. (Ed.), Contribution to the Seismotectonics of Iran, Part II. vol. 39. Geol. Surv. Iran, pp. 143–186, (in English).

Berberian, M., 1976d. On the Metamorphic Rocks of the Maku Quadrangle: First Documented Evidence of Syntectonic Metamorphism of the Caledonian Movements in Iran. Geol. Surv. Iran, Internal Report, 3 pp. (in English).

Berberian, M., 1976e. First seismotectonic map of Iran, 1:2,500,000. In: Berberian, M. (Ed.), Contribution to the Seismotectonics of Iran, Part II. vol. 39. Geol. Surv. Iran, color map (in English).

Berberian, M., 1976f. An explanatory note on the first seismotectonic map of Iran; a seismotectonic review of the country. In: Berberian, M. (Ed.), Contribution to the Seismotectonics of Iran, Part II. vol. 39. Geol. Surv. Iran, pp. 7–142, (in English).

Berberian, M., 1976g. Epicentre map of Iran (1900–1976), 1:5,000,000. In: Berberian, M. (Ed.), Contribution to the Seismotectonics of Iran, Part II. vol. 39. Geol. Surv. Iran, color map (in English).

Berberian, M., 1976h. Areas of destructive earthquakes in Iran (4th century B.C. to 1976 A.D.), 1:5000,000. In: Berberian, M. (Ed.), Contribution to the Seismotectonics of Iran, Part II. vol. 39. Geol. Surv. Iran, color map (in English).

Berberian, M., 1976i. Macroseismic epicentres of destructive and damaging earthquakes in Iran (1900–1976), 1:5,000,000. In: Contribution to the Seismotectonics of Iran, Part II. vol. 39. Geol. Surv. Iran, color map (in English).

Berberian, M., 1976j. Generalized fault map of Iran, 1:5,000,000. In: Berberian, M. (Ed.), Contribution to the Seismotectonics of Iran, Part II. vol. 39. Geol. Surv. Iran, color map (in English).

Berberian, M., 1976k. Pre-Quaternary faults in Iran. In: Berberian, M. (Ed.), Contribution to the Seismotectonics of Iran, Part II. vol. 39. Geol. Surv. Iran, pp. 259–270, (in English).

Berberian, M., 1976l. The 1962 earthquake and earlier deformations along the Ipak earthquake fault (Iran). In: Berberian, M. (Ed.), Contribution to the Seismotectonics of Iran, Part II. vol. 39. Geol. Surv. Iran, pp. 419–428, (in English).

Berberian, M., 1977a. Berberian, M. (Ed.), Contribution to the Seismotectonics of Iran (Part III), Geol. Min. Surv. Iran, vol. 40, 300 pp. (in English).

Berberian, M., 1977b. Three phases of metamorphism in the Hadji–Abad quadrangle (southeastern extremity of the Sanandaj–Sirjan Zone): a paleotectonic discussion. In: Berberian, M. (Ed.), Contribution to the Seismotectonics of Iran, Part III, Geol. Min. Surv. Iran, vol. 40, pp. 239–262, (in English).

Berberian, M., 1977c. Macroseismic epicentres of the Iranian earthquakes. In: Berberian, M. (Ed.), Contribution to the Seismotectonics of Iran, Part III, Geol. Min. Surv. Iran, vol. 40, pp. 79–100, (in English).

Berberian, M., 1977d. Historical seismicity (pre-1900) map of Iran, 1:5,000,000. In: Berberian, M. (Ed.), Contribution to the Seismotectonics of Iran, Part III, Geol. Min. Surv. Iran, vol. 40, color map (in English).

Berberian, M., 1977e. Maximum intensity, isoseismal and intensity zone maps of Iran (4th century B.C. to 1977). In: Berberian, M. (Ed.), Contribution to the Seismotectonics of Iran, Part III, Geol. Min. Surv. Iran, vol. 40, pp. 101–120, (in English).

Berberian, M., 1977f. Intensity zone map of Iran (4th century B.C. to 1900 A.D.), 1:5,000,000. In: Berberian, M. (Ed.), Contribution to the Seismotectonics of Iran, Part III, Geol. Min. Surv. Iran, vol. 40, color map (in English).

Berberian, M., 1977g. Maximum intensity of earthquakes in Iran (1900-1977), 1:5,000,000. In: Berberian, M. (Ed.), Contribution to the Seismotectonics of Iran, Part III, Geol. Min. Surv. Iran, vol. 40, color map (in English).

Berberian, M., 1977h. Isoseismal map of Iran (1900–1977), 1:5,000,000. In: Berberian, M. (Ed.), Contribution to the Seismotectonics of Iran, Part III, Geol. Min. Surv. Iran, vol. 40, color map (in English).

Berberian, M., 1977i. Intensity zone map of Iran (1900–1977), 1:5,000,000. In: Berberian, M. (Ed.), Contribution to the Seismotectonics of Iran, Part III, Geol. Min. Surv. Iran, vol. 40, color map (in English).

Berberian, M., 1977j. Intensity zone map of Iran (4th century B.C. to 1977 A.D.), 1:5,000,000. In: Berberian, M. (Ed.), Contribution to the Seismotectonics of Iran, Part III, Geol. Min. Surv. Iran, vol. 40, color map(in English).

Berberian, M., 1977k. An introduction to the seismotectonics of the Maku region (NW Iran). In: Berberian, M. (Ed.), Contribution to the Seismotectonics of Iran, Part III, Geol. Min. Surv. Iran, vol. 40, pp. 151–202, (in English).

Berberian, M., 1979a. Earthquake faulting and bedding thrust associated with the Tabas-e-Golshan (Iran) earthquake of September 16, 1978. Bull. Seismol. Soc. Am. 69 (6), 1861–1887.

Berberian, M., 1979b. Discussion of the paper A.A. Nowroozi, 1976 'Seismotectonic provinces of Iran. Bull. Seismol. Soc. Am. 69 (1), 293–297.

Berberian, M., 1979c. Evaluation of the instrumental and relocated epicentres of Iranian earthquakes. Geophys. J. R. Astron. Soc. Lond. 58, 625–630.

Berberian, M., 1979d. Tabas-e-Golshan (Iran) catastrophic earthquake of September 16, 1978; a preliminary field report. Disaster 2 (4), 207–219.

Berberian, M., 1981a. Active faulting and tectonics of Iran. In: Gupta, H.K., Delany, F.M. (Eds.), Zagros-Hindu Kush-Himalaya: Geodynamic Evolution, Am. Geophys. Union, Geodynamics Series, vol. 3, pp. 33–69.

Berberian, M., 1981b. Continental Deformation in the Iranian Plateau. Ph.D. Dissertation, University of Cambridge, Department of Earth Sciences, 300 pp.

Berberian, M., 1982. Aftershock tectonics of the 1978 Tabas-e Golshan (Iran) earthquake sequence: a documented active 'thin-and thick-skinned tectonic' case. Geophys. J. R. Astron. Soc. Lond. 68, 499–530.

Berberian, M., 1983a. The Southern Caspian: a compressional depression floored by a trapped, modified oceanic crust. Can. J. Earth Sci. 20 (2), 163–183.

Berberian, M., 1983b. Generalized tectonic map of Iran (contribution to the Seismotectonics of Iran, Part IV). 1:1,500,000. Geol. Surv. Iran 52 (in English).

Berberian, M., 1983c. Continental Deformation in the Iranian Plateau (Contribution to the Seismotectonics of Iran, Part IV). vol. 52 Geol. Surv. Iran, 625 pp. (in English plus 74 pp. in Persian).

Berberian, M., 1983d. On the Seismicity of the Baron Dam Site (Zang-e Mār, South Māku): A Preliminary Note, Water Engineer. Services (Mahāb), Ministry of Power, 15 pp.

Berberian, M., 1984a. Active tectonics of Iran. In: 27th International Geological Congress, Moscow, Proc. Vol. IV, Section 08, Geophysics: Seismicity (Abstract), p. 38.

Berberian, M., 1984b. Structural evolution and tectonics of the Iranian plateau: a plate tectonic approach. In: 27th International Geological Congress, Moscow, Proc. Vol. IV, Section 07, Tectonics of Continental Fold Belts, (Abstract), p. 140.

Berberian, M., 1988. Earthquake-Fault Hazard Study of the Ostur Dam Site (Āzarbāijān). Mahāb Qods Consulting Engineers, Tehran, Internal Report, 32 pp. (in Persian).

Berberian, M., 1989. Tectonic evolution of the Iranian mountain belts. In: 28th International Geological Congress, Washington, DC, July 9–19, 1989, Abstract, vol. 1 of 3, 1-129-130, USA.

Berberian, M., 1990. Mountains from the Perspective of the Ancient Iranians and the Geometric Pattern of the Alborz Mountains in Ancient Beliefs. Simorgh, no. 17, 18, 10–11, Los Angeles (in Persian).

Berberian, M., 1991. Is the theory of earthquake Greek or Iranian? Iranshenasi II (4), 834–845, Bethesda, MA., (in Persian) and 120 (abstract in English), Bethesda, MA.

Berberian, M., 1992. Biruni Khārazmi and the Theory of Continental Drift. Irānshenāsi IV (1), 139–147, Bethesda, MA., 17 (in Persian).

Berberian, M., 1994. Natural Hazards and the First Earthquake Catalogue of Iran. Volume 1: Historical Hazards in Iran Prior to 1900. A UNESCO/IIEES Project During the United Nations International Decade for Natural Disaster Reduction (IDNDR:1900–2000). International Institute of Earthquake Engineering and Seismology (IIEES), Tehran, 603 pp. (in English) & 66 pp. (in Persian).

Berberian, M., 1995. Master 'Blind' thrust faults hidden under the Zagros folds: active basement tectonics and surface morphotectonics. Tectonophysics 241, 193–224.

Berberian, M., 1996. The historical record of earthquakes in Persia. Encyclopaedia Iranica, VII (F.6) Drugs-Ebn al-Atir, Mazda Publishers, Costa Mesa, CA, pp. 635–640.

Berberian, M., 1997a. Seismic sources of the Transcaucasian historical earthquakes. In: Giardini, D., Balassanian, S. (Eds.), Historical and Prehistorical Earthquakes in the Caucasus. Kluwer Academic Press, The Netherlands, NATO ASI Series, 2. Environment vol. 28, pp. 233–311.

Berberian, M., 1997b. In Search of the Earth and Planetary Sciences in Irānveij. Bonyād Neyshābur Publ., Tehran, [Jostāri Dar Dānesh-e Kayhān va Zamin Dar Irānveij], 551 pp. (in Persian).

Berberian, M., 1997c. Tectonics and Fault Re-evaluation Study of the Bushehr Nuclear Power Plant Site. Khak-e-Khoob Consulting Engineers, Tehran, 98 pp.

Berberian, M., 1998. Seismicity and Seismotectonic Re-evaluation of the Bushehr Nuclear Power Plant Site. Ira-Tom/Khak-e-Khoob Consulting Engineers, Tehran, 228 pp.

Berberian, M., 2005. The 2003 Bam urban earthquake: a predictable seismotectonic pattern along the western margin of the rigid Lut Block, southeast Iran. Earthq. Spectra. 21 (S1), S35–S99. http://dx.doi.org/10.1193/1.2127909, Earthquake Engineering Research Institute.

Berberian, M., 2013. Early earthquake detection and warning alarm system in Iran by a telegraph operator: a 116-year old disaster prevention attempt. Seismol. Res. Lett. 85 (5), 816–819.

Berberian, M., Alavi-Tehrani, M., 1977a. Structural analysis of the Hamadan metamorphic tectonites; a paleotectonic discussion. In: Berberian, M. (Ed.), Contribution to the Seismotectonics of Iran, Part III. Geol. Min. Surv. Iran, 40, pp. 263–280 (in English).

Berberian, M., Alavi-Tehrani, M., 1977b. Structural pattern of the Hamadan metamorphic tectonites. In: Berberian, M. (Ed.), Contribution to the Seismotectonics of Iran, Part III. Geol. Min. Surv. Iran, vol. 40, map (in English).

Berberian, M., Arshadi, S., 1976. On the evidence of the youngest activity of the North Tabriz Fault and the seismicity of Tabriz City. In: Berberian, M. (Ed.), Contribution to the Seismotectonics of Iran, Part II. vol. 39. Geol. Surv. Iran, pp. 397–418.

Berberian, M., Arshadi, S., 1977. The Shibli rift system (Sahand region, NW Iran). In: Berberian, M. (Ed.), Contribution to the Seismotectonics of Iran, Part III. Geol. Min. Surv. Iran, vol. 40, pp. 229–238 (in English).

Berberian, M., King, G.C.P., 1981a. Towards a paleogeography and tectonic evolution of Iran. Can. J. Earth Sci. 18 (2), 210–265.

Berberian, M., Mohajer-Ashjai, A., 1977. Seismic risk map of Iran, a proposal. In: Berberian, M. (Ed.), Contribution to the Seismotectonics of Iran, Part III. Geol. Min. Surv. Iran, vol. 40, pp. 121–150 (in English).

Berberian, M., Navai, I., 1977. Naghan (Chahar Mahal Bakhtiari, High Zagros, Iran) earthquake of April 6, 1977; a preliminary field report and a seismotectonic discussion. In: Berberian, M. (Ed.), Contribution to the Seismotectonics of Iran, Part III. Geol. Min. Surv. Iran, vol. 40, pp. 51–78 (in English).

Berberian, M., Navai, I., 1978. Naghan (Chahar Mahal Bakhtiari-High Zagros, Iran) earthquake of 6 April 1977; A preliminary field report and a seismotectonic discussion. Annali Geofis. 31, 5–27.

Berberian, M., Papastamatiou, D., 1978. Kurgu (north Bandar Abbas, Iran), earthquake of March 21, 1977; a preliminary field report and a seismotectonic discussion. Bull. Seismol. Soc. Am. 68 (2), 411–428.

Berberian, M., Qorashi, M., 1984. Recent Tectonics, Seismotectonics and Earthquake-Fault Hazard Study of the Zanjan Lead-Zinc Melting Plant. Geol. Surv. Iran, Internal Report, 88 pp. (in Persian).

Berberian, M., Qorashi, M., 1985. General Geological, Tectonics, Seismotectonics and Engineering Seismology Study of the Chashm Dam Site area, Semnan. Pazhuhāb Consulting Engineers, Semnan Water Board, Internal Report, Iran, 53 pp. (in Persian).

Berberian, M., Qorashi, M., 1987. Earthquake-Fault Hazard Study of the Lake Urumiyeh Region. Geol. Surv. Iran and Ministry of Roads and Transportation, Lake Urumiyeh Highway Project, 123 pp.

Berberian, M., Qorashi, M., 1988. Seismotectonics and Earthquake-Fault Hazard Study of the Mojen Dam Site, Shahrud, Iran. Lar Consulting Engineers, Internal Report, Tehran, 239 pp. (in Persian).

Berberian, M., Qorashi, M., 1989a. Earthquake-Fault Hazard Study of the Kerman Power Plant. Moshanir Consulting Engineers, Internal Report, Tehran (in Persian).

Berberian, M., Qorashi, M., 1989b. Seismotectonic and Earthquake-Fault Hazard Study of the Neyshabur Power Plant Site. Moshanir Consulting Engineers, Internal Report, Ministry of Power, Tehran, 145 pp. (in Persian).

Berberian, M., Qorashi, M., 1994. Coseismic fault-related folding during the South Golbaf earthquake of November 20, 1989, in southeast Iran. Geology 22, 531–534.

Berberian, M., Tchalenko, J.S., 1976a. Field study and documentation of the 1930 Salmas (Shahpur-Azarbaidjan) earthquake. In: Berberian, M. (Ed.), Contribution to the Seismotectonics of Iran, Part II. vol. 39. Geol. Surv. Iran, pp. 271–342 (in English).

Berberian, M., Tchalenko, J.S., 1976b. Earthquakes of the southern Zagros (Iran): Bushehr region. In: Berberian, M. (Ed.), Contribution to the Seismotectonics of Iran, Part II. vol. 39. Geol. Surv. Iran, pp. 343–370 (in English).

Berberian, M., Tchalenko, J.S., 1976c. Earthquakes of Bandar Abbas-Hadji-Abad region, Zagros (Iran). In: Berberian, M. (Ed.), Contribution to the Seismotectonics of Iran, Part II. vol. 39. Geol. Surv. Iran, pp. 371–396 (in English).

Berberian, M., Walker, R., 2010. The Rudbar M_w 7.3 earthquake of 1990 June 20; seismotectonics, coseismic and geomorphic displacements, and historic earthquakes of the western High-Alborz, Iran. Geophys. J. Int.. 182 (3), 1577–1602. http://dx.doi.org/10.1111/j.1365-246X.2010.04705.x.

Berberian, M., Yasini, I., 1983. Different sedimentary facies and palaeogeography of the Neogene deposits in Iran (Tethys, Paratethys, Qom, and the Makran basins). In: Berberian, M. (Ed.), 1983: Continental Deformation in the Iranian Plateau (Contribution to the Seismotectonics of Iran, Part IV). Geol. Surv. Iran, vol. 52, pp. 72–74.

Berberian, M., Yeats, R.S., 1999. Patterns of historical rupture in the Iranian Plateau. Bull. Seismol. Soc. Am. 89 (1), 120–139.

Berberian, M., Yeats, R.S., 2001. Contribution of archaeological data to studies of earthquake history in the Iranian Plateau. Paul Hancock Memorial Issue. J. Struct. Geol. 23, 536–584.

Berberian, M., Yeats, R.S., 2014. Tehran, an earthquake time bomb. In: Tectonic Evolution and Seismicity of Southwest Asia; In Honor of the 40 Years of Manuel Berberian'a Research Contributions. Geol. Soc. Am, in preparation.

Berberian, M., Hushmandzadeh, A., Lotfi, M., 1973. Deformational Phases and Related Metamorphism in the Gorgan Area. Geol. Surv. Iran, Internal Report, 14 pp.

Berberian, M., Papastamatiou, D., Qoraishi, M., 1977. Khurgu (north Bandar Abbas-Iran) earthquake of March 21, 1977: a preliminary field report and a seismotectonic discussion. In: Berberian, M. (Ed.), Contribution to the Seismotectonics of Iran, Part III. Geol. Min. Surv. Iran, vol. 40, pp. 7–50 (in English).

Berberian, M., Asudeh, I., Arshadi, S., 1979a. Surface rupture and mechanism of the Bob-Tangol (southeastern Iran) earthquake of December 19, 1977. Earth Planet. Sci. Lett. 42 (3), 456–462.

Berberian, M., Asudeh, I., Bilham, R.G., Scholz, C.H., Soufleris, C., 1979b. Mechanism of the main shock and the aftershock study of the Tabas-e Golshan (Iran) earthquake of September 16, 1978; a preliminary report. Bull. Seismol. Soc. Am. 69 (6), 1851–1859.

Berberian, M., Amidi, S.M., Babakhani, A., 1981. Discovery of the Qaradagh Ophiolite Belt, the Southern Continuation of the Sevan-Akera (Little Caucasus) Ophiolite Belt in Northwestern Iran (Ahar Quadrangle): A Preliminary Field Note. Geol. Surv. Iran, Internal Report, 15 pp. (in Persian).

Berberian, M., Qorashi, M., Arzhangravesh, B., Mohajer-Ashjai, A., 1983. Recent Tectonics, Seismotectonics and Earthquake-Fault Hazard Study in the Greater Qazvin Region. vol. 57. Geol. Surv. Iran, 84 pp. (in Persian).

Berberian, M., Jackcon, J., Ghorashi, M., Kadjar, M.H., 1984. Field and teleseismic observations of the 1981 Golbaf-Sirch earthquakes in SE Iran. Geophys. J. R. Astron. Soc. Lond. 77, 809–838.

Berberian, M., Qorashi, M., Berberian, M., Qorashi, M., Arzhang-ravesh, B., Mohajer-Ashjai, A., 1985. Recent Tectonics, Seismotectonics and Earthquake-Fault Hazard Study in the Tehran Qazvin Area. vol. 56 Geol. Surv. Iran, 316 pp. (in Persian).

Berberian, M., Qorashi, M., Jackson, J.A., Priestley, K., Wallace, T., 1992. The Rudbar-Tarom earthquake of June 20, 1990 in NW Persia: preliminary field and seismological observations, and its tectonic significance. Bull. Seismol. Soc. Am. 82 (4), 1726–1755.

Berberian, M., Qorashi, M., Talebian, M., Shoja-Taheri, J., 1996. Berberian, M. (Ed.), Seismotectonic and Earthquake-Fault Hazard Investigations in the Semnan Region (Contribution to the Seismotectonics of Iran, part VII). vol. 63. Geol. Surv. Iran, 266 pp. (in Persian).

Berberian, M., Jackson, J.A., Qorashi, M., Khatib, M.M., Priestley, K., Talebian, M., Ghafuri-Ashtiani, M., 1999. The 1997 May 10 Zirkuh (Qa'enat) earthquake (Mw 7.2): faulting along the Sistan suture zone of eastern Iran. Geophys. J. Int. 136 (3), 671–694.

Berberian, M., Qorashi, M., Shoja-Taheri, J., Talebian, M., 2000a. Berberian, M. (Ed.), Seismotectonics and Earthquake-Fault Hazard Investigations in the Mashhad-Neyshabur Region. Contribution to the Seismotectonics of Iran (part VIII). vol. 72. Geol. Surv. Iran, 233 pp. (in Persian).

Berberian, M., Jackson, J.A., Qorashi, M., Talebian, M., Khatib, M., Priestley, K., 2000b. The 1994 Sefidabeh earthquakes in eastern Iran: blind thrusting and bedding-plane slip on a growing anticline, and active tectonics of the Sistan suture zone. Geophys. J. Int. 142 (2), 283–299.

Berberian, M., Jackson, J.A., Fielding, E., Parsons, B.E., Priestly, K., Qorashi, M., Talebian, M., Walker, R., Wright, T.J., Baker, E., 2001. The 1998 March 14 Fandoqa earthquake (M_w6.6) in Kerman, southeast Iran: re-rupture of the 1981 Sirch earthquake fault, triggering of slip on adjacent thrusts, and the active tectonics of the Gowk fault zone. Geophys. J. Int. 146 (2), 371–398.

Berberian, M., Malek Shahmirzādi, S., Nokandeh, J., Djamāli, M., 2012. Archaeoseismicity and environmental crises at the Sialk mounds, Central Iranian Plateau, since the Early Neolithic. J. Archaeol. Sci.. 39 (9), 2845–2858. http://dx.doi.org/10.1016/j.jas.2012.04.001.

Berberian, M., Petrie, C.A., Potts, D.T., Asgari Chaverdi, A., Dusting, A., Sardari Zarchi, A., Weeks, L., Ghāssemi, P., Noruzi, R., 2014. Archaeoseismicity of the mounds and monuments along the Kāzerun fault (western Zāgros, sw Iranian Plateau) since the Chalcolithic period. Iranica Antiqua. XLIX, 1–81. http://dx.doi.org/10.2143/JA.49.0.

Berryman, K.R., 1980. Late Quaternary movement on White Creek fault, South Island, New Zealand. N. Z. J. Geol. Geophys. 23, 93–101.

Beveridge, A.S. (Trans.), 1922. The Bābur-nāma in English (Memoirs of Bābur), Translated from the original Turki Text of Zahiru'd-din Muhammad Bābur Pādshāh Ghāzi, vol. 1. Luzac & Co., London. archive.org/stream/baburnamainengli01babuuoft#page/246/mode/2up.

Bezzeghoud, M., Buffon, E., 1999. Source parameters of the 1992 Melilla (Spain, M_w 4.8), 1994 Al Hoceima (Morocco, M_w 5.8) and 1994 Mascara (Algeria, M_w 5.7) earthquakes and seismotectonic implications. Bull. Seismol. Soc. Am. 89, 359–372.

Bharier, J., 1968. A note on the population of Iran, 1900–1966. Popul. Stud. 22 (2), 273–279.

Bharier, J., 1972. The growth of the towns and villages in Iran, 1900-1966. Middle East. Stud. 8 (1), 51–61, Repr. in J. A. Momeni, Ed., The Population of Iran. A Selection of Readings, Honolulu and Shiraz, 1977, pp. 331–341.

BHRC, 1988. Iranian Seismic Resistant Design of Buildings, National Standard No. 2800. Building and Housing Research Centre, vol. 82. Tehran (in Persian), http://www.bhrc.gov.ir.

BHRC, 1999. Iranian Code of Practice for Seismic Resistant Design of Buildings, National Standard No. 2800, second ed. Building and Housing Research Centre, Tehran. PN S 253 (in Persian), http://www.bhrc.gov.ir.

BHRC, 2003. The History of Bam Accelerograph Station; Iran Strong Motion Network (ISMN). Building and Housing Research Centre, Tehran. PN S 253 (in Persian), http://www.bhrc.gov.ir.

BHRC, 2004. Preliminary Report of the Bam Earthquake. Building and Housing Research Centre, Tehran (in Persian). http://www.bhrc.gov.ir.

Bier, L., 1986. Sarvistan, A Study in Early Iranian Architecture. The College Art Association of America, The Pennsylvania State University Press, University Park and London, 81 pp. + Illustrations.

Biggs, J., Bergman, E., Emmerson, B., Funning, G.J., Jackson, J., Parsons, B., Wright, T.J., 2006. Fault identification for buried strike-slip earthquakes using InSAR: the 1994 and 2004 Al Hoceima, Morocco earthquakes. Geophys. J. Int.. 166, 1347–1362. http://dx.doi.org/10.1111/j.1365-246X.2006.03071.x.

Bilham, R., 2009. The seismic future of cities. Bull. Earthq. Eng. Springer. 7, 839–889. http://dx.doi.org/10.1007/s10518-009-9147-0.

Bilham, R., Gaur, V., 2013. Buildings as weapon of mass destruction. Science 341, 618–619.

Biruni Khārazmi, Abu Rayhān Mohammad ebn-e Ahmad, 1025. Tahdid Nihāyāt al-Amākin Litashih Masāfāt al-Masākin [Determination of the Coordinates of Positions for Correction of Distances Between Cities; 416/1025]. Translated into Persian by Ahmad Ārām. Tehran University Publ. no. 1392, 1352/1973. English Trans. Jamil 'Ali. The American University of Beirut, Centennial Publ., Lebanon, 1967, 278 pp.

Biruni Khārazmi, Abu Rayhāan Mohammad ebn-e Ahmad, 1030. Kitāb fi Tahqiq ma li'l-Hind *men Ma'qula fi al-'Aqlom Mazdula* [Critical Study of What India Says; Whether Accepted by Reason or Refused 421/1030]. Tahqiq Mal al-Hend by Master Abu Raihan Biruni, Persian Trans. by Manuchehr Soduqi-Sahā (1362/1983), *Mo'asseseh Motāle'āt va Tahqiqāt Farhangi,* 521, vol. 1, Tehran, 260 pp. (in Persian). Also, Sachau, E.C. (Trans.), 1910, Alberuni's India. An Account of the Religion, Philosophy, Geography, Chronology, Astronomy, Customs, Laws, and Astrology of India. 2 vols. Kegan Paul, Trench Trubner & Co., London.

Bivar, A.D.H., 2000. Excavations at Ghubayra, Iran. School of Oriental and African Studies, University of London, 317 pp.

Bivar, A.D.H., Yarshater, E. (Eds.), 1978. Eastern Māzandarān I. In: Corpus Inscriptionum Iranicarum, part IV. Persian Inscriptions Down to The Early Safavid Period, vol. VI, Māzandarān Province, Portfolio I. Corpus Inscriptionum Iranicarum, School of Oriental and African Studies, London.

Black, J.A., Cunningham, G., Ebeling, J., Flückiger-Hawker, E., Robson, E., Taylor, J., Zólyomi, G., 1998–2006. The Electronic Text Corpus of Sumerian Literature (ETCSL). A project of the University of Oxford, Faculty of Oriental Sciences, http://etcsl.orinst.ox.ac.uk/.

Blair, S.S., 1985. The Madrasa at Zuzan: Islamic architecture in eastern Iran on the eve of the Mongol invasions. Muqarnas 3, 75–91.

de Bode, C.A., 1845. Travels in Luristan and Arabistan. J. Madden & Co, London, 2 vols.

Bodhi, B. (Trans.), 2000. The Connected Discourses of the Buddha: A Translation of the Samyutta Nikaya. Wisdom Publ., Boston, 2 vols.

Bolurchi, H., Hājiān, J., 1979. Geological Quadrangle Map of the Kabudarahang, 1:250,000. vol. D5. Geol. Surv. Iran.

Bolurchi, M.J., Ansari Moqaddam, F., Darzadeh, H., 2013. Preliminary report of the Bashagerd earthquake, Hormozgan province. Geol. Surv. Iran, Internal report, gsi.ir (in Persian).

Bondāri Esfahāni, 13th century. Tārikh Selseleh Saljuqi: Zobdat al-Nosra wa Nokhbat al-'Usra. Ed. Mohammad Hossein Jalili. Bonyād Farhang Irān, Tehran, 2536 (in Persian).

Bondéhéshn. See Bundahishn.

Bontemps, A., 1807. Safarnameh-ye Bontan [Bontemps] (Travels of Bontemps). Translated into Persian by M. Etehādiyeh, Tehran. From the French book, Relation d'un voyage en Turquie et en Perse pendant l'année, 1807.

Book of Tobit. The Book of Tobit. Ebible.org/kjv/Tobit.htm.

Bosworth, A.B., 1983. Arrian at the Caspian Gates: a study in methodology. Class. Q. 33 (i), 265–276.

Bosworth, C.E., 1999. See Tabari.

Bosworth, C.F., 2010. Nishapur. I. Historical Geography and History to the Beginning of the 20th Century. Encyclopaedia Iranica. Iranicaonline.org, updated September 17, 2010.

Bouchon, M., Hatzfeld, D., Jackson, J.A., Haghshenas, E., 2006. Some insight on why Bam (Iran) was destroyed by an earthquake of relatively moderate size. Geophys. Res. Lett.. 33 (L09309), 2006. http://dx.doi.org/10.1029/2006GL025906.

Bournoutian, G.A. (Trans., Ed.), 2005. The History of Vartapet Arak'el of Tabriz (Patmut'iwn Arak'el Vardapeti Dawrizhets'woy). Introduction and Annotated Translation from the Critical Text, vol. I. Mazda Publishers, Inc., Costa Mesa, CA, pp. 1–298.

Bournoutian, G.A. (Trans., Ed.), 2006. See Vārtābed Ārākel Tāvrizetsi. The History of Vartapet Arak'el of Tabriz (Patmut'iwn Arak'el Vardapeti Dawrizhets'woy). Introduction and Annotated Translation from the Critical Text, vol. II. Mazda Publishers, Inc., Costa Mesa, CA, pp. 299–618.

Boyce, M., 1967. Bibi Shahrbānū and the Lady of Pārs. Bull. Sch. Orient. Afr. Stud. 30 (1), 30–44.

Boyce, M., 1989. A History of Zoroastianism, vol. 1, The Early Period. E.J. Brill, Leiden, 350 pp.

Boyce, M., 2005. Further on the calendar of Zoroastrian feasts. Iran XLIII (43), 1–38.

Bozorgnia, F., 1966. Qum Formation stratigraphy of the central basin of Iran and its intercontinental position. Bull. Iran. Pet. Inst. 24, 69–75.

Briant, P., 1982. Alexander the Great. Encyclopaedia Iranica, vol. I, pp. 827–830, Also iranicaonline.org.

Brocker, et al., 2013. New age constraints for the geodynamic evolution of the Sistan Suture zone, eastern Iran. Lithos, 17 Quittmeyer 0–171, 17–34.

Brommer, J.J., Ambraseys, N.N., 1989. The Spitak earthquake of 7 December 1988. J. Earthq. Eng. Stud. Dyn. 18, 921–925.

Brosset, M., 1849. Rapport sur en Voyage Archéologique dans la Géorgie et dans l'Arméenie Executé en 1847-1848. 1. St. Pétersbourg, 147 f.

Brosset, M.F., 1874-6. Collection d'Historiens Armeniens, 2 vols. Impremerie de l'Academie Imperiale des Science, St. Petersburg.

Brown, T.S., 1949. Onesicritus: A Study in Hellenistic Historiography, vol. 39. Berkeley University of California Press, 196 pp.

Brown, R.D., Vedder, J.G., Wallace, R.E., Roth, E.F., Yerker, R.F., Castle, R.O., Waanonen, A.O., Page, R.W., Eaton, J.P. (Eds.), 1967. The Parkfield-Cholame, California, Earthquakes of June-August 1966—Surface Geologic Effects, Water-Resources Aspects, and Preliminary Seismic Data, vol. 579. U.S. Geol. Surv. Prof. Paper, 66 pp.

Brown, R.D., Ward, P.L., Plafker, G., 1973. Geologic and seismologic aspects of the Managua. Nicaragua earthquake of December 23, 1972, vol. 838. U.S. Geol. Surv. Prof. Paper, 34 pp.

Brunk, R., 1930. Das Erdbeben in Persien vom 6 Mai 1920. Peterm. Geogr. Mitt. 11/12, 289.

Brydges, H.J., 1834. An Account of the Transactions of His Majesty's Mission to the Court of Persia in the Years 1807-1811, vol. 1, London.

Bull, G., 1989 (Trans.). The Pilgrim; The Travels of Pietro Della Valle. Trans., Abridged & Introduction by George Bull. The Folio Society Ltd, 1990, Century Hutchinson Ltd, Middlesex House, London, 303 pp.

Bulliet, R.W., 1976. Medieval Nishapur; a topographic and demographic reconstruction. Studia Iranica 5 (1), 67–89.

Bundahishn [Bondéhéshn]: (a) in English: Pahlavi texts. The Sacred Books of the East in 5 vols. Trans. E.W. West, vol. 5. Motilal Banarsidass, Delhi, India, 1962, pp. 1–151. (b) in Persian: Bondéhésh by Farnbagh Dādagi. Translated into Persian by M. Bahār 1369/1990. Tus Publ., Tehran. (c) in Persian: Bondehesh Hendi. Translated into Persian by Roqiyeh Behzādi. Mo'asseseh Motāle'āt va Tahqiqāt Farhangi Publ., no. 593, Tehran, 1368/1989, 392 pp.

Bulloch, J., 1972. Death in a land behind God's back. N.Y. Times, E3, 16 April 1972.

Bune, V.I., Gorshkov, G.P., 1980. Seismic Zoning of the USSR. Izd. Akad. Nauk. SSSR, Moscow, 307 pp. (in Russian).

Burgess, Ch.E., 1942. Letters from Persia Written by Charles and Edward Brugess 1828-1855. In: Schwartz, B. (Ed.), New York.

Busk, H.G., 1926. The Shimbar valley Landslip dam, Bakhtiari Country, South Persia. Geol. Mag. LXIII, 355–359.

Buwalda, J.P., St. Amand, P., 1955. Geological effects of the Arvin-Tehachapi earthquake. Calif. Div. Mines Geol. Bull. 171, 41–56.

Byrne, D.E., Sykes, L.R., Davis, D.M., 1992. Great thrust earthquakes and aseismic slip along the palte boundary of the Makran subduction zone. J. Geophys. Res. 97, 449–478.

Cakir, Z., Meghraoui, M., Akoglu, A.M., Jabour, N., Belabbes, S., Ait-Brahim, L., 2006. Surface deformation associated with the Mw 6.4, February 24, 2004 Al Hoceima (Morocco) earthquake deduced from InSAR: implications for the active tectonics along North Africa. Bull. Seismol. Soc. Am.. 96 (1), 59–68. http://dx.doi.org/10.1785/0120050108.

Caldwell, J.R. (Ed.), 1967. Investigations at Tal-i-Iblis. Illinois State Museum, Preliminary Report, 9, 408 pp.

Caldwell, J.R., 1968. Tal-i-Iblis and the beginning of copper metallurgy at the fifth millennium. Archaeologia Viva 1 (1), 145–150.

Caldwell, J.R., Malek Shahmirzadi, S., 1966. Tal-e-Iblis, the Kerman range and the beginning of smelting: Preliminary Report 7. Illinois State Museum, Springfield, IL.

Camp, V.E., Griffis, R.J., 1982. Character, genesis and tectonic setting of igneous rocks in the Sistan suture zone, eastern Iran. Lithos 3, 221–239.

Campos, J., Madariaga, R., Nabelek, J., Bukchin, B.G., Deschams, A., 1994. Faulting process of the 1990 June 20 Iran earthquake from broad-band records. Geophys. J. Int. 118, 31–46.

Canltez, N., 1969. The focal mechanisms in Iran and their relations to tectonics. Pure Appl. Geophys. 75 (IV), 76–87.

Canitez, N., Ucer, B., 1967. Computer determinations for the fault-plane solutions in and near Anatolia. Tectonophysics 4 (3), 235–244.

Carbon, D., 1996. Tectonique post-obduction des montagnes d'Oman dans le cadre de la convergence Arabie-Iran. Thése, Université Montpellier II.

Carus, P., 1990. The History of the Devil and the Idea of Evil: From the Earliest Times to the Present Day. Sacred-texts.com.

Cereti, C., 1996. Zand i Wahman Yasn. A Zoroastiran Apocalypse. IsMEO, Rome.

Cereti, C., 2012. Kar-Namag i Ardashir Papagan. Encyclopaedia Iranica XV (6), 585–588. iranicaonline.org.

Chandra, U., 1981. Focal mechanism solutions and their tectonic implications for the eastern Alpine-Himalayan region. In: Gupta, H.K., Delany, F.M. (Eds.), Zagros-Hindu Kush-Himalaya Geodynamic Evolution. Am. Geophys. Union, Geodynamic Series, vol. 3, pp. 243–271.

Chandra, U., 1984. Focal mechanism solution for earthquakes in Iran. Phys. Earth Planet. Inter. 34, 9–16.

Chardin, J., 1673, 1735, 1811. Les voyages du Chevalier Jean Chardin en Perse et autres lieux de l'Orient (1686), Paris, 1811.

Chase, D.W., Fehervari, G., Caldwell, J.R., 1967. Reconnaissances in the Bard Sir Valley. In: Caldwell, J.R. (Ed.), Investigation at Tal-i-Iblis, pp. 73–107, Illinois State Museum Preliminary Report, No. 9.

Choy, G., Zednik, J., 1997. The rupture process of the Manjil, Iran earthquake of June 20 1990 and implications for intraplate strike-slip earthquakes. Stud. Geophys. Geod. 41, 45–63.

Christensen, A., 1931. Les Kaynides. AF Høst & Søn, Copenhagen, Stockholm. Persian Trans. Zabihollāh Safā 1336/1957, first ed. and 1368/1989, fifth ed., Enteshārāt 'Elmi va Farhangi Publ., 163, Tehran, 242 pp.

CHTHO, 2011. Cultural Heritage, Tourism, and Handicraft Organization of Iran. Ichhto.org, 2011.03.07.

Chu, D., Gordon, R.G., 1998. Current plate motions across the Red Sea. Geophys. J. Int. 135, 313–328.

Cisternas, A., Philip, H., Bousquet, J.C., Cara, M., Deschamps, A., Dorbath, L., Dorbath, C., Haessler, H., Jimenez, E., Nercessian, A., Rivera, L., Romanowics, B., Gvishiani, A., Shebalin, N.V., Aptekman, I., Arefiev, S., Borisov, N.A., Gorshkov, A., Graizer, V., Lander, A., Pletnev, K., Rogozhin, A.I., Tatevossian, R., 1989. The Spitak (Armenia) earthquake of 7 December 1988: field observations, seismology and tectonics. Nature 339, 675–679.

Clark, C., 1861. Notes on Persia, Khurasan and Afghanistan. J. R. Geogr. Soc. Lond. 31, 51.

Clark, M.M., 1972. Surface rupture along the Coyote Creek fault, in the Borrego Mountain earthquake of April 9, 1968. U.S. Geol. Surv. Prof. Pap. 787, 55–86.

Clavijo, Ruy Gonzalez de, 1405. Narrative of the Embassy of Roy Gonzalez de Clavijo to the Court of Timur. English Trans. C.R. Markham, Hakluyt Society, London, 1859. Safarnāmeh, Persian Trans. Historia del Gran Tamerlan, Guy le Stange, Embassy to Tamerlane, 1403–1406, London, by Mas'ud Rajabniyā, Bongāh Tarjomeh va Nashr-e-Ketāb Publ., Tehran, second ed., 1344/1965.

CNRS-INSU. Institut National des Sciences de l'Universe, Centre National de la Recherche Scientifique, France.

Code of Hammurabi. See King, 1915.

Cohen, S., 1973. Enmerkar and the Lord of Aratta. Thesis, University of Philadelphia.

Colarusso, J., 2002. Nart Sagas from the Caucasus: Myths and Legends from the Circassians, Abazas, Abkhaz, and Ubykhs. Princeton University Press, 560 pp.

Conolly, A., 1838. Journey to the North of India, Overland from England, Through Russia, Persia, and Afghanistan. Richard Bentley, London, 417 pp.

Coon, C.S., 1952. Excavations in Hotu Cave, Iran, a preliminary report. Proc. Am. Philos. Soc. 96 (3), 231–249.

Coon, C.S., 1957. The Seven Caves; Archaeological Explorations in the Middle East. Alfred A. Knopf, Inc., New York, 338 pp.

Copley, A., Jackson, J., 2006. Active tectonics of the Turkish-Iranian Plateau. Tectonics. 25, http://dx.doi.org/10.1029/2005TC001906, TC6006.

Copley, A., Faridi, M., Ghorashi, M., Hollingsworth, J., Jackson, J., Nazari, H., Oveisi, B., Talebian, M., 2013. The 2012 August 11 Ahar earthquakes: consequences for tectonics and earthquake hazard in the Turkish-Iranian Plateau. Geophys. J. Int.. http://dx.doi.org/10.1093/gji/ggt379.

Costa, C., Machette, M.N., Dart, R.L., Bastias, H.E., Paredes, J.D., Petrucca, L.P., Tello, G.E., Haller, K.M., 2000. Map and database of Quaternary faults and folds in Argentina. U.S. Geol. Surv., Open-File rep., 00-108, 76 pp.

Cowan, H.A., 1990. Late Quaternary displacements on the Hope fault at Glynn Wye, north Canterbury, New Zealand. J. Geol. Geophys. 33, 285–293.

CSEM. Centre seismologique Euro-Mediterraneen, c/o LDG, BP12, 91680 Bruyeres-le-Chatel, France. http://www.emsc-csem.org.

Cumont, F., 1931. La fin du monde selon les mages occidentaux. Revue de liHistoire des Religions 103, 78, n.2.

Curtis, V.S., 1993. Iranian Myths. British Museum Press, The Legendary Past Series, 80 pp.

Dādistān-e Dinik. Ed. Anklesaria, B.T., Datestan i Denik [Religious Decisions], 1911, Bombay; avesta.org. Also, West, E.W. (Ed., Trans.), 1882, The Dadistan-i Dinik and the epistles of Manuskihar. Sacred Books of the East, Pahlavi Texts; Ed. Max Muller. Motilal Banarsidass, Delhi, 1970. Electronic version available at: http://www.avesta.org/mp/zadspram.html.

Daneshdoust, Y., 1980. Honar-e Me'māri dar daureh-ye Taimuriān. Āthār, Sāzémān Melli Hefāzat Āthār Bāstāni, 1, Spring 1359, pp. 91–102 (in Persian).

Daneshdoust, Y., 1998. Tabas Shahri ke bud: Banāhayeh Tārikhi-ye Tabas (Tabas the Town That Was; The Monument of Tabas) Soroush Press, Tehran, 1376/1998, 596 pp. (in Persian).

Darmesteter, J., 1877. Ormazd et Ahriman, Paris.

Darmesteter, J., 1879–1887. The Zend-Avesta: i-Vendidad; ii-Sirozahs; Yashts, and Nyayish; iii-Yasna, Vispered. In: Muller, M. (Ed.), Sacred Books of the East, Delhi, 1974, 1975.

Darmesteter, J., 1898. Yashts. The Sacred Books of the East. Avesta.org.

Daryāee, T., 2002. Shrestānihā I Eranshahr; A Middle Persian Text on Late Antique Geography, Epic, and History. Mazda Publishers, Inc., CA, 90 pp.

Dashti, Gh., 1990. The responsible fault of the June 20, 1990 earthquake. In: The Free Conference on June 20, 1990 Gilan-Mazandaran Earthquake, 11–12 August 1990. Building and Housing Research Center, Tehran.

Daulatabadi, A., 1964. Zelzeleh-ha-ye Tabriz (Earthquakes of Tabriz), 1343/1964. Rev. Fac. de Letters de l'Universite de Tabriz XVI, 131–162, Summer 2 (in Persian).

Daulatshāh Samarqandi, ebn 'Ala al-Dauleh Amir, 1487. Tazkerāt al-Shu'ara (892/1487). Ed. Mahmud 'Abbāsi (n.d.), Padideh (Khavar) Publ., second ed., 1366/1987, 140 pp. Also Ed. E.G. Browne, Leiden, and London, 1901.

Davies, R.G., Jones, C.R., Hamzepour, B., Clark, G.C., 1972. Geology of the Masuleh Sheet, 1:100,000, Northwest Iran, vol. 24. Geol. Surv. Iran, 110 pp.

Davis, T.L., Namson, J.S., 1994. A balanced cross section of the 1994 Northridge earthquake, southern California. Nature 372, 167–169.

De Ballore, 1906. See Ballore, 1906.

Dehkhodā. Loghatnāmeh Dehkhodā [Dehkhodā Persian Dictionary]. University of Tehran (in Persian). loghatnameh.org. Also, vajehnameh.com.

De Martini, P.M., Hessami, K., Pantosti, D., D'Addezio, G., Alinaghi, H., Ghafory-Ashtiani, M., 1998. A geologic contribution to the evaluation of the seismic potential of the Kahrizak fault (Tehran, Iran). Tectonophysics 287 (1–4), 187–199.

De Martini, P.M., 2007. Identification et caracterisation des structures sismogenes de Apennins (Italie): Analyse conjointe (tectonique et geodesie) a diverse echelles temporelle. These, Universite Louis Pasteur-Strasbourg I, 219 pp. http://scd-theses.u-strasbg.fr/1430/01/DEMARTINI_Paolo_Marco_2007.pdf.

DeMets, C., Gordon, R.G., Argus, D.F., Stein, S., 1990. Current plate motion. Geophys. J. Int. 101, 425–478.

DeMets, C., Gordon, R.G., Argus, D.F., Stein, S., 1994. Effects of recent revisions to the geomagnetic reversal time scale on estimates of current plate motions. Geophys. Res. Lett. 21, 2191–2194.

Denisari, A., 1272. Navāder al-Tabādor le-Tohfata al-Bahādor. In: Razavi, M. (Ed.), 1971, Tehran.

Denkard. See Dinkart.

Dewey, J.W., Grantz, A., 1973. The Ohlr earthquake of 10 April 1972 in the Zagros mountains of southern Iran: seismotectonic aspects and some results of a field reconnaissance. Bull. Seismol. Soc. Am. 63 (6), 2071–2090.

al-Dhahabi (Zahabi), Mohammad ebn Ahmad ibn 'Umān, 1315. Kitāb Duwal al-Islām, 715/1315, with Zail by al-Sakhāwi to 744/1343. Ed. Hyderabad, 1337/1919. Also Les Dynasties de l'Islam, Institut Francais de Damas, 1969.

Dinkart (Acts of Religion. Madan). Ed. Peshotun Dastoor Behramjee Sanjana, 1876, English Trans. Ratanshah E. Kohiya, 1876, avesta.org. Madan, D., 1911. The complete text of the Pahlavi Dinkard. 2 vols., Bombay. Also manuscript D 10 A, Publ. Pahlavi University, Shirāz 1355/1976. Avesta.org.

Divānbaigui, R'.A., 1969. Tazalzol [Shaking]. Sālnāmeh-ye Donyā [Donyā Yearbook] 25, 182–191, Tehran, 1348/1969 (in Persian).

Djamali, M., Akhani, H., Andrieu-Ponel, V., Braconnot, P., Brewer, S., de Beaulieu, J.-L., Fleitmann, D., Fleury, J., Gasse, F., Guibal, F., Jackson, S.T., Lézine, A.-M., Médail, M., Ponel, P., Roberts, N., Stevens, L., 2010. Indian summer monsoon variations could have affected the early Holocene woodland expansion in the Near East. The Holocene 20, 813–820.

Djamour, Y., Vernant, P., Bayer, R., Nankali, H.R., Ritz, J.-F., Hinderer, J., Hatam, Y., Luck, B., Moigne, N. Le, Sedighi, M., Khorrami, F., 2010. GPS and gravity constraints on continental deformation in the Alborz mountain range, Iran. Geophys. J. Int. 183 (3), 1287–1301. http://dx.doi.org/10.1111/j.1365-246X.2010.04811.x.

Djamour, Y., Vernant, P., Nankali, H.R., Tavakoli, F., 2011. NW Iran-eastern Turkey present-day kinematics: results from the Iranian permanent GPS network. Earth Planet. Sci. Lett. 307, 27–34.

Dolan, J.F., Christofferson, S.A., Shaw, J.H., 2003. Recognition of paleoearthquakes on the Puente Hills blind thrust fault, California. Science 300, 115–118.

Dollan, J.F., Bowman, D.D., Sammis, C.G., 2007. Long-range and long-term fault interactions in southern California. Geology 35, 855–858.

Donaldson, D.M., 1933. The Shiite Religion; A History of Islam in Persia and Irak. Luzac & Co., London.

Donboli (Maftun), 'Abd al-Razzaq Baig, 1813. Tajrebat al-Ahrār va Tasliyat al-Abrār (1228/1813). Ed. Hassan Qazi Tabataba'i, 2 vols., Tabriz, 1349/1970, 1351/1972.

Drouville, G., 1825. Voyage en Perse fait en 1812 et 1813, 2 vols., Paris. Persian Trans. Safarnāmeh-ye Drouville, by Javād Mohay, Enteshārāt Gutenberg, Tehran, 2 vols., 1348/1969.

Du, Y., Aydin, A., 1993. Stress transfer and decomposition during three sequential earthquakes along the central Calaveras fault, California. J. Geophys. Res. 98, 9947–9962.

Duchesne-Guillemin, J., 1962. La religion de l'Iran ancient. Presses Universitaire de France, Paris, 359 pp.

EB., 1985, 2009. Buddhism. Encyclopaedia Britannica Online Library Edition. Iranicaonline.org.

Ebbesen, H., Kuijpers, A., Moros, M., Lloyd, J., Seidenkrantz, M.-S., Troelstra, S., 2008. The 8.2 ka cooling event related to extensive melting of the Greenland Ice Sheet. Clim. Past Discuss. 4, 1219–1235.

Eberhart-Phillips, J.S., Robinson, T.M., Hatch, D.L., Parrish, R.G., 1994. Profile of mortality from the 1989 Loma Prieta earthquake using coroner and medical examiner reports. Disasters 18 (2), 160–170.

Ebn Haukal. See Ibn Haukal.

EERI, 1998. Newsletter. Earthquake Engineering Research Institute. eeri.org.

Eftekhar-nezhad, J., Haghipur, 'A., Davudzadeh, M., 1968. Gozāreshi az bar-resiyeh zelzeleh-ye khorāsān ke dar Shahrivar 1347 rui-dād. Geol. Surv. Iran, 1347/1968, 35 pp. (in Persian).

Eghishé. See Yeghishé.

Ehrich, R.W., 1992. third ed. Chronologies in Old World Archaeology, vol. 1. The University of Chicago Press, 543 pp.

Ekstrom, G., Dziewonski, A., 1988. Evidence of bias in estimation of earthquake size. Nature 332, 319–323.

Ekstrom, G., Stein, R.S., Eaton, J.P., Eberhart-Phillips, D., 1992. Seismicity, and geometry of a 110-km-long blind thrust fault; 1. The 1985 Kettleman Hills, California, earthquake. J. Geophys. Res. 97, 4843–4864.

Empedocles, 490–430 BCE. On nature. Plato.stanford.edu/entries/Empedocles/#1.

Encyclopedia of Britannica -> See EB. Iranicaonline.org.

Engdahl, E.R., van der Hilst, R., Buland, R., 1998. Global teleseismic earthquake relocation with improved travel times and procedures for depth determination. Bull. Seismol. Soc. Am. 88, 722–743.

Engdahl, E.R., Jackson, J.A., Myers, S.C., Bergman, E.A., Priestley, K., 2006. Relocation and assessment of seismicity in the Iran region. Geophys. J. Int. 167, 761–778.

England, P.C., Jackson, J., 1989. Active deformation of the continents. Annu. Rev. Earth Planet. Sci. 17, 197–226.

England, P.C., Molnar, P., 1990. Right-lateral shear and rotation as the explanation for strike-slip faulting in eastern Tibet. Nature 344, 140–142.

Eqbāl Āshtiāni, '.A., 1986. Tārikh-e Moghol (The History of the Mongols, 1365). Amir Kabir Publ., Tehran, (in Persian).

al-Esfahani, Mohammad Mehdi ebn Mohammad Reza, 1891. Sotudeh, M. (Ed.), Nesft-e Jahān fi Ta'rif-e Esfahān, first ed. Amir Kabir Publ., Tehran, 1340/1961, second ed. 1368/1989.

Esfezāri, Khājeh Abu Hātam Mozafar ibn Esmā'il (1045–1121). Resāleh Āsār 'Alavi. Ed. Mohammad Taghi Moddares Razavi. Bonyād Farhang Irān, Tehrān, 1356/1977 (in Persian).

Esfezāri, Mu'in al-Din Muhammad al-Zamaji al-Isfizāri, 1493. Rauḑāt al-Jannāt fi Ausāf Madinat Harāt: Dar Wasf Ma'réfat al-Arḑi wa Joghrāfiyā-ye-Tārikhi, Siyāsi Ādābi-e Khorāsān va Vasf-e Shahrhā-yé Mohem-e-ān Sarzamin: Neyshābur, Harāt, Balkh, Tus-Mashhad, Marv, Shāhijān, Ghaznain va Ghayreh. Vasf-e Āb va Havā, Zerā'at, m'adan, Ahvāl-e Rejāl, Mazārāt, Abniyeh va 'Emārāt-e Mashhur-e ān Sāmān, 899/1493. vol. II (574). Ed. Sayyed Muhammad Kāzem Emām, 1339/1960, Tehran University Publ., 2 vols., 537–574.

Eshraghi, S.A., Jalali, A., 2006. Geological Map of Moaleman, 1:100,000. Geol. Surv. Iran.

Eshrāqi, E. (Ed.), 2004. Kholāsat al-Tavārikh-e Qāzi Aham Qomi [Historical Summary by Qāzi Ahamd Qomi], 2 vols. University of Tehran Publication, Tehra, 1383/2004 (in Persian).

Eslami, A.A., 1992. An illustration of the June 1990 Rudbar earthquake fault system. Geosci. Geol. Surv. Iran 2 (5), 48–53 (in Persian).

ESSN. The United System of Seismic Observations.

Estakhri, Abu Eshāq Ebrāhim ibn Muhammad Fārsi, 951. Masālek va Mamālik, 350/951. Translated into Persian by Iraj Afshār, Bongāh Tarjomeh va Nashr Ketāb Publ., 1340/1961, 1347/1968, Tehran (in Persian).

Etelā'āt, 1968. Dasht-e Bayaz and Ferdows earthquake reports. Etelā'āt evening newspaper, Tehran, 2 September 1968 (in Persian).

Etelā'āt, 1972. Qir-Kārzin Earthquake Reports. Etelā'āt evening newspaper, Tehran, 15 April 1968 (in Persian).

Etelā'āt, 1981. The Golbāf Earthquake Reports. Etelā'āt evening newspaper, Tehran, 21 June 1981 (in Persian).

Etelā'āt, 1990. The Rudbār earthquake reports. *Etelā'āt evening newspaper*, Tehran, 21 June 1990, No. 19072 (in Persian).

Etelā'āt. Evening Persian Newspaper in Tehran, Iran.

Evans, E.P., 1880. The Sacred Books of the Parsis. Unitarian Rev. Relig. Mag. XIV (6), 538.

Eznik [Yeznik Koghbātsi], 5th century ed. Venice, 1826.

Fārābi [Alpharabius], Abu Nasr Mohammad ibn, 872–951. Ahsa' al-'Ulum [Counting the Knowledge]. Translated into Persian, Hossein Khadiv-Djam, Sherkat Enteshārāt 'Elmi va Farhangi Publ., Tehran, 1364/1985.

Faridi, M., 2012. Geological Aspects of the Ahar-Varzeghan, Earthquake, 11/Aug/2012.

Faridi, M., Hagh Farshi, E., Zamanimehr, S., Sayyedie Sahebary, Allamadadie Sh, 2006. Geological Map of Khoja at a scale of 1:100,000. Geol. Surv. Iran, Tehran.

Farrokhi Sistāni, d. 1038. Divān-e Farrokhi Sistāni [Collection of Poems of Farrokhi Sistāni]. Ed. Dabir Siyāqi, S.M., 1363/1984, Tehran (in Persian).

Farooqi, M.A., 2000. The Simurgh-feather guide to the poetics; Dastan-e Amir Hamza Sahibqiran. Annu. Urdu Stud. 15, 119–167.

Farooqi, M.A. (Trans.), 2008. The Adventures of Amir Hamza. By G. Lakhnai and A, Bilgrami. Random House Modern Library Reprint Edition, 992 pp.

Faruqi, A., 1997. Iran's Response to Killer Quake: Typically Fatalistic, 'Will of God. Associated Press, 14 May 1997, USA. apr.org.

Fasihi Khafi, Fasih al-Din Ahmad ebn Jalal al-Din Mohammad Khāfi, 1442. Mojmal-e Fasihi (Mojmal al-Tavārikh, 845/1442). Ed. Mahmud Farrokh, 3 vols., Bāstān Publ., Mashhad, 1339/1960, 1341/1962.

Fatemi Aghda, S.M., 2004. The new Arg-e Bam fault. In: International Seminar on Policies and Practices in the Management of Seismic Risks in Urban Areas, Tehran, 16–18 November, 2004. Abstracts.

Fattahi, M., Walker, R., Hollingsworth, J., Bahroudi, A., Nazari, H., Talebian, M., Armitage, S., Stokes, S., 2006. Holocene slip-rate on the Sabzevar thrust fault, NE Iran, determined using optically stimulated luminescence (OSL). Earth Planet. Sci. Lett. 245 (3–4), 673–684.

Fattahi, M., Walker, R., Khatib, M.M., Dolati, A., Bahroudi, J., 2007. Slip-rate estimates and past earthquakes on the Doruneh fault, eastern Iran. Geophys. J. Int. 168, 691–709.

Fattahi, M., Nazari, H., Bateman, M.D., Meyer, B., Serbrier, M., Talebian, M., Le Dortz, K., Foroutan, M., Ahmadi Givi, F., Ghorashi, M., 2010. Refining the OSL age of the last earthquake on the Dehshir fault, Central Iran. Quat. Geochronol. 5, 286–292.

Fattahi, M., Walker, R.T., Talebian, M., Sloan, R.A., Rasheedi, A., 2011. The structure and late Quaternary slip rate of the Rafsanjan strike-slip fault, SE Iran. Geosphere. 7, 1159–1174. http://dx.doi.org/10.1130/GES00651.1.

Fāzeli, H., Wong, E.H., Potts, D., 2005. The Qazvin plain revisited: A reappraisal of the chronology of northwestern central plateau, Iran, in the 6th to the 4th Millennium BC. Anc. Near East. Stud. 42, 3–82.

Fāzeli Nashli, H., Dārābi, H., Nāseri, R., Fallāhiān, Y., 2011. Gāhnegāri-ye nesbi va motlaq-e-tapeh-ye Sagzābād-e Dasht-e Qazvin (Relative and absolute chronology of the Sagzābād mound, Qazvin Plain) , Nashriyeh 'elmi-Pazhuheshi Motāle'āt Bāstānshenāsi, Tehrān University, 3, 133–158, Day 1390 (in Persian).

Fehrevari, G., Caldwell, J.R., 1967. Islamic sites outside the Bardsir Valley: Surveys of 1964 and 1966. In: Caldwell, J. (Ed.), Investigations at Tal-I Iblis. Illinois State Museum Preliminary Report, 9, Springfield, IL, pp. 41–64.

Ferdowsi Tusi, A., 1010. Shāhnāmeh [Book of Kings]. Ed. Joneidi, F. 2008, Shāhnāmé Ferdowsi. Nashr Balkh/Bonyād Neyshābur Pul., Tehran, 6 vols. (in Persian), 1387, Tehran.

Ferrier, J.P., 1857. Caravan Journeys and Wanderings in Persia, Afghanistan, Turkistan and Beloochistan, London.

Ferrier, J.P., 1860. Voyages en Perse dans l'Afghanistan, la Beloutchistan et la Turkestan. 2 vols, Paris.

Fielding, E.J., Wright, T.J., Muller, J., Parsons, B.E., Walker, R., 2004. Aseismic deformation of a fold-and-thrust belt imaged by synthetic aperture radar interferometry near Shahdad, southeast Iran. Geology 32 (7), 577–580. http://dx.doi.org/10.1130/G20452.1.

Fielding, E.J., Talebian, M., Rosen, P.A., Nazari, H., Jackson, J.A., Ghorashi, M., Walker, R., 2005. Surface ruptures and building damage of the 2003 Bam, Iran, earthquake mapped by satellite synthetic aperture radar interferometric correlation. J. Geophys. Res.. 110http://dx.doi.org/10.1029/2004JB003299.

Fiey, J.M., 1965. Assyrie chretienne, 2 vols. Beirut.

Fischer, M., 1991. Sacred Circles: Iranian (Zoroastrian and Shi'ite Muslim) Feasting and Pilgrimage Circuits. In: Scott, J., Simpson-Housley, P. (Eds.), Sacred Places and Profane Spaces. Greenwood Press, New York, pp. 131–144.

Fitch, T.J., 1972. Plate convergence, transcurrent faults, and internal deformation adjacent to southeast Asia and the western Pacific. J. Geophys. Res. 77, 4432–4462.

Fleitmann, D., Burns, S.J., Mudelsee, M., Neff, U., Kramers, J., Mangini, A., Mater, A., 2003. Holocene forcing of the Indian monsoon recorded in stalagmites from southern Oman. Science 300, 1737–1739.

Fleitmann, D., Burns, S.J., Mangini, A., Mudelsee, M., Kramers, J., Villa, I., Neff, U., Al-Subbary, A.A., Buettner, A., Hippler, D., Mater, A., 2007. Holocene ITCZ and Indian monsoon dynamics recorded in stalagmites from Oman and Yemen (Socotra). Quat. Sci. Rev. 26 (1–2), 170–188.

Forbes, R., 1931. Conflict; Angora to Afghanistan. Frederick A. Stokes Co, New York, 333 pp.

Foroutan, M., Bolurchi, M.J., Solaymani-Azad, S., Oveisi, B., Nemati, M., Shokri, M.A., Eskandari, M., 2010. Preliminary report of Konarak earthquake 2010 Dec. 20 (Rigan, Kerman). Geol. Surv. Iran, gsi.ir.

Foroutan, M., Sebrier, M., Nazari, H., Meyer, B., Fattahi, M., Rashidi, A., Le Dortz, K., Bateman, M.-D., 2012. New evidence for large earthquakes on the Central Iran plateau: palaeoseismology of the Anar fault. Geophys. J. Int.. http://dx.doi.org/10.1111/j.1365-246X.2012.05365.x.

Foster, B.R., 2001. The Epic of Gilgamesh. W.W. Norton & Company, New York.

Foster, G., 1798. A Journey from Baghdad to England, 2 vols. London.

Fouache, E., Cosandey, C., Wormser, P., Kervran, M., Labbaf Khaniki, R.A., 2011. The river of Nishapur. Studia Iranica 40, 99–119.

Franssen, P., 1930. Correspondence of the Missions Lazaristes. Rezayeh.

Fraser, J.B., 1825. Narrative of a Journey into Churasan in the Years 1821, 2 vols. London, 1825, 1840.

Freund, R., 1970. Rotation of strike-slip faults in Sistan, southeast Iran. J. Geol. 78, 188–200.

Fu, B., Ninomiya, Y., Lei, X., Toda, S., Awata, Y., 2004. Mapping active fault associated with the 2003 Mw 6.6 Bam (SE Iran) earthquake with ASTER 3D images. Remote Sens. Environ. 92, 153–157.

Fuchs, C.W., 1886. Statistik der Erdbeben 1865–1885. Sitzungsber. Kais, Akad. Wiss. Math.-Naturwiss. Kl. 92 (1), 215–625, Vienna.

Fumal., T.E., Weldon, R.J., Biasi, G.P., Dawson, T.E., Seitz, G.G., Frost, W.T., Schwartz, D.P., 2002. Evidence for large earthquakes on the San Andreas fault at the Wrightwood, California, paleoseismic site: A.D. 500 to present. Bull. Seismol. Soc. Am. 92, 2726–2760.

Funning, G.J., Parsons, B., Wright, T.J., Fielding, E.J., Jackson, J., 2005. Surface displacements and source parameters of the 2003 Bam (Iran) earthquake from Envisat ASAR imagery. J. Geophys. Res.. 110http://dx.doi.org/10.1029/2004JB003338, B09406.

Galadini, F., Hinzen, K.-G., Stiros, S., 2006. Archaeoseismology: Methodological issues and procedure. J. Seismol. 10, 395–414.

Galanopoulos, A., 1956. The seismic risk at Athens. Praktika Akadimias Athinon 31, 464–472 (in Greek).

Galdieri, E., 1979. Esfahan, ʻĀli Qāpu; An Archaeological Survey. IsMEO, Rome, Restorations, 5, 155 pp.

Galdieri, G., 1983. ʻĀli Qāpu. Trans. ʻAbdollah Jabal ʻĀmeli, Sāzemān Melli Hefāzat Āthār Bāstāni Irān, 208 pp., 1362 (in Persian).

Gansser, A., 1969. The large earthquakes of Iran and their geological frame. Eclog. Geol. Helv. 62 (2), 443–466.

Gao, L., Wallace, T.C., 1995. The 1990 Rudbar-Tarom Iranian sequence: evidence for slip partitioning. J. Geophys. Res. 100, 15137–15332.

Gao, L., Wallace, T.C., Jackson, J.A., 1991. Aftershocks of the June 1990 Rudbar-Tarom earthquake: Evidence for slip partitioning (Abstract). EOS Trans. AGU 72 (44), Fall Meeting Suppl. 335.

Gardizi, ʻAbu Sa'id ʻAbd al-Hayy ebn al-Zahak ebn Mahmud, 1048. Tarikh-e Gardizi (Zain al-Akhbar, 440/1048). Ed. ʻAbd al-Hayy Habibi. Bonyad Farhang Iran, Tehran, 1347/1968. Donya-ye Ketab, Tehran, 1363/1984.

Gāthās. The Hymns of Zarathushtra. Avesta.org. Also see Ervad Kavasji Edalji Karga, 1997, Bombay.

Gaube, H., 1979. Bam- A provincial center. In: Gaube, H. (Ed.), Iranian Cities. Hagop Kevorkian Series on Near Eastern Art and Civilization, vol. 2. University Press, New York, pp. 99–132.

Gavin, 2010.

GCMT: The Global CMT Project, Lamont Doherty Earth Observatory (LDEO), Columbia University, 61 Route 9W, P.O. Box 1000, Palisades, NY 10964-1000, USA. http://globalcmt.org.

Geological Survey of Iran, Northwestern Regional Office, Tabriz Branch, Tabriz, 43 pp. (in Persian).

George, A., 1999. The Epic of Gilgamesh: The Babylonian Epic Poem and Other Texts in Akkadian and Sumerian. The Penguin Press/Allen Lane, Harmondsworth.

George, A., 2003. The Babylonian Gilgamesh Epic—Introduction, Critical Edition and Cuneiform Texts, 2 vols.. Oxford University Press.

Gerayeli, F., 1978. Neyshabur, Shahr-e-Firuzeh [Neyshabur, The City of Turquoise]. Daneshgah Ferdowsi Mashhad,1357, 698 pp. (in Persian).

Ghafory-Ashtiany, M., 2004. Case study on Bam earthquake of 05:26:26 of 26 December 2003, M_s 6.5. In: Education for Sustainable Development; Towards Effective Disaster Reduction and Enhancement Human Security. Annex II: Case Studies—Transforming Ignorance, Apathy, Disciplinary Boundaries, and Lack of Political Will into Enlightenment, Empowerment, Enrichment, and Enablement. The Global Blueprint for Change, Section 13, 4 pp., Global Alliance for Disaster Reduction (GADR). www.gadr.giees.uncc.edu.

Ghaffāri Kāshāni, Abu'l-Hasan, 1796. Tārikh-e Golshan-e Morād (Tārikh-e Zandiyeh). Ed. Gholam Rezā Tabatabāi-Majd, Zarrin Publ., 53, 1369/1990, Tehrān (in Persian).

Ghalamghāsh, J., Rashi, H., 2002. Geological Sheet of Jirandeh, 1:1,00,000. Geol. Surv. Iran, 5963.

Gharib, B., 2009. Hafta ('week'), history of the calendar week in Iran. Iranicaonline.org.

Ghasemi, H., Juteau, T., Bellon, H., Sabzehei, M., Whitechurch, H., Ricou, L.E., 2002. The mafic-ultramafic complex of Sikhoran (central Iran); a polygenetic ophiolite complex. C. R. Géosci. 334 (6), 431–438.

Ghazar Parpetsi, 500. The History of Ghazar Parpetsi. Trans. Robert W. Thomson, 1991. Scolar Press Atlanta, Georgia, Columbia University Program in Armenian Studies; Suren D. Fesjian Academic Publication No. 4, 304p.

al-Ghāzzi, Kamāl al-Din, 1782. Al-Tadhkirāt al-Kamāliyya. In: Taher, M.A. (Ed.), 1974, 1979.

al-Ghazzali [Algazel], Emam Abu Hamed Mohammad, 1058–1111. Tahafut al-Falāsifa [The Incoherence of the Philosophers]. Translated into Persian by 'Ali Asghar Halabi, Markaz Nashr Dāneshgahi, Tehran, 1361/1982. Also Trans. Sāhib Ahmad Kamāli. http://www.ghazali.org/books/tf/Problem_XVII.htm).

Gheitānchi, M.R., 1999. Strong earthquakes in Kerman province, south-east Iran. In: Proc. 3rd Int. Conf. Seismology & Earthq. Eng., IIEES, SEE, May 17–19, 1999, Tehran, pp. 141–147.

Ghirshman, R., 1938. Les fouilles de Chapour, Iran (deuxième campagne, décembre 1936-avril 1937). Revue des Arts Asiatiques XII (I), 12–19, Pl. IX–XIV.

Ghirshman, R., 1956. Fouilees de Chapour, II, Les Mosaïques Sassanides, Paris, 201 pp., 74 Figs, 4 Plans, XXXPl.

Ghirshman, R., 1971. Bichapour I. Librairie Orientaliste Paul Geuthner, Musé3 du Louvre, Déepartment des Antiquitée Oreintales, Série Archéologique, VI, Paris.

Gidon, M., Berthier, F., Billiault, J.P, Halbronn, B., Maurizot, P., 1974. Sur le carateres et l'ampleur du coulissement de la "Main Fault" dans la region de Broujerd-Dorud, Zagros oriental, Iran. C. R. Acad. Sci. Paris Ser. D. 278, 701–704.

Godard, A., 1937. Isfahan. 2Āthār-e Irān, 176 pp., Tehran (in French).

Godard, A., 1949. Khorāsān. Athār-é Iran 4, 113–125.

Godard, A., 1965. The Art of Iran. London. Translated from the French original: L'Art de l'Iran, 1962 by Michael Heron. George Allen & Unwin, London, 358 pp.

Golinsky, G.L., 1974. Chronological Catalog of Earthquakes Felt in Turkmenia from 2000 B.C. to 1973 A.D. Inst. Phys. Earth Atmos. Akad. Sci. TSSR, Ashkhabad.

Golinsky, G.L., 1977, 1982. Description of the major earthquakes in Western Turkmenia SSR. In: Kondorskaya, N.V., Shebalin, N.V. (Eds.), New Catalog of Strong Earthquakes in the USSR from Ancient Times Through 1977. World Data Center A., NOAA, Report SE-31, 519-524.

Golombek, L., Wilber, D., 1988. The Timurid Architecture of Iran and Turan, 2 vols.. Princeton University Press, New Jersey, 510 pp. (vol. 1), 161 Figures (vol. 2).

Gorshkov, G.P., 1947a. Tectonic earthquakes and regional seismicity in the territory of the USSR (Tektonicheskie zemletryaseniy i seysmicheskoye rayonirovanie territorii SSSR). Inst. Phys. Earth Acad. Sci. USSR, (in Russian).

Gorshkov, G.P., 1947b. Earthquakes of Turkmenia (zemletryaseniya Turkmenii). Trudy Seismol. 122Acad. Sci. USSR.

GP-DRR, 2013. Resilient people, resilient planet. Global Platform for Disaster Risk Reduction, Fourth Session, Geneva, Switzerland, 19–23 May, 2013. http://www.preventionweb.net/files/33306_finalchairssummaryoffourthsessionof.pdf.

Gronenborn, D., 2009. Climate fluctuations and trajectories to complexity in the Neolithic: towards a theory. Documenta Praehistorica 36, 97–110.

Grant, L.B., Donnellan, A., 1994. 1855 and 1991 surveys of the San Andreas fault: implications for fault mechanics. Bull. Seismol. Soc. Am. 84, 241–246.

Grant, L.B., Sieh, K., 1994. Stratigraphic evidence for several meters of dextral slip on the San Andreas fault during the 1857 earthquake in the Carrizo Plain. Bull. Seismol. Soc. Am. 83, 619–635.

Greater Bondéhéshn. See Bondéhéshn.

Griffith, R.T.H., 1896. The Rig Veda. Sacred Writings, vol. 5, Hinduism. Motilal Banarsidass Publishers, PVT. Ltd. 656 pp. Electronic version available at: http://www.sacred-texts.com/hin/rigveda/index.htm.

Grigoryan, G., Grigoryan, V., 1977. Davit of Sasu, New Narrations. *Erivan* (in Armenian). Russian Trans. David of Sasun, Armenian National Epochs. *Moskova*, 1982 (in Russian).

Grubin, D., 2010. The Buddha. PBS, 4/7/2010. http://www.pbs.org/thebuddha/?gclid=CN76nsew96ACFYwM2godgVE0uA.

GSI, 1972a. Geological Map of the Lut, 1:500,000. Geol. Surv. Iran.

GSI, 1972b. Geological Quadrangle Map of Ardakan, 1:250,000. H-8Geol. Surv. Iran.

GSI, 1986a. Geological Quadrangle Map of Bojnurd, 1:250,000. J-3Geol. Surv. Iran.

GSI, 1986b. Geological Quadrangle Map of Mashhad, 1:250,000. J-4Geol. Surv. Iran.

GSI, 1989. Geological Quadrangle Map of Gorgan, 1:250,000. H4Geol. Surv. Iran.

GSI, 1991. Geological Quadrangle Map of Qa'en, 1:250,000. K-7Geol. Surv. Iran.

GSI, 1992a. Geological Quadrangle Map of Birjand, 1:250,000. K-8Geol. Surv. Iran.

GSI, 1992b. Geological Quadrangle Map of Rafsanjan, 1:250,000. I-10Geol. Surv. Iran.

GSI, 1992c. Geological Quadrangle Map of Jahānābād, 1:250,000. K-12Geol. Surv. Iran.

GSI, 1992d. Geological Quadrangle Map of Dareh-gaz, 1:250,000. K-3Geol. Surv. Iran.

GSI, 1993. Geological Quadrangle Map of Sabzevar, 1:250,000. J-3Geol. Surv. Iran.

GSI, 1999. Geological Map of Neyshabur, 1:100,000. Sheet No. 7766Geol. Surv. Iran.

GSi, 2000. Geological Map of Mashkan, 1:100,000. Sheet No. 7563Geol. Surv. Iran.

GSI, 2005. Geological Map of Pol-e Sefid, 1:100,000. Sheet No. 6662Geol. Surv. Iran.

GSI, 2013. Preliminary Report of the 31 Ordibehesht, 1382 (M 6.2) of North Jāsk-Goharān. Seismotectonic and Seismology Department, Geol. Surv. Iran, Internal Report, Ordibehesht, 1392, gsi.ir (in Persian).

Guest, B., Axen, G.J., Lam, P.S., Hassanzadeh, J., 2006a. Late Cenozoic shortening in the west central Alborz mountains, northern Iran, by combined conjugate strike–slip and thin-skinned deformation. Geosphere. 2, 35–52. http://dx.doi.org/10.1130/GES00019.1.

Guest, B., Stockli, D., Grove, M., Axen, G., Lam, P., Hassanzadeh, J., 2006b. Thermal histories from the central Alborz Mountains, northern Iran: implications for the spatial and temporal distribution of deformation in northern Iran. Geol. Soc. Am. Bull. 118, 1507–1521.

Guidoboni, E., 1994. Catalogue of Ancient Earthquakes in the Mediterranean Area up to the 10th Century. Istitute Nazionale di Geofisica, Rome, 504 pp.

Guidoboni, E., 1997. Historical Seismology Research in the Cacasus: Methodological Aspects and Some results. In: Giardini, D., Balassanian, S. (Eds.), Historical and Prehistorical Earthquakes in the Caucasus. Kluwer Academic Press, The Netherlands, pp. 333–356, NATO ASI Series, 2. Environment, vol. 28.

Guidoboni, E., 1998. Earthquakes, Theories from Antiquity to 1600. Sciences of the Earth; An Encyclopedia of Events, People, and Phenomena. Ed. George A. Good., vol. 1. Garland Publishing, Inc., 197–214.

Guidoboni, E., Comastri, A., 2005. Catalogue of Earthquakes and Tsunamis in the Mediterranean Area from the 11th to the 15th Century. INGV Istituo Nazionale di Geofisica e Vulcanologia, 1037 pp.

Guidoboni, E., Traina, G., 1995. A new catalogue of earthquakes in the historical Armenian area from antiquity to the 12th century. Annali di Geofisica XXXVII (1), 85–147.

Gulbenkian, R., Berberian, H., 1971. La légend de David de Sassoun d'après deux voyageurs portugais du XVIe siècle. Revue des Etudes Arméniennes VIII, 175–188.

Hādi Hamédāni [Nesbat]. In Zokā' (1989).

Hāfez Abru, Shahāb al-Din 'Abdolāh ebn 'Abdol Rashid Khāfi, 1414. Joghrāfiyā-ye Hāfez-e Abru: Qesmat-e-Rob'-e Khorāsān, Harāt [Geography of Hāfez-e Abru: Khorāsān Quarter, Harāt; 817/1414]. Ed. Māyel Heravi, Bonyad Farhang Iran, 1349/1970. **Aslo Joghrāfiyā-ye Tārikhi-ye Khorāsān dar Tārikh-e-Hāfez-e-Abru (817/1414). Ed. Gholām Rezā Varahrām, Etela'at Publ., 1370/1991, 166 pp. (in Persian). Hafiz-e Abru, Djughrafiya, Ed. D. Krawulsky, Horasan zur Timuridenzeit, 2 vols. *TAVO* Beiheft B 46/1-2, 1982–1984, Wiesbaden.

Hāfez Abru, Shahāb al-Din 'Abdolāh ebn 'Abdol Rashid Khāfi, 1431. Zail-e Jāme' al-Tavārikh-e Rashidi. Ed. K. Khānbābā Bayāni, Tehran 1350/1971 (in Persian).

al-Hāfez ibn Asākir, 1176. Kitāb al-Zalāzil [The Earthquake Book, Lost manuscript, mentioned in later treatises.

Haghipour, A., Amidi, M., 1980. The November 14 to December 25, 1979 Ghaenat earthquakes of northeast Iran and their tectonic implications. Bull. Seismol. Soc. Am. 70 (5), 1751–1757.

Haghipour, A., Iranmanesh, M.H., Takin, M., 1972. The Ghir Earthquake in Southern Persia (A Field Report and Geological Discussion). Geol. Surv. Iran, Int. Rep. 52.

Haghshénās, E., 1998. Geotechnical aspects overview of the Manjil earthquake north of Iran (20 June 1990). In: Proc. 1st Iran-Japan Workshop Recent Earthquakes in Iran & Japan, May 16–18, 1998, Tehran, pp. 171–183.

Hagiwara, T., Naito, T., 1959. A Report of the Japanese Mission Sent to Iran for Investigating the Problems Related to Disastrous Earthquakes in Iran. Faculty of Science, University of Tehran, Tokyo, 100 pp.

Hajialioghli, R., Moazzen, M., Droop, G.T.R., Oberhänsli, R., Bousquet, R., Jahangiri, A., Ziemann, M., 2007a. Serpentine polymorphs and P–T evolution of meta-peridotites and serpentinites in the Takab area, NW Iran. Mineral. Mag. 71, 155–174.

Hajialioghli, R., Moazzen, M., Jahangiri, A., Droop, G.T.R., Bousquet, R., Oberhänsli, R., 2007b. Petrogenesis of meta-peridotites in the Takab area, NW Iran. In: Goldschmidt Conference Abstracts, Germany, A370.

Hajialioghli, R., Moazzen, M., Jahangiri, A., Oberhänsli, R., Mocek, B., Altenberger, U., 2011. Petrogenesis and tectonic evolution of metaluminous sub-alkaline granitoids from the Takab Complex, NW Iran. Geol. Mag.. 148 (2), 250–268. http://dx.doi.org/10.1017/S0016756810000683, Cambridge University Press 2010, 250.

Hāji Hādi Hamedāni (Nesbat). See Tabātabā'i Tabrizi (1875).

Hāji Āqā Bābā-ye Hamedāni. See Tabātabā'i Tabrizi (1875).

Hākem Neyshāburi, Abu 'Abd Allāh, 998. Tārikh-e Neyshābur [The History of Neyshābur; 388/998). Persian Trans. Mohammad ibn Hossein Khalifeh Neyshāburi, 808/1406 (Revised after 808/1406 by one of his sons Ahmad ibn Muhammad ibn Hussain ibn Ahmad). Ed. Mohammad Rezā Shafi'ee-Kadkani, 1375/1996, Āgah Publ., Tehran, 410 pp. (in Persian).

Hakhnazarian, H., 1980. Basic Chronological History. Documents of Armenian Architecture, 10, S. Stephanops. Edizioni Ares, Milano, Italy, pp. 18–19.

Hakobyan, V.A. (Ed.), 1951. Mānr Zhāmānākāgrutyunner XIII–XVIII DD (Short Chronicles, 13th-18th Centuries). In: Materials on the History of the Armenian People, vol. 1. Armenian SSR Academy of Sciences, Institute of History, Yerevān.

Hakobyan, V.A. (Ed.), 1956. Manr Zhāmānākāgrutyunner XIII–XVIII DD (Short Chronicles, 13th-18th Centuries). In: Materials on the History of the Armenian People, Book Seven, vol. 2. Armenian SSR Academy of Sciences, Institute of History, Yerevān.

von Hammer-Purgstall, J., 1827–1835. Geschichte des Osmanischen Reiches, 10 vols., Pest.

Hamilton, W., 1979. Tectonics of the Indonesian region. U.S. Geol. Surv. Prof. Pap. 1078, 345 pp.

Hamilton, H.C., Falcon, W., 1854. The Geography of Strabo. Archive.org, perseus.tuffs.edu.

Harms, J.C., Cappel, H.N., Francis, D.C., 1984. The Makran coast of Pakistan: its stratigraphy and hydrocarbon potential. In: Haq, B.U., Milliman, J.D. (Eds.), Marine Geology and Oceanography of the Arabian Sea and Coastal Pakistan. Van Nosrand Reinhold, Co., New York, pp. 3–26

Hamzeh Esfahāni, Abu 'Abdolāh ebn Hossain, 961. Tārikh Seni Moluk al-Arḍ wa'l-Anbiyā [History of Prophets and Kings; 350/961]. Translated into Persian by Ja'far Sho'aār. first ed. 1346/1967, Bonyād Farhang Irān, Publ. Amir Kabir Publ., Tehran, 1367/1988, 269 pp. (in Persian). Also Hamza ibn Hassan al-Isfahani, Tarikh Sini Muluk al-Arḍ wa'l-Anbiya, Ed. J.M.E. Gottwaldt, *Leipzig*, 1844–1848.

Hamzeloo, H., Chander, R., Sarkar, I., 1997. Probable role of the Sefidrud reservoir in the occurrence of the Rudbar earthquake of 1990. Bull. Ind. Soc. Tech. 34, 17–25.

Hanaway Jr., W.L. (Ed.), 1977. Khorasan I. Corpus Inscriptionum Iranicarum, Part IV. Persian Inscriptions Down to the Early Safavid Period, vol. II, Khorasan Province, Portfolio I. Corpus Inscriptionum Iranicarum, School of Oriental and African Studies, London.

Hansen, G., Condon, E., 1989. Denial of Disaster. Cameron and Co., San Francisco, 160 pp.

Hansman, J., 1968. The problems of Qūmis. J. R. Asiatic Soc. 111–139.

Hansman, J., 1985, 2011. Anshan. Encyclopaedia Iranica II (1), 103–107. iranicaonline.org, Also.

Harding, T.P., 1973. Newport-Inglewood trend, California—an example of wrenching style deformation. AAPG Bull. 57, 97–116.

Hāroutyunian, S., 1997. Armenian epic tradition and Kurdish folklore. Iran Caucasus 1, 85–92, Brill.

Harris, R.A., Simpson, R.W., 1996. In the shadow of 1857—effect of the Great Ft. Tejon Earthquake on subsequent earthquakes in southern California. Geophys. Res. Lett. 23, 229–232.

Harris, R.A., Simpson, R.W., 2002. The 1999 MW 7.1 Hector Mine, California earthquake—a test of the stress shadow hypothesis? Bull. Seismol. Soc. Am. 4, 1497–1512.

Harrison, J.V., 1936. Kuhgalu: south-west Iran. Geogr. J. 88, 20–36.

Harrison, J.V., 1943. The Jaz Murian Depression, Persian Baluchestan. Geogr. J. 101 (5/6), 206–225.

Harrison, J.V., 1946. South-west Persia: a survey of Pish-i-Kuh in Luristan. Geogr. J. 108, 55–71.

Harrison, J.V., Falcon, N.L., 1934. Collapse structures (Kuhgalu district, Persia). Geol. Mag. 71 (846), 529–539.

Harrison, J.V., Falcon, N.L., 1937. The Saidmarreh landslip, south-west Iran. Geogr. J. LXXXIX (1), 42–47.

Harrison, J.V., Falcon, N.L., 1938. An ancient landslip at Saidmarreh in southwestern Iran. J. Geol. 46, 296–309.

Hartzell, S.H., Mendoza, C., 1991. Application of an iterative least-squares waveform inversion of strong-motion and teleseismic records of the 1978 Tabas, Iran earthquake. Bull. Seismol. Soc. Am. 81, 305–331.

Hassanzadeh, J., Stockli, D.F., Horton, B.K., Axen, G.J., Stockli, L.D., Grove, M., Schmitt, A.K., Walker, J.D., 2008. U–Pb zircon geochronology of late Neoproterozoic–Early Cambrian granitoids in Iran: implications for paleogeography, magmatism, and exhumation history of Iranian basement. Tectonophysics 451 (1–4), 71–96.

Hātam Nāui'i, 2013. Dau'ā-yeh khair-e shohadā dalil-e kamtarin talafat-e zelzele-yeh 7.7 rishteri-ye Sarāvān shod. Kayhān evening newspaper, Tehran. www.kayhan.ir/920131/3.htm#other315, No. 20468, 31 Farvardin 1392/20 April 2013, Tehran (in Persian).

Hataria, Manekji Limji, 1865. Ezhāre Syāte Iran. Trans. Ankelsaria, 1234/1865, Bombay.

Hātef, Esfahāni, Sayyed Ahmad hosseini, 1783. In: Dastjerdi, Vahid (Ed.), Divan-e 'Ash'aār'-e Hātef-e Esfahani (Collection of poems of Hatef-e Esfahāni), Tehran 1332/1953. Also see Ghaffāri Kāshāni, 1796; Zarrābi, 1870; Nakhjavāni, 1964.

Hatzfeld, D., Tatar, M., Priestley, K., Ghafory-Ashtyany, M., 2003. Seismological constraints on the crustal structure beneath the Zagros mountain belt (Iran). Geophys. J. Int. 155, 403–410.

Haug, M., 1869. Das Achtzehnte Kapitel des Wendidad Ubersetzt und erklart (Als probe einer voll-slandegen Uebersetzung dieses werker). Akademische Buchdruck von F. Strauss, Munchen.

Hauksson, E., 1987. Seismotectonics of the Newport-Inglewood fault zone in the Los Angeles Basin, California. Bull. Seismol. Soc. Am. 77, 539–561.

Hauksson, E., Jones, L.M., Hutton, K., Eberhart-Phillips, D., 1993. The 1992 Landers earthquake sequence: seismological observations. J. Geophys. Res.. 98, 19835–19858. http://dx.doi.org/10.1029/93JB02384.

Hedāyat, Mirzā Rezā Qoli Khān, Amir al-Sho'arā', 1800–1871. Rouzat al-Safā-ye Nāseri, vol. VIII, Ed. Parviz, 1960, Addition to the Chronicle of the Rozat al-Safā of Mir Khānd.

Hedin, S., 1910. Overland to India, 2 vols.. Macmillan and Co., Ltd., London, Chapter XXXVII: The Oasis of Tebes.

Hekmat al-Ishrāq. See Sohrévardi.

Helmberger, D.V., Somerville, P.G., Garnero, E., 1992. The location and source parameters of the Lompoc, California, earthquake of 4 November 1927. Bull. Seismol. Soc. Am. 82, 1678–1709.

Helwing, B., 2005. Tappeh Sialk South Mound: operation 3. In: Sadegh Malek Shahmirzadi (Ed.), The Fishermen of Sialk; Sialk Reconsideration Project Report No. 4. Iranian Center for Archaeological Research, Deputy of Research, Archaeological Report Monograph Series, 7, 27–66, Tehrān.

Henderson, J., 1937. The west Nelson earthquake of 1929. N. Z. J. Sci. Technol. 19 (2), 66–143.

Henning, W.B., 1945. The Manichean Fasts. JRAS 146–164, repr., Acta Iranica 15, 1977, pp. 205 ff.

Herodotus, 484–425 B.C. The History of Herodotus. Trans. Macaulay, Macmillan, London, 1890. Sacred-text.com.

Herzfeld, E., 1935. Archaeological History of Iran. Oxford University Press, London.

Hessami, Kh., Pantosi, D., Tabassi, H., Shabanian, E., Abbassi, M.R., Feghhi, Kh., Solaymanian, E., 2003. Paleoearthquakes and slip rates of the North Tabriz fault, NW Iran: preliminary results. Ann. Geophys. 46 (5), 903–915.

Hessami, Kh., Tabassi, H., Abbassi, R., Azuma, T., Okumura, K., Echigo, T., Kondo, H., 2004. Surface expression of the Bam fault zone in southeastern Iran: causative fault of the 26 December 2003 Bam earthquake. J. Seismol. Earthq. Eng., Special Issue on Bam Earthquake 5 (4), 6 (1), IIEES, Tehran, 5–14.

Heuckroth, L.E., Karim, R.A., 1970. Earthquake History, Seismicity and Tectonics of the Regions of Afghanistan. Kabul University Seismological Center, 102 pp.

Hiebert, F., Dyson Jr., R.H., 2002. Prehistoric Nishapur and the frontier between Central Asia and Iran. Iranica Antiqua 37, 111–147.

Hill, K.C., 1991. Structure of the Papuan fold belt, Papua New Guinea. Am. Assoc. Pet. Geol. Bull. 75, 857–872.

Hinnells, J.R., 1973. The Zoroastrian doctrine of salvation in the Roman world: a study of the Oracle of Hystaspes. In: Hinnells, Sharpe, (Eds.), Man and His Salvation, Studies in Memory of S.G.F. Brandom. Manchester University Press, Manchester, G.B., pp. 125–148.

Hitchen, C.S., 1946. Earthquake Messages from the Mutasarrifs of Sulaimaniyah and Mosoul: Unpublished Report. Adm. Office, Baghdad.

Hodud al'Alam. Hudud al-'Alam men al-Mashreq el al-Maghreb [The Limits of the World from the East to the West], 372/982; a Persian geography written by an anonymous writer,

372/982. Ed. Manuchehr Sotudeh. Tāhuri Publ., Tehran, 1362/1983 (in Persian). Also Hudud al-'Alam, Trans. & comments by V. Minorsky, second ed. Bosworth, London, 1970.

Hofrichter, H., 1980. Historical-Illustrative Monograph. Documents of Armenian Architecture, 10, S. Stephanops. Edizioni Ares, Milano, Italy, pp. 5–9.

Hollingsworth, J., Jackson, J., Walker, R., Gheitanchi, M., Bolourchi, M., 2006. Strike-slip faulting, rotation, and along-strike elongation in the Kopeh Dagh mountains, NE Iran. Geophys. J. Int. 166, 1161–1177.

Hollingsworth, J., Jackson, J., Alarcón, J.E., Bommer, J.J., Bolourchi, M.J., 2007. The 4th February 1997 Bojnurd (Garmkhan) earthquake in NE Iran: field, teleseismic, and strong-motion evidence for rupture directivity effects on a strike-slip fault. J. Earthq. Eng. 11, 193–214.

Hollingsworth, J., Jackson, J., Walker, R., Nazari, H., 2008. Extrusion tectonics and subduction in the eastern South Caspian. J. Geol. 36 (10), 763–766.

Hollingsworth, J., Nazari, H., Ritz, J.-F., Salamati, R., Talebian, M., Bahroudi, A., Walker, R., Rizza, M., Jackson, J., 2010. Active tectonics of the east Alborz mountains, NE Iran: rupture of the left-lateral Astaneh fault system during the great 856 A.D. Qumis earthquake. J. Geophys. Res.. 115http://dx.doi.org/10.1029/2009JB007185, B12313, B12313, 1978–2012.

Homke, S., Verges, J., Garces, M., Emami, H., Karpuz, R., 2004. Magnetostratigraphy of Miocene-Pliocene Zagros foreland deposits in the front of Push-e Kuh Arc (Lurestan province, Iran). Earth Planet. Sci. Lett. 225, 397–410.

Horne, C.F., 1917. See Yadegar-e-Zariaran.

Horton, B.K., Hassanzadeh, J., Stockli, D.F., Axen, G.J., Gillis, R.J., Guest, B., Amini, A., Fakhari, M.D., Zamanzadeh, S.M., Grove, M., 2008. Detrital zircon provenance of Neoproterozoic to Cenozoic deposits in Iran: implications for chronostratigraphy and collisional tectonics. Tectonophysics 451, 97–122.

Hosseini, K., 'Asgari, A., 1997. Barresiyeh Moqadamāti-ye Larzeh-zaminsākht mantaqeyeh Bojnurd. Geol. Surv. Iran, Northeast Office, Internal Report, 37 pp. (in Persian).

Hosseini-barzi, M., Talbot, C.K., 2003. A tectonic pulse in the Makran accretionary prism recorded in Iranian coastal sediments. J. Geol. Soc. Lond. 160, 903–910.

Hourcade, B., 1994. Demography, the statistical study of characteristics of human populations in Persia since 1319/1940. Encyclopaedia Iranica XI (3), 258–276. 2011iranicaonline.org.

Houtum Schindler, A., 1881. The Persian Government telegraphs. J. Soc. Telegr. Eng. 4 (13–14), 262.

Howell Jr., B.F., 1986. History of ideas on the cause of earthquakes. EOS, Trans. Am. Geophys. Union, 67 (46), 1323–1326.

HRVD. Centroid moment tensor catalogue. Harvard University, Department of Geological Sciences, Cambridge, MA 02138, USA. globalcmt.org.

Hsu, Y.-J., Avouac, J.-P., Yu, S.-B., et al., 2009. Spatio-temporal slip, and stress level on the faults within the Western Foothills of Taiwan: implications for faults frictional properties. Pure Appl. Geophys. 166, 1853–1884.

Huckriede, R., Kursten, M., Venzlaff, H., 1962. Zur geology desgebietes zwischen Kerman und Sagand (Iran). vol. 51. Geol. Jahrb., Beihft, 197 pp.

Hudnut, K.W., Seeber, L., Pacheco, J., 1989. Cross-fault triggering in the November 1987 Superstition Hills earthquake sequence, Southern California. Geophys. Res. Lett.. 0094-8276-16http://dx.doi.org/10.1029/88GL04147.

Hume, G.W., 1967. Comments on the geology and archaeology of the Bard Sir Valley. In: Caldwell, J.R. (Ed.), Investigation at Tal-i-Iblis, pp. 108–110, Illinois State Museum Preliminary Report, No. 9.

Ibn al-'Amid. In Biruni Khārazmi, 1025.

Ibn al-Athir, 'Izz al-Din Abu al-Hassan, 1231. Al-Kāmil fi al-Tārikh [The Complete History; 630/1231]. Ed. C.J. Tornberg, Leiden, 1851–1876. Al-Muniriyya Press, Cairo, 1348–1357 [1929–1938], 9 vols.

Ibn Batuta [Ebn-e-Battuteh Maghrebi], 'Abu 'Abdolāh Mohammad ebn 'Abdolāh Tanji, 1355. Safarnāmeh-ye Ebn-e Battuteh [Travel Book of Ibn Batuta, 1325–1354], 570/1355. Translated into Persian by Mohammad 'Ali Movahed, Bongāh Tarjomeh va Nashr-e-Ketāb Publ., no. 75, Tehran, 2 vols., 1337/1958 & 1348/1969. Also Ibn Battoutah, Ed. & Trans. Defremy and Sanguinetti, Voyages d'Ibn Batoutah, 4 vols. Paris, 1914. Ed. H.A.R. Gibb. Cambridge, Hakluyt Society, 3 vols., 1958–1971.

Ibn al-Faqih al-Hamedan [ebn-e-Faqih Hamédani], Abubakr Ahmad ebn Mohammad (903). Akhbār al-Beldān [News of the Cities; ca. 290/903]. Translated into Persian by H. Mas'ud, *Manābe' Tārikh va Joghrāfiyā-ye Irān*, no. 35, Tehrān 1349/1370 (in Persian). Also Ibn al-Faqih, Ahmad ibn Muhammad, Mukhtasar Kitab al-Buldan, Ed. de Goeje, *Bibliotheca Geographorum Arabicorum, Leiden*, BGA v, 1885. Trans. H. Massé, Abrégé du livre des pays, Damascus, 1973.

Ibn Funduq, Abu'l-Hasan 'Ali ibn Zayd Bayhaqi, d. 1169. Tāritkh-e Bayhaq [The History of Baihaq; modern Sabzévār]. Ed. Ahmad Bahmanyār, with an introduction by Muhammad Qazvini, Tehran, 1317/1938, 1361/1982 (in Persian).

Ibn Hawqal (Hauqal), Abolqāsem Muhammad, 978. Surat al-Arz [The Face of the Earth, ca. 367/978]. Translated into Persian by Ja'far Sho'aār, Bonyād Farhang Irān Publ., Tehrān, 1345/1966 (in Persian). Also The Geography of Ebn Haukal; An Arabian Traveller of the Tenth Century. English Trans. Sir William Ouseley, 1800. T. Cadell, Jun. and W. Davies, Strand, Wilson & Co., London, 327 pp.

Ibn al'Imād al-'Akar, 'Abd al-Hayy ibn Ahmad, d. 1678. Shadharāt al-dhahab fi Akhbār man Dhahaba. Ed. 'Abd al-Qādir Arnā'ut and Mahmud Arnā'ut, 8 vols in 4. *Dār ibn Kathir,* Damascus and Beirut.

Ibn al-Jauzi, 'Abd al-Rahmān ibn 'Ali Jamāl al-Din, 1181. Al-Muntazam fi Tārikh al-Muluk wa al-Umam. Osmania Oriental Publ. Bureau, Hyderabad-Deccan, 1357–1361 [1938–1942], 10 vols.

Ibn Khurdādhbeh, 'Ubaydallah ibn 'Abdallah (846). Kitāb al-Masālik wa al-Mamālik. Ed. M.J. de Goeje, Leyden, BGA VI, 1889; *Bibliotheca Geographorum Arabicorum*, second ed., Leiden, 1967.

Ibn Miskawayh [Miskawaih Rāzi], Abu 'abdallāh Ahmad ibn Muhammad ibn Ya'qub, d.421/1030. Tajārub al-Umam fi al-Tārikh. Zaki al-Kurdi Press, Cairo, 1914–1915. The last portion (2 vols.) is republished in 'The Eclipse of the 'Abbasid Caliphate, vols. I. II, & III, Arabic Text, Blackwell, Oxford, 1920–1921, Trans. H.F. Amedrose, D.S. Margoliouth. Also Ed. & Trans. H.F. Amedroz, D.S. Margoliouth, Oxford, 1921. Also Maskuyeh Razi, Abu 'Ali. Tajārob al-Omam. Trans. Abolqāsem Emāmi, Sorush Publ., Tehran, 1370, 544 pp. (in Persian).

Ibn Rusteh (Rasteh), Esfahāni, Abu 'Ali Ahmad ibn 'Umar, 903. Kitāb al-'Alaq al-Nafisa. Tr. into Persian by Hossein Qarachanlu, Amir Kabir Publication, Tehran, 1365/1986. M.J. de Goeje (Ed.), Leyden, BGA VII, 1891.

Ibn Sinā (Avicenna; Pur Sina), Abu 'Ali Hussain ibn Abdallāh, 1020. Fann Samāa' Tabi'i az Ketāb-e Shafā. Translated into Persian by Mohamad 'Ali Foroughi, Vezārat Ma'āref va Owqāf va Sanāye' Mostazrafeh Publ., Tehrān, vol. 1: 1316/1937, 684 pp., vol. 2: 1319/1940, 779 pp. (in Persian). Also Fan-e Samāa'-e Tabi'i, from Shafā (*The Book of the Remedy; Sufficientia*). Translated into Persian by Mohammad 'Ali Foroughi, Chapkhaneh Majles Publ. 1316/1937, 1949; Amir Kabir Publ. 1361/1982, 675 pp.

Ibn Sinā (Avicenna; Pur Sinā], Abu 'Ali Hussain ibn Abdallāh, 980–1037. Kanz al-Masā'il fi Arba' Rasā'il' ['Treasury of Questions in Four Epistles'], Question No. 2 'Resāleh dar

Bayān-e Hoduth-e Zelzeleh va Falsafeyeh ān, az Fārābi' ['Epistle in the Occurrence of Earthquake and its phylosophy' by Fārābi; Alpharabius: 872-950]. Ed. Ziya'aldin Dari, Khayyām Publ., Tehran, 1330/1951, 70–75 (in Persian).

Idrisi, Mohammad, al-, 1154. Kitāb Nuzhat al-Mushtaq fi Ikhtirāq al-Āfāq [The Book of Pleasant Journeys into Faraway Lands]. 2 vols., Paris.

IFRC. International Federation of Red Cross and Red Crescent Societies, Switzerland. ifrc.org.

IGUT, 1963. Report on the Great Buyin-Zahra Earthquake of Sept. 1st 1962. Institute of Geophysics, Tehran University, 80 pp.

IGUT, 1969. Report on the Great Dasht-e Bayaz Earthquake of Aug. 31st and Sep. 1st 1968. Institute of Geophysics, Tehran University, 46, 89–91.

IIEES. International Institute of Earthquake Engineering and Seismology, Tehran. Iran.

Ikhwān al-Safā [Ekhvān al-Safā; *the Brethren of Purity*], 971. Rasā'il Ikhwān al-Safā wa Khullān al-Wafā wa ahl al-Hamd wa Abāa' al Majd [Epistles of the True Friends, the Faithful Comrades, the People Deserving Praise, and the Sons of Glory; 971; (Halabi, 1981)]; Matn-e Rasā'el-e Ekhvān al-Safā]. Persian translation from Arabic by 'Ali Asghar Halabi. Zavāreh Publ., Tehran, 1360/1981.

Irān. Persian newspaper, Tehrān, Irān.

Irān, 1309. Ruznāmeh-ye Irān [Iran Newspaper], Tehran, Iran (in Persian).

Irān-e Mā, 1324. Ruznāmeh-ye Irān-e Mā [Irān-e Mā Newspaper], 23 Esfand 1324, Tehran, Iran (in Persian).

Iranian (Greater) Bondéhéshn. See Bundahishn.

Iranian Red Crescent Society. rcs.ir.

Iranicaonline. Encyclopaedia Iranica online. iranicaonline.org.

IRNA. Islamic Republic News Agency, Tehrān. irna.ir.

Irvāni (Qodsi), Mirzā Moslem, (1696–1739). In Daulatābādi, 1964, Zokā', 1989.

Issawi, C., 1971. The Economic History of Iran 1800-1914. Center for Middle Eastern Studies, 8, University of Chicago Press, 405 pp.

ISC. International Seismological Centre, Pipers Lane, Thatcham, Berkshire RG19-4NS, UK. http://www.isc.ac.uk.

al-Isfezāri, Mu'in al-Din Muhammad. Rauḑāt al-Jannāt fi Ausāf Madinat Harāt. Trans. M. Barbier de Meynard, Extrits de la chronique persane d'heart, J. Asiatique, 5th ser., 1861 (Extrait no. 11) et 1862, 268–319.

Isfizāri, 899/1493. See Esfezāri, 899/1493.

Ishihara, K., Haeri, M., Moinfar, A.A., Towhata, I., Tsujino, S., 1992. Geotechnical aspects of the June 20, 1990 Manjil earthquake in Iran. Soils Found. 32 (3), 61–78, (Japanese Society of Soil Mechanics and Foundation Engineering).

Isidore of Charax, 1st century. The Parthian Stations by Isidore of Charax: The Greek Text, with a translation and commentary, Wilfred H. Schoff, 1914, London. http://www.parthia.com/doc/parthian_stations.htm.

Istakhri. See Estakhri.

Jackson, A.V.W., 1921. The location of the Faranbag Fire, the most ancient of the Zoroastrian Fires. J. Am. Orient. Soc. 41, 81–106.

Jackson, A., 1928. The cypress of Kashmar and Zoroaster. In: Zoroastrian Studies; The Iranian Religion and Various Monographs. Columbia University Indo-Iranian SeriesColumbia University Press, New York, 325 pp. http://www.Vohuman.org/Library/.

Jackson, J.A., 1992. Partitioning of strike-slip and convergent motion between Eurasia and Arabia in Eastern Turkey and the Caucasus. J. Geophys. Res. 97 (B9), 12471–12479.

Jackson, J.A., 1995. The accommodation of the Arabia–Eurasia plate convergence in Iran. J. Geophys. Res. 100, 15205–15219.

Jackson, J.A., 1980. Reactivation of basement faults and crustal shortening in orogenic belts. Nature 283, 343–346.

Jackson, J.A., 1999. Fault depth: a perspective from activity deforming regions. J. Struct. Geol. 21, 1003–1010.

Jackson, J.A., 2001. Living with earthquakes: know your faults. J. Earthq. Eng. 5 (Sp. Issue 1), 5–123.

Jackson, J.A., 2002. Using earthquakes for continental tectonic geology. In: Lee, W.H.K., Kanamori, H., Jennings, P.C., Kisslinger, C. (Eds.), International Handbook of Earthquakes and Engineering Seismology. Academic Press, pp. 491–503, Part A.

Jackson, J.A., Ambraseys, N.N., 1997. Convergence between Eurasia and Arabia in eastern Turkey and the Caucasus. In: Giardidni, D., Balasanian, S. (Eds.), Historical and Prehistorical Earthquakes in the Caucasus. Kluwer Academic Publishers, pp. 79–90.

Jackson, J.A., Fitch, T.J., 1979. Seismotectonic implications of relocated aftershock sequences in Iran and Turkey. Geophys. J. R. Astron. Soc. 54, 209–229.

Jackson, J.A., Fitch, T., 1981. Basement faulting and the focal depths of the larger earthquakes in the Zagros mountains (Iran). Geophys. J. R. Astron. Soc. 64, 561–586.

Jackson, J.A., McKenzie, D.P., 1984. Active tectonics of the Alpine-Himalayan belt between western Turkey and Pakistan. Geophys. J. R. Astron. Soc. 77 (1), 185–264.

Jackson, J.A., McKenzie, D.P., 1988. The relationship between plate motions and seismic tensors, and the rates of active deformation in the Mediterranean and Middle East. Geophys. J. R. Astron. Soc. 93, 45–73.

Jackson, J.A., Molnar, P., 1990. Active faulting and block rotation in the western Transverse Ranges, California. J. Geophys. Res. 95, 22073–22087.

Jackson, J.A., Fitch, T., McKenzie, D.P., 1981. Active thrusting and the evolution of the Zagros fold belt thrust and Nappe Tectonics. In: McClay, K., Price, N. (Eds.), Spec. Publ. Geol. Soc. Lond., vol. 9, pp. 371–379.

Jackson, J.A., Haines, A., Holt, W., 1992. The horizontal velocity field in the deforming Aegian Sea region determined from the tensors of earthquakes. J. Geophys. Res. 97, 17657–17684.

Jackson, J.A., Haines, J., Holt, W., 1995. The accommodation of Arabia-Eurasia plate convergence in Iran. J. Geophys. Res. 100 (B8), 15205–15219.

Jackson, J.A., Norris, R., Younngson, J., 1996. The structural development of active fault and fold systems in central Otago, New Zealand: evidence revealed by drainage patterns. J. Struct. Geol. 18, 217–234.

Jackson, J., Priestley, K., Allen, M., Berberian, M., 2002. Active tectonics of the South Caspian Basin. Geophys. J. Int. 148 (2), 214–245.

Jackson, J., Bouchon, M., Fielding, E., Funning, G., Ghorashi, M., Hatzfeld, D., Nazari, H., Parsons, B., Priestley, K., Talebian, M., Tatar, M., Walker, R., Wright, T., 2006. Seismotectonic, rupture-process, and earthquake-hazard aspects of the 26 December 2003 Bam, Iran, earthquake. Geophys. J. Int. 163, 90–105.

Jacob, K.H., Quittmeyer, R.L., 1979. The Makran region of Pakistan and Iran: trench-arc system with active plate subduction. In: Farah, A., DeJong, K.A. (Eds.), Geodynamics of Pakistan. Geol. Surv. Pakistan, Quetta, pp. 305–317.

Jacques, E., King, G.C.P., Tapponnier, P., Ruegg, J.C., Manighetti, I., 1996. Seismic activity triggered by stress changes after the 1978 events in the Asal rift, Djibouti. Geophys. Res. Lett. 23, 2481–2484.

Jāmāsp Nāmag, 1903. See Modi.

Jame' al-Hekmatin. See Nāser Khosrow.

James, G.A., Ghashghaie, M., 1960. The Lar Earthquake of April 24, 1960. Iranian Oil Exploration and Producing Company, Geological and Exploration Department, Report No. 950. Masjed-e Suleimān, 11 pp.

James, G.A., Wynd, J.G., 1965. Stratigraphic nomenclature of Iranian Oil Consortium Agreement area. Am. Assoc. Pet. Geol. Bull. 49 (12), 2182–2245.

Javidfakhr, B., Bellier, O., Shabanian, E., Siame, L., Léanni, L., Bourlès, D., Ahmadian, S., 2011. Fault kinematics and active tectonics at the southeastern boundary of the eastern Alborz (Abr and Khij fault zones): geodynamic implications for NNE Iran. J. Geodyn.. 52 (3-4), 290–303. http://dx.doi.org/10.1016/j.jog.2011.02.005.

Jayhāni, Abolqāsem ibn Mohammad, 976. Ashkāl al-'Alam. Trans. 'Ali ibn 'Abdolsalām Kāteb. Beh Nashr Publ., Mashhad (in Persian).

Jestin, F., Huchon, P., Gaulier, J.M., 1994. The Somalia plate and the East African rift system: present-day kinematics. Geophys. J. Int. 116, 637–654.

JICA, 2000. The study on seismic microzonation of the Greater Tehran in the Islamic Republic of Iran. Draft Final report. Japan International Cooperation Agency, Pacific Consultant International, 3 vols., in corporation with Center for Earthquake and Environmental Studies of Tehran, and Tehran Municipality.

Johnson, A.M., Fleming, R.W., Cruihshank, K.M., 1994. Shear zones formed along long straight traces of fault zones during the 28 June 1992 Landers, California, earthquake. Bull. Seismol. Soc. Am. 84 (3), 449–510.

Jomhuri Eslāmi. Persian newspaper, Tehrān, Iran.

Joneidi, F., 1987. Neyshābur. In: The Iranian Cities, Ed. Mohammad Yusef Kiyāni, 362–382. *Vezārat Farhan va Ershād Eslāmi*, 1366, Tehran (in Persian).

Joneidi, F., 2008. Shāhnāmé Ferdowsi. Nashr Balkh/Bonyād Neyshābur Publ., Tehran, 6 vols. (in Persian), 1387, Tehran (in Persian).

Jovaini, 'Alā al-Din Abolfazl 'Atā Malek, 1260. The Tārikh-I Jahān-gushā of 'Alāu'd-Din 'Atā Malik-i-Juwayni [Tārikh-e Jahān-Goshā-ye Jovaini; The History of the World Conqueror (Changiz Khān); 658/1260]. Edited from seven MSS. by Mirzā Muhammad of Qazwin [Qazvini], 3vols. Leiden, and E.F.W. Gibb Memorial Series, 16, London, 1912 (i), 1916 (ii), and 1937 (iii). Transl. J.A. Boyle, The History of the World Conquerer by Juvayni, 2 vols., Manchester 1958. Also Mohammad Qazvini (Ed.), 'Elmi Publication, Tehran (in Persian).

Justi, F., 1864. Handbuch der Zendsprache. vol. I. F.C.W. Vogel, Trubner & Co., London, p. 373

Kabābi Bandar 'Abbāsi, Mohammad 'Ali Sadid al-Saltaneh, 1963. Bandar 'Abbās va Khalij-e Fārs (A'alām al-Nās fi Ahvāl Bandar 'Abbās) [Bandar 'Abbās and the Persian Gulf]. Ed. Ahmad Eqtedāri and 'Ali Setāyesh. Ebn-e Sina Publ., Tehran, 1342/1963, 735+95 pp. (in Persian). Also Ed. Sayyed Jalāl al-Din Tehrāni (3rd v., 1314/1935), Kolāleh Khāvar Publ., Tehrān (in Persian).

Kāboli, M.'A., 2000. Barresi-hāyeh Bāstanshenāsi-e Qomrud [Archaeological Survey at Qomrud]. Iranian Cultural Heritage Organization, Deputy for Research, Archaeological Research Centre, 105 (in Persian).

Kadinsky-Cade, K., Barazangi, M., 1982. Seismotectonics of Southern Iran: the Oman Line. Tectonics 1, 389–412.

Kadinsky Cade, K., Reilinger, R., Isacks, B., 1985. Surface deformation associated with the November 23, 1977 Caucete, Argentina earthquake sequence. J. Geophys. Res. 90, 12701–12712.

Kagan, Y.Y., Jackson, D.D., 1991. Long-term earthquake clustering. Geophys. J. Int. 104, 117–133.

Kagan, Y., Knopoff, L., 1976. Statistical search for non-random features of the seismicity of strong earthquakes. Phys. Earth Planet. Inter. 12, 291–318.

Kahle, J.E., Wills, C.J., Hart, E.W., Treiman, J.A., Greenwood, R.B., Kaumeyer, R.S., 1988. Preliminary report: surface rupture, superstition Hills earthquakes of November 23 and 24, 1987, Imperial County, California. Calif. Geol. 41, 75–84.

Kambakhsh-Fard, S., 1994. The Anahita Temple Kangavar; Archaeological Excavations and Survey: The Reconstruction and Architectural Restoration of the Nahid Temple and Taghe-Gera. vol. 23. Iranian Cultural Heritage Organization, 405 pp., Tehran (in Persian).

Kanamori, H., Anderson, D.L., 1975. Theoretical basis of some empirical relations in seismology. Bull. Seismol. Soc. Am. 65 (5), 1037–1095.

Kaphadarian, K., 1946. Ohannavank and Its Inscriptions. Akademia Publishing House, Yerevan, (in Armenian).

Kaphadarian, K., 1975. Medieval Monuments and Waft Inscriptions. Akademia Publishing House, Yerevān, (in Armenian).

Karagaranbafghi, F., Foeken, J.P.T., Guest, B., Stuart, F.M., 2012. Cooling history of the Chapedony metamorphic core complex, Central Iran: implications for the Eurasia–Arabia collision. Tectonophysics 524–525, 100–107.

Karaji, Abu Bakr Mohammad ibn al-Hassan al-Hasseb, 953–1019. Estekhrāj Ābhāyeh Penhāni ['The Extraction of Hidden Waters (groundwater) to the surface']. Persian translation from Arabic by Hossein Khadiv-Djam, 1345/1966, *Enteshārāt Boyād Farhang Irān* [Iran Cultural Foundation], 8, Science in Iran, 2, Tehran. Groundwater Exploration; The Oldest Hydrology Text Book, by Aboubakr Karaji English translation from Persian by Nour Zaghi, June 2007. http://www.uscashagent.com/groundwater-exploration.htm.

Karakhanian, A.S., 1993. Active faults of the Armenian Upland. In: Proc. Scientific Meeting Seismic Protection, Venice 12–13 July 1993, Palazzo Balbi, Regione Veneto, pp. 88–92.

Karakhanian, A., Abgarian, Y., 2004. Evidence of historical seismicity and volcanism in the Armenian Highland (from Armenian and other sources). Ann. Geophys. 47 (2/3), 793–810.

Kārang, 'A.'A., 1958 (tr.). Tārikh-e Tabriz [History of Tabriz]. Translated from V. Minorsky's article. Ketāb Forushi Tehrān Publication, Tabriz, 1337/1958 (in Persian).

Karapetian, N.K., 1991. Earthquakes of the Armenian Highlands (Seismic Setting). University of Southern California, Dept. Civil Eng., Los Angeles, CA, 234 pp. (English translation of the 1986 publication in Russian).

Karimān, H., 1977. Qasrān (Kuhsārān), Mabāhes-e Tārikhi va Joghrāfiyā'i va Ejtemā'i va Mazhabi va Ātashgāh-e Mantaqeh-ye Kuhestāni-ye Ray-e Bāstān va Tehrān-e Konuni, 1356/1977. Anjoman Āthār Melli Publ., 137. In: vol. 1, Tehran, 591 pp. (in Persian).

Karimpour, M.H., Stern, C.R., Lang Farmer, G., Moradi, M., 2009. Geology and petrology of two distinct types of granitoids, Eastern Najmabad, Ghonabad, Iran. In: Geological Society of America Portland Meeting, Abstract.

Karimpour, M.H., Stern, C.R., Farmer, L., 2010. UPb-Th (zircon) geochronology, Rb-Sr & Sm-Nd isotopic composition and petrogenesis of Dehnow Kuhsangi Paleo-Tethys Diorite-Granodiorite, Mashhad, Iran. J. Asian Earth Sci. 37, 384–393.

Karimpour, M.H., Lang Farmer, G., Stern, C.R., Salati, E., 2011. U-Pb zircon geochronology and Sr-Nd isotopic characteristic of Late Neoproterozoic Bornaward granitoids (Taknar zone exotic block), Iran. J. Crystallogr. Mineral. 19 (1), 1–18, Spring 1390/2011.

Karimzadeh, S., Cakir, Z., Osmanŏglu, B., Schmalzle, G., Miyajima, M., Amiraslanzadeh, R., Djamour, Y., 2013. Interseismic strain accumulation across the North Tabriz Fault (NW Iran) deduced from InSAR time series. J. Geodyn. 66, 53–59.

Kārnāmak Artakhshir Papakan, 272–273. Book of Deeds of Ardeshir, Son of Bābag, a Pahlavi novel thought to have been written in the reign of the Sāsānian King Hormizd I, ca. AD 272/3. Avesta.rog. Sanjana, D.D.P., 1896 (English Trans.). Education Society's Team Press, Byculla, Bombay. Ed. Anklesaria, 1935; Ed. Mashkur, 1950. Also see Markus-Takeshita, 2001; Shahbāzi, 2003; Shāki, 2003; Cereti, 2012.

Karnik, V., 1971. Seismicity of the European Area. Part 2. Academia Publ. House Czechoslov. Acad. Sci., 218 pp.

Kasak, E., Veede, R., 2001. Understanding planets in ancient Mesopotamia. Folklore 16, 7–35.

Kasravi, A., 1929. Shahriyārān-e Gomnām, Tehran 1308/1929, third ed. Amir Kabir Publ., Tehran, 1353/1974.

Kasravi, A., 1948. Karvand-e Kasravi, Majmu'eh-ye 78 Resāleh va Goftār az Ahmad Kasravi. [Collection of 78 Articles of Ahmad Kasravi]. In: Yahya Zokā' (Ed.), Ketābhāyeh Jibi-Franklin Publication, Tehran, 1327/1948, 1335/1956, and 1352/1973 (in Persian).

Kayhān, 1968. Dasht-e Bayaz Earthquake Report. Kayhān evening newspaper, Tehran, 2 September 1968 (in Persian).

Kayhān, 1972. Qir-Kārzin Earthquake Reports. Kayhān evening newspaper, Tehran, 15 April 1968 (in Persian).

Kayhān. Evening Persian newspaper in Tehran, Iran.

Kayhān International. Iranian newspaper in English, Tehran, Iran.

Kenner, S.J., Simons, M., 2005. Temporal clustering of major earthquakes along individual faults due to post-seismic reloading. Geophys. J. Int. 160, 179–194.

Kervran, M., Labbaf Khaniki, R.A., 2008. New researches at Nishapur. A report on the first three seasons of excavation of the ICAR-CNRS-Louvre Museum, 2008, Tehran. Iranian Center of Archaeological Research, Tehran.

Kervran, M., Labbaf-Khaniki, M., 2008. New researches at Nishabur. A report on the first three seasons of excavation of the ICAR-CNRS-Louvre Museum, 2008. Iranian Center of Archaeological Research, Tehran, 250 pp.

Khachaturian, A., 1974. Inscriptions et histoire des églises Arméniennes. Milan.

Khāfi, Majd al-Din, 1336. Rauzeh-ye Khold, 737/1336. Mohammad Farokh (Ed.), Tehrān, 1346/1967. Abbās 'Ali Vafāee (Ed.), Sokhan Publication, Tehran (in Persian).

Khāfi, Majd al-Din, 1342. Rauzeh-ye Khold, 743/1342. Ed. M. Farokh, Tehrān, 1346/1967.

Khalaf Tabrizi, M.H., 1632. See Borhan-e Qate'.

Khalifeh Neyshāburi, 1291–1349. See Hākem Neyshāburi, Abu 'Abd Allāh, 998.

Khamiri (Mosāfer), Mollā Solaymān. See Nurbakhsh, 1990.

Khandmir, Ghiyāth al-Din ibn Hamal al-Din al-Hussaini, 1521/928. Tārikh Habib al-Sayr fi Akhbār Afrad Bashar, 828/1521. Ed. S.M. Dabir Siyaqi, Khayyām Publ., Tehran, 4 vols., 1954/1333.

Khanikoff, N., 1861 (in Abivh, H., 1861). Memoire sur la Parties Méridinale de l'Asie Centrale. Imprimerie de L. Martinet, Paris, 234 pp.

Khanikoff, N., 1858 (in Abich, H. 1858). Tremblement de terre observe' a' Tabriz en Septembre 1856, Notice physiques et geographiques de M. Khanykof sur l'Azerbeidjan, communiqees par H. Abich (lu le 17 Janvier, 1857). Bull. Cl. Phys. Math. Acad. Imp. Sci. St. Petersburg, 16(22), 337–352.

Khanlarian, L.A., 1982. Svedeniia Arakela Davrizhshogo o sovremennom emu messianskom dvizheni [Information Supplied by Arak'el of Tabriz Concerning a Messianic Movement Which Occurred in His Time]. *Kavkaz I Vizantiia* [The Caucasus and Byzantium], III, Erevan.

Khāqāni Shervāni, 1121–1190. Divān-e Khāqāni-ye Shervāni [Collection of Poems]. Ed. Sayyed Ziyā'al-Din Sajjādi, Zavvār Publ., Tehran (in Persian).

Khosravi, M.R., 1987. Joghrāfiyāyeh Tārikhiyeh Velāyat-e-Zāveh: Pazuheshi dar Joghrāfiyāyeh Tārikhiyeh Torbat Haydariyeh, Khāf, Zauzan, Rokh, va Mahvelāt [Historical Geography of Zaveh; A study of Historical Geography of Turbat-e Haydariyeh, Khāf, Zuzan, Rukh, and Mahvelāt], 1366/1987. Āstān Qods Razavi Publ., 55, 470 pp. (in Persian).

Khosravi, Kh., 1990. Enfejar-e Jam'iyat (Shahri va Rusta'ee). Āyandeh 16 (9–12), 104–113, Azar-Esfand 1369, Tehran (in Persian).

King, G., Soufleris, C., Berberian, M., 1981. The source parameters, surface deformation and tectonic setting of three recent earthquakes: Thessaloniki (Greece); Tabas-e-Golshan (Iran); and Carlisle (U.K.). Disasters 5 (1), 36–46.

King, G.C., Stein, R.S., Lin, J., 1994. Static stress changes and the triggering of earthquakes. Bull. Seismol. Soc. Am. 84, 935–953.

King, L.W., 1915. The Code of Hammurabi (1792–1750 B.C.). Trans. Leonard William King. http://www.general-intelligence.com/library/hr.pdf.

Kinnier Wilson, J.V., 1979. The Rebel Lands; An Investigation into the Origins of Early Mesopotamian Mythology. Cambridge University Press, 148 pp.

Kleiss, W., 1969. Bericht uber zwei Erkundungsfahrten in Nordwest Iran. *Archeologische Mitteilung aus Iran,* Neue Folge, Band 2, 85–119, Berlin.

Kleiss, W., 1971. The Monastery of St. Thaddeus (XII–XIXth centuries). Documents of Armenian Architecture, 4, S. Thadei Vank. Edizioni Ares, Milano, Italy, pp. 5–8.

Kondorskaya, N.V., Shebalin, N.V. (Eds.), 1977. Novii Katalog Silnikh Zemletriasenii na Territori S.S.S.R, Moscow.

Kondorskaya, N.V., Shebalin, N.V. (Eds.), 1982. New Catalog of Strong Earthquakes in the U.S.S.R. From Ancient Times Through 1977. World Data Center A for Solid Earth Geophysics, Report SE-31. NOAA, EDIS, Boulder, CO, (English Translation), 608 pp.

Kopp, C., Fruehnb, J., Flueha, E.R., Reichert, C., Kukowski, N., Bialas, J., Klaeschena, D., 2000. Structure of the Makran subduction zone from wide-angle and reflection seismic data. Tectonophysics 329, 171–191.

Kotzebue, M. von, 1819. Narrative of a Journey into Persia in the Suite of the Imperial Russian Embassy, in the Year 1817. Longman, Hurst, Rees, Orme, and Brown.

Kramer, S.N., 1944. Sumerian Mythology. Memoirs of the American Philosophical Society, Philadelphia, XXI, 125 pp.

Kreemer, C., Holt, W.E., Haines, A.J., 2003. An integrated global model of present-day plate motions and plate boundary deformation. Geophys. J. Int. 154, 8–34.

Laane, J.L., Chen, W.-P., 1989. The Makran earthquake of 1883 April 18—a possible analogue to the Puget Sound earthquake of 1965? Geophys. J. Int. 98 (1), 1–9.

Labbāf Khāniki, R., 1988. Masjed-e-Robāt-e-Ziyārat [The Robāt Ziyārat Mosque], Athar, Sazeman Mirath Farhangi Keshvar, 15 & 16, 164-172 (in Persian).

Labbāf Khāniki, R., 1997. Barresi va shenāsāee qanavāt Gonābād (Exploration of the Qanāts of Gonābād). Gozāresh-hāyeh Bāstānshenāsi 1, 271–298, Sāzemān Mirāth Farhangi Keshvar-83, Mehr 1376, Tehran (in Persian). Archaeological Reports of Iran, 1, Iranian Cultural Heritage Organization-83, Deputy Research, Archaeological Research Center, Tehran, Iran.

Labbāf Khāniki, R., 2013. Gozāresh Ejmāli Motāle'at Bāstānshenāsi dar Qā'en [Preliminary archaeological study report in Qā'en]. In: National Archaeological Conference. Birjand University, 1392/2013, 7p., Birjand, Iran (in Persian).

Labbāf Khāniki, R.A., Labaf Khaniki, M., 2012. Me'māri Kohandezh Neyshābur dar Daureh-yeh Sāsāni Sasanid Architecture of Kohandezh Niyshapur In: Fahimi, H., Alizādeh, K. (Eds.), Nāmnāmeh: Papers in Honour of Massoud Āzarnoush. Nashr Irān Negār, Tehran, pp. 337–346, (in Persian).

Landgraf, A., Ballato, P., Strecker, M.R., Friedrich, A., Tabatabaei, S.H., Shahpasandzadeh, M., 2009. Fault-kinematic and geomorphic observations along the North Tehran Thrust and Mosha Fasham Fault, Alborz mountains Iran: implications for fault system evolution and interaction in a changing tectonic regime. Geophys. J. Int. 177, 676–690.

Landgraf, A., Zielke, O., Arrowsmith, J.R., Ballato, P., Strecker, M.R., Schildgen, T.F., Friedrich, A.M., Tabatabaei, S.H., 2013. Differentiating simple and composite tectonic landscapes using numerical fault slip modeling with an example from the south central Alborz Mountains, Iran. J. Geophys. Res.. 118, 1–14. http://dx.doi.org/10.1002/jgrf.20109.

Larsson, L., 2013. Pa jakt efter Persien: 100 ar efter Sven Hedin. Natl. Geogr. Sverige 5, 20–43.

Lasemi, Y., 1995. Platform carbonates of the upper Jurassic Mozduran Formation in the Kopet Dagh basin, NE Iran—facies, paleoenvironment and sequences. Sediment. Geol. 99, 151–164.

Layard, A.H., 1846. A description of the province of Khuzistan. J. R. Geogr. Soc. Lond. 16, 1–105.

Layard, A.H., 1887. Early Adventures in Persia, Susiana and Babylonia, 2 vols.. John Murray, London.

Le Dortz, K., Meyer, B., Sebrier, M., Nazari, H., Braucher, R., Fattahi, M., Benedetti, L., Foroutan, M., Siame, L., Bourles, D., Talebian, M., Bateman, M.D., Ghoraishi, M., 2009. Holocene right-slip rate determined by cosmogenic and OSL dating on the Anar fault, central Iran. Geophys. J. Int.. 179, 700–710. http://dx.doi.org/10.1111/j.1365-246X.2009.04309.x.

Le Dortz, K., Meyer, B., Sébrier, M., Braucher, R., Bourlès, D., Benedetti, L., Nazari, H., Foroutan, M., 2012. Interpreting scattered in-situ produced cosmogenic nuclide depth-profile data. Quat. Geochronol. 11, 98–115.

Lenson, G.J., Suggate, R.P., 1968. Preliminary reports on the Inangahua earthquake, New Zealand. DSIR Res. Bull. 193, 17–36.

Le Strange, G., 1905. The Lands of the Eastern Caliphate; Mesopotamia, Persia, and Central Asia Form the Moslem Conquest to the Time of Timur. Cambridge; and Frank Gass, London. Also Translated into Persian by Mahmood Erfān, Publ. Sherkat Enteshārāt 'Elmi va Farhangi, 74, Tehran, 1337/1958 (first ed.), 1364/1985 (second ed.), 610 pp.

Lettis, W.R., Wells, D.L., Baldwin, J.N., 1997. Empirical observations regarding reverse earthquakes, blind thrust faults, and Quaternary deformation; are blind thrust faults truly blind? Bull. Seismol. Soc. Am. 87, 1171–1198.

Lézine, A.-M., Tiercelin, J.-J., Robert, C., Saliége, J.-F., Cleuziou, S., Inizan, L.-L., et al., 2007. Centennial to millennial-scale variability of the Indian monsoon during the early Holocene from a sediment, pollen and isotope record from the desert of yemen. Palaeogeogr. Palaeoclimatol. Palaeoecol. 243, 235–249.

Levy, R. (Trans.), 1967. The Epic of the Kings; Shah-Mama, the National Epic of Persia by Ferdowsi. The Royal Institute of Publication, Tehran, printed by Routledge & Kegan Paul, London, 423 pp.

Lewisohn, L., 1998. An introduction to the history of modern Persian Sufism, Part I: the Niʿmatullāhī order: persecution, revival and schism author(s). Bull. Sch. Orient. Afr. Stud. 61 (3), 437–464, University of London.

Liddall, H.S., 1961. Greek-English Lexicon. Clarendon Press, Oxford, England.

Lincoln, B., 1982. The earth becomes flat—a study of Apocalyptic imagery. Comp. Stud. Sociol. Hist. 25 (1), 136–153.

Lockhart, L., 1960. Persian Cities. Luzac, 188 pp.

Lyberis, N., Manby, G., 1999. Oblique to orthogonal convergence across the Turan block in the post-Miocene. Am. Assoc. Pet. Geol. Bull. 83 (7), 1135–1160.

Mackenzie, D.N., 1964. Zoroastrian Astrology in the Bundahišn. BSOAS 27, 511–529.

McKenzie, D.A., 1915. Myths of Babylonia and Asyria. sacret-texts.com (Chapter XV).

Madani, M., 1977. A study of the sedimentology, stratigraphy and regional geology of the Jurassic rocks of eastern Kopet-Dagh, NE Iran. Thesis, Petroleum Geology Section, Imperial College, London.

Maggi, A., Jackson, J.A., Priestley, K., Baker, C., 2000a. A re-assessment of focal depth distribution in southern Iran, the Tien Shan and northern India; do earthquakes really occur in the continental mantle? Geophys. J. Int. 143, 629–661.

Maggi, A., Jackson, J.A., McKenzie, D., Priestley, K., 2000b. Earthquake focal depths, effective elastic thickness, and strength of the continental lithosphere. Geology 28, 495–498.

Māheri, M.R., 1991. Response of large structure to Manjil earthquake of June 1990. In: Proceedings of 1st International Conference on Seismology & Earthquake Engineering, IIEES, Tehran, May 27–29, 1991. vol. 2, pp. 1019–1028.

al-Majlisi, 'Allāmah, Mullā Mohammad Bāqer Majlesi, 1668. Hilyāt ul-Muttaqin' [The Adornment of the God-Fearing']. Tehran. shiabooks.ir/1389/09/22/post-136/.

Malcom, J., 1829. History of Persia. From The Most Early Period To The Present Time, vol. 2. John Murray and Longman and Co., London.

Maleki, V., Shomali, Z.H., Hatami, M.R., 2012. Relocation of aftershocks of Mohamad Abad earthquake December 20, 1010, M_N=6.5 using nonlinear method. Iran. J. Geophys. 6 (4), 96–111.

Mallet, R., 1850, 1852, 1853, 1854, 1858. Report on the facts of earthquake phenomena. British Assoc. Advancement Sci. (BASS), London, combined edition.

Mallowan, M.E.L., 1966. Nimrud and Its Remains, 2 vols.. Dodd, Mead, London.

al-Malek (Rezā'i), Mirzā Tebyān. In Nakhjavāni (1964) and Zokā' (1989).

Mansuri, F., 1991. The Earthquakes of Tabriz. Āyandeh, 9–12, Āzar-Esfand, 1370, 770–781 (in Persian). Also The Earthquakes of Tabriz (continuation). Āyandeh, 1–6, Farvardin-Shahrivar, 1370, 164–165 (in Persian).

al-Maqrizi, Ahmad ibn 'Ali Taqi al-Din, 1426. Al-Suluk fi Ma'rifat Duwal al-Muluk, Cairo, 4 vols., 1934–1971.

Marco, S., 2008. Recognition of earthquake-related damage in archaeological sites: examples from the Dead Sea fault zone. Tectonophysics. 453, 148–156. http://dx.doi.org/10.1016/j.tecto.2007.04.011.

Marco Polo. See Polo and Pisa.

Mārdiros Khālifā. In Hakobyan, 1951.

Markaz-e Āmār-e Irān, 1956–1986. Sarshomāri-ye 'Omumi-ye Nofus va Maskan. Tehran, 1335/1956; 1345/1966; 1355/1976; 1365/1986. sci.org.ir; Amar.org.ir.

Markaz-e Āmār-e Irān, 1978. Population Growth Survey of Iran. Final Report 1973–1976, Tehran, 1357/1978. sci.org.ir; Amar.org.ir.

Markaz-e Āmār-e Irān, 1988. Sarshomāri-e Ejtemā'i-Eqteṣādi-e 'Āšāyer-e Kūchandeh.1366, Tehran, 1367/1988. sci.org.ir; Amar.org.ir.

Markus-Takeshita, K.I., 2001. From Iranian myth to folk literature: the legend of the Dragon-Slayer and the Spinning Maiden in the Persian Book of the Kings. Asian Folk. Stud. 60 (1), 203–214.

Mashkur, M.G., 1950. The Karnamak i Artakhshir i Papakan: Records of Ardashir Babakan, the Founder of the Sasanian Dynasty in Iran; Containing the Original Pahlavi Text, Glossary, Persian Translation, Comparison with Shâhnâmeh, Historical Notes and Appendices. Anjoman Āthar Melli, Tehran, 1971 (in Persian).

Mashkur, M.J., 1973. Tārikh-e Tabriz tā Pāyān-e Qarn-e Nohom-e Hejri The History of Tabriz Until the Ninth Century H. , Tehran, 1352 (in Persian).

Massé, H., 1938. Croyances et Coutumes Persanes suivies de Contes et Chansons Populaires, 2 vols.. Librarie Orientale et Améicaine, Maisonneuve, Paris.

Masson, F., Djamour, Y., van Gorp, S., Chery, J., Tatar, M., Tavakoli, F., Nankali, H., Vernant, P., 2006. Extension in NW Iran driven by the motion of the south Caspian Basin. Earth planet. Sci. Lett. 252, 180–188.

Masson, F., Anvari, M., Djamour, Y., Walpersdorf, A., Tavakoli, F., Daignieres, M., Nankali, H., Van Gorp, S., 2007. Large-scale velocity field and strain tensor in Iran inferred from GPS measurements: new insight for the present-day deformation pattern within NE Iran. Geophys. J. Int. 170, 436–440.

Massonnet, D., Rossi, M., Carmona, C., Adragna, F., Peltzer, G., Feigl, K., Rabaute, T., 1993. The displacement field of the Landers earthquake mapped by radar interferometry. Nature 364, 138–142.

al-Mas'udi, Abol Hassan 'Ali ibn Hossein, 956. Al-Tanbih wa al-Ashrāf [Notification and Review]. Persian translation by Abolqāsem Pāyandeh, Sherkat Enteshārāt 'Elmi va Farhangi, Tehran, 1365/1986. Also Ed. De Goeje, 1894, Bibliotheca Geographoium Arabicorum, Leiden. Also Beirut, 1961.

Mas'udi Marvazi, Sharaf al-Din Mohammad (second half of the 12th century AD). Resalehyeh Āthār-e 'Alavi'. Ed. Mohammad Taqi Dāneshpazhu. Farhang Irānzamin, Tehrān, 1337/1958 (in Persian).

Matheson, S.A., 1972. Persia: An Archaeological Guide, first ed. Faber and Faber Ltd., London, 330 pp.

Mattei, M., Cifelli, F., Muttoni, G., Zanchi, A., Berra, F., Mossavvari, F., Eshraghi, S.A., 2012. Neogene block rotation in central Iran: evidence from paleomagnetic data. Geol. Soc. Am. Bull.. 124, 943–956. http://dx.doi.org/10.1130/B30479.1.

McBurney, C.B.M., 1968. The cave of Ali Tappeh and the Epi-Paleolithic in NE Iran. Prehist. Soc. 34, 385–413, Liverpool.

McCaffrey, R., 1991. Slip vectors and stretching of the Sumatran fore arc. Geology 19, 881–884.

McCaffrey, R., 1992. Oblique plate convergence, slip vectors, and forearc deformation. J. Geophys. Res. 97, 8905–8915.

McCaffrey, R., Nabelek, J., 1998. Role of oblique convergence in the active deformation of the Himalayas and southern Tibet plateau. Geology 26, 691–694.

McCalpin, J.P., Nishenko, S.P., 1996. Holocene paleoseismicity, temporal clustering, and probabilities of future large ($M>7$) earthquakes on the Wasatch fault zone, Utah. J. Geophys. Res.. 0148-0227101, 6233–6253. http://dx.doi.org/10.1029/95JB02851.

McCall, G.J.H., 1985. Area Report, East Iran Project, Area No. 1. vol. 57. Geol. Surv. Iran, 634 pp.

McCall, G.J.H., 1997. The geotectonic history of the Makran and adjacent areas of southern Iran. J. Asian Sci. 15, 517–531.

McCall, G.J.H., 2003. A critique of the analogy between Archaean and Phanerozoic tectonics based on regional mapping of the Mesozoic-Cenozoic plate convergent zone in the Makran, Iran. Precambrian Res. 127, 5–17.

McCall, G.J.H., Kidd, R.G.W., 1992. The Makran, southeastern Iran: the anatomy of a convergent plate margin active from the Cretaceous to Present. In: Leggett, J.K. (Ed.), Trench-Fore-arc Geology. vol. 10. Geological Society, London, Spec. Publ., pp. 387–397

McClusky, S., Balassanian, S., Barka, A., Demir, C., Ergintav, S., Georgiev, I., Gurkan, O., Hamburger, M., Hurst, K., Kahle, H., Kastens, K., Kekelidze, G., King, R., Kotzev, V., Lenk, O., Mahmoud, S., Mishin, A., Nadariya, M., Ouzounis, A., Paradissis, D., Peter, Y., Prilepin, M., Reilinger, R., Sanli, I., Seeger, H., Tealeb, A., Toksöz, M.N., Veis, G., 2000. Global positioning system constraints on plate kinematics and dynamics in the eastern Mediterranean and Caucasus. J. Geophys. Res. 105, 5695–5719.

McClusky, S.M., Reillinger, R., Mahmoud, S., Ben Sari, D., Tealeb, A., 2003. GPS constraints on Africa (Nubia) and Arabia plate motions. Geophys. J. Int. 155 (1), 126–138.

McGuire, W.J., Griffith, D.R., Hancock, P.L., Stewart, I.S. (Eds.), 2000. The Archaeology of Geological Catastrophes. In: vol. 171. Geological Society, London, Spec. Publ., 417 pp.

McKenzie, D.P., 1972. Active tectonics of the Mediterranean region. Geophys. J. R. Astron. Soc. Lond. 30, 109–185.

McLaren, M.K., Hardebeck, J.L., van der Elst, N., et al., 2008. Complex faulting associated with 22 December 2003 M_w San Simon, California, earthquake, aftershocks, and postseismic surface deformation. Bull. Seismol. Soc. Am. 98, 1659–1680.

Meghraoui, M., Gomez, F., Sbeinati, R., Van der Woerd, J., Mounty, M., Darkal, A.N., Raswan, Y., Layyous, I., Al-Najjar, H., Darawcheh, R., Hijazi, F., Al-Ghazzi, R., Barazangi, M., 2003. Evidence for 830 years of seismic quiescence from paleoseismology, archaeoseismology, and historical seismicity along the Dead Sea fault in Syria. Earth Planet. Sci. Lett. 201, 35–52.

Mehdi Soheili. Collection of Poems. Tehran. sohaili.com/Persian.htm.

Mehryar, M., Kabiri, A., 1986. Preliminary report on the Delazian (Cheshmeh Shaikh) archaeological site. *Āthār,* Sāzeman Melli Hefazat Āthār Bāstāni Irān, 12, 13, 14, Esfand 1365, Tehran, 3–46 (in Persian).

Melville, C.P., 1980. Earthquakes in the history of Nishapur. Iran J. Br. Inst. Persian Stud. XVIII, 103–122.

Melville, C.P., 1981. Historical Monuments and Earthquakes in Tabriz, vol. XIX. Iran J. Br. Inst. Persian Stud., London, pp. 159–171

Meshkāti, N., 1970. Fehrest-e Banāha-ye Tārikhi va Amāken-e Bāstāni-e Irān Catalogue of Historical Buildings and Ancient Places of Iran Sāzemān Melli Hefāzat Āthār Bāstāni Irān, Tehran, 370 pp., 1349 (in Persian).

Messner, C.A. (Trans.), 1954. See Massé, H., 1938.

Meyer, B., LeDortz, K., 2007. Strike-slip kinematics in Central and Eastern Iran: estimating fault slip-rates averaged over the Holocene. Tectonics. 26http://dx.doi.org/10.1029/2006TC002073, TC5009.

Meyer, B., Mouthereau, F., Lacombe, O., Agard, P., 2006. Evidence of Quaternary activity along the Deshir Fault: implication for the Tertiary tectonics of Central Iran. Geophys. J. Int.. 164, 192–201. http://dx.doi.org/10.1111/j.1365-246X.2005.02784.x.

MI/DPS, 1961. National and Province Statistics of the First Census of November 1956. vol. 1. Ministry of Interior, Department of Public Statistics, Tehran.

MI/DPS, 1962. National and Province Statistics of the First Census of November 1956. vol. 2. Ministry of Interior, Department of Public Statistics, Tehran.

Mikhitār Gosh, 1130–1213. Colophon (The Aghuanian Chronicle). Rbedrosian.com.

Milkereit, C., Grosser, H., Wang, R., Wetzel, H.-U., Woith, H., Karakisa, S., Zünbül, S., Zschau, J., 2004. Implications of the 2003 Bingöl earthquake for the interaction between the North and East Anatolian faults. Bull. Seismol. Soc. Am.. 94 (6), 2400–2406. http://dx.doi.org/10.1785/0120030194v.

Miller, N.F., Abdi, K., 2003. Yeki Bud, Yeki Nabud; Essays on the Archaeology of Iran in Honor of William M. Sumner. Monograph 48, Regents of the University of California, The Cotsen Institute of Archaeology of California, Los Angeles, 340 pp.

Milne, J., 1911. Catalogue of destructive earthquakes AD 7 to 1899. Report of the 81st Meeting of the British Association for the Advancement of Science, vol. 80. BAAS, Portsmouth, UK, pp. 694–740.

Minorsky, V., 1951. Caucasica II. The Georgian Maliks of Ahar. Bull. Sch. Orient. Afr. Stud. 13 (4), 868–877, University of London.

Minorsky, V., 1956. Pūr-i Bahā's 'Mongol' ode (Mongolica, 2). Bull. Sch. Orient. Afr. Stud. 18 (2), 261–278, University of London.

Mir Baqā' Badakhshi, ca. 1664. Tazkareh-ye Mirzā Mohammad Tāher Nasrābādi Esfahāni (born in 1618), Ed. Vahid Dastgerdi, 1317/1938 (in Persian).

Moazami Goudarzi, Ghaderi-Tafreshi, 1977. Preliminary report on the Qaen, Khorassan, earthquake of November 7, 1976. In: Proc. CENTO Seminar on the Recent Advances in Earthquake Hazard Minimization, 14–17 November, 1976. vol. 70. Plan and Budget Organization, Tehran, pp. 272–278.

Moazzen, M., Oberhänsli, R., 2008. Whole rock and relict igneous clinopyroxene geochemistry of ophiolite related amphibolites from NW Iran—implications for protolith nature. Neues Jahrb. Mineral. 185 (1), 51–62.

Moazzen, M., Oberhänsli, R., Hajialioghli, R., Möller, A., Bousquet, R., Droop, G.T.R., Ahangiri, A., 2009. Peak and post-peak P–T conditions and fluid composition for scapolite–clinopyroxene–garnet calc-silicate rocks from the Takab area, NW Iran. Eur. J. Mineral. 21, 149–162.

Moazzen, M., Hajialioghli, R., Moller, A., Droop, G.T.R., Oberhansli, R., Altenberger, U., Jahangiri, A., 2013. Oligocene partial melting in the Takab metamorphic complex, NW Iran: evidence from in situ U-Pb geochronology. J. Sci. Iran 24 (3), 217–228.

Modi, J.J. (Trans.), 1903. Jamasp Namak [The Book of Jamaspi]. K.R. Cama Oriental Institute, Bombay, India. Also avesta.org.

Mogi, K., 1985. Earthquake Prediction. Academic Press, Tokyo, 355 pp.

Mohajer, G.A., Pierce, G.R., 1963. Qazvin, Iran, earthquake. Bull. Am. Pet. Geol. 47 (10), 1878–1883.

Mohajer-Ashjai, A., Behzadi, H., Berberian, M., 1975. Reflections of the rigidity of the Lut Block and recent crustal deformation in eastern Iran. Tectonophysics 25, 281–301.

Mohajer-Ashjai, A., Nowroozi, A.A., 1979. The Tabas earthquake of September 16, 1978 in east-central Iran; a preliminary filed report. Geophys. Res. J. 6 (9), 689–692.

Mohājer-Ashjai, A., Nowroozi, A.A., Taghizadeh, Gh., Zohurian Izadpanah, A.A., 1980. Gozāreshi az dau zelzeleh-ye mozdavaj dar Qā'enāt (1358). vol. 122. IAEA, 60 pp., Azar 1359 (in Persian).

Mohammad Nassir, 1641. Jong-e Dastnevis dar bareh-ye zelzeleyeh 1050 Tabriz [Earthquake Manuscript on the 1641 earthquake in Tabriz]. National Iranian Library, Tehran, (in Persian).

Mo'infar, A.A., 1978. Gozāresh-e Kutāhi az zelzeleyeh Tabas. Nashriyeh Kānun Mohandesin Irān 60, 19–23, Azar 1357, Tehran (in Persian).

Moinfar, A.A., Naderzadeh, A., 1990. An Immediate and Preliminary Report on the Manjil, Iran Earthquake of June 20, 1990. vol. 119. Building and Housing Research Center, Ministry of Housing and Urban Development, 57 pp. (in Persian w/11 pp. in English).

Mo'infar, A.A., Berberian, M., Qorashi, M., ZohuriānIzadpanāh, A.A., Nāderzādeh, A., 1987. Preliminary Seismic Zones of Iran for the Revision of the Second Iranian Code for Seismic Resistant Design of Buildings (Code No. 2800). Research Center of Buildings and Homes (Markaz Tahqiqāt Sākhtemān va Maskan), Rep. No. 74, 32 pp. with a color map. Tehran, Khordād, 1366.

Mojmal al-Tavārikh val Qesās, 1126. By anonymous writer from Asadābād, Hamédān in 520/1126. Ed. Malak al-Sho'arā Bahār, 1318/1939, Chāpkhāneh-ye Kolāleh Khāvar, Tehran, 567 pp. (in Persian).

Molinaro, M., Guezou, J.C., Leturmy, P., Eshraghi, S.A., Frizon de Lamotte, D., 2004. The origin of changes in structural style across the Bandar Abbas syntaxis, SE Zagros (Iran). Mar. Pet. Geol. 21, 735–752.

Molnar, P., Gipson, J.M., 1994. Very long baseline interferometry and active rotations of crustal blocks in the Western transverse ranges, California. Geol. Soc. Am. Bull. 106, 594–606.

Montandon, F., 1957. Les grandes catastrophes en 1956 et 1957. Revue pour l'Etude des Calamites 34, 53–54.

Moqaddasi, Shams al-Din Abu 'Abdollah Mohammad ebn Ahmad ebn Abibakr ebn Shami 985. Ahsan al-Taqāsim fi Ma'refat al-Aqālim (375/985). Translated into Persian by 'Alinaqi Monzavi. Sherkat Mo'alefān va Motarjemān Irān, Tehran, 2 vols., 1361/1982.

Moradi, A.S., Hatzfeld, D., Tatar, M., 2011. Microseismicity and seismotectonics of the North Tabriz fault (Iran). Tectonophysics 506, 22–30.

Morgan, J., 1894. Mission geographique en Perse. Etudes geographiques I–II, Paris.

Morgan, P., 1991. New thoughts on Old HormuzL Chinese ceramics in the Hormuz region in the thirteenth and fourteenth centuries. Iran J. Br. Inst. Persian Stud. XXIV, 67–83.

Mori, J., Wald, D.J., Wesson, R.L., 1995. Overlapping fault planes of the 1971 San Fernando and 1994 Northridge, California earthquakes. Geophys. Res. Lett. 22, 1033–1036.

Morier, J., 1818. A Second Journey Through Persia, Armenia, and Asia Minor, to Constantinople, Between the Years 1810 and 1816. With a Journal of the Voyage by the Brazil and Bombay to the Persian Gulf. Longman, Hurst, Rees, and Brown, London, 435 pp.

Mosanafāt. See Naraqi Kāshāni.

Mostaufi Qazvini, Hamdollāh, 1330. Tarikh-e Gozideh, 730/1330. Ed. 'Abdol Hossain Navā'i, 1387/2008. Amir kabir Publ., Tehran, fifth ed., 816 pp. (in Persian). Also Amir Kabir Publ., Tehran, 1339/1960, 816 pp. (in Persian).

Mostaufi Qazvini, Hamdollāh, 1334–1335. Zafarnāmeh, 735. Hamdallāh Mustaufi-Zafarnāma, Markaz Nashr Dāneshgāhi Irān-Verlag der Osterreichischen Akadmie der Wissenschaften, Tehran, 1377, Vienna, 1999. Also, Ed. Mansureh Sharifizādeh. Makaz Nashr Dāneshgāhi Publ., 1364, Tehran. Also Pazhuheshgāh 'Olum Ensāni Motāle'āt Farhangi Publ., Tehran, 1380, vol. 2.

Mostaufi Qazvini, Hamdollāh, 1340. The Geographical Part of the Nuzhat al-Qulub [Pleasure of the Hearts; 740/1340 AD], Composed by Hamd-Allāh Mustawfi of Qazwin in 740 (1340). English Trans. G. Le Strange, 1919. E.J.W. Gibb Memorial Series, 23, E.J. Brill Imprimerimerie Orientale, London: Leyden, Luzac & Co., 1919, 322 pp.

Mount, V.S., Suppe, J., 1987. State of stress near the San Andreas Fault: implications for wrench tectonics. Geology 15, 1143–1146.

Mouood Web site. Mouood Cultural Institute, Qom. http://mouood.org/content/view/5909/2/).

Mousavi, Z., Walpersdorf, A., Walkerc, R.T., Tavakoli, F., Pathiera, E., Nankali, H., Nilfouroushan, F., Djamour, Y., 2013. Global Positioning System constraints on the active tectonics of NE Iran and the South Caspian region. Earth Planet. Sci. Lett. 377–378, 287–298.

Movses Khorénātsi [Moses of Khoren], 466. Hāiots Bādmutiun [History of the Armenians]. Translated into Modern Armenian, Introduction and Notes by Sd. Malkhasiants. Cairo, 1953. Also, Moses Khorenatsi's History of the Armenians, Trans. and commentary on the literary sources by Robert W. Thomson, Harvard University Press, 1978, Cambridge, MA, USA, 408 pp. Alos, Tārikh-e Armanestān- Movses Khorénātsi [History of Armenia of Movses Khorenatsi]. Translated into Persian by G. Nālbāndiān, 1984. State University of Yerevān, 338 pp.

Mual, S., 1997. Auf den Spuren assyrischer Gelehrsamkeit. Ruperto Carola, Research Magazine of the University of Heidelberg. http://hsozkult.geschichte.hu-berlin.de/tagungsberichte/id=2997. See also, http://prelectur.stanford.edu/lecturers/maul/tracing.html.

Muhammad Khān ibn Musa Khān Durrāni, 1881. Tārikh Sultāni Durrāni [A History of the Dorrāni Dynasty of Afghanistan in Persian], Bombay, 1881, 292 pp.

Muhammad Zaki Khalil, 1953. Al-Resala al0Thaniha l'Abu Dulaf Mas'ar ibn al-Muhalhel alKhazraji. , Matba Jame'a al-Qahera. Rabi' al-Avval sana 1373/ 9 November 1953.

al-Muqaddasi, Muhammad ibn Ahmad, 985. Ahsan al-Taqāsim fi Ma'rifat al-Aqālim [The Best Divisions for Knowledge of the Regions; 375/985]. Ed. M. J. De Goeje, Bibliotheca Geographorum Arabicorum, Leiden, 1877; Leiden, BGA III, 1906. Also English Trans. Basil Anthony Collins, 1995, Ithaca Press, Garnet Education Publishing, UK, 498 pp. Persian Trans. by 'Alinaqi Monzavi, 1361/1982, Sherkat Mo'aléfān va Motarjémān Iran, Tehran, 2 vols., 636 pp. (in Persian).

Murdock, G.P., 1934. Our Primitive Contemporaries. The Macmillan Company, New York, 648 pp.

Musketoff (Mushketov), I.V., 1891. Zemletresenie v' Asterabad: Materialy dlya izucheniya zemletryaseniy Rossii (Materials for a Study of Earthquakes in Russia). Izv. Russ. Geogr. O-Va (Russ. Geogr. Soc.), 27, issues 1, 2; 35. St. Petersburg.

Musketoff (Mushketov), I.V., 1899. Materialy dlya izucheniya zemletryaseniy Rossii (Materials for a Study of Earthquakes in Russia). Imper. Rossii Notes, Ptilozh. Izv. Russ. Geogr. O-Va (Russ. Geogr. Soc.), 35, St. Petersburg.

Musketoff (Mushketov), I.V., and Orloff (Orlov), A., 1893. Katalog Zemletriesenii Rossiskoi Imperii (Catalogue of Earthquakes in the Russian Empire). Zapiski Imp. Russ. Geograf. Obches (Notes, Russ. Geogr. Soc.), 26, St. Petersburg, Russia.

Nabavi, S.M., 1972. Seismicity of Iran. M.Ph. Thesis, Imperial College, University of London.

Nabavi, S.M., 1977. Aspects of the seismic behaviour of Iran—especially the southern Zāgros area. Ph.D. Thesis, Department of Geology, Univ. College, London.

Nabavi, M.S., 1985. Preliminary study of the formation of the Gahar Lake (Oshtoran Kuh, Lorestan). Geol. Surv. Iran, Int. Rep., 42 pp., 1364 (in Persian).

Nāderi, B., 1980. Negāhi gozarā bar Gilān va Āthār-e Tārikhi-yeh ān [A glance on Gilān and its historical monuments]. Āthār, Sāzemān Melli Hefāzat Āthār Bāstāni Irān, 2, 3, 4, 176–184, Tehran, 1359 (in Persian).

Naderi, B., 1989. Masjed-e Jame of Qaen. Āthār, Bull. National Iranian Archaeological Service 10, 103–107, 1359 (in Persian).

Nāder Mirzā, Qājār, 1883-1885. Tārikh va Joghrāfiyā-ye Dār al-Saltaneh-ye Tabriz, be Zamimeh-ye Sharh-e Hāl-e Bozorgān [History and Geography of the Capital City Tabriz], Written by Lesān al-Molk Malek al- Movarekhin, Sephr Lith. Ed., Tehran, 1323/1905 and 1351/1972 (in Persian).

Naiieri, A., Jokar, N., Khademi, M.H., Mirzaei Alavijeh, H., Tabatabei, S.H., 1997. Preliminary report of the Sarein (SW Ardebil) earthquake of February 28, 1997. Building and Housing Research Center, 249, 59 pp., 1375 (in Persian).

Nakamura, T., Suzuki, S., Matsushima, T., Ito, Y., Hosseini, S.K., Gandomi, A.J., Sadeghi, H., Maleki, M., Aghda, S.M.F., 2004. Source fault of the 2003 Bam earthquake, southeastern Iran, inferred from aftershock distribution and its relation to the heavily damaged area: existence of the Arg-e-Bam fault proposed. http://www.gaea.kyushu-u.ac.jp/research/iran2004/paper/GRL2004.html.

Nakata, T., Otsuki, K., Khan, S.H., 1990. Active faults, stress fields, and plate motion along the Indo-Eurasian plate boundary. Tectonophysics 181, 83–95.

Nakhjavāni, H.H., 1964. Mavād al-Tavārikh [Chromatographic Verses], 1343/1964. Adabiyeh Publ., Tehrān, 836 pp. (in Persian).

Nalbant, S., Steacy, S., McCloskey, J., 2006. Stress transfer relations among the earthquakes that occurred in Kerman province, southern Iran since 1981. Geophys. J. Int. 167, 309–318.

Nankali, H.R., 2011. Slip rate of the Kazerun Fault and Main Recent Fault (Zagros, Iran) from 3D mechanical modeling. J. Asian Earth Sci. 41, 89–98.

Napier, G.C., 1876. Collections of journals and reports, 1874, London.

Narāqi, H., 1966. Tārikh-e Ejtemāi'-ye Kāshān The Social History of Kāshān. vol. 42. Tehrān University Publ., Inst. Social Studies, 387 pp. 1345/1966 (in Persian).

Narāqi Kāshāni, Afzal al-Din Muhammad, d. early 13th century. Mosanafāt. Ed. Mojtabā Minavi and Yahyā Mahdavi, Sherkat Enteshārāt Khārazmi Publ., Tehran, 1366/1987 (in Persian).

NAS, 2003. Living on an Active Earth: Perspectives on Earthquake Science. Committee on the Science of Earthquakes, Board on Earth Sciences and Resources, Division on Earth and Life Studies. National Research Council of the National Academies. The National Academic Press, Washington, DC, 418 pp.

Nāser Khosraw Qobadiyani Marvazi, 1052. Safarnāmeh va Zād al-Mosāferin [The Book of Travel/Travel Diary of Naser Khosrau/Travelogue; 444/1052]. Ed. Nāder Varzinpur. Sherkat Sahāmi Ketāb-hāyeh Jibi, Majmu'eh-ye Sokhan-e Pārsi, Tehrān 1350/1971 and 1354/1975, 198 pp. (in Persian). Ed. Mohammad Dabir Siyāqi, Tehran 1355/1976. Also ed. Nader Vazinpur, Tehran, 1350/1971, second ed., 1354 Sherkat Sahāmi Ketābhā-ye Jibi Tehrān Publ., Sokhan-e-Parsi-3, 198 pp. fourth ed., 1358/1979, Amir Kabir Publ., Tehran, 120 pp. Also Ed. & Trans. C. Schefer, as Safar Nameh: Relation du voyage de Nassiri Khosrau en Syrie, en Palestine, en Egypte, en Arabie, et en Perse pendant le annees de l'hegire 437–444/1035–1042. Publication de l'Ecole de Langues orientales vivantes, II (1), Paris, 1881. 1961–1962 reprint of 1881 edition, Berlin.

Nāser Khosraw Qobadiyani Marvazi, 1069. Kitāb-e Jāmi' al-Hekmatain [Jāme' Hekmat-e Yunān va Osul-e 'Aqā'yed-e Esmā'iliān, 1069]. Ed. Corbain, Mo'in, 1953. Tehran (in Persian).

Nasr, S.H., 1963. Nazar Motefakerān Eslāmi dar bārehye Tabi'at [Conception of Nature in Islamic Thought]. 1342/1963 (first ed.), 1359/1980 (third ed.), Kharazmi Pul. Tehran, 470 pp. (in Persian).

Nasr, S.H., 1980. 'Elm va Tamadon dar Eslām. Translated into Persian by Ahmad Aram, Kharazmi Publ., Tehran, 1359/1980 (in Persian).

Nasr, S.H., 1993. An Introduction to Islamic Cosmological Doctrine. State University of New York Press, 322 pp.

Nassir al-Din Tusi, 1256–1259. Tansukhnāmeh Ilkhāni (ca. 1256-1259). Ed. S.M.T. Modarres Razavi. Bonyād Farhang Irān Publ., Tehran, 1348/1969 (first ed.); Etela'at Publ., Tehran, 1363/1994 (second ed.), 358 pp. (in Persian).

Nāteqi Elāhi, F., Tavakoli, S., J'afarzādeh, F., Shabāniān, B., 1991. Dārāb earthquake of 15 Ābān, 1369, Furk, Fars. IIEES, Internal Report, 45 pp. (in Persian).

Navāder al-Tabādor le-Tohfat al-Bahādor. See Denisari.

Nazarevsky, N.V., 1929. Germabskoye zemletryasenie 1 Maya 1929 goda. Bull. Moskova Isp. Prirody Otd. Geol. 10, 113–122.

Nazari, H., Salamati, R., 1998. Geological Map of Rudbar, 1:100,000. Sheet No. 5863Geol. Surv. Iran.

Nazari, H., Ritz, J.-F., Salamati, R., Solaymani, S., Balescu, S., Michelot, J.-L., Ghassemi, A., Talebian, M., Lamothe, M., Ghorashi, M., 2007. Palaeoseismological analysis in Central Alborz, Iran. In: The 1957 Gobi-Altay Earthquake Commemorating Conference, 25 July–08 August, 2007, Ulaanbaatar, Mongolia, 3 pp.

Nazari, H., Ritz, J.-F., Salamati, R., Balescu, S., Michelot, J.-L., Massault, M., Ghorashi, M., 2008. Palaeoseismic Analysis in Tehran Region Along the North Tehran and Taleqan Faults, and the Rey and Kahrizak Scarps. SGF, Paris, 1 pp.

Nazari, H., Ritz, J.-F., Salamati, R., Shafei, A., Ghassemi, A., Michelot, J.-L., Massault, M., Ghorashi, M., 2009a. Morphological and palaeoseismological analysis along the Taleghan fault (Central Alborz, Iran.). Geophys. J. Int.. 178, 1028–1041. http://dx.doi.org/10.1111/j.1365-246X.2009.04173.x.

Nazari, H., Fattahi, M., Meyer, B., Sebrier, M., Talebian, M., Foroutan, M., Le Dortz, K., Bateman, D.M., Ghorashi, M., 2009b. First evidence for large earthquakes on the Deshir Fault, Central Iran Plateau. Terra Nova. 21 (6), 417–426. http://dx.doi.org/10.1111/j.1365-3121.2009.00892.x.

Nazari, H., Ritz, J.-F., Salamati, R., Shahidi, A., Habibi, H., Ghorashi, M., Karimi Bavandpur, A., 2010. Distinguishing between fault scarps and shorelines: the question of the nature of the Kahrizak, North Rey and South Rey features in the Tehran plain (Iran). Terra Nova 22, 227–237.

Nazari, H., Ritz, J.-F., Ghassemi, A., Bahar-Firouzi, K., Salamati, R., Shafei, A., Fonoudi, M., 2011. Paleoearthquakes determinations of magnitude 6.5 on the North Tehran fault, Iran. JSEE, IIEES 13 (1), 159–166.

NCC. National Cartographuc Center, Tehran, Iran.

Negahban, E.O., 1964. A preliminary report on Marlik excavation, Gohar Rud Expedition, Rudbar, 1961-1962. Offset Press, Tehran.

Negahbān, E.O., 1971. Quake destroyed 9000 year-old civilization. Keyhān International Newspaper, XIII (no. 4098; 13 March, 1971), p. 7.

Negahbān, E.O., 1973. Report of preliminary excavations for two month at Tapeh Sagzābād in the Qazvin plain. Mārlik, J. Inst. Dep. Archaeol., 1, Faculty of Letters and Humanities, University of Tehran, 10–24 (in Persian & English).

Negahbān, E.O., 1974a. Sagzabad excavations, 1970-1971. In: A Survey of Persian Art, vol. XV. Asia Institute Books, Ashiya, Japan.

Negahbān, E.O., 1974b. Survey of Excavations in Sagzabad During 1972-73, vol. XII. Iran. J. Persian Stud., London, p. 216

Negahbān, E.O., 1976. Preliminary report of the excavation of Sagzabad. In: The Memorial of the VIth International Congress of Iranian Art and Archaeology, Oxford, Tehran, pp. 247–271.

Negahbān, E.O., 1977. Report of preliminary excavations at Tapeh Sagzābād in the Qazvin plain. Mārlik, J. Inst. Dep. Archaeol., 2, Faculty of Letters and Humanities, University of Tehran, pp. 33 & 40 (in Persian & English).

Negahbān, E.O., 1984. A Preliminary Report on Marlik Excavation, Gohar Rud Expedition, Rudbar, 1961–1862. Tehran Offset Press, Tehran, Iran (in Persian).

Negahbān, E.O., 1990. Silver vessel of Marlik with gold spout and impressed gold designs. Iranica Varia 16, 144–151.

Negahbān, E.O., 1996. Marlik: the complete excavation report. Pennsylvania University Museum Monograph, 87, The University of Pennsylvania, Philadelphia.

Negahbān, E.O., 2006. Moruri bar Panjāh Sāl Bāstānshenāsi-ye Irān A Review of Fifty Years of Archaeology of Iran Sāzémān Mirāth Farhangi Keshvar, Nashr Sobhānpur, 1385/2006, p. 476.

NEIC. National Earthquake Information Center. U.S. Geological Survey Stop 967, National Earthquake Information Center, DFC, Denver, CO 80225-0046, USA. http://neic.usgs.gov/.

New Testament, The Holly Bible, King James Version. http://utile.autre.net/livre/biblekjvl.html or http://www.online-literature.com/bible/bible.php.

Nezāmi Ganjavi (Nezāmi of Ganja), 1191. Khamseh: Sharafnāmeh [lit., 'Book of Honor' the first book of Eskandarnāmeh [lit., the 'Book of Alexander'; 1191 named Eqbālnāmeh]]. Ed. Vahid Dastgerdi, 1963. Tehran.

Ngdir.ir, 2006. National Geoscience Database of Iran. Shekl 13-Naqsheh-ye hamsheddat-e zaminlarzeh-ye 11 Farvardin 1385 Chālānchulān hamrāh ba gosalhāyeh fa'āl-e mantaqeh, zaminlarzehāyeh tārikhi va dastgāhi va niz gosikhtegi-ye sathi-ye zaminlarzeh-ye 3 Bahman 1285 Silākhor. ngdir.ir. Tehran.

NGDIR, 2010. News and Reports. National Geoscience Database of Iran, 12 January 2010, Ngdir.ir.

Ni, J., Barazangi, M., 1986. Seismotectonics of the Zagros continental collision zone and a comparison with the Himalayas. J. Geophys. Res. 91, 8205–8218.

Niazi, M., 1969. Source dynamics of the Dasht-e Bayaz earthquake of August 31, 1968. Bull. Seismol. Soc. Am. 59 (5), 1843–1861.

Niazi, M., 1977. Focal Mechanism and Aftershocks of the November 7, 1976 Qayen Earthquake in NE Iran. IAEI, Joint General Assemblies, Durham, 1977.

Niãzi, M., Bozorgniã, Y., 1992. The 1990 Manjil, Iran, earthquake: geology and seismology overview, PGA attenuation, and observed damage. Bull. Seismol. Soc. Am. 82, 774–799.

Niazi, M., Kanamori, H., 1981. Source parameters of 1978 Tabas and 1979 Qainat, Iran earthquakes from period surface waves. Bull. Seismol. Soc. Am. 71 (4), 1201–1213.

Nicholson, R.A. (Ed., Trans.), 1926. The Mathnawi of Jalālu'ddin Rumi. The E.J.W. Gibb Memorial Thrust, London.

Niemi, T., 2008. Historical earthquake catalogue and archaeological data: Avoiding circular reasoning. Seismol. Res. Lett. 79, 289. See also Geol. Soc. Am. Sp. Pap. 2010, 471, 97-106.

Nikolski, N., 1909a. Letter report No. 38 of Russian Cousul at Kermanshah to E.V. Sablin, Charge d'Affairs at Imperial Mission Tehran. Dated 18th February old style. Kermanshah (unpublished; in Ambraseys, 1974b).

Nikolski, N., 1909b. Letter report No. 41 of Russian Cousul at Kermanshah to H.M the Shah, Tehran. February 24 (old style), 1909. Kermanshah.

Nikolski, N., 1909c. Letter report No. 42 of Russian Counsul at Kermanshah addressed to the First Department of the Ministry of Foreign Affairs at St. Petersburg. February 24 (old style), 1909. Kermanshah.

Nikolski, N., 1909d. Letter report of mission to Burujerd earthquake area of Russian Cousul at Kermanshah addressed to Nikolayev Central Observatory. February 1909. Kermanshah.

Nikolski, N., 1909e. Letter report of earthquake effects sent to Novoi Vremia, St. Petersburg. February 1909. Kermanshah.

Nikonov, A.A., 1991. Destructive historical earthquakes in Soviet Armenia. Tectonophysics 193, 225–229.

Nilforoushan, F., Masson, F., Vernant, P., Vigny, C., Martinod, J., Abbassi, M., Nankali, H., Hatzfeld, D., Bayer, R., Tavakoli, F., Ashtiani, A., Doerflinger, E., Daignieres, M., Collard, P., Chery, J., 2003. GPS network monitors the Arabia-Eurasi collision deformation in Iran. J. Geod. 77, 411–422.

NIOC, 1975. Geological Maps 1:250,000 Series. Nos. 20511 and 20512Oil Service Company of Iran, National Iranian Oil Company.

Nissen, E., Ghorashi, M., Jackson, J., Parsons, B., Talebian, M., 2007. The 2005 Qeshm Island earthquake (Iran)–a link between buried reverse faulting and surface folding in the Zagros Simply Folded Belt? Geophys. J. Int. 171, 326–338.

Nissen, E., Yaminifard, F., Tatar, M., Gholamzadeh, A., Bergman, E., Elliott, J.R., Jackson, J.A., Parsons, B., 2010. The vertical separation of mainshock rupture and microseismicity at Qeshm Island in the Zagros fold-and-thrust belt, Iran. Earth Planet. Sci. Lett. 296, 181–194.

North, R., 1972. A thrust event in southern Iran. Lincoln Lab., Semi-Annual Technical Summary, pp. 60–61, M.I.T.

North, R., 1973. Seismic source parameters. Ph.D. Thesis, Churchill College, Cambridge, UK, 132 pp.

Nowroozi, A.A., 1971. Seismo-tectonics of the Persian plateau, eastern Turkey, Caucasus, and Hindu-Kush regions. Bull Seismol. Soc. Am. 61 (2), 317–341.

Nowroozi, A.A., 1972. Focal mechanism of earthquakes in Persia, Turkey, West Pakistan and Afghanistan and plate tectonics of the Middle East. Bull. Seismol. Soc. Am. 62 (3), 823–850.

Nowroozi, A.A., Mohajer–Ashjai, A., Payman, M., Taqizadeh, Gh.A., 1980. Aftershocks and Damages and Field Report of the Great Tabas, East Iran Earthquake, 77 Journal Petroleum Institute, Iran, pp. 15–30.

Nowroozi, A.A., Mohajer Ashjai, A., 1980a. Faulting of Kurizan and Koli (Qaenat Iran) earthquakes of Nov. 1979: a field report. Iran. Pet. Inst. 78, 8–20.

Nowroozi, A.A., Mohajer Ashjai, A., 1980/1981b. Faulting of Kurizan and Koli (Iran) earthquakes of November 1979; a field report. Bull. B.R.G.M. (deuxième série) IV (2), 91–99.

Nozhat al-Qolub. See Mostaufi Qazvini.

Nozhat Nāmeh 'Alā'i. See Shahmardān ibn Abi al-Khair.

NSSP, 1995. The map of earthquake source origination zones in the Republic of Armenia and adjacent territory. National Survey of Seismic Protection under the government of the Republic of Armenia.

Nurbakhsh, H., 1990. Jazireh-ye Qeshm va Khalij-e Fārs The Qeshm Island and the Persian Gulf Amir Kabir Publ., pp. 88–93, 1369 (in Persian).

Oberlander, T., 1965. The Zagros Streams: A New Interpretation of Transverse Drainage in an Orogenic Zone. Syracuse Univ. Press, Syracuse Geography Ser, 168 pp.

Oberlies, T., 1998. Die Religion des Rgveda, Vienna.

Old Testament, the. King James Version. http://utile.autre.net/livre/biblekjvl.html.

'Omar Khayyām Neyshāburi, 1048–1122. Rubāiyāt [Quatrains]. The Rubāiyāt of 'Omar Khayyām, Ed. Edward Fitzgerald. Gutenberg.org/ebooks/246.

Omote, S., Nakagawa, K., Kobayashi, H., Nakaoka, E., 1965. A report on the Buyin earthquake (Iran) of Sep. 1, 1962. In: Proc. 3rd World Conf. Earthq. Eng. New Zealand. vol. III, pp. 27–44.

ORIAS. The Illustrated Life of the Buddha. ORIAS, Changhwan Park. http://orias.berkeley.edu/visuals/buddha/life.html#6; 7/8/2010).

Oskian, H., 1942. Die Armenischen Klöster In Vaspowrakan-Van. 2. Teil. 5. Die Klöster Von Arčišahovit, K'Ajbarownik' Und Arcke, 6. Das Klöster Von Artaz, 7.Die Klöster In Naxčawan, 8. Die Patriarchalschrbiben Zugunsten Des Kloösters Des Protoartyrs Małard, 9. Die Klöster in Gołt'n. Mechitaristen-Buchdruckerei, Wien, p. 458, (in Armenian).

Ouseley, W., 1823. Travels in Various Countries of the East More Particularly Persia. V., vol. III Rodwell & Martin, London, 600 pp.

Pace, E., 1968. Toll in Iran's quake now put at 12,000. N.Y. Times, p. 1, 3 September 1968.

Pacheco, J.F., Sykes, L.R., 1992. Seismic moment catalog of large shallow earthquakes, 1900 to 1989. Bull. Seismol. Soc. Am. 82 (3), 1306–1349.

Pahlavi, F., 2004. An Enduring Love. English Trans. Patricia Clancy. Miramax Books, New York, 447 pp.

Palvis, T.L., Bruhn, R.L., 1983. Deep-seated flow as a mechanism for the uplift of broad forearc ridges and its role in the exposure of high P/T metamorphic terraces. Tectonics 2 (5), 473–497.

Parham, B., 1971. Barresi va Arzeshyabi-yeh Nausazi-ye-Manateq-e-Zelzelh-zadeh-yeh Iran; Jozveh-yeh Avval-Torud (Evaluation and Assessment of the Renovations of the Earthquake-Striken Areas in Iran; First Book—Torud) vol. 28. Ministry of Development and Housing, Department of Research and Planning, 88 pp., 1350/1971 (in Persian).

Parsons, B., Wright, T., Rowe, P., Andrews, J., Jackson, J., Walker, R., Khatib, M., Talebian, M., Bergman, E., Engdahl, E.R., 2006. The 1994 Sefidabeh (eastern Iran) earthquakes revisited: new evidence from satellite radar interferometry and carbonate dating about the growth of an active fold above a blind thrust fault. Geophys. J. Int. 164, 202–217.

Pāyandeh, 1975. See Tabari.

Pérez-Lopez, R., Grutzner, C., Lario, J., Reicherter, k, Silva, P.G. (Eds.), 2009. Archaeoseismology and Palaeoseismology in the Alpine-Himalayan Collision Zone. Baelo Claudia, Cadiz, Spain.

Peronaci, F., 1958a. Characteristiques sismiques de l'Iran. Ist. Naz. Geofisica, Monte Porzio Catone, Unpublished Internal Report.

Peronaci, F., 1958a. Sismicita dell'Iran. Ann. Geofis. XI, 55–68.

Peronaci, F., 1959. Contributo alla consoscenza delle caractteristiche sismiche dell'Iran settentrionale. Ann. Geofis. XII, 523–534.

Peronaci, F., 1971. Il treemoto di Farsinich dell 13 Dicembre 1957. Ist. Naz. Geofisica, Monte Porzio Catone, Unpublished Internal Report.

Perrey, A., 1845. Liste de tremblements de terre ressentis en Europe et dans les parties adjacentes de l'Afrique et de l'Asie pendant les annees 1843 et 1844. C. R. Acad. Sci. 20, 1444.

Perrey, A., 1848. Note sur le tremblements de terre en 1847. Bull. Acad. R. Sci. Bruxelles 15, 442–454.

Perrey, A., 1850. Memoire sur le tremblements de terre ressentis dans la penisule Turco-Hellenique et en Syrie. Mem. Acad. R. Sci. Belgique 23 (1), 3–37, Bruxelles.

Perrey, A., 1851. Tremblements de terre ressentis en 1850. Bull. Acad. R. Sci. Bruxelles 18 (1), 191–308.

Perrey, A., 1852a. Treblements de terre sessentis en 1851. Bull. Acad. R. Sci. Bruxelles 19 (1), 353–396.

Perrey, A., 1852b. Supplement a la note sur les tremblements de terre ressentis en 1851. Bull. Acad. R. Sci. Bruxelles 19 (2), 21–28.

Perrey, A., 1853. Tremblements de terre en 1852. Bull. Acad. R. Sci. Bruxelles 20 (2), 39–69.

Perrey, A., 1854. Note sur les tremblements de terre en 1853. Bull. Acad. R. Sci. Bruxelles 21 (1), 457–495.

Perrey, A., 1855. Note sure les tremblements de terre en 1854 avec supplements pour les annees anterieures. Bull. Acad. R. Sci. Bruxelles 22 (1), 526–572.

Perrey, A., 1856. Note sur les tremblements de terre en 1855, avec supplements. Bull. Acad. R. Sci. Bruxelles 23, 23–68.

Perrey, A., 1857. Notes sur les tremblements de terre ressentis en 1855, avec supplements pour les annees anterieures. Bull. Acad. R. Sci. Bruxelles 24, 64–128.

Perrey, A., 1859. Notes sur les tremblements de terre en 1856 avec supplements pour les annees anterieures. Mem. Cour. Acad. R. Belgique 8 (3), 3–79.

Perrey, A., 1860. Note sur les tremblements de terre en 1857 avec supplements pour les annees anterieurs 1842. Mem. Cour. Acad. R. Belgique, Bruxelles 10 (4), 3–114.

Perrey, A., 1862a. Note sur les tremblements de terre en 1858 avec supplements pour les annees anterieurs 1843. Mem. Cour. Acad. R. Bruxelles 12 (4), 3–68.

Perrey, A., 1862b. Note sur les tremblements de terre en 1859. Mem. Cour. Acad. R. Bruxelles 13 (3), 3–78.

Perrey, A., 1862c. Note sur les tremblements de terre en 1860. Mem. Cour. Acad. R. Bruxelles 14 (3), 3–75.

Perrey, A., 1864a. Note sur les tremblements de terre en 1861. Mem. Cour. Acad. R. Bruxelles 16 (5), 2–112.

Perrey, A., 1864b. b. Note sur les tremblements de terre en 1862 avec supplements pour les annees anterieurs. Mem. Cour. Acad. R. Bruxelles 16 (6), 3–179.

Perrey, A., 1865. Note sur les tremblements de terre en 1863 avec supplements pour les annees 1843-1862. Mem. Cour. Acad. R. Bruxelles 17 (5), 1–213.

Perrey, A., 1866. Note sur les tremblements de terre en 1864 avec supplements pour les annees anterieurs 1843–1863. Mem. Cour. Acad. R. Bruxelles 18 (4), 3–98.

Perrey, A., 1867. Note sur les tremblements de terre en 1865 avec supplements pour les annees anterieurs 1843–1861. Mem. Cour. Bruxelles 19 (3).

Perrey, A., 1870. Note sur les tremblements de terre en 1866. Mem. Cour. Acad. R. Bruxelles 21 (5), 3–223.

Perrey, A., 1872a. Note sur les tremblements de terre en 1868 avec supplements pour les annees anterieurs 1843-1867. Mem. Cour. Acad. R. Bruxelles 22 (3), 1–116.

Perrey, A., 1872b. Note sur les tremblements de terre en 1869 avec supplements pour les annees anterieurs 1843-1868. Mem. Cour. Acad. R. Bruxelles 22 (4), 1–116.

Perrey, A., 1873. Supplements aux note sur les tremblements de terre ressentis de 1843 a 1868. Mem. Cour. Acad. R. Bruxelles 23 (6), 3–70.

Perrey, A., 1875a. Note sur les tremblements de terre en 1870. Mem. Cour. Acad. R. Bruxelles 24 (3), 2–146.

Perrey, A., 1875b. Note sur les tremblements de terre en 1871 avec supplements pour les annees anterieur de 1843-1870. Mem. Cour. Acad. R. Bruxelle 24 (4), 1–143.

Peterson, J.H., 2002. Digital Version of the Iranian or Greater Bundahishn. Trans. Anklesaria, 1956, http://www.avesta.org/mp/grb.htm.

Petrescu, G., Purcaru, G., 1964. The mechanism and stress pattern at the focus of the September 1, 1962, Buyin-Zara (Iran) earthquake. Am. Geophys. 20 (3), 242–247.

Petrie, C.A., Sardari Zarchi, A., Havanmard Zadeh, A., 2007. Developing societies and economies in fourth millennium B.C. Fars: further excavations at Tol-e Spid. Iran 45, 301–309.

Petrie, C.A., Asgari Chaverdi, A., Seyedin, Mo, 2009. Excavation at Tol-e Spid. In: Potts, D.T., Roustaei, K., Petrie, C.A., Weeks, L.R. (Eds.), The Mamasani Archaeological Project Stage One. A Report on the First Two Seasons of the ICAR-University of Sydney Expedition to the Mamasani District, Fars Province, Iran. Archaeological Gordon House, Oxford, England, pp. 89–132 (Chapter 4).

Peyret, M., Chery, J., Djamour, Y., Avallone, A., Sarti, F., Briole, P., Sarpoulaki, M., 2007. The source motion of 2003 Bam (Iran) earthquake constrained by satellite and ground based geodetic data. Geophys. J. Int. 163, 849–865.

Peyret, M., Rolandone, F., Dominguez, S., Djamour, Y., Meyer, B., 2008. Source model for the Mw 6.1, 31 March 2006, Chalan-Chulan Earthquake (Iran) from InSAR. Terra Nova. 20 (2), 126–133. http://dx.doi.org/10.1111/j.1365-121.2008.00797.x.

Peyret, M., Djamour, Y., Hessami, K., Regard, V., Bellier, O., Vernant, P., Daignères, M., Nankali, H., Van Gorp, S., Goudarzi, M., Chéry, J., Bayer, R., Rigoulay, M., 2009. Present-day strain distribution across the Minab-Zendan-Palami fault system from dense GPS transects. Geophys. J. Int.. 179, 751–762. http://dx.doi.org/10.1111/j.1365-246X.2009.04321.x.

Pezzo, G., Tolomei, C., Atzori, S., Salvi, S., Shabanian, E., Bellier, O., Farbod, Y., 2012. New kinematic constraints of the western Doruneh fault, northeastern Iran, from interseismic deformation analysis. Geophys. J. Int.. 190, 622–628. http://dx.doi.org/10.1111/j.1365-246X.2012.05509.x.

Philip, H., Rogozhin, E., Cisternas, A., Bousquet, J.C., Borisov, B., Karakhanian, A., 1992. The Armenian earthquake of 1988 December 7: faulting and folding, neotectonics and palaeoseismicity. Geophys. J. Int. 110, 141–158.

Pickett, D., 1984. Inscriptions by Muhammad Riḍā Al-Imāmi. Iran XXII, 91–102.

Plassard, J., Kogoj, B., 1981. third ed. Seismicité du Liban: catalogue des séismes ressentis, vol. IV Collection dés Annales-Mémoires de l'Observatoire de Ksara, Beirut.

Plato, 428–348 B.C. Works of Plato, Benjamin Jowett (Trans.), 1852, classics.mit/Browse/browse-Plato.html. Also Plato: Complete Works, John M. Cooper, D.S. Hutchinson (Eds.), 1997, Hackett Publ. Co., 1808 pp.

Platt, J.P., Leggett, J.K., Young, J., Raza, H., Alam, S., 1985. Large-scale underplating in the Makran accretionary prism, southwest Pakistan. Geology 13, 507–511.

Platt, J.P., Leggett, J.K., Alam, S., 1988. Slip vectors and fault mechanics in the Makran accretionary wedge, southwest Pakistan. J. Geophys. Res. 93, 7955–7973.

PO/ISC, 1967. Plan Organization of Iran, Statistical Center of Iran. Census Results (in Persian).

Poirier, J.P., Taher, M.A., 1980. Historical seismicity in the Near and Middle East, north Africa, and Spain from Arabic documents (VIIth–XVIII century). Bull. Seismol. Soc. Am. 70 (6), 2185–2201.

Pollitz, F.F., Sacks, I.S., 2002. Stress triggering of the 1999 Hector Mine earthquake by transient deformation following the 1992 Landers earthquake. Bull. Seismol. Soc. Am. 4, 1487–1496.

Polo, Marco and Rustichello da Pisa, 13th century. In: Henry Yule and Henri Cordier: The Travels of Marco Polo: The Complete Yule-Cordier Edition, 1903.

Pope, A.U., 1965. Persian Architecture; A Triumph of Form and Color. George Braziller, New York, 288 pp.

Popov, V.V., 1940. Seismicity of the Caspian Sea coast. Nat. Acad. Sci. USSR 29 (7), 69–70, (in Russian).

Potts, D.T., Askari Chaverdi, A., Mcrae, I.K., Alamdari, K., Dusting, A., Jaffari, J., Elliot, T.M., Setoudeh, A., Lashkari, A., Ameli, Sh., Yazdani, A., 2009. Further excavations at Qaleh Kali (MS 46) by the joint ICAR-University of Sydney Mamasani expedition: results of the 2008 season. Iran. Antiq. XLIV, 207–259.

Poure Davoud (Purdāvud), E., 1964. Visparad: A Part of the Avestā with the Last Prophet Zarthosht, Fire, Seven Countries, Swearing Formula. Tehran, 1343 (in Persian).

Poure Davoud (Purdāvud), E., 1968. Yashthā (the Holy Yashts), vol. 1, 1347/1968, Tahuri Publ., Zabān va Farhang-e Irān, 50, 626 pp. (in Persian).

Poure Davoud (Purdāvud), E. (Trans.), 1977a. The Yashts, vol. 1. Tehrān University Publ. 1637, third ed., 626 pp. (in Persian).

Poure Davoud (Purdāvud), E. (Trans.), 1977b. The Yashts, vol. 2. Tehrān University Publ. 1638, third ed., 545 pp., 1356 (in Persian).

Poure Davoud (Purdāvud), E. (Trans.), 1977a. The Yasnā, vol. 1, 1356/1977b, Tehrān University Publ., 1596 (no. 1971), third ed., 269 pp. (in Persian).

Price, N.J., 1978. Assessment of structures in cores. In: Dames, Moore, (Eds.), Supplementary Geologic and Geophysical Investigation, Proposed Nuclear Power Plant, Esfahan, vol. 1, Appendix F. Atomic Energy Organization of Iran, Tehran.

Priestley, K., Patton, H.J., Schultz, C.A., 2001. Modeling anomalous surface-wave propagation across the Southern Caspian Basin. Bull. Seismol. Soc. Am.. 91, 1924–1929. http://dx.doi.org/10.1785/0120010118.

Priestley, K., Baker, C., Jackson, J., 1994. Implications of earthquake focal mechanism data for the active tectonics of the South Caspian Basin and surrounding regions. Geophys. J. Int. 118, 111–141.

Pritchett, F.W., 1991, 2004. The Romance Tradition in Urdu; Adventures from Dastan of Amir Hamza. Columbia University Press. http://www.columbia.edu/itc/mealac/pritchett/00litlinks/hamzah.

Purbahā Jāmi, Maulānā Tāj al-Din, d. 1299. See Minorsky, 1956.

Pursinā. See Ibn Sinā.

Qatrān Tabrizi, Abu Mansur, 1009–1072. Divān-e Qatrān-e Tabrizi [Collection of Poems of Qatrān Tabrizi]. Ed. Nakhjavāni, Tabriz, 1333/1954. Ed. Hossain Ahi, Mo'asseseh Matbua'ti Khazar, Tehrān (in Persian).

Qavāmi Rāzi, Badr al-Din, d. 1164. Divān-e Ash'ār-e Sharaf al-Sho'arā Badr al-Din Qavām-e Rāzi [Collection of Poems of Badr al-Din Qvām-e Rāzi]. Ed. Mir Jalal al-Din Hossaini Armavi. Sepehr Publ., Tehrān, 1334/1955 (in Persian).

Qorashi, M., Ovaisi, B., 2002. Negāhi gozarā bar ruydād-e zaminlarzeh-ye Changureh Avaj (jonub-e Ostān-e Qazvin). Geol. Surv. Iran, 19 pp. (in Persian).

Quchāni, A., 1988. Bāzshenāsi-ye Yek Masjed-e Qadimi dar Qeshm be Estenād-e Katibeyeh Chubi-ye Mote'aleq be Qarn-e Hashtom-e Hejri. Majaleh Bāstānshenāsi va Tārikh 3 (1), 1367, Tehran (in Persian).

Quigley, M., Fattahi, M., Sohbati, R., Schmidt, A., 2011. Palaeoseismicity and pottery: investigating earthquake and archaeological chronologies on the Hajiarab alluvial fan, Iran. Quat. Int. 242 (1), 185–195.

Quittmeyer, R.C., Kafka, A.L., 1984. Constraints on plate motions in southern Pakistan and the Northern Arabian Sea from the focal mechanisms of small earthquakes. J. Geophys. Res. 89, 2444–2458.

Qurān. English Translations from: http://quran.com; Islamicity.com; and by Dr. T.B. Irving at http://www.isgkc.org/translat.htm.

al-Qurtubi, 'Arib ibn Sa'ad al-Kātib, 923. Kitāb al-Anwā [The Calendar of Cordoba for the Year 961]. Ed. R. Dozy, Le Calendarier de Cordoue de la'année 961, Leiden, Brill, 1873. Ed. C. Pellat, Leiden, Brill, 1961.

Rabino, H.L., 1928. Mazandarān va Astarābād. Persian Translation of Mazandaran and Asterabad, London (1928), by Vahid Mazandarāni, *Bongāh Tarjomeh va Nashr Ketāb*, Tehran, 1343/1964.

Rā'in, E., 1971. Daryānavardi-ye Irānian, 2 vols. Also 1358/1979 ed., *Jāvidān Publ.*, Tehrān (in Persian).

Raisi, M., 1991. The Darab, Iran Earthquake of 6 November 1990. B.Sc. Thesis, Department of Earth Sciences, College of Science, University of Shiraz, Iran.

Rajavi, Kāzem. In Zokā', 1989.

Ramazi, H.R., 1991. Seismotectonics of the 1990 Manjil-Rudbar earthauke, In Analytical report No. 1; the Manjil-Rudbar earthquake (Iran), Khordad 31, 1369 [June 20, 1990]. IIEES, Tehran, 70-91-1, 32-44 (in Persian).

Ramazi, H.R., Nayyeri, A., Khademi, M.H., Yasrebi, F., Farzanegan, E., 1997. An Immediate-Preliminary Report on the Garmkhan (North Bojnurd) Earthquake of 4 Feb. 1997. vol. 240. Building and Housing Research Center, Ministry of Housing and Urban Development, 68 pp. (in Persian).

Ramezani, J., Tucker, R.D., 2003. The Saghand region, Central Iran: U-Pb geochronology. Petrogenesis and implications for Gondwana tectonics. Am. J. Sci. 303, 622–665.

Rante, R., Collinet, A., 2013. Nishapur Revisited: Stratigraphy and Ceramics of the Qohandezh. Oxbow Books, 144 pp.

Rashid al-Din Fazlollāh Hamédāni, 1304. Jame' al-Tavārikh-e Rashidi [Complete History of Rashidi]. Bahman Karimi (Ed.), Eqbal Publication. Tehran, 2 vols, Mongols, 1338/1959 (in Persian). Esmā'ilian vol., Mohammad Taqi Dānesh Pazhuh and Mohammad Modarresi Zanjāni (Eds.), Bongāh Tarjomeh va Nashr Ketāb publication, Tehran (in Persian). Also Tr. E. Quatremére, Histoire des Mongols de la Perse, Paris, 1836. Tr. J.A. Boyle, The Successor of Genghis Khan, London, 1971.

Rashid al-Din Fazlollāh Hamédāni, 1304. Jāme' al-Tavārikh. Trans. E. Quatremére, Histoire des Mongols de la Perse, Paris, 1836. Trans. J.A. Boyle, The Successor of Genghis Khan, London, 1971.

Ravaut, P., Bayer, R., Hassani, R., Rousset, D., Al Yahya'ey, A., 1997. Structure and evolution of the Northern Oman margin: gravity and seismic Constraints over the Zagros-Makran-Oman collision zone. Tectonophysics 279, 253–280.

Al-Rāwandi [Rāvandi], 1238. Rahat al-Sodur wa Āyat al-Sorur dar Tārikh-e Āl-e Saljuq. In: Eqbāl, M. (Ed.), Amir Kabir Publ., Tehrān, 1333 (in Persian).

Razaqi-Azar, N., Farzanegan, E., Hadi Khademi, M., 1994. Gozāresh-e moqadamāti-yeh zelzele-ye Sefidābeh (ostān Sistān va Baluchestān), 4 Esfand 1372. *Markaz tahqiqat Sakhtema va Maskan*, 5, 51 pp., Esfand 1372 (in Persian).

al-Rawi, M.M., 2002. The Contribution of Ibn Sina (Avicenna) to the Development of Earth Science. vol. 4039. Foundation for Science, Technology, and Civilization, Limited. 2002–2003, 12 pp. http://www.fstc.co.uk.

Reasenberg, P.A., Simpson, R.W., 1992. Response of regional seismicity to the static stress change produced by the Loma Prieta earthquake. Science 255, 1687–1690.

Regard, V., Bellier, O., Thomas, J.C., Abbassi, M.R., Mercier, J., Shabanian, E., Feghhi, K., Soleymani, S., 2004. Accommodation of Arabia-Eurasia convergence in the Zagros-Makran transfer zone, SE Iran: a transition between collision and subduction through a young deforming system. Tectonics. 23http://dx.doi.org/10.1029/2003TC001599, TC4007.

Regard, V., Bellier, O., Thomas, J.-C., Bourles, D., Bonnet, S., Abbassi, M.R., Braucher, R., Mercier, J., Shabanian, E., Soleymani, Sh., Feghhi, Kh., 2005. Cumulative right-lateral fault slip rate across the Zagros-Makran transfer zone: role of the Minab-Zendan fault system in accommodating Arabia-Eurasia convergence in southeast Iran. Geophys. J. Int. 162, 177–203.

Regard, V., Hatzfeld, D., Molinaro, M., Aubourg, C., Bayer, R., Bellier, O., Yamini-Fard, F., Peyret, M., Abbassi, M., 2010. The transition between Makran and the Zagros collision: recent advances in its structure and active deformation. In: Leturmy, P., Robin, C. (Eds.), Tectonic and Stratigraphic Evolution of Zagros and Makran During the Mesozoic-Cenozoic. No. 330Geological Society Spec. Publ., London, pp. 43–64

Reicherer, K., Michetti, A.M., Silva, P.G., 2009. Palaeoseismology: Historical and Praehistorical records of earthquake ground effects for seismic hazard assessment. In: Leturmy, P., Robin, C. (Eds.), Tectonic and Stratigraphic Evolution of Zagros and Makran During the Mesozoic-Cenozoic. No. 316Geol. Soc. Spec. Publ., London, pp. 155–159.

reliefweb.net. United Nations Office for Coordination of Human Affairs (OCHA). reliefweb.net.

al-Resāla al-Sanjariya fi al-Kā'enāt al-Onsoriya. See Sahlān al-Sāvi.

Resāleh Āthār 'Alavi. See Esfézāri, Esmā'il.

Resāleh Āthār 'Alavi. See Mas'ud Marvazi.

Reynolds, J., 1858. See al-'Utbi, 1036.

Rezapour, M., Mohsenpur, A., 2013. The 2010 M_w Rigan, Iran, earthquake aftershock sequence. Bull. Seismol. Soc. Am. 103 (3), 1793–1899.

Rich, C.J., 1836. Narrative of a residence in Kordistan and on the site of Ancient Nineveh; with journal of a voyage down the Tigris to Baghdad; and an account of a visit to Shrizur and Persepolis. 2 vols.. James Dunca, Paternoster Row, London.

Richardson, R.K., 1922. The Geology of the Kishm Island and Adjoining Regions of the Persian Gulf. Anglo-Persian Oil Co., Ltd., Internal Report No. 119, p. 67.

Rig Vedā. See Jalali Naini, M.R., 1348/1969; Griffith, 1889.

Rishbeth, H., 2006. Ionoquakes: earthquake precursors in the ionosphere? EOS, Trans. Am. Geophys. Union 87 (32), 316.

Ritz, J.-F., 2009. Extrusion tectonics and subduction in the eastern South Caspian region since 10 Ma: Comment. Geology 37, E191. http://dx.doi.org/10.1130/G25627C.1.

Ritz, J.-F., Nazari, H., Ghassemi, A., Salamati, R., Shafei, A., Solaymani, S., Vernant, P., 2006. Active transtension inside central Alborz: a new insight into northern Iran–Southern Caspian geodynamics. Geology 34 (6), 477–480. http://dx.doi.org/10.1130/G22319.1.

Ritz, J.-F., Nazari, H., Balescu, S., Lamothe, M., Salamati, R., Chassemi, A., Shafei, A., Ghorashi, M., Saidi, A., 2012. Paleoearthquakes of the past 30,000 years along the North Tehran fault (Iran). J. Geophys. Res.. 117http://dx.doi.org/10.1029/2012JB009147, B06305.

Riyāzi Heravi, Mohammad Yusef Khān. Bahr al-Favā'ed. In: Shākeri, 1967, 1981, 2001.

Rizza, M., Vernant, P., Ritz, J.-F., Peyret, M., Nankali, H., Nazari, H., Djamour, Y., Salamati, R., Tavakoli, F., Ch´ery, J., Mahan, S.A., Masson, F., 2013. Morphotectonic and geodetic evidence for a constant slip-rate over the last 45 kyr along the Tabriz fault (Iran). Geophys. J. Int.. 2013 (193), 1083–1094. http://dx.doi.org/10.1093/gji/ggt041.

Rochberg, F., 2002. A consideration of Babylonian astronomy within the historiography of science. Stud. Hist. Philos. Sci. 33 (4), 661–684.

Rockwell, T.K., Lindvall, S., Herzberg, M., Murback, D., Dawson, T., Berger, G., 2000. Paleoseismology of the Johnson Valley, Kickapoo, and Homestead Valley faults: clustering of earthquakes in the Eastern California Shear Zone. Bull. Seismol. Soc. Am. 90, 1200–1236.

Rockwell, T., Seitz, G., Young, J., Dawson, T., 2004. Late Holocene Earthquake History of the Anza Seismic Gap, Central San Jacinto Fault Zone, Southern California. U.S. Geol. Surv, Final Technical Report, 13 pp.

Rogers, A., 1907. The Shah-Namah of Firdusi. Chapman & Hall, London, 1989 ed., Saeed International, New Delhi, 551 pp.

Rohling, E.J., Pälike, H., 2005. Centennial-scale climate cooling with a sudden cold event around 8,200 years ago. Nature 434, 975–979.

Rosenberg, M., 1979. Survey of excavation in Iran-1978; Eshkaft-e Gavi (The Malyan Project). Iran XVII, 148–149.

Rothé, J.P., 1948. Chronique séismologique. Rev. Et. Calam. 11 (26/27), 10.

Rothe, J.P., 1969. The Seismicity of the Earth 1965. UNESCO, SC.68/D.67/AF, Paris.

Roustaei, M., Nissen, E., Abbassi, M., Gholamzadeh, A., Ghorashi, M., Tatar, M., Yamini-Fard, F., Bergman, E., Jackson, J., Parsons, B., 2010. The 2006 March 25 Fin earthquakes (Iran)—insights into the vertical extents of faulting in the Zagros Simply Folded Belt. Geophys. J. Int.. 181, 1275–1291. http://dx.doi.org/10.1111/j.1365-246X.2010.04601.x.

Rubin, C.M., Sieh, K., 1997. Long dormancy, low slip rate, and similar slip-per-event for the Emerson fault, eastern California shear zone. J. Geophys. Res.. 0148-0227102, 15319–15333. http://dx.doi.org/10.1029/97JB00265.

Rubin, C.M., Sieh, K., Chen, Y.-G., et al., 2001. Surface rupture and behavior of thrust faults probed in Taiwan. EOS, Trans. Am. Geophys. Union 82, 565–569.

Rumi, Maulānā Jalāl U'ddin, 1273. Mathnavi-ye Ma'navi. Ed. & Trans. Nicholson, 1926, The Mathnawi of Jalālu'ddin Rumi. The E.J.W. Gibb Memorial Thrust, London. Persian edition, Ed. Badi' al-Zamān Foruzānfar, Amir Mosta'ān Publ., Tehrān, 1378.

Rypka, J., 1968. History of Iranian Literature. D. Reidel Publishing Company, Dordrecht.

Sa'adi, 1257. Bustān. In: Hossein Yusefi, Gholām (Ed.), third ed. Kharazmi Publ., Tehra, p. 618, 1368/1989 (in Persian).

Sabāhi Bidgoli Kāshāni, Hāji Soleymān, d. 1803. See Ghaffāri Kāshāni, 1796; Zarrābi, 1870; Nakhjavāni, 1964.

Sablin, E.V., 1909. Appendix to report no. 167 of Russian Imperial Mission to H.M. the Shah at Tehran to the First Dept. Minister of Foreign Affairs. St. Petersburg (no date), Tehran.

Saccani, E., Azimzadeh, Z., Dilek, Y., Jahangiri, A., 2013. Geochronology and petrology of the early Carboniferous Misho Mafic complex (NW Iran), and implications for the melt evolution of Paleo-Tethyan rifting in Western Cimmeria. Lithos 162–163, 264–278.

Sachau, E.C. (Trans.), 1910. Alberuni's India. An Account of the Religion, Philosophy, Geography, Chronology, Astronomy, Customs, Laws, and Astrology of India, 2 vols. Kegan Paul, Trench Trubner & Co., London.

Safar Nāmeh-ye Chardin (1664-1670/1074-1080, and 1673-1677/1084-1088 in Iran). Hossain 'Arizi (Ed.), Tehran, 1362/1983.

Sahlān Sāvi (Sāvoji), Zain al-Din 'Omar ibn (12th century AD). Resāleh Sanjariyeh ['al-Resala al-Sanjaria fi al-Kā'enāt al-O'nsoriya']. Ed. Mohammad Taqi Dānesh-Pazhuh. *Farhang Irānzamin*, Tehran, 1337/1958 (in Persian).

Saki, A., 2010. Proto-Tethyan remnants in northwest Iran: geochemistry of the gneisses and metapelitic rocks. Gondwana Res. 17, 704–714.

Saki, A., 2011. Mineralogy, geochemistry and geodynamic setting of the Mahneshan granitoids, NW Iran. Geol. J. 45, 451–466.

Saki, A., Moazzen, M., Modjtahedi, M., Oberähnsli, R., 2008. Phase relations and reaction histories of chloritoid-free and chloritoid-bearing metapelites from the Mahneshan area, NW Iran. Iran. J. Crystallogr. Mineral. 16, 622–640.

Saki, A., Moazzen, M., Modjtahedi, M., Oberähnsli, R., 2008. Determination of P-T conditions of metamorphism of Mahneshan Complex, NW Iran. Iran. J. Geosci. 68, 80–94.

Saki, A., Moazzen, M., Oberhansli, R., 2011. P-T evolution of the Precambrian metamorphic complex, NW Iran: a study of metapelitic rocks. Geol. J., 46 (1), 10–25. http://dx.doi.org/10.1002/gj.1236.

Saki, A., Moazzen, M., Oberhänsli, R., 2012. Mineral chemistry and thermobarometry of the staurolite chloritoid schists from Poshtuk, NW Iran. Geol. Mag.. 149 (6), 1077–1088. http://dx.doi.org/10.1017/S0016756812000209.

Sa'mete Borujerdi, 1909. Zelzeleh-ye Silākhor [The Earthquake of Silākhor]. In: Sa'id Esmā'ili Ketābchi, (Ed.), Collected Poems. Divān, Kermanshah (in Persian).

Sanders, C.O., 1993. Interaction of the San Jacinto and San Andreas *fault* zones, Southern California: triggered earthquake migration and coupled recurrence intervals. Science 260 (5110), 973–976.

Sani' al-Dauleh ('Etemād al-Saltaneh), Mohammad Hassan Khān, 1880–1882. Montazam-e Nāseri (Chronicles of the Nāser al-Din Shāh period). 3 vols. (1298–1300/1880–1882). Lith. Tehran (in Persian).

Sani' al-Dauleh ('Etemād al-Saltaneh), Mohammad Hassan Khān, 1883–1886. Matla' al-Shams. In: Borhan Limudehi, Taimur (Ed.), 3 vols. (1301–1303/1883–1886). Yasāveli Publ., Tehran, 1363/1984 (in Persian).

Saraby, F., Foroughi, B., 1962. A Survey of the Qazvin Earthquake. , Geol. Surv. Iran, Int. Rep., 17 pp.

Sarfaraz, A.A., 1987. Bishapur. In: Kiyani, M.Y. (Ed.), Iranian Cities, vol. 2. Ministry of Islamic Guidance Press, Tehran, pp. 22–74, 1366 (in Persian).

Sarfaraz, 1994 is a personal communication and is not a reference (Pers. Com.).

Sarfaraz, A.A., 1994b. A personal letter from 'Ali Akbar Sarfarāz to Manuel Berberian, 21 June 1994.

Sarkar, I., SriRam, V., Hamzeloo, H., Khattri, K.N., 2005. Subevent analysis for the Tabas earthquake of September 16, 1978, using near field accelerograms. Phys. Earth Planet. In. 151, 53–76.

Saroglu, F., Emre, O., Kuscu, I., 1992. Active Fault Map of Turkey. MTA, Ankara, Turkey.

Satake, K., Somerville, P.G., 1992. Location and size of the 1927 lompoc, California, earthquake from tsunami data. Bull. Seismol. Soc. Am. 82, 1710–1725.

Sauber, J., Thatcher, W., Solomon, S.C., 1986. Geodetic measurement of deformation in the central Mojave Desert, California. J. Geophys. Res. 91 (12), 12683–12693.

Sauber, J., Thatcher, W., Solomon, S.C., Lisowski, M., 1994. Geodetic slip rate for the Eastern California shear zone and the recurrence time of Mojave Desert earthquakes. Nature (Lond.) 367 (6460), 264–266.

Savage, C.D.W., 1957. Earthquake of 2nd July, and Safari Memorandum. Sir A. Gibbs and Partners, London, Internal Report B.45 dated 20 July 1957, and B.58 dated 12 August 1957.

Savarenski, E.F., Soloviev, S.L., Kharin, D.A. (Eds.), 1962. Atlas Zemletrayasenii v SSR (Earthquake in USSR), Akad. Nauk. SSR, 337p.

Sawyer, H.A., 1891. Report of a reconnaissance in the Bakhtiari country. Pub. Govt. Central Printing House, Simla.

Sawyer, H.A., 1894. The Bakhtiary Mountains and Upper Elam. Geogr. J. IV (6), 481–501.

Sayyās, R., 1366/1987. Is the core of the planet earth void? In: Proceedings of the 6th Earth Science Conference. Geological Survey of Iran, Tehran, pp. 30–31, 1366/1987, (in Persian).

Sbeinati, M.R., Darawcheh, R., Mouty, M., 2005. The historical earthquakes of Syria: an analysis of large and moderate earthquakes from 1365 BC to 1900 AD. Ann. Geophys. 48 (3), 347–435.

Schmidt, A., Fāzeli, H., 2007. Tepe Ghabristan: a Chalcolithic tell buried in alluvium. Archaeol. Prospect. 14, 38–46.

Schmidt, A., Quigley, M., Fattahi, M., Azizi, G., Maghsoudi, M., Fazeli, H., 2011. Holocene settlement shifts and palaeoenvironments on the Central Iranian Plateau: investigating linked systems. Holocene. 21, 583–595. http://dx.doi.org/10.1177/0959683610385961.

Schmidt, J.J.F., 1879. Studien uber die Erdbeben. A. Georgi, Leipzig.

Schwartz, H., 2004. Tree of Souls: The Mythology of Judaism. Oxford University Press, USA. 704 pp. http://www.umsl.edu/~schwartzh/samplemyths_2.htm.

SCI, 1967–2013. Statistical Center of Iran (sci.org.ir). Amar.org.ir.

SCI, 2004. Numbers of fatalities, casualties, and missing people as a result of the Bam earthquake (in Persian); and Statistical Data of the Iranian Cities and Provinces. Statistical Centre of Iran, Amar.sci.org.ir.

SC, P&BOGI. Statistical Center, Plan & Budget Organization of Government of Iran. Later transferred to amar.org.ir.

Secor, L., 2014. From Shah to Supreme Leader. Foreign Aff. 93 (1), 160–166.

Seilacher, A., 1969. Fault-graded beds interpreted as seismites. Sedimentology 13, 15–159.

Seilacher, A., 1984. Sedimentary structures tentatively attributed to seismic events. Mar. Geol. 55, 1–12.

Sella, G.F., Dixon, T.H., Mao, A., 2002. REVEL: A model for recent plate velocities from space geodesy. J. Geophys. Res.. 107 (B4), 2081. http://dx.doi.org/10.1029/2000JB000033, ETG11-1-ETG11-30.

Seneca, 86. Naturales Quaestiones. Encyclopedia of the Natural World. Thelatinlibrary.com. naturalesquaestiones.blogspot.com.

Setudehnia, A., 1972. Iran du sud-ouest. Lexique Stratigraphique International HI, Fascicule 9b. Centre National de la Researche Scientifique, Iran, Pads, pp. 285–376.

Shabanian, E., 2009. Active tectonic study in northeast Iran: contribution of the Kopeh Dagh and Binalud mountains to the accommodation of the Arabia-Eurasia convergence. Ph.D. Thesis, Cezanne Univ, Aix-en-Provence, France. Available at http://tel.archives-ouvertes.fr/tel-00433264/fr/.

Shabanian, E., Siame, L., Bellier, O., Benedetti, L., Abbassi, M.R., 2009. Quaternary slip-rates along the north-eastern boundary of the Arabia-Eurasia collision zone (Kopeh Dagh Mountains, north-east Iran). Geophys. J. Int.. 178, 1055–1077. http://dx.doi.org/10.1111/j.1365-246X.2009.04183.x.

Shabanian, E., Bellier, O., Siame, L., Arnaud, N., Abbassi, M.R., Cochemé, J.J., 2009. New tectonic configuration in NE Iran: active strike-slip faulting between the Kopeh Dagh and Binalud mountains. Tectonics 28http://dx.doi.org/10.1029/2008TC002444, TC5002.

Shabanian, E., Bellier, O., Abbassi, M.R., Siame, L.L., Farbod, Y., 2010. Plio-Quaternary stress states in NE Iran: Kopeh Dagh and Allah Dagh-Binalud mountains. Tectonophysics. 480, 280–304. http://dx.doi.org/10.1016/j.tecto.2009.10.022.

Shabanian, E., Bellier, O., Siame, L., Abbassi, M.R., Bourlès, D., Braucher, R., Farbod, Y., 2012. The Binalud Mountains: a key piece for the geodynamic puzzle of NE Iran. Tectonics 31http://dx.doi.org/10.1029/2012TC003183, TC6003.

Shabanian, E., Acocella, V., Gioncada, A., Ghasemi, H., Bellier, O., 2012. Structural control on volcanism in intraplate post collisional settings: Late Cenozoic to Quaternary examples of Iran and eastern Turkey. Tectonics 31http://dx.doi.org/10.1029/2011TC003042, TC3013.

Shafi'ee-Kadkani, 1996. See Hākem Neyshāburi, 998.

Shahbāzi, S., 2003a. Haftvād. Encyclopaedia Iranica, 11. www.iranicaonline.org.

Shahbāzi, S., 2003b. Iranians and Alexander. Am. J. Anc. Hist. 2, 15–38.

Shahidi, S.J., 1954. Bahthi dar Bāreh-ye Shahrbānu. Cherāgh-e Rrowshan dar Donyā-ye Tārik, Tehran, 1333/1954–1955, 186–201.

Shahmardan ebn-e Abi al-Khair, 1095. Nozhat-nāmeh-ye 'Alā'i (488/1095). Ed. Farhang Jahānpur, 1983. Mo'assesseh Motāle'at va Tahqiqāt Farhangi, Tehran, 1362/1983 (in Persian).

Shāhnāmeh, 1010. The Epic of the Kings. Ferdowsi Tusi. Ed. A. Rogers, 1907, Saeed Int., New Delhi. Ed. H. Zimmern, 1926, The Macmillan Co., New York. Ed. R. Levy, 1967, Routledge & Kegan Paul., The Royal Institute of Publication of Tehran. Ed. F. Joneidi, 2008, Nshar Balkh, Neyshābur Foundation, Tehran.

Shāhpasandzādeh, M., Zāre', M., 1996. Baresi-ye gosalesh-e jonbā dar reshteh kuhāyeh Zāgros az didgāh-e larzeh-zaminsākht; gosal-e zamilarzehi-yeh Mishān [Study of active faulting in the Zagros Mountains from the seismotectonic point of view; the Mishān earthquake fault]. IIEES Bull. 4 (5), 10–12, 1347, Tehran (in Persian).

Shahrestānhā-yé Irānshahr. See Tafazzoli, 1989; Ed. Daryāee, 2002.

Shaibāni, 'Adib, Ahmad, Mirzā, 1977. See Narāqi, 1966; Zarrābi.

Shaibāni, Z., 1986. The Shebli Tomb Tower of Damāvand. *Āsār, Sāzémān Meli Hefāzat Āsar Bāstāni*, No. 12, 13, 14, 59–71, 1365 (in Persian).

Shaibāni, Z., 1990. Reconstruction Plan for Emāmzādeh 'Abdolāh and 'Abidolāh of Damāvand. *Āsār, Sāzémān Meli Hefāzat Āsār Bāstāni*, No. 18, 19, 128–158, 1369 (in Persian).

Shākeri, R., 1967, 1981, 1986, 2001. Atraknāmeh, Tārikh-e Jāme'-e Quchān. second ed. Amir Kabir Publ, 487 pp., 1381 (in Persian).

Shākeri, R., 2002. Atraknāmeh, Tārikh-e Jāme'-e Quchān [The Book of Atrak, Complete History of Quchān], second ed. Amir Kabir Publication, Tehran, Iran, p. 487, 1381/2002 (in Persian).

Shakespeare, W., 1597. Henry IV.

Shakespeare, W., 1602. Hamlet.

Shāki, M., 2002. Haftānboxt: traditional reading of the name of a legendary warlord in southern Persia, mentioned in the Kar-nāmag I Ardasir Pābagān (The exploits of Ardasir son of Pābag). Encylopaesia Iranica XI (5), 530–531.

Shaki, M., 2003. Haftānbokht. Encyclopaedia Iranica, 11. www.iranicaonline.org.

Sharifi, H., Qasabian, N.N., 2013. The Gosht-Sarāvān, Sistān-Baluchestān province earthquake of April 22, 2013 (Farvardin 27, 1392). Geol. Surv. Iran, Int. Rep. 20p, (in Persian).

Sharp, R.V., 1975. Displacement on tectonic ruptures. Calif. Div. Mines Geol. Bull. 196, 187–194.

Sharp, R.V., 1976. Comparison of 1979 Surface Faulting with Earlier Displacements in the Imperial Valley, California, Earthquake of October 15, 1979. U.S. Geol. Surv., Prof. Pap., 1254, 213–221.

Sharp, R.V., Lienkaemper, J.J., Bonilla, M.G., Burke, D.B., Fox, B.F., Herd, D.G., Miller, D.M., Morton, D.M., Ponti, D.J., Rymer, M.J., Tinsley, J.C., Kahle, J.E., Hart, E.W., Sieh, K.E., 1982. Surface faulting in the central Imperial Valley. In The Imperial Valley California, earthquake of October 15, 1979. U.S. Geol. Surv. Prof. Pap. 1254, 119–143.

Sharp, R.V., Budding, K., Boatwright, J., Adler, M.J., Binilla, M.G., Clark, m.M., Fumal, M.M., Harms, K.K., Lienkaeper, J.J., Morton, D.M., O'Neill, B.J., Ostergren, C.L., Ponti, D.J., Pymer, M.J., Saxton, J.L., Sims, J.D., 1989. Surface faulting along the Superstition Hills fault zone and nearby faults associated with the earthquakes of 24 November 1987. Bull. Seismol. Soc. Am. 79, 252–281.

Shijādel, M., 2008. St. Thaddeus Armenian Church, Maku, Iran. Azarbāijān Department of the Cultural Heritage, Tourism, and Handicrafts, Tabriz, Iran.

Shenon, Ph., 1990. Some foreigners ousted in quake area. N.Y. Times , A10, 26 June 1990.

Shirokova, E.I., 1962. Stresses effective in earthquake foci in the Caucasus and adjacent districts. Bull. Acad. USSR, Geophys. Ser. 10, 809–815 (English Trans.).

Shirokova, E.I., 1967. General features in the orientation of principal stresses in earthquake foci in the Mediterranean-Asian seismic belt. Bull. Acad. USSR Earth Phys. 1, 22–36.

Shirvāni, Zain al-'Abedin, 1837. Hadā'eq al-Siyāha, 1857. Tehran.

Shoja-Taheri, J., Anderson, J.G., 1988. The 1978, Tabas, Iran, earthquake: an interpretation of the strong motion records. Bull. Seismol. Soc. Am. 78, 142–171.

Shoja-Taheri, J., Naserieh, S., Ghafoorian-Nasab, A.H., 2005. The 2003 Bam, Iran, Earthquake: An Interpretation of the Strong Motion Records. Earthq. Spectra. 21 (Sp. Issue. S1), S181–S206. http://dx.doi.org/10.1193/1.2101887, The Earthquake Engineering Research Institute, Oakland, Calif.

Shushtari, M.'.A.E., 1952. Tārikh-e Joghrāfiyāee-ye Khuzestān [History of Geography of Khuzestān]. Amir Kabir Publ., Tehran, 1331 (in Persian).

Siame, L.L., Bellier, O., Sebrier, M., Bourles, D.L., et al., 2002. Seismic hazard reappraisal from combined structural geology, geomorphology and cosmic ray exposure dating analyses: the Eastern Precordillera thrust system (NW Argentina). Geophys. J. Int. 150, 241–260.

Sibson, R.H., 1986. Rupture interaction with fault jogs. In: Earthquake Source Mechanics, Geophysical Monograph, vol. 37 (Maurice Ewing 6), Am. Geophys. Union, pp. 157–167.

Sibt Ibn al-Jauzi, Shams al-Din, 1256. Mir'at al-Zaman. Ed. Jewett, Chicago, 1907.

Sieberg, A., 1932. Erdbebengeographie. In: Guttenberg, B. (Ed.), Hanbuch der Geophysik, Berlin, pp. 90–126.

Sieh, K.I., 1978. Slip along the San Andreas fault associated with the great 1857 earthquake. Bull. Seismol. Soc. Am. 68, 1421–1448.

Sieh, K.L., 1978. Pre-historic large earthquakes produced by slip on the San Andreas fault at Pallet Creek, California. J. Geophys. Res. 83, 3907–3939.

Sieh, K.L., Jones, L., Hauksson, E., Hudnut, K., Eberhart-Phillips, D., Heaton, T.H., Hough, S., Hutton, K., Kanamori, H., Lilje, A., Lindvall, S., McGill, S.F., Mori, J., Rubin, C., Spotila, J.A., Stock, J., Thio, H.K., Treiman, J., Wernicke, B., Zachariasen, J., 1993. Near-field investigation of the Landers earthquake sequence, April to july, 1992. Science 260, 171–176.

Silva, P.G., Sintubin, M., Reicherter, K. (Eds.), 2011. Earthquake Archaeology and Paleoseismology. Quat. Int. 242 (1), 258.

Silver, P.G., Jordan, T.H., 1983. Total-moment spectra of fourteen large earthquakes. J. Geophys. Res. 88 (B4), 3273–3293.

Singh, G., Wasson, R.J., Agrawal, D.P., 1990. Vegetational and seasonal climatic changes since the last full glacial in the Thar Desert, northwestern India. Rev. Palaeobot. Palynol. 64, 351–358.

Siyāhat Nāmeh-ye Chardin (1664-1670/1074-1080, and 1673-1677/1084-1088 in Iran). 6 vols. Mohammad 'Abbāsi (Ed.). Amir Kabir Publication, Tehran 1338/1959 (in Persian).

Smalley, R., Pujol, J., Regnier, M., Chiu, J.-M., Chatelain, J.-L., Isacks, B.L., Araujo, M., Puebla, N., 1993. Basement seismicity beneath the Andean Precordillera thin-skinned thrust belt and implications for crustal and lithospheric behavior. Tectonics 12, 63–76.

Smith, C.B.E., 1876a. The Perso-Baluch Frontier Mission 1870, 1871, and the Perso-Afghan Mission 1871, 1872. In: Goldsmid, F.J. (Ed.), Eastern Persia, vol. 1. Macmillan & Co., London.

Smith, G., 1876b. The Chaldean Account of Genesis. Thomas Scott, London.

Smith, G.L., McNeill, L., Wang, K., He, J., 2013. Thermal structure and megathrust seismogenic potential of the Makran subduction zone. Geophys. Res. Lett.. 40, 1–6. http://dx.doi.org/10.1002/grl.50374.

Sobuti, F., 1963. Sur le mechanism au foyer dan l'arc séismique entre l'Hindou-Kouch et la Méditerranée. Ann. De Geophys. 19 (1), 51–62, Paris.

Sobouti, M., Wshghi, I., Mostaanpour, M.M., 1978. Shamil-Siahou earthquake "north-east Bandar-Abbas" of 21st March, 1977. J. Geophys. 44, 663–674.

Sohrab Sepehri, 1964. The Sound of Water's Footsteps, Kāshān, Chenār (Plane Tree) Village, Summer, 1343. Hasht Ketāb. Tahuri Publ., Tehran, 1360/1981.

Sohrevardi, Shaikh Shahāb al-Din Yahyā, 1186. Hikmat al-Ishrāq [The Theosophy of the Orient of Light/The School of Illumination (532/1186)]. Translated into Persian by S. J'afar Sohrevardi, Tehran University Publ., 1361/1982 (in Persian).

Solaymani Azad, S., 2009. Evaluation de l'aléa sismique pour les villes de Téhéran, Tabriz et Zandjan dans le NW de l'Iran. Ph.D. Thèses, Approche morphotectnique et aléosismologique [Seismic Hazard Assessment for Tehran, Tabriz and Zandjan Cities (NW Iran) Based on Morphotectonics and Palaeoseismology], vol. 2 de l'Université de Montpellier, 150 pp.

Solaymani, S., Feghhi, K., 2002. Report of Surface Faulting and Morphotectonics of "Avaj Region" Earthquake on June 22, 2002. Iiees.ac.ir, 11 pp.

Solaymani Azad, S., Philip, H., Dominguez, S., Hessami, K., Shahpasandzadeh, M., Forutan, M.-R., Tabassi, H., Lamothe, M., 2009. Paleoseismological and Morphological Evidences of Slip Variations Along the North Tabriz Fault (NW Iran). In: Solaymani Azad, S. (Ed.), submitted for publication.

Solaymani Azad, S., Ritz, J.-F., Abbassi, M.R., 2011. Left-lateral active deformation along the Mosha–North Tehran fault system (Iran): morphotectonics and paleoseismological investigations. Tectonophysics 497, 1–14.

Solecki, R.S., 1955. Shanidar Cave, a Paleolithic Site in Northern Iraq. Smithsonian Institute, Washington, DC, Smithsonian Report for 1954, 389–425.

Solecki, R.S., 1963. Prehistory in Shimbar Valley, Northern Iraq. Science 139 (3551), 179–193.

Solecki, R.S., 1964. Shanidar Cave, a Late Pleistocene Site in Northern Iraq. In: INQUA, VIth Congr., Warsaw, 1961, pp. 413–422.

Solecki, R.S., 1971. Shasnidar, The First Flower People. Alfred A. Knopf, New York.

Solecki, R.S., 1977. The implications of the Shanidar cave Neanderthal flower burial. Ann. N. Y. Acad. Sci. 293, 114–124.

Solecki, R.L., 1981. An Early Village Site at Zawi Chemi Shanida. vol. 13 Bibliotheca Mesopotamia, Malibu.

Soleki, R.S., 1997. Shanidar Cave. Encyclopedia-Near Eastern Archaeology, 15–16.

Soltāni-Moqadam, 2012. Unpublished poems by the sons of Vesāl Shirāzi regarding the 1853 earthquake in Shirāz. Payām Bahārestān 14 (2), 1421–1437, 1391, Library, Museum, and Document Center of Iran Parliament, www.ical.ir, 1391 (in Persian).

Sorush, M.R., 2007. Sounding of the old Qa'en congregational mosque, southern Khorasan: Report on the Archaeological Activities in 2006. In: Fazeli Nashli, Hassan (Ed.), Iranian Cultural Heritage, Handicrafts, and Tourism Organization, ichto.ir, 2 pp. (in Persian).

Sorushian, J.S., 1956. Farhang-e Behdinān. Ed. Manuchehr Sotudeh, Tehrān, 1335/1956.

Sotudeh, M., 1970, 1972. Az Āstārā ta Astarābād [Form Āstārā to Astarābād]. 2 vols., 1349, 1351, Tehran. Also 1357/1978 ed. (in Persian).

Sobouti, M., Eshghi, I., Hossein-Javaheri, J., 1972. The Qir earthquake of 10th April 1972. J. Earth Space Phys. Inst. Geophys. 1 (2), 17–24, Tehran Univ.

Spiegel, F., 1882. Vergleichende Grammatik der Alteranischen Sprache, Leipzig.

Sprenger, A., 1843. As-Soyuti's work on earthquakes. J. Asiatic Soc. Bengal 12 (141), 741–749, Calcutta.

Stack, E., 1882. Six Months in Persia. vol. 1 Sampson Low, Marston, Searle, and Rivington, London, 2 vols.

Stahl, A.F., 1911. Handbuch der regionalen Geologie. vol. 5. , Persien, part v.6. Heidelberg.

Stahl, P., 1962. Sur la seismicite de l'Iran. *Bul. Etud. no. 3, Stn. Seismol. de Shiraz.*

Standish, J.F., 1970. The Caspian Gates. Greece Rome 17 (1), 17–24.

Stark, F., 1934. The site of the city of Kalar in Kalardasht. Geogr. J. 83 (3), 211–217.

Stein, M.A., 1940. Old Routes of Western Iran. Macmillan and Sons, London.

Stein, R.S., Ekström, G., 1992. Seismicity and geometry of a 110-km-long blind thrust fault: 2. Synthesis of the 1982-1985 California earthquake sequence. J. Geophys. Res. 97, 4865–4883.

Stein, R.S., Lisowski, M., 1983. The 1979 Homestead valley earthquake sequence, California—control of aftershocks and postseismic deformation. J. Geophys. Res. 88, 6477–6490.

Stein, R.S., Thatcher, W., 1981. Seismic and aseismic deformation associated with the 1952 Kern County, California, earthquake and relationship to the Quaternary history of the White Wolf fault. J. Geophys. Res. 86, 4913–4928.

Stack, R.S., King, G.C.P., Lin, J., 1992. Change in failure stress on the southern San Andreas fault system caused by the 1992 magnitude = 7.4 Landers earthquakes. Science 238, 1328–1332.

Staubwasser, M., Weiss, H., 2006. Holocene climate and cultural evolution in late prehistoric West Asia. Quat. Res. 66, 372–387.

Stein, R.S., Barka, A.A., Dieterich, J.H., 1997. Progressive failure on the North Anatolian fault since 1939 by earthquake stress triggering. Geophys. J. Int. 128, 594–604.

Stepanos Orbelyan (Sisāgā'i/Swinetsi), 13th century. Bādmutiun Nāhānkin Sisāgān [History of the Province of Sisāk]. Ed. Chahnazarian, I–II, Paris, 1859 (in Armenian).

Stepanyan, V.A., 1942. Istoricheskii obzor o zemletriasenikh v Armenii i vprilegaiuschikh rayonakh; kratkaia chronologiya (Historical Survey of Earthquakes in Armenia and Surrounding Regions). In: Proc. Zakayakazskaia Konfeerentsiia po Antiseismitseskoi Struitel'stvu [Biuro Antiseisimich, Stroitel]. Akad. Nauk Izdat. Armen. F.A.N, Yerevan (in Armenian), pp. 43–72, Reprinted in 1964 as Historical Review of Earthquakes in Armenia ann Surrounding Area, 246p Yerevan, Armenia SSR (in Armenian)..

Stern, S.M., 1966. The inscription of the Kharaqān mausoleum. Iran IV, 21–27.

Stewart, C.E., 1881. The country of the Tekke Turkomans and the Tejend and Murghab Rivers. Proc. R. Geogr. Soc. 3, 513–546.

Stewart, C.E., 1911. Through Persia in Disguise with Reminiscences of the Indian Mutiny. George Routledge & Sons, Ltd, London.

Stocklin, 1961. Lagoonal formations and salt domes in East Iran. Pet. Inst. Bull. 3, 29–46, Also: Lagunare Formationen und salzdome in ostian. Eclog. Geol. Hel., 54 (1), 1–27.

Stocklin, J., 1968. Structural history and tectonics of Iran: a review. Am. Assoc. Pet. Geol. Bull. 52, 1229–1258.

Stocklin, J., 1968. Salt deposits of the Middle East. Geol. Soc. Am. Spec. Pap. 88, 157–181.

Stocklin, J., 1972. Iran Central, Septentrional et Oriental. Lexique Stratigraphique International, III. Centre National de la Recherche Scientifique, Paris, pp. 1–283, Fascicule 9b, Iran.

Stocklin, J., 1974. Possible ancient continental margins in Iran. In: Burk, C.A., Drake, C.L. (Eds.), The Geology of Continental Margins. Springer, New York, pp. 873–887.

Stocklin, J., 1977. Structural Correlation of the Alpine Ranges Between Iran and Central Asia. Mrmoire Hors-Srrie No 8 de la Socirt6 Grologique de, France, 8, 333–353.

Stocklin, J., Eftekhārnezhād, J. (Compilers), 1969. Geological Quadrangle Map of Zanjan. Geol. Surv. Iran, No. D7. The 1990 earthquake area was mapped by Eftekharnezhad, J., Nabavi, M.H., and Valeh, N.

Storey, C.A., 1927. Persian Literature. Luzac & Co, London, Archive.org.

Storey, C.A., De Blois, F., De Neufville, R., 2004. Persian Literature—A Bibliographical Survey: Poetry of the Pre-Momgol Period. vol. V Routledge Curzon.

Stover, C.W., Coffman, J.L., 1993. Seismicity of the United States, 1568-1989 (Revised). U.S. Geological Survey Professional Paper 1527. United States Government Printing Office, Washington.

Strabo. Geographica. The Geography of Strabo. The Loeb Classical Library, 1932. http://penelope.uchicago.edu/Thayer/E/Roman/Texts/Strabo/home.html.

Stronach, D., Cuyler Young Jr., T., 1966. Three Seljuq tomb towers. Iran IV, 1–20.

Suarez, G., Molnar, P., Burchfiel, C., 1983. Seismicity, fault plane solutions, depth of faulting, and active tectonics of the Andes of Peru, Ecuador and southern Columbia. J. Geophys. Res. 88, 10403–10428.

Sublett, K. Epic of Gilgamesh. Piney.com.

Sudhaus, H., Jónsson, S., 2011. Source model for the 1997 Zirkuh earthquake (MW=7.2) in Iran derived from JERS and ERS InSAR observations. Geophys. J. Int. 185 (2), 672–676.

Sultan, D.I., 1931. The Managua earthquake. Mil. Eng. 2, 354–361.

Surmeliān, L., 1966. Daredevils of Sāssoun; The Armenian National Epic. George Allen & Unwin Ltd., London, 251 pp.

al-Suyuti, Jalal al-Din, 1499. Kashf al-Salsalah ‘an Wasf al-Zalzalah [The Transmitted Expositions Concerning the Description of the Earthquake of Doomsday, 905 H/AD 1499]. Ed. ‘Abd al-Latif Sa’adāni, Fez. 1971, Cairo, Egypt. French Trans., Ed., Sa’id al-Nejjār, Traité du Tremblement de terre, *Cahiers du Centre Universitaire de la Recherche Scientifique,* Rabat, Morocco, 1975, 105 pp. Also Ed. Muhammad Kamal al-Din ‘Izz al-Din, *‘Alim al-Kutub, 1987,* Beirut.

Suzuki, S., Matsushima, T., Ito, Y., Hosseini, S.K., Nakamura, T., Gandomi, A.J., Sadeghi, H., Maleki, M., Aghda, F., 2004. Source Fault of the 2003/12/26 Bam Earthquake (Mw 6.5) in Southeastern Iran Inferred from Aftershock Observation Data by Temporal High-Sensitive Seismograph Network. EOS Trans. AGU 85 (17), Jt. Assem. Suppl., Abstract S23A-07.

Sykes, P.M., 1902. A fourth journey in Persia, 1897-1901. Geogr. J. 19 (2), 121.

Sykes, P.M., 1911. A sixth journey in Persia. Geogr. J. R. Soc. Lond. 37 (2), 149–166.

Szabo, F., Kheradpir, A., 1978. Permian and Triassic stratigraphy, Zagros Basin, southwest Iran. J. Pet. Geol. 1, 57–82.

Tabandeh, Haji Soltān Hossain, 1969. Tārikh va Joghrāfiya-ye Gonābād. Dāneshgah Tehrān Publ., Tehrān, 1348 (in Persian).

Tabari, Abu Ja’far Muhammad ibn Jarir abu al-, 915. Tārikh al-Rusul wa’l-Muluk [The History of Prophets and Kings], or Tārikh al-Tabari/Tārikh-e Tabari [The History of al-Tabari]. Cairo, *Dār al-Ma’ārif, 1960-69* (10 vols.). Persian translation by ‘Abolqāsem Pāyandeh. Bonyād Farhang Publ., Tehrān, 1354/1975. English translation by C.E. Bosworth, 1999. The History of Al-Tabari; An Annotated Translation. State University of New York Press, vol. V: The Sasanids, the Byzantines, the Lakhmidds, and Yemen), 458 pp. Also English translation & annotation by Franz Rosenthal, 1989. The History of Al-Tabari Tarikh al-Rusul Wal-Muluk). State University of New York Press, vol. I: General Introduction and From the Creation to the Flood, 458 pp.

al-Tabari, 'Ali ibn Sahl (Rabbān), 850. Firdaws al-Hikma ['Paradise of Wisdom'], ed. M. Meyerhof, 1931, Chapter 3I2, P. 5I9. Also Ed. DKI, Firdaws al-Hikma fi al-Tibb ['Paradise of Wisdom in Medicie'], Beirut, 2002 (in Arabic).

Tabātabā'i Tabrizi, Mohammad Rezā, 1875. Tārikh-e Aulād al-Athār (1292/1875). Tabriz, 1304/1887; Tehran, 1314/1896.

Tabib Ordubādi, Sayyed Najib Mir 'Ali Naqi, 18th century. Tārikh-e Zelzeleh-ye Dār al-Saltaneh (History of the earthquakes of the capital city). Central Library, Tehran University, ms. 2591:661.

Tafazzoli, A., 1989. Shahrestān-hāyeh Iran [Iranian Cities]. In: Kiyāni, M.Y. (Ed.), Shahrhāyeh Irān [Iranian Cities]. Ministry of Islamic Guidance Press, Tehran, 1368, Tehran, v. 3, 349 pp (in Persian).

Tafazzoli, A., 1996. Dezh-e Bahman. Encyclopaedia Iranica. www.iranicaonline.org.

Taher, M.A., 1974. Textes d'historiens damascenes sur les tremblements de terre du XII[e] siecle de l'Hegire (XVII–XVIIIe S). Bull. Etudes Orientales 27, 51–108, Damascus (in Arabic).

Taher, M.A., 1979. Corpus des textes Arabes relatifs aux tremblements de terre et autres catastrophes naturelles, de la conquete Arabe au XII H/XVIII JC. LLD Thesis, Univ. Paris I (in French).

Tahman al-Baihaqi, Davud ebn, 856. Poem on the Kumes/Dāmghān earthquake of 22 December 856 (242.08.18). In Abol Hassan 'Ali ebn-e Zaid Baihaqi (Ebn-e Fandoq), Tārikh-e Baihaq. Ed. Ahmad Bahmanyār, Bongāh-e Dānesh Publ., Tehran, 1317/1939.

Takin, M., 1972. Iranian geology and continental drift in the Middle East. Nature 235, 147–150.

Talā'i, H., 1998. Preliminary Report of the 10th Excavation Season (1998) at tapeh Sagzabad, Qazvin Plain. Archaeological Institute, Tehran University, unpublished internal report, 31 pp.

Talebian, M., Jackson, J., 2002. Offset on the main recent fault of the NW Iran and implications on the late Cenozoic tectonics of the Arabia–Eurasia collision zone. Geophys. J. Int. 150, 422–439.

Talebian, M., Jackson, J.J., 2004. A reappraisal of earthquake focal mechanisms and active shortening in the Zagros mountains of Iran. Geophys. J. Int. 156 (3), 506–526.

Talebian, M., Fielding, E.J., Funning, G.J., Ghorashi, M., Jackson, J., Nazari, H., Parsons, B., Priestley, K., Rosen, P.A., Walker, R., Wright, T.J., 2004. The 2003 Bam (Iran) earthquake: Rupture of a blind strikeslip fault. Geophy. Res. Lett.. 31http://dx.doi.org/10.1029/2004GL020058, 4 p.

Talebian, M., Biggs, J., Bolourchi, M., Copley, A., Gassemi, A., Ghorashi, M., Hollingsworth, J., Jackson, J., Nissen, E., Oveisi, B., Parsons, B., Priestley, K., Saiidi, A., 2006. The Dahuiyeh (Zarand) earthquake of 2005 February 22 in central Iran: Reactivation of an intramountain reverse fault. Geophys. J. Int. 164 (1), 137–148. http://dx.doi.org/10.1111/j.1365-246X.2005.02839.x.

Talebian, M., Qorashi, M., Nazari, H., Baharvand, M., 2006. Gozāresh-e moqadamāti-ye zaminlarzeh-ye Dasht-e Silākhor. Geol. Surv, Iran, internal report, gsi.ir. National Geoscience Database of Iran, ngdir.ir.

Tarasov, S.M., 1902. Earthquakes in Akhalqalak region December 1931, 1899, and August 1, 1899. News Cacasus Brach Russian Geographical Soc. XV (1), 24–43, Tiflis.

Tarbiyat, M.'A.Kh, 1935. Dāneshmandān-e Āzarbāijān. 1314, Tehran (in Persian).

Tarāyeq al-Haqāyeq of Ma'sum 'Alishāh (v.III). In Shākeri, 2001.

Tārikh Bal'ami. See Bal'ami.

Tārikh al-Rasul wal-Moluk. See Tabari, 915.

Tārikh-e Sistān, ca., 1062. History of Sistān from the 11th century with short continuation up to 1325, written by an anonymous author. In: Malek al-Sho'arā Bahār (Ed.), Kolāleh Khavar Publ., Tehran, 1314/1935 (in Persian).

Tātār, M., 2001. Etude sismotectonique de deux zones de collision continentale: le Zagros et l'Alborz (Iran). These, L'Universite Joseph Fourier-Grenoble I, France, pp. 144–258.

Tatar, M., Hatzfeld, D., 2008. Microseismic evidence of slip partitioning for the Rudbar-Tarom earthquake (Ms 7.7) of, 1990 June 20 in NW Iran. Geophys. J. Int.. http://dx.doi.org/10.1111/j.1365-246X.2008.03976.x.

Tatar, M., Yaminifard, F., Hessami Azar, K.T., Haqshenas, E., Hosseinzadeh, N., Ghafuri Zarandi, A.R., 1997. Gozāresh-e moqadamāti zaminlarzeh 16/11/75 shomāl Bojnurd. IIEES 1375, 36 pp. Tehran, Esfand (in Persian).

Tatar, M., Hatzfeld, D., Martinod, J., Walpersdorf, A., Ghafori-Ashtiani, A., Chery, J., 2002. The present-day deformation of central Zagros from GPS measurements. Geophys. Res. Lett.. 29 (19), 1927–1930. http://dx.doi.org/10.1029/2002 GLO15427.

Tatar, M., Hatzfeld, D., Ghafory-Ashtiany, M., 2004. Tectonics of the Central Zagros (Iran) deduced from microearthquake seismicity. Geophys. J. Int. 156, 255–266.

Tatar, M., Hatzfeld, D., Moradi, A.S., Paul, A., 2005. The 2003 December 26 Bam earthquake (Iran), *Mw* 6.6, aftershock Sequence. Geophys. J. Int.. 163, 90–105. http://dx.doi.org/10.1111/j.1365-246X.2005.02639.x.

Tatar, M., Jackson, J., Hatzfeld, D., Bergman, E., 2007. The 28 May 2004 Baladeh earthquake (Mw $\sim$ 6.2) in the Alborz, Iran: implications for the geology of the South Caspian Basin margin and for the seismic hazard of Tehran. Geophys. J. Int. 170, 249–261.

Tate, G.P., 1910–12. Seistan: A Memoir on the History, Topography, Ruins, and People of the Country. Roy.SVO, 2 vols. Superintendent Government Printing, Calcutta, Also, Gosha-e Adab, Quetta, Pakists, 1977, 378 pp.

Tavakoli, F., 2007. Present-Day Deformation and Kinematics of the Active Faults Observed by GPS in the Zagros and East. Thesis, Université Joseph Fouruer, Iran, 313 pp.

Tavernier, J.B., 1681. Les six voyages de Jean-Baptiste Tavernier, qu'il a fait en Turquie, en Perse, et aux Indes. 3 vols. G. Clouzier, Paris.

Tchalenko, J.S., 1973. The Kashmar (Turshiz) 1903 and Torbat-e Heidariyeh (south) 1923 earthquakes in central Khorassan (Iran). Annali di Geofisica XXVI (1), 29–40.

Tchalenko, J.S., 1974. Recent destructive earthquakes in the Central Alborz. In: Materials for the Study of Seismotectonics of Iran: North-Central Iran. vol. 29. Geol. Surv. Iran, pp. 97–116.

Tchalenko, J.S., 1975. Seismicity and structure of the Kopeh Dagh (Iran, USSR). Philos. Trans. R. Soc. Lond. A 278 (1275), 1-28.

Tchalenko, J.S., 1977. A reconnaissance of the seismicity and tectonics at the northern border of the Arabian Plate (Lake Van region). Revue de Géographie Physique et de Géologie Dynamique XIX (2), 189–208.

Tchalenko, J., Ambraseys, N.N., 1970. Structural analysis of the Dasht-e-Bayaz earthquake fractures. Bull. Geol. Soc. Am. 81, 41–60.

Tchalenko, J., Ambraseys, N.N., 1973. Earthquake destruction of adobe villages in Iran. Annali di Geofis. 26, 357–389.

Tchalenko, J.S., Berberian, M., 1974. The Salmas (Iran) earthquake of May 6, 1930. Annali di Geofisica 27 (1–2), 151–212.

Tchalenko, J.S., Berberian, M., 1975. Dasht-e-Bayaz fault, Iran: earthquake and earlier related structures in bedrock. Geol. Soc. Am. Bull. 86, 703–709.

Tchalenko, J.S., Braud, J., 1974. Seismicity and structure of the Zagros (Iran)-the Main Recent Fault between 33 and 35.N. Phil. Trans. Roy. Soc. Lond. 277, 1–25.

Tchalenko, J.S., Berberian, M., Behzadi, H., 1973. Geomorphic and seismic evidence for recent activity on the Doruneh fault, Iran. Tectonophysics 19, 333–341.

Tchalenko, J.S., Braud, J., Berberian, M., 1974. Discovery of three earthquake faults in Iran. Nature 248, 661–663.

TEH. Institute of Geophysics, University of Tehran (IGUT), Tehran. http://irsc.ut.ac.it/.

Ter-Grigorian, Kh., 1971. Brief chronological history. Documents of Armenian Architecture, U.S. Thadei Vank. Edizioni Ares, 12–13.

Theatrum Europaeum, 1617-1721. Ed. J.B. Abelin, 21 vols. Frankfurt.

Thompson, R.C., 1903. The Devils and Evil Spirits of Babylonia: being Babylonian and Assyrian incantations against the demons, ghouls, vampires, hobgoblins, ghosts, and kindred evil spirits, which attack mankind, translated from the original Cuneiform texts, with transliterations, vocabulary, notes, etc. Evil Spirits, vol. 1. Luzac & Co, London, 248 pp.

Thompson, R.C., 1904. The Reports of the Magicians and Astrologers of Nineveh and Babylon, in Assyrian and Babylonian Literature. Appleton & Company, New York, pp. 451–460. Selected Transactions, With a Critical Introduction by Robert Francis Harper, D. http://www.earth-history.com/Babylon/astrology.htm.

Thompson, R.C., 1937. A new record of an Assyrian earthquake. Br. Inst. Study Iraq IV (2), 186–189.

Thompson, R.C., Hamilton, R.W., 1932. The British Museum excavations on the temple of Ishtar at Nineveh 1930–31. Liverpool Annals Archaeol. Antropol. 19, 55–116.

Tirrul, R., Bell, I.R., Griffis, R.J., Camp, V.E., 1983. The Sistan suture zone of eastern Iran. Geol. Soc. Am. Bull. 94, 134–150.

Toksoz, M.N., Arpat, E., Saroglu, F., 1977. East Anatolian earthquake of 24 November 1976. Nature 270, 423–425.

Toksoz, M.N., Nabelek, J., Arpat, E., 1978. Source properties of the 1976 earthquake in east Turkey: a comparison of field data and teleseismic results. Tectonophysics 49, 199–205.

Tovmā Mecop'etsi. colophon in the Matenādārān, Yerevān, manuscript 9247, fol. 299r. Yerevān, Armenia.

Transparency Int. Transparency International. Transparency.org.

Tributsch, H., 1982. When the Snakes Awake: Animal and Earthquake Prediction. The MIT Press, Cambridge, MA, USA, 248 pp., Translated By Paul Langner.

Trifonov, V.G., Karakhanian, A.S., Assaturian, A.O., Ivanova, T.P., 1994. Relationship of earthquakes and active faulta in Anatolia, the Lesser Caucasus, and the Middle East. In: Bolt, B.A., Amirbekian, R. (Eds.), Proc. Int. Conf. Continental Collision Zone Earthquakes and Seismic Hazard Reduction, Yerevan-Sevan, Armenia, 1–6 October 1993, IASPE/IDNDR, 1 September 1994, pp. 41–55.

Tsukuda, T., Skai, K., Hashimoto, S., Gheitanchi, M.R., Soltanian, S., Mozaffari, P., Mozaffari, N., Akasheh, B., Javaherian, A., 1991. Aftershock distribution of the 1990 Rudbar, northwest Iran, earthquake of M 7.3 and its tectonic implications. Bull. Earthq. Res. Inst. Univ. 66, 351–381, Tokyo.

Tsutsumi, H., Yeats, R.S., 1999. Tectonic setting of the 1971 Sylmar and 1994 Northridge earthquakes in the San Fernando Valley. California. Seismol. Soc. Am. Bull. 89, 1232–1249.

Tusi, Mohammad ibn Mahmmud ibn Ahmad, 1167. Ajā'ib al-Makhluqāt va Gharā'ib al-Mawjudāt [The Wonders of Creation; 1161-1167]. In: Sotudeh, Manuchehr (Ed.), Bogāh Tarjomeh va Nashr Ketāb, Persian, Text Series, No. 27, 1345/1966, Tehran, 711 pp. (in Persian).

Ulrich, F.P., 1941. The Imperial Valley earthquake of 1940. Seismol. Soc. Am. Bull. 31, 13–31.

Uluhogiān, G., 1980. Voitive Inscriptions at the Monastery of St. Stephen the Protomartyr. Documents of Armenian Architecture, 10, S. Stephanops, Edizioni Ares, Milano, Italy, pp. 10–17.

al-'Umari, Yāsin ibn Khair Allāh al-Khatib, 1793. Al-Āthār al-Jaliyya fi al-Hawādith al-Arḍiyya. Al-Adab Press, Al-Najaf, Iraq, 1973.

Urumiyeh'i (Mohit), Mirzā 'Ali 'Asgar. In Nakhjavāni, 1964; Zokā', 1989.
USCGS. United States Coast and Geodetic Survey.
U.S. Energy Information Administration. http://tonto.eia.gov/country/index.cfm.
USGS. Significant Earthquakes of the World. http://earthquake.usgs.gov/eqcenter/eqarchives/significant.
USGS/GNIS. Geographic Names Information System, Unites States Geological Survey. Geonames. usgs.gov.
al-'Utbi, Abu Nasr Muhammad ibn 'Abd al-Jabbar, 1036. Kitab-i Yamini. Translated by J. Reynolds, Historical Memoirs of the Amir Sabaktagin and the Sultan Mahmud of Ghazna. Translated from the Persian Version of the Contemporary Chronicle of al-'Utbi, London, Oriental Translation Fund, 1858. Ed. 'Ezzat Bek Quyun, al-Yamini, l'l-'Utbi, Central Library of the University of Tehran, microfilm No. 340.
Utsu, T., 2002. A list of deadly earthquakes in the world: 1500–2000. In: Lee, William H.K., Kanamori, Hiroo, Jennings, Paul C., Kisslinger, Carl (Eds.), International Handbook of Earthquake Engineering and Seismology, Part A. Academic Press, Amsterdam, pp. 691–717.
Valle, Pietro Della, 1677. Viaggi di Pietro Della Valle il Pellegrino. Bologna. English Trans. George Bull, 1989, The Folio Society Ltd, 1990, Century Hutchinson Ltd, Middlesex House, London.
Van Dissen, R., Yeats, R.S., 1991. Hope fault, Jordan thrust, and uplift of the Seaward Kaikoura Range, New Zealand. Geology 19, 393–396.
van Zeist, W.V., Bottema, S., 1977. Palynological investigations in western Iran. Palaeohistoria XIX, 20–85.
Vanden Berghe, L., 1973. La necropolis de Hahalan, VII (1971) et VIII (1972) campagnes. Archeologia, 57, Paris.
Vārtābed Ārākel Tāvrizetzi. See Ārākel Vārtābed Tāvrizetzi.
Vendidād. Avesta.org.
Verdel, C., Wernicke, B.P., Hassanzadeh, J., Guest, B., 2011. A Paleogene extensional arc flare-up in Iran. Tectonics. 30http://dx.doi.org/10.1029/2010TC002809, TC3008.
Vernant, P., Nilforoushan, F., Hatzfeld, D., Abbassi, M., Vigny, C., Masson, F., Nankali, H., Martinod, J., Ashtiani, M., Bayer, R., Tavakoli, F., Chéry, J., 2004. Present-day crustal deformation and plate kinematics in the Middle East constrained by GPS measurements in Iran and northern Oman. Geophys. J. Int.. 157, 381–398. http://dx.doi.org/10.1111/j.1365-246X.2004.02222.x.
Vernant, P., Chery, J., 2006. Low fault friction in Iran implies localized deformation for the Arabia-Eurasia collision zone. Earth Planet. Sci. Lett. 246, 197–206.
Vesāl Shirāzi (1782–1845). Collection of Poems. Theran (in Persian).
Vigne, G.T., 1840. A personal Narrative to a Visit to Ghuzni, Kabul, and Afghanistan, and of a Residence at the Court of Dost Mohamed with Notices of Runjit Sing, Khiva, and the Russian Expedition. Whittaker & Co., London, 479 pp.
Virieux, J., Deschamps, A., Perrot, J., Campos, J., 1994. Source mechanisms and near-source wave propagation from broadband seismograms. Ann. Geofis. XXXVII (6), 1565–1590.
Volponi, F.S., 1979. Report of the Argentina-United States Binational Symposium About Caucete Earthquake of November 23rd, 1977. Universidad Nacional de San Juan, Instituto Sismológico Zonda, San Juan, Argentina, 44 pp. Univ. Nac. de San Juan, Inst. Sismol. Zonda, San Juan.
Vranes, K., Pielke Jr., R.A., 2009. Normalized earthquake damage and fatalities in the United States: 1900–2005. Nat. Hazards Rev.. 10, 84–101. http://dx.doi.org/10.1061/ASCE1527-6988200910.384.
Vrolyk, E., 1957. Mission de Protection civile en Iran. Report of the Government of Algeria, September 1957, 15 pp (in French).

Walker, F., Allen, M.B., 2012a. Offset rivers, drainage spacing and the record of strike-slip faulting: The Kuh Banan Fault, Iran. Tectonophysics. 530–531, 251–263. http://dx.doi.org/10.1016/j.tecto.2012.01.001.

Walker, R., 2003. Active Faulting and Tectonics of Iran. Ph.D. Theses, University of Cambridge, UK, 203 pp.

Walker, F., Allen, M.B., 2012b. Offset rivers, drainage spacing and the record of strike-slip faulting in Kuh Banan, Iran. Tectonophysics. 530–1, 251–263. http://dx.doi.org/10.1016/j.tecto.2012.01.001.

Walker, R., Jackson, J., 2002. Offset and evolution of the Gowk fault, SE Iran; a major intra-continental strike-slip system. J. Struct. Geol. 24, 1677–1698.

Walker, R., Jackson, J., 2004. Active tectonics and late Cenozoic strain distribution in central and eastern Iran. Tectonics. 23http://dx.doi.org/10.1029/2003TC001529, TC5010.

Walker, R., Jackson, J., Baker, C., 2003. Surface expression of thrust faulting in eastern Iran: source parameters and surface deformation of the 1978 Tabas and 1968 Ferdows earthquake sequences. Geophys. J. Int. 152, 749–765.

Walker, R., Jackson, J., Baker, C., 2004. Active faulting and seismicity of the Dasht-e-Bayaz region, eastern Iran. Geophys. J. Int. 157, 265–285.

Walker, R.T., Andalibi, M.J., Gheitanchi, M.R., Jackson, J.A., Karegar, S., Priestley, K., 2005a. Seismological and field observations from the 6 November 1990 Furg (Hormozgan) earthquake: a rare case of surface rupture in the Zagros mountains of Iran. Geophys. J. Int. 163, 567–579.

Walker, T.W., Bergman, E., Jackso, J., Ghorashi, M., Talebian, M., 2005b. The 2002 June 22 Changureh (Avaj) earthquake in Qazvin province, northwest Iran: epicentral relocation, source parameters, surface deformation and geomorphology. Geophys. J. Int. 160, 707–720.

Walker, R.T., Gans, P., Allen, M.B., Jackson, J., Khatib, M., Marsh, N., Zarrinkoub, M., 2009. Late Cenozoic volcanism and rates of active faulting in eastern Iran. Geophys. J. Int.. 177, 783–805. http://dx.doi.org/10.1111/j.1365-246X.2008.04024.x.

Walker, R.T., Bergman, E.A., Szeliga, W., Fielding, E.J., 2011. Insights into the 1968-1997 Dasht-e-Bayaz and Zirkuh earthquake sequences, eastern Iran, from calibrated relocations, InSAR and high-resolution satellite imagery. Geophys. J. Int. 187, 1577–1603.

Walker, R.T., Bergman, E.A., Elliott, J.R., Fielding, E.J., Ghods, A.R., Ghoraishi, M., Jackson, J., Nazari, H., Nemati, M., Oveisi, B., Talebian, M., Walters, R.J., 2013a. The 2010–2011 South Rigan (Baluchestan) earthquake sequence and its implications for distributed deformation and earthquake hazard in southeast Iran. Geophy. J. Int.. 193 (1), 349–374. http://dx.doi.org/10.1093/gji/ggs109.

Walker, R.T., Khatib, M.M., Bahroudi, A., Schnabel, C., Rodés, A., Fattahi, M., Talebian, M., Bergman, E., 2013b. Co-seismic, Geomorphic, and Geologic fold growth associated with the 1978 Tabas earthquake fault in eastern Iran. Geomorphology (online accessed in 2013; still in press in April 2014). http://dx.doi.org/10.1016/j.geomorph.2013.02.016.

Walker, R.T., Bergman, E.A., Elliott, J.R., Fielding, E.J., Ghods, A.-R., Ghoraishi, M., Jackson, J., Nazari, H., Nemati, M., Oveisi, B., Talebian, M., Walters, R.W., 2013c. The 2010–2011 South Rigan (Baluchestan) earthquake sequence and its implications for distributed deformation and earthquake hazard in southeast Iran. Geophys. J. Int.. http://dx.doi.org/10.1093/gji/ggs109.

Wallace, R.E., 1987. Grouping and migration of surface faulting and variations of slip rate on faults in the Great Basin province. Bull. Seismol. Soc. Am. 77, 868–876.

Walpersdorf, A., Hatzfeld, D., Nankali, H., Tavakoli, F., Nilforoushan, F., Tatar, M., Vernant, P., Chéry, J., Masson, F., 2006. Difference in the GPS deformation pattern of North and Central Zagros (Iran). Geophys. J. Int. 167, 1077–1088.

Wang, C.-Y., Yang, Z.-E., Luo, H., Mooney, W.D., 2004a. Crustal structure of the northern margin of the eastern Tien Shan and its tectonic implications for the 1906 M 7.7 Manas earthquakes. Earth Planet. Sci. Lett. 223, 187–202.

Wang, R., Xia, Y., Grosser, H., Wetzel, H.U., Kaufmann, H., Zschau, J., 2004b. The 2003 Bam (SE Iran) earthquake: precise source parameters from satellite radar interferometry. Geophys. J. Int. 159, 917–922.

Ward, W.H., 1909. The origin of the worship of Yahwe. Am. J. Semitic Lang. Lit. XXV (3), 180.

Waterman, L., 1929–31. Royal Correspondence of the Assyrian Empire, 4 vols. University of Michigan.

Watson, R.G., 1866. A History of Persian from the Beginning of the 19th Century to the Year 1858, with a Review of the Principal Events that Led to the Establishment of the Kajar Dynasty. Smith, Elder, & Co, London, 498 pp.

Watson, R.A., Wright Jr., H.E., 1969. The Saidmarreh landslide, Iran. Sp. Pap. 123, Geological Society of America Inc., pp. 115–138.

Weimer, C., 2002. See => Yadegar-e Zariaran.

Weinder, E.F., 1939. Ereškigal als Göttin des Erdbebens (Ereshkigal as goddess of earthquakes). Arch. Oriental Res. 13, 231–233.

Weiss, H., 2000. Beyound the younger Dryas: collapse as adaptation to abrupt climate changes in ancient West Asia and the Eastern Mediterranean. In: Bawden, Garth, Martin Reycraft, Richard (Eds.), Environmental disaster and the Ar0063haeology of human response. In: Maxwell Museum of Anthropology, Anthropological Paper, 7, pp. 75–98.

Weldon, R., Scharer, K., Fumal, T., Biasi, G., 2004. Wrightwood and the earthquake cycle: what a long recurrence record tells us about how faults work. GSA Today 14 (9), 4–10.

Wells, D.L., Coppersmith, K.J., 1994. New empirical relationships among magnitude, rupture length, rupture width, rupture area, and surface displacement. Bull. Seismol. Soc. Am. 84 (4), 974–1002.

Weninger, B., Alram-Stern, E., Bauer, E., Clare, L., Danzeglocke, U., Jöris, O., Kubatzki, C., Rollefson, G., Todorova, H., van Andel, T., 2006. Climate forcing due to the 8200 cal yr BP event observed at early Neolithicsites in the eastern Mediterranean. Quat. Res. 66, 401–410.

West, E.W. (Ed., Trans.), 1880. Bundahis. Sacred Books of the East, 5, Oxford University Press. Also Motilal Banarsidass, Delhi. Also see avesta.org.

West, E.W. (Ed., Trans.), 1882. The Dadistan-i Dinik and the epistles of Manuskihar. Sacred Books of the East, Pahlavi Texts; Edited by Max Muller. Motilal Banarsidass, Delhi, 1970. Electronic version available at: http://www.avesta.org/mp/zadspram.html.

West, E.W. (Ed., Trans.), 1897. Selections of Zadspram. Sacred Books of the East, 5, Oxford University Press WP. Also see Avesta.org.

Westaway, R., 1990. Seismicity and tectonic deformation rate in Soviet Armenia: implications for local earthquake hazard and evolution of adjacent regions. Tectonics 9 (3), 477–503.

White, R.S., 1982. Deformation of the Makran accretionary sediment prism in the Gulf of Oman (North-West Indian Ocean). In: Leggett, J.K. (Ed.), Trench-Forearc Geology: Sedimentation and Tectonics on Modern and Ancient Active Plate Margins. Geological Society of London, Special Publication 10, Blackwell Scientific Publications, Oxford, pp. 357–372.

White, R.S., Louden, K.E., 1982. The Makran continental margin: structure of a thickly sedimented convergent plate boundary. In: Watkins, J.S., Drake, C.L. (Eds.), Studies in Continental Margin Geology, AAPG Bull., 34, pp. 499–518.

Wicken, A.J., Hodgson, J.H., 1967. Computer re-evaluation of earthquake mechanism solutions 1922–1962. Publ. Dom. Obs. Ottawa 33, 1.

Wilber, D.N., 1955. The architecture of Islamic Iran: the Il-Khanid Period. Princeton Monographs in Art and Archaeology XXIX, Oriental Studies XVII, Department of Art and Archaeology and Oriental Languages and Literatures, Princeton University Press, Also Persian trans. 'Abdollāh Faryār, Bongāh Tarjomeh va Nashr Ketāb Publ., 256, Majmu'eh Ma'āref 'Omumi-36, Tehran, 1346/1965, 228 pp.

Wilkinson, C.K., 1975. Nishapur: Pottery of the Early Islamic Period. The Metropolitan Museum of Art, Greenwich, Connecticut, distributed by New York Graphic Society, 374 pp.

Wilkinson, C.K., 1986. Nishapur; Some Early Islamic Buildings and Their Decorations. The Metropolitan Museum of Art, New York, 328 pp.

Willis, B., 1928. Earthquakes in the Holly Land. Bull. Seismol. Soc. Am. 18, 73–113.

Wills, B., 1933. Earthquakes in the Holly Land, a correction. Science 77, 351.

Wills, C.J., 1894. Behind an Eastern Veil. A Plain Tale of Events Occurring in the Experience of a Lady who had a Unique Opportunity of Observing the Inner Life of Ladies of the Upper Class in Persia. William Blackwood & Sons, Edinburgh & London, 376 pp.

Wilmsen, M., Fürsich, F.T., Seyed-Emami, K., Majidifard, M.R., Taheri, J., 2009. The Cimmerian orogeny in northern Iran: tectono-stratigraphic evidence from the foreland. Terra Nova 21, 211–218.

Wilson, A., 1930. Earthquakes in Persia. Bull. Sch. Orient. Study (SOAS). Lond. Inst. 6 (1), 103–131.

Wilson, S.G., 1895. Persian Life and Customs; With Scenes and Incidents of Residence and Travel in the Land of the Lion and the Sun, third ed. Fleming H. Revell Co., New York, 333 pp.

Winston, D., 1966. The Iranian component in the Bible. Apocrypha, and Qumran: a review of the evidence. Hist. Relig. 5 (2), 191 ff.

Wiseman, D.J., 1979. Jonah's Nineveh. The Tyndale Biblical Archaeology Lecture, 1977. Tyndale Bull. 30, 29–52. http://www.tyndalehouse.com/tynbul/library/TynBull_1979_30_02_Wiseman_JonahsNineveh.doc.

Worldwide Aerial Surveys, Inc., 1956. Aerial photographs of Iran, Project 158, scale ~1:55,000, Cartographic Department of the Imperial Government of Iran, Joint Operations Graphics, USA, WWS.

Wright, R., 2009. The Evolution of God. Little, Brown & Co., 576 pp.

Wu, F.T., Ben-Menahem, A., 1965. Surface wave radiation pattern and source mechanism of the September 1, 1962, Iran earthquake. J. Geophys. Res. 70 (16), 3943–3949.

Wüstenfeld, F., and Mahler, E., (revised by Qureshi, H.), 1982. Comparative Chronological tables of Muslim and Christian Calendars by Ferdinand Wustenfeld and Edward Mahler. Translated and Revised by Hakeem-Uddeen Qureshi, New Delhi, India. Niavaran Cultural Centre, Tehran, 1360 (in Persian).

WWSSN. The Worldwide Standardized Seismographic Stations.

Xu, X., Deng, Q., 1996. Nonlinear characteristics of paleoseismicity in China. J. Geophys. Res. 101 (B3), 6209–6231.

Yamini Fard, F., Hatzfeld, D., Tatar, M., Mokhtari, M., 2006. Microseismicity at the intersection between the Kazerun fault and the Main Recent Fault (Zagros-Iran). Geophys. J. Int. 166, 186–196.

Yamini-Fard, F., Hatzfeld, D., Farahbod, A.M., Paul, A., Mokhtari, M., 2007. The diffuse transition between the Zagros continental collision and the Makran oceanic subduction (Iran): microearthquake seismicity and crustal structure. Geophys. J. Int. 170, 182–194.

Ya'qubi, Ahmad ibn 'Abi Ya'qub ebn Ja'far ebn Vahāb ebn Vaze', 892. Tārikh Ya'qubi, [History of Ya'qubi, 279/892]. Ed. Beirut, 1970. Also, Ed. Th. Houtsma, Leiden, 1883. Tārikh-e Ya'qubi (279/892). Translated into Persian by Ahmad Ebrāhim Āyati. Sherkat Enteshārāt 'Elmi va

Farhangi, Tehrān, first ed. 1342/1963, fifth ed. 1366/1987, ninth ed. 1382/2003. Translated into Persian by E. Āyati, Bongāh-e Tarjomeh va Nashr Ketāb, Tehran, vol. 1, 610 pp. (in Persian).

Yāqut, ibn 'Abdollah al-Rumi al Hamawi, 1225. Mu'jam al-Boldān [Dictionary of Countries, 623/1225]. Trans. C. Barbier de Meynard, Dictionaire de la Perse, Extraits de Mo'jem al-Buldan de Yaqut, Paris, 1861. W. Jwaideh, The Introductory Chapters of Yaqut's Mu'jam al-Buldan, Leyden, 1971. Also, Egypt, Beirut, mimograph, Tehran (in Arabic). Also Ed. Wustenfeld, Leipzig 1866–73. Also Asadi Bookstore Publ., 1965, Tehran (in Persian). Ed. Mohammad Ebrāhim Āyati, Bonqāh Tarjomeh va Nashr Ketāb, Tehrān, 1343/1964, 1356/1977.

Yāqut, ibn 'Abdollah al-Rumi al Hamawi, 1226. Mu'jam al-Udabā'. Ed. Rifā'i, Cairo 1936–38.

Yarshater, E., 1983. Iranian Historical Tradition: Iranian Common Beliefs and World-View. In: The Cambridge History of Iran, III. The Cambridge University Press, U.K., pp. 341–358.

Yashts. See Poure-Davud. Also avesta.org. See also Poure Davoud, 1968.

Yasnā. See Poure-Davud. Also avesta.org. See also Poure Davoud, 1977.

Yeats, R.S., 1973. Newport-Inglewood fault zone, Los Angeles Basin, California. AAPG Bull. 57, 117–135.

Yeats, R.S., 2000. The 1968 Inangahua, New Zealand, and 1994 Northridge, California, earthquakes: implications for northwest Nelson. N. Z. J. Geol. Geophys. 43, 587–599.

Yeats, R.S., 2007. Paleoseismology: why can't earthquakes keep on schedule? Geology 35, 863–864.

Yeats, R., 2012. Active Faults of the World. Cambridge University Press, 621 pp.

Yeats, R.S., Berryman, K.R., 1987. South Island, New Zealand, and Transverse Ranges, California: a seismotectonics comparison. Tectonics 6, 363–376.

Yeats, R.S., Huftile, G.J., 1995. The Oak Ridge fault system and the 1994 Northridge earthquake. Nature 373, 418–420.

Yeats, R.S., Thakur, V.C., 2008. Active faulting south of the Himalayan front: establishing a new plate boundary. Tectonophysics 453, 63–73.

Yeats, R., Clark, M., Keller, E., Rockwell, T., 1981. Active fault hazard in southern California: ground rupture versus seismic shaking. Geol. Soc. Am. Bull. 92, 189–196.

Yeats, R.S., Sieh, K.E., Allen, C.R., 1997. The Geology of Earthquakes. Oxford University Press, New York, 568 pp.

Yeghishé, Vārtāpet, 1952. History of Vartan and the Armenian War. English translation by Dikran Boyajian, The Knight of Vartan, Inc., USA.

Yeznik Koghbātsi (Yeznik of Kolb), Fifth Century. Refutation of the Sects, A Retelling of Yeznik Koghbātso's Apology (Book 1, The Nature of God). Trans. Thomas Samuelian, *St. Vartan Press*, 1986, Armenian Church Classics, NY; http://www.tertullian.org/rpearse/scanned/yeznik_refutation.htm.

Young Jr., T.C., 1968. Survey of Excavations, Godin TepeVI Iran, J. British Inst. Persian Study, London, pp. 160–161.

Young Jr., T.C., 1969. Excavation at Godin Tepe: First Progress Report. Occasional Papers no. 17, Royal Ontario Museum, Art & Archaeology, Toronto.

Young Jr., T.C., Levine, L.D., 1974. Excavations of the Godin Project: Second Progress Report. Occasional Papers no. 26, Royal Ontario Museum, Art & Archaeology, Toronto.

Zāchāriāh of Kānāker, 1627–1699. In Babayan, 2006.

Zād Spram. Rashed Mohasel, M.T., (Trans.), 1987, Gozidehā-ye Zād Esparam (Selection of Zād Esparam). Mo'aseseh Motāle'āt va Tahqiqāt Farhangi, 1366/1987, Tehrān, 162 pp. Also Selection of Zadspram, avesta.org.

Zaehner, R.C., 1972. Zurvan: A Zoroastrian Dilemma. Biblio and Tannen, New York, 495 pp.

Zāghi, N., 2007. Groundwater Exploration; The Oldest Hydrology Text Book, by Aboubakr Karaji. English translation from Persian by Nour Zaghi, June 2007. http://www.uscashagent.com/groundwater-exploration.htm.

Zakariyā Qazvini, ibn Muhammad, 1263. Ajā'ib al-Makhluqāt wa Gharā'ib al-Maujudāt [Marvels of Things Created and Miraculous Aspects of Things Existing; 661 AH/1263 AD]. Ed. Nasrallāh Sabuhi, Ketābkhāneh Markazi, Tehrān, 1361/1982 (in Persian). Also Ed. Wustenfeld, Gottingen, 1848.

Zākariā, Abbot, 1329. Inscription on a Stone Panel Regarding the 1319 Earthquake Repairs at the Armenian Monastery of Surp Thādei Vānk (St. Thāddeus) in Her (modern Māku) Signed by Abbot (Vanahāyr) Zākariā/Zechariah, (see Plate 8.2).

Zamani, A., Raisi, M., 1991. The Fork (Darab) earthquake of 6 Nov. 1990, SW Iran. Department of Earth Sciences, College of Science, University of Shiraz, Iran.

Zanchetta, S., Zanchi, A., Villa, I., Poli, S., Muttoni, G., 2009. The Shanderman eclogite: a Late Carboniferous high-pressure event in the NW Talesh Mountains (NW Iran). In: Brunet, M.-F., Wilmsen, M., Granath, J.W. (Eds.), South Caspian to Central Iran Basins. Sp. Publ. 312. The Geological Society London, pp. 57–78.

Zanchetta, S., Berra, F., Zanchi, A., Bergomi, M., Caridroit, M., Nicora, A., Heidarzadeh, G., 2013. The record of the Late Palaeozoic active margin of the Palaeotethys in NE Iran: constraints on the Cimmerian orogeny. Gondwana Res. 24, 1237–1266.

Zanchi, A., Berra, F., Mattei, M., Ghassemi, M.R., Sabouri, J., 2006. Inversion tectonics in central Alborz, Iran. J. Struct. Geol. 28, 2023–2037.

Zanchi, A., Zanchetta, S., Berra, F., Mattei, M., Garzanti, E., Molyneaux, S., Nawab, A., Aabouri, J., 2009. The Eo-Cimmerian (Late? Triassic) orogeny in North Iran. In: Brunet, M.-F., Wilmsen, M., Granath, J.W. (Eds.), South Caspian to Central Iran Basins. Sp. Publ. No. 312. The Geological Society London, pp. 31–55.

Zander, G., 1972. La réstauration de quelques monuments historigue d'Ispahan: une nouvelle lumiére sur les problémes d'histoire de l'architecture s'y rattachant. In: Kiani, M.Y., Tajvivi, A. (Eds.), The Memorial Volume. International Congress of Iranian Art & Archaeol., vol. 2. The Ministry of Culture and Arts, Tehran, pp. 246–259.

Zandjani, H., 1977. Évolution de la population iranienne d'après les recensements. Population 6, 1277–1283.

Zandjani, H., 1992. Téhéran et sa population. Deux siècles d'histoire. In: Adle, C., Hourcade, B. (Eds.), Téhéran. Capitale bicentenaire. Ifri, Paris, pp. 251–266.

Zandjani, H., Raḥmani, F., 1989. Rāhnemā-ye jamʿiyat-e šahrhā-ye Īrān 1335–1370, Tehran, 1368.

Zar'ayān, p, 1997. Moruri bar āsibhāyeh geotechniki-ye zaminlarzeh-ye akhir dar Iran (shomāl-e Bojnurd, Ardebil, va Birjand). Anjoman Mechānic Khāk va mohandesi pay Irān, Khabarnāmeh 18, 10–14, 1376 (in Persian).

Zāre', M., 1991a. The seismotectonic study of the reconstruction of the Sefid-Rud dam with special reference to the earthquake of 20th June 1990, Manji NW of Iran. In: Proceedings of the 1st International Conference on Seismology & Earthquake Engineering, SSE-1, IIEES, May 27029, 1991. 1, pp. 85–93.

Zāre', M., 1991b. The seismotectonic and neotectonic analysis of the reconstruction operation of the Sefid Rud dam, with special reference to the earthquake of 20th June 1990, Manjil, North-West of Iran. In: Proceedings of the 5th International Conference on Soil Dynamics & Earthquake Engineering, Karlsruhe, Germany. 2Computational Mechanics Publication, Southampton, UK.

Zāre', M., 1995. Nonlinear soil behavior in the near-field; a case-study on the recent Iranian accelerometric data. In: Proceedings of the 2nd International Conference on Seismology and Earthquake Engineering, IIEES, May 14–17, 1995, pp. 1283–1292.

Zare, M., 2003. Seismological aspects of the Bam, SE Iran earthquake of 26/12/2003, Mw6.5. IIEES website, http://www.iiees.ac.ir.

Zāre', M., 2002. Neotectonic study of the epicentral region of the 1962 Bu'in Zahra earthquake (M_s 7.2) in the Qazvin Plain and west of Tehrān. Pazhuheshnāmeh Zelzelehshenāsi va Mohandesi Zelzeleh, IIEEA 1 (5), 18–32, Spring 1381, (in Persian), 1381/2002.

Zaré, M., and Ansari, A., 2013. Gozaresh-e bazdid-e fanni-e shenasaee 29 Ordibehesht 1992 az zelzeleyeh Irar (Goharan-Bashagerd) dar 21 Ordibehesht 1992 ba bozorgayeh 6.2 [Technical reconnaissance report of the 11 May 2013 Irar (Goharan-Bashager) earthquake of M_w 6.2]. iiees.ac.ir.

Zaré, M., Hamzehloo, H., 2004. A study of the strong ground motions of 26 December 2003 Bam earthquake: Mw6.5. J. Seism. Earthq. Eng., Special Issue on Bam Earthquake 5 (4), 33–56, 6 (1), IIEES, Tehran.

Zāre', M., Moinfar, A.A., 1993. Earthquake fault induced landslide during the Manji, Iran (1990) earthquake. In: Proceedings of the 8th International Seminar on Earthquake Prognostics. 1. National Disaster Prevention Center of Iran, Tehran, pp. 39–40.

Zāre', M., Moinfar, A.A., 1994. Comment on "The Rudbar-Tarom earthquake of 20 June 1990 in NW Persia: a preliminary field and seismological observations and its tectonic significance. Bull Seismol. Soc. Am. 84 (2), 484–485, by M. Berberian, M. Qorashi, J.A. Jackson, K. Priestely, and T. Wallace.

Zarrābi (Sohail-e-Kāshāni), 'Abd al-Rahim Kalāntar, 1870. Tārikh-e Kāshān [History of Kashan; 1287/1870]. Ed. Iraj Afshār. Publ. Farhang Irān Zamin-2, Tehran, 1341/1962, 2535/1976, 208 pp.

Zayia, A., 1930a. Lettre de M. Abel Zaya a M. Berthounesque. 11 Mai 1930. Ann. Mission Lazarite 95, 834–837.

Zayia, A., 1930b. Les tremblements de terre. Nissions Lazaristes et Filles de la Charité. Bull. 8e année 1930, 205–220.

Zeitouniān, A.S., 1991. The Earthquakes in Armenia and Neighbouring Territories. Ac. Sc. RA, Yerevān, Armenia.

Zimern, H. (Trans.), 1926. The Epic of Kings; Heo Tales of Ancient Persia. The Macmillan Company, New York, 333 pp.

Zohoorian Izadpanah, A.A., Mohajer-Ashjai, A., Salehi Rad, M.R., Taghizadeh, Gh.A., Kabiri, A., 1981. Damage distribution and aftershock sequence of Zarand earthquake of 19 December 1977. J. Earth Space Phys. 10 (1 & 2), 25–42.

Zokā', Y., 1989. Zamin-larzehāyeh Tabriz The Earthquakes of Tabriz Ketābsarā, Tehran, 1368/1989, 216 pp. (in Persian).

Zonuzi, Mirzā Mohammad Hassan al-Hossaini, 1801. Riyāz al-Janat (1216/1801). al-Qesm al-Avval Men al-Rauzat al-Rābe'āt. Ed. Sayyed Mahmmud al-Mar'ashi, and 'Ali Fafi'i. Āyatollāh al-Ozmā al-Mar'ashi Publ., Qom, 1370/1991, 628 pp.

Index

Note: Page numbers followed by "*f*" indicate figures, "*t*" indicate tables, and "*p*" indicate plates.

C

D

F

G

H

I

J

Q

R

T

U

V

W

Z

FSC
www.fsc.org
MIX
Papier | Fördert gute Waldnutzung
FSC® C083411

Kolibri 360 GmbH
Ferdinand-Jühlke-Straße 7
99095 Erfurt, Deutschland
produktsicherheit@kolibri360.de